Fields Institute Communications

Volume 87

The Communications series features conference proceedings, surveys, and lecture notes generated from the activities at the Fields Institute for Research in the Mathematical Sciences. The publications evolve from each year's main program and conferences. Many volumes are interdisciplinary in nature, covering applications of mathematics in science, engineering, medicine, industry, and finance.

Ilia Binder • Damir Kinzebulatov •
Javad Mashreghi
Editors

Function Spaces, Theory and Applications

Springer

FIELDS

Editors
Ilia Binder
Department of Mathematics
University of Toronto
Toronto, ON, Canada

Damir Kinzebulatov
Department of Mathematics and Statistics
Laval University
Québec City, QC, Canada

Javad Mashreghi
Department of Mathematics and Statistics
Laval University
Québec City, QC, Canada

ISSN 1069-5265 ISSN 2194-1564 (electronic)
Fields Institute Communications
ISBN 978-3-031-39269-6 ISBN 978-3-031-39270-2 (eBook)
https://doi.org/10.1007/978-3-031-39270-2

This Springer imprint is published by the registered company Springer Nature Switzerland AG
The registered company address is: Gewerbestrasse 11, 6330 Cham, Switzerland

Paper in this product is recyclable.

Preface

The focus program on *Analytic Function Spaces and Their Applications* took place at Fields Institute from July 1 to December 31, 2021. Hilbert spaces of analytic functions form one of the pillars of complex analysis. These spaces have a rich structure and for more than a century have been studied by many prominent mathematicians. They also have several essential applications in other fields of mathematics and engineering, e.g., robust control engineering, signal and image processing, and theory of communication. The most important Hilbert space of analytic functions is the Hardy class H^2. However, its close cousins, e.g., the Bergman space A^2, the Dirichlet space D, the model subspaces K_Θ, and the de Branges–Rovnyak spaces H(b), have also been the center of attention in the past two decades. Studying the Hilbert spaces of analytic functions and the operators acting on them, as well as their applications in other parts of mathematics or engineering were the main subjects of this program. During the program, the world leading experts on function spaces gathered and discussed the new achievements and future venues of research on analytic function spaces, their operators, and their applications in other domains.

With more than 250 hours of lectures by prominent mathematicians, a wide variety of topics were covered. More explicitly, there were mini-courses and workshops on

1. Hardy Spaces
2. Dirichlet Spaces
3. Bergman Spaces
4. Model Spaces
5. Interpolation and Sampling
6. Riesz Bases, Frames, and Signal Processing
7. Bounded Mean Oscillation
8. de Branges-Rovnyak Spaces
9. Operators on Function Spaces
10. Truncated Toeplitz Operators
11. Blaschke Products and Inner Functions

12. Discrete and Continuous Semigroups of Composition Operators
13. The Corona Problem
14. Non-commutative Function Theory
15. Drury-Arveson Space
16. Convergence of Scattering Data and Non-linear Fourier Transform

At the end of each week, there was a high profile colloquium talk on the current topic. The program also contained two semester-long advanced courses on (i) Schramm Loewner Evolution and Lattice Models and (ii) Reproducing Kernel Hilbert Space of Analytic Functions.

The current volume features a more detailed version of some of the talks presented during the program. However, videos of almost all the Mini Courses, Advanced Courses, Plenary Talks, Colloquium Talks, and Contributed Talks are available at: http://www.fields.utoronto.ca/activities/21-22/function

Toronto, ON, Canada — Ilia Binder
Québec City, QC, Canada — Damir Kinzebulatov
Québec City, QC, Canada — Javad Mashreghi
January 2023

Contents

Contributors

Elena Afanas'eva Institute of Applied Mathematics and Mechanics of the NAS of Ukraine, Slavyansk, Ukraine

Raúl E. Curto Department of Mathematics, University of Iowa Iowa, IA, USA

Konstantin Fedorovskiy Faculty of Mechanics and Mathematics, Lomonosov Moscow State University Moscow, Russia
Moscow Center for Fundamental and Applied Mathematics, Lomonosov Moscow State University Moscow, Russia
St. Petersburg State University, St. Petersburg, Russia

Paul M. Gauthier Département de mathématiques et de statistique, Université de Montréal Montréal, QC, Canada

Soumitra Ghara Department of Mathematics and Statistics, Indian Institute of Technology Kanpur Kanpur, India

Anatoly Golberg Department of Mathematics, Holon Institute of Technology Holon, Israel

Burak Hatinoğlu School of Mathematics, Georgia Institute of Technology Atlanta, GA, USA

In Sung Hwang Department of Mathematics, Sungkyunkwan University Suwon, Korea

Michael T. Jury Department of Mathematics, University of Florida Gainesville, FL, USA

Boban Karapetrović Faculty of Mathematics, University of Belgrade Belgrade, Serbia

Paul Koosis Outremont, QC, Canada

Matthias Langer Department of Mathematics and Statistics, University of Strathclyde Glasgow, UK

Juliette Leblond INRIA, Team Factas Sophia Antipolis, France

Woo Young Lee Department of Mathematics and RIM, Seoul National University Seoul, Korea

Robert T. W. Martin Department of Mathematics, University of Manitoba Winnipeg, MB, Canada

Gadadhar Misra Department of Mathematics, Indian Institute of Technology Gandhinagar, India
Statistics and Mathematics Unit, Indian Statistical Institute Bangalore, India

Lajos Molnár Bolyai Institute, University of Szeged Szeged, Hungary
Department of Analysis, Institute of Mathematics, Budapest University of Technology and Economics Budapest, Hungary

Elodie Pozzi Department of Mathematics and Statistics, Saint Louis University Saint Louis, MO, USA

James Rovnyak Department of Mathematics, University of Virginia Charlottesville, VA, USA

Mohammad Shirazi Department of Mathematics and Statistics, McGill University Montréal, QC, Canada

Harald Woracek Institute for Analysis and Scientific Computing, Vienna University of Technology Wien, Austria

Kehe Zhu Department of Mathematics and Statistics, SUNY at Albany Albany, NY, USA

Absolute Continuity in Higher Dimensions

Elena Afanas'eva and Anatoly Golberg

2020 Mathematics Subject Classification: Primary: 26B30; Secondary: 46E35, 30C65

1 Introduction

A function φ of one variable is absolutely continuous on $[a, b]$ provided that for every $\varepsilon > 0$, there exists $\delta > 0$ such that for any finite collection $\{(a_n, b_n)\}$, $n = 1, \ldots, N$, of disjoint open subintervals in (a, b), one has

$$\sum_{n=1}^{N} |\varphi(b_n) - \varphi(a_n)| < \varepsilon ,$$

whenever $\sum_{n=1}^{N}(b_n - a_n) < \delta$.

The notion of absolute continuity has an important role in the analysis of one variable. It can be illustrated by the following chains. Over a compact subset of the real line,

$$\mathrm{AC} \subseteq \mathrm{UC} = \mathrm{C} ,$$

for a compact interval,

$$\mathrm{C}^1 \subseteq \mathrm{Lip} \subseteq \mathrm{AC} \subseteq \mathrm{BV} \subseteq \mathrm{Dif}_{a.e.} .$$

E. Afanas'eva
Institute of Applied Mathematics and Mechanics of the NAS of Ukraine, Slavyansk, Ukraine

A. Golberg (✉)
Department of Mathematics, Holon Institute of Technology, Holon, Israel
e-mail: golberga@hit.ac.il

I. Binder et al. (eds.), *Function Spaces, Theory and Applications*, Fields Institute Communications 87, https://doi.org/10.1007/978-3-031-39270-2_1

Here C, C^1, AC, UC, Lip, BV, and $\mathrm{Dif}_{a.e.}$ stand for continuous, continuously differentiable, absolutely continuous, uniformly continuous functions, Lipschitz functions, functions of bounded variations and functions differentiable almost everywhere (a.e.). More precisely, on a nondegenerate segment $[a, b]$, Lip $\subsetneq$ AC $\subsetneq$ BV and UC $\neq$ BV. Some trivial and nontrivial examples describe the sharpness of the above implications. One of such representatives is the Cantor function which is of bounded variations but fails to be absolutely continuous. A fractal curve (e.g. the Weierstrass function) is an example of a function that is continuous everywhere but differentiable nowhere. Here we refer to [12, 15, 16, 31, 34].

The following condition (Banach–Zaretsky's theorem) completely provides the relationship between AC and C, BV and the Lusin (N)-property; see, e.g. [25].

Theorem 1.1 *A function $\varphi : [a, b] \to \mathbb{R}$ is* AC *if and only if all three of the following conditions hold: (i) $\varphi \in$ C, (ii) $\varphi \in$ BV, and (iii) if E has zero Lebesgue measure then $\varphi(E)$ is so.*

Another description of absolute continuity can be given by the following way. A function $f : \mathbb{R} \to \mathbb{R}$ is absolutely continuous if there exists an integrable function $g : \mathbb{R} \to \mathbb{R}$ such that $|f(b) - f(a)| \leq \int_a^b g(x)\, dx$ for each $a < b$; see, e.g. [32].

The situation in higher dimensions is rather more complicated. The aim of this paper is to survey the corresponding relations and related notions for the multidimensional Euclidean settings. In 1999, Jan Malý [35] contributed a classical counterpart of absolute continuity in the sense of Banach or in the sense of Tonelli to $\mathbb{R}^n$. This notion has cast light on various problems including the area, coarea and degree formulas. Note that the n-absolute continuity forms a proper subclass of Sobolev class $W^{1,n}$ including validity of Lusin's (N)-property. Further investigations, mainly due to Csörnyei, Hencl and Bongiorno, essentially extended and weakened the assumptions in Malý's ideas and provided a wide spectrum of absolute continuity/bounded variation concepts. Such main results are also included in our survey.

We also intend to illustrate and connect these results for the classes of finitely bi-Lipschitz mappings and mappings of finite metric and area distortion, whose definitions rely on a metric approach and further can be easily extended to a more general setting than $\mathbb{R}^n$. Some recent results and illustrating examples are included. We also pose an open problem on the connection between the absolute continuity AC^n_Λ and absolute continuity on lines ACL.

2 Sobolev Spaces and Absolute Continuity on Lines

2.1 Sobolev Classes

Obviously the notion of absolute continuity is strongly connected with Sobolev classes in $\mathbb{R}^n$. We recall the definition of Sobolev spaces $W^{1,p}$, $p \geq 1$, following [28].

Definition 2.1 Let $\Omega \subset \mathbb{R}^n$ be open and $u \in L^1_{\text{loc}}(\Omega)$. A function $v \in L^1_{\text{loc}}(\Omega)$ is called a *weak derivative of* u if

$$\int_\Omega \varphi(x)v(x)\,dm(x) = -\int_\Omega u(x)\nabla\varphi(x)\,dm(x)$$

for every $\varphi \in \mathrm{C}^\infty_C(\Omega)$. The function v is referred to as Du. For $1 \le p \le \infty$ the Sobolev space is defined by

$$W^{1,p}(\Omega) = \{u \in L^p(\Omega) : Du \in L^p(\Omega)\}$$

with the norm

$$\|u\|_{W^{1,p}(\Omega)} = \left(\int_\Omega |u|^p + \int_\Omega |Du|^p\right)^{1/p}.$$

Here C_C denotes the collection of continuous functions with compact support.

2.2 *Absolute Continuity on Lines*

Following, e.g. [47], we recall the notions of ACL and ACL^p, $p \ge 1$. See also [20, 46].

Definition 2.2 For $j = 1, \dots, n$ let $R^n_j = \{x \in \mathbb{R}^n : x_j = 0\}$ and $T_j = \mathbb{R}^n \to R^n_j$ be an orthogonal projection $T_j x = x - x_j e_j$. Suppose that $D \subset \mathbb{R}^n$ is an open set and $u : D \to \mathbb{R}$ is a continuous function. The function u is called *absolutely continuous on lines*, abbr. ACL, if for every cube Q with $\overline{Q} \subset D$, the set $A_j \subset T_j D \subset R^n_j$ of all points $z \in T_j Q$ such that the function $t \mapsto u(z + te_j)$, $z + te_j \in Q$, is not absolutely continuous as a function of a single variable satisfies $m_{n-1}A_j = 0$ for all $j = 1, \dots, n$.

Absolutely continuous functions of a single variable on a compact interval lie between the classes of Lipschitz continuous and of differentiable a.e. functions. Thus, for absolutely continuous functions the derivative exists a.e. and are Borel-measurable. Together with Fubini's theorem this implies that an ACL function $u : D \to \mathbb{R}$ has partial derivatives with respect to every variable $x_1, \dots, x_n$ a.e. (with respect to the n-dimensional Lebesgue measure) in D.

Definition 2.3 We say that an ACL function $u : D \to \mathbb{R}$ is ACL^p, $p \ge 1$, if $\partial u(x)/\partial x_j \in L^p(K)$, $j = 1, \dots, n$, whenever $K \subset D$ is compact.

Definition 2.4 A vector-valued function is said to be ACL (ACL^p) if and only if each coordinate function belongs to this class.

For homeomorphisms (and, moreover, for general mappings from L^p) the classes $W^{1,p}$ and ACL^p, $p \geq 1$, coincide; see, e.g. [28, Thm A.15].

2.3 Mappings of Bounded and Finite Distortions

The analytic definitions of the following classes of mappings (quasiconformal, quasiregular, of finite distortion) involve mappings of Sobolev classes.

Definition 2.5 Let D be a domain in $\mathbb{R}^n$. Then a mapping $f : D \to \mathbb{R}^n$ is called *quasiregular* (or a *mapping of bounded distortion*) if $f \in W^{1,n}_{\mathrm{loc}}(D)$ and there is a constant $K \geq 1$ so that

$$||f'(x)||^n \leq K\, J_f(x) \quad \text{a.e.} \tag{1}$$

Homeomorphic quasiregular mappings are called *quasiconformal.* Here $f'(x)$ denotes the Jacobi matrix of f at x, and $J_f(x)$ is the Jacobian; $||f'(x)|| = \sup_{|h|=1} |f'(x)h|$.

Many important properties of quasiconformal/quasiregular mappings (like ACL, Hölder continuity, higher integrability of partial derivatives, openness and discreteness, etc.) can be derived from the above definition; see, e.g. [28, 42, 43, 47].

An essentially wider class of mappings where the uniform boundedness a.e. is relaxed by finiteness a.e. has the following description.

Definition 2.6 We say that a mapping $f : D \to \mathbb{R}^n$ on a domain $D \subset \mathbb{R}^n$ has *finite distortion* if $f \in W^{1,1}_{\mathrm{loc}}(D)$, $J_f \in L^1_{\mathrm{loc}}$ and there is a function $K : D \to [1, \infty]$ with $K(x) < \infty$ a.e. so that $||f'(x)||^n \leq K(x) J_f(x)$ for a.e. $x \in D$.

Clearly, the differential and topological features of mappings of finite distortion depend on appropriate restrictions on the distortion function $K(x)$; see, e.g. [28, 30] and the references therein. For homeomorphisms of finite distortion the condition $J_f \in L^1_{\mathrm{loc}}$ can be removed.

3 Lusin's (N) and (N^{-1})-Properties and Area Formulas

Definition 3.1 Let $\Omega \subset \mathbb{R}^n$ be an open set and $f : \Omega \to \mathbb{R}^n$ be a mapping. We say that f possesses the *Lusin (N)-property* on a set $\Omega' \subset \Omega$ if the implication

$$mE = 0 \quad \Longrightarrow \quad mf(E) = 0$$

holds for each subset E of Ω'.

The above definition can be extended to the k-dimensional Hausdorff measure $\mathcal{H}^k$, $k = 1, \ldots, n-1$ by replacing the n-dimensional Lebesgue measure m with $\mathcal{H}^k$. In this case we deal with the Lusin (N)-property with respect to the k-dimensional Hausdorff measure.

In literature, the Lusin (N)-property is also called *absolute continuity in measure* (or, sometimes, simply absolute continuity). And for a mapping $f : (X, \mu) \to (Y, \nu)$ between two metric spaces it means that $\nu f(E) = 0$ for all measurable sets $E \subset X$ satisfying $\mu E = 0$.

Definition 3.2 Let $\Omega \subset \mathbb{R}^n$ be an open set. We say that $f : \Omega \to \mathbb{R}^n$ satisfies the *Lusin (N^{-1})-property* if for each $E \subset f(\Omega)$ such that $mE = 0$ we have $mf^{-1}(E) = 0$.

An advanced version of the theorem on change of variables in integral is due to [19]. It states that the *area formula*

$$\int_M u(x)|J_f(x)|\,dm(x) = \int_{\mathbb{R}^n} \left(\sum_{\{x \in M : f(x) = y\}} u(x) \right) dm(y) \tag{2}$$

is valid for all non-negative measurable functions $u : \Omega \to \mathbb{R}$ if $\Omega \subset \mathbb{R}^n$ is an open set, $f : \Omega \to \mathbb{R}^n$ is approximately differentiable a.e. (e.g. a Sobolev mapping, or equivalently, $f \in W^{1,1}$) and $M \subset \Omega$ is a suitable set of full measure; see [32]. For approximately differentiable functions the area formula (2) is equivalent to the Lusin (N)-property.

The Lusin (N)-property is satisfied for general Sobolev mappings for the case $p > n$; see, e.g. [28].

Theorem 3.1 *Let $\Omega \subset \mathbb{R}^n$ and $p > n$. Suppose that $f \in W^{1,p}(\Omega)$ is continuous. Then f possesses the Lusin (N)-property.*

Remark 3.1 By Cesari's result [11] (cf. [28] and [36]) there exists a continuous mapping $f : \mathbb{R}^n \to \mathbb{R}^m$, $f \in W^{1,n}([-1,1]^n)$, $n, m \in \mathbb{N}$, such that $f([-1,1] \times \{0\}^{n-1}) = [-1,1]^m$, and hence f fails the Lusin (N)-property if $m = n$.

The limiting case $p = n$ for homeomorphisms also guarantees possessing the Lusin (N)-property.

Theorem 3.2 *Let $\Omega \subset \mathbb{R}^n$ and $f \in W^{1,n}(\Omega)$ be a homeomorphism. Then f satisfies the Lusin (N)-property.*

Remark 3.2 Let $p < n$. There exists a homeomorphism of finite distortion $f \in W^{1,p}((-1,1)^n)$ for which the Lusin (N)-property fails; see [28].

The following statement provides a sufficient condition for the mappings of finite distortion to satisfy the Lusin (N^{-1})-property; see [28].

Theorem 3.3 *Let a continuous mapping $f \in W^{1,1}(\Omega)$ be a mapping of finite distortion with $K_f^{1/(n-1)} \in L^1(\Omega)$. If the multiplicity of f is essentially bounded by a constant N and f is not constant, then $J_f(x) > 0$ a.e. in Ω, and hence f satisfies the Lusin (N^{-1})-property.*

The exponent $1/(n-1)$ is crucial and cannot be reduced even for homeomorphisms.

Remark 3.3 Let $a < 1/(n-1)$. There exists a Lipschitz homeomorphism f of finite distortion $f \in W^{1,1}((-1,1)^n)$ and $K_f^a \in L^1((-1,1)^n)$, for which the Lusin (N^{-1})-property fails; see again [28].

For $K_q(x)$-distortion function, $q \le n$, the Lusin (N^{-1})-property can be derived from the following statement.

Theorem 3.4 *Let $f \in W^{1,1}(\Omega)$ be a homeomorphism of finite distortion with $K_q \in L^{1/(q-1)}(\Omega)$ for some $q \in [1,n]$. Then f satisfies the Lusin (N^{-1})-property.*

Remark 3.4 Note that the Lusin (N^{-1})-property is equivalent to the condition that the Jacobian does not vanish a.e.; cf. [40].

4 n-Absolute Continuity and n-Bounded Variation

4.1 n-Absolute Continuity

The following kind of absolute continuity for higher dimensions was introduced in [35].

Definition 4.1 Let $\Omega \subset \mathbb{R}^n$ be an open set. A function $f : \Omega \to \mathbb{R}$ is *n-absolutely continuous* (shortly AC^n) if for each $\varepsilon > 0$ there exists $\delta > 0$ such that for every pairwise disjoint finite collection $\{B(x_j, r_j)\}$ of balls in Ω, we have

$$\sum_j mB(x_j, r_j) < \delta \qquad \Longrightarrow \qquad \sum_j (\operatorname{osc}_{B(x_j,r_j)} f)^n < \varepsilon\,, \tag{3}$$

where $\operatorname{osc}_B f = \sup\{|f(x) - f(y)| : x, y \in B\}$ for a ball B.

Clearly, any n-absolutely continuous mapping is continuous and n-absolute continuity of a mapping $f = (f_1, \ldots, f_n)$ is equivalent to n-absolute continuity of each real-valued function f_j, $j = 1, \ldots, n$.

Following [32] we recall the n-dimensional counterpart of absolute continuity involving integral upper bounds for oscillation and going back to [41].

Definition 4.2 Let $\Omega \subset \mathbb{R}^n$ be an open set. A function $f : \Omega \to \mathbb{R}$ is *generalized Lipschitz continuous of class* RR^n if there exists a function $\theta \in L^1_{\mathrm{loc}}(\Omega)$ such that for each ball B we have

$$(\operatorname{osc}_B f)^n \le \int_B \theta(x)\, dm(x).$$

The following results provide some relations between AC^n, RR^n, Sobolev spaces and (N)-property; cf. [32].

Theorem 4.1 ([35]) *Let $\Omega \subset \mathbb{R}^n$ be an open set and $f : \Omega \to \mathbb{R}^n$ be an* AC^n *mapping. Then $f \in W^{1,n}_{\mathrm{loc}}(\Omega)$ and f satisfies the Lusin (N)-property.*

Theorem 4.2 ([35]) *Let $\Omega \subset \mathbb{R}^n$ be an open set and $f : \Omega \to \mathbb{R}^n$ belong to $W^{1,p}(\Omega)$ with $p > n$. Then $f \in \mathrm{RR}^n$, and, therefore, $f \in \mathrm{AC}^n$.*

Theorem 4.3 ([13]) *Let $\Omega \subset \mathbb{R}^n$ be an open set and $f : \Omega \to \mathbb{R}^n$ be a generalized Lipschitz continuous mapping of class* RR^n*. Then f is an* AC^n *mapping.*

In [27], the following important generalization of AC^n has been defined. Instead of the implication (3), we have

$$\sum_j mB(x_j, r_j) < \delta \quad \Longrightarrow \quad \sum_j (\operatorname{osc}_{B(x_j,\lambda r_j)} f)^n < \varepsilon, \tag{4}$$

for some λ, $0 < \lambda \le 1$. Obviously, this definition coincides with Definition 4.1 only for $\lambda = 1$. Such kind of absolute continuity can be called (λ, n)-absolute continuity (abbr. AC^n_λ). One of key results of [27] says that $\mathrm{AC}^n_{\lambda_1} = \mathrm{AC}^n_{\lambda_2}$ for arbitrary λ_1, λ_2, provided $0 < \lambda_1, \lambda_2 < 1$.

It is shown in [13] that in Definitions 4.1–4.2 the balls cannot be replaced by cubes. However, the Hencl condition (4) meets both the balls and cubes.

The following statement is quite obvious. Nevertheless, we include its proof in our survey.

Theorem 4.4 *Let $\Omega \subset \mathbb{R}^n$ be an open set and $f : \Omega \to \mathbb{R}^n$ be an L-Lipschitz mapping. Then f is an* AC^n_λ *mapping.*

Proof For arbitrary L-Lipschitz mapping f on Ω, i.e.

$$|f(x) - f(y)| \le L|x - y|, \qquad \forall\, x, y \in \Omega,$$

consider a family of piecewise nonintersecting balls $\{B(x_j, r_j)\}$ such that $\sum_j mB(x_j, r_j) < \delta$. Then

$$\sum_j (\operatorname{osc}_{B(x_j,\lambda r_j)} f)^n = \sum_j \left[\sup_{B(x_j,\lambda r_j)} |f(x) - f(y)| \right]^n \le \sum_j (2\lambda L r_j)^n$$
$$= \frac{(2\lambda L)^n}{\Omega_n} \sum_j mB(x_j, \lambda r_j) < \varepsilon,$$

where $\lambda = 1/L$, and $\delta = \Omega_n \varepsilon / 2^n$. Here Ω_n stands for the volume of the unit ball $\mathbb{B}^n$ in $\mathbb{R}^n$. □

An extension of Hencl's definition of absolute continuity is given in [10]. By this approach, the constant λ is replaced by a function $\lambda : \Omega \to (0, 1]$. Denote by Λ a family of all such functions λ.

Definition 4.3 We say that a mapping $f : \Omega \to \mathbb{R}^m$ is (n, Λ)*-absolutely continuous* (shortly $\mathrm{AC}^n_\Lambda(\Omega)$) if there is $\lambda \in \Lambda$ satisfying the following condition: for each $\varepsilon > 0$ there exists $\delta > 0$ such that for every pairwise disjoint finite collection $\{B(x_j, r_j)\}$ of balls in Ω, we have

$$\sum_j mB(x_j, r_j) < \delta \quad \Longrightarrow \quad \sum_j (\operatorname{osc}_{B(x_j, \lambda(x_j) r_j)} f)^n < \varepsilon .$$

The following important properties characterize the class AC^n_Λ. First, AC^n_λ is a proper class of AC^n_Λ intersected with $W^{1,n}_{\mathrm{loc}}$. Second, $\mathrm{AC}^n_\Lambda \setminus W^{1,n}_{\mathrm{loc}} \neq \varnothing$. Note also that AC^n_Λ is stable under a large class of mappings including the quasiconformal ones; see [10].

We define a class $\mathrm{AC}^{n,p}_\Lambda$ of mappings which belong to AC^n_Λ and whose partial derivatives are integrable with the degree p, $p \geq 1$.

Another approach to the absolute continuity (relies on the classical Vitali's notation in $\mathbb{R}$) can be found in [8, 9]. Here the closed intervals in $\mathbb{R}$ are replaced by the n-dimensional intervals admitting an appropriate regularity condition. The closed interval $[\mathbf{x}, \mathbf{y}]$ of two points $\mathbf{x} = (x_1, \dots, x_n)$ and $\mathbf{y} = (y_1, \dots, y_n)$ in $\mathbb{R}^n$ such that $x_1 < y_1, \dots, x_n < y_n$ is defined by

$$[\mathbf{x}, \mathbf{y}] = \{z = (z_1, \dots, z_n) : x_i \leq z_i \leq y_i,\ i = 1, \dots, n\}.$$

Given an n-dimensional interval $[\mathbf{x}, \mathbf{y}]$ we determine a parameter of regularity by

$$r[\mathbf{x}, \mathbf{y}] = \frac{m[\mathbf{x}, \mathbf{y}]}{\left(\max_j |x_j - y_j|\right)^n} .$$

If $r[\mathbf{x}, \mathbf{y}] \geq \alpha$, for some $\alpha \in (0, 1)$, then the interval is called α*-regular.*

Definition 4.4 Let $0 < \alpha < 1$. A mapping $f : \Omega \to \mathbb{R}^m$ is called α*-absolutely continuous* (briefly $f \in \alpha\text{-}\mathrm{AC}^{(n)}$) if for any $\varepsilon > 0$ there exists $\delta > 0$ such that for each disjoint finite family of α-regular intervals $\{[\mathbf{a}_j, \mathbf{b}_j] \subset \Omega\}$ with $\sum_j m[\mathbf{a}_j, \mathbf{b}_j] < \delta$ one has

$$\sum_j |f(\mathbf{b}_j) - f(\mathbf{a}_j)|^n < \varepsilon . \tag{5}$$

Similarly to Hencl's definition of (λ, n)-absolute continuity one can consider shrunken intervals (by λ, $0 < \lambda < 1$) and insert in the bound (5). We call such a class $\alpha\text{-}\mathrm{AC}^{(n)}_\lambda$. Then the main result of [9] states

$$\mathrm{AC}^n_\lambda(\Omega) = \alpha\text{-}\mathrm{AC}^{(n)}_\lambda, \qquad \forall\, 0 < \alpha, \lambda < 1. \tag{6}$$

Recall that both AC^n_λ and α-$\mathrm{AC}^{(n)}_\lambda$ are independent on λ, and the latter does not depend on α. The relation (6) implies that the class α-$\mathrm{AC}^{(n)}_\lambda(\Omega)$ is a regular subclass of $W^{1,n}_{\mathrm{loc}}(\Omega)$ and it is stable under quasiconformal mappings. On the other hand, α-$\mathrm{AC}^{(n)}_\lambda(\Omega)$ properly lies between the Malý and Hencl classes; see [8].

For one more extension of absolute continuity with respect to real parameter p, $1 \le p < n$, we refer to [7]. It provides a generalization of Definition 4.1 in the following way.

Definition 4.5 Let $\Omega \subset \mathbb{R}^n$ be an open set and $1 \le p \le n$. A mapping $f : \Omega \to \mathbb{R}^m$ is *p-absolutely continuous* (shortly $\mathrm{AC}^p(\Omega)$) if for each $\varepsilon > 0$ there exists $\delta > 0$ such that for every pairwise disjoint finite collection $\{B(x_j, r_j)\}$ of balls in Ω, we have

$$\sum_j mB(x_j, r_j) < \delta \quad \Longrightarrow \quad \sum_j (\mathrm{osc}_{B(x_j,r_j)} f)^p r_j^{n-p} < \varepsilon. \tag{7}$$

The AC^p-absolute continuity is essentially weaker than its AC^n-counterpart. More precisely, the following statement holds.

Theorem 4.5 ([7]) *For $1 \le p < q \le n$ there is a continuous function $f \in \mathrm{AC}^p(B(0,1))$ such that $f \notin W^{1,q}(B(0,1))$.*

Another crucial difference between the AC^p-absolute continuity and the mentioned above classes is that the Lusin (N)-property need not hold.

Theorem 4.6 ([7]) *There is a homeomorphism $f : [0,1]^n \to [0,1]^n$ not satisfying Lusin's (N)-property which belongs to $\mathrm{AC}^p((0,1)^n)$, for $1 \le p < n$.*

For some additional classes of absolutely continuous mappings and their relations to the Malý, Hencl and Bongiorno definitions, we also refer to [17].

4.2 n-Bounded Variation

The bounded variation is closely related to the absolute continuity. We start with the following definition in [35].

Definition 4.6 Given a mapping $f : \Omega \to \mathbb{R}^m$ and an open set $G \subset \Omega$, we define the *n-variation* of f on G by

$$V^n(f, G) = \sup\left\{\sum_j \left(\mathrm{osc}_{B(x_j,r_j)} f\right)^n : \{B(x_j, r_j)\} \text{ is a disjoint family of balls in } G\right\}.$$

We say that f has a bounded *n-variation* in Ω if $V^n(f, \Omega) < \infty$. By $\mathrm{BV}^n(\Omega)$ we denote the class of all mappings with bounded n-variation with the seminorm $||f||_{\mathrm{BV}^n(\Omega)} = (V_n(f, \Omega))^{1/n}$.

The following deep properties characterize the mappings of BV^n class.

Theorem 4.7 ([35]) *If $n \geq 2$, then $\mathrm{BV}^n_{\mathrm{loc}}(\Omega) \subset W^{1,n}_{\mathrm{loc}}(\Omega)$.*

Theorem 4.8 ([35]) *Let $f \in \mathrm{BV}^n_{\mathrm{loc}}(\Omega)$. Then f is differentiable a.e. and $f' \in L^n(\Omega)$.*

A similar extension of the class BV^n by Malý was given in [27]; cf. implication (4).

Definition 4.7 Given a mapping $f : \Omega \to \mathbb{R}^m$ and a measurable set $G \subset \Omega$, we define the (n, λ)*-variation* of f on G by

$$V^n_\lambda(f, G) = \sup\left\{\sum_j \left(\operatorname{osc}_{B(x_j,\lambda r_j)} f\right)^n : \{B(x_j, r_j)\} \text{ is a disjoint family of balls in } G\right\},$$

where $0 < \lambda \leq 1$. We denote by $\mathrm{BV}^n_\lambda(\Omega)$ the class of all mappings f such that $V^n_\lambda(f, \Omega) < \infty$.

Note also that similar to the (n, λ)-absolute continuity the classes $\mathrm{BV}^n_{\lambda_1}(\Omega)$ and $\mathrm{BV}^n_{\lambda_2}(\Omega)$ coincide where $0 < \lambda_1 < \lambda_2 < 1$. The main results in [27] state that $\mathrm{BV}^n_{\lambda,\mathrm{loc}}(\Omega) \subset W^{1,n}_{\mathrm{loc}}(\Omega)$, any of its mapping is differentiable a.e. in Ω with n-integrable partial derivatives.

Shrinking the balls in Definition 4.2 and following [27], we say that a mapping f on Ω satisfies the RR^n_λ-condition if there is an $L^1(\Omega)$-mapping g called a *weight* such that

$$\left(\operatorname{osc}_{B(x,\lambda r)} f\right)^n \leq \int_{B(x,r)} g(y)\, dm(y)$$

for every ball $B(x, r) \subset \Omega$. We also say that a mapping f on Ω satisfies the RR^{*n}_λ-condition if there is a finite (not necessary absolutely continuous) Borel measure μ such that

$$\left(\operatorname{osc}_{B(x,\lambda r)} f\right)^n \leq \mu B(x, r)$$

for every ball $B(x, r) \subset \Omega$.

These conditions go back to [41] where a similar requirement was posed in order to provide a sufficient condition for the area formula and for differentiability a.e. The following results established in [27] show that $\mathrm{RR}^n_\lambda(\Omega) = \mathrm{AC}^n_\lambda(\Omega)$, whereas $\mathrm{RR}^{*n}_\lambda(\Omega) = \mathrm{BV}^n_\lambda(\Omega)$. Theorems 6.1–6.2 in [27] illustrate that the bounded variation by Malý is essentially stronger that the same notion by Hencl. Indeed, any mapping

f from BV^n has a continuous representative $\bar{f}$ such that $\overline{f}(x) = \lim_{y\to x} f(y)$. However, for $0 < \lambda < 1$ and $n = 2$ there is a mapping $f \in \mathrm{BV}_\lambda^2$ of the punctured ball of radius 1/8 which does not have any continuous extension to the origin.

In [7], the class of mappings of p-bounded variation is defined by the same way as of n-bounded variation replacing the sum of oscillations with the power n in (3) by the corresponding sum in (7).

Definition 4.8 We say that f has bounded *p-variation* (abbr., $f \in \mathrm{BV}^p(\Omega)$) if there exist $M > 0$ and $\eta > 0$ such that

$$\sum_j (\mathrm{osc}_{B(x_j,r_j)} f)^p r_j^{n-p} < M$$

for each disjoint system of balls $\{B(x_j, r_j)\}$ in Ω such that $r_j < \eta$.

Theorem 4.9 ([7]) *Let* $f \in \mathrm{AC}^p_{\mathrm{loc}}(\Omega)$, *then* $f \in \mathrm{BV}^p_{\mathrm{loc}}(\Omega)$. *Moreover, every mapping* $f \in \mathrm{AC}^p(\Omega)$, *is locally bounded.*

Theorem 4.10 ([7]) $\mathrm{BV}^q(\Omega) \subset \mathrm{AC}^p(\Omega)$ *for* $1 \le p < q \le n$.

Theorem 4.11 ([7]) *For* $1 < p \le n$, $\mathrm{BV}^p_{\mathrm{loc}}(\Omega) \subset W^{1,p}_{\mathrm{loc}}(\Omega)$. *Under* $p = 1$, $\mathrm{BV}^1_{\mathrm{loc}}(\Omega) \subset \mathrm{BV}(\Omega)$.

Recall that a mapping $f \in L^1(\Omega)$ is of bounded variation, $f \in \mathrm{BV}(\Omega)$, if the coordinate functions of f belong to the space $\mathrm{BV}(\Omega)$. This means that the distributional derivatives of each coordinate function f_j are measures with finite total variations in Ω; see, e.g. [14].

Theorem 4.12 ([7]) *Let* $1 \le p \le n$ *and* $f \in \mathrm{BV}^p(\Omega)$. *Then* f *is differentiable a.e.*

Remark 4.1 The proof of Theorem 4.12 holds for $0 \le p < n$.

The bounded variation BV^n_Λ involves the sum of oscillations like in Definition 4.3 and, in addition, the infimum taken over all functions $\lambda : \Omega \to (0, 1]$; see [10].

5 Modulus of Curve and Surface Families

5.1 Definition of Modulus

For a Borel function $\rho : \mathbb{R}^n \to [0, \infty]$, its *integral over a k-dimensional surface* $\mathcal{S}$ (a continuous mapping $\mathcal{S} : D_{\mathcal{S}} \to \mathbb{R}^n$, $D_{\mathcal{S}}$ is a domain in $\mathbb{R}^k$, $k = 1, \dots, n-1$) is determined by

$$\int_{\mathcal{S}} \rho \, d\mathcal{A} := \int_{\mathbb{R}^n} \rho(y)\, N(\mathcal{S}, y)\, d\mathcal{H}^k y ,$$

where $N(\mathcal{S}, y)$ stands for the multiplicity function of $\mathcal{S}$, namely, the multiplicity of covering the point y by the surface $\mathcal{S}$, $N(\mathcal{S}, y) = \operatorname{card} \mathcal{S}^{-1}(y)$, which is lower semi-continuous and, therefore, is measurable with respect to the Hausdorff measure $\mathcal{H}^k$; see [41, p. 160], cf. [37, p. 177].

A Borel function $\rho : \mathbb{R}^n \to [0, \infty]$ is called *admissible* for the family of k-dimensional surfaces Γ in $\mathbb{R}^n$, $k = 1, 2, \dots, n-1$, abbr. $\rho \in \operatorname{adm} \Gamma$, if

$$\int_S \rho^k \, d\mathcal{A} \geq 1 \qquad \forall \mathcal{S} \in \Gamma .$$

By the *k-dimensional Hausdorff area* of a Borel set B in $\mathbb{R}^n$ (or simply *area* of B in the case $k = n-1$) associated with the surface $\mathcal{S} : \omega \to \mathbb{R}^n$, we mean

$$\mathcal{A}_{\mathcal{S}}(B) = \mathcal{A}^k_{\mathcal{S}}(B) := \int_B N(\mathcal{S}, y) \, d\mathcal{H}^k y ,$$

(cf. [19, Ch. 3.2.1]). The surface $\mathcal{S}$ is called *rectifiable (quadrable)*, if $\mathcal{A}_{\mathcal{S}}(\mathbb{R}^n) < \infty$ (see, e.g. [37, Ch. 9.2]).

The *modulus of family* Γ (conformal modulus) is defined by

$$\mathcal{M}(\Gamma) := \inf_{\rho \in \operatorname{adm} \Gamma} \int_D \rho^n(x) \, dm(x) . \tag{8}$$

Replacing the exponent n in (8) by real p, $p \geq 1$, we arrive at the quantity which is called the p-modulus $\mathcal{M}_p(\Gamma)$ of the family Γ.

We say that a property P holds a.e. $\mathcal{S} \in \Gamma$, if the corresponding modulus of a subfamily Γ_* of $\mathcal{S} \in \Gamma$ for which P is not true vanishes.

5.2 Distortion of Modulus Under Quasiregular Mappings

The significance of moduli of curve/surface families follows mainly from the fact that the conformal modulus remains invariant under conformal mappings. Moreover, various inequalities for moduli form the basis for the geometric part of quasiconformality/quasiregularity. The main inequalities are called the K_O and K_I inequalities. The smallest K in (1) is called the *outer dilation* $K_O(f)$ of f.

Theorem 5.1 ([43]) *Let $f : D \to \mathbb{R}^n$ be a nonconstant quasiregular mapping. Let $G \subset D$ be a Borel set with the multiplicity function $N(f, G) < \infty$, and Γ be a family of curves in G. Then*

$$\mathcal{M}(\Gamma) \leq K_O(f) N(f, G) \mathcal{M}(f(\Gamma)) .$$

If f is quasiregular, then the following inequality

$$J_f(x) \le K\ell(f'(x))^n \tag{9}$$

holds a.e. for some $K \ge 1$ where $\ell(f'(x)) = \inf_{|h|=1} |f'(x)h|$. The smallest $K \ge 1$ in (9) is the *inner dilatation* $K_I(f)$ of f.

Theorem 5.2 ([43]) *Let $f : D \to \mathbb{R}^n$ be a nonconstant quasiregular mapping. Let Γ be a family of curves in D. Then*

$$\mathcal{M}(f(\Gamma)) \le K_I(f)\mathcal{M}(\Gamma).$$

This inequality is called the Poletskiĭ inequality. For extensions of K_I and K_O inequalities to the moduli (p-moduli) of families of surfaces, we refer to [5, 21].

6 Absolute Continuity on Paths and on Surfaces

6.1 Absolute Continuity on Paths

A path γ in $\mathbb{R}^n$ is a continuous mapping $\gamma : I \to \mathbb{R}^n$, where I is an interval in $\mathbb{R}$. Its locus $\gamma(I)$ is denoted by $|\gamma|$. If $\gamma : I \to \mathbb{R}^n$ is a locally rectifiable path then there is the unique increasing length function ℓ_γ onto a length interval $I_\gamma \subset \mathbb{R}$ with a prescribed normalization $\ell_\gamma(t_0) \in I_\gamma$, $t_0 \in I$, such that $\ell_\gamma(t)$ is equal to the length of the subpath $\gamma|_{[t_0,t]}$ of γ if $t > t_0$, and $\ell_\gamma(t)$ equals $-\ell(\gamma|_{[t,t_0]})$ otherwise; see, e.g. [37, Sect 8].

Let $f : |\gamma| \to \mathbb{R}^n$ be a continuous mapping and suppose that the path $\tilde{\gamma} = f \circ \gamma$ is also locally rectifiable. Then there is a unique increasing function $L_{\gamma,f} : I_\gamma \to I_{\tilde{\gamma}}$ such that

$$L_{\gamma,f}(\ell_\gamma(t)) = \ell_{\tilde{\gamma}}(t)$$

for all $t \in I$.

A mapping $f : D \to \mathbb{R}^n$ is *absolutely continuous on paths* (abbr. ACP) if $L_{\gamma,f}$ is absolutely continuous on closed subintervals of I_γ for a.e. path in D. Note that, in particular, absolute continuity on paths obviously implies absolute continuity on lines, so ACP $\subset$ ACL. We also say that f is *absolutely continuous on paths in the inverse direction* (abbr. ACP^{-1}), if $L^{-1}_{\gamma,f}$ is absolutely continuous on closed intervals of $I_{\tilde{\gamma}}$ for a.e. path $\tilde{\gamma}$ in $f(D)$ and for each lifting γ of $\tilde{\gamma}$.

Recall that a path γ is called a lifting of a path $\tilde{\gamma} \in \mathbb{R}^n$ under $f : D \to \mathbb{R}^n$ if $\tilde{\gamma} = f \circ \gamma$.

In [37, Ch. 8], for $f : D \to \mathbb{R}^n$ the (L)-property related to paths which lie in a domain D is defined. This condition requires that the both assertions hold:

(L_1) : For a.e. path γ in D, $\tilde{\gamma} = f \circ \gamma$ is locally rectifiable, and the function $L_{\gamma,f}$ has the Lusin (N)-property with respect to the one dimensional Lebesgue measure.

(L_2) : For a.e. path $\tilde{\gamma}$ in $f(D)$, each lifting γ of $\tilde{\gamma}$ is locally rectifiable, and the function $L_{\gamma,f}$ possesses the Lusin (N^{-1})-property with respect to the one dimensional Lebesgue measure.

Note that the (L_1)-property implies ACP, whereas the (L_2)-property guarantees for mapping to be ACP^{-1}.

6.2 Absolute Continuity on Surfaces

The following extension of the (L)-property has been introduced in [33] (cf. [37, Ch. 10]).

A mapping $f : D \to \mathbb{R}^n$ has the (A_k)-*property* if the following two conditions hold:

$(A_k^{(1)})$: For a.e. k-dimensional surface $\mathcal{S}$ in D, the restriction $f|_{\mathcal{S}}$ has the (N)-property with respect to area.

$(A_k^{(2)})$: For a.e. k-dimensional surface $\mathcal{S}'$ in $D' = f(D)$, the restriction $f|_{\mathcal{S}}$ has the (N^{-1})-property for each lifting $\mathcal{S}$ of $\mathcal{S}'$ with respect to area.

Here a surface S in D is a *lifting* of a surface $\mathcal{S}'$ in $\mathbb{R}^n$ under a mapping $f : D \to \mathbb{R}^n$ if $\mathcal{S}' = f \circ \mathcal{S}$.

6.3 Absolute Continuity of Quasiconformal Mappings

The question whether a quasiconformal mapping restricted to a smooth hypersurface always maps sets of positive Hausdorff measure to sets of positive Hausdorff measure has been posed by Gehring in the 1970s. The answer was recently obtained in [39]. In this paper, a quasiconformal mapping $f : \mathbb{R}^3 \to \mathbb{R}^3$ carries a set of positive $\mathcal{H}^2$-Hausdorff measure onto a set of Hausdorff 2-measure zero. So, this mapping fails to be absolutely continuous in measure in the inverse direction, or in our terms f does not possess the (A_2^2)-property.

It was shown in [22] that open discrete mappings of Sobolev classes $W^{1,p}_{\mathrm{loc}}$, $p > n-1$, with locally integrable inner dilatations admit ACP$_p^{-1}$-property, which means that these mappings are absolutely continuous on almost all preimage paths with respect to p-modulus. In particular, these results extend the well-known Poletskiĭ lemma for quasiregular mappings; see, e.g. [43].

7 Mappings of Finite Metric and Finite Area Distortions

We recall some definitions of mapping classes involving a purely metric approach. All of them are far advanced counterparts of Lipschitz (bi-Lipschitz) mappings.

7.1 Definitions

A fruitful extension of the bi-Lipschitz mappings provides an essentially wider mapping class. We say that $f : D \to D'$ is *finitely Lipschitz* if $L(x, f) < \infty$ for all $x \in D$ and *finitely bi-Lipschitz* if $0 < l(x, f) \leq L(x, f) < \infty$ for all $x \in D$, where

$$L(x, f) := \limsup_{y \to x} \frac{d'(f(x), f(y))}{d(x, y)},$$

and

$$l(x, f) := \liminf_{y \to x} \frac{d'(f(x), f(y))}{d(x, y)}.$$

Note that the boundedness of the above quantities does not bring any extension with respect to the standard bi-Lipschitzness; see [2, Thm 4.5]. So, these classes are equivalent and this result is quantitative.

The significance of finitely Lipschitz mappings is that by the Stepanoff theorem [44] they are differentiable a.e. However, they are far from being Sobolev mappings ($W^{1,1}_{\mathrm{loc}}$).

We say that a mapping $f : D \to D'$ is of *finite metric distortion,* abbr. $f \in$ FMD, if f admits the Lusin (N)-property with respect to the n-dimensional Hausdorff measure and $0 < l(x, f) \leq L(x, f) < \infty$ a.e.; cf. [37].

Although any FMD-mapping possesses the Lusin (N)-property with respect to the n-dimensional Hausdorff measure, for mappings of finite area distortion we add two requirements related to preserving k-dimensional areas; see [33, 37].

Recall that a mapping $f : D \to D'$ is of *finite area distortion in dimension* $k = 1, \ldots, n-1$, abbr. $f \in \mathrm{FAD}_k$, if $f \in$ FMD and has the (A_k)-property.

We also say that a mapping $f : D \to D'$ is of *finite area distortion*, abbr. $f \in$ FAD, if $f \in \mathrm{FAD}_k$ for every $k = 1, \ldots, n-1$.

In connection with the n-absolute continuity we present a recent result given in [2] for Riemannian manifolds.

Theorem 7.1 *Let D and D' be two domains in $\mathbb{R}^n$. Suppose that $f : D \to D'$ is a finitely Lipschitz homeomorphism with $L(x, f) \in L^1_{\mathrm{loc}}(D)$ then $f \in \mathrm{AC}^{n,1}_{\Lambda}(D)$.*

Corollary 7.1 *Any* FAD*-homeomorphism belongs to* BV^n_Λ.

7.2 *Metric Quasiconformality*

The quasiconformality can be defined in a metric space; see, e.g. [26].

Definition 7.1 A homeomorphism $f : X \to Y$ between metric spaces X and Y is said to be *metrically quasiconformal* if it satisfies

$$H_f(x) := \limsup_{r \to 0} H_f(x, r) \leq H < \infty$$

for all $x \in X$ for some H independent of x, where $H_f(x, r) = L(x, r, f)/l(x, r, f)$, $L(x, r, f) = \sup\{d'(f(x), f(y)) : d(x, y) \leq r\}$ and $l(x, r, f) = \inf\{d'(f(x), f(y)) : d(x, y) \geq r\}$.

In the Euclidean setting the limsup in the above definition can be replaced by liminf. In particular, the finiteness of $h_f(x) = \liminf_{r \to 0} H_f(x, r)$ implies that f is a Sobolev mapping (belongs to $W^{1,1}_{\mathrm{loc}}$), and, therefore, f is absolutely continuous on a.e. curves; see, e.g. [6].

8 Other Results and Illustrating Examples

In this section, we provide some other statements concerning mappings of finite metric and finite area distortion, Sobolev classes and several examples of FMD- and FAD-homeomorphisms in $\mathbb{R}^n$ as well.

8.1 *Absolute Continuity in* $W^{1,n-1}$

We start with some results connected to the absolute continuity, bounded variations and Sobolev class $W^{1,p}_{\mathrm{loc}}$ with the borderline exponent $p = n - 1$. This class is of special interest since any homeomorphism of $W^{1,p}_{\mathrm{loc}}$ with $p > n - 1$ is differentiable a.e.; see, e.g. [28]. Note that in the general case, p cannot be reduced without extra restriction. For recent results on differentiability a.e. for the mappings of $W^{1,n-1}_{\mathrm{loc}}$, we refer to [45]; cf. [4].

Theorem 8.1 *Suppose that* D *is a domain in* $\mathbb{R}^n$, $n \geq 2$. *Let* $f \in W^{1,n-1}_{\mathrm{loc}}(D)$ *be a continuous, discrete and open mapping satisfying* $K_O(\cdot, f) \in L^1_{\mathrm{loc}}(D)$. *Then* f *is differentiable a.e. in* D.

The second result in [45] ensuring differentiability a.e. relies on integrability of p-outer dilatation.

Theorem 8.2 *Suppose that D is a domain in $\mathbb{R}^n$, $n \geq 2$. Let $f \in W^{1,n-1}_{\mathrm{loc}}(D)$ be a continuous, discrete and open mapping satisfying $K_{O,q}(\cdot, f) \in L^1_{\mathrm{loc}}(D)$ for some $n-1 < q \leq n$. Then f is differentiable a.e. in D.*

The following three statements can be found in [14]. For mappings of finite distortion we refer again to [28].

Theorem 8.3 *Let Ω be open set and $f \in W^{1,n-1}_{\mathrm{loc}}(\Omega)$ be a homeomorphism. Then $f^{-1} \in \mathrm{BV}(f(\Omega))$.*

Theorem 8.4 *Let Ω be open set and $f \in W^{1,n-1}_{\mathrm{loc}}(\Omega)$ be a homeomorphism of finite distortion. Then $f^{-1} \in W^{1,1}_{\mathrm{loc}}(f(\Omega))$ and f^{-1} is a mapping of finite distortion.*

Theorem 8.5 *Let $f \in W^{1,n-1}_{\mathrm{loc}}((-1,1)^n)$ be a homeomorphism. Then for almost every $y \in (-1,1)$ the mapping $f|_{\{y\}\times(-1,1)^{n-1}}$ satisfies the Lusin (N)-property with respect to the $(n-1)$-dimensional Hausdorff measure.*

The following important result related to the absolute continuity on surfaces can be derived from the last theorem.

Corollary 8.1 ([14]) *Let $f \in W^{1,n-1}(B(x, r_0))$ be a homeomorphism. Then for almost every $r \in (0, r_0)$ the mapping $f : S(x,r) \to \mathbb{R}^n$ satisfies the Lusin (N)-property, i.e.*

$$\mathcal{H}^{n-1} f(A) = 0 \text{ for every } A \subset S(x,r) \text{ such that } \mathcal{H}^{n-1} A = 0\,.$$

Here $S(x,r)$ denotes the Euclidean $(n-1)$-dimensional sphere centered at x with radius r.

One more result on mappings of Sobolev class $W^{1,n-1}_{\mathrm{loc}}$ providing rich regularity properties for their inverses can be found in [29].

Theorem 8.6 *Let $f : D \to D'$ be a $W^{1,n-1}_{\mathrm{loc}}$-homeomorphism of finite inner distortion. If $K_I(\cdot, f) \in L^1_{\mathrm{loc}}(D)$ then $f^{-1} \in W^{1,n}_{\mathrm{loc}}(D')$.*

The following statements given in [38] and [18], respectively, describe some other properties of the inverses for homeomorphisms of finite inner distortion, i.e. for which $K_I(x, f)$ is finite a.e.

Theorem 8.7 *Let $f : D \to D'$ be a homeomorphism of the class $W^{1,n-1}_{\mathrm{loc}}(D)$ with finite inner distortion such that $K_I \in L^1(D)$. Then $\|(f^{-1}(y))'\| \in L^n(D')$ and $\int_{D'} \|(f^{-1}(y))'\|^n \, dm(y) = \int_D K_I(x, f)\, dm(x)$.*

Theorem 8.8 *Let $f : D \to D'$ be a homeomorphism of finite inner distortion and $f \in W^{1,n-1}(D)$. Assume that $u \in W^{1,\infty}_{\mathrm{loc}}(D)$. Then $u \circ f^{-1} \in W^{1,1}_{\mathrm{loc}}(D')$.*

8.2 Examples

Here we provide several examples of finitely bi-Lipschitz mappings and mappings of finite metric and finite area distortion.

Consider a homeomorphic mapping of the form

$$f_1(x) = \frac{1+|x|^\alpha}{|x|}\, x\,, \qquad 0 < \alpha < 1\,.$$

This homeomorphism carries the punctured unit ball $\mathbb{B}^n\backslash\{0\}$ in $\mathbb{R}^n$ onto the spherical ring $A = \{y \in \mathbb{R}^n : 1 < |y| < 2\}$. It is clear that such a mapping cannot be extended continuously to the origin, since the image of $x = 0$ is the unit sphere S^{n-1}. By a straightforward calculation,

$$L(x, f_1) = \frac{1+|x|^\alpha}{|x|}\,, \quad l(x, f_1) = \alpha|x|^{\alpha-1}, \quad J_{f_1}(x) = \frac{\alpha(1+|x|^\alpha)^{n-1}}{|x|^{n-\alpha}}\,,$$

cf. [23] and [37, Prop 6.3]. Obviously, f_1 belongs to FMD, since $L(x, f_1) < \infty$ and $l(x, f_1) > 0$ in $\mathbb{B}^n\backslash\{0\}$. Moreover, f_1 is a diffeomorphism, and, therefore, it possesses the Lusin (N)-property.

The stretching

$$f_2(x) = x(1 - \log|x|)$$

is an automorphism of the punctured unit ball $\mathbb{B}^n\backslash\{0\}$. For this homeomorphism, one gets

$$L(x, f_2) = 1+\log\frac{1}{|x|}\,, \quad l(x, f_2) = \log\frac{1}{|x|}, \quad J_{f_2}(x) = \left(1+\log\frac{1}{|x|}\right)^{n-1} \log\frac{1}{|x|},$$

and f_2 also is an FMD-homeomorphism, which has a removable singularity at the origin.

The FMD-homeomorphism

$$f_3(x) = |x|^{\alpha-1}x\,, \qquad 0 < \alpha < 1\,,$$

also has a removable singularity at the origin and can be extended to a homeomorphism of the whole unit ball $\mathbb{B}^n$. Indeed, for f_3, we have

$$L(x, f_3) = |x|^{\alpha-1}, \quad l(x, f_3) = \alpha|x|^{\alpha-1}, \quad J_{f_3}(x) = \alpha|x|^{n(\alpha-1)}\,,$$

which imply that f_3 is a diffeomorphism with $0 < l(x, f_3) < L(x, f_3) < \infty$ in $\mathbb{B}^n\backslash\{0\}$.

Consider in the punctured unit ball $\mathbb{B}^n \setminus \{0\}$ a mapping defined by

$$f_4(x) = (x_1 \cos\theta - x_2 \sin\theta, x_2 \cos\theta + x_1 \sin\theta, x_3, \ldots, x_n),$$

with $x = (x_1, \ldots, x_n)$ and $\theta = \log(x_1^2 + x_2^2)$. This mapping preserves the volume.
A straightforward calculation yields

$$L(x, f_4) = 1 + \sqrt{2}, \quad l(x, f_4) = \sqrt{2} - 1, \quad J_{f_4}(x) = 1;$$

see [24]. Thus, f_4 is L-bi-Lipschitz with $L = 1 + \sqrt{2}$, and, therefore, it belongs to FMD.

Consider a homeomorphism

$$f_5(x) = \left(x_1, \ldots, x_{n-1}, \frac{x_n^{1-c}}{1-c}\right), \qquad 0 < c < 1,$$

between the unit open cube $(0, 1)^n$ and n-dimensional interval $(0, 1)^{n-1} \times (0, 1/(1-c))$ of $\mathbb{R}^n$.

By a direct calculation, $l(x, f_5) = 1$ and $L(x, f_5) = x_n^{-c}$, and, therefore, f_5 is finitely Lipschitz with locally integrable $L(x, f_5)$ for any c, $0 < c < 1$. However,

$$\int\limits_{(0,1)^n} L^p(x, f_5)\, dm(x) = \int\limits_0^1 x_n^{-cp}\, dx_n = \infty$$

for any $p \geq 1/c$.

8.3 Absolute Continuity of Finitely Lipschitz Mappings

The validity of the Lusin (N)-property with respect to the k-dimensional Hausdorff measure for the classes of mappings mentioned above (bi-Lipschitz, of finite metric and area distortion) gives rise to the hypothesis that they have properties similar to those of the class $AC_\Lambda^{n,p}$. On the Lusin properties for such classes of mappings, see [1] (for Finsler manifolds), [2] (for Riemannian manifolds) and [37] (in the Euclidean setting).

Remark 8.1 The exponent 1 in Theorem 7.1 is sharp (w.r.t. $\mathrm{AC}_\Lambda^{n,1}$) and cannot be replaced by any larger real number.

Theorem 8.9 *There exists a finitely Lipschitz homeomorphism f of locally integrable $L(x, f)$ such that f fails to belong to $\mathrm{AC}_\Lambda^{n,1+\varepsilon}$ for any positive ε.*

Remark 8.2 Moreover, there exist finitely Lipschitz homeomorphisms which do not belong to $AC^{n,1}_\Lambda$.

Following [2], consider an automorphism of the unit ball $\mathbb{B}^n$ in $\mathbb{R}^n$ of the form

$$f_6(x) = \left(x_1 \cos\frac{1}{x_1^2+x_2^2} - x_2 \sin\frac{1}{x_1^2+x_2^2}, x_1 \sin\frac{1}{x_1^2+x_2^2}\right. \\ \left. +x_2 \cos\frac{1}{x_1^2+x_2^2}, x_3, \dots, x_n\right)$$

extended to the origin by continuity, i.e. $f_6(0) = 0$. This mapping provides a rotation in the plane x_1, x_2 with the angle $\theta = 1/(x_1^2+x_2^2)$ which tends to infinity as both x_1, x_2 approach 0. By a straightforward calculation,

$$l(x, f_6) = \sqrt{1+\theta^2} - \theta\,, \qquad L(x, f_6) = \sqrt{1+\theta^2} + \theta\,,$$

whereas the rest of stretchings are equal to 1. Hence, $J_{f_6}(x) = 1$, and f is a diffeomorphism in the punctured ball. Moreover, f_6 satisfies the Lusin (N) and (N^{-1})-properties with respect to each Hausdorff measure in $\mathbb{B}^n$. So, f_6 is an FAD-homeomorphism. However, the partial derivatives of the first two coordinate functions by x_1 and x_2 are not locally integrable in any neighborhood of points $x = (0, 0, x_3, \dots, x_n) \in \mathbb{B}^n$.

Remark 8.3 On the other hand, a finitely Lipschitz homeomorphism (with unbounded $L(x, f)$ from above) can belong to $AC^{n,p}_\Lambda$ for some $p > 1$.

Indeed, as any quasiconformal mapping f_3 belongs to $W^{1,p}$, where $p > n$ and depends only on the coefficient of quasiconformality and on n. Here the coefficient of quasiconformality is equal to $1/\alpha$. Since $L(x, f_3) = |x|^{\alpha-1}$, $l(x, f_3) = \alpha|x|^{\alpha-1}$, f_3 is finite Lipschitz in $\mathbb{B}^n$, and its inverse is $(1/\alpha)$-Lipschitz in the same ball. Note that

$$\int\limits_{\mathbb{B}^n} L^p(x, f_3)\, dm(x) = \omega_{n-1} \int\limits_0^1 r^{(\alpha-1)p+n-1}\, dr < \infty\,,$$

if $p < n/(1-\alpha)$. Here ω_{n-1} denotes the area of $(n-1)$-dimensional unit sphere in $\mathbb{R}^n$.

The above examples can serve as illustrations by investigation of the boundary correspondence under the above classes of mappings; see, e.g. [1, 2, 4].

8.4 Absolute Continuity of Sobolev Mappings

It was recently shown in [3] that homeomorphisms of $W^{1,1}_{\mathrm{loc}}$ with $K_{O,p} \in L^{\frac{n-1}{p-n+1}}_{\mathrm{loc}}$ (a p-counterpart of the outer dilatation locally integrable in the degree $(n-1)/(p-n+1)$) belong to $\mathrm{BV}^{n-\alpha}$, $\alpha = p/(p-n+1)$, $p \in [n, n+1/(n-2))$.

Theorem 8.10 *Let $f : D \to \mathbb{R}^n$ be a homeomorphism of the Sobolev class $W^{1,1}_{\mathrm{loc}}$ and $K_{O,p} \in L^{\frac{n-1}{p-n+1}}(D)$, $p \in [n, n+1/(n-2))$. Then $f \in \mathrm{BV}^{n-\alpha}(D)$, $\alpha = p/(p-n+1)$.*

Due to Remark 4.1, the last statement provides the following regularity property of Sobolev homeomorphisms of integrable p-outer dilatation.

Corollary 8.2 *Let $f : D \to \mathbb{R}^n$ be a homeomorphism of the Sobolev class $W^{1,1}_{\mathrm{loc}}$ and $K_{O,p} \in L^{\frac{n-1}{p-n+1}}(D)$, $p \in [n, n+1/(n-2))$. Then f is differentiable a.e. in D.*

8.5 Open Problem

Open problem. Does ACL follow from AC^n_Λ or BV^n_Λ?

Acknowledgments The second author cordially thanks the organizers Ilia Binder, Damir Kinzebulatov and Javad Mashreghi for an invitation to participate in the Focus Program on Analytic Function Spaces and their Applications (July 1–December 31, 2021) and to deliver a plenary lecture. The idea to extend our research to the notion of absolute continuity in the above frame has been motivated by one of excellent talks given by Javad Mashreghi throughout this distinguished event.

The authors would like to express special thanks to Stanislav Hencl and Matti Vuorinen for their help in collecting materials and careful reading of our paper, and to the referees as well.

References

1. E. Afanas'eva and A. Golberg, *Finitely bi-Lipschitz homeomorphisms between Finsler manifolds*, Anal. Math. Phys. **10** (2020), no. 4, Paper No. 48, 16 pp.
2. E. Afanas'eva and A. Golberg, *Topological mappings of finite area distortion*, Anal. Math. Phys., **12** (2022), no. 2, Paper No. 54, 29 pp.
3. E. Afanas'eva, A. Golberg and R. Salimov, *Distortion theorems for homeomorphic Sobolev mappings of integrable p-dilatations*, Stud. Univ. Babeş-Bolyai Math. **67** (2022), no. 2, 403–420.
4. E. S. Afanas'eva, V. I. Ryazanov and R. R. Salimov, *Toward the theory of Sobolev-class mappings with a critical exponent*. (Russian) Ukr. Mat. Visn. **15** (2018), no. 2, 154–176, 295; translation in J. Math. Sci. (N.Y.) **239** (2019), no. 1, 1–16.
5. C. Andreian Cazacu, *Module inequalities for quasiregular mappings*. Ann. Acad. Sci. Fenn. Ser. A I Math. **2** (1976), 17–28.

6. Z. M. Balogh, P. Koskela and S. Rogovin, *Absolute Continuity of Quasiconformal Mappings on Curves*. Geom. Funct. Anal. **17** (2007), no. 3, 645–664.
7. D. Bongiorno, *A regularity condition in Sobolev spaces* $W^{1,p}_{\mathrm{loc}}(\mathbb{R}^n)$ *with* $1 \leq p < n$. Illinois J. Math. **46** (2002), no. 2, 557–570.
8. D. Bongiorno, *Absolutely continuous functions in* $\mathbb{R}^n$. J. Math. Anal. Appl. **303** (2005), no. 1, 119–134.
9. D. Bongiorno, *On the Hencl's notion of absolute continuity*. J. Math. Anal. Appl. **350** (2009), no. 2, 562–567.
10. D. Bongiorno, *On the problem of regularity in the Sobolev space* $W^{1,n}_{\mathrm{loc}}$. Topology Appl. **156** (2009), no. 18, 2986–2995.
11. L. Cesari, *Sulle trasformazioni continue*. Ann. Math. Pura Appl. **21** (1942), 157–188.
12. K. C. Ciesielski and J. B. Seoane-Sepúlveda, *Differentiability versus continuity: restriction and extension theorems and monstrous examples*. Bull. Amer. Math. Soc. (N.S.) **56** (2019), no. 2, 211–260.
13. M. Csörnyei, *Absolutely continuous functions of Rado, Reichelderfer, and Malý*. J. Math. Anal. Appl. **252** (2000), no. 1, 147–166.
14. M. Csörnyei, S. Hencl and J. Malý, *Homeomorphisms in the Sobolev space* $W^{1,n-1}$. J. Reine Angew. Math. **644** (2010), 221–235.
15. O. Dovgoshey, O. Martio, V. Ryazanov and M. Vuorinen, *The Cantor function*. Expo. Math. **24** (2006), no. 1, 1–37.
16. S. Dubuc, *Non-differentiability and Hölder properties of self-affine functions*. Expo. Math. **36** (2018), no. 2, 119–142.
17. M. Dymond, B. Randrianantoanina and H. Xu, *On interval based generalizations of absolute continuity for functions on* $\mathbb{R}^n$. Real Anal. Exchange **42** (2017), no. 1, 49–78.
18. F. Farroni, R. Giova, G. Moscariello and R. Schiattarella, *Homeomorphisms of finite inner distortion: composition operators on Zygmund-Sobolev and Lorentz-Sobolev spaces*. Math. Scand. **116** (2015), no. 1, 34–52.
19. H. Federer, *Geometric Measure Theory*. Die Grundlehren der mathematischen Wissenschaften, Band 153. Springer-Verlag New York Inc., New York 1969 (Second edition 1996).
20. F. W. Gehring, G. J. Martin and B. P. Palka, *An introduction to the theory of higher-dimensional quasiconformal mappings*. Mathematical Surveys and Monographs, 216. American Mathematical Society, Providence, RI, 2017.
21. A. Golberg, *Generalized classes of quasiconformal homeomorphisms*. Math. Rep. (Bucur.) **7(57)** (2005), no. 4, 289–303.
22. A. Golberg and E. Sevost'yanov, *Absolute continuity on paths of spatial open discrete mappings*. Anal. Math. Phys. **8** (2018), no. 1, 25–35.
23. A. Golberg, R. Salimov and E. Sevost'yanov, *Singularities of discrete open mappings with controlled p-module*, J. Anal. Math. **127** (2015), 303–328.
24. V. Ya. Gutlyanskiĭ and A. Golberg, *On Lipschitz continuity of quasiconformal mappings in space*, J. Anal. Math. **109** (2009), 233–251.
25. C. Heil, *Absolute Continuity and the Banach–Zaretsky Theorem*. In: Hirn M., Li S., Okoudjou K.A., Saliani S., Yilmaz Ö. (eds) Excursions in Harmonic Analysis, Volume 6. Applied and Numerical Harmonic Analysis. Birkhäuser, Cham, 2015, 27–51.
26. J. Heinonen and P. Koskela, *Quasiconformal maps in metric spaces with controlled geometry*. Acta Math. **181** (1998), no. 1, 1–61.
27. S. Hencl, *On the notions of absolute continuity for functions of several variables*. Fund. Math. **173** (2002), no. 2, 175–189.
28. S. Hencl and P. Koskela, *Lectures on mappings of finite distortion*. Lecture Notes in Mathematics, 2096. Springer, Cham, 2014.
29. S. Hencl, G. Moscariello, A. Passarelli di Napoli and C. Sbordone, *Bi-Sobolev mappings and elliptic equations in the plane*. J. Math. Anal. Appl. **355** (2009), no. 1, 22–32.
30. T. Iwaniec and G. Martin, *Geometric Function Theory and Nonlinear Analysis*. Oxford University Press, 2001.

31. M. Jarnicki and P. Pflug, *Continuous nowhere differentiable functions. The monsters of analysis*. Springer Monographs in Mathematics. Springer, Cham, 2015.
32. P. Koskela, J. Malý and T. Zürcher, *Lusin's condition* (N) *and Sobolev mappings*, Rend. Lincei. Mat. Appl. **23** (2012), 455–465.
33. D. Kovtonyuk and V. Ryazanov, *On the theory of mappings with finite area distortion*. J. Anal. Math. **104** (2008), 291–306.
34. G. Leoni, *A first course in Sobolev spaces*. Graduate Studies in Mathematics, 105. American Mathematical Society, Providence, RI, 2009.
35. J. Malý, *Absolutely continuous functions of several variables*, J. Math. Anal. Appl. **231(2)** (1999), 59–61.
36. J. Malý and O. Martio, *Lusin's condition* (N) *and mappings of the class* $W^{1,n}$, J. Reine Angew. Math **458** (1995), 492–508.
37. O. Martio, V. Ryazanov, U. Srebro and E. Yakubov, *Moduli in Modern Mapping Theory*. Springer Monographs in Mathematics. Springer, New York, 2009.
38. G. Moscariello and A. Passarelli di Napoli, *The regularity of the inverses of Sobolev homeomorphisms with finite distortion*. J. Geom. Anal. **24** (2014), no. 1, 571–594.
39. D. Ntalampekos and M. Romney, *On the inverse absolute continuity of quasiconformal mappings on hypersurfaces*. American Journal of Mathematics **143** (2021), 1633–1659.
40. S. P. Ponomarev, *The* N^{-1}*-property of mappings, and Luzin's* (N) *condition*. (Russian) Mat. Zametki **58** (1995), no. 3, 411–418, 480; translation in Math. Notes **58** (1995), no. 3–4, 960–965.
41. T. Rado and P. V. Reichelderfer, *Continuous transformations in analysis*. With an introduction to algebraic topology. Die Grundlehren der mathematischen Wissenschaften in Einzeldarstellungen mit besonderer Berücksichtigung der Anwendungsgebiete, Bd. LXXV. Springer-Verlag, Berlin-Göttingen-Heidelberg, 1955.
42. Yu. G. Reshetnyak, *Space Mappings with Bounded Distortion*, Trans. of Mathematical Monographs, Amer. Math. Soc., vol. 73, 1989.
43. S. Rickman, *Quasiregular mappings*. Springer-Verlag, Berlin, 1993.
44. W. Stepanoff, *Sur les conditions de l'existence de la differentielle totale*, Rec. Math. Soc. Math. Moscou **32** (1925), 511–526.
45. V. Tengvall, *Differentiability in the Sobolev space* $W^{1,n-1}$. Calc. Var. Partial Differential Equations **51** (2014), no. 1–2, 381–399.
46. J. Väisälä, *Lectures on n-dimensional quasiconformal mappings*, Lecture Notes in Mathematics, Vol. 229. Springer-Verlag, Berlin-New York, 1971.
47. M. Vuorinen, *Conformal geometry and quasiregular mappings*, Lecture Notes in Mathematics, 1319. Springer-Verlag, Berlin, 1988.

An Indefinite Analog of Sarason's Generalized Interpolation Theorem

James Rovnyak

1 Introduction and Preliminaries

This paper is an excursion in the theory of $\mathcal{H}(B)$ spaces [6], motivated by a problem in indefinite interpolation. The main application is to spaces that are contained isometrically in the Hardy space H^2 and is based on a theorem by Alpay, Dijksma, and the author [1]. This result appears in Theorem 3.3 and may be viewed as an indefinite analog of Donald Sarason's generalized interpolation theorem [11].

For notational reasons, $\mathcal{H}(B)$ spaces are here denoted $\mathcal{H}(C)$. If C is an analytic function which is defined and bounded by one on the unit disk $\mathbb{D}$, $\mathcal{H}(C)$ is the Hilbert space of analytic functions on $\mathbb{D}$ with reproducing kernel

$$K_C(w, z) = \frac{1 - C(z)\overline{C(w)}}{1 - z\overline{w}}, \qquad w, z \in \mathbb{D}.$$

Equivalently, $\mathcal{H}(C)$ is the space of all functions h in the Hardy space H^2 such that

$$\|h\|_C^2 = \sup\left[\|h + Cg\|_2^2 - \|g\|_2^2\right] < \infty, \tag{1}$$

where the supremum is over all g in H^2 and $\|\cdot\|_2$ is the norm of H^2. We assume basic familiarity with such spaces, as may be found, for example, in [2, 4–6, 8, 13, 17]. Every space $\mathcal{H}(C)$ contains all difference quotients

The author thanks the organizers of the *Fields Focus Program on Analytic Function Spaces and their Applications* for the invitation to participate in the program.

J. Rovnyak (✉)
Department of Mathematics, University of Virginia, Charlottesville, VA, USA
e-mail: rovnyak@virginia.edu

I. Binder et al. (eds.), *Function Spaces, Theory and Applications*, Fields Institute Communications 87, https://doi.org/10.1007/978-3-031-39270-2_2

$$\frac{h(z)-h(w)}{z-w} \quad \text{and} \quad \frac{C(z)-C(w)}{z-w}$$

whenever $h(z)$ is in the space and $w \in \mathbb{D}$. The formula

$$T : h(z) \to zh(z) - C(z)\left\langle h(z), \frac{C(z)-C(0)}{z}\right\rangle_C \tag{2}$$

defines a contraction operator on $\mathcal{H}(C)$ with adjoint

$$T^* : h(z) \to \frac{h(z)-h(0)}{z}. \tag{3}$$

By the difference-quotient inequality (see (8)), the only h in $\mathcal{H}(C)$ such that $\|T^{*n}h\|_C = \|h\|_C$ for all $n \geq 1$ is $h = 0$. Therefore the operators T and T^* are completely nonunitary by [16, Ch. I, Th. 3.2]. This allows us to use the functional calculus of Schreiber [14] and Sz.-Nagy and Foias [15, 16]: for any completely nonunitary contraction operator A on a Hilbert space $\mathcal{H}$ and any H^∞ function $f(z) = \sum_{j=0}^{\infty} f_j z^j$ on the unit disk, a bounded operator $f(A)$ on $\mathcal{H}$ is defined by the formula

$$f(A) = \operatorname*{s-lim}_{r\uparrow 1} \sum_{j=0}^{\infty} f_j r^j A^j. \tag{4}$$

The operator $f(A)$ commutes with A and satisfies

$$\|f(A)\| \leq \|f\|_\infty. \tag{5}$$

The correspondence $f \to f(A)$ is an algebra homomorphism from H^∞ into $\mathcal{L}(\mathcal{H})$; see [16, Ch. III, Th. 2.1].

1.1 The Inner Case and Sarason's Theorem

When C is an inner function, $\mathcal{H}(C)$ is a model space and contained isometrically in H^2 as $H^2 \ominus CH^2$. Let S be the operator multiplication by z on H^2, and let P_C be the self-adjoint projection on H^2 with range $\mathcal{H}(C)$. Then

$$T = P_C S|_{\mathcal{H}(C)}.$$

For any H^∞ function f on $\mathbb{D}$, $f(S)$ is multiplication by $f(z)$ on H^2. One can also show that

$$f(T) = P_C f(S)|_{\mathcal{H}(C)}.$$

The operator $f(T)$ commutes with T. Sarason's generalized interpolation theorem [11, Th. 1] states that, conversely, if R is a bounded operator on $\mathcal{H}(C)$ that commutes with T, there is an $f \in H^\infty$ such that $R = f(T)$, and f can be chosen such that $\|f\|_\infty = \|R\|$. See [7, Th. 14.38] and [9, Th. 10.9] for recent accounts.

Sarason's theorem can be restated in an equivalent form for contraction operators.

Theorem 1.1 *Let C be an inner function. If R is a contraction operator on $\mathcal{H}(C)$ that commutes with T, there is a Schur function f on $\mathbb{D}$ such that $R = f(T)$.*

The **Schur class** is the set of analytic functions that are defined and bounded by one on the unit disk.

Clearly Theorem 1 in [11] implies Theorem 1.1. Conversely, assume Theorem 1.1. Let R be any nonzero bounded operator on $\mathcal{H}(C)$ that commutes with T. Then $R_1 = R/\|R\|$ is a contraction that commutes with T. By Theorem 1.1, $R_1 = g(T)$ for some Schur function g. Then

$$R = \|R\| R_1 = \|R\| g(T).$$

Thus $R = f(T)$, where $f(z) = \|R\| g(z)$ is in H^∞ and $\|f\|_\infty \leq \|R\|$. Since $R = f(T)$, the reverse inequality $\|R\| = \|f(T)\| \leq \|f\|_\infty$ follows from (5). Therefore, $\|R\| = \|f\|_\infty$, which yields Theorem 1 in [11].

1.2 An Indefinite Generalization of Theorem 1.1

Throughout this work, the letter κ denotes a nonnegative integer. If H is a self-adjoint operator on a Hilbert space, we say that H has κ **negative squares** and write $\mathrm{sq}_- H = \kappa$, if the negative spectrum of H consists of a finite number of eigenvalues of total multiplicity κ. The following result is proved by an abstract method that uses a commutant lifting theorem in [3].

Theorem 1.2 ([1, Th. 3.7]) *Let C be an inner function, and let R be a bounded operator on $\mathcal{H}(C)$ that commutes with T and satisfies*

$$\mathrm{sq}_- (1 - RR^*) = \kappa$$

for some nonnegative integer κ. Then there exist a Blaschke product B of degree κ and a Schur function f such that

$$B(T)R = f(T).$$

Conversely, if such f and B exist, $1 - RR^*$ has at most κ negative squares.

The case $\kappa = 0$ in Theorem 1.2 reduces to Theorem 1.1 since a Blaschke product of degree zero is a constant of modulus one.

In this paper we resolve a question left open in Theorem 1.2: what does the operator equation $B(T)R = f(T)$ actually say about R? The results needed to answer this question hold more generally in a broad class of spaces $\mathcal{H}(C)$. These are collected in Sect. 2, which is a study of the root subspaces of the operators T and T^* in any space $\mathcal{H}(C)$ that satisfies the identity for difference quotients.

In Sect. 3, the results on root subspaces are applied to the operator equation $B(T)R = f(T)$, yielding our main results, Theorems 3.2 and 3.3. Briefly, the operator R in Theorem 1.2 is determined on a subspace of $\mathcal{H}(C)$ of codimension at most κ, and on this subspace R is given by an explicit formula that depends on f and B.

Section 4 formulates dual results, Theorems 4.2 and 4.3, which hold for operators R that commute with T^*. The identity $B(T)R = f(T)$ is replaced by $B(T^*)R = f(T^*)$, and the condition $\mathrm{sq}_-(1 - RR^*) = \kappa$ is replaced by $\mathrm{sq}_-(1 - R^*R) = \kappa$. The dual results are derived from an interplay between the spaces $\mathcal{H}(C)$ and $\mathcal{H}(\widetilde{C})$, where $\widetilde{C}(z) = \overline{C(\overline{z})}$, which is described in Appendix 5.

2 The Root Subspaces of T and T^*

For any space $\mathcal{H}(C)$, the operator T in (2) is also written in the form

$$\begin{aligned} T\colon h(z) &\to zh(z) - C(z)\widetilde{h}(0), \\ \text{where } \widetilde{h}(0) &= \left\langle h(z), \frac{C(z)-C(0)}{z} \right\rangle_C . \end{aligned} \tag{6}$$

More generally, for each $w \in \mathbb{D}$, the difference-quotient operator

$$R(w)\colon h(z) \to \frac{h(z)-h(w)}{z-w}$$

is everywhere defined and bounded on $\mathcal{H}(C)$, and

$$\begin{aligned} R(w)^*\colon h(z) &\to \frac{zh(z) - C(z)\widetilde{h}(\overline{w})}{1-\overline{w}z}, \\ \text{where } \widetilde{h}(\overline{w}) &= \left\langle h(z), \frac{C(z)-C(w)}{z-w} \right\rangle_C . \end{aligned} \tag{7}$$

The formula (7) is given in [6, Prob. 85] and [2, p. 89]; we caution that in [2] the roles of T and T^* are reversed from their meaning here.

For any space $\mathcal{H}(C)$, the **difference-quotient inequality**

$$\left\| \frac{h(z) - h(0)}{z} \right\|_C^2 \leq \|h(z)\|_C^2 - |h(0)|^2 \tag{8}$$

holds for every h in $\mathcal{H}(C)$. The case of equality is characterized in [6, Th. 16], [12, p. 158], and [8, Cor. 25.15]:
The ***difference-quotient identity***

$$\left\| \frac{h(z) - h(0)}{z} \right\|_C^2 = \|h(z)\|_C^2 - |h(0)|^2$$

holds for every h in $\mathcal{H}(C)$ if and only if $C \notin \mathcal{H}(C)$, or equivalently, C is an extreme point of the unit ball in H^∞.

The difference-quotient identity has an inner product form by the polarization identity. Less obvious inner product identities are known. For any space $\mathcal{H}(C)$ such that $C \notin \mathcal{H}(C)$, and for all $h \in \mathcal{H}(C)$ and $\alpha, \beta \in \mathbb{D}$,

$$\begin{aligned}\overline{C(\beta)}h(\alpha) = \overline{\beta}\left\langle h(z), \frac{C(z) - C(\beta)}{z - \beta} \right\rangle_C \\ - (1 - \alpha\overline{\beta})\left\langle \frac{h(z) - h(\alpha)}{z - \alpha}, \frac{C(z) - C(\beta)}{z - \beta} \right\rangle_C\end{aligned} \tag{9}$$

and

$$\frac{1 - \overline{C(\beta)}C(\alpha)}{1 - \alpha\overline{\beta}} = \left\langle \frac{C(z) - C(\alpha)}{z - \alpha}, \frac{C(z) - C(\beta)}{z - \beta} \right\rangle_C. \tag{10}$$

See Problems 88 and 89 in [6] and Theorems 3.2.4 and 3.2.5 in [2]. We use these identities in a special case.

Lemma 2.1 *Let $\mathcal{H}(C)$ be a given space such that $C \notin \mathcal{H}(C)$. Then*

$$\left\langle \frac{h(z) - h(w)}{z - w}, \frac{C(z) - C(0)}{z} \right\rangle_C = -\overline{C(0)}h(w), \tag{11}$$

$$\left\langle \frac{C(z) - C(w)}{z - w}, \frac{C(z) - C(0)}{z} \right\rangle_C = 1 - \overline{C(0)}C(w), \tag{12}$$

for all h in $\mathcal{H}(C)$ and w in $\mathbb{D}$.

The spectrum and resolvent of T^* in the extreme point case are described by Sarason [13, pp. 41–42]. The formulas in Lemma 2.1 provide another approach.

Theorem 2.1 *Let $\mathcal{H}(C)$ be a given space such that $C \notin \mathcal{H}(C)$. If $w \in \mathbb{D}$ and $C(w) \neq 0$, then $T - w$ is invertible and*

$$(T - w)^{-1} \colon h(z) \to \frac{h(z) - h(w)C(z)/C(w)}{z - w}. \tag{13}$$

In fact, the operator

$$h(z) \to \frac{h(z) - h(w)C(z)/C(w)}{z - w} \tag{14}$$

is a left inverse of $T - w$ whether or not $C \notin \mathcal{H}(C)$. Two-sided invertibility, however, requires the assumption that $C \notin \mathcal{H}(C)$.

Proof We show that (14) is both a left and right inverse of $T - w$. For any $h(z)$ in $\mathcal{H}(C)$, the function

$$\begin{aligned} k(z) &= \frac{h(z) - h(w)C(z)/C(w)}{z - w} \\ &= \frac{h(z) - h(w)}{z - w} - \frac{h(w)}{C(w)} \frac{C(z) - C(w)}{z - w} \end{aligned}$$

is in $\mathcal{H}(C)$, and by (11) and (12),

$$\begin{aligned} (T - w)k(z) &= (z - w)k(z) - C(z)\left\langle k(z), \frac{C(z) - C(0)}{z} \right\rangle_C \\ &= (z - w)k(z) - C(z)\left\langle \frac{h(z) - h(w)}{z - w}, \frac{C(z) - C(0)}{z} \right\rangle_C \\ &\qquad + C(z)\frac{h(w)}{C(w)}\left\langle \frac{C(z) - C(w)}{z - w}, \frac{C(z) - C(0)}{z} \right\rangle_C \\ &= (z - w)k(z) - C(z)\Big[- \overline{C(0)}h(w)\Big] \\ &\qquad + C(z)\frac{h(w)}{C(w)}\Big[1 - \overline{C(0)}C(w)\Big] \\ &= h(z). \end{aligned}$$

Thus (14) is a right inverse of $T - w$. If $h(z)$ is in $\mathcal{H}(C)$ and

$$\begin{aligned} k(z) &= (T - w)h(z) \\ &= (z - w)h(z) - C(z)\left\langle h(z), \frac{C(z) - C(0)}{z} \right\rangle_C, \end{aligned}$$

then

$$\begin{aligned}\frac{k(z)-k(w)C(z)/C(w)}{z-w}\\ &= \frac{1}{z-w}\Big[(z-w)h(z) - C(z)\Big\langle h(z), \frac{C(z)-C(0)}{z}\Big\rangle_C \\ &\qquad + C(w)\Big\langle h(z), \frac{C(z)-C(0)}{z}\Big\rangle_C \frac{C(z)}{C(w)}\Big] \\ &= h(z).\end{aligned}$$

Therefore, (14) is a left inverse of $T - w$. □

The **geometric multiplicity** of an eigenvalue α of any Hilbert space operator A is the dimension of $\ker(A-\alpha)$. Any vector $f \neq 0$ such that $(A-\alpha)^n f = 0$ for some $n \geq 1$ is called a **root vector**. If the subspace consisting of all root vectors plus the zero vector has finite dimension, it is called the **root subspace**, and its dimension is the **algebraic multiplicity** of α.

In a space $\mathcal{H}(C)$ with C inner, the root subspaces for T and T^* are described by Nikol′skiĭ [10, Lecture IV]. Here we consider spaces $\mathcal{H}(C)$ that satisfy the identity for difference quotients (the extreme point case).

Let $\mathcal{H}(C)$ be a given space such that $C \notin \mathcal{H}(C)$, and suppose $\alpha \in \mathbb{D}$ is a zero of C. Then

1. α is an eigenvalue of T, and $\ker(T-\alpha) = \left[\dfrac{C(z)}{z-\alpha}\right]$.
2. $\overline{\alpha}$ is an eigenvalue of T^*, and $\ker(T^*-\overline{\alpha}) = \left[\dfrac{1}{1-\overline{\alpha}z}\right]$.

Square brackets around any set of vectors denote the linear span of the vectors. See [13, Ch. V], [8, Th. 26.1 and Cor. 26.3] and for the inner case also [9, Th. 9.22 and Cor. 9.25]. We extend these results to arbitrary root subspaces.

Theorem 2.2 *Let $\mathcal{H}(C)$ be a given space such that $C \notin \mathcal{H}(C)$, and let α be a zero of C of order n. Then α is an eigenvalue for T with geometric multiplicity one and algebraic multiplicity n. Moreover:*

(1) *The functions*

$$Q_j(z) = \frac{C(z)}{(z-\alpha)^j}, \qquad j = 1, \dots, n, \tag{15}$$

are root vectors for the operator T and eigenvalue α such that

$$\begin{aligned}(T-\alpha)Q_1 &= 0,\\ (T-\alpha)Q_j &= Q_{j-1}, \quad j = 2, \dots, n.\end{aligned} \tag{16}$$

(2) *The subspaces* $\mathcal{R}_k = \ker(T-\alpha)^k$, $k \geq 1$, *are given by*

$$\mathcal{R}_k = \begin{cases} [Q_1, \ldots, Q_k], & k = 1, \ldots, n, \\ \mathcal{R}_n, & k > n. \end{cases}$$

Thus the root subspace for T *and eigenvalue* α *is* $\mathcal{R}_n$.

Proof By (2), for all $u \in \mathcal{H}(C)$,

$$(T-\alpha)u(z) = (z-\alpha)u(z) - C(z)\left\langle u(z), \frac{C(z)-C(0)}{z}\right\rangle_C. \tag{17}$$

(1) Since C has a zero of order n at α, $C(\alpha) = C'(\alpha) = \cdots = C^{(n-1)}(\alpha) = 0$ and hence $Q_1, \cdots, Q_n$ are given recursively by

$$\begin{aligned} Q_1(z) &= \frac{C(z)-C(\alpha)}{z-\alpha}, \\ Q_j(z) &= R(\alpha)Q_{j-1}(\alpha), \qquad j = 2, \ldots, n. \end{aligned} \tag{18}$$

Therefore $Q_1, \ldots, Q_n$ belong to $\mathcal{H}(C)$. By (6) and (12),

$$\begin{aligned} (T-\alpha)Q_1(z) &= (z-\alpha)Q_1(z) - C(z)\left\langle \frac{C(z)-C(\alpha)}{z-\alpha}, \frac{C(z)-C(0)}{z}\right\rangle_C \\ &= C(z) - C(z)\big[1 - \overline{C(0)}C(\alpha)\big] \\ &= 0, \end{aligned}$$

which is the first relation in (16). For $j = 2, \ldots, n$, by (11),

$$\begin{aligned} (T-\alpha)Q_j(z) &= (z-\alpha)Q_j(z) \\ &\qquad - C(z)\left\langle \frac{Q_{j-1}(z)-Q_{j-1}(\alpha)}{z-\alpha}, \frac{C(z)-C(0)}{z}\right\rangle_C \\ &= \frac{C(z)}{(z-\alpha)^{j-1}} + C(z)\overline{C(0)}Q_{j-1}(\alpha) \\ &= Q_{j-1}(z) \end{aligned}$$

because $Q_{j-1}(\alpha) = 0$, yielding the second relation in (16). By (16), $(T-\alpha)^k Q_k = 0$ for each $k = 1, \ldots, n$. Thus $Q_1, \ldots, Q_n$ are root vectors for T, and (1) follows.

(2) Set $\mathcal{R}_k = \ker(T-\alpha)^k$, $k \geq 1$. For all $k = 1, \ldots, n$, $\mathcal{R}_k \supseteq [Q_1, \ldots, Q_k]$ by part (1). Equality holds when $k = 1$, because if $u(z)$ is in $\ker(T-\alpha)$, then

$(z-\alpha)u(z) - C(z)\widetilde{u}(0) = 0$ by (6), and hence $u(z)$ is a constant multiple of $Q_1(z) = C(z)/(z-\alpha)$. We show that equality also holds for each $k = 2, \dots, n$.

Assume that $\mathcal{R}_k = [Q_1, \dots, Q_k]$ for some $k = 1, \dots, n-1$. Suppose $u \in \mathcal{R}_{k+1}$. Then

$$(T-\alpha)^k (T-\alpha)u = 0, \tag{19}$$

and therefore $(T-\alpha)u \in \mathcal{R}_k = [Q_1, \dots, Q_k]$, say $(T-\alpha)u = \sum_{j=1}^k \gamma_j Q_j$, where $\gamma_1, \dots, \gamma_k$ are constants. Then

$$(z-\alpha)u(z) - C(z)\widetilde{u}(0) = \sum_{j=1}^k \gamma_j \frac{C(z)}{(z-\alpha)^j},$$

and hence

$$u(z) = \frac{C(z)}{z-\alpha}\widetilde{u}(0) + \sum_{j=1}^k \gamma_j \frac{C(z)}{(z-\alpha)^{j+1}}. \tag{20}$$

Thus $u \in [Q_1, \dots, Q_k, Q_{k+1}]$, and so $\mathcal{R}_{k+1} \subseteq [Q_1, \dots, Q_k, Q_{k+1}]$. The reverse inclusion holds since $k+1 \le n$, and hence equality holds. Therefore by induction, $\mathcal{R}_k = [Q_1, \dots, Q_k]$ for all $k = 1, \dots, n$.

We show that $\mathcal{R}_k = \mathcal{R}_n$ for all $k > n$. First let $k = n+1$. The inclusion $\mathcal{R}_{n+1} \supseteq \mathcal{R}_n$ is trivial. Let $u \in \mathcal{R}_{n+1}$, and repeat steps (19) through (20) with $k = n$ to obtain

$$u(z) = \frac{C(z)}{z-\alpha}\widetilde{u}(0) + \sum_{j=1}^n \gamma_j \frac{C(z)}{(z-\alpha)^{j+1}}.$$

Then $\gamma_n = 0$ because u is analytic at $z = \alpha$ and $C^{(n)}(\alpha) \neq 0$ by our assumption that C has a zero at α of order n. Therefore u is in $[Q_1, \dots, Q_n] = \mathcal{R}_n$, and hence $\mathcal{R}_{n+1} = \mathcal{R}_n$.

We proceed by induction. Assume that $\mathcal{R}_k = \mathcal{R}_n$ for some $k > n$. The inclusion $\mathcal{R}_{k+1} \supseteq \mathcal{R}_n$ is again trivial. Let $u \in \mathcal{R}_{k+1}$. Then

$$(T-\alpha)^k (T-\alpha)u = 0,$$

and so $(T-\alpha)u \in \mathcal{R}_k$. By our inductive assumption, this implies that $(T-\alpha)u \in \mathcal{R}_n$ and hence $u \in \mathcal{R}_{n+1}$. We showed above that $\mathcal{R}_{n+1} = \mathcal{R}_n$, and therefore $u \in \mathcal{R}_n$. Thus $\mathcal{R}_{k+1} \subseteq \mathcal{R}_n$ and hence $\mathcal{R}_{k+1} = \mathcal{R}_n$. This completes the inductive step. Therefore $\mathcal{R}_k = \mathcal{R}_n$ for all $k > n$. Part (2) of the theorem follows.

By (16), α is an eigenvalue for T. By part (2) its geometric multiplicity is one, and its algebraic multiplicity is n.

□

Corollary 2.1 *Let $\mathcal{H}(C)$ be a given space such that $C \notin \mathcal{H}(C)$, and let $Q_1, \ldots, Q_n$ be the functions* (15) *corresponding to some zero α of C of order n. Suppose $f \in H^\infty$ and*

$$f(T)Q_k = 0$$

for some k, $1 \le k \le n$. Then f has a zero at α of order at least k.

Proof Define $g \in H^\infty$ by

$$f(z) = f(\alpha) + f'(\alpha)(z-\alpha) + \cdots + \frac{f^{(k-1)}(\alpha)}{(k-1)!}(z-\alpha)^{k-1} + (z-\alpha)^k g(z).$$

Then

$$f(T) = f(\alpha) + f'(\alpha)(T-\alpha) + \cdots + \frac{f^{(k-1)}(\alpha)}{(k-1)!}(T-\alpha)^{k-1} + (T-\alpha)^k g(T).$$

By assumption, $f(T)Q_k = 0$. Hence by Theorem 2.2,

$$\begin{aligned} 0 &= f(\alpha)Q_k + f'(\alpha)Q_{k-1} + \cdots + \frac{f^{(k-1)}(\alpha)}{(k-1)!}Q_1 + (T-\alpha)^k g(T)Q_k \\ &= f(\alpha)Q_k + f'(\alpha)Q_{k-1} + \cdots + \frac{f^{(k-1)}(\alpha)}{(k-1)!}Q_1, \end{aligned}$$

because $(T-\alpha)^k g(T)Q_k = g(T)(T-\alpha)^k Q_k = 0$. The functions $Q_1, \ldots, Q_k$ are linearly independent, and therefore

$$f(\alpha) = f'(\alpha) = \cdots = f^{(k-1)}(\alpha) = 0.$$

Thus f has a zero at α of order at least k. □

Theorem 2.2 has a companion for T^*.

Theorem 2.3 *Let $\mathcal{H}(C)$ be a given space such that $C \notin \mathcal{H}(C)$, and let α be a zero of C of order n. Then $\overline{\alpha}$ is an eigenvalue for T^* with geometric multiplicity one and algebraic multiplicity n. Moreover:*

(1) *The functions*

$$P_j(z) = \frac{z^{j-1}}{(1-\overline{\alpha}z)^j}, \qquad j = 1, \ldots, n, \tag{21}$$

are root vectors for the operator T^* *and eigenvalue* $\overline{\alpha}$ *such that*

$$\begin{aligned}(T^* - \overline{\alpha})P_1 &= 0,\\ (T^* - \overline{\alpha})P_j &= P_{j-1}, \quad j = 2, \dots, n.\end{aligned} \tag{22}$$

(2) *The subspaces* $\widetilde{\mathcal{R}}_k = \ker(T^* - \overline{\alpha})^k$, $k \geq 1$, *are given by*

$$\widetilde{\mathcal{R}}_k = \begin{cases} [P_1, \dots, P_k], & k = 1, \dots, n, \\ \widetilde{\mathcal{R}}_n, & k > n. \end{cases}$$

Thus the root subspace for T^* *and eigenvalue* $\overline{\alpha}$ *is* $\widetilde{\mathcal{R}}_n$.

Theorem 2.3 can be deduced from Theorem 2.2 by means of the unitary equivalence in Theorem 5.1 of the Appendix 5. Instead we give a direct proof, which also has some points of interest.

Lemma 2.2 *Let* $B(z) = (z - \alpha)/(1 - \overline{\alpha}z)$ *for some* $\alpha \in \mathbb{D}$. *Then for any positive integer* k, $\dim \mathcal{H}(B^k) = k$ *and*

$$\mathcal{H}(B^k) = [P_1, \dots, P_k], \tag{23}$$

where $P_1, \dots, P_k$ *are the functions* (21). *Moreover*

$$\mathcal{H}(B^k) = \left\{ \frac{p(z)}{(1 - \overline{\alpha}z)^k} : p \text{ is a polynomial of degree } \leq k - 1 \right\}. \tag{24}$$

Lemma 2.3 *Let* $\mathcal{H}(C)$ *be a given space such that* $C \notin \mathcal{H}(C)$, *and let* α *be a zero of* C *of order* n. *Then the functions* (21) *belong to* $\mathcal{H}(C)$, *but* $z^n/(1 - \overline{\alpha}z)^{n+1}$ *is not in* $\mathcal{H}(C)$.

***Proof of Lemma* 2.2** As a preliminary, first note that the rational functions $P_1, P_2, P_3, \dots$ are linearly independent over the complex plane and hence over $\mathbb{D}$. This is clear if $\alpha = 0$. For $\alpha \neq 0$, linear independence follows from the fact that the P_js have poles of different orders at $1/\overline{\alpha}$.

We prove the first assertion by induction. When $k = 1$,

$$\frac{1 - B(z)\overline{B(w)}}{1 - z\overline{w}} = \frac{1 - |\alpha|^2}{(1 - \overline{\alpha}z)(1 - \alpha\overline{w})}.$$

Hence $\mathcal{H}(B)$ is one-dimensional and satisfies (23). Suppose $\mathcal{H}(B^k)$ has dimension k and (23) holds for some $k \geq 1$. Since $B^{k+1} = B^k B$, by [6, Problems 48, 52],

$$\begin{aligned}\mathcal{H}(B^{k+1}) &= \mathcal{H}(B^k) + B^k \mathcal{H}(B) \\ &= [P_1, \dots, P_k] + B^k [P_1]\end{aligned} \tag{25}$$

as linear spaces. Therefore $\dim \mathcal{H}(B^{k+1}) \leq k+1$ and the function $(z-\alpha)^k/(1-\overline{\alpha}z)^{k+1}$ belongs to $\mathcal{H}(B^{k+1})$. Since $\mathcal{H}(B^{k+1})$ is invariant under $R(\alpha)$, it contains the $k+1$ functions

$$\frac{1}{(1-\overline{\alpha}z)^{k+1}}, \frac{z-\alpha}{(1-\overline{\alpha}z)^{k+1}}, \dots, \frac{(z-\alpha)^k}{(1-\overline{\alpha}z)^{k+1}}.$$

The linear span of these functions includes P_{k+1}, and hence

$$\mathcal{H}(B^{k+1}) \supseteq [P_1, \dots, P_k, P_{k+1}].$$

Since $P_1, \dots, P_k, P_{k+1}$ are linearly independent, $\dim \mathcal{H}(B^{k+1}) = k+1$ and $\mathcal{H}(B^{k+1}) = [P_1, \dots, P_k, P_{k+1}]$. The first assertion of the lemma follows by induction.

Let $\mathcal{M}$ denote the right side of (24). Clearly $\dim \mathcal{M} = k$ and $\mathcal{M}$ contains $P_1, \dots, P_k$. Hence $\mathcal{H}(B^k) \subseteq \mathcal{M}$ by (23). Since $\mathcal{H}(B^k)$ and $\mathcal{M}$ both have dimension k, (24) follows. □

***Proof of Lemma* 2.3** Set $B(z) = (z-\alpha)/(1-\overline{\alpha}z)$. Since C has a zero of order n at α,

$$C(z) = B(z)^n C_1(z),$$

where a space $\mathcal{H}(C_1)$ exists and $C_1(\alpha) \neq 0$. By [6, Problems 48, 52],

$$\mathcal{H}(C) = \mathcal{H}(B^n) + B^n \mathcal{H}(C_1) \tag{26}$$

as linear spaces. Therefore $P_1, \dots, P_n$ belong to $\mathcal{H}(C)$ by Lemma 2.2.

To see that $z^n/(1-\overline{\alpha}z)^{n+1} \notin \mathcal{H}(C)$, we argue by contradiction. Assume that $z^n/(1-\overline{\alpha}z)^{n+1} \in \mathcal{H}(C)$. Then $z^j/(1-\overline{\alpha}z)^{n+1} \in \mathcal{H}(C)$ for all $j = 0, \dots, n$ because $\mathcal{H}(C)$ is invariant under the formation of difference quotients at the origin. Hence

$$\frac{(z-\alpha)^n}{(1-\overline{\alpha}z)^{n+1}} \in \mathcal{H}(C).$$

By (26),

$$\frac{(z-\alpha)^n}{(1-\overline{\alpha}z)^{n+1}} = h(z) + B(z)^n k(z). \tag{27}$$

for some $h(z)$ in $\mathcal{H}(B^n)$ and $k(z) \in \mathcal{H}(C_1)$. By Lemma 2.2,

$$h(z) = \frac{p(z)}{(1-\overline{\alpha}z)^n},$$

where $p(z)$ is a polynomial of degree at most $n-1$. If $p(z)$ is not identically zero, it has a zero at α of order at least n, which is impossible because $\deg p \le n-1$. Therefore $p(z) \equiv 0$, so $h(z) \equiv 0$ and

$$k(z) = \frac{1}{1-\overline{\alpha} z} \in \mathcal{H}(C_1).$$

Let T_1 and T_1^* be the operators (2) and (3) on $\mathcal{H}(C_1)$. Then $k(z)$ is an eigenfunction for T_1^* and eigenvalue $\overline{\alpha}$. Hence $T_1 - \alpha$ is not invertible. By Theorem 2.1, this implies $C_1(\alpha) = 0$, a contradiction. Therefore $z^n/(1-\overline{\alpha} z)^{n+1} \notin \mathcal{H}(C)$. □

Proof of Theorem 2.3 For every $u \in \mathcal{H}(C)$,

$$(T^* - \overline{\alpha})u(z) = \frac{(1-\overline{\alpha} z)u(z) - u(0)}{z}, \tag{28}$$

by (3).

(1) The functions $P_1, \ldots, P_n$ belong to $\mathcal{H}(C)$ by Lemma 2.3, and (22) follows by routine algebra using (28). The relations (22) imply that $(T^* - \overline{\alpha})^k P_k = 0$ for all $k = 1, \ldots, n$, and so $P_1, \ldots, P_n$ are root vectors for T^*.

(2a) Set $\widetilde{\mathcal{R}}_k = \ker(T^* - \overline{\alpha})^k$, $k = 1, 2, 3, \ldots$. We show that $\widetilde{\mathcal{R}}_k = [P_1, \ldots, P_k]$ for $k = 1, \ldots, n$. By part (1),

$$\widetilde{\mathcal{R}}_k \supseteq [P_1, \ldots, P_k], \qquad k = 1, \ldots, n, \tag{29}$$

Equality holds in (29) when $k = 1$. For if $u \in \widetilde{\mathcal{R}}$, then $(T^* - \overline{\alpha})u = 0$, so by (28), $u(z) = u(0)/(1-\overline{\alpha} z) = u(0)P_1(z)$. We show that equality also holds for $k = 2, \ldots, n$.

Suppose $\widetilde{\mathcal{R}}_k = [P_1, \ldots, P_k]$ for some $k = 1, \ldots, n-1$. By (29) with k replaced by $k+1$, $\widetilde{\mathcal{R}}_{k+1} \supseteq [P_1, \ldots, P_k, P_{k+1}]$. For any $u \in \mathcal{R}_{k+1}$,

$$(T^* - \overline{\alpha})^k (T^* - \overline{\alpha})u = 0.$$

Therefore $(T^* - \overline{\alpha})u \in \widetilde{\mathcal{R}}_k = [P_1, \ldots, P_k]$, say $(T^* - \overline{\alpha})u = \sum_{j=1}^k \gamma_j P_j$, where $\gamma_1, \ldots, \gamma_k$ are scalars. Then by (28),

$$\frac{(1-\overline{\alpha} z)u(z) - u(0)}{z} = \sum_{j=1}^k \gamma_j P_j(z),$$

and hence

$$u(z) = \frac{u(0)}{1-\overline{\alpha} z} + \sum_{j=1}^k \gamma_j \frac{zP_j(z)}{1-\overline{\alpha} z}$$

$$= u(0)P_1(z) + \sum_{j=1}^{k} \gamma_j P_{j+1}(z).$$

Therefore $u \in [P_1, \ldots, P_k, P_{k+1}]$, and so $\widetilde{\mathcal{R}}_{k+1} = [P_1, \ldots, P_k, P_{k+1}]$. Hence by induction, $\widetilde{\mathcal{R}}_k = [P_1, \ldots, P_k]$ for all $k = 1, \ldots, n$.

(2b) We show by induction that $\widetilde{\mathcal{R}}_{n+k} = \widetilde{\mathcal{R}}_n$ for all $k \geq 1$. For the case $k = 1$, the inclusion $\widetilde{\mathcal{R}}_{n+1} \supseteq \widetilde{\mathcal{R}}_n$ is obvious. Let $u \in \widetilde{\mathcal{R}}_{n+1}$. Then

$$(T^* - \overline{\alpha})^n (T^* - \overline{\alpha})u = 0.$$

Hence $(T^* - \overline{\alpha})u \in \widetilde{\mathcal{R}}_n$, and so $(T^* - \overline{\alpha})u = \sum_{j=1}^{n} \gamma_j P_j$, for some constants $\gamma_1, \ldots, \gamma_n$ by part (1). By (28),

$$\frac{(1 - \overline{\alpha}z)u(z) - u(0)}{z} = \sum_{j=1}^{n} \gamma_j \frac{z^{j-1}}{(1 - \overline{\alpha}z)^j},$$

hence

$$\begin{aligned} u(z) &= \frac{u(0)}{1 - \overline{\alpha}z} + \sum_{j=1}^{n} \gamma_j \frac{z^j}{(1 - \overline{\alpha}z)^{j+1}} \\ &= u(0)P_1(z) + \sum_{j=1}^{n-1} \gamma_j P_{j+1}(z) + \gamma_n \frac{z^n}{(1 - \overline{\alpha}z)^{n+1}}. \end{aligned}$$

Here $\gamma_n = 0$ because otherwise $z^n/(1 - \overline{\alpha}z)^{n+1} \in \mathcal{H}(C)$, contradicting Lemma 2.3. Therefore u is in the span of $P_1, \ldots, P_n$, that is, $u \in \widetilde{\mathcal{R}}_n$, and thus $\widetilde{\mathcal{R}}_{n+1} = \widetilde{\mathcal{R}}_n$.

For the inductive step, assume that $\widetilde{\mathcal{R}}_{n+k} = \widetilde{\mathcal{R}}_n$ for some $k \geq 1$. Trivially, $\widetilde{\mathcal{R}}_{n+k+1} \supseteq \widetilde{\mathcal{R}}_n$. Let $u \in \widetilde{\mathcal{R}}_{n+k+1}$. Then

$$(T^* - \overline{\alpha})^{n+k} (T^* - \overline{\alpha})u = 0,$$

so $(T^* - \overline{\alpha})u \in \widetilde{\mathcal{R}}_{n+k}$. By our inductive assumption, $(T^* - \overline{\alpha})u \in \widetilde{\mathcal{R}}_n$. Then $(T^* - \overline{\alpha})^{n+1}u = (T^* - \overline{\alpha})^n (T^* - \overline{\alpha})u = 0$, and hence $u \in \widetilde{\mathcal{R}}_{n+1}$. Since we already showed that $\widetilde{\mathcal{R}}_{n+1} = \widetilde{\mathcal{R}}_n$, $u \in \widetilde{\mathcal{R}}_n$. Thus $\widetilde{\mathcal{R}}_{n+k} = \widetilde{\mathcal{R}}_n$. Therefore by induction, $\widetilde{\mathcal{R}}_{n+k} = \widetilde{\mathcal{R}}_n$ for all $k \geq 1$.

By (22), $\overline{\alpha}$ is an eigenvalue for T^*. By part (2) its geometric multiplicity is one, and its algebraic multiplicity is n. □

Theorem 2.4 *Let $\mathcal{H}(C)$ be a given space such that $C \notin \mathcal{H}(C)$, and let α be a zero of C of order n. The functions $P_1, \ldots, P_n$ in Theorem 2.3 are also given by*

$$P_k(z) = R(\alpha)^{*k-1} K_C(\alpha, z), \qquad k = 1, \ldots, n. \tag{30}$$

Hence for each $k = 1, \ldots, n$,

$$\begin{aligned}&\left[\ker(T^* - \overline{\alpha})^k\right]^{\perp} \\ &\qquad = \left\{h \in \mathcal{H}(C)\colon h(\alpha) = h'(\alpha) = \cdots = h^{k-1}(\alpha) = 0\right\}.\end{aligned} \tag{31}$$

Proof Since $C(\alpha) = 0$, $K_C(\alpha, z) = 1/(1 - \overline{\alpha} z)$, and so (30) holds for $k = 1$. Assume (30) holds for some $k = 1, \ldots, n - 1$. Then by (7),

$$\begin{aligned}R(\alpha)^{*k} K_C(\alpha, z) &= R(\alpha)^* P_k(z) \\ &= \frac{z}{1 - \overline{\alpha} z} P_k(z) - \frac{C(z)}{1 - \overline{\alpha} z} \widetilde{P}_k(\overline{\alpha}).\end{aligned}$$

By our inductive assumption and (18),

$$\begin{aligned}\widetilde{P}_k(\overline{\alpha}) &= \left\langle P_k(z), \frac{C(z) - C(\alpha)}{z - \alpha} \right\rangle_C \\ &= \left\langle R(\alpha)^{*k-1} K_C(\alpha, z), \frac{C(z) - C(\alpha)}{z - \alpha} \right\rangle_C \\ &= \left\langle K_C(\alpha, z), R(\alpha)^{k-1} Q_1(z) \right\rangle_C \\ &= \langle K_C(\alpha, z), Q_k(z) \rangle_C \\ &= \overline{Q_k(\alpha)} \\ &= 0.\end{aligned}$$

The last equality holds since $C(z)$ has a zero of order n at α and $Q_k(z) = C(z)/(z - \alpha)^k$ where $k \le n - 1$. Thus

$$R(\alpha)^{*k} K_C(\alpha, z) = \frac{z}{1 - \overline{\alpha} z} P_k(z) = P_{k+1}(z).$$

The identity (30) follows by induction.

To prove (31), fix some $k = 1, \ldots, n$. By Theorem 2.3, $\ker(T^* - \overline{\alpha})^k$ is spanned by $P_1, \ldots, P_k$. Therefore by (30) a function h in $\mathcal{H}(C)$ is orthogonal to $\ker(T^* - \overline{\alpha})^k$ if and only if

$$\left\langle K_C(\alpha, z), R(\alpha)^{k-1} h(z) \right\rangle_C = 0, \qquad j = 1, \ldots, n,$$

that is, $h(\alpha) = h'(\alpha) = \cdots = h^{k-1}(\alpha) = 0$. □

Theorem 2.5 *Let $\mathcal{H}(C)$ be a given space such that $C \notin \mathcal{H}(C)$. Let $\alpha_1, \ldots, \alpha_r$ be distinct zeros of C. For any positive integers $m_1, \ldots, m_r$,*

$$\ker\Big[(T^* - \overline{\alpha}_1)^{m_1} \cdots (T^* - \overline{\alpha}_r)^{m_r}\Big] = \sum_{j=1}^{r} \ker(T^* - \overline{\alpha}_j)^{m_j}. \tag{32}$$

Lemma 2.4 *For any space $\mathcal{H}(C)$ and $\alpha \in \mathbb{D}$, if $u \in \mathcal{H}(C)$ and*

$$(T^* - \overline{\alpha})u(z) = v(z),$$

then

$$u(z) = \frac{u(0)}{1 - \overline{\alpha}z} + \frac{z}{1 - \overline{\alpha}z}\, v(z).$$

Lemma 2.5 *Let $\alpha, \beta \in \mathbb{D}$, $\alpha \neq \beta$. Then for all $m \geq 1$,*

$$\begin{aligned} &\frac{z}{1 - \overline{\alpha}z}\left[\frac{1}{1 - \overline{\beta}z}, \frac{z}{(1 - \overline{\beta}z)^2}, \ldots, \frac{z^{m-1}}{(1 - \overline{\beta}z)^m}\right] \\ &\quad \subseteq \left[\frac{1}{1 - \overline{\alpha}z}\right] + \left[\frac{1}{1 - \overline{\beta}z}, \frac{z}{(1 - \overline{\beta}z)^2}, \ldots, \frac{z^{m-1}}{(1 - \overline{\beta}z)^m}\right]. \end{aligned} \tag{33}$$

***Proof of Lemma* 2.4** Solve $[u(z) - u(0)]/z - \overline{\alpha}u(z) = v(z)$ for $u(z)$. □

***Proof of Lemma* 2.5** When $m = 1$, this follows from the identity

$$\frac{z}{1 - \overline{\alpha}z}\, \frac{1}{1 - \overline{\beta}z} = \frac{1}{\overline{\alpha} - \overline{\beta}}\left[\frac{1}{1 - \overline{\alpha}z} - \frac{1}{1 - \overline{\beta}z}\right].$$

Assume (33) is known for some $m \geq 1$. Then to show that it holds with m replaced by $m + 1$, it is sufficient to show that

$$\frac{z}{1 - \overline{\alpha}z}\, \frac{z^m}{(1 - \overline{\beta}z)^{m+1}} \in \left[\frac{1}{1 - \overline{\alpha}z}\right] + \left[\frac{1}{1 - \overline{\beta}z}, \ldots, \frac{z^{m-1}}{(1 - \overline{\beta}z)^m}, \frac{z^m}{(1 - \overline{\beta}z)^{m+1}}\right].$$

In fact,

$$\begin{aligned} \frac{z}{1 - \overline{\alpha}z}\, \frac{z^m}{(1 - \overline{\beta}z)^{m+1}} &= \frac{z}{(1 - \overline{\alpha}z)(1 - \overline{\beta}z)}\, \frac{z^m}{(1 - \overline{\beta}z)^m} \\ &= \frac{1}{\overline{\alpha} - \overline{\beta}}\left[\frac{1}{1 - \overline{\alpha}z} - \frac{1}{1 - \overline{\beta}z}\right] \frac{z^m}{(1 - \overline{\beta}z)^m} \end{aligned}$$

$$= \frac{1}{\overline{\alpha}-\overline{\beta}} \frac{z}{1-\overline{\alpha}z} \frac{z^{m-1}}{(1-\overline{\beta}z)^m} - \frac{1}{\overline{\alpha}-\overline{\beta}} \frac{z^m}{(1-\overline{\beta}z)^{m+1}}.$$

Both terms on the right side belong to the required set. □

***Proof of Theorem* 2.5** In the proof we denote by $n_1, \ldots, n_r$ the orders of $\alpha_1, \ldots, \alpha_r$ as zeros of C. The numbers $m_1, \ldots, m_r$ are arbitrary positive integers. Set

$$k_j = \min(m_j, n_j), \qquad j = 1, \ldots, r.$$

By Theorem 2.3(2), for all $j = 1, \ldots, r$,

$$\begin{aligned} \ker(T^* - \overline{\alpha}_j)^{m_j} &= \ker(T^* - \overline{\alpha}_j)^{k_j} \\ &= \left[\frac{1}{1-\overline{\alpha}_j z}, \frac{z}{(1-\overline{\alpha}_j z)^2}, \ldots, \frac{z^{k_j-1}}{(1-\overline{\alpha}_j z)^{k_j}}\right]. \end{aligned} \tag{34}$$

We prove (32) by induction on the number $M = m_1 + \cdots + m_r$ of factors in the product. The equality is obvious when $M = 1$. Assume (32) holds for some $\alpha_1, \ldots, \alpha_r$ and $m_1, \ldots, m_r$. We show that (32) also holds if one new factor is added to the product.

Case 1. *We add a new zero α_{r+1} of C to the list $\alpha_1, \ldots, \alpha_r$.*

We must show that

$$\begin{aligned} &\ker\left[(T^* - \overline{\alpha}_1)^{m_1} \cdots (T^* - \overline{\alpha}_r)^{m_r} (T^* - \overline{\alpha}_{r+1})\right] \\ &\qquad = \sum_{j=1}^{r} \ker(T^* - \overline{\alpha}_j)^{m_j} + \ker(T^* - \overline{\alpha}_{r+1}). \end{aligned} \tag{35}$$

The inclusion $\supseteq$ is clear. Let

$$u \in \ker\left[(T^* - \overline{\alpha}_1)^{m_1} \cdots (T^* - \overline{\alpha}_r)^{m_r} (T^* - \overline{\alpha}_{r+1})\right].$$

Then

$$(T^* - \overline{\alpha}_{r+1})u \in \ker\left[(T^* - \overline{\alpha}_1)^{m_1} \cdots (T^* - \overline{\alpha}_r)^{m_r}\right].$$

By our inductive assumption,

$$(T^* - \overline{\alpha}_{r+1})u = \sum_{j=1}^{r} v_j(z),$$

where $v_j \in \ker(T^* - \overline{\alpha}_j)^{m_j}$, $j = 1, \dots, r$. By (34),

$$v_j(z) \in \left[\frac{1}{1-\overline{\alpha}_j z}, \frac{z}{(1-\overline{\alpha}_j z)^2}, \dots, \frac{z^{k_j-1}}{(1-\overline{\alpha}_j z)^{k_j}}\right], \qquad j = 1, \dots, r.$$

By Lemma 2.4,

$$u(z) = \frac{u(0)}{1-\overline{\alpha}_{r+1} z} + \sum_{j=1}^{r} \frac{z}{1-\overline{\alpha}_{r+1} z} v_j(z).$$

By Lemma 2.5, for all $j = 1, \dots, r$,

$$\frac{z}{1-\overline{\alpha}_{r+1} z} v_j(z) \in \left[\frac{1}{1-\overline{\alpha}_{r+1} z}\right] + \left[\frac{1}{1-\overline{\alpha}_j z}, \frac{z}{(1-\overline{\alpha}_j z)^2}, \dots, \frac{z^{k_j-1}}{(1-\overline{\alpha}_j z)^{k_j}}\right].$$

Therefore

$$u(z) \in \left[\frac{1}{1-\overline{\alpha}_{r+1} z}\right] + \sum_{j=1}^{r} \left[\frac{1}{1-\overline{\alpha}_j z}, \frac{z}{(1-\overline{\alpha}_j z)^2}, \dots, \frac{z^{k_j-1}}{(1-\overline{\alpha}_j z)^{k_j}}\right].$$

Hence

$$u \in \ker(T^* - \overline{\alpha}_{r+1}) + \sum_{j=1}^{r} \ker(T^* - \overline{\alpha}_j)^{m_j}$$

by (34), and this proves (35).

Case 2A. *For some $\ell \in \{1, \dots, r\}$ such that $m_\ell < n_\ell$, we increase m_ℓ by one.*

Without loss of generality, we can suppose that $\ell = r$. Thus $m_r < n_r$, and we must show that

$$\begin{aligned} \ker\Big[(T^* - \overline{\alpha}_1)^{m_1} \cdots (T^* - \overline{\alpha}_{r-1})^{m_{r-1}} (T^* - \overline{\alpha}_r)^{m_r+1}\Big] \\ = \sum_{j=1}^{r-1} \ker(T^* - \overline{\alpha}_j)^{m_j} + \ker(T^* - \overline{\alpha}_r)^{m_r+1}. \end{aligned} \tag{36}$$

The inclusion $\supseteq$ is obvious. Suppose

$$u \in \ker\Big[(T^* - \overline{\alpha}_1)^{m_1} \cdots (T^* - \overline{\alpha}_{r-1})^{m_{r-1}} (T^* - \overline{\alpha}_r)^{m_r+1}\Big].$$

Then

$$(T^* - \overline{\alpha}_r)u \in \left[(T^* - \overline{\alpha}_1)^{m_1} \cdots (T^* - \overline{\alpha}_r)^{m_r}\right].$$

By our inductive assumption, then

$$(T^* - \overline{\alpha}_r)u = \sum_{j=1}^{r} v_j(z),$$

where by (34), for each $j = 1, \ldots, r$,

$$v_j(z) \in \ker(T^* - \overline{\alpha}_j)^{m_j} = \left[\frac{1}{1 - \overline{\alpha}_j z}, \frac{z}{(1 - \overline{\alpha}_j z)^2}, \ldots, \frac{z^{k_j - 1}}{(1 - \overline{\alpha}_j z)^{k_j}}\right].$$

Here $k_r = m_r$ because $m_r < n_r$ by assumption. Also $m_r + 1 \leq n_r$, and therefore

$$\ker(T^* - \overline{\alpha}_r)^{m_r+1} = \left[\frac{1}{1 - \overline{\alpha}_j z}, \frac{z}{(1 - \overline{\alpha}_j z)^2}, \ldots, \frac{z^{m_r}}{(1 - \overline{\alpha}_j z)^{m_r+1}}\right].$$

By Lemma 2.4,

$$\begin{aligned} u(z) &= \frac{u(0)}{1 - \overline{\alpha}_r z} + \sum_{j=1}^{r} \frac{z}{1 - \overline{\alpha}_r z} v_j(z) \\ &= \frac{u(0)}{1 - \overline{\alpha}_r z} + \sum_{j=1}^{r-1} \frac{z}{1 - \overline{\alpha}_r z} v_j(z) + \frac{z}{1 - \overline{\alpha}_r z} v_r(z). \end{aligned}$$

The first and third terms on the right are in $\ker(T^* - \overline{\alpha}_r)^{m_r+1}$. Since $\alpha_1, \ldots, \alpha_r$ are distinct, by Lemma 2.5 the second term belongs to

$$\begin{aligned} &\left[\frac{1}{1 - \overline{\alpha}_r z}\right] + \sum_{j=1}^{r-1} \left[\frac{1}{1 - \overline{\alpha}_j z}, \frac{z}{(1 - \overline{\alpha}_j z)^2}, \ldots, \frac{z^{k_j - 1}}{(1 - \overline{\alpha}_j z)^{k_j}}\right] \\ &\qquad = \ker(T^* - \overline{\alpha}_r) + \sum_{j=1}^{r-1} \ker(T^* - \overline{\alpha}_j)^{m_j}. \end{aligned}$$

Therefore $u(z) \in \sum_{j=1}^{r-1} \ker(T^* - \overline{\alpha}_j)^{m_j} + \ker(T^* - \overline{\alpha}_r)^{m_r+1}$, which proves (36).

Case 2B. *For some $\ell \in \{1, \ldots, r\}$ such that $m_\ell \geq n_\ell$, we increase m_ℓ by one.*

Suppose first that $\ell = 1$, so $m_1 \geq n_1$. By Theorem 2.3,

$$\ker(T^* - \overline{\alpha}_1)^{m_1+1} = \ker(T^* - \overline{\alpha}_1)^{m_1}.$$

Therefore both the left side and the right side of (32) are unchanged if we replace m_1 by m_1+1. For the general case, by commutivity we can move any factor $(T^*-\overline{\alpha}_\ell)^{m_\ell}$ to the left in the product, and the same argument applies.

We have shown in all cases that (32) holds for $M+1$ factors if it holds for M factors. The theorem follows by induction. □

3 The Operator Equation $B(T)R = f(T)$

We now consider the operator equation $B(T)R = f(T)$ in an arbitrary space $\mathcal{H}(C)$ that satisfies the identity for difference quotients, assuming that R commutes with T. In Theorem 3.2 we solve the operator equation for R, given any Blaschke product B of degree κ and any Schur function f. When specialized to the inner case and combined with Theorem 1.2, this result implies Theorem 3.3, which may be viewed as an indefinite analog of Sarason's generalized interpolation theorem.

Theorem 3.1 *Let R be a bounded operator on a space $\mathcal{H}(C)$ such that $C \notin \mathcal{H}(C)$. Assume that R commutes with T and satisfies $B(T)R = f(T)$, where f is a Schur function and B is a Blaschke product of degree κ. If B and C have no common zero, then $B(T)$ is invertible, and*

$$R = B(T)^{-1} f(T).$$

Proof The problem is to show that $B(T)$ is invertible. Since no zero of B is a zero of C, $B(T)$ is a finite product of factors $(1-\overline{\gamma}T)^{-1}(T-\gamma)$ such that $C(\gamma) \neq 0$. Each such factor is invertible by Theorem 2.1, and hence $B(T)$ is invertible. □

For the general case, let {Zeros of B} $\cap$ {Zeros of C} $= \{\alpha_1, \ldots, \alpha_r\}$, where $\alpha_1, \ldots, \alpha_r$ are distinct points of $\mathbb{D}$. Set

$$m_1, \ldots, m_r = \text{orders of } \alpha_1, \ldots, \alpha_r \text{ as zeros of } B,$$

$$n_1, \ldots, n_r = \text{orders of } \alpha_1, \ldots, \alpha_r \text{ as zeros of } C.$$

Factor $B(z)$ in the form

$$B(z) = B_1(z)B_0(z)B_2(z), \tag{37}$$

where

$$B_1(z) = \prod_{B(\gamma)=0,\, C(\gamma)\neq 0} \frac{z-\gamma}{1-\overline{\gamma}z},$$

$$B_0(z) = \prod_{j=1}^{r} \left(\frac{z - \alpha_j}{1 - \overline{\alpha}_j z} \right)^{k_j}, \qquad k_j = \min(m_j, n_j),$$

$$B_2(z) = \prod_{m_j > n_j} \left(\frac{z - \alpha_j}{1 - \overline{\alpha}_j z} \right)^{m_j - n_j}.$$

An empty product may be viewed as 1 or simply not present.

Theorem 3.2 *Let R be a bounded operator on a space $\mathcal{H}(C)$ such that $C \notin \mathcal{H}(C)$. Assume that R commutes with T and satisfies*

$$B(T)R = f(T), \tag{38}$$

where f is a Schur function and B is a Blaschke product of degree κ. Let B be factored as in (37)*. Then the subspace $\mathcal{K}$ of all h in $\mathcal{H}(C)$ such that*

$$h(\alpha_j) = h'(\alpha_j) = \cdots = h^{(k_j - 1)}(\alpha_j) = 0, \qquad j = 1, \ldots, r, \tag{39}$$

has codimension at most κ in $\mathcal{H}(C)$, and $\mathcal{K}$ is invariant under T and R. Moreover, $g = f/B_0$ is a Schur function, and

$$R|_{\mathcal{K}} = B_1\big(T|_{\mathcal{K}}\big)^{-1} g\big(T|_{\mathcal{K}}\big) B_2\big(T|_{\mathcal{K}}\big)^{-1}. \tag{40}$$

The invertibility of $B_1\big(T|_{\mathcal{K}}\big)$ and $B_2\big(T|_{\mathcal{K}}\big)$ is shown in the proof.

Lemma 3.1 *Let $A \in \mathcal{L}(\mathcal{H})$ be an invertible operator on a Hilbert space $\mathcal{H}$. Let $\mathcal{K}$ be a closed invariant subspace of A, and set $A_{\mathcal{K}} = A|_{\mathcal{K}}$. Then $A_{\mathcal{K}}$ is invertible in $\mathcal{L}(\mathcal{K})$ if and only if $\mathcal{K}$ is invariant under A^{-1}. In this case, $A_{\mathcal{K}}^{-1} = A^{-1}|_{\mathcal{K}}$.*

***Proof of Lemma* 3.1** Assume $A^{-1}\mathcal{K} \subseteq \mathcal{K}$. For any $v \in \mathcal{K}$, $A^{-1}v \in \mathcal{K}$ and

$$A_{\mathcal{K}}(A^{-1}v) = A(A^{-1}v) = v. \tag{41}$$

So $A_{\mathcal{K}}$ maps $\mathcal{K}$ onto itself. Since A is one-to-one, so is $A_{\mathcal{K}} = A|_{\mathcal{K}}$. Therefore $A_{\mathcal{K}}$ is invertible in $\mathcal{L}(\mathcal{K})$. By (41),

$$A_{\mathcal{K}}^{-1} v = A^{-1} v, \qquad v \in \mathcal{K},$$

hence $A_{\mathcal{K}}^{-1} = A^{-1}|_{\mathcal{K}}$. Conversely, let $A_{\mathcal{K}}$ be invertible in $\mathcal{L}(\mathcal{K})$. Given $v \in \mathcal{K}$, let $u = A_{\mathcal{K}}^{-1} v$. Then $u \in \mathcal{K}$ and $Au = A_{\mathcal{K}} A_{\mathcal{K}}^{-1} v = v$. Hence $A^{-1} v = u$ belongs to $\mathcal{K}$. Thus $A^{-1}\mathcal{K} \subseteq \mathcal{K}$. □

***Proof of Theorem* 3.2** Since the functional calculus is an algebra homomorphism, $B(T) = B_1(T)B_0(T)B_2(T)$ and

$$B_1(T) = \prod_{C(\gamma)\neq 0} (1-\overline{\gamma}T)^{-1}(T-\gamma), \tag{42}$$

$$B_0(T) = \prod_{j=1}^{r} \Big[(1-\overline{\alpha}_j T)^{-1}(T-\alpha_j)\Big]^{k_j}, \tag{43}$$

$$B_2(T) = \prod_{m_j>n_j} \Big[(1-\overline{\alpha}_j T)^{-1}(T-\alpha_j)\Big]^{m_j-n_j}. \tag{44}$$

All factors in (42), (43), and (44) commute with R and with each other. Each factor $(1-\overline{\gamma}T)^{-1}(T-\gamma)$ in $B_1(T)$ is invertible by Theorem 2.1, and hence $B_1(T)$ is invertible. Therefore by (38),

$$RB_2(T)B_0(T) = B_1(T)^{-1}f(T). \tag{45}$$

Claim 1: *$f(T) = g(T)B_0(T)$ where g is a Schur function.*

By (45), the kernel of $f(T)$ contains the kernel of $B_0(T)$. For each $j = 1, \dots, r$, the kernel of $B_0(T)$ contains

$$\ker(T-\alpha_j)^{k_j} = [Q_1, \dots, Q_{k_j}]$$

by (43) and Theorem 2.2(2). Thus $B_0(T)Q_{k_j} = 0$, and hence

$$f(T)Q_{k_j} = 0.$$

By Corollary 2.1, f has a zero at α_j of order at least k_j. Since this holds for all $j = 1, \dots, r$, by standard function theory $f(z)/B_0(z)$ is the restriction of a Schur function. Claim 1 follows.

Define $\mathcal{K}$ as the closure of the range of $B_0(T)$. By (45) and Claim 1,

$$RB_2(T)B_0(T) = B_1(T)^{-1}g(T)B_0(T),$$

and hence

$$RB_2(T)|_{\mathcal{K}} = B_1(T)^{-1}g(T)\big|\mathcal{K}. \tag{46}$$

Clearly $\mathcal{K}$ is invariant under T. We show that $\mathcal{K}$ has codimension at most κ and is characterized by (39). By (43) and Theorem 2.5,

$$\mathcal{K} = \Big[\ker B_0(T)^*\Big]^{\perp}$$

$$= \Big[\ker\big[(T^*-\overline{\alpha}_1)^{k_1}\cdots(T^*-\overline{\alpha}_r)^{k_r}\big]\Big]^{\perp}$$

$$= \Big[\sum_{j=1}^{r} \ker(T^* - \overline{\alpha}_j)^{k_j}\Big]^{\perp}.$$

Therefore by Theorem 2.4, $\mathcal{K}$ is the set of all h in $\mathcal{H}(C)$ such that

$$h(\alpha_j) = h'(\alpha_j) = \cdots = h^{(k_j-1)}(\alpha_j) = 0$$

for all $j = 1, \ldots, r$. It also follows $\mathcal{K}^{\perp}$ is spanned by the sum of the kernels $\ker(T^* - \overline{\alpha}_j)^{k_j}$, $j = 1, \ldots, r$. Hence by Theorem 2.2, $\mathcal{K}^{\perp}$ is spanned by $k_1 + \cdots + k_r$ linearly independent functions, and so

$$\operatorname{codim} \mathcal{K} = k_1 + \cdots + k_r \le \kappa.$$

Claim 2: *$B_2(T)|_{\mathcal{K}} = B_2\big(T|_{\mathcal{K}}\big)$ is an invertible operator on $\mathcal{K}$.*

It is enough to prove this when $B_2(z)$ consists of a single linear fractional factor. For definiteness, let this factor correspond to α_1, so

$$B_2(T) = (1 - \overline{\alpha}_1 T)^{-1}(T - \alpha_1). \tag{47}$$

By (44), $m_1 > n_1$. Therefore $k_1 = \min(m_1, n_1) = n_1$, and hence

$$C(\alpha_1) = C'(\alpha_1) = \cdots = C^{(k_1-1)}(\alpha_1) = 0 \quad \text{and} \quad C^{k_1}(\alpha_1) \neq 0. \tag{48}$$

We treat the two factors in (47) separately.

1°) *$(T - \alpha_1)|_{\mathcal{K}} = T|_{\mathcal{K}} - \alpha_1$ is an invertible operator on $\mathcal{K}$.*

We show that $(T - \alpha_1)|_{\mathcal{K}}$ is a one-to-one mapping of $\mathcal{K}$ onto itself. If $(T - \alpha_1)u = 0$ for some u in $\mathcal{K}$, then by (6),

$$u(z) = \frac{C(z)}{z - \alpha_1}\widetilde{u}(0). \tag{49}$$

Since $u \in \mathcal{K}$, $u^{(k_1-1)}(\alpha_1) = 0$. Then by (49), $C^{(k_1)}(\alpha_1)\widetilde{u}(0) = 0$. Hence by (48), $\widetilde{u}(0) = 0$ and so $u = 0$. Thus $(T - \alpha_1)|_{\mathcal{K}}$ is one-to-one.

To see that $(T - \alpha_1)\mathcal{K} = \mathcal{K}$, consider an arbitrary $v \in \mathcal{K}$. We seek a solution $u \in \mathcal{K}$ of

$$(T - \alpha_1)u = v. \tag{50}$$

We first find a solution u_0 of (50) in $\mathcal{H}(C)$. Since $v \in \mathcal{K}$,

$$v(\alpha_j) = v'(\alpha_j) = \cdots = v^{(k_j-1)}(\alpha_j) = 0, \qquad j = 1, \ldots, r. \tag{51}$$

In particular, $v(\alpha_1) = 0$. Define $u_0 \in \mathcal{H}(C)$ by

$$u_0(z) = \frac{v(z) - v(\alpha_1)}{z - \alpha_1} = \frac{v(z)}{z - \alpha_1}.$$

By (6),

$$(T - \alpha_1)u_0 = (z - \alpha_1)\,\frac{v(z)}{z - \alpha_1} - C(z)\widetilde{u}_0(0) = v(z) - C(z)\widetilde{u}_0(0),$$

where by (11),

$$\widetilde{u}_0(0) = \left\langle \frac{v(z) - v(\alpha_1)}{z - \alpha_1}, \frac{C(z) - C(0)}{z} \right\rangle_C = -\overline{C(0)}\, v(\alpha_1) = 0.$$

Thus $(T - \alpha_1)u_0 = v$ and u_0 is a solution of (50) in $\mathcal{H}(C)$. To find a solution u of (50) in $\mathcal{K}$, we modify u_0 by writing

$$u(z) = \frac{v(z)}{z - \alpha_1} + \gamma\,\frac{C(z)}{z - \alpha_1}, \tag{52}$$

where γ is a constant to be determined. No matter how γ is chosen, u satisfies (50) because $C(z)/(z - \alpha_1)$ is in the kernel of $T - \alpha_1$ by Theorem 2.2. For arbitrary γ and $2 \le j \le r$, the function (52) satisfies

$$u(\alpha_j) = u'(\alpha_j) = \cdots = u^{(k_j - 1)}(\alpha_j) = 0, \tag{53}$$

because both $v(z)$ and $C(z)$ have a zero at α_j of order at least k_j and $\alpha_j \neq \alpha_1$. It remains to choose γ so that (53) holds for $j = 1$. We use the Taylor expansions of $v(z)$ and $C(z)$ at α_1. By (51) and (48),

$$u(z) = \left[\frac{v^{(k_1)}(\alpha_1)}{k!} + \gamma\,\frac{C^{(k_1)}(\alpha_1)}{k!}\right](z - \alpha_1)^{k_1 - 1} + \text{higher powers}.$$

Since $C^{(k_1)}(\alpha_1) \neq 0$ by (48), we can choose γ such that

$$\frac{v^{(k_1)}(\alpha_1)}{k!} + \gamma\,\frac{C^{(k_1)}(\alpha_1)}{k!} = 0.$$

Then u satisfies (53) for $j = 1$. Hence u is a solution of (50) belonging to $\mathcal{K}$, and so $(T - \alpha_1)\mathcal{K} = \mathcal{K}$. We have shown that $(T - \alpha_1)|_{\mathcal{K}}$ maps $\mathcal{K}$ one-to-one onto itself, and hence it is an invertible operator in $\mathcal{L}(\mathcal{K})$. The relation $(T - \alpha_1)|_{\mathcal{K}} = T|_{\mathcal{K}} - \alpha_1$ is clear, and so 1°) follows.

2°) $(1 - \overline{\alpha}_1 T)^{-1}|_{\mathcal{K}} = (1 - \overline{\alpha}_1 T|_{\mathcal{K}})^{-1}$ *is an invertible operator on* $\mathcal{K}$.

For any $u \in \mathcal{K}$,

$$(1-\overline{\alpha}_1 T)^{-1}u = \sum_{j=0}^{\infty} \overline{\alpha}_1^j T^j u = \sum_{j=0}^{\infty} \overline{\alpha}_1^j T|_{\mathcal{K}}^j u = (1-\overline{\alpha}_1 T|_{\mathcal{K}})^{-1}u,$$

so $(1-\overline{\alpha}_1 T)^{-1}|_{\mathcal{K}} = (1-\overline{\alpha}_1 T|_{\mathcal{K}})^{-1}$. Obviously $(1-\overline{\alpha}_1 T|_{\mathcal{K}})^{-1}$ is an invertible operator in $\mathcal{L}(\mathcal{K})$. This proves 2°), and Claim 2 follows.

It was shown above that $B_1(T)$ is an invertible operator on $\mathcal{H}(C)$.

Claim 3: $B_1\big(T|_{\mathcal{K}}\big)$ is invertible and $B_1\big(T|_{\mathcal{K}}\big)^{-1} = B_1(T)^{-1}|_{\mathcal{K}}$.

It is enough to prove this when $B_1(T)$ consists of just one linear fractional factor, say

$$B_1(T) = (1-\overline{\gamma}T)^{-1}(T-\gamma),$$

where $\gamma \in \mathbb{D}$ and $C(\gamma) \neq 0$. We shall apply Lemma 3.1 to the operator $B_1(T)$ on $\mathcal{H}(C)$ and invariant subspace $\mathcal{K}$. We show that

$$B_1(T)^{-1}\mathcal{K} \subseteq \mathcal{K}. \tag{54}$$

This reduces to showing that $(T-\gamma)^{-1}\mathcal{K} \subseteq \mathcal{K}$. Suppose $u \in \mathcal{K}$. Since $C(\gamma) \neq 0$, by Theorem 2.1,

$$(T-\gamma)^{-1}u(z) = \frac{u(z) - u(\gamma)C(z)/C(\gamma)}{z-\gamma}.$$

For any α_j, $j = 1, \ldots, r$, the first k_j Taylor coefficients of $u(z)$ at α_j are zero by the characterization (39) of $\mathcal{K}$, which was proved above. The first k_j Taylor coefficients of $C(z)$ at α_j are zero because $C(z)$ has a zero at α_j of order at least k_j by assumption. Since $1/(z-\gamma)$ is analytic at α_j, the same is true for $(T-\gamma)^{-1}u(z)$, and hence $(T-\gamma)^{-1}u(z)$ belongs to $\mathcal{K}$. Thus (54) follows. By Lemma 3.1, $B_1(T)|_{\mathcal{K}}$ is invertible and

$$\big(B_1(T)|_{\mathcal{K}}\big)^{-1} = B_1(T)^{-1}|_{\mathcal{K}}.$$

Claim 3 follows because $B_1(T)|_{\mathcal{K}} = B_1\big(T|_{\mathcal{K}}\big)$ by the functional calculus.

We can now complete the proof. In (46) we showed that

$$RB_2(T)|_{\mathcal{K}} = B_1(T)^{-1}g(T)\big|\mathcal{K}$$

as operators from $\mathcal{K}$ into $\mathcal{H}(C)$. Here since $\mathcal{K}$ is invariant under T, by the functional calculus (4),

$$B_2(T)|_{\mathcal{K}} = B_2(T|_{\mathcal{K}}) \quad \text{and} \quad g(T)|_{\mathcal{K}} = g(T|_{\mathcal{K}}).$$

Thus

$$RB_2(T|_{\mathcal{K}}) = B_1(T)^{-1} g(T|_{\mathcal{K}}).$$

By Claim 2, $B_2(T|_{\mathcal{K}})$ is invertible in $\mathcal{L}(\mathcal{K})$, and so

$$R|_{\mathcal{K}} = B_1(T)^{-1} g(T|_{\mathcal{K}}) B_2(T|_{\mathcal{K}})^{-1}.$$

By Claim 3, $B_1(T)^{-1}|_{\mathcal{K}} = B_1(T|_{\mathcal{K}})^{-1}$. Therefore $R\mathcal{K} \subseteq \mathcal{K}$ and

$$R|_{\mathcal{K}} = B_1(T|_{\mathcal{K}})^{-1} g(T|_{\mathcal{K}}) B_2(T|_{\mathcal{K}})^{-1},$$

which is (40).

By the functional calculus, $B_1(T|_{\mathcal{K}}), g(T|_{\mathcal{K}}), B_2(T|_{\mathcal{K}})$ commute with $T|_{\mathcal{K}}$ and with each other. By Claims 2 and 3, $B_1(T|_{\mathcal{K}})$ and $B_2(T|_{\mathcal{K}})$ are invertible. Hence $B_1(T|_{\mathcal{K}})^{-1}, g(T|_{\mathcal{K}}), B_2(T|_{\mathcal{K}})^{-1}$ commute with $T|_{\mathcal{K}}$ and with each other. □

We can now state the form of Theorem 1.2 promised in Sect. 1.

Theorem 3.3 *Let C be an inner function, and let R be a bounded operator on $\mathcal{H}(C)$ that commutes with T and satisfies*

$$\text{sq}_-(1 - RR^*) = \kappa$$

for some nonnegative integer κ. Then there exist a Schur function f and a Blaschke product B of degree κ such that $B(T)R = f(T)$. Let B be factored as in (37). Then the subspace $\mathcal{K}$ of all h in $\mathcal{H}(C)$ such that

$$h(\alpha_j) = h'(\alpha_j) = \cdots = h^{(k_j-1)}(\alpha_j) = 0, \qquad j = 1, \dots, r,$$

has codimension at most κ in $\mathcal{H}(C)$, and $\mathcal{K}$ is invariant under T and R. Moreover, $g = f/B_0$ is a Schur function, and

$$R|_{\mathcal{K}} = B_1(T|_{\mathcal{K}})^{-1} g(T|_{\mathcal{K}}) B_2(T|_{\mathcal{K}})^{-1}.$$

Proof By Theorem 1.2, there exist a Schur function f and a Blaschke product B of degree κ such that $B(T)R = f(T)$. The remaining assertions in the theorem then follow from Theorem 3.2. □

4 Dual Results

The isomorphism in Theorem 5.1 of Appendix 5 allows us to formulate dual versions of our main theorems with T replaced by T^*. Theorem 1.2 has the following dual version.

Theorem 4.1 *Let C be an inner function, and let R be a bounded operator on $\mathcal{H}(C)$ that commutes with T^* and satisfies*

$$\mathrm{sq}_-\,(1 - R^*R) = \kappa$$

for some nonnegative integer κ. Then there exist a Blaschke product B of degree κ and a Schur function f such that

$$B(T^*)R = f(T^*).$$

Conversely, if such f and B exist, $1 - R^*R$ has at most κ negative squares.

Proof Set $R_\times = R^*$. Since R commutes with T^* and $\mathrm{sq}_-\,(1 - R^*R) = \kappa$, $R_\times$ commutes with T and

$$\mathrm{sq}_-\,(1 - R_\times R_\times^*) = \kappa.$$

By Theorem 1.2, there exist a Schur function $f_\times$ and Blaschke product $B_\times$ of degree κ such that $B_\times(T)R_\times = f_\times(T)$, and hence

$$B_\times(T)^*R = f_\times(T)^*.$$

By the functional calculus (4),

$$B_\times(T)^* = \widetilde{B}_\times(T^*) \qquad \text{and} \qquad f_\times(T)^* = \widetilde{f}_\times(T^*),$$

where $\widetilde{B}_\times(z) = \overline{B_\times(\overline{z})}$ is a Blaschke product of degree κ and $\widetilde{f}_\times(z) = \overline{f_\times(\overline{z})}$ is a Schur function. Thus

$$\widetilde{B}_\times(T^*)R = \widetilde{f}_\times(T^*).$$

Setting $f = \widetilde{f}_\times$ and $B = \widetilde{B}_\times$, we obtain $B(T^*)R = f(T^*)$, as required.

The converse direction is similarly deduced from the converse part of Theorem 1.2. □

The dual version of Theorem 3.2 uses an analog of the factorization (37). Assume given a space $\mathcal{H}(C)$, $C \notin \mathcal{H}(C)$, and a Blaschke product B of degree κ. Set

$$\widetilde{C}(z) = \overline{C(\overline{z})}.$$

Let {Zeros of B} $\cap$ {Zeros of $\widetilde{C}$} $= \{\beta_1, \ldots, \beta_s\}$, and

$$m_1, \ldots, m_s = \text{orders of } \beta_1, \ldots, \beta_s \text{ as zeros of } B,$$
$$n_1, \ldots, n_s = \text{orders of } \beta_1, \ldots, \beta_s \text{ as zeros of } \widetilde{C}.$$

Write

$$B(z) = B_1(z) B_0(z) B_2(z), \tag{55}$$

where

$$B_1(z) = \prod_{B(\gamma)=0,\, \widetilde{C}(\gamma)\neq 0} \frac{z-\gamma}{1-\overline{\gamma}z},$$

$$B_0(z) = \prod_{j=1}^{s} \left(\frac{z-\beta_j}{1-\overline{\beta}_j z}\right)^{k_j}, \qquad k_j = \min(m_j, n_j),$$

$$B_2(z) = \prod_{m_j > n_j} \left(\frac{z-\beta_j}{1-\overline{\beta}_j z}\right)^{m_j - n_j}.$$

For any h in $\mathcal{H}(C)$, in accordance with (7) we set

$$\widetilde{h}(w) = \left\langle h(z), \frac{C(z) - C(\overline{w})}{z - \overline{w}} \right\rangle_C, \qquad w \in \mathbb{D}.$$

Theorem 4.2 *Let $\mathcal{H}(C)$ be a given space such that $C \notin \mathcal{H}(C)$, and let R be a bounded operator on $\mathcal{H}(C)$ that commutes with T^*. Assume that*

$$B(T^*)R = f(T^*),$$

where f is a Schur function and B is a Blaschke product of degree κ. Let B be factored as in (55). *Then the subspace $\mathcal{K}$ of all h in $\mathcal{H}(C)$ such that*

$$\widetilde{h}(\beta_j) = \widetilde{h}'(\beta_j) = \cdots = \widetilde{h}^{(k_j-1)}(\beta_j) = 0, \qquad j = 1, \ldots, s, \tag{56}$$

has codimension at most κ in $\mathcal{H}(C)$, and $\mathcal{K}$ is invariant under T^* and R. Moreover, $g = f/B_0$ is a Schur function, and

$$R|_{\mathcal{K}} = B_1\big(T^*|_{\mathcal{K}}\big)^{-1} g\big(T^*|_{\mathcal{K}}\big) B_2\big(T^*|_{\mathcal{K}}\big)^{-1}. \tag{57}$$

Proof The mapping $U : h \to \widetilde{h}$ is a unitary operator from $\mathcal{H}(C)$ onto $\mathcal{H}(\widetilde{C})$ by Theorem 5.1. Let $\widetilde{T}$, $\widetilde{T}^*$ be the operators on $\mathcal{H}(\widetilde{C})$ corresponding to T, T^* on $\mathcal{H}(C)$. By Theorem 5.1,

$$\widetilde{T} = UT^*U^{-1}.$$

Set $\widetilde{R} = URU^{-1}$. Then since $T^*R = RT^*$,

$$\widetilde{T}\widetilde{R} = UT^*RU^{-1} = URT^*U^{-1} = \widetilde{R}\widetilde{T}. \tag{58}$$

Since $B(T^*)R = f(T^*)$,

$$UB(T^*)U^{-1}URU^{-1} = Uf(T^*)U^{-1}.$$

By the functional calculus (4),

$$UB(T^*)U^{-1} = B(UT^*U^{-1}) = B(\widetilde{T}),$$
$$Uf(T^*)U^{-1} = f(UT^*U^{-1}) = f(\widetilde{T}).$$

Hence

$$B(\widetilde{T})\widetilde{R} = f(\widetilde{T}). \tag{59}$$

By (58) and (59), we can apply Theorem 3.2 with the same f, B but with $\mathcal{H}(C)$, R replaced by $\mathcal{H}(\widetilde{C})$, $\widetilde{R}$. Let $\widetilde{\mathcal{K}}$ be the set of all $\widetilde{h}$ in $\mathcal{H}(\widetilde{C})$ that satisfy (56). By Theorem 3.2, $\widetilde{\mathcal{K}}$ has codimension at most κ and is invariant under $\widetilde{T}$. Moreover, $\widetilde{\mathcal{K}}$ is invariant under $\widetilde{R}$, and

$$\widetilde{R}|_{\widetilde{\mathcal{K}}} = B_1\big(\widetilde{T}|_{\widetilde{\mathcal{K}}}\big)^{-1} g\big(\widetilde{T}|_{\widetilde{\mathcal{K}}}\big) B_2\big(\widetilde{T}|_{\widetilde{\mathcal{K}}}\big)^{-1} \tag{60}$$

where $g(z) = f(z)/B_0(z)$ is a Schur function. Let $U_{\mathcal{K}}$ be the restriction of U to the subspace $\mathcal{K} = U^{-1}\widetilde{\mathcal{K}}$ of $\mathcal{H}(C)$, so $U_{\mathcal{K}}$ is a unitary operator from $\mathcal{K}$ onto $\widetilde{\mathcal{K}}$. By (4), for any φ in H^∞,

$$\varphi\big(\widetilde{T}|_{\widetilde{\mathcal{K}}}\big) = U_{\mathcal{K}}\varphi\big(T^*|_{\mathcal{K}}\big)U_{\mathcal{K}}^{-1}.$$

Since also $\widetilde{R}|_{\widetilde{\mathcal{K}}} = U_{\mathcal{K}} R|_{\mathcal{K}} U_{\mathcal{K}}^{-1}$, the relation (57) follows from (60). The theorem follows. □

Theorem 3.3 also has a dual version.

Theorem 4.3 *Let C be an inner function, and let R be a bounded operator on $\mathcal{H}(C)$ that commutes with T^* and satisfies*

$$\mathrm{sq}_-\,(1 - R^*R) = \kappa$$

for some nonnegative integer κ. Then there exist a Schur function f and a Blaschke product B of degree κ such that $B(T^*)R = f(T^*)$. Let B be factored as in (55). Then the subspace $\mathcal{K}$ of all h in $\mathcal{H}(C)$ such that

$$\widetilde{h}(\beta_j) = \widetilde{h}'(\beta_j) = \cdots = \widetilde{h}^{(k_j-1)}(\beta_j) = 0, \qquad j = 1, \ldots, s, \tag{61}$$

has codimension at most κ in $\mathcal{H}(C)$, and $\mathcal{K}$ is invariant under T^* and R. Moreover, $g = f/B_0$ is a Schur function, and

$$R|_{\mathcal{K}} = B_1\big(T^*|_{\mathcal{K}}\big)^{-1} g\big(T^*|_{\mathcal{K}}\big) B_2\big(T^*|_{\mathcal{K}}\big)^{-1}. \tag{62}$$

Proof By Theorem 4.1, there exist a Schur function f and a Blaschke product B of degree κ such that $B(T^*)R = f(T^*)$. Hence the result follows from Theorem 4.2. □

5 Appendix

Let $\mathcal{H}(C)$ be a given space such that $C \notin \mathcal{H}(C)$. Set

$$\widetilde{C}(z) = \overline{C(\overline{z})}.$$

Then a space $\mathcal{H}(\widetilde{C})$ exists, and $\widetilde{C} \notin \mathcal{H}(\widetilde{C})$ by (1). We denote by $\widetilde{T}$ and $\widetilde{T}^*$ the operators on $\mathcal{H}(\widetilde{C})$ corresponding to (2) and (3).

Theorem 5.1 *The mapping $U : h(z) \to \widetilde{h}(z)$ defined by*

$$\widetilde{h}(w) = \left\langle h(z), \frac{C(z) - C(\overline{w})}{z - \overline{w}} \right\rangle_C, \qquad w \in \mathbb{D}, \tag{63}$$

is a unitary operator from $\mathcal{H}(C)$ onto $\mathcal{H}(\widetilde{C})$ such that

$$UTU^{-1} = \widetilde{T}^*. \tag{64}$$

The inverse of U is the corresponding mapping from $\mathcal{H}(\widetilde{C})$ onto $\mathcal{H}(C)$, that is, if h and $\widetilde{h}$ are connected by (63), then also

$$h(w) = \left\langle \widetilde{h}(z), \frac{\widetilde{C}(z) - \widetilde{C}(\overline{w})}{z - \overline{w}} \right\rangle_{\widetilde{C}}, \qquad w \in \mathbb{D}. \tag{65}$$

This result is given in [6, Th. 18] (see also [2, Th. 3.4.2(C)]). We sketch a proof by reproducing kernel methods.

Proof If $h(z) = K_C(\alpha, z)$, $\alpha \in \mathbb{D}$, is any kernel function in $\mathcal{H}(C)$, then

$$\widetilde{h}(w) = \left\langle K_C(\alpha, z), \frac{C(z) - C(\overline{w})}{z - \overline{w}} \right\rangle_C = \frac{\widetilde{C}(w) - \widetilde{C}(\overline{\alpha})}{w - \overline{\alpha}},$$

which belongs to $\mathcal{H}(\widetilde{C})$. By (10) applied to $\mathcal{H}(\widetilde{C})$, for any $\alpha, \beta \in \mathbb{D}$,

$$\begin{aligned}\left\langle \frac{\widetilde{C}(z) - \widetilde{C}(\overline{\alpha})}{z - \overline{\alpha}}, \frac{\widetilde{C}(z) - \widetilde{C}(\overline{\beta})}{z - \overline{\beta}} \right\rangle_{\widetilde{C}} &= K_C(\alpha, \beta) \\ &= \langle K_C(\alpha, z), K_C(\beta, z)\rangle_C.\end{aligned}$$

Therefore the restriction of U to the span of kernel functions is an isometric mapping from a dense subspace of $\mathcal{H}(C)$ into $\mathcal{H}(\widetilde{C})$. The restriction has a unique extension by continuity to an isometry from $\mathcal{H}(C)$ into $\mathcal{H}(\widetilde{C})$. A short argument shows that the extension coincides with the mapping U defined by (63). Therefore U is an isometry from $\mathcal{H}(C)$ into $\mathcal{H}(\widetilde{C})$. To see that U is onto, we use (10) again, to show that

$$U\colon \frac{C(z) - C(\overline{\alpha})}{z - \overline{\alpha}} \to K_{\widetilde{C}}(\alpha, z), \qquad \alpha \in \mathbb{D}. \tag{66}$$

Thus the range of U contains all kernel functions in $\mathcal{H}(\widetilde{C})$ and hence is dense in $\mathcal{H}(\widetilde{C})$. Therefore U is unitary.

To prove (64), consider any h in $\mathcal{H}(C)$, and compute: for $w \in \mathbb{D} \setminus \{0\}$,

$$\begin{aligned}(UTh)(w) &= \left\langle Th(z), \frac{C(z) - C(\overline{w})}{z - \overline{w}} \right\rangle_C \\ &= \left\langle h(z), T^* \frac{C(z) - C(\overline{w})}{z - \overline{w}} \right\rangle_C \\ &= \left\langle h(z), \frac{1}{z}\left[\frac{C(z) - C(\overline{w})}{z - \overline{w}} - \frac{C(0) - C(\overline{w})}{-\overline{w}} \right] \right\rangle_C \\ &= \left\langle h(z), \frac{1}{\overline{w}}\left[\frac{C(z) - C(\overline{w})}{z - \overline{w}} - \frac{C(z) - C(0)}{z} \right] \right\rangle_C \\ &= \frac{\widetilde{h}(w) - \widetilde{h}(0)}{\overline{w}}.\end{aligned}$$

Therefore $UT = \widetilde{T}^* U$, which is equivalent to (64).

For the last statement, it is enough to check (65) when $\widetilde{h}(z)$ is a kernel function for $\mathcal{H}(\widetilde{C})$. By (66) we may take

$$h(z) = \frac{C(z) - C(\overline{\alpha})}{z - \overline{\alpha}} \qquad \text{and} \qquad \widetilde{h}(z) = K_{\widetilde{C}}(\alpha, z)$$

in (66). Then (65) reduces to the identity

$$\left\langle K_{\widetilde{C}}(\alpha, z), \frac{\widetilde{C}(z) - \widetilde{C}(\overline{w})}{z - \overline{w}} \right\rangle_{\widetilde{C}} = \frac{C(w) - C(\overline{\alpha})}{w - \overline{\alpha}}, \tag{67}$$

and the result follows. □

References

1. Daniel Alpay, Aad Dijksma, and James Rovnyak, *On Nudel′man's problem and indefinite interpolation in the generalized Schur and Nevanlinna classes*, Complex Anal. Oper. Theory **14** (2020), no. 1, Art. 25, 30. MR 4060499
2. Daniel Alpay, Aad Dijksma, James Rovnyak, and Hendrik de Snoo, *Schur functions, operator colligations, and reproducing kernel Pontryagin spaces*, Operator Theory: Advances and Applications, vol. 96, Birkhäuser Verlag, Basel, 1997. MR 1465432
3. Rodrigo Arocena, Tomas Ya. Azizov, Aad Dijksma, and Stefania A. M. Marcantognini, *On commutant lifting with finite defect. II*, J. Funct. Anal. **144** (1997), no. 1, 105–116. MR 1430717
4. J. A. Ball and V. Bolotnikov, *de Branges-Rovnyak spaces: basics and theory*, Operator theory (D. Alpay, ed.), Springer, Basel, 2015, pp. 631–679.
5. Louis de Branges and James Rovnyak, *Canonical models in quantum scattering theory*, Perturbation Theory and its Applications in Quantum Mechanics (Proc. Adv. Sem. Math. Res. Center, U.S. Army, Theoret. Chem. Inst., Univ. of Wisconsin, Madison, Wis., 1965), Wiley, New York, 1966, pp. 295–392. MR 0244795
6. ——, *Square summable power series*, Holt, Rinehart and Winston, New York-Toronto, Ont.-London, 1966. MR 0215065
7. Emmanuel Fricain and Javad Mashreghi, *The theory of $\mathcal{H}(b)$ spaces. Vol. 1*, New Mathematical Monographs, vol. 20, Cambridge University Press, Cambridge, 2016. MR 3497010
8. ——, *The theory of $\mathcal{H}(b)$ spaces. Vol. 2*, New Mathematical Monographs, vol. 21, Cambridge University Press, Cambridge, 2016. MR 3617311
9. Stephan Ramon Garcia, Javad Mashreghi, and William T. Ross, *Introduction to model spaces and their operators*, Cambridge Studies in Advanced Mathematics, vol. 148, Cambridge University Press, Cambridge, 2016. MR 3526203
10. N. K. Nikol′skiĭ, *Treatise on the shift operator*, Grundlehren der Mathematischen Wissenschaften [Fundamental Principles of Mathematical Sciences], vol. 273, Springer-Verlag, Berlin, 1986, Spectral function theory, With an appendix by S. V. Hruščev [S. V. Khrushchëv] and V. V. Peller, Translated from the Russian by Jaak Peetre. MR 827223
11. Donald Sarason, *Generalized interpolation in H^∞*, Trans. Amer. Math. Soc. **127** (1967), 179–203. MR 208383
12. ——, *Shift-invariant spaces from the Brangesian point of view*, The Bieberbach conjecture (West Lafayette, Ind., 1985), Math. Surveys Monogr., vol. 21, Amer. Math. Soc., Providence, RI, 1986, pp. 153–166. MR 875239
13. ——, *Sub-Hardy Hilbert spaces in the unit disk*, University of Arkansas Lecture Notes in the Mathematical Sciences, vol. 10, John Wiley & Sons, Inc., New York, 1994, A Wiley-Interscience Publication. MR 1289670
14. Morris Schreiber, *A functional calculus for general operators in Hilbert space*, Trans. Amer. Math. Soc. **87** (1958), 108–118. MR 99601

15. Béla Sz.-Nagy and Ciprian Foiaş, *Sur les contractions de l'espace de Hilbert. III*, Acta Sci. Math. (Szeged) **19** (1958), 26–45. MR 103418
16. ——, *Harmonic analysis of operators on Hilbert space*, North-Holland Publishing Co., Amsterdam-London; American Elsevier Publishing Co., Inc., New York; Akadémiai Kiadó, Budapest, 1970, Translated from the French and revised. MR 0275190
17. Dan Timotin, *A short introduction to de Branges–Rovnyak spaces*, Invariant subspaces of the shift operator, Contemp. Math., vol. 638, Amer. Math. Soc., Providence, RI, 2015, pp. 21–38. MR 3309347

An Operator Theoretical Approach of Some Inverse Problems

Juliette Leblond and Elodie Pozzi

1 Introduction

This paper aims at giving an overview of the solution of inverse problem with boundary data. The inverse problems we consider involve elliptic partial differential equations (PDE) of the form $\mathrm{div}(\sigma \nabla u) = 0$ where σ is a fixed function in a Sobolev space $W^{1,r}(\Omega)$ in a simply connected bounded domain Ω of the complex plane. The goal is to describe the existing results enhancing the connections between PDE, function spaces and operator theory for a deep understanding of inverse problems.

In [7], the authors established results in the Hardy spaces of the unit disk $\mathbb{D}$ (for $\sigma = 1$, i.e. $\Delta u = 0$) for partial boundary data in $L^p(\mathbb{T})$. This work built the foundations of an approach to resolve a class of inverse problems which led to several other instances of these ideas applied in other, more complicated settings. For example, the case where $\Omega = \mathbb{D}$ with $\sigma \in W^{1,\infty}(\mathbb{D})$ has been investigated in [8], and with $\sigma \in W^{1,r}(\mathbb{D})$, $r \geq 2$, in [2]. When the domain Ω is an annulus, the inverse problem for the Laplace equation ($\sigma = 1$) has been solved in [13, 23]; for σ in the space $W^{1,\infty}(\Omega)$, see [5, 18].

We consider the case where $\sigma \in W^{1,r}(\Omega)$, $r \geq 2$, on simply connected Dini-smooth domains Ω of the complex plane; some remarks in conclusion will specify the validity of the results in less smooth domains and for multi-connected domains. The regularity of the boundary $\partial\Omega$ of the domain plays a fundamental role in the solution of inverse problem when the boundary data is in $L^p(\partial\Omega)$, for $1 < p < \infty$. Indeed, the conformal map from $\mathbb{D}$ onto a simply connected Dini-smooth bounded

J. Leblond
INRIA, Team Factas, Sophia Antipolis, France
e-mail: juliette.Leblond@inria.fr

E. Pozzi (✉)
Department of Mathematics and Statistics, Saint Louis University, Saint Louis, MO, USA
e-mail: elodie.pozzi@slu.edu

I. Binder et al. (eds.), *Function Spaces, Theory and Applications*, Fields Institute Communications 87, https://doi.org/10.1007/978-3-031-39270-2_3

domain Ω has a derivative that can be extended continuously to the unit circle $\mathbb{T}$: it permits an extension of the results on the unit disk to the case of simply connected Dini-smooth bounded domain using the conformal map.

The critical cases for such a study correspond to $r = 2$ (when the exponent r is equal to the dimension of the domain) for Dini-smooth domains Ω or $r \geq 2$ and Ω is only rectifiable (the above conformal map has a derivative that admits an $L^1(\mathbb{T})$ extension on $\mathbb{T}$), and the results cannot be obtained directly using conformal equivalence between $\mathbb{D}$ and the rectifiable domain Ω. We will thus focus on $r > 2$ and a simply connected Dini-smooth bounded domain Ω.

The paper is organized as follows. Section 2 installs the applied context of the PDE and the operators related to the considered inverse problems, together with notation and definitions. Section 3 is devoted to the treatment of the forward and the inverse problems; Sect. 3.1 describes the methods and the solutions for the Laplace equation (i.e. in the case $\sigma = 1$), while Sect. 3.2 is concerned with their generalizations to the conductivity equation. We finally provide some concluding remarks in Sect. 4.

2 Framework, Notation, Definitions

2.1 With PDE, Examples

We consider linear elliptic PDEs from a family of conductivity equations:

$$\begin{cases} A\,u = \operatorname{div}(\sigma \nabla u) = f \text{ in } \Omega\,, & \\ B\,u = b \text{ on } I \subset \partial\Omega\,, & B\,u = u_{|_I} \text{ or } B\,u = (\sigma\,\partial_n u)_{|_I}\,, \end{cases} \tag{1}$$

for smooth enough domains $\Omega \subset \mathbb{R}^2$, conductivity coefficient σ with positive values, source term f and boundary data b. The particular case where $\sigma = 1$ corresponds to Poisson (Laplace) equation, in the inhomogeneous situation when $f \neq 0$.

When settled in dimension 3, such PDEs are involved in many application areas in physics, among which those related to electromagnetism on Maxwell's equation under the quasi-static assumption (neglecting the time derivative of the electromagnetic fields) but also to gravimetry issues and Newton equation. Two-dimensional situations arise for instance whenever the 3D geometry admits symmetry properties while the solution is invariant in some direction (like for plasma in a toroidal *tokamak* where a 2D conductivity PDE arises from a cylindrical change of variable in Poisson-Laplace equation), but are also of interest by themselves since their solution can be handled with tools from complex analysis and analytic function spaces.

We have the following application in mind, related to electromagnetism, where u stands for a potential.

Coming from neurosciences and brain imaging, a model for electroencephalography (EEG) is given by (1) with Ω made of (disjoint union), say, three homogeneous nested layers Ω_i, $i = 0, 1, 2$ (brain, skull, scalp), each of constant conductivity σ_i, and $f = \operatorname{div} \mu$ (current source) supported in the innermost layer Ω_0 is to be estimated from the Cauchy data (electric potential, current flux) that are available on a part $I \subset \partial\Omega$ of the outer boundary. Due to continuity properties at interfaces, this can be separated into a pair of data completion (boundary value) Cauchy problems, and an inverse source estimation problem, for Poisson Laplace equations; see [1, 10] in $\mathbb{R}^2$, and [15] in $\mathbb{R}^3$. For $i = 2, 1$, we wish to find u such that

$$\begin{cases} \Delta\, u = 0 \text{ in } \Omega_i\,, \\ u_{|_I}\,,\ (\partial_n u)_{|_I} \text{ given on } I \subset \partial\Omega_i\,. \end{cases}$$

If the conductivity is allowed to vary, which is a reasonable assumption in particular within the intermediate layer Ω_1, we are back to the general conductivity PDE:

$$\begin{cases} \operatorname{div}(\sigma\, \nabla\, u) = 0 \text{ in } \Omega_i\,, \\ u_{|_I}\,,\ (\sigma\, \partial_n u)_{|_I} \text{ given on } I \subset \partial\Omega_i\,. \end{cases} \tag{2}$$

This solves the so-called cortical mapping step and provides data on the cortex $\partial\Omega_0$ from which we state the source estimation issue: find μ supported within Ω_0 (satisfying some assumptions, like being a linear combination of Dirac masses) such that

$$\begin{cases} \Delta\, u = \operatorname{div} \mu \text{ in } \Omega_0\,, \\ u_{|_{\partial\Omega_0}}\,,\ (\partial_n u)_{|_{\partial\Omega_0}} \text{ given on } \partial\Omega_0\,. \end{cases}$$

In this work, we will discuss the above boundary value problems (2), though not the EEG source estimation one.

2.2 *With Banach Spaces and Their Operators*

Let V and W be two Banach spaces. Let $T : V \longrightarrow W$ be a bounded linear map. We denote by V^* and W^* the respective dual spaces of V and W and we will write $\langle L, f\rangle_{V^*,V} = L(f)$ for $f \in V$, $L \in V^*$. The adjoint T^* of T is a bounded linear operator defined from W^* to V^* such that $\langle g, Tf\rangle_{W^*,W} = \langle T^*g, f\rangle_{V^*,V}$ for $f \in V$ and $g \in W^*$. In the case of Hilbert spaces, $V = \overline{\operatorname{Ran} T^*} \oplus \ker T$ and $W = \ker T^* \oplus \overline{\operatorname{Ran} T}$; the direct sums are orthogonal for the respective inner products on V and W; see [12, Rmk 17] or [9, Prop. 3].

Consider two elements $b \in W$ and $m \in V$, and T a linear and bounded operator from V to W such that $T(m) = b$. Given $b \in W$ and the linear bounded operator T, is it possible to find $m \in V$?

The problem is *well-posed in the Hadamard sense* [21] if the three following conditions are satisfied:

- A solution exists.
- It is unique.
- The solution m depends continuously on the initial data b.

Assume that V and W are reflexive Banach spaces. This question is trivial when T is invertible: there are existence and uniqueness of a solution and the solution depends continuously on b since T^{-1} is invertible. The most interesting situation occurs when T is not invertible. Two situations happen: $R = \operatorname{Ran} T$ is closed or not closed. If the range R of T is closed, one can write $T = P_R \widetilde{T}$ where P_R denotes the (orthogonal?) projection from W onto R and $\widetilde{T}$ is an invertible linear bounded operator from $V \backslash \ker T$ onto $W \backslash \ker T^*$. Let $Q = \overline{\operatorname{Ran} T^*} \subset V$. Then, the linear bounded operator $T^\dagger = P_Q \widetilde{T}^{-1}$ is called the pseudo-inverse of T, and whenever T is injective, $T^\dagger b$ is the unique element in V such that

$$\min_{m \in V} \|T(m) - b\|_W = \|T^\dagger b - m\|_V;$$

see [11, Thms 1, 5]. In the case of Hilbert spaces, the pseudo-inverse is given by $T^\dagger = T^*(T T^*)^{-1}$. The minimization problem is well-posed in the Hadamard sense.

If $\ker T \neq \{0\}$, the problem is said to have silent sources as elements m_S of V that are not detected by T. They lead to non-uniqueness properties, as $T(m + m_S) = T(m)$ for $m_s \in \ker T$.

Lemma 2.1 *When R is dense in W, for any $b \notin R$ there is a sequence $(m_k)_{k \geq 1}$ of elements in V such that $\|T(m_k) - b\|_W \underset{k \to +\infty}{\longrightarrow} 0$. As a consequence,* $\lim_{k \to +\infty} \|m_k\|_V = +\infty$.

Proof Assume that $\sup_{k \in \mathbb{Z}^+} \|m_k\|_V < \infty$ and there exists a subsequence $(m_{n_k})_{k \geq 1}$ converging to an element $m \in V$, and by continuity of T, $(T(m_k))_{k \geq 1}$ converges to $T(m) = b$ by uniqueness of a limit, which leads to a contradiction. ■

The limit of the norm of m_k has an impact on the error in the reconstruction: indeed, if for $k \geq 1$, $\eta_k = m_k - m$ then $\|\eta_k\|_V \geq \|m_k\|_V - \|m\|_V \underset{k \to +\infty}{\longrightarrow} +\infty$. To avoid this phenomenon, we express a Tykhonov-like regularization process for classes of (ill-posed) inverse problems related to some elliptic partial differential equations in appropriate framework called bounded extremal problems denoted by BEP. The BEP is an approximation problem with a constraint in norm to avoid an arbitrary large norm of the element m in V such that $\|T(m) - b\|_W$ is small. It can be stated as follows:

$$\min_{\substack{g\in V\\ \|g\|_V\le M}} \|T(g)-b\|_W\,. \tag{BEP}$$

The BEP is formulated in normed function spaces linked to the partial differential equation and using operators related the initial data b.

We will use the terminology "solving an inverse problem" throughout whenever a Tychonov-like regularization method leads to its well-posedness, hence to existence and uniqueness of a solution. In the Hilbertian case, the solution is constructively given by an implicit equation.

In the sequel, we will write $\widetilde{T}$ instead of T to specify that the operator acts on a real Banach space V. For $I \subsetneq \partial\Omega$, we will write $J = \partial\Omega\backslash I$.

2.3 Notation, Definitions

Let Ω be a domain of $\mathbb{C} \simeq \mathbb{R}^2$. We will write $\mathcal{H}^1$ for the Hausdorff measure on $\partial\Omega$.

A domain in $\mathbb{C}$ is said to be Dini-smooth (see [28]) if its boundary $\partial\Omega$ has a parametrization ψ such that its derivative ψ' is Dini-continuous: its modulus of continuity $\omega_{\psi'}$ satisfies $\int_0^\varepsilon \frac{\omega_{\psi'}(t)}{t}dt < \infty$ for any $\varepsilon > 0$. If Ω is a simply connected Dini-smooth domain, a conformal map ψ from $\mathbb{D}$ onto Ω is such that ψ' has a continuous extension to $\overline{\mathbb{D}}$ (see [28, Thm 3.5]). This result permits to extend most of the results from $\mathbb{D}$ to a simply connected Dini-smooth domain and is still true for a conformal map ψ between two multi-connected bounded domains (see [5]). Throughout this paper, Ω is a Dini-smooth domain of $\mathbb{C}$.

We will write $\mathcal{D}(\Omega)$ for the space of C^∞-smooth complex-valued functions with compact support in Ω, equipped with the usual topology. Its dual is the space of distributions $\mathcal{D}'(\Omega)$ on Ω.

For $1 \le p < \infty$, $L^p(\Omega)$ denotes the space of complex-valued measurable functions f on Ω such that the norm $\|f\|_{L^p(\Omega)} := \left(\int_\Omega |f(z)|^p dz\right)^{1/p}$ is finite; $L^\infty(\Omega)$ is the space of essentially bounded functions on Ω, equipped with the norm defined by the essential sup of the modulus. The Sobolev space $W^{1,p}(\Omega)$ is the set of complex-valued functions $f \in L^p(\Omega)$ with distributional derivatives $\partial f := \partial_z f = \frac{1}{2}(\partial_x - i\partial_y)f, \overline{\partial} f := \partial_{\overline{z}} f = \frac{1}{2}(\partial_x + i\partial_y) \in L^p(\Omega)$ for $z = x+iy \in \mathbb{C}$. We equip $W^{1,p}(\Omega)$ with the norm

$$\|f\|_{W^{1,p}(\Omega)} := \left(\|f\|^p_{L^p(\Omega)} + \|\partial f\|^p_{L^p(\Omega)} + \|\overline{\partial} f\|^p_{L^p(\Omega)}\right)^{1/p}.$$

The space $W^{1,\infty}(\Omega)$ is defined similarly, with obvious modifications of the norm. For $1 \le p \le \infty$, the space $L^p_{\mathbb{R}}(\Omega)$ (respectively $W^{1,p}_{\mathbb{R}}(\Omega)$) denotes the space of real-valued functions in $L^p(\Omega)$ (in $W^{1,p}(\Omega)$ respectively). For p strictly greater than the dimension of Ω, $W^{1,p}(\Omega)$ embeds into $L^\infty(\Omega)$ [12, Cor. 9.14] and in the space of Hölder continuous functions. As a consequence, $W^{1,p}(\Omega)$-functions extend continuously to $\partial\Omega$. For $p > 2$, a function in $W^{1,p}(\Omega)$ can be extended continuously to $\overline{\Omega}$: the extension to $\partial\Omega$ of a function $f \in W^{1,p}(\Omega)$ is denoted by $\mathrm{tr}_{\partial\Omega}(f)$.

The space $W^{1-1/p,p}(\partial\Omega)$ is a real interpolation between the spaces $L^p(\partial\Omega)$ and $W^{1,p}(\partial\Omega)$; $W^{1-1/p,p}(\partial\Omega)$ is equipped with the norm:

$$\|f\|_{W^{1-1/p,p}(\partial\Omega)} := \|f\|_{L^p(\partial\Omega)} + \left(\int_{\partial\Omega\times\partial\Omega} \frac{|f(t)-f(u)|^p}{(\mathcal{H}^1(t,u))^p} d\mathcal{H}^1(t)d\mathcal{H}^1(u)\right)^{1/p},$$

where $\mathcal{H}^1(t,u)$ denotes the length of the arc (t,u) on $\partial\Omega$. We will write $L^p_{loc}(\Omega)$ ($W^{1,p}_{loc}(\Omega)$ respectively) for the distributions on Ω such that their restriction to a compact subset Ω_0 of Ω belongs to $L^p(\Omega_0)$ ($W^{1,p}(\Omega_0)$ respectively).

The distribution $\nu\overline{\partial f}$ with $\nu \in W^{1,r}_{\mathbb{R}}(\Omega)$ is defined using the Leibniz rule:

$$\langle \nu\overline{\partial f}, \phi\rangle = -\int_\Omega (\nu\overline{f}\,\overline{\partial}\phi + \overline{\partial}\nu\overline{f}\phi), \qquad \text{for } \phi \in \mathcal{D}(\Omega),$$

where $\langle , \rangle$ denotes the duality product between $\mathcal{D}'(\Omega)$ and $\mathcal{D}(\Omega)$. For $r > 2$, if $\sigma \in W^{1,r}(\Omega)$ with $0 < c < \sigma < C$ and $\mathrm{div}(\sigma\nabla u) = 0$ for $u \in W^{1,p}_{\mathbb{R}}(\Omega)$, then the normal derivative $\partial_n u$ is the unique element of $W^{-1/p,p}(\partial\Omega) = (W^{1-1/q,q}(\partial\Omega))^*$ with $1/p + 1/q = 1$

$$\langle \sigma\partial_n u, \phi\rangle = \int_\Omega \sigma\nabla u\nabla g, \quad \mathrm{tr}_{\partial\Omega}\phi = g.$$

3 Inverse Boundary Value Problems

Let Ω be a Dini-smooth simply connected domain of $\mathbb{C}$ and $I \subset \partial\Omega$, $J\partial\Omega \setminus I$, with $\mathcal{H}^1(I) > 0$ and $\mathcal{H}^1(J) > 0$. We will denote by ϕ a conformal map from $\mathbb{D}$ onto Ω. Let $1 < p < \infty$ and $b \in L^p_{\mathbb{R}}(I)$. Let $\sigma \in W^{1,r}_{\mathbb{R}}(\Omega)$ for $r > 2$ satisfying $0 < c_\sigma < \sigma < C_\sigma < 1$ with $c_\sigma, C_\sigma > 0$.

The inverse data extension problem can be stated as follows, with $I \subset \partial\Omega$: given $b = (b_D, b_N) \in \left(L^p_{\mathbb{R}}(I)\right)^2$, is there m such that

$$\begin{cases} \mathrm{div}\,(\sigma\nabla m) = 0 & \text{on } \Omega\,, \\ m = b_D \text{ and/or } \sigma\,\partial_n m = b_N & \text{on } I? \end{cases} \tag{3}$$

Note that for $I = \partial\Omega$, only one of the two boundary conditions can be imposed. The operator $A = \mathrm{div}\,(\sigma\nabla\cdot)$ is an elliptic operator of the form $A = \mathrm{div}(M\nabla\cdot)$ in the isotropic case where $M = \sigma I_2$ where I_2 is the identity matrix of size 2.

Remark 3.1 The existence of a solution m to Equation (3) can be written in terms of operators as follows (see BEP): for which space V of functions $m : \Omega \longrightarrow \mathbb{R}$ satisfying $\mathrm{div}\,(\sigma \nabla m) = 0$ on Ω such that m can be extended to $\partial\Omega$ as a function in $L^p(\partial\Omega)$ is the operator $\widetilde{T} : V \longrightarrow L^p_{\mathbb{R}}(I)$, $\widetilde{T}(m) = b_D$ for $I = \partial\Omega$, $\widetilde{T} : V \longrightarrow \left(L^p_{\mathbb{R}}(I)\right)^2$, $\widetilde{T}(m) = (b_D, b_N)$ for $I \subset \partial\Omega$, continuous and surjective? See also Remark 3.2.

3.1 Harmonic Solutions, Hardy Spaces

In the homogeneous situation where $\sigma = 1$, (3) rewrites as a Laplace equation:

$$\begin{cases} \Delta u = 0 & \text{on } \Omega\,, \\ u = b_D \text{ and/or } \partial_n u = b_N & \text{on } I\,. \end{cases} \tag{4}$$

The condition $u = b_D$ on I refers to a Dirichlet boundary condition (D) and the condition $\partial_n u = b_N$ on I is a Neumann boundary condition (N). Coupling the two as in (4) leads to a Cauchy-type problem, with overdetermined partial data. In what follows, we assume that $b_D \in L^p_{\mathbb{R}}(I)$, $b_N \in W^{1,p}_{\mathbb{R}}(I)$ for $1 < p < \infty$.

3.1.1 Forward Problems

The forward problem consists from a given harmonic function u on Ω in finding $\mathrm{tr}_{\partial\Omega} u = b_D$ ((D) condition) and $\partial_n u = b_N$ ((N) condition) on $I \subset \partial\Omega$.

In the case of the unit disk $\mathbb{D}$, it is known that a harmonic function $u : \mathbb{D} \longrightarrow \mathbb{R}$ has a trace on $\mathbb{T}$ almost everywhere that belongs to $L^p_{\mathbb{R}}(\mathbb{T})$ if and only if u satisfies

$$\sup_{0<r<1} \int_0^{2\pi} |u(re^{it})|^p dt < \infty\,. \tag{5}$$

For any harmonic function u in $\mathbb{D}$ satisfying (5), the restriction to I of $\mathrm{tr}_{\mathbb{T}} u$ is in $L^p(I)$.

For a bounded Dini-smooth domain Ω, if $u : \Omega \longrightarrow \mathbb{C}$ is a harmonic function such that

$$\sup_{0<r<1} \int_{\Gamma_r} |u(\zeta)|^p d\mathcal{H}^1(\zeta) < \infty, \tag{6}$$

where $\Gamma_r = \psi(r\mathbb{T})$, then u has a trace on $\partial\Omega$ denoted by $\mathrm{tr}_{\partial\Omega} u \in L^p_{\mathbb{R}}(\partial\Omega)$ (see [5, Thm]) and $(\mathrm{tr}_{\partial\Omega} u)_{|_I} \in L^p_{\mathbb{R}}(I)$ which solves the part (D) of the forward problem.

Remark 3.2 The real Banach space V in Remark 3.1 is defined by

$$V = \mathrm{tr}_{\partial\Omega}\left(\left\{u : \Omega \longrightarrow \mathbb{R}\text{harmonic such that } \|u\|^p := \sup_{0<r<1} \int_{\Gamma_r} |u(\zeta)|^p d\mathcal{H}^1(\zeta) < \infty\right\}\right).$$

Another way to find a solution to the forward problem in Dini-smooth domain is to reformulate it using complex-valued functions. Since the function u is harmonic in Ω, it is possible to construct a harmonic function $\widetilde{u}$ on Ω such that $f = u + i\widetilde{u}$ is analytic on Ω (i.e. $\overline{\partial} f = 0$ in Ω). The function $\widetilde{u}$ is called the harmonic conjugate of u. The construction of $\widetilde{u}$ is as follows: if Ω is simply connected, given a harmonic function $u : \Omega \longrightarrow \mathbb{R}$, there is a harmonic function $\widetilde{u}$ on Ω such that u and $\widetilde{u}$ satisfy the Cauchy-Riemann equations

$$\begin{cases} \partial_x \widetilde{u} = -\partial_y u\,, \\ \partial_y \widetilde{u} \ = \partial_x u. \end{cases}$$

The harmonic conjugate is unique up to an additive constant. Whenever u is harmonic in Ω and satisfies (6), its harmonic conjugate $\widetilde{u}$ is such that $\mathrm{tr}_{\partial\Omega}\widetilde{u} \in L^p_{\mathbb{R}}(\partial\Omega)$. To get lighter notations, we still write u and $\widetilde{u}$ instead of $\mathrm{tr}_{\partial\Omega}\, u$ and $\mathrm{tr}_{\partial\Omega}\, \widetilde{u}$. This result is known as Riesz theorem (see [20, Ch. III, Thm. 2.3]) and can be stated as follows:

$$\begin{aligned} \mathfrak{H} : L^p_{\mathbb{R}}(\partial\Omega) &\longrightarrow L^p_{\mathbb{R}}(\partial\Omega) \\ u &\longmapsto \widetilde{u}, \text{ with the normalization } \int_{\partial\Omega} \widetilde{u}\, d\mathcal{H}^1 = 0, \end{aligned}$$

is a bounded linear map on $L^p_{\mathbb{R}}(\partial\Omega)$. The operator $\mathfrak{H}$ is called the conjugation map.

The function $f = u + i\widetilde{u}$ with u and $\widetilde{u}$ satisfying the Cauchy-Riemann equations is analytic since $\overline{\partial} f = \partial_x(u + i\widetilde{u}) + i\partial_y(u + i\widetilde{u}) = 0$ in Ω. It is known that if an analytic function f on Ω is such that

$$\|f\|_p^p := \sup_{0<r<1} \|f\|^p_{L^p(\Gamma_r)} < \infty, \tag{7}$$

where $\Gamma_r = \phi(r\mathbb{T})$, then f has a non-tangential limit on $\partial\Omega$ called trace of f and denoted by $\mathrm{tr}_{\partial\Omega}\, f$ (see [2, 16, 20]).

Definition 3.1 An analytic function on Ω satisfying (7) belongs to the Hardy space $H^p(\Omega)$.

The space $H^p(\Omega)$ equipped with the norm (7) is a Banach space. The space of $\mathrm{tr}_{\partial\Omega} f$ for $f \in H^p(\Omega)$ is a closed subspace of $L^p(\partial\Omega)$ denoted by $H^p(\partial\Omega)$ (see [16, Thm 3.3] for the case of the unit disk).

Proposition 3.1 *The linear map*

$$H^p(\Omega) \longrightarrow H^p(\partial\Omega)$$
$$f \longmapsto tr_{\partial\Omega} f$$

is an isometric isomorphism. As a consequence, a function $f \in H^p(\Omega)$ *can be identified to its trace* $tr_{\partial\Omega}(f)$.

When $f = u + i\widetilde{u} \in H^p(\Omega)$, $\mathrm{Re}\,(\mathrm{tr}_{\partial\Omega} f) = u \in L^p_{\mathbb{R}}(\partial\Omega)$. Thus, the forward problem is resolved since for any $f \in H^p(\Omega)$, the restriction to I of $\mathrm{Re}(\mathrm{tr}_{\partial\Omega} f)$ is in $L^p_{\mathbb{R}}(I)$. Moreover, one has $\mathrm{Re}\, H^p(\partial\Omega) = L^p_{\mathbb{R}}(\partial\Omega)$.

Remark 3.3 The forward problem (3) with the Dirichlet boundary condition is formulated in the complex plane in terms of the operator T introduced in Sect. 2: for $I \subset \partial\Omega$, the operator $T : V = H^p(\partial\Omega) \longrightarrow L^p_{\mathbb{R}}(I)$ with $T(f) = \mathrm{Re}\,(f)_{|_I}$ is well defined, bounded and surjective.

3.1.2 Inverse Problems

We saw above that any harmonic function u real-valued and satisfying (5) (for $\Omega = \mathbb{D}$) has a trace in $L^p_{\mathbb{R}}(\Omega)$. Conversely, given a function $b_D \in L^p_{\mathbb{R}}(\partial\Omega)$, it is possible to construct a function u harmonic on Ω such that u has a trace on $L^p_{\mathbb{R}}(\partial\Omega)$ which coincides with b_D on $\partial\Omega$; for $\Omega = \mathbb{D}$, u is given by

$$u(z) = \frac{1}{2\pi}\int_0^{2\pi} \frac{1 - |z|^2}{|e^{it} - z|^2} b_D(e^{it})dt;$$

see [16, Chap. 1] or [20, Chap. 1.3]. This result gives a solution to the problem (3) for the (D) condition on $\partial\Omega$ and similarly with the (N) condition there.
Whenever $I \subsetneq \mathbb{T} = \partial\mathbb{D}$ with $\mathcal{H}^1(I) > 0$ and $\mathcal{H}^1(\mathbb{T} \setminus I) > 0$, if $b = b_D + i \int b_N$ on I ($b \in L^p(I)$), then there exists a sequence of $H^p(\mathbb{D})$ functions $(g_\lambda)_\lambda$ defined by

$$g_\lambda(z) = \lambda h_\lambda(z) \frac{1}{2i\pi} \int_I \frac{\overline{h_\lambda(w)} b(w)}{w - z} d\mathcal{H}^1(w)$$

where

$$h_\lambda(z) = \exp\left(\frac{-1}{4\pi} \ln(1 + \lambda) \frac{e^{i\theta} + z}{e^{i\theta} - z}\right)$$

such that the restriction to I of $(\mathrm{tr}_{\mathbb{T}} g_\lambda)$ converges in $L^p(I)$ to b, showing that $H^p(\mathbb{D})_{|_I}$ is dense in $L^p(I)$; see [27, Thm. 1]. In [7, Prop. 1], the authors prove the same result for $\Omega = \mathbb{T}$ and $p = 2$ using the property of $H^2(\mathbb{T})$. Precisely, let $g \in L^2(I)$ such that for all $f \in H^2(\mathbb{T})$,

$$
\begin{aligned}
0 &= \langle g, f \rangle_{L^2(I)} \\
&= \langle g \vee 0, f \rangle_{L^2(\mathbb{T})} \\
&= \langle P_+(g \vee 0), f \rangle_{L^2(\mathbb{T})}
\end{aligned}
$$

where $g \vee 0$ is the extension of g to J by 0 and P_+ denotes the orthogonal projection from $L^2(\mathbb{T})$ onto $H^2(\mathbb{T})$. By taking $f = e^{in\theta} \in H^2(\mathbb{T})$, $n \geq 0$, one has $P_+(g \vee 0) = 0$ so $g \vee 0 \in (H^2(\mathbb{T}))^\perp$ which is impossible since a function in $(H^2(\mathbb{T}))^\perp$ cannot vanish on a subset of $\mathbb{T}$ (it is a consequence of [16, Thm 2.2]). Thus, $g = 0$ and $H^2(\mathbb{T})_{|_I}$ is dense in $L^2(I)$.

For a simply connected Dini-smooth domain Ω and $I = \partial\Omega$, for a given function $b_D \in L^p_{\mathbb{R}}(\partial\Omega)$, there is $f \in H^p(\mathbb{D})$ such that $\mathrm{Re}(\mathrm{tr}_{\mathbb{T}} f) = b_D \circ \psi$ and $g = f \circ \psi^{-1} \in H^p(\Omega)$ is such that $\mathrm{Re}(\mathrm{tr}_{\partial\Omega} g) = b_D$ on Ω. Now if $I \subsetneq \partial\Omega$ with $\mathcal{H}^1(I) > 0$ and $\mathcal{H}^1(J) > 0$, the functions $\widetilde{g_\lambda} = g_\lambda \circ \psi^{-1}$ are in $H^p(\Omega)$ and $\|\mathrm{Re}(\mathrm{tr}_{\partial\Omega} g_\lambda \circ \psi^{-1})_{|_I} - b_D\|_{L^p(I)} \simeq \|\mathrm{Re}(\mathrm{tr}_{\partial\Omega} g_\lambda)_{|_{\psi^{-1}(I)}} - b_D \circ \psi\|_{L^p(\psi^{-1}(I))}$ converges to 0 as λ goes to $+\infty$. Thus, we get the following result:

Proposition 3.2 *The space $H^p(\partial\Omega)_{|_I}$ is dense in $L^p(I)$. However, Re$(H^p(\partial\Omega))_{|_I} = L^p_{\mathbb{R}}(I)$. Moreover,*

(i) If $I \subset \partial\Omega$, $T : H^p(\partial\Omega) \longrightarrow L^p_{\mathbb{R}}(I)$, $g \mapsto Re\, g_{|_I}$ is surjective.
(ii) If $I \subsetneq \partial\Omega$ with $\mathcal{H}^1(I) > 0$ and $\mathcal{H}^1(J) > 0$, the range of $T : H^p(\partial\Omega) \longrightarrow L^p(I)$, $g \mapsto g_{|_I}$ is dense in $L^p(I)$ and the inverse problem is ill-posed.

Remark 3.4

- For $I = \partial\Omega$, for $b_D \in L^p_{\mathbb{R}}(\partial\Omega)$ a function $f \in H^p(\partial\Omega)$ such that $T(f) = \mathrm{Re}\, f = b_D$ is unique up to the condition $\int_{\partial\Omega} \mathrm{Im} f \, d\mathcal{H}^1 = 0$.
- If $I \subsetneq \partial\Omega$ with $\mathcal{H}^1(I) > 0$ and $\mathcal{H}^1(J) > 0$, if $b = b_D + i \int b_N \in L^p(I)$, $b \notin \mathrm{Ran}\, T$, $T(f) = f_{|_I}$, a sequence of functions $(f_k)_{k\geq 1}$ in $H^p(\partial\Omega)$ such that $\|f_k - b\|_{L^p(I)} \longrightarrow 0$ will be such that $\|f_k\|_{L^p(J)} \longrightarrow +\infty$ leading to instability in the reconstruction.

It is possible to solve the inverse problem using the following bounded extremal problem: let $b_D \in L^p_{\mathbb{R}}(I) = W$, let $M > 0$, and the BEP is

$$
\min_{g \in B_M} \|(\mathrm{Re}\, g)_{|_I} - b_D\|_{L^p_{\mathbb{R}}(I)},
$$

where:

$$
B_M = \{g \in H^p(\partial\Omega)\,,\ \|(\mathrm{Re}\, g)_{|_J}\|_{L^p(J)} \leq M\}\,.
$$

Proposition 3.3 *Let $b_D \in L^p_{\mathbb{R}}(I)$, $M > 0$. There is a unique $g_0 \in B_M$ which achieves the above minimum among such functions. Moreover, if $b_D \notin Re\,(B_M)_{|_I}$, then g_0 saturates the constraint in the sense that $\|(Re\, g_0)_{|_J}\|_{L^p(J)} = M$.*

If $p = 2$, g_0 is the solution of the implicit equation

$$P_+ (\chi_I g_0) + \gamma P_+ (\chi_J Re g_0) = P_+(b_D \vee 0), \text{ for some } \gamma > 0, \tag{8}$$

where $b_D \vee 0$ coincides with b on I and vanishes on $J = \partial\Omega\backslash I$ and P_+ denotes the projection from $L^2(\partial\Omega)$ onto $H^2(\partial\Omega)$ defined by

$$P_+(f) = \frac{1}{2}(I + i\mathfrak{H})f + \frac{1}{2}L(f), \quad L(f) = \int_{\partial\Omega} f(t)d\mathcal{H}^1(t).$$

Remark 3.5 For $\Gamma \subset \partial\Omega$, the operator $\mathcal{T}_\Gamma$ acting on $H^2(\partial\Omega)$ and defined by $\mathcal{T}_\Gamma(f) = P_+(\chi_\Gamma f)$ is the Toeplitz operator with symbol $\phi = \chi_\Gamma$.

Actually, the above proposition is a corollary of the following one, from [7]. Let $b = b_D + i \int b_N \in L^p(I) = W$, and let $M > 0$. Another version of the BEP is

$$\min_{g \in B_M} \|g_{|_I} - b\|_{L^p(I)},$$

where:

$$B_M = \{g \in H^p(\partial\Omega), \ \|g_{|_J}\|_{L^p(J)} \leq M\}.$$

Proposition 3.4 *Let $b \in L^p(I)$, $M > 0$. There is a unique $g_0 \in B_M$ which achieves the above minimum among such functions. Moreover, if $b \notin (B_M)_{|_I}$, then g_0 saturates the constraint in the sense that $\|(g_0)_{|_J}\|_{L^p(J)} = M$.*

If $p = 2$, g_0 is the solution of the implicit equation

$$P_+ (\chi_I g_0) + \gamma P_+ (\chi_J g_0) = P_+(b \vee 0), \text{ for some } \gamma > 0,$$

where $b \vee 0$ coincides with b on I and vanishes on J.

The existence of a solution for the bounded extremal problem follows from the use of a standard weak compactness argument; see [14, Lem 2.1]. The saturation of the constraint by g_0 is a consequence of the density of the range of T. The uniqueness of g_0 relies on the strict convexity of the norm. In the case of $p = 2$, the implicit equation comes from the minimization of the form $\psi(g) = \|g_{|_I} - b\|^2_{L^2(I)}$ under the constraint $\phi(g) = \|(g_0)_{|_J}\|^2_{L^2(J)} = M$ using a Lagrange multiplier; see [19, Prop.2] or [14, Thm.2.1].

Proposition 3.3 is generalized to the elliptic operator $\text{div}(\sigma\nabla\cdot)$ in Sect. 3.2.2.

3.2 Conductivity PDE, Generalized Hardy Spaces

In this section, we consider the case when $\sigma \in W^{1,r}_{\mathbb{R}}(\Omega)$, $r > 2$ with $\sigma \neq 1$ for the initial problem (3). To solve the forward and the inverse problems, we transpose them in the complex plane. When $\sigma = 1$, the key ingredient to rewrite the Laplace equation as a complex partial differential equation is the existence of a harmonic conjugate. Precisely, for $\Omega = \mathbb{D}$ and $\sigma = 1$ and u solution to $\Delta u = 0$ there is a real-valued function v defined on $\mathbb{D}$ such that $u + iv$ is analytic in $\mathbb{D}$ (i.e. $\overline{\partial} f = 0$). This result can be extended to a simply connected Dini-smooth domain by the use of a conformal map. The existence of a harmonic conjugate remains true in an annular domain and thus to a finitely connected Dini-smooth domain by the use of a conformal map.

When σ is no longer constant and Ω is a simply connected domain, if $u \in L^{\gamma}_{\mathbb{R}}(\Omega)$, and $\gamma > \frac{r}{r-1}$ satisfies $\operatorname{div}(\sigma \nabla u) = 0$ in Ω, the differential form $a = (\sigma \partial_y u)dx + (\sigma \partial_x u)dy$ is closed in a simply connected domain which implies that a is exact. Thus, there exists $v \in L^{\gamma}_{\mathbb{R}}$ such that $a = (\partial_x v)dx + (\partial_y v)dy$. Then, u and v satisfy the generalized Cauchy-Riemann equations in Ω

$$\begin{cases} \partial_x v = -\sigma \partial_y u \\ \partial_y v \ = \sigma \partial_x u \end{cases}$$

and v is a solution to $\operatorname{div}\left(\frac{1}{\sigma} \nabla v\right) = 0$. For such u and v satisfying the generalized Cauchy-Riemann equations, the complex-valued function $f = u + iv$ satisfies in the distributional sense

$$\begin{aligned} \overline{\partial} f &= \partial_x u + i \partial_x v + i \partial_y u - \partial_y v \\ &= (1 - \sigma)(\partial_x u + i \partial_y u) \end{aligned}$$

while

$$\begin{aligned} \overline{\partial f} &= \overline{\partial_x u - i \partial_x v + i \partial_y u + \partial_y v} \\ &= (1 + \sigma)(\partial_x u + i \partial_y u), \end{aligned}$$

which implies that $\overline{\partial} f = \nu \overline{\partial f}$ with $\nu = \frac{1+\sigma}{1-\sigma}$. Now, if u is solution to $\operatorname{div}(\sigma \nabla u) = 0$ and

$$\operatorname*{ess\ sup}_{0<r<1} \|u\|_{L^p(\Gamma_r)} < \infty, \text{ where } \Gamma_r = \phi(r\mathbb{T}), \tag{9}$$

then u has a trace on $\partial\Omega$ and $\operatorname{tr}_{\partial\Omega} u \in L^p_{\mathbb{R}}(\partial\Omega)$ and its harmonic conjugate v has also a trace in $L^p_{\mathbb{R}}(\partial\Omega)$ making the linear operator

$$\begin{aligned} \mathfrak{H}_\nu : L^p_{\mathbb{R}}(\partial\Omega) &\longrightarrow L^p_{\mathbb{R}}(\partial\Omega) \\ u &\longmapsto v, \text{ with the normalization } \int_{\partial\Omega} v \, d\mathcal{H}^1 = 0. \end{aligned}$$

bounded on $L^p_{\mathbb{R}}(\partial\Omega)$; see [8, Cor.4.4.2.1(2)] and [18, Cor.1]. The operator $\mathfrak{H}_\nu$ is called the generalized conjugation map. For a Dini-smooth finitely connected domain Ω, a function u satisfying a condition like (9) ($\phi(r\mathbb{T})$ being replaced by $\partial\Delta_n$ which is a finite union of rectifiable Jordan curves of uniformly bounded length) has a harmonic conjugate if and only if $\int_\Gamma \sigma\partial_n u = 0$ for any smooth-curve $\Gamma \subset \Omega$; see [5, Lem.5.2].

The authors in [5, 18] proved the following result for a Dini-smooth multi-connected domain Ω, $\sigma \in W^{1,r}_{\mathbb{R}}(\Omega)$, $r > 2$. Let $\gamma > \frac{r}{r-1}$.

Proposition 3.5 *Let* $\nu = \frac{1-\sigma}{1+\sigma} \in W^{1,r}_{\mathbb{R}}(\Omega)$ *with* $\|\nu\|_{L^\infty(\Omega)} \leq \kappa < 1$. *Then, the following assertions are equivalent*

(i) $f = u + iv \in L^\gamma(\Omega)$ *satisfies the Beltrami conjugate equation*

$$\overline{\partial} f = \nu\overline{\partial f}. \tag{10}$$

(ii) $u, v \in L^\gamma_{\mathbb{R}}(\Omega)$ *satisfy respectively* $div(\sigma\nabla u) = 0$ *and* $div\left(\frac{1}{\sigma}\nabla v\right) = 0$.

(iii) $u, v \in L^\gamma_{\mathbb{R}}(\Omega)$ *are solutions of the generalized Cauchy-Riemann equations*

$$\begin{cases} \partial_x v = -\sigma\partial_y u \\ \partial_y v \ = \sigma\partial_x u. \end{cases}$$

(iv) $w = \mathcal{J}_{\nu,\alpha} f = \frac{f-\nu\overline{f}}{\sqrt{1-\nu^2}} = \sigma^{1/2}u + i\sigma^{-1/2}v$ *is the solution of* $\overline{\partial} w = \alpha\overline{w}$ *for* $\alpha = -\frac{\overline{\partial}\nu}{1-\nu^2} = \overline{\partial}\log\sigma^{1/2}$.

Definition 3.2 The generalized Hardy space of functions on Ω denoted by $H^p_\nu(\Omega)$ is the set of functions f satisfying (10) such that

$$\|f\|^p_{p,\nu} = \operatorname{ess}\sup_{0<r<1} \|f\|^p_{L^p(\Gamma_r)} < \infty,$$

where $\Gamma_r = \psi(r\mathbb{T})$. In the same way, $G^p_\alpha(\Omega)$ denotes the space of functions w satisfying $\overline{\partial} w = \alpha\overline{w}$ and $\|w\|^p_{p,\alpha} = \operatorname{ess}\sup_{0<r<1} \|w\|^p_{L^p(\Gamma_r)} < \infty$.

Remark 3.6

- When $\sigma = 1$ ($\nu = 0$, $\alpha = 0$), the space $H^p_0(\Omega) = G^p_0(\Omega)$ corresponds to the Hardy space $H^p(\Omega)$ and the norms coincide.
- The spaces H^p_ν and $G^p_\alpha(\Omega)$ equipped respectively with the norm $\|\cdot\|_{p,\nu}$ and $\|\cdot\|_{p,\alpha}$ are Banach spaces. The map $\mathcal{J}_{\nu,\alpha}$ in Proposition 3.5 is an isomorphism between the two Banach spaces.

The spaces $G^p_\alpha(\Omega)$ also called generalized Hardy spaces are related to the spaces $H^p(\Omega)$ through the following factorization theorem called Bers similarity principle; see [2, Lem. 3.1].

Proposition 3.6 *Let $\alpha \in L^r(\Omega)$ and $w \in L^{\gamma}_{loc}(\Omega)$, with $\gamma > \frac{r}{r-1}$. Then, w can be factorized as $w = e^s F$, where F is analytic in Ω and $s \in W^{1,r}(\Omega)$ with*

$$\|s\|_{W^{1,r}(\Omega)} \leq C\|\alpha\|_{L^r(\Omega)}, \tag{11}$$

the constant C depending only on r and Ω.

Remark 3.7 The Bers similarity principle remains true if Ω is a bounded domain of $\mathbb{C}$ without specific regularity on the boundary.

A consequence of this result is that $w = e^s F \in G^p_\alpha(\Omega)$ if and only if $F \in H^p(\Omega)$. The factorization combined with Equation (11) permits to establish properties of $G^p_\alpha(\Omega)$ functions from the spaces $H^p(\Omega)$. Using the isomorphism $\mathcal{J}_{\nu,\alpha}$ between $(H^p_\nu(\Omega), \|\cdot\|_{p,\nu})$ and $(G^p_\alpha(\Omega), \|\cdot\|_{p,\alpha})$, the properties will be extended to $H^p_\nu(\Omega)$-functions.

The authors in [2] give a converse of the similarity principle in the case of the unit disk for $r = 2$. We give a version of this result for $r > 2$ and a Dini-smooth domain Ω:

Proposition 3.7 *Let $\alpha \in L^r(\Omega)$, $r > 2$ and F be an analytic function on Ω. Fix $\psi \in W^{1-1/r,r}(\partial\omega)$ and $\lambda \in \mathbb{R}$. There is a unique $s \in W^{1,r}(\Omega)$ such that $Re(tr_{\partial\Omega}s) = \psi$ and $\int_{\partial\Omega} Im(s) d\mathcal{H}^1 = \lambda$ and s satisfies* (11).

Let $F \in H^p(\Omega)$ and $\alpha \in L^r(\Omega)$, $r > 2$. There is $s \in W^{1,r}(\Omega)$ such that $w = e^s F$. The map $s \longmapsto e^s$ is continuous from $W^{1,r}(\Omega)$ to $W^{1,r}(\Omega)$; see [2, Section 2].

3.2.1 Forward Problem

The following proposition gives a condition on the existence of a boundary value on $\partial\Omega$ for a function satisfying (10).

Proposition 3.8

- *A function $f \in H^p_\nu(\Omega)$ (resp. $w \in G^p_\alpha(\Omega)$) has a trace on $\partial\Omega$, $tr_{\partial\Omega} f \in L^p(\partial\Omega)$ (resp. $tr_{\partial\Omega} w \in L^p(\partial\Omega)$); moreover, there exist $C_{i,\nu}$, $C_{i,\alpha} > 0$, $i = 1, 2$, such that:*

$$C_{1,\nu}\|tr_{\partial\Omega} f\|_{L^p(\partial\Omega)} \leq \|f\|_{H^p_\nu(\partial\Omega)} \leq C_{2,\nu}\|tr_{\partial\Omega} f\|_{L^p(\partial\Omega)},$$

and

$$C_{1,\alpha}\|tr_{\partial\Omega} w\|_{L^p(\partial\Omega)} \leq \|w\|_{G^p_\alpha(\partial\Omega)} \leq C_{2,\alpha}\|tr_{\partial\Omega} w\|_{L^p(\partial\Omega)}.$$

- *The space $tr_{\partial\Omega}(H^p_\nu(\Omega))$ and $tr_{\partial\Omega}(G^p_\alpha(\Omega))$ are closed subspaces of $L^p(\partial\Omega)$ denoted by $H^p_\nu(\partial\Omega)$ and $G^p_\alpha(\partial\Omega)$.*

- *The spaces* $(H_\nu^p(\Omega), \|\cdot\|_{p,\nu})$ *and* $(H_\nu^p(\partial\Omega), \|\cdot\|_p)$ *are isomorphic; in the same way, the spaces* $(G_\alpha^p(\Omega), \|\cdot\|_{p,\alpha})$ *and* $(G_\alpha^p(\partial\Omega), \|\cdot\|_p)$ *are isomorphic.*

The forward problem in the complex plane that consists in the existence of a trace for a function in $H_\nu^p(\Omega)$ is solved: the operator $T : H_\nu^p(\Omega) \longrightarrow L_\mathbb{R}^p(\partial\Omega)$ given by $T(f) = \text{Re}(tr_{\partial\Omega} f)$ is well defined, linear and bounded. It implies that the forward problem for $\text{div}(\sigma\nabla u) = 0$ on Ω with $\text{tr}_{\partial\Omega}\, u = b_D \in L_\mathbb{R}^p(\partial\Omega)$ and the operator $\widetilde{T} : \text{Re}\,(H_\nu^p(\Omega)) \longrightarrow L_\mathbb{R}^p(\partial\Omega)$ given by $\widetilde{T}(u) = \text{tr}_{\partial\Omega} u$ are well defined, linear and bounded.

For the Neumann boundary condition (N), if $u \in W_\mathbb{R}^{1,p}(\mathbb{D})$ is such that $\text{div}(\sigma\nabla u) = 0$ with ∇u satisfying the condition (5) has a trace $\text{tr}_\mathbb{T} u \in W^{1,p}(\mathbb{T})$ and $\partial_n u \in L_\mathbb{R}^p(\mathbb{T})$ with $\int_\mathbb{T} \sigma\partial_n u = 0$, see [8, Cor.4.4.3.1] which is a generalization of [17].

3.2.2 Inverse Problems

Assume that $I = \Omega$. Given a function $b \in L_\mathbb{R}^p(\partial\Omega)$, the inverse problem asks the existence of a function $f \in H_\nu^p(\Omega)$ such that $\text{Re}(\text{tr}_{\partial\Omega} f) = b$ on $\partial\Omega$. This question is solved in [8] in the case of the unit disk for $r = \infty$, [5] in the case of a multi-connected Dini-smooth domain for $r > 2$ and [2] for $r \geq 2$ in the case of the unit disk. We collect the results in the following theorem:

Theorem 3.1 *Let* $\alpha \in W^{1,r}(\Omega)$ *and* $b \in L_\mathbb{R}^p(\partial\Omega)$. *There exists* $f \in H_\nu^p(\Omega)$ *such that* $Re(tr_{\partial\Omega} f) = b$ *on* $\partial\Omega$ *and* $\int_{\partial\Omega} Im(tr_{\partial\Omega} f) d\mathcal{H}^1 = 0$. *Moreover, there is* $C_{p,\alpha,\nu} > 0$ *such that:*

$$\|tr_{\partial\Omega} f\|_p \leq C_{p,\alpha,\nu}\|b\|_p. \tag{12}$$

Remark 3.8 Under the same hypothesis, there is a unique function $u \in \text{Re}(H_\nu^p(\Omega))$ defined on Ω such that $\text{tr}_{\partial\Omega} u = b$.

Thus, T is surjective and bijective when restricted to:

$$H_\nu^p(\Omega) / \left\{ f \in H_\nu^p(\Omega) : \int_{\partial\Omega} \text{tr}_{\partial\Omega}\, f d\mathcal{H}^1 = 0 \right\}.$$

Given a function $g \in L_\mathbb{R}^p(\mathbb{T})$ such that $\int_\mathbb{T} \sigma\partial_n(u) = 0$, there exists $u \in W_\mathbb{R}^{1,p}(\mathbb{D})$ unique up to an additive constant satisfying $\text{div}(\sigma\nabla u) = 0$ in $\mathbb{D}$ with ∇u satisfying (5) with $\partial_n(u) = g$ on $\mathbb{T}$.

Assume now that $I \subsetneq \partial\Omega$ with $\mathcal{H}^1(I) > 0$ and $\mathcal{H}^1(J) > 0$, $J = \partial\Omega \backslash I$.

Proposition 3.9 *The space* $(H_\nu^p(\partial\Omega))_{|_I}$ *is dense in* $L^p(I)$. *Moreover, the range of* T *is dense in* $L_\mathbb{R}^p(\partial\Omega)$ *and when* $b_D \notin Ran\ T$ *the inverse problem is ill-posed.*

In particular, $(\mathrm{Re}(H^p_\nu(\partial\Omega)))_{|_I}$ is dense in $L^p_{\mathbb{R}}(I)$. Because of the ill-posedness of the inverse problem, one can use a bounded extremal problem (BEP) formulated in the more general setting of generalized Hardy spaces:

$$\min_{g\in B_M} \|\mathrm{Re}(g)_{|_I} - b\|_{L^p_{\mathbb{R}}(I)}, \tag{13}$$

where:

$$B_M = \{g \in H^p_\nu(\partial\Omega),\ \|\mathrm{Re}(g)_{|_J}\|_{L^p_{\mathbb{R}}(J)} \le M\}.$$

Proposition 3.10 *Let $b \in L^p_{\mathbb{R}}(I)$, $M > 0$. There exists a unique $g_0 \in B_M$ such that g_0 solves* (13). *Moreover, if $b \notin (B_M)_{|_I}$, then g_0 saturates the constraint: $\|Re(g_0)\|_{L^p_{\mathbb{R}}(I)} = M$.*

If $p = 2$ and $\Omega = \mathbb{D}$, g_0 satisfies the implicit equation

$$\mathcal{P}_\nu(\chi_I g_0) + \gamma\mathcal{P}_\nu(\chi_J Re(g_0)) = \mathcal{P}_\nu(b \vee 0),$$

where $\mathcal{P}_\nu : L^2(\mathbb{T}) \longrightarrow H^p_\nu(\mathbb{T})$ defined by

$$\mathcal{P}_\nu(f) = \frac{1}{2}(I + i\mathfrak{H}_\nu)f + \frac{1}{2}L(f), \quad L(f) = \frac{1}{2\pi}\int_0^{2\pi} f(e^{it})dt.$$

Remark 3.9 The operator $\mathcal{P}_\nu$ generalizes P_+ (see Remark 3.5) and thus extends the definition of a Toeplitz operator with symbol $\phi = \chi_I$ on generalized Hardy spaces: $\mathcal{T}_\Gamma : L^p(\partial\Omega) \longrightarrow H^p_\nu(\partial\Omega)$, with $\mathcal{T}_\Gamma(f) = P_\nu(\chi_I f)$, $f \in L^p(\partial\Omega)$.

4 Conclusion

The results for the forward and the inverse problems can be extended to finitely connected bounded Dini-smooth domains [5], with some variations of the BEP in terms of the constraint on the imaginary part of the function g; see [22, 27–29].

We focused on inverse problems with partial boundary data: this situation occurs when physical quantities are measured on the boundary of a bounded domain and the source of such measurements is located inside the domain, like the EEG source estimation problem described in Sect. 1. In some situations, the measures are taken inside the domain and the source is defined on the boundary. For example, in geosciences and planetary sciences, the inverse magnetization problem is to estimate the magnetization μ from measurements of the normal component of the magnetic field measured within a domain Ω, far from the support of μ (see [26] in $\mathbb{R}^2$, [3, 6] in $\mathbb{R}^3$): find μ supported on $S \subset \partial\Omega$ (model for a magnetized thin rock sample),

$$\begin{cases} \Delta u = \operatorname{div} \mu \,, \\ (\partial_n u)_{|K} \text{ given on } K \subset \Omega \,. \end{cases}$$

Such problems have been studied in [26] where $\Omega = \Pi_+$ the upper-half plane that corresponds to the two-dimensional analogous of the physical three-dimensional situation (see [3, 4, 6]) and to the Hilbertian setting where $p = 2$. This problem and related issues of course make sense in more general domains Ω and for $p \neq 2$. It is our belief that the corresponding properties and results remain valid for $1 < p < \infty$ and in Dini-smooth domains Ω.

References

1. L. Baratchart, A. Ben Abda, F. Ben Hassen, J. Leblond, *Recovery of pointwise sources or small inclusions in 2D domains and rational approximation*, Inverse problems, 21, 51–74, 2005.
2. L. Baratchart, A. Borichev, S. Chaabi, *Pseudo-holomorphic functions at the critical exponent*, J. Eur. Math. Soc. (JEMS), 18 (9), 1919–1960, 2016.
3. L. Baratchart, S. Chevillard, D. Hardin, J. Leblond, E. A. Lima, J.-P. Marmorat, *Magnetic moment estimation and bounded extremal problems*, Inverse Problems & Imaging, 13 (1), 2019.
4. L. Baratchart, S. Chevillard, J. Leblond, *Silent and equivalent magnetic distributions on thin plates*, Theta Series in Advanced Mathematics, Proceedings of the Conference "Analyse Harmonique et Fonctionnelle, Théorie des Opérateurs et Applications" 2015 (in honor of Jean Esterle), eds. Ph. Jaming, A. Hartmann, K. Kellay, S. Kupin, G. Pisier, D. Timotin, Theta Series in Advanced Mathematics, 11–28, 2017.
5. L. Baratchart, Y. Fischer, J. Leblond, *Dirichlet/Neumann problems and Hardy classes for the planar conductivity equation*, Complex Variable & Elliptic Equations, 59 (4), 504–538, 2014.
6. L. Baratchart, D. P. Hardin, E. A. Lima, E. B. Saff, B. P. Weiss, *Characterizing kernels of operators related to thin-plate magnetizations via generalizations of Hodge decompositions*, Inverse Problems, 29 (1), 2013.
7. L. Baratchart, J. Leblond, *Hardy approximation to L^p functions on subsets of the circle with* $1 \leq p < \infty$, Constr. Approx., 14 (1), 1998.
8. L. Baratchart, J. Leblond, S. Rigat, E. Russ, *Hardy spaces of the conjugate Beltrami equation*, Journal of Functional Analysis, 259, 384–427, 2010.
9. B. Beauzamy, *Introduction to Banach spaces and their geometry*, North-Holland Mathematics Studies, 68, 1985.
10. A. Ben Abda, F. Ben Hassen, J. Leblond, M. Mahjoub, *Sources recovery from boundary data: a model related to electroencephalography*, Mathematical and Computer Modelling, 49, 2213–2223, 2009.
11. F. J. Beutler,*The operator theory of the pseudo-inverse. II. Unbounded operators with arbitrary range*, Journal of Mathematical Analysis and Applications, 10, 471–493, 1965.
12. H. Brezis, *Functional Analysis, Sobolev Spaces and Partial Differential Equations*, Springer, 2011.
13. I. Chalendar, J. Leblond, J. R. Partington, *Approximation problems in some holomorphic spaces, with applications*, Systems, Approximation, Singular Integral Operators, and Related Topics, Proceedings of IWOTA 2000, eds. A.A. Borichev, N.K. Nikolski, Integral Equations and Operator Theory 129, 143–169, 2001.
14. I. Chalendar, J. R. Partington, *Constrained approximation and invariant subspaces*, J. Math. Anal. Appl., 280, 176–187, 2003.

15. M. Clerc, J. Leblond, J.-P. Marmorat, T. Papadopoulo, *Source localization in EEG using rational approximation on plane sections*, Inverse Problems, 28, 2012.
16. P. L. Duren, *Theory of H^p spaces*, Pure and Applied Mathematics, 38, 1970.
17. E. Fabes, M. Jodeit, N. Rivière,*Potential techniques for boundary value problems on C^1 domains*, Acta Math., 141,165–186, 1978.
18. Y. Fischer, J. Leblond, *Solutions to conjugate Beltrami equations and approximation in generalized Hardy spaces*, Advances in Pure and Applied Mathematics, 2 (1), 2011.
19. Y. Fischer, J. Leblond, J. R. Partington, E. Sincich, *Bounded extremal problems in Hardy spaces for the conjugate Beltrami equation in simply-connected domains*, Appl. Comput. Harmon. Anal., 31 (2), 264–285, 2011.
20. J. B. Garnett, *Bounded analytic functions*, Graduate Texts in Mathematics, 236, Springer, 2007.
21. M. S. Gockenbach, *Linear inverse problems and Tikhonov regularization*, Carus Mathematical Monographs, 32, Mathematical Association of America, Washington, DC, 2016.
22. B. Jacob, J. Leblond, J.-P. Marmorat, J. R. Partington, *A constrained approximation problem arising in parameter identification*, Linear Algebra and its Applications, 351–352, 487–500, 2002.
23. J. Leblond, J. R. Partington, E. Pozzi, *Best approximation problems in Hardy spaces and by polynomials, with norm constraints*, Integral Equations & Operator Theory, 75 (4), 491–516, 2013.
24. J. Leblond, D. Ponomarev, *On some extremal problems for analytic functions with constraints on real or imaginary parts*, Festschrift in Honor of Daniel Alpay's 60th Birthday, F. Colombo & al. (eds.), Advances in Complex Analysis and Operator Theory, Trends in Math., Birkhäuser, 219–236, 2017.
25. J. Leblond, D. Ponomarev, *Recovery of analytic functions with prescribed pointwise values on the disk from partial boundary data*, Journal of Inverse and Ill-Posed Problems, 25 (2), 157–174, 2017.
26. J. Leblond, E. Pozzi, *Solutions to inverse moment estimation problems in dimension 2, using best constrained approximation*, Journal of Approximation Theory, 264, 2021.
27. D. J. Patil, *Representation of H^p-functions*, Bulletin of the American Mathematical Society, 78, 1972.
28. Ch. Pommerenke, *Boundary behaviour of conformal maps*, Grundlehren der mathematischen Wissenschaften [Fundamental Principles of Mathematical Sciences], 299, 1992.
29. D. Ponomarev, *Some inverse problems with partial data*, PhD thesis, Université Nice Sophia Antipolis, 2016.

Applications of the Automatic Additivity of Positive Homogeneous Order Isomorphisms Between Positive Definite Cones in C^*-Algebras

Lajos Molnár

Dedicated to my little Elza Janka

1 Introduction

For a C^*-algebra $\mathcal{A}$ (in this paper all C^*-algebras are supposed to be unital), we denote by $\mathcal{A}_s, \mathcal{A}^+, \mathcal{A}^{++}$ the sets of all self-adjoint, all positive semidefinite (i.e., self-adjoint with nonnegative spectrum), and all positive definite (i.e., invertible positive semidefinite) elements of $\mathcal{A}$, respectively. For apparent reasons, we call $\mathcal{A}^+$ the positive semidefinite cone and $\mathcal{A}^{++}$ the positive definite cone of $\mathcal{A}$. The usual order among the elements of $\mathcal{A}_s$ is defined in the following way. For any $A, B \in \mathcal{A}_s$, we write $A \leq B$ if and only if $B - A \in \mathcal{A}^+$.

In what follows, we survey results on the descriptions of the structures of certain isomorphisms, isometries, or, in general, symmetries of positive definite cones in C^*-algebras in the proof of which a former observation of ours asserting that homogeneous order isomorphisms are automatically additive plays a fundamental role.

In fact, the key concept here is that of the Jordan *-isomorphisms. A bijective linear map $J : \mathcal{A} \to \mathcal{B}$ between C^*-algebras $\mathcal{A}, \mathcal{B}$ is called a Jordan *-isomorphism if it preserves the Jordan product as well as the involution, that is, if it satisfies

$$J(AB + BA) = J(A)J(B) + J(B)J(A) \quad \text{and} \quad J(A^*) = J(A)^*, \quad A, B \in \mathcal{A}.$$

Those maps represent one of the most important classes of symmetries of C^*-algebras.

For our present purposes, notice the following obvious facts. Jordan *-isomorphisms between C^*-algebras map positive definite cones onto positive

L. Molnár (✉)
Bolyai Institute, University of Szeged, Szeged, Hungary
Department of Analysis, Institute of Mathematics, Budapest University of Technology and Economics, Budapest, Hungary
e-mail: molnarl@math.u-szeged.hu

I. Binder et al. (eds.), *Function Spaces, Theory and Applications*, Fields Institute Communications 87, https://doi.org/10.1007/978-3-031-39270-2_4

definite cones and they are positive homogeneous order isomorphisms (when restricted to such cones). A bijective map ϕ between the positive definite cones $\mathcal{A}^{++}, \mathcal{B}^{++}$ is called an order isomorphism if for any $A, B \in \mathcal{A}^{++}$ we have

$$A \leq B \Longleftrightarrow \phi(A) \leq \phi(B).$$

Positive homogeneity means that $\phi(\lambda A) = \lambda\phi(A)$ holds for all $A \in \mathcal{A}^{++}$ and positive real number λ. Apparently, besides Jordan *-isomorphisms, all congruence transformations, i.e., maps of the form $A \mapsto TAT^*$, where T is any fixed invertible element of $\mathcal{A}$, also have those properties; they map positive definite cones onto positive definite cones and are positive homogeneous order isomorphisms (when restricted to such cones). Surprisingly, a kind of converse statement is also true. Namely, all positive homogeneous order isomorphisms between positive definite cones can be obtained as compositions of those two types of transformations. Namely, we have the following statement which is of central importance for the proofs of the results to be presented in this paper.

Proposition K *Let $\mathcal{A}, \mathcal{B}$ be C^*-algebras. Assume that $\phi : \mathcal{A}^{++} \to \mathcal{B}^{++}$ is a positive homogeneous order isomorphism. Then ϕ is necessarily of the form*

$$\phi(A) = TJ(A)T, \quad A \in \mathcal{A}^{++}$$

*with some $T \in \mathcal{B}^{++}$ and Jordan *-isomorphism $J : \mathcal{A} \to \mathcal{B}$.*

(The reason behind that T above is not a general invertible element but a positive one is that, by polar decomposition, any invertible element $T \in \mathcal{A}$ can be written as $T^* = U^*|T^*|$ with a unitary $U \in \mathcal{A}$, and the map $A \mapsto UAU^*$ is an algebra *-automorphism and hence also a Jordan *-automorphism of $\mathcal{A}$.)

The trivial consequence of Proposition K is that the additivity of positive homogeneous order isomorphisms between positive definite cones in C^*-algebras is automatic, we get it for free.

Our key proposition is in fact a simple consequence of our former result Theorem 9 in [18] in which we described the structure of all surjective Thompson isometries between positive definite cones. We will discuss that theorem in the next section.

Observe that, obviously, the condition of positive homogeneity in Proposition K cannot be dropped as one can easily see it by considering the algebra of complex numbers, or any other commutative C^*-algebra. The question of when do we have a similar conclusion (we mean automatic additivity) for merely order isomorphisms (i.e., without the assumption of positive homogeneity) has recently been closely investigated. In this respect we mention our paper [23] the results in which motivated the large part of those investigations and then the deep results in [39] (in the setting of the full Hilbert space operator algebra) and in [37] (in the context of von Neumann algebras without commutative direct summands) and the paper [20] (where important general results are presented which can be employed in the particular case of so-called standard operator algebras; see, e.g., Proposition 17 in [33]).

In the past few years, we found several interesting (at least we believe so) applications of Proposition K, some of which we surveyed in the paper [31]. In the present article we summarize our most recent results whose proofs rest also on our key proposition formulated above.

2 Thompson Isometries

Let us briefly discuss the background of our Proposition K. As already mentioned, that statement is a simple corollary of our former result Theorem 9 in [18] on the description of the structure of the so-called surjective Thompson isometries.

The Thompson metric (or Thompson part metric) is an important distance measure on positive definite cones. In fact, it is originally defined in a setting which is far more general than that of the C^*-algebras; see [40]. However, here we restrict our attention only to the latter structures. The Thompson metric d_T is defined on the positive definite cone $\mathcal{A}^{++}$ of the C^*-algebra $\mathcal{A}$ as follows:

$$d_T(A, B) = \log \max\{M(A/B), M(B/A)\}, \quad A, B \in \mathcal{A}^{++}, \tag{1}$$

where $M(X/Y) = \inf\{t > 0 : X \le tY\}$, $X, Y \in \mathcal{A}^{++}$. It is quite easy to see that d_T is a true metric and it is just a triviality that all positive homogeneous order isomorphisms between positive definite cones are surjective Thompson isometries.

Actually, in the current setting of C^*-algebras, d_T can be computed in a more explicit way. Indeed, since for $t, s > 0$, the inequalities $A \le tB$, $B \le sA$ are equivalent to

$$-(\log t)I \le \log A^{-1/2}BA^{-1/2} \le (\log s)I,$$

it is apparent that d_T can also be rewritten as

$$d_T(A, B) = \left\| \log \left(A^{-1/2}BA^{-1/2} \right) \right\|, \quad A, B \in \mathcal{A}^{++}. \tag{2}$$

Observe that, although the triangle inequality for d_T is quite easily deducible using (1), it is very difficult to see it from (2) (cf. the remarks given at the end of this section). Let us also mention the important fact that d_T shows up also as a geodesic distance in a Finsler-type geometry on the positive definite cones of C^*-algebras which was introduced and studied by Corach, Porta, and Recht in the 1990s; see [8–10], and also [11].

One of the most important properties of the Thompson metric which makes it so useful is that, with respect to d_T, the positive definite cone is a complete metric space and its topology coincides with the topology of the C^*-norm (these facts can be verified with not much efforts).

Let us now turn to the description of the structure of surjective Thompson isometries between positive definite cones. First, observe the following. The Jordan *-isomorphisms, the congruence transformations defined above, as well as the map of the inverse operation are all surjective Thompson isometries when restricted to positive definite cones (these follow easily from the original definition of the Thompson metric (1)). Quite surprisingly, it turns out that, to a certain extent, those are the prototypes of the surjective Thompson isometries. Indeed, Theorem 9 in [18] reads as follows. An element of an algebra is called central if it commutes with all elements of the algebra.

Theorem 2.1 *Let $\mathcal{A}, \mathcal{B}$ be C^*-algebras. The surjective map $\phi : \mathcal{A}^{++} \to \mathcal{B}^{++}$ is a Thompson isometry if and only if there is a central projection P in $\mathcal{B}$ and a Jordan *-isomorphism $J : \mathcal{A} \to \mathcal{B}$ such that ϕ is of the form*

$$\phi(A) = \phi(I)^{1/2} \left(PJ(A) + (I - P)J(A^{-1})\right) \phi(I)^{1/2}, \quad A \in \mathcal{A}^{++}. \tag{3}$$

The statement in Proposition K immediately follows from this theorem. Indeed, as we have already pointed out, any positive homogeneous order isomorphism between positive definite cones is a surjective Thompson isometry, so Theorem 2.1 applies and the positive homogeneity clearly implies that the central projection P above is necessarily the identity (because of the positive homogeneity, the part $(I - P)J(A^{-1})$ in (3) must be missing).

The original proof of Theorem 2.1 was given in [18]. In [31] we presented a somewhat different argument based on the interesting observation that surjective Thompson isometries necessarily preserve the geometric mean. In the rest of the section we discuss this observation.

The geometric mean (first considered by Pusz and Woronowicz in [38]) of the elements $A, B \in \mathcal{A}^{++}$ is

$$A\#B = A^{1/2}(A^{-1/2}BA^{-1/2})^{1/2}A^{1/2}.$$

(This is an operator mean in the Kubo-Ando sense; see the next section for some very basic facts of the Kubo-Ando theory).

The abovementioned mean-preserving property of surjective Thompson isometries is in a very nice analogy with the main observation in the famous Mazur-Ulam theorem [22] which is that surjective isometries between normed linear spaces preserve the arithmetic mean of elements (this immediately implies that those isometries are necessarily affine maps). In [31] we verified the geometric mean-preserving property of surjective Thompson isometries by referring to our former paper [25] where we actually used some basic ideas from the ingenious proof of the Mazur-Ulam theorem given by Väisälä [42].

In what follows we demonstrate that the geometric mean-preserving property can in fact be obtained by employing Mazur and Ulam's original ideas, too. Indeed, we are going to present a metric characterization of the geometric mean in terms of the Thompson metric. (Since the Thompson metric is expressed by the order and the operation of positive scalar multiplication, we also obtain that the geometric mean can, implicitly, be characterized by them as well.)

We begin with some notation. Let $\mathcal{S}$ be a bounded subset of a metric space $\mathcal{X}$ with metric d. Define

$$\kappa(\mathcal{S}) = \{X \in \mathcal{S} : d(X, Y) \le (1/2)\mathrm{diam}(\mathcal{S}), Y \in \mathcal{S}\}.$$

Apparently, $\kappa(\mathcal{S}) \subset \mathcal{S}$ and $\mathrm{diam}(\kappa(\mathcal{S})) \le (1/2)\mathrm{diam}(\mathcal{S})$. Therefore, defining recursively $\kappa_0(\mathcal{S}) = \mathcal{S}$ and $\kappa_n(\mathcal{S}) = \kappa(\kappa_{n-1}(\mathcal{S}))$ for every positive integer n, we see that $(\kappa_n(\mathcal{S}))$ is a decreasing sequence of subsets of $\mathcal{X}$ whose intersection $C(\mathcal{S})$ is either empty or a singleton. In the latter case we call the unique element of $C(\mathcal{S})$ the core of $\mathcal{S}$. For example, if the space $\mathcal{X}$ is complete and $\mathcal{S}$ is closed, then we necessarily have that $\mathcal{S}$ has a core.

For any $A, B \in \mathcal{X}$ define

$$\mathcal{E}(A, B) = \{X \in \mathcal{X} : d(X, A) = d(X, B) = (1/2)d(A, B)\}.$$

This is the set of all points which are in equal distance to A and B and this distance is the half of the distance between A and B. One can call the elements of $\mathcal{E}(A, B)$ the metric midpoints between A and B. We will prove that for the positive definite cone $\mathcal{A}^{++}$ of the C^*-algebra $\mathcal{A}$ equipped with the Thompson metric d_T, the core of $\mathcal{E}(A, B)$ is exactly $A\#B$ which gives the desired metric characterization of the geometric mean.

Proposition 2.1 *If $\mathcal{A}$ is a C^*-algebra and $\mathcal{A}^{++}$ is equipped with the Thompson metric, then for any $A, B \in \mathcal{A}^{++}$ we have*

$$C(\mathcal{E}(A, B)) = \{A\#B\}.$$

To verify this statement we first prove the following lemma which says that the core of a kind of a symmetric bounded set in the positive definite cone (namely, one which is invariant under the inverse operation) that contains the identity is the identity.

Lemma 2.1 *Let $\mathcal{S}$ be a bounded subset of $\mathcal{A}^{++}$ such that $I \in \mathcal{S}$ and for any $A \in \mathcal{S}$ we have $A^{-1} \in \mathcal{S}$. Then $C(\mathcal{S}) = \{I\}$.*

Proof We first prove that $\kappa(\mathcal{S})$ contains I and it is invariant under taking inverses. It will follow that I belongs to all $\kappa_n(\mathcal{S})$ and we will be done.

Recall that the map of the inverse operation is a Thompson isometry. If $X \in \kappa(\mathcal{S})$, then for any $Y \in \mathcal{S}$ we have $Y^{-1} \in \mathcal{S}$ and

$$d_T(X^{-1}, Y) = d_T(X, Y^{-1}) \le (1/2)\mathrm{diam}(\mathcal{S}).$$

This implies that $X^{-1} \in \kappa(\mathcal{S})$. To see that $I \in \kappa(\mathcal{S})$, pick any $Y \in \mathcal{S}$. Using the formula (2), we easily compute

$$d_T(I, Y) = (1/2)d_T(Y, Y^{-1}) \leq (1/2)\text{diam}(\mathcal{S})$$

and this completes the proof. □

Let now $\mathcal{X}, \mathcal{Y}$ be metric spaces and $\phi : \mathcal{X} \to \mathcal{Y}$ be a surjective isometry. Then for any bounded subset $\mathcal{S}$ of $\mathcal{X}$ we have $C(\phi(\mathcal{S})) = \phi(C(\mathcal{S}))$. Indeed, it is easy to see that

$$\phi(\kappa(\mathcal{S})) = \kappa(\phi(\mathcal{S}))$$

and then that

$$\phi(\kappa_n(\mathcal{S})) = \kappa_n(\phi(\mathcal{S}))$$

holds for every positive integer n which implies $C(\phi(\mathcal{S})) = \phi(C(\mathcal{S}))$. Therefore, any surjective isometry maps the core of a set to the core of the image of that set.

***Proof (Proof of Proposition* 2.1)** Assume first that $B = A^{-1}$. Then the set $\mathcal{E}(A, A^{-1})$ contains I and it is invariant under taking inverses. Indeed,

$$d_T(I, A) = d_T(I, A^{-1}) = (1/2)d_T(A, A^{-1}),$$

hence $I \in \mathcal{E}(A, A^{-1})$. For any $X \in \mathcal{E}(A, A^{-1})$ we have

$$d_T(X, A) = d_T(X, A^{-1}) = (1/2)d_T(A, A^{-1})$$

from which it follows that

$$d_T(X^{-1}, A) = d_T(X, A^{-1}) = d_T(X, A) = d_T(X^{-1}, A^{-1}) = (1/2)d_T(A, A^{-1}).$$

This implies that $X^{-1} \in \mathcal{E}(A, A^{-1})$. Hence, by the previous lemma, we obtain that $C(\mathcal{E}(A, A^{-1})) = \{I\}$.

Now pick arbitrary $A, B \in \mathcal{A}^{++}$. Let $C = (A\#B)^{-1/2}A(A\#B)^{-1/2}$ and consider the congruence transformation $X \mapsto \psi(X) = (A\#B)^{1/2}X(A\#B)^{1/2}$ on $\mathcal{A}^{++}$ which is a surjective Thompson isometry. We can see that

$$\psi(C) = A, \quad \psi(C^{-1}) = (A\#B)A^{-1}(A\#B) = B, \quad \psi(I) = A\#B.$$

Applying the transformation ψ on

$$C(\mathcal{E}(C, C^{-1})) = \{I\},$$

we obtain the desired equality $C(\mathcal{E}(A, B)) = \{A\#B\}$. □

From the above results and discussion we then obtain the key point of the proof of Theorem 2.1.

Proposition 2.2 *Any surjective Thompson isometry $\phi : \mathcal{A}^{++} \to \mathcal{B}^{++}$ between positive definite cones in C^*-algebras preserves the geometric mean, i.e., it satisfies*

$$\phi(A\#B) = \phi(A)\#\phi(B), \quad A, B \in \mathcal{A}^{++}. \tag{4}$$

This means that ϕ is a isomorphism under the operation of the geometric mean.

We finish this section with some remarks. Above we have considered the Thompson metric

$$d_T(A, B) = \left\| \log \left(A^{-1/2} B A^{-1/2} \right) \right\|, \quad A, B \in \mathcal{A}^{++}$$

and presented the description of the structure of the corresponding surjective isometries.

Following the work [15] by Fujii which concerned matrix algebras, in Theorem 8 in [17] Hatori proved the following statement concerning operator algebras: If H is a complex Hilbert space and $|||.|||$ is a complete unitarily invariant uniform norm on the algebra $B(H)$ of all bounded linear operators acting on H, then

$$(A, B) \to \left|\left|\left| \log A^{-1/2} B A^{-1/2} \right|\right|\right| \tag{5}$$

is a metric on the positive definite cone of $B(H)$. Let us emphasize that this is a really difficult statement whose proof requires deep arguments (which is in contrast to the case of the C^*-norm where it follows easily from (1)).

Next, observe that the map of the inverse operation is an isometry with respect to any metric of the form (5), this is just a triviality. However, to see that the congruence transformations are also isometries is no longer as easy as in the case of the Thompson metric. In fact, it follows from the observation that, given a C^*-algebra $\mathcal{A}$, for any $A, B \in \mathcal{A}^{++}$ and invertible $T \in \mathcal{A}$, the element $(TAT^*)^{-1/2} TBT^* (TAT^*)^{-1/2}$ is unitarily equivalent to $A^{-1/2} B A^{-1/2}$. Namely, with the polar decomposition $X = V|X|$ of the element $X = A^{-1/2} BT^* (TAT^*)^{-1/2}$ we have

$$(TAT^*)^{-1/2} TBT^* (TAT^*)^{-1/2} = V^* A^{-1/2} B A^{-1/2} V$$

(the details are given on pp. 153–154 in [36]). As an immediate consequence of these facts, one can readily verify that the metric characterization Proposition 2.1 of the geometric mean remains valid also for any metrics that appear in (5).

In view of the result Theorem 2.1 it is now a natural question to ask if one can precisely describe all the surjective isometries between positive definite cones corresponding to the metrics in (5). The fact is that in the general case the problem seems quite hopeless but in the finite-dimensional case, for matrix algebras, we have an acceptable result. We refer to Theorem 3 in [26] which is a sharp statement in the following sense. Depending on the dimension and depending on whether the considered unitarily invariant norm (in finite dimension such norms are known to be automatically complete and uniform) $|||.|||$ is a scalar multiple of the Hilbert-Schmidt norm (in another word Frobenius norm) or not, that result gives the precise list of possible surjective isometries (of the positive definite cone) for the collection of all possible norms $|||.|||$.

Our last remark in this section concerns Proposition 2.1. Denote by $\mathbb{M}_n$ the algebra of all n by n complex matrices and by $\mathbb{P}_n$ the corresponding positive definite cone. If $|||.|||$ is the Hilbert-Schmidt norm on $\mathbb{M}_n$, i.e., we consider the metric

$$(A, B) \mapsto (\mathrm{Tr}(\log A^{-1/2}BA^{-1/2})^2)^{1/2} \tag{6}$$

on $\mathbb{P}_n$, then we actually have the equality

$$\mathcal{E}(A, B) = \{A\#B\} \quad A, B \in \mathbb{P}_n.$$

This means that there is a unique metric midpoint between A and B and it is exactly the geometric mean. This observation follows from the so-called semiparallelogram law; see 6.1.9 Theorem in [2]. Actually, the special metric (6) is particularly important since behind that there is a Riemannian structure on $\mathbb{P}_n$ which has wide-ranging applications. It is a very notable fact that in that structure, for any two points $A, B \in \mathbb{P}_n$, there is a unique geodesic curve connecting A and B, namely,

$$t \to A^{1/2}(A^{-1/2}BA^{-1/2})^t A^{1/2}, \quad t \in [0, 1],$$

and its midpoint is the geometric mean $A\#B$ (the general points of that curve are the weighted geometric means of A and B) and the length of that curve is just the value in (6). For details, see Chapter 6 in Bhatia's book [2]. Similar statement is true in the context of general C^*-algebras relating the Thompson metric; see [9].

3 Preservers of Means

In the previous section we have seen that surjective Thompson isometries between positive definite cones necessarily preserve the geometric mean; they are algebraic isomorphisms with respect to that operation. The natural question arises whether we can precisely describe the structure of those transformations. The fact is that we do not have such a description concerning general C^*-algebras; we have a satisfactory result only for factor von Neumann algebras. The necessity parts of the following statement appeared as Theorem 4 in [27]. Recall that on any finite von Neumann algebra there is a unique center-valued positive linear transformation which is tracial (i.e., vanishes on commutators) and acts as the identity on the center. In the case of a finite factor, this functional is called normalized trace, and we denote it by Tr.

Theorem 3.1 *Assume that $\mathcal{A}, \mathcal{B}$ are factor von Neumann algebras, and $\phi : \mathcal{A}^{++} \to \mathcal{B}^{++}$ is a continuous bijective map.*

*Suppose that $\mathcal{A}$ is of infinite type. Then ϕ preserves the geometric mean (satisfies (4)) if and only if there is either an algebra *-isomorphism or an algebra *-antiisomorphism $\theta : \mathcal{A} \to \mathcal{B}$, $c \in \{-1, 1\}$, and an element $T \in \mathcal{B}^{++}$ such that*

$$\phi(A) = T\theta(A^c)T, \qquad A \in \mathcal{A}^{++}.$$

*Assume that $\mathcal{A}$ is of finite type. Then ϕ preserves the geometric mean if and only if there is either an algebra *-isomorphism or an algebra *-antiisomorphism $\theta : \mathcal{A} \to \mathcal{B}$, $c \in \{-1, 1\}$, a real number d with $d \neq -c$, and an element $T \in \mathcal{B}^{++}$ such that*

$$\phi(A) = e^{d\mathrm{Tr}(\log A)} T\theta(A^c)T, \qquad A \in \mathcal{A}^{++}.$$

Above we have obtained the precise descriptions of the structures of continuous geometric mean-preserving bijections between the positive definite cones of von Neumann factors. Concerning general C^*-algebras we do not have any result; indeed, we consider that a difficult problem.

As for the continuity assumption in Theorem 3.1, it is not difficult to show that it cannot be dropped. To see this, let $\alpha : \mathbb{R} \to \mathbb{R}$ be a discontinuous bijective additive function. Pick a positive integer n and set $a(t) = (\alpha(t) - t)/n$, $t \in \mathbb{R}$. Then the transformation

$$\phi(A) = e^{a(\log \det A)} A, \quad A \in \mathbb{P}_n$$

is a bijective selfmap of $\mathbb{P}_n$ which preserves the geometric mean but not continuous, hence not of the form that appears in Theorem 3.1.

Generally speaking, means play very important roles in a number of areas of mathematics including operator theory. So, in view of the previous theorem, it is a natural problem to investigate isomorphisms or symmetries with respect not only to the geometric but also to other means.

We recall that there is a general theory of operator means due to Kubo and Ando [19]. Their kinds of means are originally defined for the positive semidefinite cone of the full operator algebra $B(H)$ over an (infinite-dimensional) Hilbert space H as two-variable operations satisfying certain conditions. One of the fundamental results of the celebrated Kubo-Ando theory is that those types of means are in one-to-one correspondence with operator monotone functions on the positive reals. This very important observation makes it possible to extend the concept of Kubo-Ando means to the positive definite(!) cones of abstract C^*-algebras in a straightforward, simple way as follows. Let $f :]0, \infty[\to]0, \infty[$ be an operator monotone function with the property $f(1) = 1$. The corresponding Kubo-Ando mean σ on the positive definite cone of any C^*-algebra $\mathcal{A}$ is defined by the formula

$$A\sigma B = A^{1/2} f(A^{-1/2} B A^{-1/2}) A^{1/2}, \quad A, B \in \mathcal{A}^{++}. \tag{7}$$

The most important properties of Kubo-Ando means are the monotonicity in their variables and the transfer equality/property, i.e., the invariance of those means under all congruence transformations. If we look at the definition (7) above given for general positive definite cones, then we notice that the monotonicity of $A\sigma B$ in the variable B is obvious. To see the monotonicity in the other variable A, one can use the equality

$$A^{1/2}f(A^{-1/2}BA^{-1/2})A^{1/2} = B^{1/2}g(B^{-1/2}AB^{-1/2})B^{1/2}, \tag{8}$$

where $g(t) = tf(1/t)$, $t > 0$, and this function g is known to be also operator monotone (in fact, g is the operator monotone function corresponding to the mean $(A, B) \mapsto B\sigma A$ called the transpose of σ). Behind equality (8) stands the identity $Xf(X^*X) = f(XX^*)X$ which is valid for all continuous functions f on the positive reals and for all invertible elements $X \in \mathcal{A}$. The invariance under the congruence transformation $T(.)T^*$ for an arbitrary invertible element $T \in \mathcal{A}$, i.e., the identity

$$TA^{1/2}f(A^{-1/2}BA^{-1/2})A^{1/2}T^*$$
$$= (TAT^*)^{1/2}f((TAT^*)^{-1/2}(TBT^*)(TAT^*)^{-1/2})(TAT^*)^{1/2}, \quad A, B \in \mathcal{A}^{++}$$

can be calculated quite easily using the polar decomposition $A^{1/2}T^* = V|A^{1/2}T^*|$ of any element $A \in \mathcal{A}^{++}$.

Let us now make a comment regarding the properties of Jordan *-isomorphisms. (For a more detailed discussion, see [31].) If $J : \mathcal{A} \to \mathcal{B}$ is a Jordan *-isomorphism, then it preserves the Jordan triple product, i.e., it satisfies $J(ABA) = J(A)J(B)J(A)$ for all $A, B \in \mathcal{A}$. For every $A \in \mathcal{A}$, A is invertible if and only if $J(A)$ is invertible and we also have $J(A)^{-1} = J(A^{-1})$. In particular, J preserves the spectra of the elements. The transformation J is compatible with the continuous functional calculus, i.e., for an arbitrary self-adjoint element $A \in \mathcal{A}$ and continuous real function f on the spectrum of A, we have $J(f(A)) = f(J(A))$. As a consequence of these properties, it is trivial to check that any Jordan *-isomorphism J between C^*-algebras, when restricted to the positive definite cone, preserves all Kubo-Ando means, i.e., it is an isomorphism with respect to all such operations. Therefore, transformations of the form $A \mapsto TJ(A)T$, where J is a Jordan *-isomorphism and T is a positive definite element, are bijections between positive definite cones which preserve any Kubo-Ando mean. The reader is asked to recall this fact when studying the next results.

Having seen Theorem 3.1 and the general notion of Kubo-Ando means, it seems to be a natural and interesting problem to study the isomorphisms of positive definite cones with respect to Kubo-Ando means which are different from the geometric mean. In full generality we do not have any result which in fact should not be a big surprise. Hence, in what follows we investigate the question for a certain important subcollection of Kubo-Ando means.

First of all, let us mention that the most distinguished Kubo-Ando means are naturally the arithmetic mean, the geometric mean, and the harmonic mean with representing operator monotone functions $t \mapsto (1+t)/2, t \mapsto \sqrt{t}, t \mapsto (2t)/(1+t)$, respectively. For $A, B \in \mathcal{A}^{++}$, we denote

$$A \nabla B = \frac{A+B}{2},$$
$$A \# B = A^{1/2}(A^{-1/2}BA^{-1/2})^{1/2}A^{1/2},$$
$$A!B = 2(A^{-1}+B^{-1})^{-1}.$$

Moreover, there is a natural "path" in the collection of Kubo-Ando means connecting those three particular means which is the parametric family of power means. For any $p \in [-1, 1]$, $p \neq 0$, the Kubo-Ando pth power mean is defined by

$$A\mathfrak{m}_p B = A^{1/2}\left(\frac{I+(A^{-1/2}BA^{-1/2})^p}{2}\right)^{1/p} A^{1/2}, \quad A, B \in \mathcal{A}^{++}. \tag{9}$$

This is the Kubo-Ando mean corresponding to the operator monotone function $t \mapsto ((1+t^p)/2)^{1/p}$ (it is known that this function is operator monotone exactly for all nonzero ps from the interval $[-1, 1]$). Clearly, $\mathfrak{m}_1 = \nabla$, $\mathfrak{m}_{-1} =!$ and we will point out later that the limiting case at $p = 0$ is the geometric mean, i.e., we have $\mathfrak{m}_p \to \#$ as $p \to 0$, and hence we can also write $\mathfrak{m}_0 = \#$.

We can now formulate our next result which describes the precise structure of the isomorphisms between positive definite cones with respect to the power mean $\mathfrak{m}_p$ for any nonzero $p \in [-1, 1]$. The proof rests on an essential application of our key result Proposition K. Actually, the result below was presented as Theorem 2 in [32] under the additional condition of continuity of the transformation. Here we provide a more careful argument yielding that the continuity assumption can in fact be dropped (we do not present all details of the proof, but hope that the argument is clear).

Theorem 3.2 *Let $\mathcal{A}, \mathcal{B}$ be C^*-algebras, let $p \in [-1, 1]$ be a nonzero real number, and let $\phi : \mathcal{A}^{++} \to \mathcal{B}^{++}$ be a bijective map. Then ϕ satisfies*

$$\phi(A\mathfrak{m}_p B) = \phi(A)\mathfrak{m}_p\phi(B), \quad A, B \in \mathcal{A}^{++}$$

*if and only if there is a Jordan *-isomorphism $J : \mathcal{A} \to \mathcal{B}$ and an element $D \in \mathcal{B}^{++}$ such that*

$$\phi(A) = DJ(A)D, \quad A \in \mathcal{A}^{++}.$$

Proof Assume that $\phi : \mathcal{A}^{++} \to \mathcal{B}^{++}$ is an isomorphism with respect to the mean $\mathfrak{m}_p$. We can and do assume that p is positive. Indeed, it is easy to see that $A\mathfrak{m}_{-p}B = (A^{-1}\mathfrak{m}_p B^{-1})^{-1}$ holds for any $A, B \in \mathcal{A}^{++}$ and then it is apparent that the bijective map $\psi : \mathcal{A}^{++} \to \mathcal{B}^{++}$ defined by $\psi(A) = \phi(A^{-1})^{-1}$, $A \in \mathcal{A}^{++}$ satisfies

$$\psi(A\mathfrak{m}_{-p}B) = \psi(A)\mathfrak{m}_{-p}\psi(B), \quad A, B \in \mathcal{A}^{++}.$$

We will prove that ϕ is a positive homogeneous order isomorphism. Observe that for a given $A \in \mathcal{A}^{++}$, as X runs through $\mathcal{A}^{++}$, the element

$$2^{1/p}A\mathfrak{m}_pX = A^{1/2}\left(I + (A^{-1/2}XA^{-1/2})^p\right)^{1/p}A^{1/2}$$

runs through the set of all elements of $\mathcal{A}^{++}$ which are strictly greater than A (an element $Y \in \mathcal{A}^{++}$ is said to be strictly greater than A if $Y - A \in \mathcal{A}^{++}$). It now easily follows that the order relation $\leq$ can be characterized in the following way: for any $A, B \in \mathcal{A}^{++}$ we have $A \leq B$ if and only if

$$\{B\mathfrak{m}_pX : X \in \mathcal{A}^{++}\} \subset \{A\mathfrak{m}_pX : X \in \mathcal{A}^{++}\}.$$

From this we obtain that ϕ is an order isomorphism.

We next prove that ϕ is positive homogeneous. For any positive real number t we have that

$$\frac{t}{2^{1/p}}I \leq (tI)\mathfrak{m}_pX$$

holds for every $X \in \mathcal{A}^{++}$. Since ϕ is an order and also an $\mathfrak{m}_p$-isomorphism, we infer that

$$\phi\left(\frac{t}{2^{1/p}}I\right) \leq \phi(tI)\mathfrak{m}_pX$$

for every $X \in \mathcal{B}^{++}$. In particular, substituting X with $\epsilon I \in \mathcal{B}^{++}$ and taking the limit $\epsilon \to 0$, we obtain that

$$\phi\left(\frac{t}{2^{1/p}}I\right) \leq \frac{\phi(tI)}{2^{1/p}}.$$

It readily follows that

$$\phi\left(\frac{I}{2^{n/p}}\right) \leq \frac{\phi(I)}{2^{n/p}}$$

is valid for every integer $n \geq 1$. Note that the right-hand side and thus the left-hand side, too, tend to zero as $n \to \infty$. Pick $A \in \mathcal{A}^{++}$. Observe that $A/2^{1/p}$ is the greatest lower bound of the decreasing sequence

$$A\mathfrak{m}_p\frac{I}{2^{n/p}} = \frac{1}{2^{1/p}}\left(A^p + \frac{I}{2^n}\right)^{1/p}.$$

Since ϕ is an order isomorphism, we see that $\phi(A/2^{1/p})$ is the greatest lower bound of the decreasing sequence

$$\phi\left(A\mathfrak{m}_p\frac{I}{2^{n/p}}\right) = \phi(A)\mathfrak{m}_p\phi\left(\frac{I}{2^{n/p}}\right)$$

which tends to $\phi(A)/2^{1/p}$ as $n \to \infty$. This implies that $\phi(A)/2^{1/p}$ is also the greatest lower bound of that sequence and hence we obtain $\phi(A/2^{1/p}) = \phi(A)/2^{1/p}$ for all $A \in \mathcal{A}^{++}$.

Observe that for commuting positive definite elements $C, D \in \mathcal{A}^{++}$ we clearly have $C\mathfrak{m}_p D = ((C^p + D^p)/2)^{1/p}$. Assume now that, for some given positive numbers α, β, the equalities $\phi(\alpha^{1/p}A) = \alpha^{1/p}\phi(A)$ and $\phi(\beta^{1/p}A) = \beta^{1/p}\phi(A)$ hold for all $A \in \mathcal{A}^{++}$. We easily compute

$$\begin{aligned} \phi((\alpha+\beta)^{1/p}A)/2^{1/p} &= \phi(((\alpha+\beta)^{1/p}A)/2^{1/p}) \\ &= \phi((\alpha^{1/p}A)\mathfrak{m}_p(\beta^{1/p}A)) = \phi(\alpha^{1/p}A)\mathfrak{m}_p\phi(\beta^{1/p}A) \\ &= (\alpha^{1/p}\phi(A))\mathfrak{m}_p(\beta^{1/p}\phi(A)) = (\alpha+\beta)^{1/p}\phi(A)/2^{1/p}. \end{aligned}$$

This gives us that

$$\phi((\alpha+\beta)^{1/p}A) = (\alpha+\beta)^{1/p}\phi(A)$$

holds for all $A \in \mathcal{A}^{++}$. From this observation we can easily deduce that $\phi(r^{1/p}A) = r^{1/p}\phi(A)$ is valid for any positive rational number r and element $A \in \mathcal{A}^{++}$. By the order-preserving property of ϕ, it readily implies that ϕ is positive homogeneous. We then apply Proposition K and complete the proof of the necessity part of the statement.

As for sufficiency, it is just trivial since, as we have already mentioned above, any Jordan *-isomorphism and any congruence transformation preserve any Kubo-Ando mean. □

The result above tells us that the automorphism groups of positive definite cones with respect to any Kubo-Ando pth power mean with $0 \neq p \in [-1, 1]$ are as small as possible (equal the groups generated by the Jordan *-automorphisms and the congruence transformations). However, recall Theorem 3.1 to see how different the situation is with the geometric mean (which represents the case $p = 0$). Moreover, do not forget that we have Theorem 3.1 only for von Neumann factors and not for all C^*-algebras. These facts may convince the reader to share our opinion that the description of the structures of the isomorphisms of positive definite cones with respect to general Kubo-Ando means is most likely a really difficult problem.

There are also non-Kubo-Ando means. For example, many people would rather define power means not as in (9), but "more naturally", by the following formula:

$$A\mathsf{m}_p B = \left(\frac{A^p + B^p}{2}\right)^{1/p}, \quad A, B \in \mathcal{A}^{++}.$$

(Obviously, for commuting A, B, we have $A\mathsf{m}_p B = A\mathfrak{m}_p B$.) The operation m_p is called the conventional pth power mean and its formula looks definitely simpler than the one in (9), but, in general, the resulted sort of mean lacks some important properties, most notably the monotonicity in its variables. In this regard, let us mention that from Propositions 9 and 14 in [35] one can immediately obtain the following. If $\mathcal{A}$ is a C^*-algebra, p is any nonzero real number different from $1, -1$, and for any $A \in \mathcal{A}^{++}$ we have that $A\mathsf{m}_p B \leq A\mathsf{m}_p C$ holds for all $B, C \in \mathcal{A}^{++}$ with $B \leq C$, then the algebra $\mathcal{A}$ is necessarily commutative.

Nevertheless m_p can certainly be considered as a kind of mean and hence a kind of operation and the description of the structure of the corresponding isomorphisms is not really difficult. The following result appeared as Theorem 1 in [32].

Theorem 3.3 *Let $\mathcal{A}, \mathcal{B}$ be C^*-algebras, let p be a nonzero real number, and let $\phi : \mathcal{A}^{++} \to \mathcal{B}^{++}$ be a bijective map. Then ϕ satisfies*

$$\phi(A\mathsf{m}_p B) = \phi(A)\mathsf{m}_p\phi(B), \quad A, B \in \mathcal{A}^{++}$$

*if and only if there is a Jordan *-isomorphism $J : \mathcal{A} \to \mathcal{B}$ and an element $D \in \mathcal{B}^{++}$ such that*

$$\phi(A) = (DJ(A)^p D)^{1/p}, \quad A \in \mathcal{A}^{++}. \tag{10}$$

(Observe that here the exponent p shows up in the decomposition (10) not like in Theorem 3.2 concerning the Kubo-Ando power means.)

Next we examine the limiting cases $p = 0$. What happens if we take the limit $p \to 0$ in the above two variants of power means? The observation that the convergence

$$A\mathsf{m}_p B = \left(\frac{A^p + B^p}{2}\right)^{1/p} \to \exp\left((1/2)(\log A + \log B)\right) \tag{11}$$

holds in norm for all positive definite Hilbert space operators A, B is attributed to Bhagwat and Subramanian [1] (see, e.g., [3]). Therefore, we write

$$A\mathsf{m}_0 B = \exp\left((1/2)(\log A + \log B)\right), \quad A, B \in \mathcal{A}^{++}$$

which is usually called the log-Euclidean mean of A and B (see, e.g., [3]). If, in (11), we write I in the place of A and $A^{-1/2}BA^{-1/2}$ in the place of B, we easily deduce that, as $p \to 0$, we have the convergence

$$A\mathfrak{m}_p B \to A^{1/2}(A^{-1/2}BA^{-1/2})^{1/2}A^{1/2} = A\#B, \quad A, B \in \mathcal{A}^{++}$$

in norm. Therefore, the limit of the Kubo-Ando pth power mean as $p \to 0$ is the Kubo-Ando geometric mean, and hence we can write $\mathfrak{m}_0 = \#$. This justifies our former statement that $\mathfrak{m}_p$, $p \in [-1, 1]$ is a parametric family of Kubo-Ando means connecting the three fundamental means $!, \#, \nabla$.

What concerns the isomorphisms with respect to the "conventional analogue" of the geometric mean, i.e., the log-Euclidean mean, is the case is very easy. Namely, the bijective map $\phi : \mathcal{A}^{++} \to \mathcal{B}^{++}$ between positive definite cones of C^*-algebras satisfies $\phi(A \mathsf{m}_0 B) = \phi(A) \mathsf{m}_0 \phi(B)$, $A, B \in \mathcal{A}^{++}$ if and only if the bijective map $\log \circ \phi \circ \exp$ from $\mathcal{A}_s$ to $\mathcal{B}_s$ preserves the arithmetic mean. Consequently, we have a huge collection of such isomorphisms very differently from the case of the geometric mean.

Above we have got a picture about the isomorphisms with respect to any power means (let they be either the Kubo-Ando power means or the conventional power means). We have noted dissimilarities which obviously imply, among others, that, in general, the Kubo-Ando pth power mean is different from the conventional pth power mean. As we have already mentioned, in commutative algebras we trivially have the equality of those two kinds of means. One can suspect that for $p \neq -1, 1$ the converse is also true. Indeed, actually much more is true, in the noncommutative case those two means cannot even be transformed onto each other. Namely, we have the following result which is an improvement of Theorem 3 in [32] in the sense that the additional condition of the continuity of the maps was assumed there which is dropped here.

Theorem 3.4 *Let $p \in]-1, 1[$ and let $\mathcal{A}, \mathcal{B}$ be C^*-algebras. Assume that $\phi : \mathcal{A}^{++} \to \mathcal{B}^{++}$ is a bijective map such that*

$$\phi(A \mathfrak{m}_p B) = \phi(A) \mathsf{m}_p \phi(B), \quad A, B \in \mathcal{A}^{++}. \tag{12}$$

If $p \neq 0$, then the algebras $\mathcal{A}, \mathcal{B}$ are necessarily commutative, while in the case where $p = 0$, it follows that $\mathcal{A}$ is necessarily commutative.

Proof We only sketch the proof and omit the details. Assume that $p \neq 0$. Following the basic idea of the proof of Theorem 3.2, one can first prove that p can be assumed to be positive. Next, one can verify that the bijection $\psi : A \mapsto \phi(A)^p$ between $\mathcal{A}^{++}$ and $\mathcal{B}^{++}$ is an order isomorphism, and then show that ϕ is positive homogeneous.

From these we deduce that

$$A \leq tB \iff \psi(A) \leq t^p \psi(B)$$

holds for any $A, B \in \mathcal{A}^{++}$ and positive real number t. Then we obtain that

$$d_T(\psi(A), \psi(B)) = p d_T(A, B), \quad A, B \in \mathcal{A}^{++},$$

(see the definition of the Thompson metric given in (1)). This means that ψ is a dilation (or, in another word, homothety) between the positive definite cones $\mathcal{A}^{++}$ and $\mathcal{B}^{++}$. But we proved in Theorem 18 in [30] the interesting fact that the existence of a non-isometric dilation with respect to the Thompson metric between the positive definite cones of C^*-algebras implies that the underlying algebras are necessarily commutative. This completes the proof of the statement in the case $p \neq 0$.

Assume now that $p = 0$. Then $\psi = \log \circ \phi : \mathcal{A}^{++} \to \mathcal{B}_s$ is a bijective map which satisfies

$$\psi(A\#B) = \frac{\psi(A) + \psi(B)}{2}, \quad A, B \in \mathcal{A}^{++}.$$

As a special case of Proposition 7 in [27] we obtain that $\mathcal{A}$ is necessarily commutative. □

We need to make a comment concerning the case $p = 0$ in which we have asserted only that $\mathcal{A}$ is necessarily commutative. Indeed, this is all what can be stated. To see this, consider the commutative C^*-algebra $\mathbb{C}^{n^2}$ and the noncommutative C^*-algebra $\mathbb{M}_n$ (n is an arbitrary positive integer greater than 1). For any bijective linear transformation ψ between the linear spaces of all self-adjoint elements of $\mathbb{C}^{n^2}$ and $\mathbb{M}_n$, we have that $\exp \circ \psi \circ \log$ is a bijective map between the positive definite cones of $\mathbb{C}^{n^2}$ and $\mathbb{M}_n$ that satisfies (12).

Besides the conventional power means, let us mention a few additional important non-Kubo-Ando means. First, there is a variant of the geometric mean due to Fiedler and Pták [14] which is usually called the spectral geometric mean. This concept was originally introduced and studied for positive definite matrices. Its most outstanding property is that the square of the spectral geometric mean of any $A, B \in \mathbb{P}_n$ is similar to AB; hence the eigenvalues of the spectral geometric mean of A and B coincide with the square roots of the eigenvalues of AB. The definition given by Fiedler and Pták can be straightforwardly extended for the elements of the positive definite cone in any C^*-algebra $\mathcal{A}$ by setting

$$A \natural B = (A^{-1}\#B)^{1/2} A (A^{-1}\#B)^{1/2}, \quad A, B \in \mathcal{A}^{++}.$$

Just as in [14], it requires some algebraic manipulations to verify that $\natural$ (which is not a Kubo-Ando type mean) is symmetric, i.e., $A \natural B = B \natural A$ holds for all $A, B \in \mathcal{A}^{++}$.

Concerning the isomorphisms of the spectral geometric mean, little is known; we do not have their precise description even in the finite-dimensional case, in matrix algebras. The information that we are currently having along with several open questions can be found in [21].

Let us mention another non-Kubo-Ando mean, called Wasserstein mean, which has recently been introduced by Bhatia, Jain, and Lim [5] (also see [4, 6]). Just as with the Fiedler-Pták mean, the original definition of the Wasserstein mean was created for positive definite matrices but the defining formula can trivially be extended to the positive definite cone of an arbitrary C^*-algebra $\mathcal{A}$ as follows:

$$A \sigma_w B = \frac{1}{4}\left(A + B + A(A^{-1}\#B) + (A^{-1}\#B)A\right), \quad A, B \in \mathcal{A}^{++}. \tag{13}$$

In the last paragraph in Sect. 2 we mentioned that behind the geometric mean there is a Riemannian structure on $\mathbb{P}_n$ which has several applications. Bhatia, Jain, and Lim observed the nice, important, and useful fact that there is a similar background also behind the Wasserstein mean. The short explanation is the following. Recall that besides the metric (6), there is another important distance on the positive definite cone of $\mathbb{M}_n$ which was called in [5] the Bures-Wasserstein metric. It is defined as

$$d_{BW}(A, B) = \left(\mathrm{Tr}A + \mathrm{Tr}B - 2\mathrm{Tr}(A^{1/2}BA^{1/2})^{1/2}\right)^{1/2} \tag{14}$$

for any $A, B \in \mathbb{P}_n$, where Tr is the usual trace on $\mathbb{M}_n$. (In quantum information theory this metric is usually called Bures metric but it has connections to statistics and optimal transport, too, where they call it Wasserstein metric.) Bhatia, Jain, and Lim observed that there is a Riemannian geometry on the positive definite cone of $\mathbb{M}_n$ whose geodesic distance is exactly the Bures-Wasserstein metric and the geodesic curve connecting the points $A, B \in \mathbb{P}_n$ is

$$t \mapsto (1-t)^2 A + t^2 B + t(1-t)(A(A^{-1}\#B) + (A^{-1}\#B)A), \quad t \in [0, 1].$$

We see that the midpoint of this curve is just the Wasserstein mean $A\sigma_w B$ of A and B.

In the next theorem we describe the structure of the isomorphisms of positive definite cones with respect to the Wasserstein mean. Let us mention that if one calculates the expression in (13), one obtains the following not easily manageable formula:

$$A\sigma_w B = \frac{1}{4}\Big(A + B + A^{1/2}(A^{1/2}BA^{1/2})^{1/2}A^{-1/2} + A^{-1/2}(A^{1/2}BA^{1/2})^{1/2}A^{1/2}\Big).$$

The next result provides the complete description of the structure of the isomorphisms with respect to that complicated operation. Again, surprisingly, the proof strongly uses our key statement, Proposition K. The following result has appeared as Theorem 1 in [34].

Theorem 3.5 *Let $\mathcal{A}, \mathcal{B}$ be C^*-algebras and $\phi : \mathcal{A}^{++} \to \mathcal{B}^{++}$ be a bijective map. Then ϕ preserves the Wasserstein mean, i.e., it satisfies*

$$\phi(A\sigma_w B) = \phi(A)\sigma_w\phi(B), \quad A, B \in \mathcal{A}^{++}$$

*if and only if there is a Jordan *-isomorphism $J : \mathcal{A} \to \mathcal{B}$ and a central element $C \in \mathcal{B}^{++}$ such that*

$$\phi(A) = CJ(A), \quad A \in \mathcal{A}^{++}.$$

Let us point out the essential difference in the conclusions of Theorems 3.2 and 3.5 which is the centrality of the elements inducing the congruence transformations in the decomposition of the isomorphisms in the latter case.

To better understand the content of the previous result, observe that by the already mentioned properties of Jordan *-isomorphisms, it is easy to see that any such transformation, when restricted to a positive definite cone, preserves the Wasserstein mean (just as it preserves all Kubo-Ando means). One can also easily check that the congruence transformations implemented by central elements of positive definite cones also have that property. The statement above tells that, in fact, all isomorphisms with respect to the Wasserstein mean can be obtained as compositions of those two types of transformations.

The strategy of the proof of Theorem 3.5 is very similar to that of the proof of Theorem 3.2. Namely, in a very similar way we prove that any bijective map $\phi : \mathcal{A}^{++} \to \mathcal{B}^{++}$ between positive definite cones which preserves the Wasserstein mean is necessarily a positive homogeneous order isomorphism. To do that, we need a characterization of the order expressed by the operation σ_w. The following holds. For any $A, B \in \mathcal{A}^{++}$, we have

$$A \leq B \quad \Longleftrightarrow \quad \{B\sigma_w X \, : \, X \in \mathcal{A}^{++}\} \subset \{A\sigma_w X \, : \, X \in \mathcal{A}^{++}\}.$$

This is the same sort of characterization as the one we have seen before concerning the Kubo-Ando power means. However the proof of this latter characterization is definitely more complicated.

In the second part of the proof we apply our key Proposition K and have that ϕ is necessarily the composition of a Jordan *-isomorphism J and a congruence transformation implemented by an element $T \in \mathcal{B}^{++}$:

$$\phi(A) = TJ(A)T, \quad A \in \mathcal{A}^{++}.$$

It remains to verify that T is a central element which is equivalent to that the element $\phi(I)$ is central. Actually, this follows from the observation that ϕ in general maps the central elements of $\mathcal{A}^{++}$ to the central elements of $\mathcal{B}^{++}$. In fact, this can be inferred from the following interesting characterization of centrality. An element $A \in \mathcal{A}^{++}$ is central if and only if A is a lower bound of the set $4\{A\sigma_w X \, : \, X \in \mathcal{A}^{++}\}$.

For the details of the above outlined proof, see the paper [34].

Finally, in finishing this section, we mention that in commutative algebras the Wasserstein mean, the Kubo-Ando power mean, and the conventional power mean both with exponent $p = 1/2$ all coincide. One can formulate and verify variants of Theorem 3.4 concerning the pairs $(\mathfrak{m}_{1/2}, \sigma_w)$ and $(\mathsf{m}_{1/2}, \sigma_w)$ of means.

4 Preservers of the Norm of Means

Above we have discussed various mean-preserving transformations and demonstrated how important the role our key result Proposition K plays in the descriptions of their structures.

We also recall that the application of Proposition K was equally essential in the solutions of problems where we studied the structures of certain isometries and sorts of relative entropy-preserving maps between positive definite cones or so-called density spaces in C^*-algebras. We mention a few of our corresponding results. To begin with, we point out that the definition of the Bures-Wasserstein metric (14) was extended by Farenick and Rahaman to the setting of C^*-algebras with faithful traces.

If $\mathcal{A}$ is a C^*-algebra and τ is a trace (a positive linear functional on $\mathcal{A}$ which is tracial, i.e., it vanishes on commutators), then τ is called faithful if $\tau(A) = 0$, $A \in \mathcal{A}^+$ implies $A = 0$. Examples for such algebras include UHF-algebras, finite factor von Neumann algebras, and irrational rotation algebras.

Let $\mathcal{A}$ be a C^*-algebra and τ be a faithful trace on $\mathcal{A}$. For any $A, B \in \mathcal{A}^+$, we set

$$F_\tau(A, B) = \tau\left((A^{1/2}BA^{1/2})^{1/2}\right)$$

and

$$d_{BW}^\tau(A, B) = (\tau(A) + \tau(B) - 2F_\tau(A, B))^{1/2}.$$

Similarly to the case of matrix algebras, the first quantity is called the fidelity of A and B; the second one is said to be the Bures-Wasserstein distance between A and B. Indeed, it can be shown (highly nontrivially) that d_{BW}^τ is a true metric on $\mathcal{A}^+$; see Section 2.2 in [13].

The next theorem describes the surjective Bures-Wasserstein isometries between positive definite cones of C^*-algebras. It appeared as Theorem 2 in [29]; for a somewhat more general statement, see Theorem 12 in [31].

Theorem 4.1 *Let $\mathcal{A}, \mathcal{B}$ be C^*-algebras with faithful traces τ, τ', respectively, and let $\phi : \mathcal{A}^{++} \to \mathcal{B}^{++}$ be a surjective map. Then ϕ is a Bures-Wasserstein isometry, i.e., it satisfies*

$$d_{BW}^{\tau'}(\phi(A), \phi(B)) = d_{BW}^\tau(A, B), \quad A, B \in \mathcal{A}^{++}$$

*if and only if there is a Jordan *-isomorphism $\phi : \mathcal{A} \to \mathcal{B}$ and a central element $C \in \mathcal{B}^{++}$ such that $\phi(A) = CJ(A)$ holds for all $A \in \mathcal{A}^{++}$ and $\tau'(CJ(X)) = \tau(X)$ holds for all $X \in \mathcal{A}$.*

The key point of the proof is where we verify that ϕ is necessarily an order isomorphism. In fact, that follows from the following observation. For any $A, B \in \mathcal{A}^{++}$, the set

$$\{(d_{BW}^\tau(B, X))^2 - (d_{BW}^\tau(A, X))^2 \,:\, X \in \mathcal{A}^{++}\}$$

is bounded from above if and only if $A \leq B$. To the positive homogeneity of ϕ we prove that it respects the traces τ, τ' and then it preserves the fidelity from which homogeneity follows easily. We next apply Proposition K.

As for the appearance of the centrality in the statement above, we note that since ϕ respects the traces τ, τ', concerning the Jordan *-isomorphism $J : \mathcal{A} \to \mathcal{B}$ and the congruence transformation implemented by $T \in \mathcal{B}^{++}$ which appear in the decomposition of ϕ, we have

$$\tau'(TJ(A)T) = \tau(A), \quad A \in \mathcal{A}^{++}.$$

This identity is proved to imply that the element $T \in \mathcal{B}^{++}$ is necessarily central. For details, see the proof of Theorem 12 in [31].

Let us make the following comment. Having a look at Theorems 3.5 and 4.1 we can notice that the structure of the surjective Bures-Wasserstein isometries is very close to that of the corresponding mean preservers, namely, the preservers of the Wasserstein mean (actually, the only difference is the preservation of the trace, i.e., the last property in Theorem 4.1). As for the structures of the surjective Thompson isometries and the Kubo-Ando geometric mean preservers, see Theorems 2.1 and 3.1; it is obvious how much larger the dissimilarity is between them.

We remark further that the statement in Theorem 4.1 was verified also for maps between so-called density spaces in C^*-algebras, which are the convex sets $\mathcal{D}_\tau(\mathcal{A}) = \{A \in \mathcal{A}^+ : \tau(A) = 1\}$ of positive semidefinite elements with unit trace. See Theorem 1 in [29]. The structure of surjective fidelity preserving maps was also determined in the same paper and proved to be the same as that of the surjective Bures-Wasserstein isometries.

Finally, we make a comment regarding the finite-dimensional case. In fact, we conjecture that on the positive definite cone of a matrix algebra, the Bures-Wasserstein isometries as selfmaps are automatically surjective. We believe that it is a quite nontrivial problem. Concerning fidelity preserving maps the relevant statement was verified in [24]; see Theorem 2 there. Also, for distance measures of the form (5), the analogue result was obtained in [28]. What motivates the question is the well-known fact that any isometry of a finite-dimensional normed space into itself is necessarily surjective.

Besides the symmetries of positive definite cones or density spaces in C^*-algebras with respect to certain distances, we also obtained similar results concerning the preservers of certain kinds of relative entropies, especially Rényi relative entropies and Umegaki and Belavkin-Staszewski relative entropies; see [30]. In fact, our key statement Proposition K played an essential role in the proofs of the corresponding results as well.

We now turn to the last part of our paper where we consider maps between positive definite cones which preserve the norm of means. The different kinds of Rényi relative entropies are functions of the trace of certain particular variants of a weighted geometric mean; hence the related symmetries can be viewed as transformations which preserve a certain norm of a certain operator mean. On the other hand, norm additive maps (transformations which preserve the norm of the sum of any two elements) on function algebras have extensively been studied in a search for criteria that would guarantee that an a priori nonlinear transformation be a weighted composition operator (or at least be close to such a map). The probably most cited paper on this topic is [41]. Norm additivity is equivalent to the preservation of the norm of the arithmetic mean. The above discussion has given the motivation to study transformations between positive definite cones of C^*-algebras which preserve the norm of some given operator mean.

Because of their special importance, one may be forced to consider first the case of Kubo-Ando means. And, of course, the first of those means to start with should be the arithmetic mean. Surprisingly, the problem is very much nontrivial even for that particularly simple mean. Our corresponding result is the following one which is to appear as Theorem 2.4 in [12] (it actually solves Problem 26 in [31]).

Theorem 4.2 *Let $\mathcal{A}, \mathcal{B}$ be C^*-algebras. Assume that $\phi : \mathcal{A}^{++} \to \mathcal{B}^{++}$ is a bijective map. Then ϕ satisfies*

$$\|\phi(A) + \phi(B)\| = \|A + B\|, \quad A, B \in \mathcal{A}^{++}, \tag{15}$$

*if and only if there is a Jordan *-isomorphism $J : \mathcal{A} \to \mathcal{B}$ which extends ϕ, i.e., $\phi(A) = J(A)$ holds for all $A \in \mathcal{A}^{++}$.*

In the following we give some details of the proof of the theorem. First we consider the problem on positive semidefinite cones. So, let $\mathcal{A}, \mathcal{B}$ be C^*-algebras and $\phi : \mathcal{A}^+ \to \mathcal{B}^+$ be a bijective map such that

$$\|\phi(A) + \phi(B)\| = \|A + B\|$$

holds for all $A, B \in \mathcal{A}^+$.

In the current situation we easily obtain that ϕ is an order isomorphism. Indeed, the order can be almost trivially characterized by the norm of sums as follows. For any $A, B \in \mathcal{A}^+$ we have $A \leq B$ if and only if $\|A + X\| \leq \|B + X\|$ holds for all $X \in \mathcal{A}^+$. For the nontrivial part of the claimed equivalence, assume that $\|A+X\| \leq \|B+X\|$ holds for all $X \in \mathcal{A}^+$. Pick a positive scalar t such that $X = tI - B \in \mathcal{A}^+$. We have $\|A+(tI-B)\| \leq t$ which implies that $A+(tI-B) \leq tI$, and thus $A \leq B$.

The difficult part of the proof of Theorem 4.2 is to show that ϕ is positive homogeneous. This is done in a series of steps. In the first one, we verify that ϕ preserves orthogonality, i.e., zero product of elements. This can be proved using the following characterization of the orthogonality relation; see Lemma 2.7 in [12].

Lemma 4.1 *For any $A, B \in \mathcal{A}^+$ we have $AB = 0$ if and only if $\|C + D\| = \max\{\|C\|, \|D\|\}$ holds for any $C, D \in \mathcal{A}^+$ with $C \leq A$ and $D \leq B$.*

It then follows that ϕ preserves the norm, the order, and the orthogonality of the positive semidefinite elements in both directions. It turns out that these imply that the map is necessarily positive homogeneous. In fact, we have the following interesting statement which provides a nonlinear characterization of Jordan *-isomorphisms on positive definite cones.

Lemma 4.2 *Let $\phi : \mathcal{A}^+ \to \mathcal{B}^+$ be a bijective map. It extends to a Jordan *-isomorphism $J : \mathcal{A} \to \mathcal{B}$ exactly when the following hold: ϕ preserves the norm, the order, and the orthogonality in both directions, i.e., if we have:*

(i) $\|\phi(A)\| = \|A\|$, $A \in \mathcal{A}^{++}$.
(ii) $A \leq B \Leftrightarrow \phi(A) \leq \phi(B)$, $A, B \in \mathcal{A}^{++}$.
(iii) $AB = 0 \Leftrightarrow \phi(A)\phi(B) = 0$, $A, B \in \mathcal{A}^{++}$.

In the proof we use the following quite complicated characterization of the equality of elements; see Lemma 2.2 in [12].

Lemma 4.3 *For any $A, B \in \mathcal{A}^{++}$, the following assertions are equivalent:*

(i) $A = B$.
(ii) *For any $C \in \mathcal{A}^+$ and positive real number δ, we have*

$$\{X \in \mathcal{A}^+ : CX = 0, \ \|X\| \leq \delta\} = \{X \in \mathcal{A}^+ : CX = 0, \ \|X\| \leq \delta, \ X \leq A\}$$

if and only if

$$\{X \in \mathcal{A}^+ : CX = 0, \ \|X\| \leq \delta\} = \{X \in \mathcal{A}^+ : CX = 0, \ \|X\| \leq \delta, \ X \leq B\}.$$

After the above characterization is verified, the proof of Lemma 4.2 goes as follows. We consider only the sufficiency part of the statement, so assume that $(i) - (iii)$ hold for ϕ. From the norm-preserving property of ϕ we deduce that $\phi(tI) = tI$ holds for every positive real number t. Indeed, we have $\phi(A) \leq tI$ if and only if $\|\phi(A)\| \leq t$, which is equivalent to $\|A\| \leq t$. The latter holds if and only if $A \leq tI$, and this is equivalent to $\phi(A) \leq \phi(tI)$ since ϕ is an order isomorphism. We then obtain the desired equality $\phi(tI) = tI$. Using the order-preserving property and this equality, it follows that ϕ preserves the positive definite elements in both directions.

Now, select any $A \in \mathcal{A}^{++}$ and positive real number t. Then for any $D \in \mathcal{B}^+$ and $\delta > 0$ we have the following equivalences:

$$\{Y \in \mathcal{B}^+ : DY = 0, \|Y\| \leq \delta\} = \{Y \in \mathcal{B}^+ : DY = 0, \ \|Y\| \leq \delta, \ Y \leq \phi(tA)\}$$
$$\Longleftrightarrow \text{for any } Y \in \mathcal{B}^+ \text{ we have } DY = 0, \ \|Y\| \leq \delta \Rightarrow Y \leq \phi(tA)$$
$$\Longleftrightarrow \text{for any } X \in \mathcal{A}^+ \text{ we have } \phi^{-1}(D)X = 0, \ \|X\| \leq \delta \Rightarrow X \leq tA$$
$$\Longleftrightarrow \text{for any } Z \in \mathcal{A}^+ \text{ we have } \phi^{-1}(D)Z = 0, \ \|Z\| \leq \delta/t \Rightarrow Z \leq A$$
$$\Longleftrightarrow \text{for any } Y \in \mathcal{B}^+ \text{ we have } DY = 0, \ \|Y\| \leq \delta/t \Rightarrow Y \leq \phi(A)$$
$$\Longleftrightarrow \{W \in \mathcal{B}^+ : DW = 0, \|W\| \leq \delta\} = \{W \in \mathcal{B}^+ : DW = 0, \ \|W\| \leq \delta, \ W \leq t\phi(A)\}.$$

By Lemma 4.3, this yields $\phi(tA) = t\phi(A)$. It follows that ϕ, when restricted to $\mathcal{A}^{++}$, is a bijection onto $\mathcal{B}^{++}$ which preserves the order in both directions and is positive homogeneous. By Proposition K, there exist a Jordan $*$-isomorphism $J : \mathcal{A} \to \mathcal{B}$ and a positive definite element $C \in \mathcal{B}$ such that $\phi(A) = CJ(A)C$ holds for all $A \in \mathcal{A}^{++}$. Since $\phi(tI) = tI$, it follows that $C = I$. This gives us that $\phi(A) = J(A)$ holds for all $A \in \mathcal{A}^{++}$.

To show that this former equality is valid also for any $A \in \mathcal{A}^{+}$, observe that

$$\phi(A + tI) = J(A + tI) = J(A) + tI, \quad t > 0.$$

Since ϕ is an order isomorphism and $A = \inf\{A + tI : t > 0\}$, we have

$$\phi(A) = \inf\{\phi(A + tI) : t > 0\} = \inf\{J(A + tI) : t > 0\} = J(A), \quad A \in \mathcal{A}^{+}.$$

Hence we are done concerning the structure of norm additive maps between positive semidefinite cones. For the proof of Theorem 4.2 we need the following simple lemma; see Lemma 2.8 in [12].

Lemma 4.4 *Let $\mathcal{A}, \mathcal{B}$ be unital C^*-algebras. Let $\phi : \mathcal{A}^{++} \to \mathcal{B}^{++}$ be an order isomorphism fixing all positive scalar multiples of the identity. For any $\epsilon > 0$, define $\psi : \mathcal{A}^{+} \to \mathcal{B}^{+}$ by $\psi(A) = \phi(A + \epsilon I) - \epsilon I$. Then ψ is an order isomorphism from $\mathcal{A}^{+}$ onto $\mathcal{B}^{+}$. Moreover, if ϕ preserves the norm, then so does ψ.*

After this, we are in a position to verify our original statement, Theorem 4.2.

Proof (Proof of Theorem* 4.2*) Assume that $\phi : \mathcal{A}^{++} \to \mathcal{B}^{++}$ is a bijective map satisfying (15). We can show just as in the proof of Theorem 3.2 that ϕ is an order isomorphism and, following the argument given in the first paragraph of the proof of Lemma 4.2, that ϕ fixes all positive scalar multiples of the identity. By Lemma 4.4, we define an order isomorphism $\psi : \mathcal{A}^{+} \to \mathcal{B}^{+}$ by setting $\psi(A) = \phi(\epsilon I + A) - \epsilon I$ for any $A \in \mathcal{A}^{+}$. We also have

$$\begin{aligned}
\|\psi(A) + \psi(B)\| &= \|\phi(\epsilon I + A) + \phi(\epsilon I + B) - 2\epsilon I\| \\
&= \|\phi(\epsilon I + A) + \phi(\epsilon I + B)\| - 2\epsilon \\
&= \|(\epsilon I + A) + (\epsilon I + B)\| - 2\epsilon \\
&= \|(\epsilon I + A) + (\epsilon I + B) - 2\epsilon I\| = \|A + B\|
\end{aligned}$$

for any $A, B \in \mathcal{A}^{+}$. This means that ψ is a norm additive bijection between positive semidefinite cones which, as we already know, extends to a Jordan $*$-isomorphism between $\mathcal{A}$ and $\mathcal{B}$. In particular, ψ is affine. It follows trivially that ϕ is also affine on the set of all elements of $\mathcal{A}^{++}$ which are bounded from below by ϵI. Since this holds for any positive ϵ, we infer that ϕ is affine. The structure of affine bijections between positive definite cones of C^*-algebras is known. In fact, we can apply Theorem 3.3 and obtain that $\phi(A) = CJ(A)C$, $A \in \mathcal{A}^{++}$ holds with some

Jordan $*$-isomorphism $J : \mathcal{A} \to \mathcal{B}$ and invertible positive element $C \in \mathcal{B}^{++}$. Since $\phi(I) = I$, we have $C = I$ and this completes the proof. □

We point out that bijective norm additive maps between positive definite cones in C^*-algebras were studied also in the paper [16]. However, not the C^*-norm but, given a faithful trace on the algebra, the corresponding Schatten p-norm was considered there for any $1 < p < \infty$. The argument was also based on our Proposition K but the proof of the positive homogeneity was far much easier than it is in the case of Theorem 4.2.

Because of some technical issues, concerning the preservers of the norm of the harmonic mean, we do not have a result in the full generality of C^*-algebras; we leave that as an open problem. But we do have a statement for so-called AW^*-algebras which reads as follows; see Theorem 2.16 in [12].

Theorem 4.3 *Let $\mathcal{A}, \mathcal{B}$ be AW^*-algebras. Assume that $\phi : \mathcal{A}^{++} \to \mathcal{B}^{++}$ is a bijective map. Then ϕ satisfies*

$$\|\phi(A)!\phi(B)\| = \|A!B\|, \quad A, B \in \mathcal{A}^{++},$$

if and only if it extends to a Jordan $*$-isomorphism from $\mathcal{A}$ onto $\mathcal{B}$.

Concerning the preservers of the norm of the geometric mean, we have the same description. In fact, chronologically, the following statement was obtained prior to the previous two results. See Theorem 1 in [7].

Theorem 4.4 *Let $\mathcal{A}, \mathcal{B}$ be C^*-algebras and assume that $\phi : \mathcal{A}^{++} \to \mathcal{B}^{++}$ is a bijective map. It satisfies*

$$\|\phi(A)\#\phi(B)\| = \|A\#B\|, \quad A, B \in \mathcal{A}^{++} \tag{16}$$

*if and only if there is a Jordan *-isomorphism $J : \mathcal{A} \to \mathcal{B}$ which extends ϕ.*

In the cases of the related results for the arithmetic and the harmonic means, the characterization of the order via the norm of the mean in question was easy. With the geometric mean the situation is very much different. Formally, we have the same type of characterization saying that for any $A, B \in \mathcal{A}^{++}$, the inequality $A \leq B$ holds if and only if $\|A\#X\| \leq \|B\#X\|$, $X \in \mathcal{A}^{++}$. But its proof is quite complicated; see Lemma 9 in [7]. After having this characterization, it is clear that any bijective map $\phi : \mathcal{A}^{++} \to \mathcal{B}^{++}$ satisfying (16) is an order isomorphism. Next, the verification of the homogeneity of ϕ is just easy due to the homogeneity (of order 1/2) of the geometric mean in its variables. Finally, we can use our Proposition K again and finish the proof of the necessity part of Theorem 4.4 in a few lines.

How could we get any further from here? How could we describe the structures of preservers of the norm of other means? That is a quite difficult question. Keeping the above arguments in mind, one could try to prove in the way as above (i.e., through the use of a characterization of the order by the norm of the mean in question) that those transformations are necessarily order isomorphisms. Therefore, it is tempting to define and study the following property that we could call the order-determining property, for a general mean.

Definition 4.1 The mean σ is said to have the (**OD**) property (or, in short, to be (**OD**)) on a C^*-algebra $\mathcal{A}$ if the following holds. For any $A, B \in \mathcal{A}^{++}$, we have

$$A \leq B \Longleftrightarrow \|A\sigma X\| \leq \|B\sigma X\|, \quad X \in \mathcal{A}^{++}.$$

When σ has the (**OD**) property on all C^*-algebras, we simply say that it has the (**OD**) property or that it is (**OD**).

Concerning this property, in the recent paper [33] we presented several results and formulated a number of open problems. Indeed, the knowledge that we have is actually quite limited. We finish the paper with a short survey of the most relevant results which appeared in [33].

Theorem 3 in [33] tells that if $f :]0, \infty[\to]0, \infty[$ is a nontrivial (i.e., not affine) operator monotone function such that $f(0+) = 0$, then the corresponding Kubo-Ando mean is necessarily (**OD**). (The geometric mean and the harmonic mean are examples of such means.)

In the case where $f(0+) > 0$, we only have a partial result, namely, Proposition 3 in [33], which says that for any Kubo-Ando mean which is the affine combination of a (nontrivial) weighted arithmetic mean and some weighted geometric means, it necessarily has the (**OD**) property. In particular, that statement covers the case of power means with exponents $p = 1/n$, where n is any positive integer. We suspect that the same conclusion holds for all $0 < p < 1$ (for any negative p we have $f(0+) = 0$ and hence the former general result applies), i.e., the Kubo-Ando pth power mean has the (**OD**) property. This is a conjecture, we do not have a proof.

However, we do have a result for the case of full operator algebras. To formulate it, we need the concept of the transpose f° of the operator monotone function $f :]0, \infty[\to]0, \infty[$ which is defined by $f^\circ(t) = tf(1/t)$, $t > 0$; see the explanation right after (8). Proposition 9 in [33] reads as follows. If $f(0+), f^\circ(0+) > 0$, and, in addition, $(f^\circ)'(0+) = \infty$ also holds, then for any $A, B \in B(H)^{++}$ we have

$$A \leq B \Longleftrightarrow \|A\sigma X\| \leq \|B\sigma X\|, \quad X \in B(H)^{++}.$$

Similar characterization can be verified in the case where $(f^\circ)'(0+) = f(0+)$ holds.

Observe that this result covers all Kubo-Ando pth power means with $0 < p \leq 1$. Therefore, all such means have the (**OD**) property on $B(H)$. Unfortunately, the statement guarantees such a positive conclusion only for two possible extreme values of $(f^\circ)'(0+)$, namely, for $f(0+)$ and ∞. What happens when we have $f(0+) < (f^\circ)'(0+) < \infty$? We do not know; we consider that a challenging problem. For example, as a very particular case, we do not have an answer even to the following question. Does the norm of the sum of the arithmetic mean and harmonic mean has the (**OD**) property at least on $B(H)$?

Several other results can also be found in [33]. For example, concerning the conventional power means we have the following statement. If p is any real number different from $-1, 1$ and $\mathcal{A}$ is a C^*-algebra, then given an $A \in \mathcal{A}^{++}$, we have the equivalence

$$A \leq B \iff \|A \mathsf{m}_p X\| \leq \|B \mathsf{m}_p X\|, \quad X \in \mathcal{A}^{++}$$

if and only if A is a central element of $\mathcal{A}$. In particular, such a mean is **(OD)** exactly on commutative C^*-algebras.

As other non-Kubo-Ando means, we mention that the Fiedler-Pták spectral geometric mean is **(OD)** whose assertion is just easy; see Proposition 8 in [33].

We stop here and hope that the above survey can convince the reader about the usefulness of the observation in Proposition K and also demonstrates that there is big room for further research on preservers relating means on positive definite cones.

Acknowledgments This paper was completed while the author was a visiting researcher at the Alfréd Rényi Institute of Mathematics. The research was supported by the Ministry of Innovation and Technology of Hungary from the National Research, Development and Innovation Fund, project no. TKP2021-NVA-09, and also by the National Research, Development and Innovation Office of Hungary, NKFIH, grant no. K134944.

References

1. K.V. Bhagwat and R. Subramanian, *Inequalities between means of positive operators,* Math. Proc. Cambridge Philos. Soc. **83** (1978), 393–401.
2. R. Bhatia, *Positive Definite Matrices,* Princeton University Press, Princeton Oxford, 2007.
3. R. Bhatia and P. Grover, *Norm inequalities related to the matrix geometric mean,* Linear Algebra Appl. **437** (2012), 726–733.
4. R. Bhatia, S. Gaubert and T. Jain, *Matrix versions of the Hellinger distance,* Lett. Math. Phys. **109** (2019), 1777–1804.
5. R. Bhatia, T. Jain and Y. Lim, *On the Bures-Wasserstein distance between positive definite matrices,* Expo. Math. **37** (2019), 165–191.
6. R. Bhatia, T. Jain and Y. Lim, *Inequalities for the Wasserstein mean of positive definite matrices,* Linear Algebra Appl. **576** (2019), 108–123.
7. F. Chabbabi, M. Mbekhta and L. Molnár, *Characterizations of Jordan *-isomorphisms of C^*-algebras by weighted geometric mean related operations and quantities,* Linear Algebra Appl. **588** (2020), 364–390.
8. G. Corach, H. Porta and L. Recht, *A geometric interpretation of Segal's inequality* $\|e^{X+Y}\| \leq \|e^{X/2}e^Y e^{X/2}\|$, Proc. Amer. Math. Soc. **115** (1992), 229–231.
9. G. Corach, H. Porta and L. Recht, *Geodesics and operator means in the space of positive operators,* Internat. J. Math. **4** (1993), 193–202.
10. G. Corach, H. Porta and L. Recht, *Convexity of the geodesic distance on spaces of positive operators,* Illinois J. Math. **38** (1994), 87–94.
11. G. Corach and A.L. Maestripieri, *Differential and metrical structure of positive operators,* Positivity **3** (1999), 297–315.
12. Y. Dong, L. Li, L. Molnár and N.-C. Wong, *Transformations preserving the norm of means between positive cones of general and commutative C^*-algebras,* J. Operator Theory **88** (2022), 365–406.

13. D. Farenick and M. Rahaman, *Bures contractive channels on operator algebras,* New York J. Math. **23** (2017), 1369–1393.
14. M. Fiedler and V. Pták, *A new positive definite geometric mean of two positive definite matrices,* Linear Algebra Appl. **251** (1997), 1–20.
15. J.I. Fujii, *Differential geometries and their metrics on the positive operators,* RIMS Kôkyûroku No. 1632 (2009), 28–37.
16. M. Gaál, *Norm-additive maps on the positive definite cone of a* C^**-algebra,* Results Math. **73**:151 (2018).
17. O. Hatori, *Examples and applications of generalized gyrovector spaces,* Results Math. **71** (2017), 295–317.
18. O. Hatori and L. Molnár, *Isometries of the unitary groups and Thompson isometries of the spaces of invertible positive elements in* C^**-algebras,* J. Math. Anal. Appl. **409** (2014), 158–167.
19. F. Kubo and T. Ando, *Means of positive linear operators,* Math. Ann. **246** (1980), 205–224.
20. B. Lemmens, O. van Gaans and H. van Imhoff, *On the linearity of order-isomorphisms,* Canad. J. Math. **73** (2021), 399–416.
21. L. Li, L. Molnár and L. Wang, *On preservers related to the spectral geometric mean,* Linear Algebra Appl. **610** (2021), 647–672.
22. S. Mazur and S. Ulam, *Sur les transformations isométriques d'espaces vectoriels normés,* C. R. Acad. Sci. Paris **194** (1932), 946–948.
23. L. Molnár, *Order-automorphisms of the set of bounded observables,* J. Math. Phys. **42** (2001), 5904–5909.
24. L. Molnár, *Fidelity preserving maps on density operators*, Rep. Math. Phys. **48** (2001), 299–303.
25. L. Molnár, *Thompson isometries of the space of invertible positive operators,* Proc. Amer. Math. Soc. **137** (2009), 3849–3859.
26. L. Molnár, *Jordan triple endomorphisms and isometries of spaces of positive definite matrices,* Linear Multilinear Alg. **63** (2015), 12–33.
27. L. Molnár, *The arithmetic, geometric and harmonic means in operator algebras and transformations among them*, Recent Methods and Research Advances in Operator Theory, Ed. F. Botelho, R. King, and T.S.S.R.K. Rao, 182–193, Contemp. Math., 687, Amer. Math. Soc., Providence, RI, 2017.
28. L. Molnár, *On the surjectivity of generalized isometries on the positive definite cone of matrices,* Mediterr. J. Math. **14**:161 (2017).
29. L. Molnár, *Bures isometries between density spaces of* C^**-algebras,* Linear Algebra Appl. **557** (2018), 22–33.
30. L. Molnár, *Quantum Rényi relative entropies: their symmetries and their essential difference,* J. Funct. Anal. **277** (2019), 3098–3130.
31. L. Molnár, *Jordan isomorphisms as preservers,* Linear and multilinear algebra and function spaces, 19–41, Contemp. Math., 750, Centre Rech. Math. Proc., Amer. Math. Soc., Providence, RI, 2020.
32. L. Molnár, *Maps on positive cones in operator algebras preserving power means,* Aequationes Math. **94** (2020), 703–722.
33. L. Molnár, *On the order determining property of the norm of a Kubo-Ando mean in operator algebras,* Integral Equations Operator Theory **93**, Article number: 53 (2021).
34. L. Molnár, *Maps on positive definite cones of* C^**-algebras preserving the Wasserstein mean,* Proc. Amer. Math. Soc. **150** (2022), 1209–1221.
35. L. Molnár, *On certain order properties of non Kubo-Ando means in operator algebras,* Integral Equations Operator Theory **94** Article number: 25 (2022).
36. L. Molnár and P. Szokol, *Transformations on positive definite matrices preserving generalized distance measures,* Linear Algebra Appl. **466** (2015), 141–159.
37. M. Mori, *Order isomorphisms of operator intervals in von Neumann algebras.* Integral Equations Operator Theory **91** (2019), Paper No. 11, 26 pp.

38. W. Pusz and S.L. Woronowicz, *Functional calculus for sesquilinear forms and the purification map*. Rep. Math. Phys. **8** (1975), 159–170.
39. P. Šemrl, *Order isomorphisms of operator intervals,* Integral Equations Operator Theory **89** (2017), 1–42.
40. A. C. Thompson, *On certain contraction mappings in a partially ordered vector space,* Proc. Amer. Math. Soc. **14** (1963), 438–443.
41. T. Tonev and R. Yates, *Norm-linear and norm-additive operators between uniform algebras*, J. Math. Anal. Appl. **357** (2009), 45–53.
42. J. Väisälä, *A proof of the Mazur-Ulam Theorem,* Amer. Math. Monthly **110** (2003), 633–635.

Direct and Inverse Spectral Theorems for a Class of Canonical Systems with Two Singular Endpoints

Matthias Langer and Harald Woracek

AMS MSC 2010: 34B05, 34L40, 34B20, 34A55, 47B50, 47B32

1 Introduction

By a *Hamiltonian*, we understand a function H defined on a (possibly unbounded) interval (a, b), which takes real and non-negative 2×2-matrices as values, is locally integrable and does not vanish on any set of positive measure. Throughout this paper, we assume that Weyl's limit point case prevails at the endpoint b; this means that for one (and hence for all) $x_0 \in (a, b)$, we have $\int_{x_0}^{b} \operatorname{tr} H(x)\, \mathrm{d}x = \infty$.

The *canonical system* associated with H is the differential equation

$$y'(x) = zJH(x)y(x), \qquad x \in (a, b), \tag{1.1}$$

where z is a complex parameter (the eigenvalue parameter), J is the signature matrix $J := \begin{pmatrix} 0 & -1 \\ 1 & 0 \end{pmatrix}$ and y is a 2-vector-valued function. Canonical systems appear frequently in natural sciences, for example, in Hamiltonian mechanics or as generalizations of Sturm–Liouville problems, e.g. in the study of a vibrating string with non-homogeneous mass distribution. They provide a unifying approach to Schrödinger operators, Jacobi operators and Krein strings. Some selected references are [1, 4, 33, 85] for relevance in physics and [6, 53, 54, 92] for the relation to scalar second-order differential or difference equations.

The theory of canonical systems was developed in works of Stieltjes, Weyl, Markov, Krein, Kac and de Branges. There is a vast literature, especially on spectral

M. Langer
Department of Mathematics and Statistics, University of Strathclyde, Glasgow, UK
e-mail: m.langer@strath.ac.uk

H. Woracek (✉)
Institute for Analysis and Scientific Computing, Vienna University of Technology, Wien, Austria
e-mail: harald.woracek@tuwien.ac.at

I. Binder et al. (eds.), *Function Spaces, Theory and Applications*, Fields Institute Communications 87, https://doi.org/10.1007/978-3-031-39270-2_5

theory, ranging from classical papers to very recent work. As examples we mention [5, 15, 41–43, 49, 52, 60, 74, 90, 93–95, 100–102]. Our standard reference is [42], where the spectral theory of canonical systems is developed in a modern operator-theoretic language.

With a Hamiltonian H, one can associate a Hilbert space $L^2(H)$ and a (minimal) differential operator $S(H)$; see Sect. 3.2. The spectral theory of $S(H)$ changes drastically depending on the left endpoint a being in Weyl's limit circle case (LC) or Weyl's limit point case (LP), i.e. whether for one (and hence for all) $x_0 \in (a, b)$

$$\text{(LC)}: \quad \int_a^{x_0} \operatorname{tr} H(x)\,\mathrm{d}x < \infty \qquad \text{or} \qquad \text{(LP)}: \quad \int_a^{x_0} \operatorname{tr} H(x)\,\mathrm{d}x = \infty.$$

Note that because of the non-negativity of H, the Hamiltonian H is in the limit circle case if and only if all entries of H are integrable at a.

Limit Circle Case

Assume that H is in the limit circle case at its left endpoint (and, as always in this paper, in the limit point case at its right endpoint). Then the operator $S(H)$ is symmetric with deficiency index $(1, 1)$. A complex-valued function q_H, the *Weyl coefficient* of H, can be constructed as follows. Let $\theta(\cdot\,; z)$ and $\varphi(\cdot\,; z)$ be the solutions of (1.1) that satisfy the initial conditions $\theta(a; z) = \binom{1}{0}$ and $\varphi(a; z) = \binom{0}{1}$, respectively; note that H is integrable at a. The Weyl coefficient q_H is defined by

$$q_H(z) := \lim_{x \nearrow b} \frac{\theta_1(x; z)\tau + \theta_2(x; z)}{\varphi_1(x; z)\tau + \varphi_2(x; z)}, \qquad z \in \mathbb{C} \setminus \mathbb{R}, \tag{1.2}$$

with $\tau \in \mathbb{R} \cup \{\infty\}$; the limit is independent of τ since H is in the limit point case at b. The function q_H belongs to the *Nevanlinna class* $\mathcal{N}_0$, i.e. it is analytic in $\mathbb{C} \setminus \mathbb{R}$, symmetric with respect to the real line in the sense that $q_H(\overline{z}) = \overline{q_H(z)}$, $z \in \mathbb{C} \setminus \mathbb{R}$, and maps the open upper half-plane $\mathbb{C}^+$ into $\mathbb{C}^+ \cup \mathbb{R}$.

The Weyl coefficient q_H can be used to construct a spectral measure and a Fourier transform. Let μ_H be the measure in the Herglotz integral representation of q_H (see (3.1) below) appropriately including a possible point mass at ∞, and define an integral transformation Θ_H by

$$(\Theta_H f)(t) := \int_a^b \varphi(x; t)^T H(x) f(x)\,\mathrm{d}x, \quad f \in L^2(H), \quad \sup(\operatorname{supp} f) < b.$$

Then a direct spectral theorem holds; more precisely, the following is true.

(1) The map Θ_H extends to an isometric isomorphism from $L^2(H)$ onto $L^2(\mu_H)$, where we tacitly understand that the space $L^2(\mu_H)$ appropriately includes a possible point mass at ∞.
(2) This extension of Θ_H establishes a unitary equivalence between the self-adjoint extension of $S(H)$ that is determined by the boundary condition $y_1(a) = 0$ and

the operator M_{μ_H} of multiplication by the independent variable in the space $L^2(\mu_H)$.

This direct theorem shows, in particular, that the mentioned self-adjoint extension of $S(H)$ has simple spectrum.
An inverse spectral theorem was proved by L. de Branges in [11–14], in particular [12, Theorem XII] and [14, Theorem VII]; see also [102] for an explicit treatment. These results include the following statements.

(1) Let a function q in the Nevanlinna class $\mathcal{N}_0$ be given. Then there exists a Hamiltonian H that is in the limit circle case at its left endpoint (and in the limit point case at its right endpoint) such that $q_H = q$.
(2) Let a positive scalar measure μ with $\int_{\mathbb{R}}(1+t^2)^{-1}\mathrm{d}\mu(t) < \infty$ be given (plus a possible point mass at ∞). Then there exists a Hamiltonian H that is in the limit circle case at its left endpoint (and in the limit point case at its right endpoint) such that $\mu = \mu_H$ (and possible point masses at ∞ coincide).
(3) Let two Hamiltonians H_1 and H_2 be given, both being in the limit circle case at their left endpoints (and in the limit point case at their right endpoints). Then we have $q_{H_1} = q_{H_2}$ if and only if H_1 and H_2 are reparameterizations of each other; the latter means that $H_2(x) = H_1(\gamma(x))\gamma'(x)$ with some increasing bijection γ such that γ and γ^{-1} are absolutely continuous.
(4) Let two Hamiltonians H_1 and H_2 be given, both being in the limit circle case at their left endpoints (and in the limit point case at their right endpoints). Then we have $\mu_{H_1} = \mu_{H_2}$ (and possible point masses at ∞ coincide) if and only if there exists a real constant α such that the Hamiltonians

$$H_1, \qquad \begin{pmatrix} 1 & \alpha \\ 0 & 1 \end{pmatrix} H_2 \begin{pmatrix} 1 & 0 \\ \alpha & 1 \end{pmatrix}$$

are reparameterizations of each other.

Limit Point Case
If the limit point case prevails (also) at the left endpoint, much less can be said in general. The operator $S(H)$ is self-adjoint, and its spectral multiplicity cannot exceed 2. A 2×2-matrix-valued Weyl coefficient can be defined. Via the Titchmarsh–Kodaira formula, this leads to a Fourier transform onto an L^2-space with respect to a 2×2-matrix-valued measure; see, e.g. [42], and [63] or [40, §2] for Schrödinger equations.

For Hamiltonians being in the limit point case, non-simple spectrum can appear; and this is not an exceptional case. The class of all Hamiltonians that have simple spectrum—despite being in the limit point case at both endpoints—can be characterized based on a theorem of I. S. Kac from the 1960s; see [50, Fundamental Theorem].[1] However, given the Hamiltonian H, the condition given in Kac's theorem is hardly accessible to computation. To the best of our knowledge, an

[1] A proof is given in [51] (in Russian).

explicit characterization of simplicity of the spectrum is not known. An easy-to-check sufficient condition for $S(H)$ having simple spectrum follows from a result of L. de Branges; see [15, Theorems 40 and 41].

In the study of limit point Hamiltonians with simple spectrum, there remain some major drawbacks compared with the limit circle situation. Even in the situation of de Branges' theorem, there is neither a canonical way to choose a scalar-valued spectral measure μ nor further information on properties of μ can be obtained. In view of this fact, naturally, there are no inverse statements asserting existence or uniqueness of a Hamiltonian which would lead to a given measure.

The Main Results of the Present Paper

We specify a class $\mathbb{H}$ of Hamiltonians, which are in the limit point case at both endpoints and for which a Weyl theory analogous to the limit circle case can be developed. This class $\mathbb{H}$ is a proper subclass of the one familiar from de Branges' theorem mentioned above, but it is still sufficiently large to cover many cases of interest.

For each Hamiltonian $H \in \mathbb{H}$, we prove the following direct spectral results.

(1) Every solution of equation (1.1) attains regularized boundary values at a in the sense that (at most) finitely many divergent terms are discarded in a well-defined way (Theorem 4.2); the regularization depends on one free parameter $x_0 \in (a, b)$. One can then define solutions θ and φ by prescribing the regularized boundary values. Hence, an analogue of the Weyl coefficient, which we call *singular Weyl coefficient*, can be defined with the help of θ and φ as in (1.2); this singular Weyl coefficient depends on the parameter x_0, but the dependence shows only in an additive real polynomial (Theorem 4.5).

(2) A Fourier transform onto an L^2-space generated by a scalar measure exists. One measure with this property can be constructed in a canonical way via the singular Weyl coefficient, and this measure is independent of the parameter x_0 (Theorem 4.8). The corresponding Fourier transform and its inverse can be written as integral transforms (Theorem 5.1).

Concerning the, now meaningfully posed, inverse spectral problem, we

(3) Characterize the class of measures occurring via the mentioned construction (Theorem 6.1)

(4) Establish global and local uniqueness results (Theorems 6.2 and 6.3)

(5) Establish a one-to-one correspondence between the growth of the Hamiltonian H at a and the growth of the spectral measure μ_H at infinity, measured by a positive integer Δ (Theorem 4.8)

Sturm–Liouville Equations

Recently, Sturm–Liouville equations, and in particular Schrödinger equations, with two singular endpoints attracted a lot of attention; for example, let us mention [30, 32, 35–37, 39, 64–68]. Sturm–Liouville equations for which the corresponding operator is bounded from below can be transformed into canonical systems of the form (1.1); see Remark 9.14. We consider two classes of Sturm–Liouville equations in detail: first, equations without potential, i.e.

$$-\left(py'\right)' = \lambda wy \tag{1.3}$$

with $p(x)$, $w(x) > 0$ a.e., $1/p$, w locally integrable and either $1/p$ or w integrable at a. Such equations, which are treated in Sects. 7 and 8, have many applications (see, e.g. [17] and [88]) and include equations in impedance form, i.e. where $p = w$; see, e.g. [2]. Second, we consider one-dimensional Schrödinger equations, i.e.

$$-y'' + qy = \lambda y \tag{1.4}$$

with q locally integrable. The class of equations we can treat includes radial equations for Schrödinger equations with spherically symmetric potentials; the corresponding operators are also called perturbed Bessel operators. We apply our results on canonical systems to the Sturm–Liouville equations (1.3) and (1.4); in particular, we construct singular Titchmarsh–Weyl coefficients, spectral measures and Fourier transforms, and we prove inverse spectral theorems.

Methods Employed

In order to establish our present results, we utilize the theory of indefinite inner product spaces. Our approach proceeds via Pontryagin space theory, i.e. the theory of indefinite inner product spaces with a finite-dimensional negative part. In some sense, our approach reaches as far as Pontryagin space models possibly can. One key idea is to extend the Hamiltonian H to the left by a so-called indivisible interval so that the original left endpoint a becomes an interior point where H is singular. We can then apply the theory of generalized Hamiltonians, developed in [59–61] and also [82], for which corresponding operator models act in Pontryagin spaces (in general, a generalized Hamiltonian can have a finite number of interior singularities).

We use operator-theoretic tools like the spectral theory of self-adjoint relations, models for generalized Nevanlinna functions and for generalized Hamiltonians, and the theory of de Branges Pontryagin spaces of entire functions. In particular, proofs rely heavily on the theory developed in [82] and [84] and in [59–61]. We would like to mention that the underlying relation in the Pontryagin space is of the most intriguing (but also most difficult to handle) kind: it is a proper relation having infinity as a singular critical point with a neutral algebraic eigenspace.

Organization of the Manuscript

The paper is divided into sections according to the following table.

Table of contents

In Sect. 2 we introduce the class $\mathbb{H}$ of Hamiltonians that is treated in our paper. The definition involves a certain growth condition of the Hamiltonian at the left endpoint a. We associate a positive integer, $\Delta(H)$, with each $H \in \mathbb{H}$, which measures the growth of H at a. Further, we define a class $\mathbb{M}$ of Borel measures on $\mathbb{R}$ that satisfy a certain growth condition at infinity; this class will turn out to be the set of spectral measures of Hamiltonians from $\mathbb{H}$. In Sect. 3 we recall the definition and certain properties of generalized Nevanlinna functions and the operator that is connected with equation (1.1). Moreover, we recall the notion of generalized Hamiltonians, a certain subclass of generalized Hamiltonians that have only one interior singularity and corresponding operator models. In Sect. 4 we show that solutions of (1.1) attain regularized boundary values at a (Theorem 4.2), and we construct singular Weyl coefficients (Theorem 4.5) and construct a spectral measure via a Stieltjes-type inversion formula (Theorem 4.8). The Fourier transform is constructed in Sect. 5 (Theorem 5.1); this shows, in particular, that the spectrum is simple. Inverse spectral theorems (existence and global and local uniqueness theorem) are proved in Sect. 6 (Theorems 6.1, 6.2 and 6.3).

In the second part of the paper, we consider Sturm–Liouville equations. First, we consider equations of the form (1.3). The case when $1/p$ is not integrable at a is considered in Sect. 7; the case when w is not integrable at a is studied in Sect. 8. Finally, Schrödinger equations of the form (1.4) are investigated in Sect. 9.

PART I: General Theory

In the first part, which comprises Sects. 2–6, the direct and inverse spectral theory of canonical systems with two singular endpoints is developed.

2 The Two Basic Classes

We start with the definition and a brief discussion of the two major objects of our investigation. These are a class $\mathbb{H}$ of Hamiltonians and a class $\mathbb{M}$ of measures, which will turn out to correspond to each other.

2.1 The Class $\mathbb{H}$ of Hamiltonians

Let us state the definition of Hamiltonians again explicitly: by a *Hamiltonian* $H = (h_{ij})_{i,j=1}^2$, we understand a function defined on some (non-empty and possibly unbounded) interval (a, b) whose values are real, non-negative 2×2-matrices, which is locally integrable and which does not vanish on any set of positive measure. In the rest of the paper, we shall also write $\mathrm{dom}(H){:=}(a,b)$ if H is defined on (a,b).

We say that two Hamiltonians H_1 and H_2 defined on intervals (a_1,b_1) and (a_2,b_2), respectively, are *reparameterizations* of each other if there exists an increasing bijection $\gamma : (a_2,b_2) \to (a_1,b_1)$ such that γ and γ^{-1} are both absolutely continuous and

$$H_2(x) = H_1\big(\gamma(x)\big) \cdot \gamma'(x), \qquad x \in (a_2,b_2) \text{ a.e.} \tag{2.1}$$

Note that in this situation, y is a solution of (1.1) with $H = H_1$ if and only if $\tilde{y}$, where $\tilde{y}(x) = y(\gamma(x))$, is a solution of (1.1) with $H = H_2$.

Remark 2.1 As a rule of thumb, Hamiltonians which are reparameterizations of each other share all their essential properties. For a detailed and explicit exposition of reparameterizations in an up-to-date language, see [104] (in particular, Theorem 3.8 therein).

We also recall the notion of indivisible intervals. An interval $(\alpha,\beta) \subseteq (a,b)$ is called *H-indivisible* (or just *indivisible*) *of type* ϕ if

$$H(x) = h(x)\xi_\phi\xi_\phi^T, \qquad x \in (\alpha,\beta), \tag{2.2}$$

where $\xi_\phi = (\cos\phi, \sin\phi)^T$ and h is a locally integrable function that is positive almost everywhere; see, e.g. [52]. An indivisible interval (α,β) is called *maximal* if it is not contained in any larger indivisible interval.

Definition 2.2 Let $H = (h_{ij})_{i,j=1}^2$ be a Hamiltonian defined on (a,b). We say that H belongs to the class $\mathbb{H}$ if H is in the limit point case at both endpoints, the interval (a,b) is neither one indivisible interval nor the union of two indivisible intervals and H satisfies the following conditions (I), (HS) and (Δ).

(I) For one (and hence for all) $x_0 \in (a,b)$,

$$\int_a^{x_0} h_{22}(x)\,\mathrm{d}x < \infty.$$

(HS) For one (and hence for all) $x_0 \in (a,b)$,

$$\int_a^{x_0}\int_a^{x} h_{22}(t)\,\mathrm{d}t\, h_{11}(x)\,\mathrm{d}x < \infty.$$

(Δ) Let $x_0 \in (a,b)$ and define functions $X_k : (a,x_0] \to \mathbb{C}^2$ recursively by

$$X_0(x) := \begin{pmatrix} 1 \\ 0 \end{pmatrix}, \qquad x \in (a, x_0],$$

$$X_k(x) := \int_{x_0}^{x} JH(t) X_{k-1}(t)\, \mathrm{d}t, \qquad x \in (a, x_0],\ k \in \mathbb{N}.$$

There exists a number $N \in \mathbb{N}_0$ such that

$$L^2\big(H|_{(a,x_0)}\big) \cap \operatorname{span}\big\{X_k : k \leq N\big\} \neq \{0\}. \tag{2.3}$$

If $H \in \mathbb{H}$, we denote by $\Delta(H)$ the smallest non-negative integer N such that (2.3) holds.

It is proved in [59, Lemma 3.12] that this definition is justified, namely, that the validity of (Δ) and the number $\Delta(H)$ do not depend on the choice of x_0 (for (I) and (HS), this is trivial to check).

Notice that, for $H \in \mathbb{H}$, we always have $\Delta(H) > 0$. This follows since we assume limit point case at a. Namely, for each $x_0 \in (a, b)$, the constant function $(0,\ 1)^T$ belongs to $L^2(H|_{(a,x_0)})$ by (I), and hence the constant $(1,\ 0)^T$ cannot be in this space.

Remark 2.3 We assume that (a, b) is neither one indivisible interval nor the union of two indivisible intervals since; otherwise, the corresponding space $L^2(H)$ (defined in Sect. 3.2) and hence also the Fourier transform would be trivial.

Remark 2.4 The conditions (I) and (HS) are, up to a normalization and exchanging upper and lower rows, precisely the conditions of de Branges' theorem [15, Theorem 41]. Note that under the conditions (I) and (HS), any self-adjoint realization corresponding to $H|_{(a,x_0)}$ has a Hilbert–Schmidt resolvent. The additional condition (Δ) arose only recently in the context of indefinite canonical systems; we recall more details in Sect. 3.2.

In general, it is difficult to decide whether a given Hamiltonian satisfies (Δ). Contrasting (I) and (HS), the condition (Δ) is of recursive nature and not accessible by simple computation. An easier-to-handle (though still recursive) criterion for the validity of (Δ) is available for Hamiltonians of diagonal form, cf. [105, Theorem 3.7] and Sect. 7. Using this criterion, various examples can be constructed. The following two examples are taken from [105, Corollary 3.14 and Example 3.15].

Example 2.5 Let $\alpha \in \mathbb{R}$ and set

$$H_\alpha(x) := \begin{pmatrix} x^{-\alpha} & 0 \\ 0 & 1 \end{pmatrix}, \qquad x \in (0, \infty).$$

Then H_α is in the limit point case at ∞ and satisfies (I) at 0. Depending on the value of α, the following conditions hold:

value of α	(LP)/(LC) at 0	(HS) and (Δ)
$\alpha < 1$	(LC)	both hold (trivially)
$1 \leq \alpha < 2$	(LP)	both hold
$\alpha \geq 2$	(LP)	none holds

Hence, we have $H_\alpha \in \mathbb{H}$ for each $\alpha \in [1, 2)$ but not for other values of α.

The number $\Delta(H_\alpha)$ can be computed, namely,

$$\Delta(H_\alpha) = n, \qquad \text{when } \alpha \in \Big(2 - \frac{1}{n}, 2 - \frac{1}{n+1}\Big) \text{ with } n \in \mathbb{N}.$$

This shows that, for a Hamiltonian of class $\mathbb{H}$, there are no a priori restrictions on the value of the number $\Delta(H)$. Computing $\Delta(H_\alpha)$ for $\alpha = 2 - \frac{1}{n}$ with $n \in \mathbb{N}$ is equally well possible, but requires more elaborate computations. These have not been carried out in [105] but will be made available elsewhere.

Example 2.6 Consider the Hamiltonian

$$H(x) := \begin{pmatrix} (x \ln x)^{-2} & 0 \\ 0 & 1 \end{pmatrix}, \qquad x \in (0, 1).$$

This Hamiltonian is in the limit point case at 0 and at 1, satisfies (I) and (HS) at 0, but does not satisfy (Δ).

This example shows that the presently considered class $\mathbb{H}$ is a proper subclass of the one treated in [15, Theorem 41].

2.2 *The Class* $\mathbb{M}$ *of Measures*

By a *positive Borel measure on* $\mathbb{R}$, we understand a (not necessarily finite) positive measure defined on the σ-algebra of all Borel subsets of $\mathbb{R}$ which takes finite values on compact sets.

Definition 2.7 Let μ be a positive Borel measure on $\mathbb{R}$. We say that μ belongs to the class $\mathbb{M}$ if there exists a number $N \in \mathbb{N}_0$ such that

$$\int_{\mathbb{R}} \frac{d\mu(t)}{(1 + t^2)^{N+1}} < \infty. \tag{2.4}$$

If $\mu \in \mathbb{M}$, we denote by $\Delta(\mu)$ the smallest non-negative integer N such that (2.4) holds.

This class of measures is known from Pontryagin space theory. A measure μ belongs to $\mathbb{M}$ if and only if it is the measure in the distributional representation of some generalized Nevanlinna function with a certain spectral behaviour, cf. [72] and [84, Theorems 2.8 and 3.9]. In the classical (positive definite) setting, this corresponds to the fact that a positive Borel measure μ satisfies $\int_{\mathbb{R}}(1+t^2)^{-1}\,d\mu(t)$ if and only if it is the measure in the Herglotz integral representation of some Nevanlinna function. We recall details in Sect. 3.1.

3 Preliminaries from Indefinite Theory

Our approach to direct and inverse spectral theory for Hamiltonians of class $\mathbb{H}$ is based on the theory of indefinite canonical systems and their Pontryagin space operator models as developed in [59–61] and further in [80, 82]. In this preliminary section, we recall the relevant notions and theorems. For the theory of Pontryagin spaces, we refer the reader, e.g. to [10].

3.1 *Generalized Nevanlinna Functions and the Class* $\mathcal{N}_{<\infty}^{(\infty)}$

As we already mentioned in the introduction, a function q is said to be a *Nevanlinna function* if it is analytic in $\mathbb{C}\setminus\mathbb{R}$ and satisfies $q(\overline{z}) = \overline{q(z)}$ for $z \in \mathbb{C}\setminus\mathbb{R}$ and $\operatorname{Im} q(z) \geq 0$ for $z \in \mathbb{C}^+$. We denote the set of all Nevanlinna functions by $\mathcal{N}_0$.

In Pontryagin space theory, an indefinite analogue of this notion appears and plays a significant role; see, e.g. [71, 72].

Definition 3.1 A function q is called a *generalized Nevanlinna function* if it is meromorphic in $\mathbb{C}\setminus\mathbb{R}$ and has the following properties (i) and (ii):

(i) $q(\overline{z}) = \overline{q(z)}$ for $z \in \rho(q)$, where $\rho(q)$ denotes the domain of analyticity of q in $\mathbb{C}\setminus\mathbb{R}$.
(ii) The reproducing kernel

$$K_q(w,z) := \frac{q(z)-\overline{q(w)}}{z-\overline{w}}, \qquad z, w \in \rho(q),$$

has a finite number of negative squares; the latter means that there exists a $\kappa \in \mathbb{N}_0$ so that for every choice of $n \in \mathbb{N}$ and $z_1, \dots, z_n \in \rho(q)$, the matrices $(K_q(z_i, z_j))_{i,j=1}^n$ have at most κ negative eigenvalues.

We denote the set of all generalized Nevanlinna functions by $\mathcal{N}_{<\infty}$. Moreover, if $q \in \mathcal{N}_{<\infty}$, we denote the actual number of negative squares of the kernel K_q (i.e. the minimal κ in the above definition) by $\operatorname{ind}_- q$. Further, we set $\mathcal{N}_\kappa := \{q \in \mathcal{N}_{<\infty} : \operatorname{ind}_- q = \kappa\}$ for $\kappa \in \mathbb{N}_0$.

That this definition is indeed an extension of the definition of $\mathcal{N}_0$, i.e. that the class $\mathcal{N}_0$ in Definition 3.1 coincides with the class defined before Definition 3.1 is a classical result, which can be traced back to as far as [44] or [91].

Let $q \in \mathcal{N}_0$. Using the representation of the positive harmonic function $\operatorname{Im} q$ as a Poisson integral, one easily obtains a representation of q with a Cauchy-type integral.

3.2 *Herglotz integral representation of $\mathcal{N}_0$-functions.* A function q belongs to the class $\mathcal{N}_0$ if and only if it can be represented in the form

$$q(z) = a + bz + \int_{\mathbb{R}} \Big(\frac{1}{t-z} - \frac{t}{1+t^2}\Big)\, d\mu(t), \qquad z \in \mathbb{C} \setminus \mathbb{R}, \tag{3.1}$$

with $a \in \mathbb{R}$, $b \geq 0$ and a positive Borel measure μ satisfying $\int_{\mathbb{R}} (1+t^2)^{-1}\, d\mu(t) < \infty$.

The analogue of this integral representation in the indefinite setting is a distributional representation of a generalized Nevanlinna function. In essence this is shown in [72], where an integral representation of $q \in \mathcal{N}_{<\infty}$ was given without using the language of distributions. The distributional viewpoint was first mentioned in [48, Introduction, p. 253], established more thoroughly in [57, Corollary 3.5] and refined in [55, Proposition 5.4]. The formulation given below is taken from our paper [84]. This paper contains several results which are crucial for the present discussion and is our standard reference in the context of distributional representations.

Before we can provide the actual statement, we need to introduce some notation. First, we denote by $\mathbb{R}(z)$ the set of all rational functions with real coefficients. Second, we denote by $\overline{\mathbb{R}}$ the one-point compactification of the real line considered as a C^∞-manifold in the usual way. Moreover, for each $z \in \mathbb{C} \setminus \mathbb{R}$, let $\beta_z : \overline{\mathbb{R}} \to \mathbb{C}$ be defined by

$$\beta_z(t) := \begin{cases} \dfrac{1+xz}{x-z}, & x \in \mathbb{R}, \\[2ex] z, & x = \infty. \end{cases}$$

Third, for a function f, set $f^\#(z) := \overline{f(\overline{z})}$ whenever $\overline{z} \in \mathbb{C}$ is in the domain of f. Further, we denote by $\widetilde{\mathcal{D}}'(\overline{\mathbb{R}})$ the set of all distributional densities on $\overline{\mathbb{R}}$; see, e.g. [45] or [84]. With each $\phi \in \widetilde{\mathcal{D}}'(\overline{\mathbb{R}})$, one can associate a linear functional on $C^\infty(\overline{\mathbb{R}})$, which is again denoted by ϕ; see [84, (2.2)]. Next, we denote by $\mathcal{F}(\overline{\mathbb{R}})$ the set of all $\phi \in \widetilde{\mathcal{D}}'(\overline{\mathbb{R}})$ for which there exists a finite subset F of $\overline{\mathbb{R}}$ such that ϕ acts as a positive measure on $\overline{\mathbb{R}} \setminus F$; for details see [84, Definitions 2.1 and 2.3]. Finally, $\mathcal{F}_{\{\infty\}}$ is the set of $\phi \in \mathcal{F}(\overline{\mathbb{R}})$ that act as a positive measure on $\mathbb{R}$. For $\phi \in \mathcal{F}_{\{\infty\}}$, we denote by μ_ϕ the unique positive Borel measure on $\mathbb{R}$ such that

$$\phi(f) = \int_{\mathbb{R}} f(x) \frac{d\mu_\phi(x)}{1+x^2}, \qquad f \in C^\infty(\overline{\mathbb{R}}),\ \operatorname{supp} f \subseteq \mathbb{R}; \tag{3.2}$$

see [84, Definition 2.4].

3.3 Distributional representation of $\mathcal{N}_{<\infty}$-functions [55, Proposition 5.4]. Let $\phi \in \mathcal{F}(\overline{\mathbb{R}})$ and $r \in \mathbb{R}(z)$. Then the function

$$q(z) := r(z) + \phi(\beta_z) \tag{3.3}$$

belongs to $\mathcal{N}_{<\infty}$.

Conversely, let $q \in \mathcal{N}_{<\infty}$ be given. Then there exist unique $\phi \in \mathcal{F}(\overline{\mathbb{R}})$ and $r \in \mathbb{R}(z)$ such that

(i) the representation (3.3) holds;
(ii) r is analytic on $\mathbb{R}$ and remains bounded for $|z| \to \infty$.

In the present paper, the following subclass of $\mathcal{N}_{<\infty}$ plays a central role.

Definition 3.4 We denote by $\mathcal{N}_\kappa^{(\infty)}$, $\kappa \in \mathbb{N}_0$, the set of all functions $q \in \mathcal{N}_\kappa$ such that

$$\lim_{z \overset{\measuredangle}{\longrightarrow} i\infty} \frac{q(z)}{z^{2\kappa-1}} \in (-\infty, 0) \qquad \text{or} \qquad \lim_{z \overset{\measuredangle}{\longrightarrow} i\infty} \left| \frac{q(z)}{z^{2\kappa-1}} \right| = \infty, \tag{3.4}$$

where $\overset{\measuredangle}{\longrightarrow}$ denotes the non-tangential limit, i.e. $z \to \infty$ inside some Stolz angle $\{z \in \mathbb{C} : \varepsilon \leq \arg z \leq \pi - \varepsilon\}$ for one (and hence for all) $\varepsilon \in (0, \frac{\pi}{2})$. Moreover, set

$$\mathcal{N}_{<\infty}^{(\infty)} := \bigcup_{\kappa \in \mathbb{N}_0} \mathcal{N}_\kappa^{(\infty)}.$$

Let us stress that the significance of the condition in this definition is not that (3.4) holds *for some* κ, but that it holds *exactly for* $\kappa = \operatorname{ind}_- q$. Note that $\mathcal{N}_0^{(\infty)} = \mathcal{N}_0$.

The classes $\mathcal{N}_\kappa^{(\infty)}$ appeared often in the recent literature, where they are also denoted by $\mathcal{N}_\kappa^\infty$. Let us mention, for instance, [24, 28, 36, 75], where Sturm–Liouville equations with singular endpoints or singular perturbations of self-adjoint operators were studied, and [23] in connection with rank one perturbations at infinite coupling, and [25–27] where operator models of such functions were investigated. The class $\mathcal{N}_{<\infty}^{(\infty)}$ has an operator-theoretic interpretation, namely, the self-adjoint relation in the operator/relation representation[2] of q has ∞ has its only spectral point of non-positive type; for details see also [84, §5].

For our present considerations, it is essential that the distributional representation of a generalized Nevanlinna function q takes a simple form if $q \in \mathcal{N}_{<\infty}^{(\infty)}$. The following result is contained in [84, Theorem 3.9 (i), (ii)].

3.5 *Distributional representation of $\mathcal{N}_{<\infty}^{(\infty)}$-functions.* A function q belongs to the class $\mathcal{N}_{<\infty}^{(\infty)}$ if and only if it can be represented as

$$q(z) := r + \phi(\beta_z) \tag{3.5}$$

with a real constant r and a distributional density $\phi \in \mathcal{F}_{\{\infty\}}$. Denote by μ_q the measure in (3.2) that is connected with the distributional density ϕ in (3.5), i.e. $\mu_q := \mu_{\phi_q}$ with notation from (3.2). Then a Stieltjes inversion formula is valid:

$$\mu_q\big([a,b]\big) = \frac{1}{\pi} \lim_{\varepsilon \searrow 0} \lim_{\delta \searrow 0} \int_{a-\varepsilon}^{b+\varepsilon} \operatorname{Im} q(t+i\delta)\mathrm{d}t, \qquad [a,b] \subseteq \mathbb{R}; \tag{3.6}$$

see [84, Theorem 3.9 (ii)].

3.6 *Operator model.* With a distributional density $\phi \in \mathcal{F}_{\{\infty\}}$, one can associate a Pontryagin space $\Pi(\phi)$, which is the completion of $C^\infty(\overline{\mathbb{R}})$ with respect to the inner product

$$[f,g]_\phi := \phi(f\overline{g}), \qquad f,g \in C^\infty(\overline{\mathbb{R}}),$$

and a self-adjoint relation A_ϕ in $\Pi(\phi)$; see [48, 57] or [84, §5]. The space $\Pi(\phi)$ contains the following set:

$$\Big\{ f \in L^2\Big(\frac{\mu_\phi(x)}{1+x^2}\Big) : \operatorname{supp} f \text{ is compact} \Big\}. \tag{3.7}$$

Let $\mathcal{E}_{A_\phi}(\infty)$ be the algebraic eigenspace at infinity of A_ϕ. By [84, Theorem 5.3] there exists an isometric, continuous, surjective map

$$\psi(\phi) : \mathcal{E}_{A_\phi}(\infty)^{[\perp]} \to L^2\Big(\frac{\mu_\phi(x)}{1+x^2}\Big),$$

which acts as the identity on functions from the set in (3.7).

[2] A detailed account on the operator representation of scalar-valued generalized Nevanlinna functions can be found in [72, §1]. For a slightly different viewpoint and results for operator- (or matrix-) valued functions, see [18, 71, §3], or some of the vast more recent literature on operator models.

Recall from [22] that every function $q \in \mathcal{N}_\kappa^{(\infty)}$ can be written as $q(z) = p(z)q_0(z)$ where $q_0 \in \mathcal{N}_0$ and p is a monic real polynomial of degree 2κ. Let μ_q be the measure associated with q as in 3.5 and let μ_0 be the measure in the integral representation (3.1) of q_0. Then the Stieltjes inversion formula (3.6) implies that

$$\mu_q([a,b]) = \int_{[a,b]} p(t)\mathrm{d}\mu_0(t)$$

for every finite interval $[a,b]$ with $\mu_q(\{a\}) = \mu_q(\{b\}) = 0$. This, together with [79, Corollary 3.1], immediately yields the following lemma.

Lemma 3.7 *Let $\kappa \in \mathbb{N}$, let $q_n \in \mathcal{N}_\kappa^{(\infty)}$ for $n \in \mathbb{N}$ and let $q \in \mathcal{N}_\kappa^{(\infty)}$ such that $q_n(z) \to q(z)$ locally uniformly on $\mathbb{C}\setminus\mathbb{R}$. Moreover, let μ_{q_n} and μ_q be the measures that are associated with q_n and q, respectively, as in 3.5. Then, for every interval $[a,b]$ with $\mu_q(\{a\}) = \mu_q(\{b\}) = 0$, we have*

$$\lim_{n\to\infty} \mu_{q_n}([a,b]) = \mu_q([a,b]).$$

In this lemma, the assumption that $\operatorname{ind}_- q = \operatorname{ind}_- q_n$ is crucial.

3.2 The Operator Associated with a Canonical System

We recall the definition of the space $L^2(H)$ and the corresponding operator. Note that the notion of indivisible intervals and the vector ξ_ϕ were defined in Sect. 2. The space $L^2(H)$ is the space of measurable functions f defined on (a,b) with values in $\mathbb{C}^2$ which satisfy $\int_a^b f^*Hf < \infty$ and have the property that $\xi_\phi^T f$ is constant on every indivisible interval of type ϕ, factorized with respect to the equivalence relation $=_H$ where

$$f =_H g \qquad \Longleftrightarrow \qquad H(f-g) = 0 \quad \text{a.e.}$$

In the space $L^2(H)$, the *operator* $T(H)$ is defined via its graph as

$$\begin{aligned} T(H) := \Big\{ (f;g) \in \big(L^2(H)\big)^2 : \exists \text{ representatives } \hat{f},\ \hat{g} \text{ of } f,\ g \text{ such that} \\ \hat{f} \text{ is locally absolutely continuous and } \hat{f}' = JH\hat{g} \text{ a.e. on } (a,b) \Big\}. \end{aligned} \tag{3.8}$$

Since H is in the limit point case at both endpoints, the operator $T(H)$ is self-adjoint; see, e.g. [42, §6]. If H is in the limit circle case at the left endpoint, then $T(H)$ is the maximal operator (or relation); its adjoint, the minimal operator $S(H)$, is a symmetric operator.

3.3 General Hamiltonians

The definition of the Pontryagin space analogue of a Hamiltonian and the description of its associated canonical system and the operator model are quite long and involved. Here we give only an intuitive picture; for a complete and logically sound formulation, we refer to [59, §8] or [82, Definitions 2.16–2.18]. The latter paper is our standard reference in the context of general Hamiltonians.

A *general Hamiltonian* $\mathfrak{h}$ is a collection of data:

$$\mathfrak{h}: \qquad n \in \mathbb{N}_0, \quad -\infty \leq \sigma_0 < \sigma_1 < \ldots < \sigma_{n+1} \leq \infty;$$

$$\text{Hamiltonians } H_i : (\sigma_i, \sigma_{i+1}) \to \mathbb{R}^{2\times 2}, \quad i = 1, \ldots, n;$$

$$\ddot{o}_i \in \mathbb{N}_0, \quad b_{i,1}, \ldots, b_{i,\ddot{o}+1} \in \mathbb{R}, \quad d_{i,0}, \ldots, d_{i,2\Delta_i - 1} \in \mathbb{R},$$
$$i = 1, \ldots, n;$$

$$E \subseteq \{\sigma_0, \sigma_{n+1}\} \cup \bigcup_{i=0}^{n} (\sigma_i, \sigma_{i+1});$$

where, among others, the following conditions have to be satisfied.

- H_0 is in the limit circle case at σ_0; if $n \geq 1$, then H_{i-1} is in the limit point case at σ_i and H_i is in the limit point case at σ_i for $i = 1, \ldots, n$.
- The growth of the Hamiltonians H_i towards the points $\sigma_1, \ldots, \sigma_n$ is restricted; the number $\Delta_i \in \mathbb{N}$ is a certain measure for this growth.
- Two adjacent Hamiltonians satisfy an interface condition at their common endpoint.

The general Hamiltonian is called *singular* if H_n is in the limit point case at σ_{n+1}; it is called *regular* if H_n is in the limit circle case at σ_{n+1}. Moreover, let H be the function defined on $\bigcup_{i=0}^{n}(\sigma_i, \sigma_{i+1})$ such that $H|_{(\sigma_i,\sigma_{i+1})} = H_i$ for $i = 0, \ldots, n$.

The following visualization of the canonical system associated with a general Hamiltonian may be helpful:

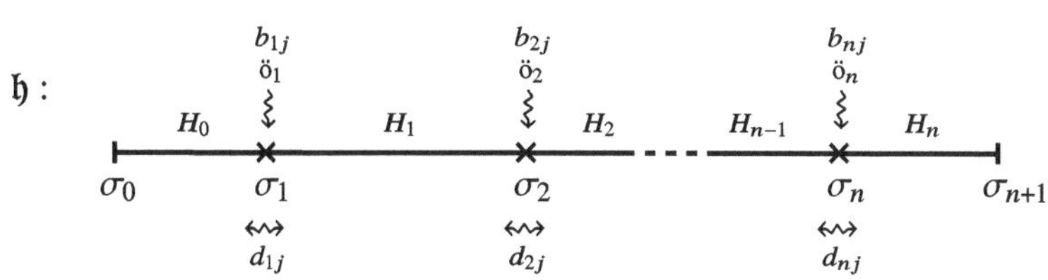

At σ_0 an initial condition can be prescribed. The points $\sigma_1, \ldots, \sigma_n$ are inner singularities: the Hamiltonian function is in the limit point case from both sides. On the interval (σ_i, σ_{i+1}) a solution of the system behaves according to the canonical differential equation $y'(x) = zJH_i(x)y(x)$. The data $\ddot{o}_i, b_{ij}, d_{ij}$ describe what happens to a solution when passing through the singularity σ_i. Thereby, $\ddot{o}_i, b_{ij}$ correspond to a point interaction inside the singularity σ_i, whereas d_{ij} correspond to a local interaction of the Hamiltonians to the left and to the right of the singularity σ_i. The set E (which we did not indicate in the picture) is used to quantitatively describe the influence of the interface conditions manifested by the data part d_{ij}. At the points from the set E the interval (σ_0, σ_{n+1}) is split into smaller pieces that contain at most one singularity.

It can be proved that the spectral theory of singular general Hamiltonians defined in this way is the full Pontryagin space analogue of the theory of classical Hamiltonians (being in the limit circle at their left and in the limit point case at their right endpoint).

(1) With a singular general Hamiltonian $\mathfrak{h}$ a boundary triple $(\mathcal{P}(\mathfrak{h}), T(\mathfrak{h}), \Gamma(\mathfrak{h}))$ can be associated and a Weyl coefficient $q_{\mathfrak{h}}$ can be constructed; see [59, Definition 8.5 and Theorem 8.7] and [62, Theorem 5.1 and Definition 5.2].
(2) The Weyl coefficient $q_{\mathfrak{h}}$ (see 3.8 below) belongs to the class $\mathcal{N}_{<\infty}$ and can be interpreted as a Q-function of the minimal operator $S(\mathfrak{h}){:=}T(\mathfrak{h})^*$, which is a completely non-self-adjoint symmetry with deficiency index $(1, 1)$; see [59, Theorem 8.7] and [62, Proposition 5.19 and Corollary 6.5].
(3) An inverse spectral theorem holds, which states that each generalized Nevanlinna function is the Weyl coefficient of some singular general Hamiltonian and that the general Hamiltonian is, up to reparameterization, uniquely determined by its Weyl coefficients; see [61, Theorem 1.4] and [62, Remark 3.38].

    A local uniqueness theorem holds, which states that beginning sections of Hamiltonians are uniquely determined (up to reparameterization) by the asymptotic behaviour of the Weyl function towards $i\infty$; see [81, Theorem 1.2 and Remark 1.3].

We should point out that the term 'reparameterization', which we used without further notice in item (3), actually requires some explanation. Not only the Hamiltonians H_i may be reparameterized (in the sense of the classical theory, see (2.1)) but also the data d_{ij} and E may be changed according to certain rules; see [62, Remark 3.38].

3.8 The fundamental solution. A crucial concept in the theory of general Hamiltonians is the *fundamental solution* associated with a general Hamiltonian $\mathfrak{h}$. This is the indefinite analogue of the fundamental matrix solution of the system (1.1) in the positive definite case.

It is shown in [62, §5a–c] that, with a general Hamiltonian $\mathfrak{h}$, a chain $\omega_{\mathfrak{h}}$ of entire 2×2-matrix functions is associated, namely, $\omega_{\mathfrak{h}} = \omega_{\mathfrak{h}}(x; z)$, $x \in [\sigma_0, \sigma_{n+1}) \setminus \{\sigma_1, \dots, \sigma_n\}$, $z \in \mathbb{C}$, so that, for fixed x, the function $\omega_{\mathfrak{h}}$ is entire in z and, for fixed z, it satisfies

$$\frac{\partial}{\partial x}\omega_{\mathfrak{h}}(x; z)J = z\omega_{\mathfrak{h}}(x; z)H(x), \qquad x \in (\sigma_0, \sigma_{n+1}) \setminus \{\sigma_1, \dots, \sigma_n\}. \tag{3.9}$$

Note that the rows of $\omega_{\mathfrak{h}}$ satisfy the differential equation (1.1). Moreover,

$$\omega_{\mathfrak{h}}(x;0)=I, \quad \det\omega_{\mathfrak{h}}(x;z)=1, \qquad x\in[\sigma_0,\sigma_{n+1})\setminus\{\sigma_1,\dots,\sigma_n\},\ z\in\mathbb{C} \tag{3.10}$$

and $\omega_{\mathfrak{h}}(\sigma_0;z)=I$, $z\in\mathbb{C}$. The chain $\omega_{\mathfrak{h}}$ is used to construct the Weyl coefficient $q_{\mathfrak{h}}$ using a similar limiting procedure as in the positive definite case: if $\omega_{\mathfrak{h}}=(\omega_{\mathfrak{h},ij})_{i,j=1}^2$, then

$$q_{\mathfrak{h}}(z)=\lim_{x\nearrow\sigma_{n+1}}\frac{\omega_{\mathfrak{h},11}(x;z)\tau+\omega_{\mathfrak{h},12}(x;z)}{\omega_{\mathfrak{h},21}(x;z)\tau+\omega_{\mathfrak{h},22}(x;z)}$$

for $\tau\in\mathbb{R}\cup\{\infty\}$ and z in the domain of holomorphy of $q_{\mathfrak{h}}$ (which is $\mathbb{C}\setminus\mathbb{R}$ with at most a finite number of points removed); the limit is locally uniform in z and independent of τ; see [57, Lemma 8.2].

In the present paper we use some specific properties of fundamental solutions; detailed references are provided at the appropriate places. Here we only would like to mention that two general Hamiltonians that are reparameterizations of each other give rise to the same fundamental solutions (up to reparameterization in the sense of [62, Definition 3.4]) and hence to the same Weyl coefficients; this is shown in [61, Theorem 1.6].

3.9 *Splitting of general Hamiltonians.* Let $\mathfrak{h}$ be a general Hamiltonian, let $s\in\bigcup_{i=0}^n(\sigma_i,\sigma_{i+1})$ and assume that s is not inner point of an indivisible interval. Then 'restrictions' of $\mathfrak{h}$ to the intervals (σ_0,s) and (s,σ_{n+1}) can be defined, which are denoted by $\mathfrak{h}_{\lhd s}$ and $\mathfrak{h}_{\rhd s}$, respectively; see [62, Definition 3.47] and also [82, §2.19].

3.4 *The Class $\mathfrak{H}_0$*

In the present paper those general Hamiltonians are of interest whose Weyl coefficients belong to the class $\mathcal{N}_{<\infty}^{(\infty)}$. They can be characterized in a neat way; see [82, Theorem 3.1]. For the notion of indivisible intervals see Sect. 2, (2.2).

Definition 3.10 We say that a singular general Hamiltonian $\mathfrak{h}$ belongs to the class $\mathfrak{H}_0$ if

(i) $\mathfrak{h}$ has exactly one singularity, i.e. H is defined on a set of the form $(\sigma_0,\sigma_1)\cup(\sigma_1,\sigma_2)$;
(ii) the interval (σ_0,σ_1) is indivisible of type 0, i.e. the Hamiltonian function H_0 of $\mathfrak{h}$ on (σ_0,σ_1) is of the form $H_0(x)=h_0(x)\begin{pmatrix}1&0\\0&0\end{pmatrix}$ with some scalar function h_0.

It follows from the definition of a general Hamiltonian and the form of H_0 that the component $(H_1)_{22}$ is integrable on (σ_1,x_0) for some (and hence all) $x_0\in(\sigma_1,\sigma_2)$, i.e. H_1 satisfies condition (I) in Definition 2.2.

3.11 The relation $\mathfrak{H}_0 \leftrightsquigarrow \mathcal{N}^{(\infty)}_{<\infty}$. The content of [82, Theorem 3.1] is the following: a general Hamiltonian $\mathfrak{h}$ belongs to $\mathfrak{H}_0$ if and only if its Weyl coefficient $q_{\mathfrak{h}}$ belongs to $\mathcal{N}^{(\infty)}_{<\infty} \setminus \mathcal{N}_0$.

Thereby, the negative index of $q_{\mathfrak{h}}$ can be expressed in terms of the general Hamiltonian $\mathfrak{h}$:

$$\operatorname{ind}_- q_{\mathfrak{h}} = \Delta_1 + \left\lfloor \frac{\ddot{o}_1}{2} \right\rfloor + \begin{cases} 1, & \ddot{o}_1 \text{ odd}, \ b_{1,1} > 0, \\ 0, & \text{otherwise}. \end{cases} \tag{3.11}$$

This is a particular instance of a general formula shown in [61, Theorem 1.4], namely, that $\operatorname{ind}_- q_{\mathfrak{h}} = \operatorname{ind}_- \mathfrak{h}$ where $\operatorname{ind}_- \mathfrak{h}$ is given by the formula [82, (2.13)].

Often it is convenient to use a particular form of a general Hamiltonian from the class $\mathfrak{H}_0$, namely

$$\mathfrak{h}: \quad \sigma_0 = -1, \ \sigma_1 = 0, \ \sigma_2 = \infty; \qquad E = \{-1, x_0, \infty\};$$

$$H_0(x) = x^{-2}\begin{pmatrix} 1 & 0 \\ 0 & 0 \end{pmatrix}, \quad x \in (-1, 0); \qquad H_1(x), \quad x \in (0, \infty); \tag{3.12}$$

$$\ddot{o}_1 \in \mathbb{N}_0, \quad b_{1,1}, \dots, b_{1,\ddot{o}_1+1} \in \mathbb{R}, \quad d_{1,0}, \dots, d_{1,2\Delta_1-1} \in \mathbb{R},$$

where $x_0 \in (0, \infty)$; one can choose $b_{1,\ddot{o}_1+1} = 0$ if no interval of the form $(0, \varepsilon)$ with $\varepsilon > 0$ is indivisible. For every given general Hamiltonian $\mathfrak{g} \in \mathfrak{H}_0$ and given $x_0 \in (0, \infty)$ there exists an $\mathfrak{h}$ as in (3.12) which is a reparameterization of $\mathfrak{g}$.

3.5 The Operator Model

The original definition of the boundary triple associated with a general Hamiltonian given in [59] is involved and quite abstract (using a completion procedure). The boundary triple associated with a general Hamiltonian $\mathfrak{h}$ that has only one singularity can be described isomorphically in a more concrete way; see [80, Definition 2.14 and Theorem 2.15]. Since we deal with the operator model in some depth, we recall its concrete description for a general Hamiltonian of the form (3.12). In the above mentioned reference the component h_{11} is integrable around σ_1 instead of the component h_{22}. One has to apply a rotation isomorphism as defined in [62, Definition 2.4] and also discussed in [82, §2.g] to transform the model from [80] to the current situation.

First we need the following fact, which was shown in [59, Lemma 3.10]; also here one has to apply a rotation isomorphism.

3.12 The functions $\mathfrak{w}_k$. Let $H \in \mathbb{H}$ with $\operatorname{dom}(H) = (a, b)$. For each $x_0 \in (a, b)$, there exists a unique sequence $(\mathfrak{w}_k)_{k\in\mathbb{N}_0}$ of absolutely continuous real 2-vector functions on (a, b) such that

$$\begin{aligned} \mathfrak{w}_0 &= \begin{pmatrix} 1 \\ 0 \end{pmatrix}, \\ \mathfrak{w}_{l+1}' &= JH\mathfrak{w}_l, && l \geq 0, \\ \mathfrak{w}_l|_{(a,x_0)} &\in L^2\big(H|_{(a,x_0)}\big), && l \geq \Delta(H), \\ \mathfrak{w}_l(x_0) &\in \operatorname{span}\left\{\begin{pmatrix} 1 \\ 0 \end{pmatrix}\right\}, && l \geq 0. \end{aligned} \tag{3.13}$$

Let $H_1, H_2 \in \mathbb{H}$, and assume that H_1 and H_2 are reparameterizations of each other, say, $H_2(x) = H_1(\gamma(x))\gamma'(x)$, where γ is an increasing bijection such that γ and γ^{-1} are absolutely continuous. Let $x_1 \in \operatorname{dom}(H_1)$, set $x_2{:=}\gamma^{-1}(x_1)$ and let $\mathfrak{w}_{1;l}$ and $\mathfrak{w}_{2;l}$ be the corresponding sequences of functions for H_1 and H_2 respectively. with x_0 replaced by x_1 and x_2 respectively. Then, as a simple calculation shows, one has

$$\mathfrak{w}_{2;l} = \mathfrak{w}_{1;l} \circ \gamma, \qquad l \geq 0.$$

In the following item 3.13, we recall the above-mentioned isomorphic form of the operator model. We restrict ourselves to the case that is needed in the present paper (this leads to a significant simplification of the formulae).

3.13 *The boundary triple* $(\mathcal{P}(\mathfrak{h}), T(\mathfrak{h}), \Gamma(\mathfrak{h}))$. Let $\mathfrak{h} \in \mathfrak{H}_0$ be given by the data as in (3.12), and assume, in addition, that $\ddot{o}_1 = 0$ and that no interval $(0, \varepsilon)$ with $\varepsilon > 0$ is indivisible. Due to the growth restriction imposed on the Hamiltonian functions of a general Hamiltonian in its definition (cf. [82, Definitions 2.16–2.18]) the Hamiltonian function H_1 of $\mathfrak{h}$ satisfies (I), (HS) and (Δ), i.e. it belongs to the class $\mathbb{H}$. Let $\mathfrak{w}_l$ be the corresponding functions (3.13) and denote by $\mathbb{1}_{\urcorner x_0}$ the indicator function of the interval $(0, x_0]$.

First, we define the base space $\mathcal{P}(\mathfrak{h})$ of the boundary triple $(\mathcal{P}(\mathfrak{h}), T(\mathfrak{h}), \Gamma(\mathfrak{h}))$. Set $\Delta{:=}\Delta(H_1) = \Delta_1$ and

$$L^2_\Delta(H_1){:=}L^2(H_1) \dotplus \operatorname{span}\big\{\mathfrak{w}_k \mathbb{1}_{\urcorner x_0} : k = 0, \ldots, \Delta - 1\big\}.$$

Then $\mathcal{P}(\mathfrak{h})$ is the linear space

$$\mathcal{P}(\mathfrak{h}){:=}L^2_\Delta(H_1) \times \mathbb{C}^\Delta$$

endowed with an inner product as follows. Let $F = (f; \xi), G = (g; \eta) \in \mathcal{P}(\mathfrak{h})$, where $\xi = (\xi_k)_{k=0}^{\Delta-1}$, $\eta = (\eta_k)_{k=0}^{\Delta-1}$, and denote by $\lambda = (\lambda_k)_{k=0}^{\Delta-1}$ and $\mu = (\mu_k)_{k=0}^{\Delta-1}$ the unique coefficients such that

$$
\begin{aligned}
\tilde{f} &:= f - \sum_{l=0}^{\Delta-1} \lambda_l \mathfrak{w}_l \mathbb{1}_{\Lsh x_0} \in L^2(H_1), \\
\tilde{g} &:= g - \sum_{l=0}^{\Delta-1} \mu_l \mathfrak{w}_l \mathbb{1}_{\Lsh x_0} \in L^2(H_1).
\end{aligned}
\tag{3.14}
$$

Then

$$[F, G] := (\tilde{f}, \tilde{g})_{L^2(H_1)} + \sum_{k=0}^{\Delta-1} \lambda_k \overline{\eta_k} + \sum_{k=0}^{\Delta-1} \xi_k \overline{\mu_k}.$$

Second, we define the maximal relation $T(\mathfrak{h})$. Set

$$T_{\Delta,\max}(H_1) := \big\{(f; g) \in L^2_\Delta(H_1) \times L^2_\Delta(H_1) : \exists \hat{f} \text{ absolutely continuous representative of } f \text{ s.t. } \hat{f}' = JH_1 g\big\}.$$

Then a pair $(F; G)$ of elements $F = (f; \xi)$, $G = (g; \eta) \in \mathcal{P}(\mathfrak{h})$ belongs to $T(\mathfrak{h})$ if and only if (with λ and μ again as in (3.14))

(i) $(f; g) \in T_{\Delta,\max}(H_1)$;
(ii) for each $k \in \{0, \dots, \Delta - 2\}$,

$$\xi_k = \eta_{k+1} + \frac{1}{2}\mu_{\Delta-1} d_{\Delta+k} + \frac{1}{2}\lambda_0 d_k - \mathfrak{w}_{k+1}(x_0)_1 f(x_0)_2;$$

(iii) $$\xi_{\Delta-1} = \int_0^{x_0} \mathfrak{w}_\Delta^* H_1 \tilde{g} + \frac{1}{2}\sum_{l=0}^{\Delta-1} \lambda_l d_{l+\Delta-1} + \mu_{\Delta-1} d_{2\Delta-1} - \mathfrak{w}_\Delta(x_0)_1 f(x_0)_2.$$

Here $\mathfrak{w}_k(x_0)_2$ denotes the lower component of the vector $\mathfrak{w}_k(x_0)$ and $f(x_0) = (f(x_0)_1, f(x_0)_2)^T$ denotes the value at x_0 of the unique absolutely continuous representative $\hat{f}$ with $\hat{f}' = JH_1 g$ (uniqueness of this representative follows since H_1 does not end indivisibly towards 0, cf. [42, Lemma 3.5]).

Finally, we define the boundary relation $\Gamma(\mathfrak{h})$: for $(F; G) \in T(\mathfrak{h})$, we set

$$\Gamma(\mathfrak{h})(F; G) := \begin{pmatrix} -\lambda_0 \\ \eta_0 - f(x_0)_2 + \frac{1}{2}\sum_{l=0}^{\Delta-1} \mu_l d_l \end{pmatrix}.$$

The space $\mathcal{P}(\mathfrak{h})$ and the relation $T(\mathfrak{h})$ are related to the space $L^2(H_1)$ and the maximal relation $T_{\max}(H_1)$ therein as follows.

3.14 *The map* $\psi(\mathfrak{h})$. The original definition of the map $\psi(\mathfrak{h})$ that establishes this relation is again implicit, cf. [59, Definitions 8.5 and 4.10]. However, based on [59, (4.12)], it is easy to obtain the following description in the concrete model space $\mathcal{P}(\mathfrak{h})$ introduced above.

Let $\mathfrak{h}$ be a general Hamiltonian as in 3.13. Then we denote by $\psi(\mathfrak{h}) : \mathcal{P}(\mathfrak{h}) \to L^2_\Delta(H_1)$ the projection onto the first component of $\mathcal{P}(\mathfrak{h})$, i.e.

$$\psi(\mathfrak{h})(f;\xi) := f, \qquad (f;\xi) \in \mathcal{P}(\mathfrak{h}). \tag{3.15}$$

This map satisfies

$$\big(\psi(\mathfrak{h}) \times \psi(\mathfrak{h})\big)T(\mathfrak{h}) = T_{\Delta,\max}(H_1);$$

see [80, Remark 2.11]. Moreover, it is obvious that $\psi(\mathfrak{h})$ maps $L^2(H_1) \times \mathbb{C}^\Delta$ isometrically and surjectively onto $L^2(H_1)$; note that the elements in $\mathbb{C}^\Delta$ are neutral. Using that

$$L^2(H_1) \times \mathbb{C}^\Delta = \big(\{0\} \times \mathbb{C}^\Delta\big)^{[\perp]}$$

we can deduce that

$$\begin{aligned} \psi(\mathfrak{h})\big((\{0\} \times \mathbb{C}^\Delta)^{[\perp]}\big) &= L^2(H_1), \\ \big(\psi(\mathfrak{h}) \times \psi(\mathfrak{h})\big)\Big(\big(T(\mathfrak{h}) \cap (\{0\} \times \mathbb{C}^\Delta)^{[\perp]}\big)^2\Big) &= T_{\max}(H_1); \end{aligned} \tag{3.16}$$

here $[\perp]$ denotes the orthogonal companion with respect to the inner product $[\,\cdot,\cdot\,]$, i.e. $\mathcal{M}^{[\perp]} = \{x : [x,y] = 0 \text{ for all } y \in \mathcal{M}\}$.

3.6 The Basic Identification

The class $\mathbb{H}$ of Hamiltonians can be identified with the class $\mathfrak{H}_0$ of general Hamiltonians up to the parameters $\ddot{o}_1$, $b_{1,j}$, $d_{1,j}$. This is nearly obvious, but is a crucial observation for our approach. Hence we point it out in this prominent way.

3.15 *The relation* $\mathfrak{H}_0 \rightsquigarrow \mathbb{H}$. Let $\mathfrak{h} \in \mathfrak{H}_0$ be given by the data (3.12). Then $H_1 \in \mathbb{H}$ and $\Delta(H_1) = \Delta_1$.

3.16 *The relation* $\mathbb{H} \rightsquigarrow \mathfrak{H}_0$. Let $H \in \mathbb{H}$ be given, assume that H is defined on $(0,\infty)$ and choose $x_0 \in (0,\infty)$. Then we associate with H a general Hamiltonian $\mathfrak{h} \in \mathfrak{H}_0$ as in (3.12) with $H_1 = H$ and $\ddot{o}_1$, $b_{1,j}$, $d_{1,j}$ arbitrary. Again one has $\Delta_1 = \Delta(H_1)$, and the negative index of $q_\mathfrak{h}$ is given by (3.11).

4 Construction of the Spectral Measure

Let a Hamiltonian $H \in \mathbb{H}$ be given. In this section we complete the following tasks: (1) we show that each solution of the canonical system (1.1) attains regularized boundary values; (2) we construct a family of functions from the class $\mathcal{N}_{<\infty}^{(\infty)}$ to which we refer as *singular Weyl coefficients* of H; and (3) we construct a positive Borel measure of class $\mathbb{M}$ to which we refer as the *spectral measure* of H. Most of these facts follow relatively easily by using the basic identification 3.16 and previous results from [82, 84].

In order to formulate the theorems, one more notation is needed.

4.1 *The defect spaces* $\mathfrak{N}_z$. Let $H \in \mathbb{H}$ and $z \in \mathbb{C}$. We denote the set of all locally absolutely continuous solutions of the differential equation (1.1) by $\mathfrak{N}_z$ and speak of the *defect space of H at the point z*. Clearly, $\mathfrak{N}_z$ is a linear space of dimension 2. Note that, for $z = 0$, this space is trivial in the sense that it consists of all constant functions.

Let $H_1, H_2 \in \mathbb{H}$ and assume that H_1 and H_2 are reparameterizations of each other: $H_2(x) = H_1(\gamma(x))\gamma'(x)$ where γ is an increasing bijection such that γ and γ^{-1} are absolutely continuous. Then a simple calculation shows that the mapping $\psi \mapsto \psi \circ \gamma$ is a bijection between the corresponding defect spaces $\mathfrak{N}_{1,z}$ and $\mathfrak{N}_{2,z}$.

In the next theorem we show that each solution of (1.1) assumes regularized boundary values at the left endpoint. These regularized boundary values will be used later to fix a fundamental system of solutions.

Theorem 4.2 (Regularized boundary values) *Let* $H \in \mathbb{H}$ *with* $\operatorname{dom}(H) = (a, b)$. *Then, for each fixed* $x_0 \in (a, b)$, *the following statements hold (the functions* $\mathfrak{w}_k$ *are as in 3.12).*

(i) *For each* $z \in \mathbb{C}$ *and each solution* $\psi = (\psi_1, \psi_2)^T \in \mathfrak{N}_z$ *the boundary value*

$$\operatorname{rbv}_{z,1}\psi := \lim_{x \searrow a} \psi_1(x)$$

and the regularized boundary value

$$\operatorname{rbv}_{z,2}\psi := -\lim_{x \searrow a}\left[\sum_{l=0}^{\Delta(H)} z^l\big(\mathfrak{w}_l(x)\big)^* J\left(\psi(x) - \lim_{t \searrow a}\psi_1(t)\sum_{k=\Delta(H)+1}^{2\Delta(H)-l} z^k \mathfrak{w}_k(x)\right)\right] \tag{4.1}$$

exist.

(ii) *For* $z \in \mathbb{C}$ *define*

$$\operatorname{rbv}_z : \begin{cases} \mathfrak{N}_z \to \mathbb{C}^2, \\ \psi \mapsto (\operatorname{rbv}_{z,1}\psi,\ \operatorname{rbv}_{z,2}\psi)^T. \end{cases}$$

Then rbv_z *is a bijection from* $\mathfrak{N}_z$ *onto* $\mathbb{C}^2$.

(iii) *For each* $z \in \mathbb{C} \setminus \{0\}$ *there exists an (up to scalar multiples) unique solution* $\psi = (\psi_1, \psi_2) \in \mathfrak{N}_z \setminus \{0\}$ *such that* $\lim_{x \searrow a} \psi_2(x)$ *exists.*
This solution is characterized by the property that $\psi|_{(a,x_0)} \in L^2(H|_{(a,x_0)})$, *and also by the property that* $\mathrm{rbv}_{z,1}\psi = 0$ *(and* $\psi \neq 0$*).*
If ψ *is such that* $\lim_{x \searrow a} \psi_2(x)$ *exists, then*

$$\mathrm{rbv}_{z,2}\psi = \lim_{x \searrow a} \psi_2(x).$$

In contrast to $\mathrm{rbv}_{z,1}\psi$, *the regularized boundary value* $\mathrm{rbv}_{z,2}\psi$ *depends on the choice of* x_0 *since the* $\mathfrak{w}_k$ *depend on* x_0. *This dependence is controlled as follows.*

(iv) *Let* $x_0, \hat{x}_0 \in (a, b)$, *and let* rbv_z *and* $\widehat{\mathrm{rbv}}_z$ *be the correspondingly defined regularized boundary value mappings. Then there exists a polynomial* $p(z)$ *with real coefficients which has no constant term and whose degree does not exceed* $2\Delta(H)$ *such that*

$$\widehat{\mathrm{rbv}}_{z,2}\psi = \mathrm{rbv}_{z,2}\psi + p(z)\mathrm{rbv}_{z,1}\psi, \qquad \psi \in \mathfrak{N}_z, \ \ z \in \mathbb{C}.$$

Remark 4.3 For $z = 0$, solutions ψ of (1.1) are constant, and for such ψ the relation

$$\mathrm{rbv}_z\psi = \psi(x), \qquad x \in (a, b),$$

holds.

Proof (Proof of Theorem 4.2) There is no loss of generality in assuming that H is defined on $(0, \infty)$. This follows since the functions $\mathfrak{w}_k$ transform naturally by composition when performing a reparameterization, cf. 3.12.

Let $\mathfrak{h}$ be the general Hamiltonian given by the data (3.12) with $H_1 = H$ as in the basic identification 3.16 with $\ddot{o}_1$, $b_{1,j}$, $d_{1,j}$ all equal to 0. Items (i) and (ii) follow immediately from [82, Theorem 5.1]; we just need to match notation. Comparing the respective definitions we can deduce that

$$\mathrm{rbv}_{z,1}\psi \mathrel{\widehat{=}} \mathrm{rbv}_{\mathrm{r}}(z)\psi, \quad \mathrm{rbv}_{z,2}\psi \mathrel{\widehat{=}} -\mathrm{rbv}_{\mathrm{s}}(z)\psi, \quad \mathrm{rbv}_z\psi \mathrel{\widehat{=}} \mathrm{rbv}(z)\psi,$$

where the expressions on the right-hand sides are generalized boundary values corresponding to the general Hamiltonian $\mathfrak{h}$ as in [82]. Item (iii) follows directly from [82, Theorem 5.2]. Only the proof of item (iv) requires an argument.

Let $\widehat{\mathfrak{h}}$ be the general Hamiltonian which is constituted by the same data as $\mathfrak{h}$ with the exception that we take $\widehat{E}{:=}\{-1, \hat{x}_0, \infty\}$ instead of E. Then $\widehat{\mathrm{rbv}}_z\psi = \mathrm{rbv}(\widehat{\mathfrak{h}}, z)\psi$, where $\mathrm{rbv}(\widehat{\mathfrak{h}}, z)$ denotes the regularized boundary value map defined for $\widehat{\mathfrak{h}}$ as in [82, Theorem 5.1].

By [59, Proposition 8.11] there exist numbers $d_0, \ldots, d_{2\Delta(H)-1} \in \mathbb{R}$ such that the general Hamiltonian $\mathfrak{g}$ defined as in (3.12) with $H_1 = H$, $\ddot{o}_1 = 0$ and $b_{1,1} = 0$

is a reparameterization of $\widehat{\mathfrak{h}}$. Thereby, the increasing bijection between the domains of $\widehat{\mathfrak{h}}$ and $\mathfrak{g}$ is the identity map. Clearly, the defect spaces $\mathfrak{N}_z$ and the functions $\mathfrak{w}_l$ in (3.13) built with the base point x_0 for $\mathfrak{h}$ and for $\mathfrak{g}$, respectively, coincide.

The fundamental solutions $\omega_{\widehat{\mathfrak{h}}}$ and $\omega_{\mathfrak{g}}$ coincide; see 3.8. Let $\operatorname{rbv}(\mathfrak{g}, z)$ be the regularized boundary value map defined for $\mathfrak{g}$ and let $\operatorname{rbv}_{\mathrm{r}}(\mathfrak{g}, z), \operatorname{rbv}_{\mathrm{s}}(\mathfrak{g}, z)$ be its components as in [82, Theorem 5.1]. By [82, Remark 5.8], equality of fundamental solutions implies equality of regularized boundary values, i.e.

$$\operatorname{rbv}_{\mathrm{r}}(\widehat{\mathfrak{h}}, z) = \operatorname{rbv}_{\mathrm{r}}(\mathfrak{g}, z), \qquad \operatorname{rbv}_{\mathrm{s}}(\widehat{\mathfrak{h}}, z) = \operatorname{rbv}_{\mathrm{s}}(\mathfrak{g}, z).$$

The first of these equalities is of course trivial; both sides, applied to a solution $\psi = (\psi_1, \psi_2)^T \in \mathfrak{N}_z$, are equal to $\lim_{x \searrow a} \psi_1(x)$. The second equality tells us that

$$\widehat{\operatorname{rbv}}_{z,2}\psi = -\operatorname{rbv}_{\mathrm{s}}(\mathfrak{g}, z)\psi, \quad \psi \in \mathfrak{N}_z.$$

Comparing the definition of $\operatorname{rbv}_{\mathrm{s}}(\mathfrak{g}, z)$ in [82, (5.3)] with the definition of $\operatorname{rbv}_{z,2}$ in (4.1) we obtain that

$$\operatorname{rbv}_{\mathrm{s}}(\mathfrak{g}, z)\psi = -\operatorname{rbv}_{z,2}\psi + \operatorname{rbv}_{z,1}\psi \cdot \sum_{l=1}^{2\Delta(H)} z^l d_{l-1}, \quad \psi \in \mathfrak{N}_z,\ z \in \mathbb{C}.$$

The assertion in item (iv) thus follows with the polynomial

$$p(z) := -\sum_{l=1}^{2\Delta(H)} z^l d_{l-1}.$$

□

Remark 4.4 In the above proof we have defined the general Hamiltonian $\mathfrak{h}$ via the basic identification using $\ddot{o}_1, b_{1,j}, d_{1,j}$ all equal to 0. This may seem artificial, and thus requires an explanation. To this end, revisit [82, (5.3)]. If we had used other values for $\ddot{o}_1, b_{1,j}, d_{1,j}$, then the regularized boundary values of $\mathfrak{h}$ as defined in [82] would have changed by the summand

$$\Big(\lim_{t \searrow a} \psi_1(t)\Big)\bigg(\sum_{l=1}^{2\Delta(H)} z^l d_{1,l-1} - \sum_{l=0}^{\ddot{o}_1} z^{2\Delta(H)+l} b_{1,\ddot{o}_1+1-l}\bigg). \tag{4.2}$$

This summand is independent of x and hence contains no information about the asymptotic behaviour of ψ. We regard the inclusion of a summand (4.2) as a distracting complication from the point of our presentation and hence use the choice of vanishing $\ddot{o}_1, b_{1,j}, d_{1,j}$.

Of course, notions intrinsic for H must not depend on the choice of parameters in the basic identification. Thus we shall keep track of the influence of $\ddot{o}_1, b_{1,j}, d_{1,j}$.

In the next theorem a fundamental system of solutions of (1.1) is constructed. Since H is not integrable at the left endpoint, this is a non-trivial task. We fix solutions with the help of the regularized boundary values from Theorem 4.2. With this fundamental system of solutions we then construct a singular Weyl coefficient, which will be used later to obtain a spectral measure. For the definition of the class $\mathcal{N}_\kappa^{(\infty)}$ see Definition 3.4.

Theorem 4.5 (Singular Weyl coefficients) *Let $H \in \mathbb{H}$ with* $\operatorname{dom}(H) = (a, b)$. *Then, for each fixed $x_0 \in (a, b)$, the following statements hold.*

(i) *For each $z \in \mathbb{C}$ denote by $\theta(\cdot\,; z) = (\theta_1(\cdot\,; z), \theta_2(\cdot\,; z))^T$ and $\varphi(\cdot\,; z) = (\varphi_1(\cdot\,; z), \varphi_2(\cdot\,; z))^T$ the unique elements of $\mathfrak{N}_z$ such that*

$$\operatorname{rbv}_z\theta(\cdot\,; z) = (1,\ 0)^T, \qquad \operatorname{rbv}_z\varphi(\cdot\,; z) = (0,\ 1)^T. \tag{4.3}$$

Then, for each $x \in (a, b)$, the functions $\theta(x; \cdot)$ and $\varphi(x; \cdot)$ are entire of finite exponential type[3]

$$\int_a^x \sqrt{\det H(t)}\,\mathrm{d}t, \tag{4.4}$$

and they satisfy $\theta_1(x; z)\varphi_2(x; z) - \theta_2(x; z)\varphi_1(x; z) = 1$ for $z \in \mathbb{C}$.

Moreover, let $a_+ = \inf\{x \in (a, b) : \int_a^x h_{22}(t)\mathrm{d}t > 0\}$. Then, for each $z \in \mathbb{C}$, the following relations hold:

$$\begin{aligned} &\lim_{x\searrow a} \theta_1(x; z) = 1, && \lim_{x\searrow a} \frac{\theta_1(x; z)}{\int_x^{x_0} h_{11}(t)\mathrm{d}t} = -z, \\ &\lim_{x\searrow a_+} \frac{\varphi_1(x; z)}{\int_a^x h_{22}(t)\mathrm{d}t} = -z, && \lim_{x\searrow a} \varphi_2(x; z) = 1. \end{aligned} \tag{4.5}$$

(ii) *For each $\tau \in \mathbb{R} \cup \{\infty\}$, the limit*

$$q_H(z) := \lim_{x\nearrow b} \frac{\theta_1(x; z)\tau + \theta_2(x; z)}{\varphi_1(x; z)\tau + \varphi_2(x; z)}, \qquad z \in \mathbb{C} \setminus \mathbb{R}, \tag{4.6}$$

exists locally uniformly on $\mathbb{C} \setminus \mathbb{R}$, defines an analytic function in z on $\mathbb{C} \setminus \mathbb{R}$ and does not depend on τ (here the fraction on the right-hand side of (4.6) is interpreted as $\frac{\theta_1(x;z)}{\varphi_1(x;z)}$ if $\tau = \infty$). The function q_H belongs to the class $\mathcal{N}_{\Delta(H)}^{(\infty)}$.

(iii) *We have*

$$\theta(\cdot\,; z) - q_H(z)\varphi(\cdot\,; z) \in L^2\big(H|_{(x_0,b)}\big), \qquad z \in \mathbb{C} \setminus \mathbb{R}, \tag{4.7}$$

[3] If the integral in (4.4) is 0, then $\theta(x; \cdot)$ and $\varphi(x; \cdot)$ are either of minimal exponential type or of order less than 1.

and this property characterizes the value $q_H(z)$ for each $z \in \mathbb{C} \setminus \mathbb{R}$.

The function q_H depends on the choice of x_0, which is controlled as follows.

(iv) *Let $x_0, \hat{x}_0 \in (a, b)$ and let q_H and $\widehat{q}_H$ be the correspondingly defined functions (4.6). Then there exists a polynomial p with real coefficients which has no constant term and whose degree does not exceed $2\Delta(H)$ such that*

$$\widehat{q}_H(z) = q_H(z) + p(z).$$

Proof Again we may assume without loss of generality that H is defined on $(0, \infty)$. Let $\mathfrak{h}$ be the general Hamiltonian (3.12) with $H_1 = H$ and $\ddot{o}_1, b_{1,j}, d_{1,j}$ all equal to 0, so that $\mathrm{rbv}_z\psi = \mathrm{rbv}(z)\psi$, where $\mathrm{rbv}(z)\psi$ is as in [82, Theorem 5.1]. It follows from [82, Corollary 5.7] that the fundamental solution $\omega_{\mathfrak{h}}$ from 3.8 associated with $\mathfrak{h}$ is given by

$$\omega_{\mathfrak{h}}(x; z) = \begin{pmatrix} \theta_1(x; z) & \theta_2(x; z) \\ \varphi_1(x; z) & \varphi_2(x; z) \end{pmatrix}. \tag{4.8}$$

The first properties of θ and φ mentioned in (i) are immediate; see 3.8. The formula for the exponential type follows from [83, Theorem 4.1] if we observe that $\det H_0(x) = 0$ for $x \in (-1, 0)$ with H_0 from (3.12). The limit relations in (4.5) is a consequence of [82, Theorem 4.1 (with $\alpha = 0$), Remark 4.2 (iii) and Lemma 4.14].

Corollary 5.7 in [82] also implies that the limit in (4.6) exists locally uniformly, that q_H is characterized by (4.7) and that q_H coincides with the Weyl coefficient $q_{\mathfrak{h}}$ of the general Hamiltonian $\mathfrak{h}$. In particular, this shows that q_H belongs to the class $\mathcal{N}^{(\infty)}_{\Delta(H)}$; see 3.11 and note that $\Delta(H) = \Delta_1$ and $\ddot{o}_1 = 0$.

For the proof of item (iv), consider again the general Hamiltonians $\widehat{\mathfrak{h}}$ and $\mathfrak{g}$ as in the proof of Theorem 4.2 (iv). Since they are reparameterizations of each other, their Weyl coefficients coincide. An application of [82, Corollary 5.9] with $\mathfrak{h}$ and $\mathfrak{g}$ gives

$$\widehat{q}_H(z) - q_H(z) = q_{\widehat{\mathfrak{h}}}(z) - q_{\mathfrak{h}}(z) = q_{\mathfrak{g}}(z) - q_{\mathfrak{h}}(z) = -\sum_{l=1}^{2\Delta(H)} z^l d_{l-1},$$

which shows (iv). □

Note that for $z = 0$ one has

$$\theta(x; 0) = \begin{pmatrix} 1 \\ 0 \end{pmatrix}, \quad \varphi(x; 0) = \begin{pmatrix} 0 \\ 1 \end{pmatrix}, \qquad x \in (a, b), \tag{4.9}$$

which follows from (4.3).

Remark 4.6 Let us study the influence of the parameters $\ddot{o}_1, b_{1,j}, d_{1,j}$ in 3.16 on the above proof. If we chose other parameters than all equal to 0 and hence alter the regularized boundary values by a polynomial summand, then the same would happen to the function q_H. In fact, revisiting [82, Corollary 5.9] we would pass from q_H to $q_H + p$ where p is a polynomial with real coefficients with $p(0) = 0$.

Notice that, conversely, each summand $p \in \mathbb{R}[z]$ with $p(0) = 0$ can be produced by a proper choice of $\ddot{o}_1, b_{1,j}, d_{1,j}$. Moreover, changing the base point x_0 (which is the second arbitrariness in our basic identification) also manifests only in adding a polynomial summand $p \in \mathbb{R}[z]$ with $p(0) = 0$.

In order to handle the arbitrariness in the basic identification in a structurally clean way, we introduce an equivalence relation on the set of Weyl coefficients. Namely, we set

$$q_1 \sim q_2 \quad :\Longleftrightarrow \quad q_1 - q_2 \in \mathbb{R}[z],\ (q_1 - q_2)(0) = 0. \tag{4.10}$$

Clearly, this is an equivalence relation on $\mathcal{N}_{<\infty}^{(\infty)}$. In this context, remember that $q \in \mathcal{N}_{<\infty}^{(\infty)}$ implies that $q + p \in \mathcal{N}_{<\infty}^{(\infty)}$ for all $p \in \mathbb{R}[z]$.

Definition 4.7 Let $H \in \mathbb{H}$ be given. Then we denote by $[q]_H$ the equivalence class modulo the relation (4.10) which contains some (and hence any) function q_H constructed in Theorem 4.5.

We speak of $[q]_H$ as *the singular Weyl coefficient of H*. Each representative q_H of $[q]_H$ is called *a* (!) singular Weyl coefficient of H.

By this definition we achieve that the singular Weyl coefficient $[q]_H$ of $H \in \mathbb{H}$ is nothing but the equivalence class which consists of all Weyl coefficients of general Hamiltonians associated with H by the basic identification.

In the following theorem a measure is constructed with the help of the singular Weyl coefficient and a Stieltjes-type inversion formula.

Theorem 4.8 (The spectral measure) *Let $H \in \mathbb{H}$ be given. Then there exists a unique positive Borel measure μ_H with*

$$\mu_H\big([s_1, s_2]\big) = \frac{1}{\pi} \lim_{\varepsilon \searrow 0} \lim_{\delta \searrow 0} \int_{s_1-\varepsilon}^{s_2+\varepsilon} \operatorname{Im} q_H(t + i\delta)\, dt, \quad -\infty < s_1 < s_2 < \infty, \tag{4.11}$$

where $q_H \in [q]_H$ is any singular Weyl coefficient of H. We have $\mu_H \in \mathbb{M}$ and $\Delta(\mu_H) = \Delta(H)$.

Proof Since $q_H \in \mathcal{N}_{<\infty}^{(\infty)}$, it has a representation $q_H(z) = r + \phi(\beta_z)$ with $r \in \mathbb{R}$ and a distributional density $\phi \in \mathcal{F}_{\{\infty\}}$, i.e. ϕ coincides with a measure μ_ϕ on $\mathbb{R}$; see 3.5. It follows from (3.6) (see also [84, Theorem 3.9 (ii)]) that the measure μ_ϕ is given by the right-hand side of (4.11) on the set of closed intervals; in particular, the double limit exists. Moreover, [84, Theorem 2.8 (ii)] implies that $\mu_H \in \mathbb{M}$.

The fact that μ_H does not depend on the choice of $q_H \in [q]_H$ is clear since a summand that is a real polynomial yields no contribution in the Stieltjes inversion formula.

Finally, we show that $\Delta(\mu_H) = \Delta(H)$. It follows from [84, Theorem 3.9 (v)] that

$$\Delta(\mu_H) = \min\big\{\operatorname{ind}_-(q_H + p) : p \in \mathbb{R}[z]\big\}$$
$$= \min\big\{\operatorname{ind}_-(q_H + p) : p \in \mathbb{R}[z],\ p(0) = 0\big\}.$$

For every $p \in \mathbb{R}[z]$ with $p(0) = 0$ there exist $\ddot{o}_1$, $b_{1,j}$, $d_{1,j}$ such that the Weyl coefficient of the general Hamiltonian $\mathfrak{h}$ in (3.12) with $H_1 = H$ is equal to $q_H + p$; and for each choice of $\ddot{o}_1$, $b_{1,j}$, $d_{1,j}$ the Weyl coefficient is of this form; see 4.6. It follows from this and (3.11) that

$$\Delta(\mu_H) = \min\big\{\operatorname{ind}_-\mathfrak{h} : \mathfrak{h} \text{ as in (3.12) with } H_1 = H \text{ and } \ddot{o}_1, b_{1,j}, d_{1,j} \text{ arbitrary}\big\}$$
$$= \Delta_1 = \Delta(H).$$

□

Definition 4.9 Let $H \in \mathbb{H}$ be given. Then we call the unique positive Borel measure μ_H defined by (4.11) the *spectral measure of H*.

The choice of this terminology is justified by Theorem 5.1 below where we construct a Fourier transform into the space $L^2(\mu_H)$. Before we establish this Fourier transform, let us mention one simple observation. Namely, it is almost immediate that Hamiltonians which are reparameterizations of each other give rise to the same singular Weyl coefficients and the same spectral measures. We provide a slightly more exhaustive variant of this fact.

Proposition 4.10 *Let $H \in \mathbb{H}$ and $\alpha \in \mathbb{R}$. Then the Hamiltonian*

$$H_\alpha := \begin{pmatrix} 1 & \alpha \\ 0 & 1 \end{pmatrix} H \begin{pmatrix} 1 & 0 \\ \alpha & 1 \end{pmatrix}$$

belongs to $\mathbb{H}$.

Let, in addition, $\tilde{H} \in \mathbb{H}$ be given and assume that $\tilde{H}$ and H_α are reparameterizations of each other. Then

(i) *for each pair of singular Weyl coefficients $q_{\tilde{H}}$ and q_H of $\tilde{H}$ and H, respectively, the difference $q_{\tilde{H}} - q_H$ is a real polynomial whose constant term equals α;*
(ii) $\mu_{\tilde{H}} = \mu_H$.

Proof Let $\mathfrak{h}$ be the general Hamiltonian as in (3.12) with $H_1 = H$ and let $\omega_{\mathfrak{h}}$ be its fundamental solution. It follows from [57, Lemma 10.2] that

$$\omega_\alpha := \begin{pmatrix} 1 & \alpha \\ 0 & 1 \end{pmatrix} \omega_{\mathfrak{h}} \begin{pmatrix} 1 & -\alpha \\ 0 & 1 \end{pmatrix}$$

is the fundamental solution of some general Hamiltonian $\mathfrak{h}_\alpha$. This factorization immediately yields

$$q_{\mathfrak{h}_\alpha} = q_{\mathfrak{h}} + \alpha, \tag{4.12}$$

which implies that $q_{\mathfrak{h}_\alpha} \in \mathcal{N}_{<\infty}^{(\infty)}$, and hence $\mathfrak{h}_\alpha \in \mathfrak{H}_0$. A short computation, based on [62, Corollary 5.6], shows that the Hamiltonian function of $\mathfrak{h}_\alpha$ on $(0, \infty)$ is equal to H_α, and therefore $H_\alpha \in \mathbb{H}$. The functions $q_{\mathfrak{h}_\alpha}$ and $q_{\mathfrak{h}}$ are singular Weyl coefficients of H_α and H, respectively. Relation (4.12), together with Theorem 4.5 (iv) implies that $q_{H_\alpha} - q_H$ is a real polynomial whose constant term is equal to α.

Assume now that $\tilde{H}$ is a reparameterizations of H_α, say (2.1) holds with $H_1 = H_\alpha$, $H_2 = \tilde{H}$ and some $\gamma : (0, \infty) \to (0, \infty)$. We can build general Hamiltonians $\tilde{\mathfrak{h}}$ and $\mathfrak{h}_\alpha$ via 3.16 with $\ddot{o}_1 = \tilde{\ddot{o}}_1 = 0$, $b_{1,1} = \tilde{b}_{1,1} = 0$, $d_{1,j} = \tilde{d}_{1,j} = 0$ and some $x_0, \tilde{x}_0 \in (0, \infty)$ such that $\gamma(\tilde{x}_0) = x_0$. Then $\tilde{\mathfrak{h}}$ and $\mathfrak{h}_\alpha$ are reparameterizations of each other; see [62, Remark 3.38] and [59, Proposition 8.13]. Hence $\tilde{\mathfrak{h}}$ and $\mathfrak{h}_\alpha$ have the same Weyl coefficients; see [61, Theorem 1.4]. This shows that $q_{\tilde{H}} = q_{H_\alpha}$, which in turn implies (i).

For (ii) observe that an entire summand yields no contribution in the Stieltjes inversion formula. □

5 The Fourier Transform

For a positive Borel measure μ on $\mathbb{R}$ we denote by M_μ the operator of multiplication by the independent variable in $L^2(\mu)$. In this section we prove that for each $H \in \mathbb{H}$ there exists a unitary operator Θ_H from $L^2(H)$ onto $L^2(\mu_H)$, the Fourier transform connected with H, which establishes unitary equivalence of $T(H)$ and M_{μ_H}. Both Θ_H and its inverse act as integral transformations. These results are the most involved ones in the paper. Their proofs require to go deeply into the operator model of a general Hamiltonian $\mathfrak{h}$. The essential ingredients are the following:

(1) the spectral theory of the model relation in the Pontryagin space $\mathcal{P}(\mathfrak{h})$, in particular, the spectral decomposition of a self-adjoint relation in a Pontryagin space,
(2) Q-function theory to relate the model relation connected with $\mathfrak{h}$ to the model relation of a distributional density,
(3) the interpretation of a fundamental solution matrix as a generalized u-resolvent matrix.

Theorem 5.1 (The Fourier transform) *Let $H \in \mathbb{H}$ with* $\operatorname{dom}(H) = (a, b)$ *be given, let $T(H)$ be the self-adjoint operator as defined in Sect. 3.2 and let μ_H be the spectral measure associated with H via* (4.11)*. Moreover, let $\varphi(\cdot\,;z) = (\varphi_1(\cdot\,;z), \varphi_2(\cdot\,;z))^T$ be the unique element of $\mathfrak{N}_z$ with* $\operatorname{rbv}_z\varphi(\cdot\,;z) = (0, 1)^T$ *as in Theorem 4.5. Then the following statements hold.*

(i) *The map defined by*

$$(\Theta_H f)(t) := \int_a^b \varphi(x;t)^T H(x) f(x)\,\mathrm{d}x, \qquad t \in \mathbb{R},$$
$$f \in L^2(H), \quad \sup(\operatorname{supp} f) < b, \tag{5.1}$$

extends to an isometric isomorphism from $L^2(H)$ *onto* $L^2(\mu_H)$*, which we again denote by* Θ_H*.*

(ii) *The operator* Θ_H *establishes a unitary equivalence between* $T(H)$ *and* M_{μ_H}*, i.e. we have*

$$\Theta_H \circ T(H) = M_{\mu_H} \circ \Theta_H.$$

(iii) *On the subspace of compactly supported functions, also the inverse of* Θ_H *acts as an integral transformation, namely,*

$$(\Theta_H^{-1} g)(x) = \int_{-\infty}^{\infty} g(t)\varphi(x;t)\,\mathrm{d}\mu_H(t), \qquad x \in (a,b),$$
$$g \in L^2(\mu_H), \quad \operatorname{supp} g \text{ compact.} \tag{5.2}$$

Remark 5.2 Note that the integrals in (5.1) and (5.2) are always well defined. For the latter this is obvious since $\varphi(x;t)$ is continuous in t; for the former it follows from Theorem 4.2 (iii).

As an additional result we prove a connection between the point mass at 0 of the spectral measure and the behaviour of H at b.

Proposition 5.3 *Let* $H = (h_{ij})_{i,j=1}^2 \in \mathbb{H}$*, defined on* (a,b)*, and let* μ_H *be the spectral measure associated with* H *via* (4.11)*. Then* $\mu_H(\{0\}) > 0$ *if and only if*

$$\int_a^b h_{22}(x)\mathrm{d}x < \infty. \tag{5.3}$$

If (5.3) *is satisfied, then*

$$\mu_H\big(\{0\}\big) = \left[\int_a^b h_{22}(x)\mathrm{d}x\right]^{-1}. \tag{5.4}$$

Note that in any case,

$$\mu_H\big(\{0\}\big) = -\lim_{y \searrow 0} iy q_H(iy) \tag{5.5}$$

by [84, (3.8)].

The rest of this section is devoted to the proof of Theorem 5.1 and Proposition 5.3. We split the proof of Theorem 5.1 into three parts, which are contained in three separate subsections. First, in Sect. 5.1, we construct a Fourier transform Θ_H from $L^2(H)$ onto $L^2(\mu_H)$ in an abstract way. In Sect. 5.2 we show that this map acts as asserted in (5.1). The formula for Θ_H^{-1} is proved in Sect. 5.3. Finally, we prove Proposition 5.3 in Sect. 5.4.

Since all statements in Theorem 5.1 and Proposition 5.3 are invariant under reparameterizations, we can assume without loss of generality that $\operatorname{dom} H = (0, \infty)$. Throughout these four subsections, keep $H \in \mathbb{H}$ fixed and let $\mathfrak{h}$ be the general Hamiltonian defined in the basic identification 3.16 with $\ddot{o}_1$, $b_{1,j}$, $d_{1,j}$ all equal to 0.

5.1 Construction of a Fourier Transform

Let us first consider the case when $(0, c)$ is a maximal H-indivisible interval of type 0. Then the space $L^2(H)$ can be identified with $L^2(H|_{(c,\infty)})$ and $\varphi(c; z) = (0,\ 1)^T$; see [82, proof of Theorem 5.1, p. 541]. Now items (i) and (ii) in Theorem 5.1 follow from [13, Theorem III]. For the rest of Sects. 5.1 and 5.2 we assume that H does not start with an indivisible interval at 0. Hence we can use the boundary triple $(\mathcal{P}(\mathfrak{h}), T(\mathfrak{h}), \Gamma(\mathfrak{h}))$ associated with the general Hamiltonian $\mathfrak{h}$ in the form in 3.13.

The Fourier transform Θ_H is constructed by combining several mappings. We provide a comprehensive summary in Fig. 1 on page 139. The reader may find it helpful to visit this summary already while going through the construction.

It is shown in [62, Proposition 5.19] that the Weyl coefficient $q_{\mathfrak{h}}$ is a Q-function of the minimal relation $S(\mathfrak{h}) := T(\mathfrak{h})^*$. In fact, denote by $\pi_{l,1}$ the projection from $\mathbb{C}^2 \times \mathbb{C}^2$ onto the upper entry of the first vector component,[4] set

$$A := \ker(\pi_{l,1} \circ \Gamma(\mathfrak{h})), \tag{5.6}$$

and let ε_z be the defect elements with

$$\big(\pi_l \circ \Gamma(\mathfrak{h})\big)(\varepsilon_z; z\varepsilon_z) = \begin{pmatrix} 1 \\ -q_{\mathfrak{h}}(z) \end{pmatrix}. \tag{5.7}$$

Then $q_{\mathfrak{h}}$ is the Q-function of $S(\mathfrak{h})$ induced by A and $(\varepsilon_z)_{z\in\rho(A)}$. In particular, since $q_{\mathfrak{h}}$ is analytic in $\mathbb{C}\backslash\mathbb{R}$ and $S(\mathfrak{h})$ is completely non-self-adjoint, we have $\mathbb{C}\backslash\mathbb{R} \subseteq \rho(A)$.

[4] Here 'l, 1' stands for 'left vector, first entry'. This is a generic notation: for example, π_l is the projection from $\mathbb{C}^2 \times \mathbb{C}^2$ onto the first vector component, $\pi_{r,2}$ onto the lower entry of the second vector component, etc. The use of 'left' and 'right' is motivated by the fact that the vectors correspond to boundary values at the left and right endpoint, respectively.

Let $\phi_{\mathfrak{h}}$ be the distributional density in the representation (3.3) of $q_{\mathfrak{h}}$, and let $\Pi(\phi_{\mathfrak{h}})$, $A_{\phi_{\mathfrak{h}}}$ and $\psi(\phi_{\mathfrak{h}})$ be as in 3.6. By [57, Proposition 3.4, Proof of Corollary 3.5], there exists an isometric isomorphism (where we use the base point $z_0 = i$)

$$\Theta_{\mathfrak{h}} : \mathcal{P}(\mathfrak{h}) \to \Pi(\phi_{\mathfrak{h}}) \quad \text{with} \quad (\Theta_{\mathfrak{h}} \times \Theta_{\mathfrak{h}})(A) = A_{\phi_{\mathfrak{h}}}. \tag{5.8}$$

This isomorphism is determined by its action on defect elements, namely,

$$\Theta_{\mathfrak{h}}(\varepsilon_z) = \hat{\varepsilon}_z, \qquad z \in \rho(A), \tag{5.9}$$

where $\hat{\varepsilon}_z \in \Pi(\phi_{\mathfrak{h}})$ is defined by

$$\hat{\varepsilon}_z(t) := \begin{cases} \dfrac{t-i}{t-z}, & t \in \mathbb{R}, \\ 1, & t = \infty. \end{cases} \tag{5.10}$$

Denote by $\mathcal{E}_A(\infty)$ and $\mathcal{E}_{\phi_{\mathfrak{h}}}(\infty)$ the algebraic eigenspaces at infinity of the relations A and $A_{\phi_{\mathfrak{h}}}$, respectively. By [82, Lemma 3.2 (d)] we have $\mathcal{E}_A(\infty) = \{0\} \times \mathbb{C}^{\Delta}$; in particular, $\mathcal{E}_A(\infty)$ is neutral. It follows from (5.8) that

$$\Theta_{\mathfrak{h}}(\mathcal{E}_A(\infty)) = \mathcal{E}_{\phi_{\mathfrak{h}}}(\infty),$$

and

$$\Theta_{\mathfrak{h}}\big(\mathcal{E}_A(\infty)^{[\perp]}\big) = \mathcal{E}_{\phi_{\mathfrak{h}}}(\infty)^{[\perp]}. \tag{5.11}$$

These relations imply that also $\mathcal{E}_{\phi_{\mathfrak{h}}}(\infty)$ is neutral and that the restriction $\Theta_{\mathfrak{h}}|_{\mathcal{E}_A(\infty)^{[\perp]}}$ is an isometric bijection from $\mathcal{E}_A(\infty)^{[\perp]}$ onto $\mathcal{E}_{\phi_{\mathfrak{h}}}(\infty)^{[\perp]}$. Let π_A and $\pi_{\phi_{\mathfrak{h}}}$ be the following canonical projections:

$$\pi_A : \mathcal{E}_A(\infty)^{[\perp]} \to \mathcal{E}_A(\infty)^{[\perp]} / \mathcal{E}_A(\infty),$$
$$\pi_{\phi_{\mathfrak{h}}} : \mathcal{E}_{\phi_{\mathfrak{h}}}(\infty)^{[\perp]} \to \mathcal{E}_{\phi_{\mathfrak{h}}}(\infty)^{[\perp]} / \mathcal{E}_{\phi_{\mathfrak{h}}}(\infty).$$

Then there exists an isometric isomorphism

$$\Lambda_{\mathfrak{h}} : \mathcal{E}_A(\infty)^{[\perp]} / \mathcal{E}_A(\infty) \to \mathcal{E}_{\phi_{\mathfrak{h}}}(\infty)^{[\perp]} / \mathcal{E}_{\phi_{\mathfrak{h}}}(\infty)$$

such that

$$\pi_{\phi_{\mathfrak{h}}} \circ \Theta_{\mathfrak{h}} = \Lambda_{\mathfrak{h}} \circ \pi_A.$$

By (3.15) we have

$$\ker \pi_A = \mathcal{E}_A(\infty) = \{0\} \times \mathbb{C}^{\Delta} = \ker \psi(\mathfrak{h}).$$

Since $\psi(\mathfrak{h})\big(\mathcal{E}_A(\infty)^{[\perp]}\big) = L^2(H)$ by the first relation in (3.16), we can deduce that there exists an isometric isomorphism

$$\Psi_A : \mathcal{E}_A(\infty)^{[\perp]}/\mathcal{E}_A(\infty) \to L^2(H) \quad \text{with} \quad \Psi_A \circ \pi_A = \psi(\mathfrak{h})|_{\mathcal{E}_A(\infty)^{[\perp]}}.$$

By [84, Theorem 5.3], $\psi(\phi_{\mathfrak{h}})$ maps $\mathcal{E}_{\phi_{\mathfrak{h}}}(\infty)^{[\perp]}$ isometrically onto $L^2(\frac{\mu_{\phi_{\mathfrak{h}}}(x)}{1+x^2})$. Since $L^2(\frac{\mu_{\phi_{\mathfrak{h}}}(x)}{1+x^2})$ is non-degenerate, we have

$$\ker \psi(\phi_{\mathfrak{h}}) = \big(\mathcal{E}_{\phi_{\mathfrak{h}}}(\infty)^{[\perp]}\big)^{\circ} = \mathcal{E}_{\phi_{\mathfrak{h}}}(\infty) = \ker \pi_{\phi_{\mathfrak{h}}},$$

and we obtain an isometric isomorphism

$$\Psi_{\phi_{\mathfrak{h}}} : \mathcal{E}_{\phi_{\mathfrak{h}}}(\infty)^{[\perp]}/\mathcal{E}_{\phi_{\mathfrak{h}}}(\infty) \to L^2\Big(\frac{\mu_{\phi_{\mathfrak{h}}}(x)}{1+x^2}\Big)$$

such that

$$\Psi_{\phi_{\mathfrak{h}}} \circ \pi_{\phi_{\mathfrak{h}}} = \psi(\phi_{\mathfrak{h}})|_{\mathcal{E}_{\phi_{\mathfrak{h}}}(\infty)^{[\perp]}}.$$

Finally, let

$$U : L^2\Big(\frac{\mu_{\phi_{\mathfrak{h}}}(x)}{1+x^2}\Big) \to L^2(\mu_{\phi_{\mathfrak{h}}})$$

be the operator of multiplication by $\frac{1}{x-i}$, which is an isomorphism from $L^2(\frac{\mu_{\phi_{\mathfrak{h}}}(x)}{1+x^2})$ onto $L^2(\mu_{\phi_{\mathfrak{h}}})$.

Now we define Θ_H as the composition of the constructed isometric isomorphisms, namely

$$\Theta_H := U \circ \Psi_{\phi_{\mathfrak{h}}} \circ \Lambda_{\mathfrak{h}} \circ \Psi_A^{-1}.$$

Then Θ_H is an isometric isomorphism from $L^2(H)$ onto $L^2(\mu_{\phi_{\mathfrak{h}}})$.

In order to see how Θ_H is related to $T(H)$, it is enough to trace back the defining procedure.[5] Using (5.8) and (5.11) we obtain

$$(\Lambda_{\mathfrak{h}} \times \Lambda_{\mathfrak{h}})\Big[(\pi_A \times \pi_A)\big(A \cap (\mathcal{E}_A(\infty)^{[\perp]})^2\big)\Big] = (\pi_{\phi_{\mathfrak{h}}} \times \pi_{\phi_{\mathfrak{h}}})\big(A_{\phi_{\mathfrak{h}}} \cap (\mathcal{E}_{\phi_{\mathfrak{h}}}(\infty)^{[\perp]})^2\big).$$

By [59, Proposition 4.17 (iii)] we have

$$(\psi(\mathfrak{h}) \times \psi(\mathfrak{h}))\big(T(\mathfrak{h}) \cap (\mathcal{E}_A(\infty)^{[\perp]})^2\big) = T(H),$$

and hence

$$(\Psi_A \times \Psi_A)\Big[(\pi_A \times \pi_A)\big(A \cap (\mathcal{E}_A(\infty)^{[\perp]})^2\big)\Big] \subseteq T(H).$$

It follows from [84, Theorem 5.3] that

$$(\psi(\phi_{\mathfrak{h}}) \times \psi(\phi_{\mathfrak{h}}))\big(A_{\phi_{\mathfrak{h}}} \cap (\mathcal{E}_{\phi_{\mathfrak{h}}}(\infty)^{[\perp]})^2\big) = M_{(1+x^2)^{-1}d\mu_{\phi_{\mathfrak{h}}}(x)},$$

and therefore

$$(\Psi_{\phi_{\mathfrak{h}}} \times \Psi_{\phi_{\mathfrak{h}}})\Big[(\pi_{\phi_{\mathfrak{h}}} \times \pi_{\phi_{\mathfrak{h}}})\big(A_{\phi_{\mathfrak{h}}} \cap (\mathcal{E}_{\phi_{\mathfrak{h}}}(\infty)^{[\perp]})^2\big)\Big] = M_{(1+x^2)^{-1}d\mu_{\phi_{\mathfrak{h}}}(x)}.$$

Putting these relations together we obtain

$$\begin{aligned}
\big(\Theta_H^{-1} \times \Theta_H^{-1}\big)M_{\mu_{\phi_{\mathfrak{h}}}} &= \big(\Psi_A\Lambda_{\mathfrak{h}}^{-1}\Psi_{\phi_{\mathfrak{h}}}^{-1} \times \Psi_A\Lambda_{\mathfrak{h}}^{-1}\Psi_{\phi_{\mathfrak{h}}}^{-1}\big)M_{(1+x^2)^{-1}d\mu_{\phi_{\mathfrak{h}}}(x)} \\
&= \big(\Psi_A\Lambda_{\mathfrak{h}}^{-1} \times \Psi_A\Lambda_{\mathfrak{h}}^{-1}\big)\Big[(\pi_{\phi_{\mathfrak{h}}} \times \pi_{\phi_{\mathfrak{h}}})\big(A_{\phi_{\mathfrak{h}}} \cap (\mathcal{E}_{\phi_{\mathfrak{h}}}(\infty)^{[\perp]})^2\big)\Big] \\
&= (\Psi_A \times \Psi_A)\Big[(\pi_A \times \pi_A)\big(A \cap (\mathcal{E}_A(\infty)^{[\perp]})^2\big)\Big] \\
&\subseteq T(H).
\end{aligned}$$

Strict inclusion cannot occur since both $M_{\mu_{\phi_{\mathfrak{h}}}}$ and $T(H)$ are self-adjoint. We conclude that

$$(\Theta_H \times \Theta_H)T(H) = M_{\mu_{\phi_{\mathfrak{h}}}}.$$

It remains to remember (from the proof of Theorem 4.8) that $\mu_H = \mu_{\phi_{\mathfrak{h}}}$.

[5] Remember in the following that $T(H)$ is self-adjoint, and hence $T(H) = T_{\max}(H) = S(H)$ in the notation of several previous papers like [59].

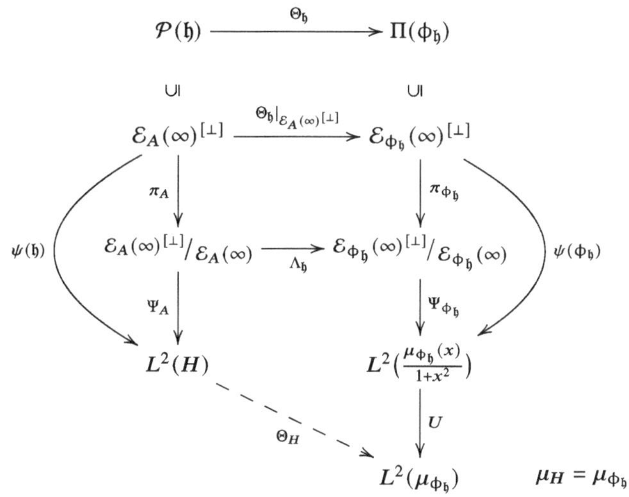

Fig. 1 Construction of Θ_H in Sect. 5.1

5.2 Computation of Θ_H as an Integral Transform

Since $\mathfrak{h}$ starts with an indivisible interval of type 0, [62, Theorem 6.4] is not applicable; a computation of the full Fourier transform $\Theta_{\mathfrak{h}}$ in the spirit of this result is not possible. However, we are only interested in Θ_H, which is essentially a restriction of $\Theta_{\mathfrak{h}}$. And it turns out that the action of this restriction can be computed. The argument is based on a refined variant of [57, Proposition 4.6]. It requires to go into the details of the constructions made in [57] and [62].

Let us lay out the operator-theoretic setup (in five parts). Thereby we use, without much further notice, terminology and results from [56]. In particular, we ask the reader to recall definitions and usage of spaces, like $\mathcal{P}_-$, dualities $[\cdot,\cdot]_\pm$ and resolvent-like operators $R_z^\pm$, as introduced and studied in [56, §3]. Moreover, we repeatedly employ terminology and results from [62]. A comprehensive summary of the involved spaces and relations can be found in Fig. 2 on page 145. We advice the reader to visit this summary already on going through the construction.

☆ ☆ ☆ *Part 1* ☆ ☆ ☆

Let A and ε_z be as in the previous section, cf. (5.6), (5.7), so that $q_{\mathfrak{h}}$ is a Q-function of $S(\mathfrak{h})$ induced by A and $(\varepsilon_z)_{z\in\rho(A)}$. Let $u \in \mathcal{P}(\mathfrak{h})_- := (S(\mathfrak{h})^*)'$ be the element defined by

$$\big[u, (f;g)\big]_\pm := \big(\pi_{l,2}\circ\Gamma(\mathfrak{h})\big)(f;g), \qquad (f;g)\in S(\mathfrak{h})^*.$$

Moreover, set

$$R_z:=(A-z)^{-1}, \qquad R_z^+ : \begin{cases} \mathcal{P}(\mathfrak{h}) \to S(\mathfrak{h})^* \\ \quad f \mapsto (R_z f; I + zR_z f) \end{cases}$$

for $z \in \rho(A)$ and let $R_z^- : \mathcal{P}(\mathfrak{h})_- \to \mathcal{P}(\mathfrak{h})$ be the dual of $R_{\overline{z}}^+$, i.e. the unique map with

$$[R_z^- v, f] = [v, R_{\overline{z}}^+ f]_\pm, \qquad v \in \mathcal{P}(\mathfrak{h})_-, \ f \in \mathcal{P}(\mathfrak{h}).$$

Then, by [62, Theorem 4.24], we have $\varepsilon_z = R_z^- u$, $z \in \rho(A)$.

☆ ☆ ☆ *Part 2* ☆ ☆ ☆

Since $q_\mathfrak{h} \in \mathcal{N}_{<\infty}^{(\infty)}$, the only critical point of A is the point ∞ and A has no finite spectral points of non-positive type, cf. [82, Lemma 2.5] and the paragraph preceding it. This implies that for each bounded Borel set Δ the spectral projection $E(\Delta)$ of A is well defined and its range is a Hilbert space. Moreover, $E_\Delta : \Delta' \mapsto E(\Delta \cap \Delta')$ is the spectral measure of the bounded self-adjoint operator $A_\Delta:=A \cap (\operatorname{ran} E(\Delta) \times \operatorname{ran} E(\Delta))$ in the Hilbert space $\operatorname{ran} E(\Delta)$, and $\sigma(A_\Delta) \subseteq \overline{\Delta}$, cf. [76, Theorem II.3.1, p. 34].[6]

For a bounded Borel set Δ and elements $f, g \in \mathcal{P}(\mathfrak{h})$ we thus have a complex measure $E_{\Delta;f,g}$ on $\mathbb{R}$ defined by

$$E_{\Delta;f,g}(\Delta'):=\big[E(\Delta \cap \Delta')f, g\big], \qquad \Delta' \text{ a Borel set on } \mathbb{R}.$$

For later use let us list some simple properties of these objects.

(1) We have

$$E_{\Delta;f,g} = E_{\Delta;E(\Delta)f,g} = E_{\Delta;f,E(\Delta)g}, \qquad f, g \in \mathcal{P}(\mathfrak{h}), \tag{5.12}$$

and

$$E_{\Delta;f,g}(\Delta') = \overline{E_{\Delta;g,f}(\Delta')}, \qquad f, g \in \mathcal{P}(\mathfrak{h}),\ \Delta' \text{ a Borel set on } \overline{\mathbb{R}}. \tag{5.13}$$

(2) Denote by $\|\cdot\|_{\mathcal{P}(\mathfrak{h})}$ some norm which induces the Pontryagin space topology of $\mathcal{P}(\mathfrak{h})$, $\|E(\Delta)\|$ the corresponding operator norm, and let $\|E_{\Delta;f,g}\|$ denote the total variation of the measure $E_{\Delta;f,g}$. Then

[6] In [76] bounded operators are treated. The extension to the case of relations is provided in [21]. In [76] results are formulated for Δ being an interval whose endpoints are not critical points. In our case, the only critical point is ∞. Moreover, with the usual measure-theoretic extension process one can define $E(\Delta)$ for each bounded Borel set; cf. the first paragraph in the proof of Lemma 5.4 below. We tacitly use this fact and often formulate results for bounded Borel sets although in the original references only intervals were considered.

$$\|E_{\Delta;f,g}\| \leq \|E(\Delta)\| \, \|f\|_{\mathcal{P}(\mathfrak{h})} \|g\|_{\mathcal{P}(\mathfrak{h})}.$$

(3) Let $T > 0$, let $F : [-T, T] \to \mathbb{C}$ be continuous and let $f, g \in \mathcal{P}(\mathfrak{h})$. Then the map

$$\Delta \mapsto \int_{[-T,T]} F \, \mathrm{d}E_{\Delta;f,g}, \qquad \Delta \text{ a Borel subset of } [-T, T],$$

is a complex Borel measure on $[-T, T]$.

(4) For each bounded Borel set Δ we have

$$E(\Delta)R_z|_{\operatorname{ran} E(\Delta)} = (A_\Delta - z)^{-1} = \int_{\mathbb{R}} \frac{1}{t-z} \, \mathrm{d}E_\Delta(t), \qquad z \in \rho(A).$$

(5) If Δ, Δ' are bounded Borel sets with $\Delta \subseteq \Delta'$ and $f, g \in \mathcal{P}(\mathfrak{h})$, then

$$E_{\Delta;f,g} \ll E_{\Delta';f,g} \quad \text{with} \quad \frac{\mathrm{d}E_{\Delta;f,g}}{\mathrm{d}E_{\Delta';f,g}} = \mathbb{1}_\Delta.$$

☆ ☆ ☆ *Part 3* ☆ ☆ ☆

Introduce the set

$$I_{\mathrm{reg}} := \big\{ t \in (0, \infty) : t \text{ is not inner point of an indivisible interval} \big\}. \tag{5.14}$$

For $t \in I_{\mathrm{reg}}$ the boundary triple $(\mathcal{P}(\mathfrak{h}), T(\mathfrak{h}), \Gamma(\mathfrak{h}))$ is isomorphic to the boundary triple that is obtained by pasting the boundary triples corresponding to $\mathfrak{h}_{\lrcorner t}$ and $\mathfrak{h}_{t\ulcorner}$; for the notation of pasting boundary triples and the mentioned result see [62, Definition 3.47] and [62, Remark 3.51] (or [80, 2.2, 2.3 and Lemma 2.5]); for the definition of $\mathfrak{h}_{\lrcorner t}$ and $\mathfrak{h}_{\ulcorner t}$ see 3.9. In particular, the space $\mathcal{P}(\mathfrak{h})$ can be decomposed as follows:

$$\mathcal{P}(\mathfrak{h}) = \mathcal{P}(\mathfrak{h}_{\lrcorner t}) \, [\dot{+}] \, \mathcal{P}(\mathfrak{h}_{t\ulcorner}),$$

and we may consider $\mathcal{P}(\mathfrak{h}_{\lrcorner t})$ naturally as a subspace of $\mathcal{P}(\mathfrak{h})$. Denote by P_t, $t \in I_{\mathrm{reg}}$, the orthogonal projection in $\mathcal{P}(\mathfrak{h})$ onto $\mathcal{P}(\mathfrak{h}_{\lrcorner t})$. Note that this projection acts as

$$P_t\big((f;\xi)\big) = (\mathbb{1}_{\lrcorner t} f; \xi), \qquad (f;\xi) \in \mathcal{P}(\mathfrak{h}). \tag{5.15}$$

Let us consider the relation

$$S_1(\mathfrak{h}_{\lrcorner t}) := \ker\big[(\pi_{l,1} \times \pi_r)\Gamma(\mathfrak{h}_{\lrcorner t})\big] \subseteq \mathcal{P}(\mathfrak{h}_{\lrcorner t})^2.$$

This relation is symmetric and has deficiency index $(1, 1)$. Let $\psi_t(z) \in \mathcal{P}(\mathfrak{h}_{\lrcorner t})$, $z \in \mathbb{C}$, be the defect elements of $S(\mathfrak{h}_{\lrcorner t})$ that satisfy

$$\big(\pi_l \circ \Gamma(\mathfrak{h}_{\restriction t})\big)\big(\psi_t(z); z\psi_t(z)\big) = \begin{pmatrix} 0 \\ 1 \end{pmatrix}, \qquad z \in \mathbb{C}; \tag{5.16}$$

then $\psi_t(z)$ are also defect elements of $S_1(\mathfrak{h}_{\restriction t})$, i.e.

$$\operatorname{ran}\big(S_1(\mathfrak{h}_{\restriction t}) - \overline{z}\big)^{\perp} = \operatorname{span}\big\{\psi_t(z)\big\}, \qquad z \in \mathbb{C}.$$

Considering $S_1(\mathfrak{h}_{\restriction t})$ as a linear relation in $\mathcal{P}(\mathfrak{h})$, we have $S_1(\mathfrak{h}_{\restriction t}) \subseteq A$ and hence $A \subseteq S_1(\mathfrak{h}_{\restriction t})^*$, where $S_1(\mathfrak{h}_{\restriction t})^*$ denotes the adjoint of $S_1(\mathfrak{h}_{\restriction t})$ as a relation in $\mathcal{P}(\mathfrak{h})$. The adjoint $S_1(\mathfrak{h}_{\restriction t})^{*_t}$ of $S_1(\mathfrak{h}_{\restriction t})$ in $\mathcal{P}(\mathfrak{h}_{\restriction t})$ is given by $S_1(\mathfrak{h}_{\restriction t})^{*_t} = \ker(\pi_{l,1} \circ \Gamma(\mathfrak{h}_{\restriction t}))$, and it follows that

$$S_1(\mathfrak{h}_{\restriction t})^{*_t} = (P_t \times P_t)(A). \tag{5.17}$$

Let $u_t \in \mathcal{P}(\mathfrak{h}_{\restriction t})_- := (S_1(\mathfrak{h}_{\restriction t})^{*_t})'$ be the unique element with

$$[u_t, (f; g)]_{\pm_t} := \big(\pi_{l,2} \circ \Gamma(\mathfrak{h}_{\restriction t})\big)(f; g), \qquad (f; g) \in S_1(\mathfrak{h}_{\restriction t})^{*_t}, \tag{5.18}$$

where $[\,\cdot\,,\cdot\,]_{\pm_t}$ denotes the duality between $S_1(\mathfrak{h}_{\restriction t})^{*_t}$ and $\mathcal{P}(\mathfrak{h}_{\restriction t})_-$. Since $(P_t \times P_t)(S(\mathfrak{h})^*) = S(\mathfrak{h}_{\restriction t})^{*_t}$ and

$$(\pi_l \circ \Gamma(\mathfrak{h}))(f; g) = (\pi_l \circ \Gamma(\mathfrak{h}_{\restriction t}))(P_t f; P_t g), \qquad (f; g) \in S(\mathfrak{h})^*,$$

we have

$$[u, (f; g)]_{\pm} = [u_t, (P_t f; P_t g)]_{\pm_t}, \qquad (f; g) \in A, \tag{5.19}$$

where $[\,\cdot\,,\cdot\,]_{\pm}$ denotes the duality between $S(\mathfrak{h})^*$ and $\mathcal{P}(\mathfrak{h})_-$.

☆ ☆ ☆ *Part 4* ☆ ☆ ☆

Let ψ_t be as in (5.16). For $t \in I_{\text{reg}}$ and $w \in \mathbb{C} \setminus \mathbb{R}$, define elements $\psi_{t,w}(z) \in \mathcal{P}(\mathfrak{h})$ by

$$\psi_{t,w}(z) := \big(I + (z - w)R_z\big)\psi_t(w), \qquad z \in \rho(A);$$

remember here that $\mathbb{C} \setminus \mathbb{R} \subseteq \rho(A)$. It follows from the resolvent identity that

$$\psi_{t,w}(z_1) = \big(I + (z_1 - z_2)R_{z_1}\big)\psi_{t,w}(z_2), \qquad z_1, z_2 \in \rho(A). \tag{5.20}$$

Moreover, since $\psi_t(w) \perp \operatorname{ran}(S_1(\mathfrak{h}_{\restriction t}) - \overline{w})$ and A is a self-adjoint extension of $S_1(\mathfrak{h}_{\restriction t})$, we have

$$\psi_{t,w}(z) \perp \operatorname{ran}\big(S_1(\mathfrak{h}_{\restriction t}) - \overline{z}\big), \qquad z \in \rho(A). \tag{5.21}$$

Hence, there exists a scalar function $\lambda_{t,w}$, which is analytic on $\rho(A)$, such that

$$P_t\psi_{t,w}(z) = \lambda_{t,w}(z)\cdot\psi_t(z), \qquad z\in\rho(A). \tag{5.22}$$

Clearly, $\lambda_{t,w}(w) = 1$, and the zeros of $\lambda_{t,w}$ form a discrete subset of $\rho(A)$. Set

$$S_{t,w}:=\big\{(f;g)\in A : g-\overline{z}f \perp \psi_{t,w}(z),\ z\in\rho(A)\big\}\subseteq \mathcal{P}(\mathfrak{h})^2.$$

Relation (5.21) implies that $S_{t,w}$ is a symmetric extension of $S_1(\mathfrak{h}_{\upharpoonleft t})\subseteq\mathcal{P}(\mathfrak{h}_{\upharpoonleft t})^2$, which acts in the larger Pontryagin space $\mathcal{P}(\mathfrak{h})$ and has deficiency index (1, 1). It follows that $(P_t\times P_t)((S_{t,w})^*)\subseteq S_1(\mathfrak{h}_{\upharpoonleft t})^{*_t}$. In fact, due to (5.17), equality must hold.

Set $\mathcal{P}(\mathfrak{h})_-^{t,w}:=(S_{t,w}^*)'$ and let $P_{t,w}:\mathcal{P}(\mathfrak{h}_{\upharpoonleft t})_-\to\mathcal{P}(\mathfrak{h})_-^{t,w}$ be the adjoint of the map $P_t\times P_t: S_{t,w}^*\to S_1(\mathfrak{h}_{\upharpoonleft t})^{*_t}$, i.e.

$$\big[P_{t,w}v,(f;g)\big]_{\pm_{t,w}} = \big[v,(P_tf;P_tg)\big]_{\pm_t}, \qquad v\in\mathcal{P}(\mathfrak{h}_{\upharpoonleft t})_-,\ (f;g)\in(S_{t,w})^*,$$

where $[\,\cdot\,,\cdot\,]_{\pm_{t,w}}$ is the duality between $S_{t,w}^*$ and $\mathcal{P}(\mathfrak{h})_-^{t,w}$. Set $u_{t,w}:=P_{t,w}u_t$, so that

$$\big[u_{t,w},(f;g)\big]_{\pm_{t,w}} = \big[u_t,(P_tf;P_tg)\big]_{\pm_t}, \qquad (f;g)\in S_{t,w}^*. \tag{5.23}$$

Remembering (5.19) we have, in particular,

$$[u_{t,w},(f;g)]_{\pm_{t,w}} = [u,(f;g)]_{\pm}, \qquad (f;g)\in A. \tag{5.24}$$

Note here that A is contained in both $S(\mathfrak{h})^*$ and $S_{t,w}^*$.

Denote by $R_{t,w;z}^-$ the dual of $R_{\overline{z}}^+:\mathcal{P}(\mathfrak{h})\to A\subseteq S_{t,w}^*$ corresponding to the duality $[\,\cdot\,,\cdot\,]_{\pm_{t,w}}$. Then, by (5.24), we have

$$[R_{t,w;z}^-u_{t,w},f] = [u_{t,w},R_{\overline{z}}^+f]_{\pm_{t,w}} = [u,R_{\overline{z}}^+f]_{\pm} = [R_z^-u,f], \qquad f\in\mathcal{P}(\mathfrak{h}),$$

and hence

$$R_{t,w;z}^-u_{t,w} = R_z^-u = \varepsilon_z, \qquad z\in\rho(A). \tag{5.25}$$

☆ ☆ ☆ *Part 5* ☆ ☆ ☆

For $t\in I_{\text{reg}}$ let $\phi_t(z)$, $z\in\mathbb{C}$, be the defect elements of $S(\mathfrak{h}_{\upharpoonleft t})$ with

$$\big(\pi_l\circ\Gamma(\mathfrak{h}_{\upharpoonleft t})\big)\big(\phi_t(z);z\phi_t(z)\big) = \begin{pmatrix}1\\0\end{pmatrix}, \qquad z\in\mathbb{C}. \tag{5.26}$$

By [62, Theorem 4.19], the map Ξ_t, which is defined by

$$(\Xi_t f)(z) = \begin{pmatrix} [f, \phi_t(\overline{z})] \\ [f, \psi_t(\overline{z})] \end{pmatrix}, \qquad f \in \mathcal{P}(\mathfrak{h}_{\restriction t}), \tag{5.27}$$

is an isomorphism from $\mathcal{P}(\mathfrak{h}_{\restriction t})$ onto the reproducing kernel space $\mathfrak{K}(\omega_{\mathfrak{h}}(t))$ with the kernel (here $\omega_{\mathfrak{h}}$ is the fundamental solution of $\mathfrak{h}$)

$$H_{\omega_{\mathfrak{h}}(t)}(w, z) = \frac{\omega_{\mathfrak{h}}(t; z) J \omega_{\mathfrak{h}}(t; w)^* - J}{z - \overline{w}}.$$

By [62, Proposition 4.4] the kernel $H_{\omega_{\mathfrak{h}}(t)}$ can be written as

$$H_{\omega_{\mathfrak{h}}(t)}(w, z) = \begin{pmatrix} [\phi_t(z), \phi_t(w)] & [\phi_t(z), \psi_t(w)] \\ [\psi_t(z), \phi_t(w)] & [\psi_t(z), \psi_t(w)] \end{pmatrix}.$$

On the dense subspace $\operatorname{span}\big(\{\phi_t(z) : z \in \mathbb{C}\} \cup \{\psi_t(z) : z \in \mathbb{C}\}\big)$ of $\mathcal{P}(\mathfrak{h}_{\restriction t})$, the action of Ξ_t is determined by linearity and the formula

$$\Xi_t\big(\lambda\phi_t(z) + \mu\psi_t(z)\big) = H_{\omega_{\mathfrak{h}}(t)}(\overline{z}, \cdot)\begin{pmatrix} \lambda \\ \mu \end{pmatrix}, \qquad \lambda, \mu \in \mathbb{C},\ z \in \mathbb{C}. \tag{5.28}$$

This relation is seen as follows ($\zeta \in \mathbb{C}$):

$$\begin{aligned} \Xi_t\big(\lambda\phi_t(z) + \mu\psi_t(z)\big)(\zeta) &= \lambda \begin{pmatrix} [\phi_t(z), \phi_t(\overline{\zeta})] \\ [\phi_t(z), \psi_t(\overline{\zeta})] \end{pmatrix} + \mu \begin{pmatrix} [\psi_t(z), \phi_t(\overline{\zeta})] \\ [\psi_t(z), \psi_t(\overline{\zeta})] \end{pmatrix} \\ &= \big(H_{\omega_{\mathfrak{h}}(t)}(\overline{\zeta}, z)\big)^T \begin{pmatrix} \lambda \\ \mu \end{pmatrix} = \big(H_{\omega_{\mathfrak{h}}(t)}(\zeta, \overline{z})\big)^* \begin{pmatrix} \lambda \\ \mu \end{pmatrix} = H_{\omega_{\mathfrak{h}}(t)}(\overline{z}, \zeta)\begin{pmatrix} \lambda \\ \mu \end{pmatrix}. \end{aligned}$$

☆ ☆ ☆ ☆ ☆ ☆ ☆ ☆ ☆

In order to prove (5.1), we proceed in several steps and compute various inner products and the action of several maps. The desired result will then follow by putting these formulae together. A comprehensive overview of the involved maps is provided in the diagram on page 155.

For $t \in I_{\text{reg}}$ denote by Θ_t the map Ξ_t followed by projection onto the second component, i.e.

$$(\Theta_t f)(z) := [f, \psi_t(\overline{z})], \qquad f \in \mathcal{P}(\mathfrak{h}_{\restriction t}). \tag{5.29}$$

Recall that the entries of $\omega_{\mathfrak{h}}(t; z)$ are by definition (see [62, Definitions 5.3, 4.3]) just the right-hand boundary values of the elements $\phi_t(z)$ and $\psi_t(z)$:

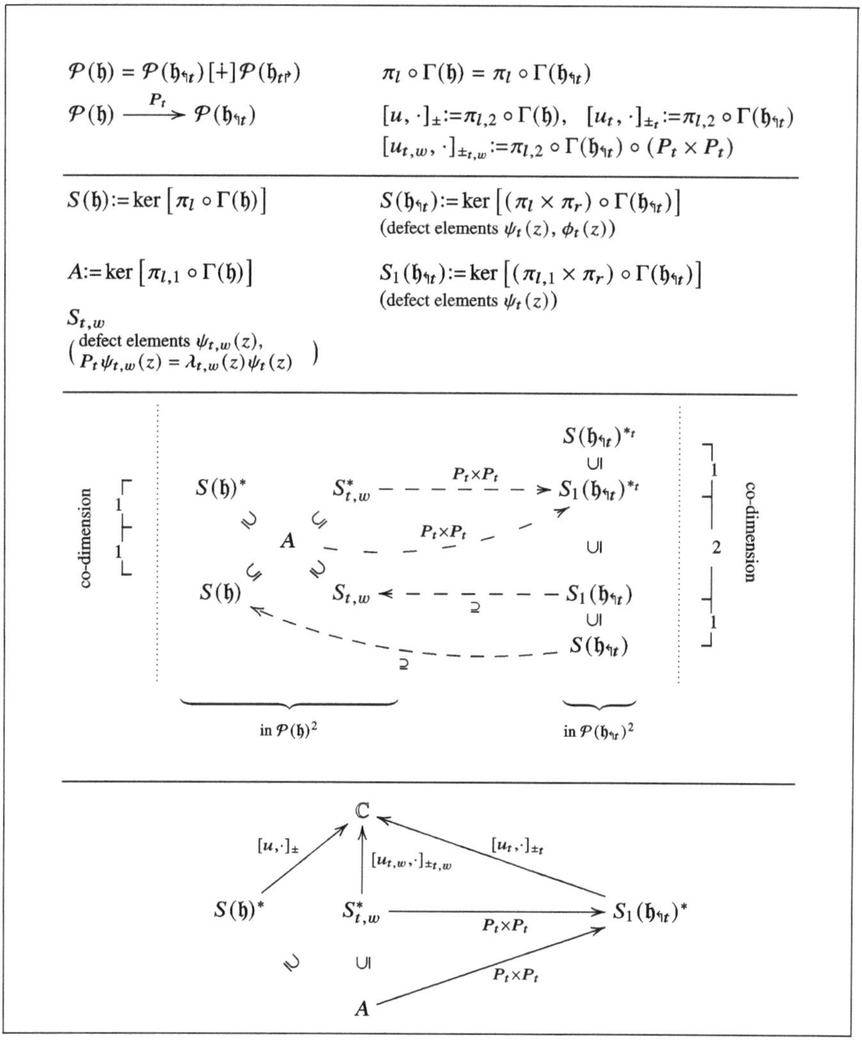

Fig. 2 Computation of Θ_H in Sect. 5.2

$$\big(\pi_r \circ \Gamma(\mathfrak{h}_{\dashv t})\big)\big(\phi_t(z); z\phi_t(z)\big) = \omega_{\mathfrak{h}}(t; z)^T \begin{pmatrix} 1 \\ 0 \end{pmatrix},$$

$$\big(\pi_r \circ \Gamma(\mathfrak{h}_{\dashv t})\big)\big(\psi_t(z); z\psi_t(z)\big) = \omega_{\mathfrak{h}}(t; z)^T \begin{pmatrix} 0 \\ 1 \end{pmatrix}.$$

With the same method that was used in the proof of [57, Proposition 4.1], we obtain the following statement.

Lemma 5.4 *Let Δ be a bounded Borel set, $z_0 \in \mathbb{C} \setminus \mathbb{R}$ and $t \in I_{\mathrm{reg}}$. Then*

$$\big[E(\Delta)f, \psi_{t,w}(\zeta)\big] = \int_{-\infty}^{\infty} (\Theta_t f)(x) \cdot (x - z_0)\, \mathrm{d}E_{\Delta;\varepsilon_{z_0},\psi_{t,w}(\zeta)}(x),$$
$$f \in \mathcal{P}(\mathfrak{h}_{\urcorner t}),\ \ \zeta, w \in \mathbb{C} \setminus \mathbb{R}. \tag{5.30}$$

Proof The algebra of all bounded Borel sets is the union $\bigcup_{T>0} \mathcal{B}_{[-T,T]}$ of the σ-algebras $\mathcal{B}_{[-T,T]}$ of all Borel subsets of $[-T, T]$. Since $\mathcal{B}_{[-T,T]}$, as a σ-algebra, is generated by the set of all closed intervals whose endpoints do not carry a point mass of E, it is enough to establish (5.30) for such intervals.

Throughout the proof fix $f \in \mathcal{P}(\mathfrak{h}_{\urcorner t})$ and an interval $[a_-, a_+]$ with $E(\{a_-\}) = E(\{a_+\}) = 0$. Moreover, choose a bounded open interval Δ_0 which contains $[a_-, a_+]$.

Step 1: some computations. To start with, we compute (indicating which equations are used)

$$\begin{aligned}\big[u_{t,w}, \big(\psi_{t,w}(\overline{z}); \overline{z}\psi_{t,w}(\overline{z})\big)\big]_{\pm_{t,w}} &\overset{(5.23)}{=} \big[u_t, \big(P_t\psi_{t,w}(\overline{z}); \overline{z}P_t\psi_{t,w}(\overline{z})\big)\big]_{\pm_t} \\ &\overset{(5.22)}{=} \overline{\lambda_{t,w}(\overline{z})}\big[u_t, \big(\psi_t(\overline{z}); \overline{z}\psi_t(\overline{z})\big)\big]_{\pm_t} \\ &\overset{(5.16,\, 5.18)}{=} \overline{\lambda_{t,w}(\overline{z})},\end{aligned}$$

which yields

$$[f, \psi_{t,w}(\overline{z})] \overset{(5.22)}{=} \overline{\lambda_{t,w}(\overline{z})}\,[f, \psi_t(\overline{z})] = (\Theta_t f)(z)\cdot\big[u_{t,w}, \big(\psi_{t,w}(\overline{z}); \overline{z}\psi_{t,w}(\overline{z})\big)\big]_{\pm_{t,w}}. \tag{5.31}$$

Since

$$(\overline{z} - \overline{z}_0)R_{\overline{z}_0}\psi_{t,w}(\overline{z}) \overset{(5.20)}{=} \psi_{t,w}(\overline{z}) - \psi_{t,w}(\overline{z}_0),$$

we have

$$\begin{aligned}(z - z_0)[\varepsilon_{z_0}, \psi_{t,w}(\overline{z})] &\overset{(5.25)}{=} (z - z_0)\big[R^-_{t,w;z_0}u_{t,w}, \psi_{t,w}(\overline{z})\big] \\ &= \big[u_{t,w}, (\overline{z} - \overline{z}_0)R^{\pm}_{\overline{z}_0}\psi_{t,w}(\overline{z})\big]_{\pm_{t,w}} \\ &= \Big[u_{t,w}, \Big((\overline{z} - \overline{z}_0)R_{\overline{z}_0}\psi_{t,w}(\overline{z}); (\overline{z} - \overline{z}_0)\big(I + \overline{z}_0R_{\overline{z}_0}\big)\psi_{t,w}(\overline{z})\Big)\Big]_{\pm_{t,w}} \\ &= \Big[u_{t,w}, \Big(\psi_{t,w}(\overline{z}) - \psi_{t,w}(\overline{z}_0); (\overline{z} - \overline{z}_0)\psi_{t,w}(\overline{z}) + \overline{z}_0\big(\psi_{t,w}(\overline{z}) - \psi_{t,w}(\overline{z}_0)\big)\Big)\Big]_{\pm_{t,w}}\end{aligned}$$

$$= \big[u_{t,w}, (\psi_{t,w}(\overline{z}); \overline{z}\psi_{t,w}(\overline{z}))\big]_{\pm_{t,w}} - \big[u_{t,w}, (\psi_{t,w}(\overline{z_0}); \overline{z_0}\psi_{t,w}(\overline{z_0}))\big]_{\pm_{t,w}}.$$

Bringing the very last summand to the left-hand side and substituting the first term on the right-hand side into (5.31), we obtain

$$[f, \psi_{t,w}(\overline{z})] = (\Theta_t f)(z) \cdot \Big((z - z_0)\big[\varepsilon_{z_0}, \psi_{t,w}(\overline{z})\big] + \big[u_{t,w}, \big(\psi_{t,w}(\overline{z_0}); \overline{z_0}\psi_{t,w}(\overline{z_0})\big)\big]_{\pm_{t,w}}\Big). \tag{5.32}$$

For each $z, \zeta \in \mathbb{C} \setminus \mathbb{R}$, we have

$$\big[(I + (z - \overline{\zeta})R_z)f, \psi_{t,w}(\zeta)\big] = \big[f, (I + (\overline{z} - \zeta)R_{\overline{z}})\psi_{t,w}(\zeta)\big] \overset{(5.20)}{=} [f, \psi_{t,w}(\overline{z})] \tag{5.33}$$

and

$$\begin{aligned}\big[\varepsilon_{z_0}, \psi_{t,w}(\overline{z})\big] &\overset{(5.20)}{=} \big[\varepsilon_{z_0}, (I + (\overline{z} - \zeta)R_{\overline{z}})\psi_{t,w}(\zeta)\big]\\ &= \big[(I + (z - \overline{\zeta})R_z)\varepsilon_{z_0}, \psi_{t,w}(\zeta)\big]\\ &= \big[\varepsilon_{z_0}, \psi_{t,w}(\zeta)\big] + (z - \overline{\zeta})\big[R_z E(\Delta_0)\varepsilon_{z_0}, \psi_{t,w}(\zeta)\big]\\ &\quad + (z - \overline{\zeta})\big[R_z E(\Delta_0^c)\varepsilon_{z_0}, \psi_{t,w}(\zeta)\big].\end{aligned} \tag{5.34}$$

If $z \neq \overline{\zeta}$, then

$$\frac{1}{z - \overline{\zeta}}[f, \psi_{t,w}(\zeta)] + [R_z f, \psi_{t,w}(\zeta)] \overset{(5.33)}{=} \frac{1}{z - \overline{\zeta}}[f, \psi_{t,w}(\overline{z})] \tag{5.35}$$

$$\overset{(5.32)}{=} \frac{1}{z-\overline{\zeta}}(\Theta_t f)(z)\cdot\Big((z-z_0)\big[\varepsilon_{z_0}, \psi_{t,w}(\overline{z})\big]+\big[u_{t,w}, (\psi_{t,w}(\overline{z_0}); \overline{z_0}\psi_{t,w}(\overline{z_0}))\big]_{\pm_{t,w}}\Big)$$

$$\overset{(5.34)}{=} (\Theta_t f)(z) \cdot \Bigg(\frac{z - z_0}{z - \overline{\zeta}}[\varepsilon_{z_0}, \psi_{t,w}(\zeta)] \tag{5.36}$$

$$+ (z - z_0)\big[R_z E(\Delta_0)\varepsilon_{z_0}, \psi_{t,w}(\zeta)\big] + (z - z_0)\big[R_z E(\Delta_0^c)\varepsilon_{z_0}, \psi_{t,w}(\zeta)\big] \tag{5.37}$$

$$+ \frac{1}{z - \overline{\zeta}}\big[u_{t,w}, \big(\psi_{t,w}(\overline{z_0}); \overline{z_0}\psi_{t,w}(\overline{z_0})\big)\big]_{\pm_{t,w}}\Bigg). \tag{5.38}$$

Step 2: use of the Stieltjes–Lifšic inversion formula. We shall apply the Stieltjes–Lifšic inversion formula as stated in [76, p. 24, Corollary II.2 (second formula)] with two minor modifications. First, since we assume that the endpoints a_+

and a_- carry no point mass of E, the limit '$\varepsilon \searrow 0$' is not needed. Second, by analyticity we may apply Cauchy's theorem to replace the path of integration used in [76] by the path γ_δ composed of the two oriented line segments connecting the points $a_- - i\delta, a_+ - i\delta$, and $a_+ + i\delta, a_- + i\delta$, respectively. Then, for each function g that is analytic in some open neighbourhood of $[a_-, a_+]$ and $u, v \in \operatorname{ran} E(\Delta_0)$,

$$\lim_{\delta\searrow 0} \frac{-1}{2\pi i} \int\limits_{\gamma_\delta} g(z)[R_z v, w]\mathrm{d}z = \int_{[a_-,a_+]} g(x)\,\mathrm{d}E_{\Delta_0;v,w}(x). \tag{5.39}$$

Since $\Theta_t f$ is entire and $\zeta \notin \mathbb{R}$, the first term on the left-hand side of (5.35), the terms in (5.36) and (5.38) and the second term in (5.37) are holomorphic in a neighbourhood of $[a_-, a_+]$. Hence,

$$\lim_{\delta\searrow 0} \int\limits_{\gamma_\delta} [R_z f, \psi_{t,w}(\zeta)]\mathrm{d}z = \lim_{\delta\searrow 0} \int\limits_{\gamma_\delta} (\Theta_t f)(z) \cdot (z - z_0)\big[R_z E(\Delta_0)\varepsilon_{z_0}, \psi_{t,w}(\zeta)\big]\mathrm{d}z. \tag{5.40}$$

Applying (5.39) twice, namely, with the entire functions $g(z) = 1$ and $g(z) = (\Theta_t f)(z) \cdot (z - z_0)$, we obtain

$$\begin{aligned}
&\big[E([a_-, a_+])f, \psi_{t,w}(\zeta)\big] \\
&\quad = \big[E([a_-, a_+])E(\Delta_0)f, E(\Delta_0)\psi_{t,w}(\zeta)\big] = \int\limits_{[a_-,a_+]} \mathrm{d}E_{\Delta_0;E(\Delta_0)f,E(\Delta_0)\psi_{t,w}(\zeta)}(x) \\
&\quad \overset{(5.39)}{=} \lim_{\delta\searrow 0} \frac{-1}{2\pi i} \int\limits_{\gamma_\delta} \big[R_z E(\Delta_0)f, E(\Delta_0)\psi_{t,w}(\zeta)\big]\,\mathrm{d}z \\
&\quad = \lim_{\delta\searrow 0} \frac{-1}{2\pi i} \int\limits_{\gamma_\delta} \big[R_z f, \psi_{t,w}(\zeta)\big]\,\mathrm{d}z \\
&\quad \overset{(5.40)}{=} \lim_{\delta\searrow 0} \frac{-1}{2\pi i} \int\limits_{\gamma_\delta} (\Theta_t f)(z) \cdot (z - z_0)\big[R_z E(\Delta_0)\varepsilon_{z_0}, E(\Delta_0)\psi_{t,w}(\zeta)\big]\mathrm{d}z \\
&\quad \overset{(5.39)}{=} \int\limits_{[a_-,a_+]} (\Theta_t f)(x) \cdot (x - z_0)\,\mathrm{d}E_{\Delta_0;E(\Delta_0)\varepsilon_{z_0},E(\Delta_0)\psi_{t,w}(\zeta)}(x) \\
&\quad = \int\limits_{[a_-,a_+]} (\Theta_t f)(x) \cdot (x - z_0)\,\mathrm{d}E_{\Delta_0;\varepsilon_{z_0},\psi_{t,w}(\zeta)}(x).
\end{aligned}$$

It remains to remember that $\frac{\mathrm{d}E_{[a_-,a_+];u,v}}{\mathrm{d}E_{\Delta_0;u,v}} = \mathbb{1}_{[a_-,a_+]}$. □

The next statement is the key lemma.

Lemma 5.5 *Assume that H does not end with an indivisible interval towards ∞. Then, for each $s_0 \in (0, \infty)$, we have*

$$c.l.s.\big\{\psi_{t,w}(z) : t \in I_{\text{reg}},\ t \geq s_0,\ z, w \in \mathbb{C} \setminus \mathbb{R}\big\} = \mathcal{P}(\mathfrak{h})\,[-]\,\mathcal{E}_A(\infty), \tag{5.41}$$

where $c.l.s.$ stands for 'closed linear span' and I_{reg} is defined in (5.14).

Proof

Step 1. Fix a point $s \in I_{\text{reg}}$ and let $M \subseteq \mathbb{C}$ be a set with a finite accumulation point. Let $\delta_j \in \mathcal{P}(\mathfrak{h}_{\upharpoonright s})$, $j = 0, \ldots, \Delta(H)-1$, be the elements $\delta_j := (0, (\delta_{jk})_{k=0}^{\Delta(H)-1})$ where δ_{jk} is the Kronecker delta. Then $(0; \delta_0), (\delta_0; \delta_1), \ldots, (\delta_{\Delta(H)-2}; \delta_{\Delta(H)-1}) \in T(\mathfrak{h}_{\upharpoonright s})$, and the boundary values of these elements vanish. Repeatedly applying the abstract Green identity [59, (2.6) and Proposition 5.2], we obtain, for $k = 0, \ldots, \Delta(H) - 1$ and $z \in C$,

$$[\delta_k, \psi_s(z)] = [\delta_{k-1}, z\psi_s(z)] = \ldots = [\delta_0, z^k\psi_s(z)] = [0, z^{k+1}\psi_s(z)] = 0.$$

In particular, we have

$$\mathcal{P}(\mathfrak{h}_{\upharpoonright s})\,[-]\,c.l.s.\big\{\psi_s(z) : z \in M\big\} \supseteq \operatorname{span}\{\delta_0, \ldots, \delta_{\Delta(H)-1}\}. \tag{5.42}$$

Applying the isomorphism $\Xi_s : \mathcal{P}(\mathfrak{h}_{\upharpoonright s}) \to \mathfrak{K}(\omega_{\mathfrak{h}}(s))$ from (5.27) and using (5.28), we can deduce that

$$\begin{aligned}&\Xi_s\Big(\mathcal{P}(\mathfrak{h}_{\upharpoonright s})\,[-]\,c.l.s.\{\psi_s(z) : z \in M\}\Big)\\&= \Big(c.l.s.\Big\{H_{\omega_{\mathfrak{h}}(s)}(\overline{z}, \cdot)\begin{pmatrix}0\\1\end{pmatrix} : z \in M\Big\}\Big)^{[\perp]}.\end{aligned}$$

By analyticity, the space on the right-hand side equals $\ker \pi_-$, where π_- denotes the projection onto the second component in $\mathfrak{K}(\omega_{\mathfrak{h}}(s))$. As shown in the proof of [106, Lemma 6.3 (subcase 3b)], we have $\dim\ker\pi_- = \Delta(H)$. Thus, equality must hold in (5.42).

Step 2. First note that $P_s\delta_k = \delta_k$ where P_s is as in (5.15), and hence

$$[\psi_{s,w}(z), \delta_k] = [P_s\psi_{s,w}(z), \delta_k] \overset{(5.22)}{=} \lambda_{s,w}(z)[\psi_s(z), \delta_k] = 0,$$

$$k = 0, \ldots, \Delta(H) - 1,\ z, w \in \rho(A).$$

Together with the fact that $\mathcal{E}_A(\infty) = \operatorname{span}\{\delta_0, \dots, \delta_{\Delta(H)-1}\}$ (see [82, Lemma 3.2 (d)]), this gives

$$\mathcal{E}_A(\infty) \subseteq \{\psi_{t,w}(z) : t \in I_{\mathrm{reg}}, t \geq s_0, z, w \in \mathbb{C} \setminus \mathbb{R}\}^{[\perp]},$$

i.e. the inclusion '$\subseteq$' in (5.41). To show the reverse inclusion, let $f \in \mathcal{P}(\mathfrak{h})$ be given and assume that $f\,[\perp]\,\psi_{s,w}(z)$, $s \in I_{\mathrm{reg}}$, $s \geq s_0$, $z, w \in \mathbb{C} \setminus \mathbb{R}$. Then, for each fixed $s \geq s_0$ and $z \in \mathbb{C} \setminus \mathbb{R}$,

$$[P_s f, \psi_s(z)] = [f, \psi_{s,z}(z)] = 0, \qquad z \in \mathbb{C} \setminus \mathbb{R}.$$

By Step 1 we therefore have $P_s f \in \operatorname{span}\{\delta_0, \dots, \delta_{\Delta(H)-1}\}$. This tells us that $\psi(\mathfrak{h}_{\upharpoonleft s}) P_s f = 0$. Since

$$\psi(\mathfrak{h}) f|_{(-1,0)\cup(0,s)} = \psi(\mathfrak{h}_{\upharpoonleft s}) P_s f,$$

and s may be chosen arbitrarily large by our hypothesis that H does not end indivisibly, it follows that $\psi(\mathfrak{h}) f = 0$. Hence,

$$f \in \operatorname{span}\{\delta_0, \dots, \delta_{\Delta(H)-1}\} = \mathcal{E}_A(\infty),$$

which proves the reverse inclusion. □

If $s_1, s_2 \in I_{\mathrm{reg}}$, $s_1 < s_2$, then clearly $P_{s_1} \psi_{s_2}(z) = \psi_{s_1}(z)$, and hence

$$\Theta_{s_1} f = \Theta_{s_2} f, \qquad f \in \mathcal{P}(\mathfrak{h}_{\upharpoonleft s_1}),$$

where P_s and Θ_s are defined in (5.15) and (5.29). Thus, a map Θ_∞ on $\bigcup_{s \in I_{\mathrm{reg}}} \mathcal{P}(\mathfrak{h}_{\upharpoonleft s})$ is well defined by setting

$$\Theta_\infty f := \Theta_{s_f} f \qquad \text{with } s_f \text{ sufficiently large so that } f \in \mathcal{P}(\mathfrak{h}_{\upharpoonleft s_f}). \tag{5.43}$$

Using Lemma 5.5 we can extend Lemma 5.4.

Proposition 5.6 *Let Δ be a bounded Borel set and let $z_0 \in \mathbb{C} \setminus \mathbb{R}$. Then*

$$[E(\Delta) f, g] = \int_{-\infty}^{\infty} (\Theta_\infty f)(x) \cdot (x - z_0)\, \mathrm{d}E_{\Delta;\varepsilon_{z_0},g}(x), \qquad f \in \bigcup_{s \in I_{\mathrm{reg}}} \mathcal{P}(\mathfrak{h}_{\upharpoonleft s}),\ g \in \mathcal{P}(\mathfrak{h}). \tag{5.44}$$

Proof Fix $f \in \bigcup_{s \in I_{\mathrm{reg}}} \mathcal{P}(\mathfrak{h}_{\upharpoonleft s})$ and let s_f be so large that $f \in \mathcal{P}(\mathfrak{h}_{\upharpoonleft s_f})$. By Lemma 5.4 the asserted relation holds for all $g \in \operatorname{span}\{\psi_{s,w}(z) : s \in I_{\mathrm{reg}}, s \geq s_f, z, w \in \mathbb{C} \setminus \mathbb{R}\}$.

Both sides of (5.44) depend continuously on g. For the left-hand side, this is obvious, and for the right-hand side, remember that E_Δ is compactly supported, and hence, for each continuous function F on $\mathbb{R}$, the integral $\int_{-\infty}^{\infty} F\,\mathrm{d}E_\Delta$ exists in the strong operator topology. We obtain from Lemma 5.5 that (5.44) holds for all $g \in \mathcal{E}_A(\infty)^{[\perp]}$.

Finally, let an arbitrary element $g \in \mathcal{P}(\mathfrak{h})$ be given. Since $E(\Delta)g \in \mathcal{E}_A(\infty)^{[\perp]}$, we may apply what we have already shown and obtain

$$[E(\Delta)f, g] = [E(\Delta)f, E(\Delta)g] = \int_{-\infty}^{\infty} (\Theta_\infty f)(x)\cdot(x - z_0)\,\mathrm{d}E_{\Delta;\varepsilon_{z_0},E(\Delta)g}(x)$$

$$\overset{(5.12)}{=} \int_{-\infty}^{\infty} (\Theta_\infty f)(x)\cdot(x - z_0)\,\mathrm{d}E_{\Delta;\varepsilon_{z_0},g}.$$

□

Lemma 5.7 *Let Δ be a bounded Borel set and $g \in \mathcal{P}(\mathfrak{h})$. Then*

$$(x - z_0)\,\mathrm{d}E_{\Delta;\varepsilon_{z_0},g} = (x - z_1)\,\mathrm{d}E_{\Delta;\varepsilon_{z_1},g}, \qquad z_0, z_1 \in \mathbb{C}\setminus\mathbb{R}. \tag{5.45}$$

Proof To see this, note that

$$I + (z_0 - z_1)(A_\Delta - z_0)^{-1} = \int_{\mathbb{R}} \frac{x - z_1}{x - z_0}\,\mathrm{d}E_\Delta, \qquad z_0, z_1 \in \mathbb{C}\setminus\mathbb{R}.$$

From the identity [56, (3.2)], we obtain

$$\big(I + (z_0 - z_1)(A - z_0)^{-1}\big)E(\Delta)R_{z_1}^- = R_{z_0}^-.$$

Hence, for each Borel set Δ' and $g \in \mathcal{P}(\mathfrak{h})$,

$$\begin{aligned}
E_{\Delta;\varepsilon_{z_0},g}(\Delta') &\overset{(5.25)}{=} [E(\Delta\cap\Delta')R_{z_0}^- u, g] = [E(\Delta')E(\Delta)R_{z_0}^- u, E(\Delta)g]\\
&= \big[E(\Delta')E(\Delta)\big(I + (z_0 - z_1)(A - z_0)^{-1}\big)R_{z_1}^- u, E(\Delta)g\big]\\
&= \big[E(\Delta')\big(I + (z_0 - z_1)(A_\Delta - z_0)^{-1}\big)E(\Delta)R_{z_1}^- u, E(\Delta)g\big]\\
&\overset{(5.25)}{=} \Bigg[\int_{\Delta'} \frac{x - z_1}{x - z_0}\,\mathrm{d}E_\Delta(x)\cdot E(\Delta)\varepsilon_{z_1}, E(\Delta)g\Bigg]\\
&= \int_{\Delta'} \frac{x - z_1}{x - z_0}\,\mathrm{d}E_{\Delta;E(\Delta)\varepsilon_{z_1},E(\Delta)g}(x) \overset{(5.12)}{=} \int_{\Delta'} \frac{x - z_1}{x - z_0}\,\mathrm{d}E_{\Delta;\varepsilon_{z_1},g}(x).
\end{aligned}$$

It follows that $E_{\Delta;\varepsilon_{z_0},g} \ll E_{\Delta;\varepsilon_{z_1},g}$ and $(x-z_0)\,\mathrm{d}E_{\Delta;\varepsilon_{z_0},g} = (x-z_1)\,\mathrm{d}E_{\Delta;\varepsilon_{z_1},g}$. □

Now we are in position to relate the maps Θ_∞ and $\Theta_{\mathfrak{h}}$.

Lemma 5.8 *Let $\phi_{\mathfrak{h}}$ be the distributional density in the representation* (3.5) *of the Weyl coefficient $q_{\mathfrak{h}}$ of $\mathfrak{h}$, let $\psi(\phi_{\mathfrak{h}})$ be as in 3.6, let $\Theta_{\mathfrak{h}}$ be the isomorphism acting as in* (5.9) *and let Θ_∞ be as in* (5.43). *Then*

$$(\Theta_\infty f)(x) = \frac{1}{x-i}\big(\big[\psi(\phi_{\mathfrak{h}}) \circ \Theta_{\mathfrak{h}}\big] f\big)(x) \qquad \mu_H\text{–a.e.},$$

$$f \in \bigcup_{s \in I_{\text{reg}}} \mathcal{P}(\mathfrak{h}_{\dashv s}) \cap \mathcal{E}_A(\infty)^{[\perp]}.$$

In particular, Θ_∞ maps $\bigcup_{s\in I_{\text{reg}}} \mathcal{P}(\mathfrak{h}_{\dashv s})$ into $L^2(\mu_H)$.

Proof Let $f \in \bigcup_{s\in I_{\text{reg}}} \mathcal{P}(\mathfrak{h}_{\dashv s}) \cap \mathcal{E}_A(\infty)^{[\perp]}$ be given. For a bounded open interval Δ and $w \in \mathbb{C}\setminus\mathbb{R}$, we compute the inner product $[E(\Delta)f, \varepsilon_w]$ in two ways.

On the one hand, we have

$$\begin{aligned}
[E(\Delta)f, \varepsilon_w] &\overset{(5.44)}{=} \int_{\mathbb{R}} (\Theta_\infty f)(x)(x-i)\,\mathrm{d}E_{\Delta;\varepsilon_i,\varepsilon_w}(x) \\
&\overset{(5.13,5.45)}{=} \int_{\mathbb{R}} (\Theta_\infty f)(x)(x-i)\frac{x+i}{x-\overline{w}}\,\mathrm{d}E_{\Delta;\varepsilon_i,\varepsilon_i}(x).
\end{aligned} \tag{5.46}$$

Using a standard argument, we now relate $E_{\Delta;\varepsilon_i,\varepsilon_i}$ to μ_H. Since q_H is the Q-function induced by A and the family $(\varepsilon_z)_{z\in\rho(A)}$, we have the representation

$$\begin{aligned}
q_{\mathfrak{h}}(z) &= \overline{q_{\mathfrak{h}}(i)} + (z+i)[\varepsilon_i, \varepsilon_i] + (z^2+1)[R_z\varepsilon_i, \varepsilon_i] \\
&= \overline{q_{\mathfrak{h}}(i)} + (z+i)[\varepsilon_i, \varepsilon_i] + (z^2+1)\big[R_z E(\Delta^c)\varepsilon_i, E(\Delta^c)\varepsilon_i\big] \\
&\quad + (z^2+1)\big[(A_\Delta - z)^{-1}E(\Delta)\varepsilon_i, E(\Delta)\varepsilon_i\big].
\end{aligned} \tag{5.47}$$

Let $[a_-, a_+] \subseteq \Delta$ be such that $\mu_H(\{a_-\}) = \mu_H(\{a_+\}) = 0$ and $E(\{a_-\}) = E(\{a_+\}) = 0$. Observing that all summands on the right-hand side of (5.47) apart from the last one are analytic across Δ, we compute (where γ_δ is as in the proof of Lemma 5.4)

$$\mu_H\big([a_-, a_+]\big) \overset{(4.11)}{=} \frac{1}{\pi}\lim_{\delta\searrow 0} \int_{[a_-,a_+]} \operatorname{Im} q_H(x+i\delta)\,\mathrm{d}x$$

$$\overset{q_{\mathfrak{h}}(\overline{z})=\overline{q_{\mathfrak{h}}(z)}}{=} \frac{1}{\pi}\lim_{\delta\searrow 0}\frac{-1}{2i}\int_{\gamma_\delta} q_H(z)\,\mathrm{d}z$$

$$\overset{(5.47)}{=} \lim_{\delta\searrow 0}\frac{-1}{2\pi i}\int_{\gamma_\delta}(z^2+1)\big[(A_\Delta - z)^{-1}E(\Delta)\varepsilon_i, E(\Delta)\varepsilon_i\big]\,\mathrm{d}z$$

$$\overset{(5.39)}{=} \int_{[a_-,a_+]}(x^2+1)\,\mathrm{d}E_{\Delta;E(\Delta)\varepsilon_i,E(\Delta)\varepsilon_i} \overset{(5.12)}{=} \int_{[a_-,a_+]}(x^2+1)\,\mathrm{d}E_{\Delta;\varepsilon_i,\varepsilon_i}.$$

From this it follows that $\mathbb{1}_\Delta(x)\,\mathrm{d}\mu_H(x) \ll \mathrm{d}E_{\Delta;\varepsilon_i,\varepsilon_i}$ and

$$\frac{\mathbb{1}_\Delta(x)\,\mathrm{d}\mu_H(x)}{\mathrm{d}E_{\Delta;\varepsilon_i,\varepsilon_i}} = x^2+1.$$

Hence, we can further rewrite the last integral in (5.46) and obtain

$$[E(\Delta)f, \varepsilon_w] = \int_{\mathbb{R}}(\Theta_\infty f)(x)(x-i)\frac{x+i}{x-\overline{w}}\,\frac{\mathbb{1}_\Delta(x)\mathrm{d}\mu_H(x)}{1+x^2}. \tag{5.48}$$

On the other hand, let $\mathcal{E}_{\phi_{\mathfrak{h}}}(\infty)^{[\perp]}$ be the algebraic eigenspace at infinity of $A_{\phi_{\mathfrak{h}}}$, let $E_{\phi_{\mathfrak{h}}}$ be the spectral measure of $A_{\phi_{\mathfrak{h}}}$ and let $\hat{\varepsilon}_w$ be as in (5.10). Since $\Theta_{\mathfrak{h}}$ is isometric, $\operatorname{ran} E_{\phi_{\mathfrak{h}}}(\Delta) \subseteq \mathcal{E}_{\phi_{\mathfrak{h}}}(\infty)^{[\perp]}$ and $\psi(\phi_{\mathfrak{h}})$ is isometric on $\mathcal{E}_{\phi_{\mathfrak{h}}}(\infty)^{[\perp]}$ (by [84, Theorem 5.3]), we have

$$\begin{aligned}
[E(\Delta)f, \varepsilon_w] &= [E(\Delta)f,\ E(\Delta)\varepsilon_w] \\
&= \big[\Theta_{\mathfrak{h}}E(\Delta)f,\ \Theta_{\mathfrak{h}}E(\Delta)\varepsilon_w\big]_{\Pi(\phi_{\mathfrak{h}})} \\
&\overset{(5.8)}{=} \big[E_{\phi_{\mathfrak{h}}}(\Delta)\Theta_{\mathfrak{h}}f,\ E_{\phi_{\mathfrak{h}}}(\Delta)\Theta_{\mathfrak{h}}\varepsilon_w\big]_{\Pi(\phi_{\mathfrak{h}})} \\
&\overset{(5.9)}{=} \int_{\mathbb{R}}\Big(\psi(\phi_{\mathfrak{h}})E_{\phi_{\mathfrak{h}}}(\Delta)\Theta_{\mathfrak{h}}f\Big)(x)\overline{\Big(\psi(\phi_{\mathfrak{h}})E_{\phi_{\mathfrak{h}}}(\Delta)\hat{\varepsilon}_w\Big)(x)}\,\frac{\mathrm{d}\mu_H(x)}{1+x^2}.
\end{aligned} \tag{5.49}$$

The mappings $\psi(\phi_{\mathfrak{h}})\circ E_{\phi_{\mathfrak{h}}}(\Delta)|_{\mathcal{E}_{\phi_{\mathfrak{h}}}(\infty)^{[\perp]}}$ and $(\cdot\,\mathbb{1}_\Delta)\circ\psi(\phi_{\mathfrak{h}})$ are continuous on $\mathcal{E}_{\phi_{\mathfrak{h}}}(\infty)^{[\perp]}$. By the definition of $\psi(\phi_{\mathfrak{h}})$ and [57, Proposition 3.1], they coincide on all compactly supported functions of $\mathcal{B}_2(\phi_{\mathfrak{h}})$ (for this notation, see [84, §5]). Their continuity implies that they coincide on $\mathcal{E}_{\phi_{\mathfrak{h}}}(\infty)^{[\perp]}$. The assumption $f \in \mathcal{E}_A(\infty)^{[\perp]}$ and the relation (5.11) imply that $\Theta_{\mathfrak{h}}f \in \mathcal{E}_{\phi_{\mathfrak{h}}}(\infty)^{[\perp]}$. Hence,

$$\Big(\psi(\phi_{\mathfrak{h}})E_{\phi_{\mathfrak{h}}}(\Delta)\Theta_{\mathfrak{h}}f\Big)(x)=\mathbb{1}_\Delta(x)\Big(\psi(\phi_{\mathfrak{h}})\Theta_{\mathfrak{h}}f\Big)(x)\qquad x\in\mathbb{R}\quad \mu_H\text{-a.e.}$$

Using [57, Proposition 3.1] and the fact that $\psi(\phi_{\mathfrak{h}})$ acts as the identity on compactly supported functions, we obtain

$$\Big(\psi(\phi_{\mathfrak{h}})E_{\phi_{\mathfrak{h}}}(\Delta)\hat{\varepsilon}_w\Big)(x)=\Big(\psi(\phi_{\mathfrak{h}})\big(\mathbb{1}_\Delta\hat{\varepsilon}_w\big)\Big)(x)=\mathbb{1}_\Delta(x)\cdot\frac{x-i}{x-w}.$$

The integral on the right-hand side of (5.49) thus equals

$$\int_{\mathbb{R}}\Big(\psi(\phi_{\mathfrak{h}})\Theta_{\mathfrak{h}}f\Big)(x)\cdot\frac{x+i}{x-\overline{w}}\mathbb{1}_\Delta(x)\,\frac{d\mu_H(x)}{1+x^2}. \tag{5.50}$$

Since the linear span of the functions $x\mapsto\frac{x+i}{x-\overline{w}}$, $w\in\rho(A)$, is dense in $L^2\big(\frac{\mathbb{1}_\Delta(x)}{1+x^2}\mu_H(x)\big)$, we conclude from (5.48), (5.49) and (5.50) that

$$(\Theta_\infty f)(x)(x-i)=\Big(\psi(\phi_{\mathfrak{h}})\Theta_{\mathfrak{h}}f\Big)(x),\qquad x\in\Delta\quad\mu_H\text{-a.e.}$$

Since Δ was an arbitrary bounded open interval, the assertion follows. □

To finish the proof of (5.1), let $[a,b]\subseteq(0,\infty)$ and denote by Ω the map defined on $L^2(H|_{[a,b]})$ as

$$(\Omega f)(z):=\begin{pmatrix}\int_{[a,b]}\big[\psi(\mathfrak{h}_{\urcorner b})\phi_b(\overline{z})\big](t)^*H(t)f(t)\,dt\\ \int_{[a,b]}\big[\psi(\mathfrak{h}_{\urcorner b})\psi_b(\overline{z})\big](t)^*H(t)f(t)\,dt\end{pmatrix},$$

where ϕ_b and ψ_b are as in (5.26) and (5.16), respectively. The function $[\psi(\mathfrak{h}_{\urcorner b})\psi_b(\overline{z})](t)$ is a solution of (1.1) with z replaced by $\overline{z}$, which assumes boundary values at 0, namely, $(0,\ 1)^T$. The function $\phi(t;\overline{z})$ shares these properties, and hence we have $[\psi(\mathfrak{h}_{\urcorner b})\psi_b(\overline{z})](t)=\phi(t;\overline{z})$. Thus, the second component of Ω can be rewritten as

$$\int_{[a,b]}\phi(t,\overline{z})^*H(t)f(t)\,dt.$$

The asserted formula (5.1) for the action of Θ_H is now obtained by putting together the so far collected knowledge. Consider the following diagram (here π_- denotes the projection onto the second component, and references between # are for the proof of the commutativity of the corresponding part of the diagram):

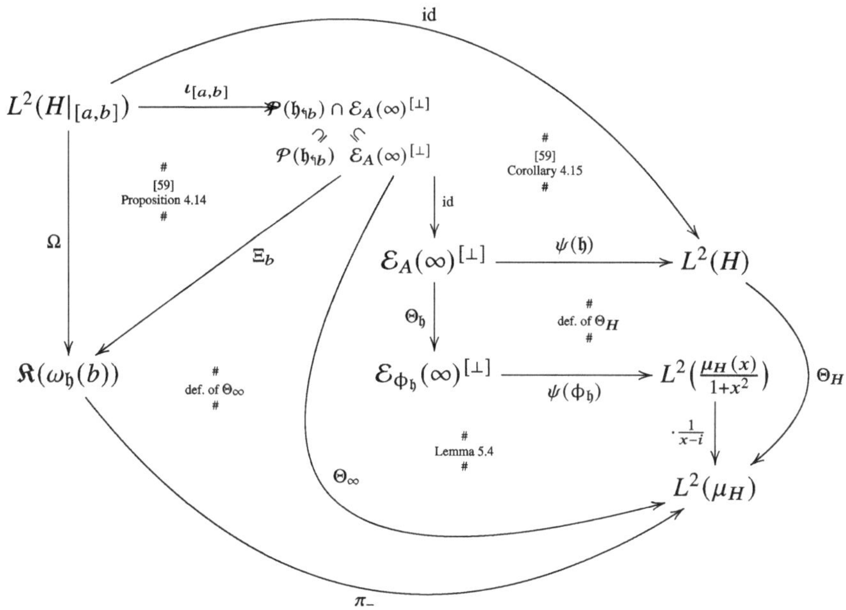

We see that (5.1) holds for each $f \in L^2(H)$ supported on $[a, b]$. For each fixed $T > 0$, both sides of (5.1) depend continuously on f when f varies in the set of all elements of $L^2(H)$ whose support is bounded above by T. Remember here that $\phi(\cdot\,; z) \in L^2(0, T)$. Hence, (5.1) holds for all $f \in L^2(H)$ whose support is bounded above by T.

5.3 Computation of Θ_H^{-1} as an Integral Transform

The final task in the proof of Theorem 5.1 is to establish the formula (5.2) for the action of Θ_H^{-1}.

Denote by E the spectral family associated with $T(H)$. Let $f \in L^2(H)$ with $\sup(\operatorname{supp} f) < b$, let $h \in L^2(H)$ be bounded and with $\sup(\operatorname{supp} h) < b$ and let $\Delta \subseteq \mathbb{R}$ be a finite interval. Using (i), (ii) of Theorem 5.1 and Fubini's theorem, we obtain

$$\begin{aligned}
\big(E(\Delta)f, h\big)_{L^2(H)} &= \int_{\mathbb{R}} \mathbb{1}_\Delta(t)\big(\Theta_H f\big)(t)\overline{\big(\Theta_H h\big)(t)}\,\mathrm{d}\mu_H(t) \\
&= \int_\Delta \big(\Theta_H f\big)(t) \int_a^b h(x)^* H(x)\varphi(x; t)\,\mathrm{d}x\,\mathrm{d}\mu_H(t) \\
&= \int_a^b h(x)^* H(x) \int_\Delta \big(\Theta_H f\big)(t)\varphi(x; t)\,\mathrm{d}\mu_H(t)\,\mathrm{d}x.
\end{aligned}$$

Since the set $\{h \in L^2(H) : \sup(\operatorname{supp} h) < b,\ h \text{ bounded}\}$ is dense in $L^2(H)$, it follows that

$$\big(E(\Delta)f\big)(x) = \int_\Delta \big(\Theta_H f\big)(t)\varphi(x;t)\,\mathrm{d}\mu_H(t) \qquad H\text{-a.e.} \tag{5.51}$$

Both sides of this equality depend continuously on f, and therefore this relation is valid for arbitrary $f \in L^2(H)$.

To complete the proof, let $g \in L^2(\mu_H)$ with compact support be given. Choose a finite interval Δ which contains supp g. Since Θ_H intertwines $T(H)$ with the multiplication operator in $L^2(\mu_H)$, we have

$$E(\Delta) \circ \Theta_H^{-1} = \Theta_H^{-1} \circ \big(\cdot\, \mathbb{1}_\Delta\big). \tag{5.52}$$

Thus,

$$\begin{aligned}\big(\Theta_H^{-1}g\big)(x) &= \big(\Theta_H^{-1}(\mathbb{1}_\Delta g)\big)(x) \overset{(5.52)}{=} \big(E(\Delta)(\Theta_H^{-1}g)\big)(x)\\ &\overset{(5.51)}{=} \int_\Delta g(t)\varphi(x;t)\,\mathrm{d}\mu_H(t) = \int_{\mathbb{R}} g(t)\varphi(x;t)\,\mathrm{d}\mu_H(t) \qquad H\text{-a.e.}\end{aligned}$$

This finishes the proof of Theorem 5.1.

5.4 The Connection Between the Point Mass at 0 and the Behaviour of H

Before we prove Proposition 5.3, we need a lemma.

Lemma 5.9 *Let $H \in \mathbb{H}$ with* $\operatorname{dom} H = (0, \infty)$ *and let $\mathfrak{h}$ be the general Hamiltonian as in 3.15. Moreover, let $\omega_{\mathfrak{h}}$ be the chain of matrices as in 3.8. Then, for each $x \in (0, \infty)$,*

$$\left[\frac{\partial}{\partial z}\omega_{\mathfrak{h},21}(x;z)\right]\bigg|_{z=0} = -\int_0^x h_{22}(t)\mathrm{d}t. \tag{5.53}$$

Proof Let $x_1 \in (0, x)$. Integrating (3.9) we obtain

$$\omega_{\mathfrak{h}}(x;z) - \omega_{\mathfrak{h}}(x_1;z) = z\int_{x_1}^x \omega_{\mathfrak{h}}(t;z)H(t)J\,\mathrm{d}t$$

for $z \in \mathbb{C}$. If we differentiate both sides with respect to z, set $z = 0$ and use (3.10), we arrive at

$$\left[\frac{\partial}{\partial z}\omega_{\mathfrak{h}}(x;z)\right]\bigg|_{z=0} - \left[\frac{\partial}{\partial z}\omega_{\mathfrak{h}}(x_1;z)\right]\bigg|_{z=0} = \int_{x_1}^{x} \omega_{\mathfrak{h}}(t;0)H(t)J\,\mathrm{d}t = \int_{x_1}^{x} H(t)J\,\mathrm{d}t. \tag{5.54}$$

It follows from [82, Theorem 4.1] that

$$\lim_{x_1 \searrow 0}\left[\frac{\partial}{\partial z}\omega_{\mathfrak{h},21}(x_1;z)\right]\bigg|_{z=0} = 0,$$

which, together with (5.54), implies (5.53). □

Proof (Proof of Proposition 5.3) Assume first that $\mu_H(\{0\}) > 0$. Define $g \in L^2(\mu_H)$ by

$$g(t) = \begin{cases} 1, & t = 0, \\ 0, & t \neq 0. \end{cases}$$

Then

$$(\Theta_H^{-1}g)(x) = \int_{-\infty}^{\infty} g(t)\varphi(x;t)\mathrm{d}\mu_H(t) = \mu_H\big(\{0\}\big)\varphi(x;0) = \mu_H\big(\{0\}\big)\begin{pmatrix}0\\1\end{pmatrix}.$$

Since Θ_H is an isometric isomorphism, we obtain

$$\mu_H\big(\{0\}\big) = \|g\|^2_{L^2(\mu_H)} = \big\|\Theta_H^{-1}g\big\|^2_{L^2(H)} = \big[\mu_H\big(\{0\}\big)\big]^2 \int_0^{\infty} \begin{pmatrix}0\\1\end{pmatrix}^* H(x)\begin{pmatrix}0\\1\end{pmatrix}\mathrm{d}x$$
$$= \big[\mu_H\big(\{0\}\big)\big]^2 \int_0^{\infty} h_{22}(x)\mathrm{d}x,$$

which implies (5.3) and (5.4).

Now assume that (5.3) is satisfied. Let $c > 0$ be large enough such that $(0, c)$ is not an indivisible interval and introduce the Hamiltonian function

$$H_c(x) := \begin{cases} H(x), & x \in (0, c], \\ \left(\begin{smallmatrix}1&0\\0&0\end{smallmatrix}\right), & x \in (c, \infty), \end{cases}$$

which belongs to $\mathbb{H}$, satisfies $\Delta(H_c) = \Delta(H)$ and is in the limit point case at infinity. Moreover, let $\mathfrak{h}_c$ be the corresponding singular general Hamiltonian as in 3.15 and let $\omega_{\mathfrak{h}_c}$ be the chain of matrices as in 3.8. The singular Weyl coefficient q_{H_c} is given by

$$q_{H_c}(z) = q_{\mathfrak{h}_c}(z) = \omega_{\mathfrak{h}_c}(c; z) \star \infty = \frac{\omega_{\mathfrak{h}_c,11}(c; z)}{\omega_{\mathfrak{h}_c,21}(c; z)}, \qquad z \in \mathbb{C} \setminus \mathbb{R};$$

it belongs to $\mathcal{N}_{\Delta(H)}^{(\infty)}$ and is meromorphic in $\mathbb{C}$. Let μ_{H_c} be the spectral measure associated with H_c via (4.11). Then Lemma 5.9 and (3.10) imply that, for every $\varepsilon > 0$,

$$\mu_{H_c}\big((-\varepsilon, \varepsilon)\big) \geq \mu_{H_c}\big(\{0\}\big) = \operatorname{Res}(q_{H_c}; 0) = -\frac{\omega_{\mathfrak{h}_c,11}(c; 0)}{\big[\frac{\partial}{\partial z}\omega_{\mathfrak{h}_c,21}(c; z)\big]\big|_{z=0}}$$

$$= \left[\int_0^c h_{22}(x)\mathrm{d}x\right]^{-1} \geq \left[\int_0^\infty h_{22}(x)\mathrm{d}x\right]^{-1} =: M.$$

By Theorem 4.5 (ii) we have $q_H(z) = \lim_{c\to\infty} q_{H_c}(z)$ locally uniformly in $\mathbb{C} \setminus \mathbb{R}$. Since $\operatorname{ind}_- q_{H_c} = \Delta(H_c) = \Delta(H) = \operatorname{ind}_- q_H$, we can apply Lemma 3.7, which implies that, for every $\varepsilon > 0$ such that $\mu_H(\{-\varepsilon\}) = \mu_H(\{\varepsilon\}) = 0$,

$$\mu_H\big((-\varepsilon, \varepsilon)\big) = \lim_{c\to\infty} \mu_{H_c}\big((-\varepsilon, \varepsilon)\big) \geq M.$$

Hence, $\mu_H(\{0\}) \geq M > 0$. □

6 Inverse Theorems

By the procedure described in Theorem 4.8, a map from $\mathbb{H}$ to $\mathbb{M}$, namely, $H \mapsto \mu_H$, is well defined. Hence, it is meaningful to pose inverse problems. Concisely formulated, we face the task to determine the range and kernel of the mapping $H \mapsto \mu_H$.

In this section, we complete this task. In fact, we provide somewhat more detailed results. They include singular Weyl functions and a local version of the uniqueness theorem. Proofs are again relatively simple; they are carried out in the same manner as in Sect. 4, using the basic identifications 3.15, 3.16, and some results taken from the literature. Recall the notation from Definition 4.7. Theorems 6.1 and 6.2 are the analogues for our class $\mathbb{H}$ of Hamiltonians with two singular endpoints to de Branges' celebrated inverse spectral theorem.

Theorem 6.1 (Existence Theorem) *The following statements hold.*

(i) *Let $q \in \mathcal{N}_{<\infty}^{(\infty)}$ with $\operatorname{ind}_- q > 0$. Then there exists a Hamiltonian $H \in \mathbb{H}$ with $q \in [q]_H$.*
(ii) *Let $\mu \in \mathbb{M}$ with $\Delta(\mu) > 0$. Then there exists a Hamiltonian $H \in \mathbb{H}$ with $\mu_H = \mu$.*

Proof To show (i) let $q \in \mathcal{N}_{<\infty}^{(\infty)}$ be given. According to [61, Theorem 1.4], there exists a general Hamiltonian $\mathfrak{h}_0$ with $q_{\mathfrak{h}_0} = q$. Moreover, [82, Theorem 3.1] implies that $\mathfrak{h}_0 \in \mathfrak{H}_0$. Applying an appropriate reparameterization, we may assume that:

- $\mathfrak{h}_0$ is defined on $(-1, 0) \cup (0, \infty)$
- The Hamiltonian function of $\mathfrak{h}$ equals $x^{-2}\begin{pmatrix} 1 & 0 \\ 0 & 0 \end{pmatrix}$ for $x \in (-1, 0)$
- $b_{\ddot{o}+1} = 0$
- $E = \{-1, 1, \infty\}$

Let H be the Hamiltonian function of $\mathfrak{h}_0$ on the interval $(0, \infty)$. Then, by our basic identification 3.15, we have $H \in \mathbb{H}$.

Let $\mathfrak{h}$ be the general Hamiltonian built from H in the basic identification 3.16 with $\ddot{o}_1, b_{1,j}, d_{1,j}$ all equal to 0. Then $\mathfrak{h}$ and $\mathfrak{h}_0$ differ only in the data part $\ddot{o}_1, b_{1,j}, d_{1,j}$. From [82, Corollary 5.9] we obtain that

$$q_H(z) = q_{\mathfrak{h}} = q_{\mathfrak{h}_0} - \sum_{l=1}^{2\Delta(H)} z^l d_{1,l-1} + \sum_{l=1}^{\ddot{o}_1} z^{2\Delta(H)+l} b_{1,\ddot{o}_1+1-l},$$

i.e. q_H and q differ only by a polynomial with real coefficients and vanishing constant term. Hence, $q \in [q]_H$.

For the proof of (ii), let $\mu \in \mathbb{M}$ be given. Choose $q \in \mathcal{N}_{<\infty}^{(\infty)}$ with $\mu_q = \mu$; this is possible by [84, Theorem 3.9 (v)].[7] An application of the already proved item (i) provides us with a Hamiltonian $H \in \mathbb{H}$ such that $q_H - q$ is a real polynomial. By the Stieltjes inversion formula (3.6) and the definition of μ_H, it follows that $\mu_H = \mu_{q_H} = \mu$. □

As we have seen in Proposition 4.10, Hamiltonians which are—essentially—reparameterizations of each other have, essentially, the same singular Weyl coefficients and have the same spectral measures. The converse of this fact is an important result.

Theorem 6.2 (Global Uniqueness Theorem) *Let $H_1, H_2 \in \mathbb{H}$ be given.*

(i) *Assume that there exist singular Weyl coefficients q_{H_1} and q_{H_2} of H_1 and H_2, respectively, such that $q_{H_1} - q_{H_2}$ is a real polynomial, and set $\alpha := (q_{H_1} - q_{H_2})(0)$. Then the Hamiltonians*

$$H_1 \quad \textit{and} \quad \begin{pmatrix} 1 & \alpha \\ 0 & 1 \end{pmatrix} H_2 \begin{pmatrix} 1 & 0 \\ \alpha & 1 \end{pmatrix}, \tag{6.1}$$

are reparameterizations of each other.

In particular, if $[q]_{H_1} = [q]_{H_2}$, then H_1 and H_2 are reparameterizations of each other.

[7] The set on the right-hand side of [84, Theorem 3.9 (v)] is certainly non-empty.

(ii) *If* $\mu_{H_1} = \mu_{H_2}$, *then there exists a real constant* α *such that the Hamiltonians in* (6.1) *are reparameterizations of each other.*

Proof Let $H_1, H_2 \in \mathbb{H}$ be given. Assume that both Hamiltonians are defined on $(0, \infty)$. This is no loss in generality since it can always be achieved by a reparameterization, and reparameterizations change neither singular Weyl coefficients nor spectral measures; see Proposition 4.10.

First, we consider the case when $[q]_{H_1} = [q]_{H_2}$. Let $\mathfrak{h}_1$ and $\mathfrak{h}_2$ be the general Hamiltonians defined for H_1 and H_2 by the basic identification 3.16 with $\ddot{o}_1$, $b_{1,j}$, $d_{1,j}$ all equal to 0 with some base points x_1 and x_2. Then

$$q_{\mathfrak{h}_1} \in [q]_{H_1} \quad \text{and} \quad q_{\mathfrak{h}_2} \in [q]_{H_2}.$$

Applying a reparameterization to $\mathfrak{h}_2$, we can achieve that $x_1 = x_2$. By [82, Corollary 5.9] there exist numbers $d_{1,0}, \dots, d_{1,2\Delta(H_2)-1} \in \mathbb{R}$ and $\ddot{o}_1 \in \mathbb{N}_0$, $b_{1,1}, \dots, b_{1,\ddot{o}_1} \in \mathbb{R}$ such that the Weyl coefficient of the general Hamiltonian $\tilde{\mathfrak{h}}_2$ constituted by the same data as $\mathfrak{h}_2$ with exception of $d_{1,j}, \ddot{o}_1, b_{1,j}$ is equal to $q_{\mathfrak{h}_1}$. By the uniqueness part in [61, Theorem 1.4], thus, $\mathfrak{h}_1$ and $\tilde{\mathfrak{h}}_2$ are reparameterizations of each other. In particular, their Hamiltonian functions on $(0, \infty)$ are reparameterizations of each other. However, the Hamiltonian function of $\mathfrak{h}_1$ on this interval is H_1 and the one of $\tilde{\mathfrak{h}}_2$ is H_2.

Now assume that some singular Weyl coefficients q_{H_1} and q_{H_2} differ by a real constant, say α. Consider the Hamiltonian $H_0 := \begin{pmatrix} 1 & \alpha \\ 0 & 1 \end{pmatrix} H_2 \begin{pmatrix} 1 & 0 \\ \alpha & 1 \end{pmatrix}$. We know from Proposition 4.10 (and its proof) that $H_0 \in \mathbb{H}$ and that $q_{H_0} - q_{H_2} = \alpha$ when we choose the same base point in the construction of q_{H_1} and q_{H_2}, respectively. We thus have $q_{H_1} = q_{H_0}$. By what we proved in the previous paragraph, this implies that H_1 and H_0 are reparameterizations of each other; hence (i) is shown.

For the proof of (ii), assume that $\mu_{H_1} = \mu_{H_2}$. Choose $x_0 \in (0, \infty)$, and let q_{H_1} and q_{H_2} be singular Weyl coefficients of H_1 and H_2, respectively, built with the base point x_0. By [84, Theorem 3.9 (iv)] the difference $q_{H_1} - q_{H_2}$ is a real polynomial. Now the already proved item (i) yields (6.1). □

Our last result in this section is a refined version of the above uniqueness theorem. It asserts that certain beginning sections of two Hamiltonians $H_1, H_2 \in \mathbb{H}$ coincide if (and only if) some of their singular Weyl coefficients are exponentially close. Local uniqueness theorems for one-dimensional Schrödinger operators were first proved by B. Simon in [99, Theorem 1.2]. For canonical systems with a regular left endpoint, a local uniqueness theorem was proved in [81, Theorem 1.2]; see also [78, Section 4] for a formulation in terms of transfer functions.

Theorem 6.3 (Local Uniqueness Theorem) *Let* $H_1, H_2 \in \mathbb{H}$ *with* $\mathrm{dom}(H_i) = (a_i, b_i)$, $i = 1, 2$, *be given and set*

$$s_i(\tau):=\sup\left\{x\in(a_i,b_i):\int_{a_i}^{x}\sqrt{\det H_i(\xi)}\,\mathrm{d}\xi<\tau\right\},\qquad \tau>0,\ i=1,2.\tag{6.2}$$

Then, for each $\tau > 0$, the following statements are equivalent.

(i) *The Hamiltonian $H_1|_{(a_1,s_1(\tau))}$ is a reparameterization of $H_2|_{(a_2,s_2(\tau))}$.*
(ii) *There exist singular Weyl coefficients q_{H_1} and q_{H_2} of H_1 and H_2, respectively, and there exists a $\beta \in (0, \pi)$ such that for each $\varepsilon > 0$,*

$$q_{H_1}(re^{i\beta})-q_{H_2}(re^{i\beta})=\mathrm{O}\big(e^{(-2\tau+\varepsilon)r\sin\beta}\big),\quad r\to\infty.$$

(iii) *There exist singular Weyl coefficients q_{H_1} and q_{H_2} of H_1 and H_2, respectively, and there exists a $k \geq 0$ such that for each $\delta \in (0, \frac{\pi}{2})$,*

$$q_{H_1}(z)-q_{H_2}(z)=\mathrm{O}\big((\operatorname{Im}z)^k e^{-2\tau\operatorname{Im}z}\big),\quad |z|\to\infty,\ z\in\Gamma_\delta,\tag{6.3}$$

where Γ_δ is the Stolz angle $\Gamma_\delta:=\{z\in\mathbb{C}:\delta\leq\arg z\leq\pi-\delta\}$.

Note that the integral in (6.2) is always finite. This is a consequence of [83, Theorem 4.1], which also implies that this integral is equal to the exponential type of each entry of $\theta(x;\cdot)$ and $\varphi(x;\cdot)$.

Proof (Proof of Theorem 6.3) This theorem is a consequence of the indefinite version of [81, Theorem 1.2] indicated in [81, Remark 1.3]. Let $H_1, H_2 \in \mathbb{H}$ be given, and assume w.l.o.g. that both are defined on $(0,\infty)$. The proof again proceeds via considering general Hamiltonians $\mathfrak{h}_1$ and $\mathfrak{h}_2$ built in our basic identification 3.16 from H_1 and H_2, respectively.

Assuming (i) we choose $ö_1, b_{1,j}, d_{1,j}$ all equal to 0 and the same base point x_0 in the definition of $\mathfrak{h}_1$ and $\mathfrak{h}_2$. Then, by [81, Theorem 1.2 (indefinite variant)], it follows that $q_{\mathfrak{h}_1}$ and $q_{\mathfrak{h}_2}$ are exponentially close in the sense of (ii) and (iii).

Conversely, assume (ii) or (iii), and choose $ö_1^{(1)}, b_{1,j}^{(1)}, d_{1,j}^{(1)}$ and x_1 and $ö_1^{(2)}, b_{1,j}^{(2)}$, $d_{1,j}^{(2)}$ and x_2 in the definition of $\mathfrak{h}_1$ and $\mathfrak{h}_2$ so that $q_{H_i}=q_{\mathfrak{h}_i}$. This is possible; cf. Remark 4.6. Then, again by [81, Theorem 1.2 (indefinite variant)], $\mathfrak{h}_{1,\upharpoonright s_1(\tau)}$ and $\mathfrak{h}_{2,\upharpoonright s_2(\tau)}$ are reparameterizations of each other. In particular, (i) holds. □

Remark 6.4 If one (and hence all) of the equivalent conditions of Theorem 6.3 holds, then (6.3) holds with

$$k:=8\max\{\Delta(H_1),\Delta(H_2)\}+3.\tag{6.4}$$

This can be seen by tracing the proof of [81, Theorem 1.2 (indefinite version)] as indicated in the footnotes in this paper.

The value (6.4) of the constant k in (6.3) is probably not the best possible. However, it is noteworthy that (6.4) depends only on $\Delta(H_1)$ and $\Delta(H_2)$.

Remark 6.5 The dependence of these results on the choices of q_{H_1} and q_{H_2} is not essential. If (ii/iii) holds with some pair $(q_{H_1}, q_{H_2}) \in [q]_{H_1} \times [q]_{H_2}$, then for each $q_1 \in [q]_{H_1}$, there exists a unique element $q_2 \in [q]_{H_2}$ such that (ii/iii) holds for (q_1, q_2). This corresponding function q_2 can be determined by starting with some $q \in [q]_{H_2}$, computing the polynomial asymptotics of $q - q_{H_1}$ at $i\infty$, and subtracting this polynomial from q.

Viewing the above remark from a slightly different perspective leads to the following more effective test for (ii/iii) to hold, which removes the dependence on the choice of q_{H_1} and q_{H_2}.

Corollary 6.6 *Assume that we are in the situation of Theorem 6.3. Pick some singular Weyl coefficients $q_1 \in [q]_{H_1}$ and $q_2 \in [q]_{H_2}$ and let*

$$q_2(iy) - q_1(iy) = \alpha_n y^n + \alpha_{n-1} y^{n-1} + \ldots + \alpha_1 y + o(y), \quad y \to \infty.$$

Then (ii/iii) *of Theorem 6.3 hold if and only if the conditions stated in* (ii/iii) *hold with $q_{H_1} = q_1$ and*

$$q_{H_2}(z) = q_2(z) - \sum_{l=1}^{n} \alpha_l (-i)^l z^l.$$

The following corollary of Theorem 6.3 is also worth mentioning. It says that under the a priori hypothesis of finite exponential type, the global uniqueness result Theorem 6.2 can be strengthened; and it may be much easier to establish exponential closeness of singular Weyl coefficients than their actual equality.

Corollary 6.7 *Let $H_1, H_2 \in \mathbb{H}$ with* $\operatorname{dom}(H_i) = (a_i, b_i)$, $i = 1, 2$, *be given and assume that*

$$\int_{a_i}^{b_i} \sqrt{\det H_i(y)}\, dy < \infty, \qquad i = 1, 2.$$

If there exist singular Weyl coefficients q_{H_1} and q_{H_2} of H_1 and H_2, respectively, and there exist $\beta \in (0, \pi)$ and $\tau > \max\left\{ \int_{a_i}^{b_i} \sqrt{\det H_i(y)}\, dy : i = 1, 2 \right\}$ such that

$$q_{H_1}(re^{i\beta}) - q_{H_2}(re^{i\beta}) = O\big(e^{-2\tau r \sin\beta}\big), \qquad r \to \infty,$$

then H_1 and H_2 are reparameterizations of each other.

PART II: Applications to Sturm–Liouville Equations

In Sects. 7–9, we study scalar second-order differential equations. Under certain assumptions, such equations can be transformed to canonical systems of the form (1.1) so that the results from Sects. 4–6 can be applied. In Sects. 7 and 8, we consider Sturm–Liouville equations of the form (7.1) below where either $1/p$ or w is not integrable at a. In Sect. 9 we study one-dimensional Schrödinger operators with a singular potential.

7 Sturm–Liouville Equations Without Potential: Singular $1/p$

In this section, we consider Sturm–Liouville equations of the form

$$-\left(py'\right)' = \lambda w y \tag{7.1}$$

on an interval (a, b) with $-\infty \le a < b \le \infty$ where $\lambda \in \mathbb{C}$ and the functions p and w satisfy the conditions

$$p(x) > 0,\ w(x) > 0 \ \text{a.e.}, \qquad \frac{1}{p},\, w \in L^1_{\text{loc}}(a, b). \tag{7.2}$$

In the following, we write $\operatorname{dom}(p; w) := (a, b)$. Moreover, let $L^2(w)$ be the weighted L^2-space with inner product $(f, g) = \int_a^b f \overline{g} w$.

We consider the following class of coefficients.

Definition 7.1 We say that $(p; w) \in \mathbb{K}_{\text{SL}}$ if p and w are defined on some interval (a, b) and they satisfy (7.2) and the following conditions.

(i) For one (and hence for all) $x_0 \in (a, b)$,

$$\int_a^{x_0} \frac{1}{p(x)} \mathrm{d}x = \infty \qquad \text{and} \qquad \int_a^{x_0} w(x) \mathrm{d}x < \infty. \tag{7.3}$$

(ii) For one (and hence for all) $x_0 \in (a, b)$,

$$\int_a^{x_0} \int_a^{x} w(t) \mathrm{d}t \, \frac{1}{p(x)} \mathrm{d}x < \infty. \tag{7.4}$$

(iii) Let $x_0 \in (a, b)$ and define functions w_l, $l = 0, 1, \ldots$, recursively by

$$w_0(x) = 1,$$

$$w_l(x) = \begin{cases} \displaystyle\int_x^{x_0} \frac{1}{p(t)} w_{l-1}(t)\mathrm{d}t & \text{if } l \text{ is odd,} \\ \displaystyle\int_a^x w(t) w_{l-1}(t)\mathrm{d}t & \text{if } l \text{ is even.} \end{cases} \tag{7.5}$$

There exists an $n \in \mathbb{N}_0$ such that

$$w_n\big|_{(a,x_0)} \in \begin{cases} L^2\big(\frac{1}{p}\big|_{(a,x_0)}\big) & \text{if } n \text{ is even,} \\ L^2\big(w\big|_{(a,x_0)}\big) & \text{if } n \text{ is odd.} \end{cases} \tag{7.6}$$

(iv) Equation (7.1) is in the limit point case at b, i.e. for $\lambda \in \mathbb{C} \setminus \mathbb{R}$, equation (7.1) has (up to a scalar multiple) only one solution in $L^2(w|_{(x_0,b)})$ for $x_0 \in (a, b)$.

If $(p; w) \in \mathbb{K}_{\mathrm{SL}}$, we denote by $\Delta_{\mathrm{SL}}(p, w)$ the minimal $n \in \mathbb{N}_0$ such that (7.6) holds.

Remark 7.2

(i) Under the assumption of (7.3), condition (7.4) is equivalent to

$$\int_a^{x_0} \int_x^{x_0} \frac{1}{p(t)} \mathrm{d}t \, w(x)\mathrm{d}x < \infty;$$

see, e.g. [82, Lemma 4.3].

(ii) Assume that (7.3) holds. Then (7.4) and (7.6) with $n = 1$ are satisfied if and only if equation (7.1) is in the limit circle case at a; this is true because the solutions of (7.1) with $\lambda = 0$ are $y(x) = c_1 w_1(x) + c_2$ with $c_1, c_2 \in \mathbb{C}$ and the limit circle case prevails at a if and only if all these solutions are in $L^2(w|_{(a,x_0)})$.

(iii) The functions w_0 and w_1 are solutions of (7.1) with $\lambda = 0$. Since $w_1(x) \to \infty$ as $x \searrow a$, the function w_0 is a principal solution and w_1 is a non-principal solution, i.e. $w_0(x) = \mathrm{o}(w_1(x))$ as $x \searrow 0$; for the notions of principal and non-principal solutions, see, e.g. [89]. Moreover, one can easily verify that

$$-\frac{1}{w}\big(p w_{l+2}'\big)' = w_l \qquad \text{when } l \in \mathbb{N} \text{ is odd.}$$

For given p and w satisfying (7.2), define the Hamiltonian

$$H(x) := \begin{pmatrix} \dfrac{1}{p(x)} & 0 \\ 0 & w(x) \end{pmatrix}, \qquad x \in (a,b). \tag{7.7}$$

If $\psi = (\psi_1, \psi_2)^T$ is a solution of equation (1.1) with H as in (7.7), then

$$\psi_1' = -zw\psi_2, \qquad \psi_2' = z\frac{1}{p}\psi_1,$$

and hence ψ_2 is a solution of (7.1) with $\lambda = z^2$. Conversely, if ψ is a solution of (7.1) and $z \in \mathbb{C}$ is such that $z^2 = \lambda$, then

$$\psi(x) = \begin{pmatrix} p(x)\psi'(x) \\ z\psi(x) \end{pmatrix} \tag{7.8}$$

satisfies (1.1) with H as in (7.7).

In the following, assume that $(p; w) \in \mathbb{K}_{\mathrm{SL}}$. The first relation in (7.3) implies that H is in the limit point case at a. Since (7.1) is in the limit point case at b, the Hamiltonian H is also in the limit point case at b because $\psi \in L^2(H|_{(x_0,b)})$ with ψ as in (7.8), $z \neq 0$ and $x_0 \in (a,b)$ implies that $\psi \in L^2(w|_{(x_0,b)})$. Therefore, the operator $T(H)$, which is defined in (3.8) and acts in the space $L^2(H) = L^2(\frac{1}{p}) \oplus L^2(w)$, is self-adjoint. Since $H(x)$ is invertible for a.e. $x \in (a,b)$, the operator $T(H)$ can be written as

$$T(H)f = H^{-1}J^{-1}f' = \begin{pmatrix} pf_2' \\ -\dfrac{1}{w}f_1' \end{pmatrix}$$

with maximal domain

$$\operatorname{dom}(T(H)) = \left\{ \begin{pmatrix} f_1 \\ f_2 \end{pmatrix} : f_1, f_2 \text{ abs. cont.}, \; pf_2' \in L^2\Big(\frac{1}{p}\Big), \; \frac{1}{w}f_1' \in L^2(w) \right\}$$

Hence, $(T(H))^2$ acts as follows:

$$\big(T(H)\big)^2 \begin{pmatrix} f_1 \\ f_2 \end{pmatrix} = \begin{pmatrix} -p\Big(\dfrac{1}{w}f_1'\Big)' \\ -\dfrac{1}{w}\big(pf_2'\big)' \end{pmatrix}. \tag{7.9}$$

With the mappings

$$\iota_2 : \begin{cases} L^2(w) \to L^2(H) \\ g \mapsto \begin{pmatrix} 0 \\ g \end{pmatrix}, \end{cases} \qquad P_2 : \begin{cases} L^2(H) \to L^2(w) \\ \begin{pmatrix} f \\ g \end{pmatrix} \mapsto g, \end{cases} \tag{7.10}$$

we define the self-adjoint operator

$$A_{p,w} := P_2 \big(T(H)\big)^2 \iota_2. \tag{7.11}$$

This operator acts like

$$A_{p,w} y = -\frac{1}{w}(py')'$$

$$\operatorname{dom}(A_{p,w}) = \Big\{ y \in L^2(w) : y,\, py' \text{ locally absolutely continuous,}$$

$$\int_a^b p(x)\big|y'(x)\big|^2 \mathrm{d}x < \infty,\ \frac{1}{w}(py')' \in L^2(w) \Big\} \tag{7.12}$$

and is the Friedrichs extension of the minimal operator associated with (7.1) since all functions in $\operatorname{dom}(A_{p,w})$ are in the form domain; note that $A_{p,w}$ is non-negative. If (7.1) is also in the limit point case at a (i.e. when $\Delta_{\mathrm{SL}}(p, w) \geq 2$), then $A_{p,w}$ coincides with the maximal operator, i.e. the condition $\int_a^b p|y'|^2 < \infty$ is automatically satisfied. If (7.1) is in the limit circle case at a, one can replace the condition $\int_a^b p|y'|^2 < \infty$ in (7.12) by any of the two boundary conditions

$$\lim_{x \searrow a} \frac{y(x)}{w_1(x)} = 0, \qquad \lim_{x \searrow a} p(x) y'(x) = 0; \tag{7.13}$$

see, e.g. [89, Theorem 4.3].

Remark 7.3 One can also treat the situation when (7.1) is either regular or in the limit circle case at b. In the former case, one extends H by an indivisible interval of infinite length; in the latter case, H is in the limit point case. In both cases, elements in the domain of $A_{p,w}$ defined via (7.11) satisfy some boundary condition at b.

Assume that $(p; w) \in \mathbb{K}_{\mathrm{SL}}$ and let H be as in (7.7). It follows from [105, Theorem 3.7] that $H \in \mathbb{H}$, that the functions $\mathfrak{w}_l$, $l \in \mathbb{N}_0$, defined in (3.13) are given by

$$\mathfrak{w}_l(x) = \begin{cases} \begin{pmatrix} w_l(x) \\ 0 \end{pmatrix} & \text{if } l \text{ is even,} \\ \begin{pmatrix} 0 \\ -w_l(x) \end{pmatrix} & \text{if } l \text{ is odd,} \end{cases} \tag{7.14}$$

and that

$$\Delta_{\mathrm{SL}}(p, w) = \Delta(H). \tag{7.15}$$

Therefore, we can apply the results from Part I to the Hamiltonian H. Using the connection between (7.1) and (1.1), we can show that regularized boundary values of solutions of (7.1) exist at a. Denote by $\mathfrak{N}_\lambda^{\mathrm{SL}}$ the set of all solutions of the Sturm–Liouville equation (7.1).

Theorem 7.4 (Regularized boundary values) *Let* $(p; w) \in \mathbb{K}_{\mathrm{SL}}$ *with* $\mathrm{dom}(p; w) = (a, b)$ *and set* $\Delta := \Delta_{\mathrm{SL}}(p, w)$*. Then, for* $x_0 \in (a, b)$*, the following statements hold.*

(i) *For each* $\lambda \in \mathbb{C}$ *and each solution* $\psi \in \mathfrak{N}_\lambda^{\mathrm{SL}}$*, the boundary value*

$$\mathrm{rbv}_{\lambda,1}^{\mathrm{SL}}\psi := \lim_{x \searrow a} p(x)\psi'(x) \tag{7.16}$$

and the regularized boundary value

$$\mathrm{rbv}_{\lambda,2}^{\mathrm{SL}}\psi := \lim_{x \searrow a} \Bigg[\sum_{k=0}^{\lfloor \frac{\Delta-1}{2} \rfloor} \lambda^k \Big(w_{2k}(x)\psi(x) + w_{2k+1}(x)p(x)\psi'(x) \Big) \tag{7.17}$$
$$+ \begin{Bmatrix} \lambda^{\frac{\Delta}{2}} w_\Delta(x)\psi(x) & \text{if } \Delta \text{ is even} \\ 0 & \text{if } \Delta \text{ is odd} \end{Bmatrix}$$
$$+ \Big(\lim_{t \searrow a} p(t)\psi'(t)\Big) \sum_{k=\lfloor \frac{\Delta+1}{2} \rfloor}^{\Delta-1} \sum_{l=0}^{2k-\Delta} (-1)^l \lambda^k w_l(x) w_{2k-l+1}(x) \Bigg]$$

exist.

(ii) *For each* $\lambda \in \mathbb{C}$*, we define*

$$\mathrm{rbv}_\lambda^{\mathrm{SL}} : \begin{cases} \mathfrak{N}_\lambda^{\mathrm{SL}} \to \mathbb{C}^2 \\ \psi \mapsto \big(\mathrm{rbv}_{\lambda,1}^{\mathrm{SL}}\psi,\ \mathrm{rbv}_{\lambda,2}^{\mathrm{SL}}\psi\big)^T. \end{cases}$$

Then $\mathrm{rbv}_\lambda^{\mathrm{SL}}$ *is a bijection from* $\mathfrak{N}_\lambda^{\mathrm{SL}}$ *onto* $\mathbb{C}^2$*.*

(iii) *For each* $\lambda \in \mathbb{C}$*, there exists an (up to scalar multiples) unique solution* $\psi \in \mathfrak{N}_\lambda^{\mathrm{SL}} \setminus \{0\}$ *such that* $\lim_{x \searrow a} \psi(x)$ *exists.*
This solution is characterized by the property that $\int_a^{x_0} p|\psi'|^2 < \infty$ *and also by the property that* $\mathrm{rbv}_{\lambda,1}^{\mathrm{SL}}\psi = 0$ *(and* $\psi \not\equiv 0$*).*
If ψ *is a solution such that* $\lim_{x \searrow a} \psi(x)$ *exists, then* $\mathrm{rbv}_{\lambda,2}^{\mathrm{SL}}\psi = \lim_{x \searrow a} \psi(x)$*.*

The regularized boundary value $\mathrm{rbv}^{\mathrm{SL}}_{\lambda,2}$ *depends on the choice of* x_0 *in the following way.*

(iv) *Let* $x_0, \hat{x}_0 \in (a,b)$*, and let* $\mathrm{rbv}^{\mathrm{SL}}_{\lambda}$ *and* $\widehat{\mathrm{rbv}}^{\mathrm{SL}}_{\lambda}$ *be the correspondingly defined regularized boundary value mappings. Then there exists a polynomial* $p_{x_0,\hat{x}_0}$ *with real coefficients whose degree does not exceed* $\Delta - 1$ *such that*

$$\widehat{\mathrm{rbv}}^{\mathrm{SL}}_{\lambda,2}\psi = \mathrm{rbv}^{\mathrm{SL}}_{\lambda,2}\psi + p_{x_0,\hat{x}_0}(\lambda)\mathrm{rbv}^{\mathrm{SL}}_{\lambda,1}\psi, \qquad \psi \in \mathfrak{N}^{\mathrm{SL}}_{\lambda}, \quad \lambda \in \mathbb{C}.$$

Moreover, clearly, $\widehat{\mathrm{rbv}}^{\mathrm{SL}}_{\lambda,1} = \mathrm{rbv}^{\mathrm{SL}}_{\lambda,1}$.

Remark 7.5

(i) Let

$$W_p(y_1, y_2)(x) := p(x)\Big(y_1(x)y_2'(x) - y_1'(x)y_2(x)\Big) \tag{7.18}$$

be the weighted Wronskian with weight p. Using (7.5) we can rewrite the expression that appears within the round brackets in (7.17) as follows:

$$w_{2k}\psi + w_{2k+1}p\psi' = W_p(w_{2k+1}, \psi).$$

(ii) When $\Delta_{\mathrm{SL}}(p,w) = 1$, i.e. when (7.1) is in the limit circle case at a, then $\mathrm{rbv}^{\mathrm{SL}}_{\lambda,2}\psi$ does not depend on λ explicitly, and it takes the form

$$\mathrm{rbv}^{\mathrm{SL}}_{\lambda,2}\psi = \lim_{x\searrow a}\Big(\psi(x) + p(x)w_1(x)\psi'(x)\Big) = \lim_{x\searrow a} W_p(w_1,\psi)(x).$$

Note also that $\mathrm{rbv}^{\mathrm{SL}}_{\lambda,1}\psi = \lim_{x\searrow a} W_p(1,\psi)(x)$.

(iii) Instead of the functions w_l, one can use functions $\check{w}_l$ that are defined by the recurrence relation $\check{w}_0 \equiv 1$ and

$$\check{w}_l(x) = \begin{cases} \displaystyle\int_x^{x_0} \frac{1}{p(t)}\check{w}_{l-1}(t)\mathrm{d}t + c_l & \text{if } l \text{ is odd,} \\ \displaystyle\int_a^x w(t)\check{w}_{l-1}(t)\mathrm{d}t & \text{if } l \text{ is even,} \end{cases}$$

with arbitrary real numbers c_l for odd l. To add the extra constants c_l is useful for practical calculations, in particular, when w_l has an asymptotic expansion (for $x \searrow a$) in which a constant term can be removed by adjusting c_l. One can show that the corresponding regularized boundary value $\check{\mathrm{rbv}}^{\mathrm{SL}}_{\lambda,2}$ satisfies

$$\check{\mathrm{rbv}}^{\mathrm{SL}}_{\lambda,2}\psi = \mathrm{rbv}^{\mathrm{SL}}_{\lambda,2}\psi + \check{p}(\lambda)\mathrm{rbv}^{\mathrm{SL}}_{\lambda,1}\psi$$

with

$$\check{p}(\lambda) = \sum_{k=0}^{\Delta-1} \lambda^k \sum_{i=0}^{k} c_{2k+1-2i} \lim_{t\searrow a} v_{2i}(t) \tag{7.19}$$

where $v_0 \equiv 1$ and

$$v_l(x) = \int_x^{x_0}\int_a^x w(t)v_{l-2}(t)\mathrm{d}t \,\frac{1}{p(x)}\mathrm{d}x, \qquad l \text{ even.}$$

The limit $\lim_{t\searrow a} v_l(t)$ exists because of condition (7.4).

Before we prove Theorem 7.4, we show the following lemma, where $\mathfrak{N}_z$ denotes the set of all solutions of (1.1); see Sect. 4.1.

Lemma 7.6 *Let $z \in \mathbb{C}$ and let ψ be a solution of* (7.1) *with $\lambda:=z^2$. Moreover, set*

$$\boldsymbol{\psi}(x):=\begin{pmatrix} p(x)\psi'(x) \\ z\psi(x) \end{pmatrix}, \qquad \widehat{\boldsymbol{\psi}}(x):=\begin{pmatrix} p(x)\psi'(x) \\ -z\psi(x) \end{pmatrix}. \tag{7.20}$$

Then

$$\boldsymbol{\psi} \in \mathfrak{N}_z, \qquad \widehat{\boldsymbol{\psi}} \in \mathfrak{N}_{-z}, \qquad \mathrm{rbv}_{z,1}\boldsymbol{\psi} = \mathrm{rbv}_{-z,1}\widehat{\boldsymbol{\psi}} \tag{7.21}$$

and

$$\begin{aligned}
\mathrm{rbv}_{z,2}\boldsymbol{\psi} &= -\mathrm{rbv}_{-z,2}\widehat{\boldsymbol{\psi}} \\
&= z\lim_{x\searrow a}\Bigg[\sum_{\substack{l:\,0\le l\le \Delta-1\\ l\ \mathit{even}}} z^l\Big(w_l(x)\psi(x) + w_{l+1}(x)p(x)\psi'(x)\Big) \\
&\quad + \begin{Bmatrix} z^\Delta w_\Delta(x)\psi(x) & \mathit{if}\ \Delta\ \mathit{is\ even} \\ 0 & \mathit{if}\ \Delta\ \mathit{is\ odd} \end{Bmatrix} \\
&\quad + \Big(\lim_{t\searrow a} p(t)\psi'(t)\Big)\sum_{k=\lfloor\frac{\Delta+1}{2}\rfloor}^{\Delta-1}\ \sum_{l=0}^{2k-\Delta}(-1)^l z^{2k} w_l(x)w_{2k-l+1}(x)\Bigg].
\end{aligned}$$

Proof Let H be as in (7.7) and set $\Delta:=\Delta(H)$. The relations in (7.21) are clear from the considerations around equation (7.8) and the fact that (7.1) does not change when we replace z by $-z$. From (4.1) and (7.14), we obtain

$$\operatorname{rbv}_{z,2}\psi = -\lim_{x\searrow a}\Bigg[\sum_{l=0}^{\Delta} z^l\big(\mathfrak{w}_l(x)\big)^* J\Bigg(\psi(x) - (\operatorname{rbv}_{z,1}\psi)\sum_{k=\Delta+1}^{2\Delta-l} z^k\mathfrak{w}_k(x)\Bigg)\Bigg]$$

$$= \lim_{x\searrow a}\Bigg[\sum_{\substack{l:\,0\le l\le\Delta\\ l\text{ even}}} z^l w_l(x)\Bigg(\psi(x) - (\operatorname{rbv}_{z,1}\psi)\sum_{k=\Delta+1}^{2\Delta-l} z^k\mathfrak{w}_k(x)\Bigg)_2$$

$$+ \sum_{\substack{l:\,1\le l\le\Delta\\ l\text{ odd}}} z^l w_l(x)\Bigg(\psi(x) - (\operatorname{rbv}_{z,1}\psi)\sum_{k=\Delta+1}^{2\Delta-l} z^k\mathfrak{w}_k(x)\Bigg)_1\Bigg]$$

$$= \lim_{x\searrow a}\Bigg[\sum_{\substack{l:\,0\le l\le\Delta\\ l\text{ even}}} z^l w_l(x)\Bigg(z\psi(x) + \Big(\lim_{t\searrow a} p(t)\psi'(t)\Big)\sum_{\substack{k:\,\Delta+1\le k\le 2\Delta-l\\ k\text{ odd}}} z^k w_k(x)\Bigg)$$

$$+ \sum_{\substack{l:\,1\le l\le\Delta\\ l\text{ odd}}} z^l w_l(x)\Bigg(p(x)\psi'(x) - \Big(\lim_{t\searrow a} p(t)\psi'(t)\Big)\sum_{\substack{k:\,\Delta+1\le k\le 2\Delta-l\\ k\text{ even}}} z^k w_k(x)\Bigg)\Bigg]$$

$$= z\lim_{x\searrow a}\Bigg[\sum_{\substack{l:\,0\le l\le\Delta\\ l\text{ even}}} z^l w_l(x)\psi(x) + \sum_{\substack{l:\,0\le l\le\Delta-1\\ l\text{ even}}} z^l w_{l+1}(x)p(x)\psi'(x)$$

$$+ \Big(\lim_{t\searrow a} p(t)\psi'(t)\Big)\Bigg(\sum_{\substack{l:\,0\le l\le\Delta\\ l\text{ even}}}\ \sum_{\substack{k:\,\Delta+1\le k\le 2\Delta-l\\ k\text{ odd}}} z^{l+k-1} w_l(x)w_k(x)$$

$$- \sum_{\substack{l:\,1\le l\le\Delta\\ l\text{ odd}}}\ \sum_{\substack{k:\,\Delta+1\le k\le 2\Delta-l\\ k\text{ even}}} z^{l+k-1} w_l(x)w_k(x)\Bigg)\Bigg]$$

$$= z\lim_{x\searrow a}\Bigg[\sum_{\substack{l:\,0\le l\le\Delta\\ l\text{ even}}} z^l w_l(x)\psi(x) + \sum_{\substack{l:\,0\le l\le\Delta-1\\ l\text{ even}}} z^l w_{l+1}(x)p(x)\psi'(x)$$

$$+ \Big(\lim_{t\searrow a} p(t)\psi'(t)\Big)\sum_{l=0}^{\Delta}\ \sum_{\substack{k:\,\Delta+1\le k\le 2\Delta-l\\ l+k\text{ odd}}} (-1)^l z^{l+k-1} w_l(x)w_k(x)\Bigg]$$

$$= z \lim_{x \searrow a} \Bigg[\sum_{\substack{l:\, 0 \le l \le \Delta-1 \\ l \text{ even}}} z^l \Big(w_l(x)\psi(x) + w_{l+1}(x)p(x)\psi'(x) \Big)$$

$$+ \begin{Bmatrix} z^\Delta w_\Delta(x)\psi(x) & \text{if } \Delta \text{ is even} \\ 0 & \text{if } \Delta \text{ is odd} \end{Bmatrix}$$

$$+ \Big(\lim_{t \searrow a} p(t)\psi'(t) \Big) \sum_{\substack{m:\, \Delta \le m \le 2\Delta-2 \\ m \text{ even}}} \sum_{l=0}^{m-\Delta} (-1)^l z^m w_l(x) w_{m-l+1}(x) \Bigg],$$

which proves the statement for $\mathrm{rbv}_{z,2}\psi$. Inside the limit, only even powers of z appear, and hence, as $\mathrm{rbv}_{-z,2}\widehat{\psi}$ is obtained from $\mathrm{rbv}_{z,2}\psi$ by replacing z by $-z$, the equality $\mathrm{rbv}_{z,2}\psi = -\mathrm{rbv}_{-z,2}\widehat{\psi}$ follows. □

Proof (Proof of Theorem 7.4) First, we settle the case $\lambda = 0$. The solutions of (7.1) with $\lambda = 0$ are of the form

$$\psi(x) = c_1 \int_{x_0}^{x} \frac{\mathrm{d}t}{p(t)} + c_2$$

with $c_1, c_2 \in \mathbb{C}$. For such a solution, the limits in (7.16) and (7.17) exist and $\mathrm{rbv}^{\mathrm{SL}}_{0,1}\psi = c_1$ and

$$\mathrm{rbv}^{\mathrm{SL}}_{0,2}\psi = \lim_{x \searrow a} \big(w_0(x)\psi(x) + w_1(x)p(x)\psi'(x) \big)$$

$$= \lim_{x \searrow a} \left(c_1 \int_{x_0}^{x} \frac{\mathrm{d}t}{p(t)} + c_2 + \int_{x}^{x_0} \frac{\mathrm{d}t}{p(t)} \cdot p(x) \cdot \frac{c_1}{p(x)} \right) = c_2.$$

This shows that $\mathrm{rbv}^{\mathrm{SL}}_0 : \mathfrak{N}^{\mathrm{SL}}_0 \to \mathbb{C}^2$ is a bijective mapping. Moreover, the conditions in (iii) are all equivalent to $c_1 = 0$ since $\frac{1}{p}$ is not integrable at a.

For the rest of the proof, assume that $\lambda \neq 0$.

(i) Let ψ be a solution of (7.1), let $z \in \mathbb{C}$ with $z^2 = \lambda$ and define ψ as in (7.20). The existence of the limit in (7.16) and the equality

$$\mathrm{rbv}^{\mathrm{SL}}_{\lambda,1}\psi = \mathrm{rbv}_{z,1}\psi \tag{7.22}$$

are immediate. The existence of the limit in (7.17) and the relation

$$\mathrm{rbv}^{\mathrm{SL}}_{\lambda,2}\psi = \frac{1}{z}\mathrm{rbv}_{z,2}\psi \tag{7.23}$$

follows from Lemma 7.6 by observing that $z^2 = \lambda$.

(ii) Theorem 4.2 (ii) and the relations in (7.22) and (7.23) show that the mapping $\mathrm{rbv}_\lambda^{\mathrm{SL}} : \mathfrak{N}_\lambda^{\mathrm{SL}} \to \mathbb{C}^2$ is bijective.

(iii) The first and the last assertions follow immediately from Theorem 4.2 (iii). For the second statement, note that there is (up to a scalar multiple) a unique solution ψ_{reg} such that $\psi_{\mathrm{reg}}|_{(a,x_0)} \in L^2(H|_{(a,x_0)})$. Hence, $p\psi'_{\mathrm{reg}}|_{(a,x_0)} = \psi_{\mathrm{reg},1}|_{(a,x_0)} \in L^2(\frac{1}{p}|_{(a,x_0)})$. Any other solution ψ is such that $\lim_{x\searrow} p(x)\psi'(x) \neq 0$ according to the already proved third statement of (iii). Since $\frac{1}{p}$ is not integrable at a by assumption, such a ψ satisfies $p\psi'|_{(a,x_0)} \notin L^2(\frac{1}{p}|_{(a,x_0)})$. Now the claim follows because $p\psi'|_{(a,x_0)} \in L^2(\frac{1}{p}|_{(a,x_0)})$ if and only if $\int_a^{x_0} p|\psi'|^2 < \infty$.

(iv) It follows from Theorem 4.2 (iv) that there exists a polynomial $\hat{p}$ of degree at most 2Δ with real coefficients and no constant term such that

$$\widehat{\mathrm{rbv}}_{z,2}\psi = \mathrm{rbv}_{z,2}\psi + \hat{p}(z)\mathrm{rbv}_{z,1}\psi \tag{7.24}$$

for all $\psi \in \mathfrak{N}_z$. If we choose ψ as in (7.20) for $\psi \in \mathfrak{N}_{z^2}^{\mathrm{SL}}$, then, by (7.22) and (7.23), we have

$$\begin{aligned}\widehat{\mathrm{rbv}}^{\mathrm{SL}}_{z^2,2}\psi &= \frac{1}{z}\widehat{\mathrm{rbv}}_{z,2}\psi = \frac{1}{z}\Big(\mathrm{rbv}_{z,2}\psi + \hat{p}(z)\mathrm{rbv}_{z,1}\psi\Big)\\ &= \mathrm{rbv}^{\mathrm{SL}}_{z^2,2}\psi + \frac{\hat{p}(z)}{z}\mathrm{rbv}^{\mathrm{SL}}_{z^2,1}\psi.\end{aligned} \tag{7.25}$$

Since this relation must be true for all $z \in \mathbb{C} \setminus \{0\}$ and all $\psi \in \mathfrak{N}_{z^2}^{\mathrm{SL}}$, it follows by replacing z by $-z$ that $\hat{p}$ is an odd polynomial. Hence, one can define a polynomial $p_{x_0,\hat{x}_0}$ by the relation $p_{x_0,\hat{x}_0}(z^2) = \frac{\hat{p}(z)}{z}$, which is a real polynomial of degree at most $\Delta - 1$. Now the assertion follows from (7.25). □

In the next theorem, we establish the existence of a singular Titchmarsh–Weyl coefficient, which is used in Theorem 7.11 to obtain a spectral measure. Recall from Definition 3.1 that $\mathcal{N}_\kappa$, $\kappa \in \mathbb{N}_0$, is the set of all generalized Nevanlinna functions with κ negative squares and that $\mathcal{N}_{<\infty} = \bigcup_{\kappa\in\mathbb{N}_0} \mathcal{N}_\kappa$. Further, denote by $\mathcal{N}_\kappa^{(\infty)}$, $\kappa \in \mathbb{N}_0$, the set of functions from $\mathcal{N}_\kappa$ whose only generalized pole of non-positive type is infinity and set $\mathcal{N}_{<\infty}^{(\infty)} = \bigcup_{\kappa\in\mathbb{N}_0} \mathcal{N}_\kappa^{(\infty)}$; see Definition 3.4. Note that a function q, defined on $\mathbb{C} \setminus \mathbb{R}$, belongs to $\mathcal{N}_\kappa^{(\infty)}$ if and only if

$$q(z) = p_{2\kappa}(z)q_0(z)$$

where $p_{2\kappa}$ is a monic real polynomial of degree 2κ and $q_0 \in \mathcal{N}_0$, i.e. $q_0(\overline{z}) = \overline{q_0(z)}$ and $\operatorname{Im} q_0(z) \geq 0$ for $z \in \mathbb{C}^+ = \{z \in \mathbb{C} : \operatorname{Im} z > 0\}$.

Theorem 7.7 (Singular Titchmarsh–Weyl coefficients) *Let* $(p; w) \in \mathbb{K}_{\mathrm{SL}}$ *with* $\mathrm{dom}(p; w) = (a, b)$ *be given. Then, for each fixed* $x_0 \in (a, b)$, *the following statements hold.*

(i) *For* $\lambda \in \mathbb{C}$ *let* $\theta(\cdot\,; \lambda)$ *and* $\varphi(\cdot\,; \lambda)$ *be the unique solutions of* (7.1) *such that*

$$\mathrm{rbv}_\lambda^{\mathrm{SL}}\theta(\cdot\,; \lambda) = \begin{pmatrix} 1 \\ 0 \end{pmatrix}, \qquad \mathrm{rbv}_\lambda^{\mathrm{SL}}\varphi(\cdot\,; \lambda) = \begin{pmatrix} 0 \\ 1 \end{pmatrix}. \tag{7.26}$$

Then, for each $x \in (a, b)$, *the functions* $\theta(x; \cdot)$ *and* $\varphi(x; \cdot)$ *are entire of order* $\frac{1}{2}$ *and finite (positive) type*

$$\int_a^x \sqrt{\frac{w(t)}{p(t)}}\,\mathrm{d}t.$$

Moreover, for each $\lambda \in \mathbb{C}$, *one has* $W_p\big(\varphi(\cdot\,; \lambda), \theta(\cdot\,; \lambda)\big) \equiv 1$ *where the weighted Wronskian* W_p *is as in* (7.18), *and the following relations hold:*

$$\begin{aligned} &\lim_{x \searrow a} \varphi(x; \lambda) = 1, && \lim_{x \searrow a} \frac{p(x)\varphi'(x; \lambda)}{\int_a^x w(t)\mathrm{d}t} = -\lambda, \\ &\lim_{x \searrow a} \frac{\theta(x; \lambda)}{w_1(x)} = -1, && \lim_{x \searrow a} p(x)\theta'(x; \lambda) = 1. \end{aligned} \tag{7.27}$$

(ii) *The limit*

$$m_{p,w}(\lambda) := \lim_{x \nearrow b} \frac{\theta(x; \lambda)}{\varphi(x; \lambda)}, \qquad \lambda \in \mathbb{C} \setminus [0, \infty), \tag{7.28}$$

exists locally uniformly on $\mathbb{C} \setminus [0, \infty)$ *and defines an analytic function in* λ. *The function* $m_{p,w}$ *belongs to the class* $\mathcal{N}_\kappa^{(\infty)}$ *with* $\kappa = \big\lfloor \frac{\Delta_{\mathrm{SL}}(p,w)}{2} \big\rfloor$.

(iii) *We have*

$$\theta(\cdot\,; \lambda) - m_{p,w}(\lambda)\varphi(\cdot\,; \lambda) \in L^2\big(w|_{(x_0,b)}\big), \qquad \lambda \in \mathbb{C} \setminus [0, \infty),$$

and this property characterizes the value $m_{p,w}(\lambda)$ *for each* $\lambda \in \mathbb{C} \setminus [0, \infty)$.

(iv) *For* $\lambda \in \mathbb{C}\setminus[0, \infty)$ *let* ψ *be any non-trivial solution of* (7.1) *such that* $\psi|_{(x_0,b)} \in L^2(w|_{(x_0,b)})$. *Then*

$$m_{p,w}(\lambda) = -\frac{\mathrm{rbv}_{\lambda,2}^{\mathrm{SL}}\psi}{\mathrm{rbv}_{\lambda,1}^{\mathrm{SL}}\psi}.$$

The function $m_{p,w}$ depends on the choice of x_0. This dependence is controlled as follows:

(iv) *Let $\hat{x}_0 \in (a, b)$, and let $\widehat{m}_{p,w}$ be the correspondingly defined singular Titchmarsh–Weyl coefficient. Then there exists a polynomial $p_{x_0,\hat{x}_0}$ with real coefficients whose degree does not exceed $\Delta_{\mathrm{SL}}(p, w) - 1$ such that*

$$\widehat{m}_{p,w}(\lambda) = m_{p,w}(\lambda) - p_{x_0,\hat{x}_0}(\lambda). \tag{7.29}$$

Before we prove Theorem 7.7, let us introduce some notation.

Definition 7.8 We refer to each function $m_{p,w}$ constructed as in Theorem 7.7 as *a singular Titchmarsh–Weyl coefficient* associated with the Sturm–Liouville equation (7.1) when $(p; w) \in \mathbb{K}_{\mathrm{SL}}$. We denote by $[m]_{p,w}$ the equivalence class of $\mathcal{N}_{<\infty}^{(\infty)}$-functions modulo the relation

$$m_1 \mathrel{\hat{\sim}} m_2 \quad :\Longleftrightarrow \quad m_1 - m_2 \in \mathbb{R}[z] \tag{7.30}$$

which contains some (and hence any) function $m_{p,w}$ in Theorem 7.7; we call $[m]_{p,w}$ *the* singular Titchmarsh–Weyl coefficient.

Remark 7.9

(ii) If $\Delta_{\mathrm{SL}}(p, w) = 1$, then $m_{p,w} \in \mathcal{N}_0$ by Theorem 7.7 (ii), which is in accordance with the classical theory since (7.1) is in the limit circle case at a; see, e.g. [9, Corollary 8.1] and [66, Corollary A.9].

(ii) According to Theorem 7.4 (iii), $\varphi(\cdot\,; \lambda)$ is the—up to a multiplicative scalar—unique solution of (7.1) that satisfies $\int_a^{x_0} p(x)|\varphi'(x; \lambda)|^2 \mathrm{d}x < \infty$ for some x_0. If (7.1) is in the limit circle case at a (i.e. if $\Delta_{\mathrm{SL}}(p, w) = 1$), then $\varphi(\cdot\,; \lambda)$ is the only solution of (7.1) that satisfies the boundary condition (7.13); if (7.1) is in the limit point case, then $\varphi(\cdot\,; \lambda)$ is the only solution in $L^2(w|_{(a,x_0)})$.

For the proof of Theorem 7.7 let $\theta(\cdot\,; \lambda)$ and $\varphi(\cdot\,; \lambda)$ be as in the statement of the theorem, i.e. the unique solutions of (7.1) that satisfy (7.26). Let $z \in \mathbb{C} \setminus \{0\}$ be such that $\lambda = z^2$ and let H be as in (7.7). Then the functions $\theta(\cdot\,; z), \varphi(\cdot\,; z) \in \mathfrak{N}_z$ that satisfy (4.3) have the following form:

$$\theta(x; z) = \begin{pmatrix} p(x)\theta'(x; z^2) \\ z\theta(x; z^2) \end{pmatrix}, \qquad \varphi(x; z) = \begin{pmatrix} \frac{1}{z} p(x)\varphi'(x; z^2) \\ \varphi(x; z^2) \end{pmatrix}, \tag{7.31}$$

where $'$ denotes the derivative with respect to x; cf. (7.22) and (7.23). It follows from Lemma 7.6 that

$$\theta(x;-z) = \begin{pmatrix} p(x)\theta'(x;z^2) \\ -z\theta(x;z^2) \end{pmatrix}, \qquad \varphi(x;-z) = \begin{pmatrix} -\frac{1}{z}p(x)\varphi'(x;z^2) \\ \varphi(x;z^2) \end{pmatrix}. \tag{7.32}$$

Moreover, let q_H be a singular Weyl coefficient of H as in (4.6).

Proof (Proof of Theorem 7.7) Let θ and φ be as in (7.31).

Item (i) follows directly from Theorem 4.5 (i). Note that $a_+ = a$ in Theorem 4.5 since $w(x) > 0$, $x \in (a, b)$ a.e.

(ii) It follows from Theorem 4.5 (ii) that, for $z \in \mathbb{C} \setminus \mathbb{R}$,

$$\frac{\theta(x;z^2)}{\varphi(x;z^2)} = \frac{\theta_2(x;z)}{z\varphi_2(x;z)} \to \frac{1}{z}q_H(z) \qquad \text{as } x \nearrow b.$$

The function q_H is an odd function as the following calculation shows (where we use (7.31) and (7.32)):

$$q_H(-z) = \lim_{x\nearrow b} \frac{\theta_2(x;-z)}{\varphi_2(x;-z)} = \lim_{x\nearrow b} \frac{-\theta_2(x;z)}{\varphi_2(x;z)} = -q_H(z). \tag{7.33}$$

Hence,

$$\lim_{x\nearrow b} \frac{\theta(x;\lambda)}{\varphi(x;\lambda)} = \lim_{x\nearrow b} \frac{\theta(x;z)}{z\varphi(x;z)} = \frac{1}{z}q_H(z) \tag{7.34}$$

is independent of the choice of z such that $z^2 = \lambda$, and therefore $m_{p,w}$ is well defined as a function of λ and analytic on $\mathbb{C} \setminus [0, \infty)$. It follows from [84, Theorem 4.4] that $m_{p,w} \in \mathcal{N}_{<\infty}^{(\infty)}$. Set $m_+(\lambda){:=}\lambda m_{p,w}(\lambda)$; then $m_+ \in \mathcal{N}_{<\infty}^{(\infty)}$ again by [84, Theorem 4.4]. Now [55, Proposition 4.8] implies that $\operatorname{ind}_- m_{p,w} + \operatorname{ind}_- m_+ = \operatorname{ind}_- q_H$. Moreover, $m_{p,w}$ and m_+ have infinity as only generalized pole of non-positive type and therefore, by the definition of m_+ and by (3.4), we have $\operatorname{ind}_- m_+ - 1 \le \operatorname{ind}_- m_{p,w} \le \operatorname{ind}_- m_+$. This, together with the fact that $\operatorname{ind}_- q_H = \Delta(H) = \Delta_{\mathrm{SL}}(p, w)$ by Theorem 4.5 (ii) and (7.15), yields $\operatorname{ind}_- m_{p,w} = \left\lfloor \frac{\Delta_{\mathrm{SL}}(p,w)}{2} \right\rfloor$.

(iii) We can write

$$\theta(\cdot\,;z^2) - m_{p,w}(z^2)\varphi(\cdot\,;z^2) = \frac{1}{z}\Big(\theta_2(\cdot\,;z) - q_H(z)\varphi_2(\cdot\,;z)\Big).$$

By Theorem 4.5 (iii) the right-hand side of this equality is in $L^2(w|_{(x_0,b)})$. Since (up to a scalar multiple) only one solution is in $L^2(w)$ at b (because of the limit point assumption at b), the value of $m_{p,w}(z^2)$ is uniquely determined by the L^2 property.

(iv) The formula follows from item (iii) since any such ψ is a multiple of $\psi_0 := \theta(\cdot\,;\lambda) - m_{p,w}(\lambda)\varphi(\cdot\,;\lambda)$ and $\mathrm{rbv}^{\mathrm{SL}}_{\lambda,1}\psi_0 = 1$, $\mathrm{rbv}^{\mathrm{SL}}_{\lambda,2}\psi_0 = -m_{p,w}(\lambda)$.

(v) Let $\psi \in \mathfrak{N}^{\mathrm{SL}}_{\lambda}$ be such that $\psi|_{(x_0,b)} \in L^2(w|_{(x_0,b)})$. It follows from item (iv) and Theorem 7.4 (iv) with the notation used there that

$$\widehat{m}_{p,w}(\lambda) = -\frac{\widehat{\mathrm{rbv}}^{\mathrm{SL}}_{\lambda,2}\psi}{\widehat{\mathrm{rbv}}^{\mathrm{SL}}_{\lambda,1}\psi} = -\frac{\mathrm{rbv}^{\mathrm{SL}}_{\lambda,2}\psi + p_{x_0,\hat{x}_0}(\lambda)\mathrm{rbv}^{\mathrm{SL}}_{\lambda,1}\psi}{\mathrm{rbv}^{\mathrm{SL}}_{\lambda,1}\psi} = m_{p,w}(\lambda) - p_{x_0,\hat{x}_0}(\lambda).$$

Since $p_{x_0,\hat{x}_0}$ is the polynomial from Theorem 7.4 (iv), it has the properties stated there.

□

From (7.34) we obtain the following relation between $m_{p,w}$ and q_H if the same base point x_0 is chosen:

$$m_{p,w}(z^2) = \frac{1}{z} q_H(z), \qquad z \in \mathbb{C} \setminus \mathbb{R}. \tag{7.35}$$

Next, we construct a measure with the help of the Stieltjes inversion formula and the singular Titchmarsh–Weyl coefficient. Before we formulate the theorem, we introduce the following class of measures.

Definition 7.10 Let ν be a Borel measure on $\mathbb{R}$. We say that $\nu \in \mathbb{M}^-$ if there exists an $n \in \mathbb{N}_0$ such that

$$\nu\big((-\infty, 0)\big) = 0 \qquad \text{and} \qquad \int_{[0,\infty)} \frac{d\nu(t)}{(1+t)^{n+1}} < \infty. \tag{7.36}$$

If $\nu \in \mathbb{M}^-$, we denote by $\Delta^-(\nu)$ the minimal $n \in \mathbb{N}_0$ such that (7.36) holds.

The reason for the use of the minus sign in the notation will become clearer in the next section; see, in particular, item 3 in 8.2.

In the next theorem, a measure is constructed, which will turn out to be a spectral measure for the Sturm–Liouville equation (7.1).

Theorem 7.11 (The spectral measure) *Let $(p; w) \in \mathbb{K}_{\mathrm{SL}}$ with $\mathrm{dom}(p; w) = (a, b)$ be given. Then there exists a unique Borel measure $\mu_{p,w}$ that satisfies*

$$\mu_{p,w}\big([s_1, s_2]\big) = \frac{1}{\pi} \lim_{\varepsilon \searrow 0} \lim_{\delta \searrow 0} \int_{s_1-\varepsilon}^{s_2+\varepsilon} \operatorname{Im} m_{p,w}(t + i\delta)\, dt, \quad -\infty < s_1 < s_2 < \infty, \tag{7.37}$$

where $m_{p,w} \in [m]_{p,w}$ is any singular Titchmarsh–Weyl coefficient associated with (7.1). We have $\mu_{p,w} \in \mathbb{M}^-$ and $\Delta^-(\mu_{p,w}) = \Delta_{\mathrm{SL}}(p, w)$.

Moreover, $\mu_{p,w}(\{0\}) > 0$ if and only if

$$\int_a^b w(x)\mathrm{d}x < \infty. \tag{7.38}$$

If (7.38) *is satisfied, then*

$$\mu_{p,w}(\{0\}) = \left[\int_a^b w(x)\mathrm{d}x\right]^{-1}. \tag{7.39}$$

We refer to the measure $\mu_{p,w}$ given by (7.37) as the *spectral measure* associated with the Sturm–Liouville equation (7.1). This is justified by Theorem 7.13.

Remark 7.12

(i) The number $\Delta^-(\mu_{p,w})$, which describes the behaviour of the spectral measure $\mu_{p,w}$ at infinity, gives a finer measure of the growth of the coefficients p and w at the endpoint a than $\mathrm{ind}_- m_{p,w}$, the negative index of the singular Titchmarsh–Weyl coefficient.

(ii) If $\Delta_{\mathrm{SL}}(p, w) = 1$, then $\int_0^\infty \frac{\mathrm{d}\mu_{p,w}(t)}{(1+t)^2} < \infty$, which is the classical growth condition for a spectral measure corresponding to a Nevanlinna function. This is in accordance with the fact that in this case (7.1) is in the limit circle case and therefore classical Hilbert space theory is sufficient to obtain a scalar spectral measure; see, e.g. [9, Theorem 11.1].

(iii) If one uses the more general functions $\breve{w}_l$ instead of w_l as in Remark 7.5 (iii), then—similarly to (7.29)—the singular Titchmarsh–Weyl coefficient changes only by the real polynomial p from (7.19), which is of degree at most $\Delta - 1$, and hence the spectral measure is unchanged.

Proof (Proof of Theorem 7.11) Let us first note that (3.6), (4.11) and (7.37) imply that

$$\mu_H = \mu_{q_H} \qquad \text{and} \qquad \mu_{p,w} = \mu_{m_{p,w}}$$

with the notation from Sect. 3.5. Define $\tau(s) := s^2$, $s \in \mathbb{R}$, and denote by μ^τ the corresponding push-forward measure for a Borel measure μ, i.e. $\mu^\tau(B) = \mu(\tau^{-1}(B))$ for Borel sets $B \subseteq \mathbb{R}$. It follows from [84, Theorem 4.4] and (7.35) that

$$\mu_{p,w} \ll \mu_H^\tau \qquad \text{and} \qquad \frac{\mathrm{d}\mu_{p,w}}{\mathrm{d}\mu_H^\tau}(t) = \mathbb{1}_{[0,\infty)}(t), \tag{7.40}$$

where $\mathbb{1}_{[0,\infty)}$ denotes the characteristic function of the interval $[0, \infty)$.

For $m \in \mathbb{N}_0$, we obtain from (7.40) that

$$\int_{\mathbb{R}} \frac{\mathrm{d}\mu_H(t)}{(1+t^2)^{m+1}} = \mu_H(\{0\}) + \int_{(0,\infty)} \frac{\mathrm{d}\mu_H^\tau(s)}{(1+s)^{m+1}}$$

$$= \mu_H(\{0\}) + \int_{(0,\infty)} \frac{\mathrm{d}\mu_{p,w}(s)}{(1+s)^{m+1}},$$

which shows that $\mu_{p,w} \in \mathbb{M}^-$ and $\Delta^-(\mu_{p,w}) = \Delta(\mu_H) = \Delta(H) = \Delta_{\mathrm{SL}}(p, w)$, where the last equality follows from (7.15). The fact that a real polynomial makes no contribution in the Stieltjes inversion formula implies that the measure $\mu_{p,w}$ does not depend on the choice of the representative from $[m]_{p,w}$.

Since $\mu_{p,w}(\{0\}) = \mu_H(\{0\})$ by (7.40), the equivalence of $\mu_{p,w}(\{0\}) > 0$ and (7.38) and the relation (7.39) follow from Proposition 5.3. □

In the following theorem, a Fourier transform is constructed, which yields the unitary equivalence of the operator $A_{p,w}$ and the multiplication operator in the space $L^2(\mu_{p,w})$. In particular, this implies that the spectrum of $A_{p,w}$ is simple. Note that the function φ that appears in the Fourier transform is the (up to a scalar multiple) unique solution of (7.1) that is 'regular' at a; cf. Remark 7.9 (ii).

Theorem 7.13 (The Fourier transform) *Let* $(p; w) \in \mathbb{K}_{\mathrm{SL}}$ *with* $\mathrm{dom}(p; w) = (a, b)$ *be given, and let* $\mu_{p,w}$ *be the spectral measure associated with* (7.1) *via* (7.37)*. Then the following statements hold.*

(i) *The map defined by*

$$(\Theta_{p,w} f)(t) := \int_a^b \varphi(x;t) f(x) w(x)\,\mathrm{d}x, \qquad t \in \mathbb{R},$$
$$f \in L^2(w), \quad \sup(\operatorname{supp} f) < b, \tag{7.41}$$

extends to an isometric isomorphism from $L^2(w)$ *onto* $L^2(\mu_{p,w})$.

(ii) *The operator* $\Theta_{p,w}$ *establishes a unitary equivalence between* $A_{p,w}$ *and the operator* $M_{\mu_{p,w}}$ *of multiplication by the independent variable in* $L^2(\mu_{p,w})$*, i.e. we have*

$$\Theta_{p,w} A_{p,w} = M_{\mu_{p,w}} \Theta_{p,w}.$$

(iii) *For compactly supported functions, the inverse of* $\Theta_{p,w}$ *acts as an integral transformation, namely,*

$$(\Theta_{p,w}^{-1} g)(x) = \int_0^\infty \varphi(x;t) g(t)\,\mathrm{d}\mu_{p,w}(t), \qquad x \in (a, b),$$
$$g \in L^2(\mu_{p,w}), \quad \operatorname{supp} g \textit{ compact}. \tag{7.42}$$

First, we need a lemma.

Lemma 7.14 *Set*

$$L^2_{\mathrm{odd}}(\mu_H) := u\big\{g \in L^2(\mu_H) : g \textit{ is odd}\big\},$$

$$L^2_{\text{even}}(\mu_H):=u\{g \in L^2(\mu_H) : g \textit{ is even}\}.$$

The Fourier transform Θ_H *from Theorem 5.1 maps*

$$\left\{\begin{pmatrix} f_1 \\ f_2 \end{pmatrix} \in L^2(H) : f_2 \equiv 0\right\} \quad \textit{bijectively onto } L^2_{\text{odd}}(\mu_H)$$

and

$$\left\{\begin{pmatrix} f_1 \\ f_2 \end{pmatrix} \in L^2(H) : f_1 \equiv 0\right\} \quad \textit{bijectively onto } L^2_{\text{even}}(\mu_H).$$

Proof Since $\varphi_1(x; -z) = -\varphi_1(x; z)$ and $\varphi_2(x; -z) = \varphi_2(x; z)$ by (7.32), we have, for $f_1 \in L^2(\frac{1}{p})$ with $\sup(\operatorname{supp} f_1) < b$, that

$$\left[\Theta_H \begin{pmatrix} f_1 \\ 0 \end{pmatrix}\right](-t) = \int_a^b \varphi(x; -t)^* H(x) \begin{pmatrix} f_1(x) \\ 0 \end{pmatrix} \mathrm{d}x = \int_a^b \varphi_1(x; -t) \frac{1}{p(x)} f_1(x) \mathrm{d}x$$

$$= -\int_a^b \varphi_1(x; t) \frac{1}{p(x)} f_1(x) \mathrm{d}x = -\left[\Theta_H \begin{pmatrix} f_1 \\ 0 \end{pmatrix}\right](t),$$

and similarly,

$$\left[\Theta_H \begin{pmatrix} 0 \\ f_2 \end{pmatrix}\right](-t) = \left[\Theta_H \begin{pmatrix} 0 \\ f_2 \end{pmatrix}\right](t)$$

for $f_2 \in L^2(w)$. Now the result follows from the bijectivity of Θ_H. □

Proof (Proof of Theorem 7.13)

(i) The operator

$$U : \begin{cases} L^2_{\text{even}}(\mu_H) \to L^2(\mu_{p,w}) \\ f \mapsto g \quad \text{with } g(s) = f\left(\sqrt{s}\right),\ s \in [0, \infty), \end{cases}$$

is well defined since $\mu_{p,w}((-\infty, 0)) = 0$. It is isometric because (7.40) implies that, for $f \in L^2_{\text{even}}(\mu_H)$,

$$\|Uf\|^2_{L^2(\mu_{p,w})} = \int_{[0,\infty)} \left|f\left(\sqrt{s}\right)\right|^2 \mathrm{d}\mu_{p,w}(s) = \int_{[0,\infty)} \left|f\left(\sqrt{s}\right)\right|^2 \mathrm{d}\mu_H^\tau(s)$$

$$= \int_{\mathbb{R}} |f(t)|^2 \mathrm{d}\mu_H(t) = \|f\|^2_{L^2(\mu_H)},$$

where μ_H^τ is as at the beginning of the proof of Theorem 7.11. Moreover, U is surjective and the inverse is given by $(U^{-1}g)(t) = g(t^2)$.

Let ι_2 and P_2 be as (7.10). By Lemma 7.14 the operator $U\Theta_H\iota_2$ is well defined and isometric from $L^2(w)$ onto $L^2(\mu_{p,w})$. Moreover,

$$\begin{aligned}\big(U\Theta_H\iota_2 f\big)(t) &= U\bigg[\int_a^b \varphi_2(x;\,\cdot)w(x)f(x)\mathrm{d}x\bigg](t)\\ &= \int_a^b \varphi_2\big(x;\sqrt{t}\,\big)w(x)f(x)\mathrm{d}x = \big(\Theta_{p,w}f\big)(t)\end{aligned}$$

by (7.31), which shows that

$$\Theta_{p,w} = U\Theta_H\iota_2. \tag{7.43}$$

(ii) Let M_{μ_H} be the multiplication operator by the independent variable in $L^2(\mu_H)$ as in Theorem 5.1 (ii). With Theorem 5.1 (ii), the definition of $A_{p,w}$ and (7.43) we obtain

$$\begin{aligned}\Theta_{p,w}A_{p,w} &= U\Theta_H\iota_2P_2\big(T(H)\big)^2\iota_2 = U\Theta_H\big(T(H)\big)^2\iota_2\\ &= UM_{\mu_H}^2\Theta_H\iota_2 = UM_{\mu_H}^2U^{-1}\Theta_{p,w} = M_{\mu_{p,w}}\Theta_{p,w}.\end{aligned} \tag{7.44}$$

(iii) It follows from (7.43) that $\Theta_{p,w}^{-1} = P_2\Theta_H^{-1}U^{-1}$. If $g \in L^2(\mu_{p,w})$ with compact support, then, by (7.31) and (7.40), we have

$$\begin{aligned}(\Theta_{p,w}^{-1}g)(x) &= P_2\int_{\mathbb{R}} g(t^2)\varphi(x;t)\mathrm{d}\mu_H(t) = \int_{\mathbb{R}} g(t^2)\varphi(x;t^2)\mathrm{d}\mu_H(t)\\ &= \int_{[0,\infty)} g(s)\varphi(x;s)\mathrm{d}\mu_H^\tau(s) = \int_{[0,\infty)} g(s)\varphi(x;s)\mathrm{d}\mu_{p,w}(s),\end{aligned}$$

which shows the desired representation for $\Theta_{p,w}^{-1}$.

□

From Theorem 6.2, we obtain the following uniqueness result, which says that equality of singular Titchmarsh–Weyl coefficients or spectral measures implies equality of the coefficients up to a reparameterization of the independent variable. We cannot prove an existence result since we cannot characterize those spectral measures that lead to diagonal Hamiltonians with non-vanishing determinant. If we considered strings and the corresponding Krein–Feller operators, we would also obtain an existence result: namely, every measure from the class $\mathbb{M}^-$ is the spectral measure of a certain string with two singular endpoints.

Theorem 7.15 (Global Uniqueness Theorem) *Let $(p_1; w_1), (p_2; w_2) \in \mathbb{K}_{SL}$ be given with* $\mathrm{dom}(p_i; w_i) = (a_i, b_i)$, *$i = 1, 2$. Then the following statements are equivalent:*

(i) *There exists an increasing bijection $\gamma : (a_2, b_2) \to (a_1, b_1)$ such that γ and γ^{-1} are locally absolutely continuous and*

$$p_2(x) = \frac{1}{\gamma'(x)} p_1\big(\gamma(x)\big), \qquad w_2(x) = \gamma'(x) w_1\big(\gamma(x)\big) \tag{7.45}$$

for $x \in (a_2, b_2)$ a.e.;

(ii) $[m]_{p_1,w_1} = [m]_{p_2,w_2}$;

(iii) $\mu_{p_1,w_1} = \mu_{p_2,w_2}$.

Proof The theorem follows from Theorem 6.2 and Proposition 4.10 if we recall (2.1) and observe that

$$q_{H_1}(z) - q_{H_2}(z) = z\big(m_{p_1,w_1}(z^2) - m_{p_2,w_2}(z^2)\big),$$

which vanishes at 0 and that hence $\alpha = 0$ in Theorem 6.2. □

Remark 7.16 If one considers Sturm–Liouville equations in impedance form, i.e. when $p = w$, and assumes that the left endpoints coincide, then equality of spectral measures is equivalent to the equality of coefficients a.e. because in this case, one has $\gamma' = 1$ if (7.45) is satisfied. We refer to [2] and the references therein for other types of inverse spectral theorems for Sturm–Liouville equations in impedance form.

We also obtain a local version of the uniqueness result, which is an extension of [81, Theorem 1.5] to the case of two singular endpoints. See also [78, §4.4] for a result on strings with regular left endpoint. The next theorem follows immediately from Theorem 6.3.

Theorem 7.17 (Local Inverse Spectral Theorem) *Let $(p_1; w_1), (p_2; w_2) \in \mathbb{K}_{SL}$ be given with* $\mathrm{dom}(p_i; w_i) = (a_i, b_i)$, *$i = 1, 2$, and let $\tau > 0$. Moreover, for $i = 1, 2$, let $s_i(\tau)$ be the unique value s_i such that*

$$\int_{a_i}^{s_i} \sqrt{\frac{w_i(\xi)}{p_i(\xi)}}\, \mathrm{d}\xi = \tau$$

if $\int_{a_i}^{b_i} \sqrt{\frac{w_i(\xi)}{p_i(\xi)}}\, \mathrm{d}\xi > \tau$ and set $s_i(\tau) := b_i$ otherwise. Then the following statements are equivalent.

(i) *There exists an increasing bijection $\gamma : (a_2, s_2(\tau)) \to (a_1, s_1(\tau))$ such that γ and γ^{-1} are locally absolutely continuous and (7.45) holds for $x \in (a_2, s_2(\tau))$ a.e.*

(ii) *There exist singular Titchmarsh–Weyl coefficients* m_{p_1,w_1} *and* m_{p_2,w_2}, *and there exists a* $\beta \in (0, 2\pi)$ *such that, for each* $\varepsilon > 0$,

$$m_{p_1,w_1}\big(re^{i\beta}\big) - m_{p_2,w_2}\big(re^{i\beta}\big) = \mathrm{O}\Big(e^{(-2\tau+\varepsilon)\sqrt{r}\sin\frac{\beta}{2}}\Big), \qquad r \to \infty.$$

(iii) *There exist singular Titchmarsh–Weyl coefficients* m_{p_1,w_1} *and* m_{p_2,w_2}, *and there exists a* $k \geq 0$ *such that, for each* $\delta \in (0, \pi)$,

$$m_{p_1,w_1}(\lambda) - m_{p_2,w_2}(\lambda) = \mathrm{O}\Big(|\lambda|^k e^{-2\tau \operatorname{Im}\sqrt{\lambda}}\Big),$$

$$|\lambda| \to \infty, \quad \lambda \in \big\{z \in \mathbb{C} : \delta \leq \arg z \leq 2\pi - \delta\big\},$$

where $\sqrt{\lambda}$ *is chosen so that* $\operatorname{Im}\sqrt{\lambda} > 0$.

In the next proposition, we provide a sufficient condition for $(p; w) \in \mathbb{K}_{\mathrm{SL}}$. This result is also used in Sect. 9. Let us recall the following notation: we write $f(x) \asymp g(x)$ as $x \searrow 0$ if there exist $c, C > 0$ and $x_0 > 0$ such that $cg(x) \leq f(x) \leq Cg(x)$ for all $x \in (0, x_0)$.

Proposition 7.18 *Let* $\alpha \geq 1$ *and assume that* p *and* w, *defined on* $(0, b)$ *with* $b > 0$ *or* $b = \infty$ *satisfy* (7.2) *with* $a = 0$ *and*

$$p(x) \asymp x^{\alpha}, \quad w(x) \asymp x^{\alpha} \qquad \text{as } x \searrow 0. \tag{7.46}$$

If (7.1) *is in the limit point case at* b, *then* $(p; w) \in \mathbb{K}_{\mathrm{SL}}$ *and* $\Delta_{\mathrm{SL}}(p, w) = \big\lfloor \frac{\alpha+1}{2} \big\rfloor$.

Before we prove Proposition 7.18, we need a lemma.

Lemma 7.19 *Let* $\alpha > 1$ *and assume that* p *and* w, *defined on* $(0, b)$ *with* $b > 0$ *or* $b = \infty$, *satisfy* (7.46). *Moreover, choose* $x_0 \in (0, b)$. *Then*

$$w_l(x) \asymp \begin{cases} x^l & \text{if } l \text{ is even and } l < \alpha + 1, \\ x^{-\alpha+l} & \text{if } l \text{ is odd and } l < \alpha, \end{cases} \tag{7.47}$$

as $x \searrow 0$.

Proof We prove the lemma by induction. For $l = 0$, the statement is clear from the definition of w_0. Now assume that (7.47) is true for $l \in \mathbb{N}_0$. If l is even and $l + 1 < \alpha$, then

$$w_{l+1}(x) \asymp \int_x^{x_0} t^{-\alpha} t^l \mathrm{d}t = \frac{1}{\alpha - l - 1}\big(x^{-\alpha+l+1} - x_0^{-\alpha+l+1}\big) \asymp x^{-\alpha+l+1}.$$

If l is odd and $l + 1 < \alpha + 1$, then

$$w_{l+1}(x) \asymp \int_0^x t^{\alpha} t^{-\alpha+l} \mathrm{d}t = \frac{1}{l+1} x^{l+1} \asymp x^{l+1}.$$

In both cases, it follows that (7.47) is true for $l+1$ instead of l. Hence, the statement follows by induction. □

Proof (Proof of Proposition 7.18) The conditions (i) and (ii) in Definition 7.1 are easy to check. For (iii) let us first consider the case $\alpha = 1$. Then $w_1(x) \asymp -\ln x$ and hence $w_1 \in L^2(w|_{(0,x_0)})$, which shows that $\Delta_{\mathrm{SL}}(p, w) = 1 = \lfloor \frac{1+1}{2} \rfloor$.

Now let $\alpha > 1$. If l is even and $l < \alpha + 1$, then

$$w_l \in L^2\Big(\frac{1}{p}\Big|_{(0,x_0)}\Big) \quad \Longleftrightarrow \quad \int_0^{x_0} x^{-\alpha} x^{2l} \mathrm{d}x < \infty \quad \Longleftrightarrow \quad l > \frac{\alpha - 1}{2}. \tag{7.48}$$

If l is odd and $l < \alpha$, then

$$w_l \in L^2(w|_{(0,x_0)}) \quad \Longleftrightarrow \quad \int_0^{x_0} x^{\alpha} x^{2(-\alpha+l)} \mathrm{d}x < \infty \quad \Longleftrightarrow \quad l > \frac{\alpha - 1}{2}. \tag{7.49}$$

The minimal integer l that satisfies $l > \frac{\alpha-1}{2}$ is $\lfloor \frac{\alpha+1}{2} \rfloor$. Since $\lfloor \frac{\alpha+1}{2} \rfloor < \alpha$ for $\alpha > 1$, the asymptotic relations (7.47) are valid for w_l, $l \leq \lfloor \frac{\alpha+1}{2} \rfloor$. □

Example 7.20 Equations of the form

$$-a_2 y'' - a_1 y' = \lambda y,$$

where a_1 and a_2 are continuous functions on (a, b) and $a_2(x) > 0$ for $x \in (a, b)$, can be written in the form (7.1) with

$$p(x) = \exp\left(\int_{x_0}^x \frac{a_1(t)}{a_2(t)} \mathrm{d}t\right), \qquad w(x) = \frac{1}{a_2(x)} \exp\left(\int_{x_0}^x \frac{a_1(t)}{a_2(t)} \mathrm{d}t\right)$$

with some $x_0 \in [a, b]$. As an example, we consider the associated Laguerre equation

$$-xy''(x) - (1 + \alpha - x)y'(x) = \lambda y(x), \qquad x \in (0, \infty),$$

with $\alpha \geq 0$. For p and w, one obtains

$$p(x) = x^{\alpha+1} e^{-x}, \qquad w(x) = x^{\alpha} e^{-x}.$$

It can be shown in a similar way as in Proposition 7.18 that $(p; w) \in \mathbb{K}_{\mathrm{SL}}$ with $\Delta = \lfloor \alpha + 1 \rfloor$. Hence, the singular Titchmarsh–Weyl coefficient belongs to $\mathcal{N}_\kappa^{(\infty)}$

with $\kappa = \lfloor \frac{\alpha+1}{2} \rfloor$. This is in agreement with [29] where a model for this singular Titchmarsh–Weyl coefficient was constructed. For $\alpha < -1$, the associated Laguerre equation was studied with the help of Pontryagin spaces in [20, 69, 70]; in this case, the results of the next subsection can be applied.

8 Sturm–Liouville Equations Without Potential: Singular *w*

In this section, we consider the case when w is not integrable at a but $\frac{1}{p}$ is. In Definition 7.1 and most theorems, one just has to swap the roles of $\frac{1}{p}$ and w. Let us state the definition of the class of coefficients explicitly.

Definition 8.1 We say that $(p; w) \in \mathbb{K}_{\mathrm{SL}}^{+}$ if p and w are defined on some interval (a, b) and they satisfy (7.2) and the following conditions.

(i) For one (and hence for all) $x_0 \in (a, b)$,

$$\int_a^{x_0} w(x)\mathrm{d}x = \infty \qquad \text{and} \qquad \int_a^{x_0} \frac{1}{p(x)}\mathrm{d}x < \infty.$$

(ii) For one (and hence for all) $x_0 \in (a, b)$,

$$\int_a^{x_0}\int_a^{x} \frac{1}{p(t)}\mathrm{d}t\, w(x)\mathrm{d}x < \infty.$$

(iii) Let $x_0 \in (a, b)$ and define functions w_l, $l = 0, 1, \dots$, recursively by

$$w_0(x) = 1,$$

$$w_l(x) = \begin{cases} \displaystyle\int_x^{x_0} w(t)w_{l-1}(t)\mathrm{d}t & \text{if } l \text{ is odd,} \\ \displaystyle\int_a^{x} \frac{1}{p(t)} w_{l-1}(t)\mathrm{d}t & \text{if } l \text{ is even.} \end{cases}$$

There exists an $n \in \mathbb{N}_0$ such that

$$w_n\big|_{(a,x_0)} \in \begin{cases} L^2\big(w\big|_{(a,x_0)}\big) & \text{if } n \text{ is even,} \\ L^2\big(\frac{1}{p}\big|_{(a,x_0)}\big) & \text{if } n \text{ is odd.} \end{cases} \tag{8.1}$$

(iv) Equation (7.1) is in the limit point case at b, i.e. for $\lambda \in \mathbb{C} \setminus \mathbb{R}$, equation (7.1) has (up to a scalar multiple) only one solution in $L^2(w|_{(x_0,b)})$ for $x_0 \in (a, b)$.

If $(p; w) \in \mathbb{K}_{\mathrm{SL}}^+$, we denote by $\Delta_{\mathrm{SL}}^+(p, w)$ the minimal $n \in \mathbb{N}_0$ such that (8.1) holds.

8.2 Differences between the classes $\mathbb{K}_{\mathrm{SL}}$ and $\mathbb{K}_{\mathrm{SL}}^+$ In the following list, we mention the major differences that occur in theorems and other important statements corresponding to coefficients in $\mathbb{K}_{\mathrm{SL}}$ and $\mathbb{K}_{\mathrm{SL}}^+$, respectively.

1. The analogue of Remark 7.2 (ii) is not true; the equation (7.1) is always in the limit point case at a if $\Delta_{\mathrm{SL}}^+(p, w) \geq 1$. This follows from the fact that $1 \notin L^2(w|_{(a,x_0)})$.
2. The regularized boundary values have the following form (with $\Delta = \Delta_{\mathrm{SL}}^+(p, w)$):

$$\mathrm{rbv}_{\lambda,1}^{\mathrm{SL},+}\psi = \lim_{x \searrow a} \psi(x),$$

$$\mathrm{rbv}_{\lambda,2}^{\mathrm{SL},+}\psi = \lim_{x \searrow a} \Bigg[p(x)\psi'(x) + \sum_{k=1}^{\lfloor \frac{\Delta}{2} \rfloor} \lambda^k \Big(w_{2k}(x)p(x)\psi'(x) - w_{2k-1}(x)\psi(x) \Big)$$

$$- \begin{Bmatrix} \lambda^{\frac{\Delta+1}{2}} w_\Delta(x)\psi(x) & \text{if } \Delta \text{ is odd} \\ 0 & \text{if } \Delta \text{ is even} \end{Bmatrix}$$

$$+ \Big(\lim_{t \searrow a} \psi(t)\Big) \Bigg(\sum_{k=\lfloor \frac{\Delta+3}{2} \rfloor}^{\Delta} \sum_{l=0}^{2k-\Delta-2} (-1)^{l+1} \lambda^k w_l(x) w_{2k-l-1}(x) \Bigg) \Bigg].$$

3. The singular Titchmarsh–Weyl coefficient $m_{p,w}^+$, which is defined as in (7.28), is connected with the singular Weyl coefficient of the corresponding canonical system via

$$m_{p,w}^+(z^2) = z q_H(z). \tag{8.2}$$

This relation explains the use of the notation with + as this was used, e.g. in [55] and [84]. The singular Titchmarsh–Weyl coefficient $m_{p,w}^+$ belongs to $\mathcal{N}_\kappa^{(\infty)}$ where $\kappa = \lfloor \frac{\Delta_{\mathrm{SL}}^+(p,w)+1}{2} \rfloor$. The equivalence classes $[m]_{p,w}^+$ are defined not with respect to the equivalence relation $\hat{\sim}$ defined in (7.30) but with the equivalence relation $\sim$ defined in (4.10), which is

$$m_1 \sim m_2 \quad :\Longleftrightarrow \quad m_1 - m_2 \in \mathbb{R}[z], \ (m_1 - m_2)(0) = 0.$$

4. The spectral measure $\mu_{p,w}^+$ belongs to the class $\mathbb{M}^+$, which is the set of Borel measures on $\mathbb{R}$ such that

$$\nu\big((-\infty,0]\big)=0 \qquad \text{and} \qquad \int_{(0,\infty)} \frac{d\nu(t)}{t(1+t)^{n+1}} < \infty. \tag{8.3}$$

If $\nu \in \mathbb{M}^+$, we denote by $\Delta^+(\nu)$ the minimal $n \in \mathbb{N}_0$ such that (8.3) holds. Then $\Delta^+(\mu^+_{p,w}) = \Delta^+_{\mathrm{SL}}(p,w)$.

5. Instead of (7.39), one has

$$-\lim_{\lambda \nearrow 0} m^+_{p,w}(\lambda) = \left[\int_a^b \frac{1}{p(x)}\,dx\right]^{-1}, \tag{8.4}$$

and that the left-hand side is zero if and only if the integral on the right-hand side is infinite. Relation (8.4) follows from (5.4), (5.5) and (8.2).

The global uniqueness result is different from the one in the previous subsection since adding a constant to the singular Titchmarsh–Weyl coefficient corresponds to a more complicated transformation; cf. also [31, Corollary 3.6] for the case when the equations are in impedance form.

Theorem 8.3 (Global Uniqueness Theorem)

(i) *Let* $(p_1; w_1), (p_2; w_2) \in \mathbb{K}^+_{\mathrm{SL}}$ *be given with* $\mathrm{dom}(p_i; w_i) = (a_i, b_i)$, $i = 1,2$. *Assume that there exist singular Titchmarsh–Weyl coefficients* $m^+_{p_i,w_i}$ *corresponding to* $(p_i; w_i)$ *for* $i = 1,2$ *such that*

$$m^+_{p_1,w_1}(\lambda) - m^+_{p_2,w_2}(\lambda) = c_n\lambda^n + \ldots + c_1\lambda + c_0 \tag{8.5}$$

with $c_0, \ldots, c_n \in \mathbb{R}$. *Then there exists an increasing bijection* $\gamma : (a_2, b_2) \to (a_1, b_1)$ *such that* γ *and* γ^{-1} *are locally absolutely continuous and*

$$\begin{aligned} p_2(x) &= \frac{1}{\gamma'(x)}\left(1 + c_0\int_{a_1}^{\gamma(x)} \frac{1}{p_1(t)}\,dt\right)^2 p_1\big(\gamma(x)\big), \\ w_2(x) &= \gamma'(x)\left(1 + c_0\int_{a_1}^{\gamma(x)} \frac{1}{p_1(t)}\,dt\right)^2 w_1\big(\gamma(x)\big) \end{aligned} \tag{8.6}$$

for $x \in (a_2, b_2)$*; for all* $x \in (a_2, b_2)$, *one has*

$$1 + c_0\int_{a_1}^{\gamma(x)} \frac{1}{p_1(t)}dt > 0. \tag{8.7}$$

Moreover, $\Delta^+_{\mathrm{SL}}(p_1, w_1) = \Delta^+_{\mathrm{SL}}(p_2, w_2)$.

(ii) *Let* $(p_1; w_1) \in \mathbb{K}^+_{\mathrm{SL}}$ *be given with* $\mathrm{dom}(p_1; w_1) = (a_1, b_1)$. *Let* $(a_2, b_2) \subseteq \mathbb{R}$ *be an open interval,* $\gamma : (a_2, b_2) \to (a_1, b_1)$ *an increasing bijection such that* γ *and* γ^{-1} *are locally absolutely continuous, and let* $c_0 \in \mathbb{R}$ *such that*

$$1 + c_0 \int_{a_1}^{b_1} \frac{1}{p_1(t)}\,\mathrm{d}t \geq 0. \tag{8.8}$$

Define functions p_2, w_2 *by* (8.6). *Then* $(p_2; w_2) \in \mathbb{K}_{\mathrm{SL}}^{+}$ *with* $\Delta_{\mathrm{SL}}^{+}(p_1, w_1) = \Delta_{\mathrm{SL}}^{+}(p_2, w_2)$, *and there exist singular Titchmarsh–Weyl coefficients* m_{p_i,w_i}^{+}, $i = 1, 2$, *such that*

$$m_{p_1,w_1}^{+}(\lambda) - m_{p_2,w_2}^{+}(\lambda) = c_0.$$

(iii) *Let* $(p_1; w_1), (p_2; w_2) \in \mathbb{K}_{\mathrm{SL}}^{+}$ *be given with* $\operatorname{dom}(p_i; w_i) = (a_i, b_i)$, $i = 1, 2$. *Then* $\mu_{p_1,w_1}^{+} = \mu_{p_2,w_2}^{+}$ *if and only if there exists* γ *as above and* $c_0 \in \mathbb{R}$ *such that* (8.6) *and* (8.8) *hold.*

Before we can prove the theorem, we need a lemma about a transformation of diagonal Hamiltonians.

Lemma 8.4 *Let* $H \in \mathbb{H}$ *be a diagonal Hamiltonian with* $\operatorname{dom} H = (a, b)$ *of the form*

$$H(x) = \begin{pmatrix} h_{11}(x) & 0 \\ 0 & h_{22}(x) \end{pmatrix}.$$

Assume that $h_{22}(x) > 0$ *for almost all* $x \in (a, b)$ *and let* q_H *be a singular Weyl coefficient and* μ_H *the corresponding spectral measure. Moreover, let* $c \in \mathbb{R}$ *and define the functions*

$$\alpha(x) := 1 + c \int_a^x h_{22}(t)\,\mathrm{d}t, \qquad x \in (a, b), \tag{8.9}$$

$$\widetilde{q}(z) := q_H(z) - \frac{c}{z}. \tag{8.10}$$

Then the following statements are equivalent:

(i) $\alpha(x) > 0$ *for all* $x \in (a, b)$.
(ii) $c + \mu_H(\{0\}) \geq 0$.
(iii) $\widetilde{q} \in \mathcal{N}_{<\infty}^{(\infty)}$.

If these conditions are satisfied, then

$$\widetilde{H}(x) := \begin{pmatrix} (\alpha(x))^2 h_{11}(x) & 0 \\ 0 & \dfrac{h_{22}(x)}{(\alpha(x))^2} \end{pmatrix}, \qquad x \in (a, b), \tag{8.11}$$

belongs to $\mathbb{H}$ *and* $\widetilde{q}$ *is a singular Weyl coefficient for* $\widetilde{H}$.

Proof First note that the integral in (8.9) is positive for all $x \in (a, b)$ and strictly increasing in x. Hence, (i) is equivalent to

$$c + \left[\int_a^b h_{22}(t)\mathrm{d}t \right]^{-1} \geq 0,$$

where we use $1/\infty = 0$ in the case when the integral is infinite. Since $\mu_H(\{0\}) = \big[\int_a^b h_{22}(t)\mathrm{d}t\big]^{-1}$ by Proposition 5.3, the equivalence of (i) and (ii) follows. The function $\widetilde{q}$ is in $\mathcal{N}_{<\infty}$. The only possible finite generalized pole of non-positive type of $\widetilde{q}$ is 0. Hence, $\widetilde{q} \in \mathcal{N}_{<\infty}^{(\infty)}$ if and only if

$$\lim_{\varepsilon \searrow 0} i\varepsilon \widetilde{q}(i\varepsilon) \leq 0;$$

see, e.g. [77, Theorem 3.1]. It follows from [84, Theorem 3.9 (ii)] that

$$c + \mu_H\big(\{0\}\big) = c - \lim_{\varepsilon \searrow 0} i\varepsilon q_H(i\varepsilon) = - \lim_{\varepsilon \searrow 0} i\varepsilon \widetilde{q}(i\varepsilon),$$

which implies the equivalence of (ii) and (iii).

For the rest of the proof, assume that $\alpha(x) > 0$ for all $x \in (a, b)$. One can easily show that the assertions of the lemma are unaffected by reparameterizations. So we can assume that H is defined on $(0, \infty)$. Let $\mathfrak{h}$ be the indefinite Hamiltonian associated with H as in Sect. 3.16 with Weyl coefficient $q_{\mathfrak{h}}$ such that $q_{\mathfrak{h}} = q_H$, and let $\omega_{\mathfrak{h}}$ be the corresponding maximal chain of matrices as in Sect. 3.8. It follows from Lemma 5.9 that

$$\widetilde{\alpha}(x) := 1 - c \frac{\partial}{\partial z} \omega_{\mathfrak{h},21}(x; z) \Big|_{z=0} = \begin{cases} 1, & x \in [-1, 0), \\ \alpha(x), & x \in (0, \infty). \end{cases}$$

The transformation $\mathfrak{T}_c$ from [58, Definition 4.1] applied to $\omega_{\mathfrak{h}}$ yields a maximal chain of matrices $\widetilde{W}$, where

$$\widetilde{W}(x; z) := \big(\mathfrak{T}_c \omega_{\mathfrak{h}}\big)(x; z) = \begin{pmatrix} 1 & -\frac{c}{z} \\ 0 & 1 \end{pmatrix} \omega_{\mathfrak{h}}(x; z) \begin{pmatrix} \frac{1}{\widetilde{\alpha}(x)} & \frac{c}{z} \\ 0 & \widetilde{\alpha}(x) \end{pmatrix},$$

so that $\widetilde{W}$ corresponds to an indefinite Hamiltonian $\widetilde{\mathfrak{h}}$ whose Weyl coefficient is

$$q_{\widetilde{\mathfrak{h}}}(z) = q_{\mathfrak{h}}(z) - \frac{c}{z};$$

see [58, Theorem 4.4]. Differentiating $\widetilde{W}$ with respect to x and using the differential equation (3.9), i.e.

$$\frac{\partial}{\partial x}\widetilde{W}(x;z)J = z\widetilde{W}(x;z)\widetilde{H}(x),$$

one can easily show that the Hamiltonian function $\widetilde{H}$ that corresponds to $\widetilde{\mathfrak{h}}$ is given by (8.11); cf. [103, Rule 4]. □

Proof (Proof of Theorem 8.3) Throughout the proof, let H_1 and H_2 be the corresponding Hamiltonians

$$H_i(x) = \begin{pmatrix} w_i(x) & 0 \\ 0 & \dfrac{1}{p_i(x)} \end{pmatrix}, \qquad i = 1, 2,$$

and let q_{H_1}, q_{H_2} be the singular Weyl coefficients such that $m^+_{p_i,w_i}(z^2) = zq_{H_i}(z)$, $i = 1, 2$ as in (8.2).

(i) Suppose that (8.5) holds. Define $\widetilde{q}$ as in (8.10) with $H = H_1$ and $c = c_0$, i.e.

$$\widetilde{q}(z) = q_{H_1}(z) - \frac{c_0}{z} = q_{H_2}(z) + c_n z^{2n-1} + c_{n-1}z^{2n-3} + \ldots + c_1 z. \qquad (8.12)$$

The equality of the first and the last expression in (8.12) implies that $\widetilde{q} \in \mathcal{N}^{(\infty)}_{<\infty}$, i.e. condition (iii) in Lemma 8.4 is satisfied. Hence, we can apply Lemma 8.4, which yields a Hamiltonian $\widetilde{H}$ with corresponding singular Weyl coefficient $\widetilde{q}$. Since $\widetilde{q}$ and q_{H_2} differ only by a real polynomial without constant term, Theorem 6.1 (iii) implies that H_2 is a reparameterization of $\widetilde{H}$. This, together with (8.11) and (8.9), shows (8.6). Relation (8.7) follows from Lemma 8.4 (i). Since $\mu^+_{p_1,w_1} = \mu^+_{p_2,w_2}$, we obtain

$$\Delta^+_{\mathrm{SL}}(p_1, w_1) = \Delta^+(\mu^+_{p_1,w_1}) = \Delta^+(\mu^+_{p_2,w_2}) = \Delta^+_{\mathrm{SL}}(p_2, w_2).$$

(ii) Condition (8.8) implies that (i) in Lemma 8.4 is satisfied with $H = H_1$ and $c = c_0$. Hence, we can apply Lemma 8.4, which yields $\widetilde{H}$. The assertion follows since H_2 is a reparameterization of $\widetilde{H}$.

(iii) It follows from [84, Theorem 3.9 (iv)] that $\mu^+_{p_1,w_1} = \mu^+_{p_2,w_2}$ if and only if $m^+_{p_1,w_1}$ and $m^+_{p_2,w_2}$ differ by a real polynomial. Now the claim follows from (i) and (ii). □

Let us finally point out that Proposition 7.18 remains valid for the situation in this section if (7.46) is replaced by

$$p(x) \asymp x^{-\alpha}, \quad w(x) \asymp x^{-\alpha} \qquad \text{as } x \searrow 0$$

and $\mathbb{K}_{\mathrm{SL}}$ is replaced by $\mathbb{K}^+_{\mathrm{SL}}$.

9 Schrödinger Equations

Let $V \in L^1_{\mathrm{loc}}(0, b)$ with $b > 0$ or $b = \infty$ and consider the one-dimensional Schrödinger equation

$$-u''(x) + V(x)u(x) = \lambda u(x). \tag{9.1}$$

In this section, the left endpoint needs to be finite, which without loss of generality we assume to be 0. In the following, we write $\mathrm{dom}(V){:=}(0, b)$.

Assume that, for $\lambda = 0$, equation (9.1) has a solution ϕ (i.e. $V = \frac{\phi''}{\phi}$) in $W^{2,1}_{\mathrm{loc}}(0, b)$ that satisfies

$$\begin{aligned} &\phi(x) > 0 \quad \text{for all} \quad x \in (0, b), \\ &\phi\big|_{(0,x_0)} \in L^2(0, x_0), \ \frac{1}{\phi}\Big|_{(0,x_0)} \notin L^2(0, x_0) \quad \text{for some} \quad x_0 \in (0, b). \end{aligned} \tag{9.2}$$

A similar approach, namely, to assume the existence of a particular solution instead of explicit conditions on the coefficients, was used in [19].

Note that ϕ with the above properties is determined only up to a multiplicative positive constant; see Remark 9.11 for a further discussion of this non-uniqueness.

Let $x_0 \in (0, b)$ and define functions $\widetilde{w}_l$, $l = 0, 1, \dots,$ recursively by

$$\begin{aligned} \widetilde{w}_0(x) &= \frac{1}{\phi(x)}, \\ \widetilde{w}_k(x) &= \begin{cases} \phi(x) \displaystyle\int_x^{x_0} \frac{1}{\phi(t)} \widetilde{w}_{k-1}(t)\mathrm{d}t & \text{if } k \text{ is odd,} \\ \dfrac{1}{\phi(x)} \displaystyle\int_0^x \phi(t)\widetilde{w}_{k-1}(t)\mathrm{d}t & \text{if } k \text{ is even.} \end{cases} \end{aligned} \tag{9.3}$$

Remark 9.1 It follows from the last condition in (9.2) and [89, Theorem 2.2] that ϕ is a principal solution of (9.1) with $\lambda = 0$. The function $\widetilde{w}_1$ is a non-principal solution of (9.1) with $\lambda = 0$, and one has

$$-\widetilde{w}''_{k+2} + V\widetilde{w}_{k+2} = \widetilde{w}_k \qquad \text{when } k \in \mathbb{N} \text{ is odd.} \tag{9.4}$$

In [75] a sequence of functions g_k was used which satisfy the relations

$$-g''_{k+1} + Vg_{k+1} - \mu_k g_{k+1} = g_k$$

with pairwise distinct numbers μ_k.

Remark 9.2 Instead of $\widetilde{w}_k$ one can use more general functions $\check{w}_k$ that are defined as in (9.3) but with the relation

$$\check{w}_k(x) = \phi(x)\left[\int_x^{x_0} \frac{1}{\phi(t)}\check{w}_{k-1}(t)\mathrm{d}t + c_k\right], \qquad k \text{ odd},$$

with arbitrary constants $c_k \in \mathbb{R}$; cf. Remarks 7.5 (iii) and 7.12 (iii). The spectral measure that is constructed below remains the same.

In this section, we consider the following class of potentials.

Definition 9.3 We say that $V \in \mathbb{K}_{\text{Schr}}$ if $V \in L^1_{\text{loc}}(0, b)$, there exists a ϕ so that ϕ is a solution of (9.1) with $\lambda = 0$ that (9.2) holds and that the following conditions are satisfied.

(i) For one (and hence for all) $x_0 \in (0, b)$,

$$\int_0^{x_0} \phi(x)\widetilde{w}_1(x)\mathrm{d}x < \infty.$$

(ii) There exists an $n \in \mathbb{N}$ such that

$$\widetilde{w}_n\big|_{(0,x_0)} \in L^2(0, x_0). \tag{9.5}$$

(iii) Equation (9.1) is in the limit point case at b.

If $V \in \mathbb{K}_{\text{Schr}}$, we denote by $\Delta_{\text{Schr}}(V)$ the minimal $n \in \mathbb{N}$ such that (9.5) holds.

Remark 9.4

(i) Since the function $\widetilde{w}_1$ is a non-principal solution of (9.1) with $\lambda = 0$, we have $\Delta(V) = 1$ if and only if equation (9.1) is regular or in the limit circle case at the left endpoint.
(ii) One can also consider the case when (9.1) is regular or in the limit circle case at the right endpoint b. In this case, one has to impose a fixed self-adjoint boundary condition at b; cf. Remark 7.3. This situation was considered, e.g. in [98] in connection with n-entire operators.

In order to apply the results from Sect. 7, we set

$$p(x) = w(x) := \big(\phi(x)\big)^2, \qquad x \in (0, b). \tag{9.6}$$

With w_k defined as in (7.5), we have

$$\widetilde{w}_k(x) = \begin{cases} \dfrac{w_k(x)}{\phi(x)} & \text{if } k \text{ is even,} \\ \phi(x) w_k(x) & \text{if } k \text{ is odd.} \end{cases} \tag{9.7}$$

It is easy to see that $V \in \mathbb{K}_{\text{Schr}}$ if and only if $(p; w) \in \mathbb{K}_{\text{SL}}$ with p, w from (9.6); for the equivalence of the limit point property at b, see (9.15). Moreover, if $V \in \mathbb{K}_{\text{Schr}}$, then $\Delta_{\text{Schr}}(V) = \Delta_{\text{SL}}(p, w)$.

Example 9.5

(i) A large subclass of $\mathbb{K}_{\text{Schr}}$ is the following. Let $b > 0$ or $b = \infty$, let $V_0 \in L^1_{\text{loc}}(0, b)$ and let $l \in \left[-\frac{1}{2}, \infty\right)$. Moreover, set

$$V(x) = \frac{l(l+1)}{x^2} + V_0(x) \tag{9.8}$$

and assume that

$$\begin{aligned} x V_0(x)\big|_{(0,x_0)} &\in L^1(0, x_0) && \text{if } l > -\frac{1}{2}, \\ (\ln x) x V_0(x)\big|_{(0,x_0)} &\in L^1(0, x_0) && \text{if } l = -\frac{1}{2} \end{aligned} \tag{9.9}$$

with some $x_0 \in (0, b)$. Moreover, suppose that the minimal operator associated with (9.2) is bounded from below and that (9.1) is in the limit point case at b. Under the assumption that (9.9) is valid, it follows from [67, Lemma 3.2] that there exists a solution ϕ of (9.1) with $\lambda = 0$ such that

$$\phi(x) = x^{l+1}\big(1 + \mathrm{o}(x)\big), \qquad x \searrow 0. \tag{9.10}$$

Assume that $\phi(x) > 0$ for $x \in (0, b)$, which is satisfied, e.g. if the minimal operator is uniformly positive, which can be achieved by a shift of the spectral parameter. Now it follows from Proposition 7.18 that $(p; w) \in \mathbb{K}_{\text{SL}}$, and hence $V \in \mathbb{K}_{\text{Schr}}$ and $\Delta_{\text{Schr}}(V) = \Delta_{\text{SL}}(p, w) = \lfloor l + \frac{3}{2} \rfloor$.

Since $l = 0$ is allowed in (9.8), the class $\mathbb{K}_{\text{Schr}}$ contains potentials where 0 is a regular endpoint. If $l \in \left[-\frac{1}{2}, \frac{1}{2}\right) \setminus \{0\}$, then (9.1) is in the limit circle case at 0 and $\Delta_{\text{Schr}}(V) = 1$.

Potentials of the form (9.8) have been studied in many papers; see, e.g. [3, 30, 35–37, 47, 64, 65, 67, 68, 75, 86, 97, 98].

(ii) The class $\mathbb{K}_{\text{Schr}}$ contains also potentials that have a stronger singularity at the left endpoint than those considered in (i). If

$$V(x) = \frac{\phi''(x)}{\phi(x)} \qquad \text{where} \qquad \phi(x) \asymp x^{\beta}, \quad x \searrow 0 \tag{9.11}$$

with $\beta \geq \frac{1}{2}$, $\phi(x) > 0$ for $x \in (0, b)$ and (9.1) is in the limit point case at b, then $V \in \mathbb{K}_{\mathrm{Schr}}$ with $\Delta_{\mathrm{Schr}}(V) = \lfloor \beta + \frac{1}{2} \rfloor$; cf. Proposition 7.18. For instance, functions of the form

$$\phi(x) = x^{\beta}\Big(2 + \sin\frac{1}{x}\Big), \qquad x \in (0, x_0),$$

with $\beta > 0$ lead to oscillatory potentials that do not satisfy (9.9), namely,

$$V(x) = -\frac{1}{x^4} \cdot \frac{\sin\frac{1}{x}}{2 + \sin\frac{1}{x}} + \mathrm{O}\Big(\frac{1}{x^3}\Big), \qquad x \searrow 0.$$

It follows from Lemma 7.19 and (9.7) that if V is of the form in (9.11), then

$$\widetilde{w}_k(x) \asymp x^{-\beta+k}, \qquad x \searrow 0,\ k \in \mathbb{N}_0,\ k < 2\beta. \tag{9.12}$$

In particular, the relation in (9.12) is valid for $k = 0, 1, \ldots, 2\Delta - 1$ if β is not an odd integer multiple of $\frac{1}{2}$, and it is valid for $k = 0, 1, \ldots, 2\Delta - 2$, otherwise.

(iii) The function $V(x) = \frac{1}{x^4}$ does not belong to $\mathbb{K}_{\mathrm{Schr}}$. It can easily be checked that the only possible choice for ϕ (up to scalar multiples) is $\phi(x) = xe^{-1/x}$. Moreover, one can show that

$$\widetilde{w}_n(x) \sim C_n x^{\alpha_n} e^{\frac{1}{x}}, \qquad x \searrow 0,$$

with some $C_n > 0$, $n \in \mathbb{N}$, and $\alpha_n = \frac{3n-2}{2}$ when n is even and $\alpha_n = \frac{3n-1}{2}$ when n is odd. Hence, condition (i) in Definition 9.3 is satisfied, but there is no $n \in \mathbb{N}$ such that (9.5) holds. This potential was also studied in [87], where it was shown that the approach with super-singular perturbations, as developed in [75] and [86], cannot be applied to this potential.

(iv) Potentials from the class $H^{-1}_{\mathrm{loc}}(0, b)$ could also be treated by our method if we relaxed the assumption $V \in L^1_{\mathrm{loc}}(0, b)$. In this case, one would only have $\phi \in H^1_{\mathrm{loc}}(0, b)$. Operators with such potentials were considered, e.g. in [31, 46, 96]. Note that the class $H^{-1}_{\mathrm{loc}}(0, b)$ includes measure coefficients.

Let us introduce the unitary operator

$$U : \begin{cases} L^2(0, b) \to L^2(w), \\ \qquad u \mapsto \dfrac{u}{\phi} \end{cases} \tag{9.13}$$

and define the self-adjoint operator

$$A_V := U^{-1} A_{p,w} U \tag{9.14}$$

with p and w as in (9.6) and $A_{p,w}$ from (7.11). For $u \in W^{2,1}(0,b)$ with compact support, we have

$$\begin{aligned} A_V u &= -\phi \frac{1}{w}\Big(p\Big(\frac{u}{\phi}\Big)'\Big)' = -\frac{1}{\phi}\Big(\phi^2 \frac{\phi u' - \phi' u}{\phi^2}\Big)' \\ &= -\frac{\phi u'' - \phi'' u}{\phi} = -u'' + Vu. \end{aligned} \tag{9.15}$$

Therefore, A_V is the Friedrichs extension of the minimal operator connected with the equation (9.1); cf. the discussion below (7.12). In particular, if $\Delta_{\mathrm{Schr}}(V) = 1$, then a possible boundary condition at 0 to characterize A_V is

$$\lim_{x \searrow 0} \frac{u(x)}{\widetilde{w}_1(x)} = 0;$$

see [89, Theorem 4.3]. As mentioned above, if $\Delta_{\mathrm{Schr}}(V) \geq 2$, then (9.1) is in the limit point case at 0 and hence no boundary condition is needed there.

We can apply all theorems from Sect. 7. In order to rewrite these results in a more intrinsic form, we define regularized boundary values by $\mathrm{rbv}_\lambda^{\mathrm{Schr}} u := \mathrm{rbv}_\lambda^{\mathrm{SL}} U u$ for $\lambda \in \mathbb{C}$ and u a solution of (9.1). Then Theorem 7.4, together with a straightforward calculation, yields the following theorem.

Theorem 9.6 (Regularized boundary values) *Let* $V \in \mathbb{K}_{\mathrm{Schr}}$ *with* $\mathrm{dom}(V) = (0,b)$, *set* $\Delta := \Delta_{\mathrm{Schr}}(V)$ *and let* $\mathfrak{N}_\lambda^{\mathrm{Schr}}$ *be the set of all solutions of* (9.1). *Then, for* $x_0 \in (0,b)$, *the following statements hold.*

(i) *For each* $\lambda \in \mathbb{C}$ *and each solution* $u \in \mathfrak{N}_\lambda^{\mathrm{Schr}}$, *the boundary value*

$$\mathrm{rbv}_{\lambda,1}^{\mathrm{Schr}} u = \lim_{x \searrow 0} \big(\phi(x) u'(x) - \phi'(x) u(x)\big),$$

and the regularized boundary value

$$\begin{aligned} \mathrm{rbv}_{\lambda,2}^{\mathrm{Schr}} u = \lim_{x \searrow 0} \Bigg[&\sum_{k=0}^{\lfloor \frac{\Delta-1}{2} \rfloor} \lambda^k \Big(\widetilde{w}_{2k+1}(x) u'(x) - \widetilde{w}'_{2k+1}(x) u(x)\Big) \\ &+ \begin{cases} \lambda^{\frac{\Delta}{2}} \widetilde{w}_\Delta(x) u(x) & \text{if } \Delta \text{ is even} \\ 0 & \text{if } \Delta \text{ is odd} \end{cases} \\ &+ \big(\mathrm{rbv}_{\lambda,1}^{\mathrm{Schr}} u\big) \Bigg(\sum_{k=\lfloor \frac{\Delta+1}{2} \rfloor}^{\Delta-1} \sum_{l=0}^{2k-\Delta} (-1)^l \lambda^k \widetilde{w}_l(x) \widetilde{w}_{2k-l+1}(x) \Bigg) \Bigg]. \end{aligned}$$

exist.

(ii) *For each* $\lambda \in \mathbb{C}$, *we define*

$$\mathrm{rbv}_\lambda^{\mathrm{Schr}} : \begin{cases} \mathfrak{N}_\lambda^{\mathrm{Schr}} & \to \mathbb{C}^2 \\ u & \mapsto \left(\mathrm{rbv}_{\lambda,1}^{\mathrm{Schr}} u,\ \mathrm{rbv}_{\lambda,2}^{\mathrm{Schr}} u\right)^T. \end{cases}$$

Then $\mathrm{rbv}_\lambda^{\mathrm{Schr}}$ *is a bijection from* $\mathfrak{N}_\lambda^{\mathrm{Schr}}$ *onto* $\mathbb{C}^2$.

(iii) *For each* $\lambda \in \mathbb{C}$, *there exists an (up to scalar multiples) unique solution* $u \in \mathfrak{N}_\lambda^{\mathrm{Schr}} \setminus \{0\}$ *such that* $\lim_{x \searrow 0} \frac{u(x)}{\phi(x)}$ *exists.*

This solution is characterized by the property that $\int_0^{x_0} \phi^2 \left|\left(\frac{u}{\phi}\right)'\right|^2 < \infty$ *and also by the property that* $\mathrm{rbv}_{\lambda,1}^{\mathrm{Schr}} u = 0$ *(and* $u \not\equiv 0$*).*

If u *is a solution such that* $\lim\limits_{x \searrow 0} \frac{u(x)}{\phi(x)}$ *exists, then* $\mathrm{rbv}_{\lambda,2}^{\mathrm{Schr}} u = \lim\limits_{x \searrow 0} \frac{u(x)}{\phi(x)}$.

The regularized boundary value $\mathrm{rbv}_{\lambda,2}^{\mathrm{Schr}}$ *depends on the choice of* x_0 *in the following way.*

(iv) *Let* $x_0, \hat{x}_0 \in (0, b)$, *and let* $\mathrm{rbv}_\lambda^{\mathrm{Schr}}$ *and* $\widehat{\mathrm{rbv}}_\lambda^{\mathrm{Schr}}$ *be the correspondingly defined regularized boundary value mappings. Then there exists a polynomial* $p_{x_0,\hat{x}_0}(z)$ *with real coefficients whose degree does not exceed* $\Delta - 1$ *such that*

$$\widehat{\mathrm{rbv}}_{\lambda,2}^{\mathrm{Schr}} u = \mathrm{rbv}_{\lambda,2}^{\mathrm{Schr}} u + p_{x_0,\hat{x}_0}(\lambda)\mathrm{rbv}_{\lambda,1}^{\mathrm{Schr}} u, \qquad u \in \mathfrak{N}_\lambda^{\mathrm{Schr}}, \quad \lambda \in \mathbb{C}.$$

Moreover, clearly, $\widehat{\mathrm{rbv}}_{\lambda,1}^{\mathrm{Schr}} = \mathrm{rbv}_{\lambda,1}^{\mathrm{Schr}}$.

The next theorem about a fundamental system of solutions of (9.1) and the existence of a singular Titchmarsh–Weyl coefficient follows from Theorem 7.7 (i) with the help of the unitary operator U from (9.13).

Theorem 9.7 (Singular Titchmarsh–Weyl coefficients) *Let* $V \in \mathbb{K}_{\mathrm{Schr}}$ *with* $\mathrm{dom}(V) = (0, b)$ *be given. Then, for each fixed* $x_0 \in (0, b)$, *the following statements hold.*

(i) *For* $\lambda \in \mathbb{C}$, *let* $\widetilde{\theta}(\cdot\,;\lambda)$ *and* $\widetilde{\varphi}(\cdot\,;\lambda)$ *be the unique solutions of* (9.1) *such that*

$$\mathrm{rbv}_\lambda^{\mathrm{Schr}} \widetilde{\theta}(\cdot\,;\lambda) = \begin{pmatrix} 1 \\ 0 \end{pmatrix}, \qquad \mathrm{rbv}_\lambda^{\mathrm{Schr}} \widetilde{\varphi}(\cdot\,;\lambda) = \begin{pmatrix} 0 \\ 1 \end{pmatrix}.$$

Then, for each $x \in (0, b)$, *the functions* $\widetilde{\theta}(x;\cdot)$ *and* $\widetilde{\varphi}(x;\cdot)$ *are entire of order* $\frac{1}{2}$ *and finite type* x. *Moreover, for each* $\lambda \in \mathbb{C}$, *one has* $W\left(\widetilde{\varphi}(\cdot\,;\lambda), \widetilde{\theta}(\cdot\,;\lambda)\right) \equiv 1$ *where* $W{:=}W_1$ *denotes the Wronskian as in* (7.18) *with* $p \equiv 1$, *and the following relations hold:*

$$\lim_{x \searrow 0} \frac{\widetilde{\varphi}(x;\lambda)}{\phi(x)} = 1, \qquad \lim_{x \searrow 0} \frac{\phi(x)\widetilde{\varphi}'(x;\lambda) - \phi'(x)\widetilde{\varphi}(x;\lambda)}{\int_a^x \left(\phi(t)\right)^2 \mathrm{d}t} = -\lambda,$$

$$\lim_{x\searrow 0}\frac{\widetilde{\theta}(x;\lambda)}{\widetilde{w}_1(x)}=-1,\qquad \lim_{x\searrow 0}\big(\phi(x)\widetilde{\theta}'(x;\lambda)-\phi'(x)\widetilde{\theta}(x;\lambda)\big)=1.$$

Further, one has $\widetilde{\varphi}(x;0)=\phi(x)$ *and* $\widetilde{\theta}(x;0)=-\widetilde{w}_1(x)$, $x\in(0,b)$.

(ii) *The limit*

$$\widetilde{m}_V(\lambda):=\lim_{x\nearrow b}\frac{\widetilde{\theta}(x;\lambda)}{\widetilde{\varphi}(x;\lambda)},\qquad \lambda\in\mathbb{C}\setminus[0,\infty),$$

exists locally uniformly on $\mathbb{C}\setminus[0,\infty)$ *and defines an analytic function in* λ. *The function* $\widetilde{m}_V$ *belongs to the class* $\mathcal{N}_\kappa^{(\infty)}$ *with* $\kappa=\lfloor\frac{\Delta_{\text{Schr}}(V)}{2}\rfloor$.

(iii) *We have*

$$\widetilde{\theta}(\cdot\,;\lambda)-\widetilde{m}_V(\lambda)\widetilde{\varphi}(\cdot\,;\lambda)\in L^2(x_0,b),\qquad \lambda\in\mathbb{C}\setminus[0,\infty),$$

and this property characterizes the value $\widetilde{m}_V(\lambda)$ *for each* $\lambda\in\mathbb{C}\setminus[0,\infty)$.

(iv) *For* $\lambda\in\mathbb{C}\setminus[0,\infty)$, *let* u *be any non-trivial solution of* (9.1) *such that* $u|_{(x_0,b)}\in L^2(x_0,b)$. *Then*

$$\widetilde{m}_V(\lambda)=-\frac{\operatorname{rbv}^{\text{Schr}}_{\lambda,2}u}{\operatorname{rbv}^{\text{Schr}}_{\lambda,1}u}.$$

The function $\widetilde{m}_V$ *depends on the choice of* x_0. *This dependence is controlled as follows.*

(v) *Let* $\hat{x}_0\in(a,b)$, *and let* $\widehat{\widetilde{m}}_V$ *be the correspondingly defined singular Titchmarsh–Weyl coefficient. Then there exists a polynomial* $p_{x_0,\hat{x}_0}$ *with real coefficients whose degree does not exceed* $\Delta_{\text{Schr}}(V)-1$ *such that*

$$\widehat{\widetilde{m}}_V(\lambda)=\widetilde{m}_V(\lambda)-p_{x_0,\hat{x}_0}(\lambda).$$

The functions $\widetilde{\theta}$ and $\widetilde{\varphi}$ are related to the functions θ and φ corresponding to (7.1) with p and w as in (9.6) as follows:

$$\widetilde{\theta}(x;\lambda)=\phi(x)\theta(x;\lambda),\qquad \widetilde{\varphi}(x;\lambda)=\phi(x)\varphi(x;\lambda),\qquad x\in(0,b). \tag{9.16}$$

The function $\widetilde{m}_V$ is called *singular Titchmarsh–Weyl coefficient.* It follows from (7.28) and (9.16) that $\widetilde{m}_V=m_{p,w}$. As in Sect. 7, one defines equivalence classes $[\widetilde{m}]_V$ with respect to the equivalence relation $\hat{\sim}$ defined in (7.30).

The next theorem about the existence of a spectral measure follows from Theorem 7.11. For the definition of the class $\mathbb{M}^-$, see Definition 7.10.

Theorem 9.8 (The spectral measure) *Let* $V\in\mathbb{K}_{\text{Schr}}$ *with* $\operatorname{dom}(V)=(0,b)$ *be given. Then there exists a unique Borel measure* $\widetilde{\mu}_V$ *that satisfies*

$$\widetilde{\mu}_V\big([s_1,s_2]\big)=\frac{1}{\pi}\lim_{\varepsilon\searrow 0}\lim_{\delta\searrow 0}\int\limits_{s_1-\varepsilon}^{s_2+\varepsilon}\operatorname{Im}\widetilde{m}_V(t+i\delta)\,\mathrm{d}t,\qquad -\infty<s_1<s_2<\infty,$$

where $\widetilde{m}_V \in [\widetilde{m}]_V$ *is any singular Titchmarsh–Weyl coefficient associated with* (9.1)*. We have* $\widetilde{\mu}_V \in \mathbb{M}^-$ *and* $\Delta^-(\widetilde{\mu}_V)=\Delta_{\mathrm{Schr}}(V)$*.*

Moreover, $\widetilde{\mu}_V(\{0\})>0$ *if and only if*

$$\int_0^b \big(\phi(x)\big)^2\mathrm{d}x<\infty. \tag{9.17}$$

If (9.17) *is satisfied, then*

$$\widetilde{\mu}_V\big(\{0\}\big)=\left[\int_0^b\big(\phi(x)\big)^2\mathrm{d}x\right]^{-1}.$$

Clearly, we have $\widetilde{\mu}_V=\mu_{p,w}$ where p and w are as in (9.6).

Example 9.9 Consider V as in Example 9.5 (i). Since $\Delta_{\mathrm{Schr}}(V)=\lfloor l+\frac{3}{2}\rfloor$, we obtain from Theorem 9.7 that $\widetilde{m}_V\in\mathcal{N}_\kappa^{(\infty)}$ with $\kappa=\lfloor\frac{l}{2}+\frac{3}{4}\rfloor$. Moreover, Theorem 9.8 yields that $\widetilde{\mu}_V\in\mathbb{M}^-$ with $\Delta^-(\widetilde{\mu}_V)=\Delta_{\mathrm{Schr}}(V)=\lfloor l+\frac{3}{2}\rfloor$.

In the next theorem, we consider the corresponding Fourier transform and its inverse. This theorem follows from Theorem 7.13.

Theorem 9.10 (The Fourier transform) *Let* $V\in\mathbb{K}_{\mathrm{Schr}}$ *with* $\mathrm{dom}(V)=(0,b)$ *be given, and let* $\widetilde{\mu}_V$ *be the spectral measure associated with* (9.1) *as in Theorem 9.8. Then the following statements hold.*

(i) *The map defined by*

$$\big(\widetilde{\Theta}_V f\big)(t):=\int_0^b\widetilde{\varphi}(x;t)f(x)\,\mathrm{d}x,\qquad t\in\mathbb{R},$$
$$f\in L^2(0,b),\quad \sup(\operatorname{supp}f)<b, \tag{9.18}$$

extends to an isometric isomorphism from $L^2(0,b)$ *onto* $L^2(\widetilde{\mu}_V)$

(ii) *The operator* $\widetilde{\Theta}_V$ *establishes a unitary equivalence between* A_V *and the operator* $M_{\widetilde{\mu}_V}$ *of multiplication by the independent variable in* $L^2(\widetilde{\mu}_V)$*, i.e. we have*

$$\widetilde{\Theta}_V A_V=M_{\widetilde{\mu}_V}\widetilde{\Theta}_V.$$

(iii) *For compactly supported functions, the inverse of* $\widetilde{\Theta}_V$ *acts as an integral transformation, namely,*

$$\big(\widetilde{\Theta}_V^{-1}g\big)(x)=\int_0^\infty \widetilde{\varphi}(x;t)g(t)\,\mathrm{d}\widetilde{\mu}_V(t), \qquad x\in(a,b),$$
$$g\in L^2(\widetilde{\mu}_V),\ \ \operatorname{supp} g \ \textit{compact}.$$

The existence of a Fourier transform into a scalar L^2-space shows in particular that the spectrum of A_V is simple.

Remark 9.11 Recall that the solution ϕ is not unique. If ϕ is multiplied by a positive constant r, then $\widetilde{w}_l$ and $\widetilde{\theta}$ are divided by r, $\widetilde{\varphi}$ is multiplied by r and $\widetilde{m}_V$ and $\widetilde{\mu}_V$ are divided by r^2. However, in the situation of Example 9.5 (i), one can normalize ϕ such that (9.10) holds.

Finally, let us state global and local uniqueness theorems. For the case of Bessel-type potentials as in Example 9.5 (i), see, e.g. [30, Theorem 5.1].

Theorem 9.12 (Global Uniqueness Theorem) *Let $V_1, V_2 \in \mathbb{K}_{\mathrm{Schr}}$ be given with* $\operatorname{dom}(V_i)=(0,b_i)$, $i=1,2$. *Then the following statements are equivalent:*

(i) $b_1=b_2$ *and* $V_1(x)=V_2(x)$, $x\in(0,b_1)$ *a.e.*
(ii) *There exists a* $c>0$ *such that* $[\widetilde{m}]_{V_1}=c[\widetilde{m}]_{V_2}$.
(iii) *There exists a* $c>0$ *such that* $\widetilde{\mu}_{V_1}=c\widetilde{\mu}_{V_2}$.

Proof For the implication (i)$\Rightarrow$(ii), see Remark 9.11. The equivalence of (ii) and (iii) is clear from the definition of $\widetilde{\mu}_V$. Now suppose that (ii) holds and let $p_i = w_i=\phi_i^2$ be as in (9.6). By rescaling ϕ_2, we may assume that $c=1$. Then we have $[m]_{p_1,w_1}=[m]_{p_2,w_2}$. It follows from Theorem 7.15 that there exists $\gamma:(0,b_2)\to(0,b_1)$ such that (7.45) holds. However, this implies that $\gamma'(x)=1$ a.e., and hence $b_1=b_2$ and $\phi_1=\phi_2$. This shows that $V_1=V_2$, i.e. (i) is satisfied. □

Local uniqueness theorems for Schrödinger equations have attracted a lot of attention recently. For the case of a regular left endpoint, B. Simon proved the first version of such a theorem in [99, Theorem 1.2]; alternative proofs were given in [7, 38] and [78]. For Bessel-type operators with potentials as in Example 9.5 (i), a local uniqueness theorem was proved in [68, Theorem 4.1].

Theorem 9.13 (Local Uniqueness Theorem) *Let $V_1, V_2\in\mathbb{K}_{\mathrm{Schr}}$ be given with* $\operatorname{dom}(V_i)=(0,b_i)$, $i=1,2$. *Then, for $\tau>0$, the following statements are equivalent:*

(i) *One has* $V_1(x)=V_2(x)$, $x\in\big(0,\min\{\tau,b_1,b_2\}\big)$ *a.e.*
(ii) *There exist singular Titchmarsh–Weyl coefficients $\widetilde{m}_{V_1}$ and $\widetilde{m}_{V_2}$, and there exist $c>0$ and $\beta\in(0,2\pi)$ such that, for each $\varepsilon>0$,*

$$\widetilde{m}_{V_1}\big(re^{i\beta}\big)-c\widetilde{m}_{V_2}\big(re^{i\beta}\big)=\mathrm{O}\Big(e^{(-2\tau+\varepsilon)\sqrt{r}\sin\frac{\beta}{2}}\Big), \qquad r\to\infty.$$

(iii) *There exist singular Titchmarsh–Weyl coefficients $\widetilde{m}_{V_1}$ and $\widetilde{m}_{V_2}$, and there exist $c>0$ and $k\geq 0$ such that, for each $\delta\in(0,\pi)$,*

$$\widetilde{m}_{V_1}(\lambda) - c\widetilde{m}_{V_2}(\lambda) = \mathrm{O}\Big(|\lambda|^k e^{-2\tau \operatorname{Im}\sqrt{\lambda}}\Big),$$
$$|\lambda| \to \infty, \quad \lambda \in \big\{z \in \mathbb{C} : \delta \le \arg z \le 2\pi - \delta\big\},$$

where $\sqrt{\lambda}$ *is chosen so that* $\operatorname{Im}\sqrt{\lambda} > 0$.

Proof This theorem follows from Theorem 7.17; we only have to observe that $s_i(\tau) = \min\{\tau, b_i\}$ and that the validity of (7.45) with $p_i = w_i$ implies that $\gamma(x) = x$ for $x \in (0, \tau)$. □

Let us conclude this section with two remarks about possible extensions.

Remark 9.14 With a similar method, one can also treat general Sturm–Liouville equations of the form

$$-\big(Py'\big)' + Qy = \lambda W y \tag{9.19}$$

where $1/P$, Q and W are locally integrable. If a positive solution ϕ of (9.19) with $\lambda = 0$ exists such that $\phi \in L^2(W|_{(a,x_0)})$, then one can use the mapping $u \mapsto \frac{u}{\phi}$ to transform (9.19) to an equation of the form (7.1) with

$$p{:=}P\phi^2, \qquad w{:=}W\phi^2;$$

cf. [89, Lemma 3.2]. Using Theorem 7.15, one can show that if the spectral measures corresponding two equations of the form (9.19) coincide, then the coefficients are related via a Liouville transform; see [31, Theorem 3.4] for a related result and [8, Theorem 4.2] for the case when $W \equiv 1$ and the left endpoint is regular.

Remark 9.15 One can also apply the results of the first part of the paper to Dirac systems of the form

$$-Ju' + Vu = zu \tag{9.20}$$

on an interval (a, b), where V is a real-valued, symmetric and locally integrable 2×2-matrix function, $z \in \mathbb{C}$ is the spectral parameter and u is a 2-vector function. Assume that there exists a solution ϕ of (9.20) with $z = 0$ (i.e. $J\phi' = V\phi$) which is in $(L^2(a, x_0))^2$ for some $x_0 \in (a, b)$. Under this assumption, we can transform (9.20) into a canonical system (1.1) as it was done in [73, Section 4.1, pp. 336, 337]. Let Φ be a 2×2-matrix solution of $J\Phi' = V\Phi$ (i.e. columns of Φ are solutions of (9.20) with $z = 0$) such that

$$\begin{pmatrix}\Phi_{12}\\ \Phi_{22}\end{pmatrix} = \phi \qquad \text{and} \qquad \det\Phi(x_0) = 1.$$

From the second relation, it follows that $\Phi(x_0)^T J\Phi(x_0) = J$. Since $\frac{\mathrm{d}}{\mathrm{d}x}(\Phi^T J\Phi) = 0$, we have

$$\Phi^T J \Phi = J \qquad \text{on } (a, b), \tag{9.21}$$

and hence $\det \Phi(x) = 1$, $x \in (a, b)$. Set

$$H{:=}\Phi^T \Phi,$$

which is clearly symmetric and non-negative and does not vanish on any set of positive measure. It is easy to see that y is a solution of (1.1) if and only if $u{:=}\Phi y$ is a solution of (9.20). Since $h_{22} = \Phi_{12}^2 + \Phi_{22}^2 = \phi_1^2 + \phi_2^2$, condition (I) in Definition 2.2 is satisfied. If $H \in \mathbb{H}$, i.e. also (HS) and (Δ) are fulfilled, then one can apply the results from Sects. 4–6.

In order to write the results in a more intrinsic form, one can use the unitary transformation

$$U : \begin{cases} L^2(H) \to (L^2(a, b))^2, \\ y \mapsto \Phi u, \end{cases}$$

whose inverse acts like $U^{-1}u = \Phi^{-1}u = -J\Phi^T Ju$. For instance, one can define regularized boundary values by $\mathrm{rbv}_z^{\mathrm{Dir}} u{:=}\mathrm{rbv}_z U^{-1} u$ as in Sect. 9. Details are left to the reader. See also, e.g. [16, 34] for different approaches to Dirac operators.

Acknowledgments The first author gratefully acknowledges the support of the Nuffield Foundation, grant no. NAL/01159/G, and the Engineering and Physical Sciences Research Council (EPSRC), grant no. EP/E037844/1. The second author was supported by the joint project I 4600 of the Austrian Science Fund (FWF) and the Russian Foundation for Basic Research (RFBR).

References

1. S. Albeverio, F. Gesztesy, R. Høegh-Krohn and H. Holden, *Solvable Models in Quantum Mechanics*. Second edition. With an appendix by Pavel Exner. AMS Chelsea Publishing, Providence, RI, 2005.
2. S. Albeverio, R. Hryniv and Ya. Mykytyuk, Inverse spectral problems for Sturm–Liouville operators in impedance form. *J. Funct. Anal.* **222** (2005), 143–177.
3. S. Albeverio, R. Hryniv and Ya. Mykytyuk, Scattering theory for Schrödinger operators with Bessel-type potentials. *J. Reine Angew. Math.* **666** (2012), 83–113.
4. V.I. Arnol′d, *Mathematical Methods of Classical Mechanics*, Graduate Texts in Mathematics, vol. 60, Springer-Verlag, New York, 1989.
5. D.Z. Arov and H. Dym, *J-Contractive Matrix Valued Functions and Related Topics. Encyclopedia of Mathematics and its Applications*, vol. 116. Cambridge University Press, Cambridge, 2008.
6. F.V. Atkinson, *Discrete and Continuous Boundary Problems*. Mathematics in Science and Engineering, vol. 8. Academic Press, New York, 1964.
7. C. Bennewitz, A proof of the local Borg–Marchenko theorem. *Comm. Math. Phys.* **218** (2001), 131–132.

8. C. Bennewitz, A Paley–Wiener theorem with applications to inverse spectral theory. In: *Advances in Differential Equations and Mathematical Physics (Birmingham, AL, 2002)*. Contemp. Math., vol. 327, Amer. Math. Soc., Providence, RI, 2003, pp. 21–31.
9. C. Bennewitz and W.N. Everitt, The Titchmarsh–Weyl eigenfunction expansion theorem for Sturm–Liouville differential equations. In: *Sturm–Liouville Theory*. Birkhäuser, Basel, 2005, pp. 137–171.
10. J. Bognár, *Indefinite Inner Product Spaces*. Ergebnisse der Mathematik und ihrer Grenzgebiete, vol. 78. Springer-Verlag, New York, 1974.
11. L. de Branges, Some Hilbert spaces of entire functions. *Trans. Amer. Math. Soc.* **96** (1960), 259–295.
12. L. de Branges, Some Hilbert spaces of entire functions. II. *Trans. Amer. Math. Soc.* **99** (1961), 118–152.
13. L. de Branges, Some Hilbert spaces of entire functions. III. *Trans. Amer. Math. Soc.* **100** (1961), 73–115.
14. L. de Branges, Some Hilbert spaces of entire functions. IV. *Trans. Amer. Math. Soc.* **105** (1962), 43–83.
15. L. de Branges, *Hilbert Spaces of Entire Functions*. Prentice-Hall Inc., Englewood Cliffs, N.J., 1968.
16. R. Brunnhuber, J. Eckhardt, A. Kostenko and G. Teschl, Singular Weyl–Titchmarsh–Kodaira theory for one-dimensional Dirac operators. *Monatsh. Math.* **174** (2014), 515–547.
17. K.P. Bube and R. Burridge, The one-dimensional inverse problem of reflection seismology. *SIAM Rev.* **25** (1983), 497–559.
18. K. Daho and H. Langer, Matrix functions of the class N_κ. *Math. Nachr.* **120** (1985), 275–294.
19. E.B. Davies, Singular Schrödinger operators in one dimension. *Mathematika* **59** (2013), 141–159.
20. V.A. Derkach, On extensions of the Laguerre operator in spaces with an indefinite metric. *Mat. Zametki* **63** (1998), 509–521.
21. A. Dijksma and H. de Snoo, Symmetric and selfadjoint relations in Kreĭn spaces. II. *Ann. Acad. Sci. Fenn. Ser. A I Math.* **12** (1987), 199–216.
22. A. Dijksma, H. Langer, A. Luger and Y. Shondin, A factorization result for generalized Nevanlinna functions of the class $\mathcal{N}_\kappa$. *Integral Equations Operator Theory* **36** (2000), 121–125.
23. A. Dijksma, H. Langer and Y. Shondin, Rank one perturbations at infinite coupling in Pontryagin spaces. *J. Funct. Anal.* **209** (2004), 206–246.
24. A. Dijksma, H. Langer, Y. Shondin and C. Zeinstra, Self-adjoint operators with inner singularities and Pontryagin spaces. In: *Operator Theory and Related Topics, vol. II (Odessa, 1997)*. Oper. Theory Adv. Appl., vol. 118. Birkhäuser, Basel, 2000, pp. 105–175.
25. A. Dijksma, A. Luger and Y. Shondin, Minimal models for $\mathcal{N}_\kappa^\infty$-functions. In: *Operator Theory and Indefinite Inner Product Spaces*. Oper. Theory Adv. Appl., vol. 163. Birkhäuser, Basel, 2006, pp. 97–134.
26. A. Dijksma, A. Luger and Y. Shondin, Approximation of $\mathcal{N}_\kappa^\infty$-functions. I: Models and regularization. In: *Spectral Theory in Inner Product Spaces and Applications*. Oper. Theory Adv. Appl., vol. 188. Birkhäuser Verlag, Basel, 2009, pp. 87–112.
27. A. Dijksma, A. Luger and Y. Shondin, Approximation of $\mathcal{N}_\kappa^\infty$-functions II: Convergence of models. In: *Recent Advances in Operator Theory in Hilbert and Krein Spaces*. Oper. Theory Adv. Appl., vol. 198. Birkhäuser Verlag, Basel, 2010, pp. 125–169.
28. A. Dijksma and Y. Shondin, Singular point-like perturbations of the Bessel operator in a Pontryagin space. *J. Differential Equations* **164** (2000), 49–91.
29. A. Dijksma and Y. Shondin, Singular point-like perturbations of the Laguerre operator in a Pontryagin space. In: *Operator Methods in Ordinary and Partial Differential Equations (Stockholm, 2000)*. Oper. Theory Adv. Appl., vol. 132. Birkhäuser, Basel, 2002, pp. 141–181.
30. J. Eckhardt, Inverse uniqueness results for Schrödinger operators using de Branges theory. *Complex Anal. Oper. Theory* **8** (2014), 37–50.

31. J. Eckhardt, F. Gesztesy, R. Nichols and G. Teschl, Inverse spectral theory for Sturm–Liouville operators with distributional potentials. *J. Lond. Math. Soc. (2)* **88** (2013), 801–828.
32. J. Eckhardt and G. Teschl, Uniqueness results for one-dimensional Schrödinger operators with purely discrete spectra. *Trans. Amer. Math. Soc.* **365** (2013), 3923–3942.
33. H. Flanders, *Differential Forms with Applications to the Physical Sciences*. Second edition. Dover Books on Advanced Mathematics. Dover Publications Inc., New York, 1989.
34. B. Fritzsche, B. Kirstein and A.L. Sakhnovich, Weyl functions of generalized Dirac systems: integral representation, the inverse problem and discrete interpolation. *J. Anal. Math.* **116** (2012), 17–51.
35. C. Fulton, Titchmarsh–Weyl m-functions for second-order Sturm–Liouville problems with two singular endpoints. *Math. Nachr.* **281** (2008), 1418–1475.
36. C. Fulton and H. Langer, Sturm–Liouville operators with singularities and generalized Nevanlinna functions. *Complex Anal. Oper. Theory* **4** (2010), 179–243.
37. C. Fulton, H. Langer and A. Luger, Mark Krein's method of directing functionals and singular potentials. *Math. Nachr.* **285** (2012), 1791–1798.
38. F. Gesztesy and B. Simon, On local Borg–Marchenko uniqueness results. *Comm. Math. Phys.* **211** (2000), 273–287.
39. F. Gesztesy and M. Zinchenko, On spectral theory for Schrödinger operators with strongly singular potentials. *Math. Nachr.* **279** (2006), 1041–1082.
40. D.J. Gilbert, On subordinacy and spectral multiplicity for a class of singular differential operators. *Proc. Roy. Soc. Edinburgh Sect. A* **128** (1998), 549–584.
41. I.C. Gohberg and M.G. Krein, *Theory and Applications of Volterra Operators in Hilbert Space*, Translations of Mathematical Monographs, vol. 24. American Mathematical Society, Providence, R.I., 1970.
42. S. Hassi, H. de Snoo and H. Winkler, Boundary-value problems for two-dimensional canonical systems. *Integral Equations Operator Theory* **36** (2000), 445–479.
43. S. Hassi, C. Remling and H. de Snoo, Subordinate solutions and spectral measures of canonical systems. *Integral Equations Operator Theory* **37** (2000), 48–63.
44. G. Herglotz, Über Potenzreihen mit positivem, reellem Teil im Einheitskreis [German]. *Leipz. Ber.* **63** (1911), 501–511.
45. L. Hörmander, *The Analysis of Linear Partial Differential Operators. I. Distribution Theory and Fourier Analysis.* Second edition. Grundlehren der Mathematischen Wissenschaften [Fundamental Principles of Mathematical Sciences], vol. 256. Springer-Verlag, Berlin, 1990.
46. R.O. Hryniv and Y.V. Mykytyuk, Self-adjointness of Schrödinger operators with singular potentials. *Methods Funct. Anal. Topology* **18** (2012), 152–159.
47. R. Hryniv and P. Sacks, Numerical solution of the inverse spectral problem for Bessel operators. *J. Comput. Appl. Math.* **235** (2010), 120–136.
48. P. Jonas, H. Langer and B. Textorius, Models and unitary equivalence of cyclic selfadjoint operators in Pontrjagin spaces. In: *Operator Theory and Complex Analysis (Sapporo, 1991)*. Oper. Theory Adv. Appl., vol. 59. Birkhäuser, Basel, 1992, pp. 252–284.
49. I. Kac, On Hilbert spaces generated by monotone Hermitian matrix-functions. *Har′ kov Gos. Univ. Uč. Zap. 34 = Zap. Mat. Otd. Fiz.-Mat. Fak. i Har′ kov. Mat. Obšč. (4)* **22** (1951), 95–113. 1950.
50. I.S. Kac, On the spectral multiplicity of a second-order differential operator. *Dokl. Akad. Nauk SSSR* **145** (1962), 510–513.
51. I.S. Kac, Spectral multiplicity of a second-order differential operator and expansion in eigenfunction. *Izv. Akad. Nauk SSSR Ser. Mat.* **27** (1963), 1081–1112.
52. I.S. Kac, Linear relations, generated by a canonical differential equation on an interval with a regular endpoint, and expansibility in eigenfunctions. *VINITI Deponirovannye Nauchnye Raboty* **195** (1985), 50 pp., b.o. 720, Deposited in Ukr NIINTI, no. 1453, 1984.
53. I.S. Kac, Inclusion of the Hamburger power moment problem in the spectral theory of canonical systems. *Zap. Nauchn. Sem. S.-Peterburg. Otdel. Mat. Inst. Steklov. (POMI)* **262** (1999) (Issled. po Linein. Oper. i Teor. Funkts. **27**), 147–171, 234. English translation: *J. Math. Sci. (New York)* **110** (2002), 2991–3004.

54. I.S. Kac and M.G. Krein, *On spectral functions of a string*, pp. 648–737. Izdat. 'Mir', Moscow, 1968. Addition II in F.V. Atkinson, Diskretnye i nepreryvnye granichnye zadachi (Russian translation of 'Discrete and Continuous Boundary Problems'). English translation: *Amer. Math. Soc. Transl. (2)* **103** (1974), 19–102.
55. M. Kaltenbäck, H. Winkler and H. Woracek, Generalized Nevanlinna functions with essentially positive spectrum. *J. Operator Theory* **55** (2006), 17–48.
56. M. Kaltenbäck and H. Woracek, Generalized resolvent matrices and spaces of analytic functions. *Integral Equations Operator Theory* **32** (1998), 282–318.
57. M. Kaltenbäck and H. Woracek, Pontryagin spaces of entire functions. II. *Integral Equations Operator Theory* **33** (1999), 305–380.
58. M. Kaltenbäck and H. Woracek, Pontryagin spaces of entire functions. III. *Acta Sci. Math. (Szeged)* **69** (2003), 241–310.
59. M. Kaltenbäck and H. Woracek, Pontryagin spaces of entire functions. IV. *Acta Sci. Math. (Szeged)* **72** (2006), 709–835.
60. M. Kaltenbäck and H. Woracek, Canonical differential equations of Hilbert–Schmidt type. In: *Operator Theory in Inner Product Spaces*. Oper. Theory Adv. Appl., vol. 175. Birkhäuser, Basel, 2007, pp. 159–168.
61. M. Kaltenbäck and H. Woracek, Pontryagin spaces of entire functions. VI. *Acta Sci. Math. (Szeged)* **76** (2010), 511–560.
62. M. Kaltenbäck and H. Woracek, Pontryagin spaces of entire functions. V. *Acta Sci. Math. (Szeged)* **77** (2011), 223–336.
63. K. Kodaira, The eigenvalue problem for ordinary differential equations of the second order and Heisenberg's theory of S-matrices. *Amer. J. Math.* **71** (1949), 921–945.
64. A. Kostenko, A. Sakhnovich and G. Teschl, Inverse eigenvalue problems for perturbed spherical Schrödinger operators. *Inverse Problems* **26** (2010), 105013, 14 pp.
65. A. Kostenko, A. Sakhnovich and G. Teschl, Commutation methods for Schrödinger operators with strongly singular potentials. *Math. Nachr.* **285** (2012), 392–410.
66. A. Kostenko, A. Sakhnovich and G. Teschl, Weyl–Titchmarsh theory for Schrödinger operators with strongly singular potentials. *Int. Math. Res. Not. IMRN* **2012** (2012), 1699–1747.
67. A. Kostenko and G. Teschl, On the singular Weyl–Titchmarsh function of perturbed spherical Schrödinger operators. *J. Differential Equations* **250** (2011), 3701–3739.
68. A. Kostenko and G. Teschl, Spectral asymptotics for perturbed spherical Schrödinger operators and applications to quantum scattering. *Comm. Math. Phys.* **322** (2013), 255–275.
69. A.M. Krall, Laguerre polynomial expansions in indefinite inner product spaces. *J. Math. Anal. Appl.* **70** (1979), 267–279.
70. A.M. Krall, On boundary values for the Laguerre operator in indefinite inner product spaces. *J. Math. Anal. Appl.* **85** (1982), 406–408.
71. M.G. Krein and H. Langer, Über die Q-Funktion eines π-hermiteschen Operators im Raume Π_κ [German]. *Acta Sci. Math. (Szeged)* **34** (1973), 191–230.
72. M.G. Krein and H. Langer, Über einige Fortsetzungsprobleme, die eng mit der Theorie hermitescher Operatoren im Raume Π_κ zusammenhängen. I. Einige Funktionenklassen und ihre Darstellungen [German]. *Math. Nachr.* **77** (1977), 187–236.
73. M.G. Krein and H. Langer, On some continuation problems which are closely related to the theory of operators in spaces Π_κ. IV. Continuous analogues of orthogonal polynomials on the unit circle with respect to an indefinite weight and related continuation problems for some classes of functions. *J. Operator Theory* **13** (1985), 299–417.
74. M.G. Krein and H. Langer, Continuation of Hermitian positive definite functions and related questions. *Integral Equations Operator Theory* **78** (2014), 1–69.
75. P. Kurasov and A. Luger, An operator theoretic interpretation of the generalized Titchmarsh–Weyl coefficient for a singular Sturm–Liouville problem. *Math. Phys. Anal. Geom.* **14** (2011), 115–151.
76. H. Langer, Spectral functions of definitizable operators in Krein spaces. In: *Functional Analysis (Dubrovnik, 1981)*. Lecture Notes in Math., vol. 948. Springer, Berlin, 1982, pp. 1–46.

77. H. Langer, A characterization of generalized zeros of negative type of functions of the class N_κ. In: *Advances in Invariant Subspaces and other Results of Operator Theory (Timişoara and Herculane, 1984)*. Oper. Theory Adv. Appl., vol. 17. Birkhäuser, Basel, 1986, pp. 201–212.
78. H. Langer, Transfer functions and local spectral uniqueness for Sturm–Liouville operators, canonical systems and strings. *Integral Equations Operator Theory* **85** (2016), 1–23.
79. H. Langer, A. Luger and V. Matsaev, Convergence of generalized Nevanlinna functions. *Acta Sci. Math. (Szeged)* **77** (2011), 425–437.
80. M. Langer and H. Woracek, A function space model for canonical systems with an inner singularity. *Acta Sci. Math. (Szeged)* **77** (2011), 101–165.
81. M. Langer and H. Woracek, A local inverse spectral theorem for Hamiltonian systems. *Inverse Problems* **27** (2011), 055002, 17 pp.
82. M. Langer and H. Woracek, Indefinite Hamiltonian systems whose Titchmarsh–Weyl coefficients have no finite generalized poles of non-positive type. *Oper. Matrices* **7** (2013), 477–555.
83. M. Langer and H. Woracek, The exponential type of the fundamental solution of an indefinite Hamiltonian system. *Complex Anal. Oper. Theory* **7** (2013), 285–312.
84. M. Langer and H. Woracek, Distributional representations of generalized Nevanlinna functions. *Math. Nachr.* **288** (2015), 1127–1149.
85. A.E. Lifschitz, *Magnetohydrodynamics and Spectral Theory*. Developments in Electromagnetic Theory and Applications, vol. 4. Kluwer Academic Publishers Group, Dordrecht, 1989.
86. A. Luger and C. Neuner, An operator theoretic interpretation of the generalized Titchmarsh–Weyl function for perturbed spherical Schrödinger operators. *Complex Anal. Oper. Theory* **9** (2015), 1391–1410.
87. A. Luger and C. Neuner, On the Weyl solution of the 1-dim Schrödinger operator with inverse fourth power potential. *Monatsh. Math.* **180** (2016), 295–303.
88. J.R. McLaughlin, Analytical methods for recovering coefficients in differential equations from spectral data. *SIAM Rev.* **28** (1986), 53–72.
89. H.-D. Niessen and A. Zettl, Singular Sturm–Liouville problems: the Friedrichs extension and comparison of eigenvalues. *Proc. London Math. Soc. (3)* **64** (1992), 545–578.
90. B.C. Orcutt, *Canonical Differential Equations*. ProQuest LLC, Ann Arbor, MI, 1969. PhD Thesis, University of Virginia.
91. G. Pick, Über die Beschränkungen analytischer Funktionen, welche durch vorgegebene Funktionswerte bewirkt werden [German]. *Math. Ann.* **77** (1915), 7–23.
92. C. Remling, Schrödinger operators and de Branges spaces. *J. Funct. Anal.* **196** (2002), 323–394.
93. C. Remling, Spectral Theory of Canonical Systems. De Gruyter Studies in Mathematics, vol. 70. De Gruyter, Berlin, 2018.
94. R. Romanov, Canonical systems and de Branges spaces, arXiv:1408.6022v1, 2014.
95. L.A. Sakhnovich, *Spectral Theory of Canonical Differential Systems. Method of Operator Identities*. Translated from the Russian manuscript by E. Melnichenko. Oper. Theory Adv. Appl., vol. 107. Birkhäuser Verlag, Basel, 1999.
96. A.M. Savchuk and A.A. Shkalikov, Sturm–Liouville operators with singular potentials. *Mat. Zametki* **66** (1999), 897–912.
97. L.O. Silva, G. Teschl and J.H. Toloza, Singular Schrödinger operators as self-adjoint extensions of N-entire operators. *Proc. Amer. Math. Soc.* **143** (2015), 2103–2115.
98. L.O. Silva and J.H. Toloza, A class of n-entire Schrödinger operators. *Complex Anal. Oper. Theory* **8** (2014), 1581–1599.
99. B. Simon, A new approach to inverse spectral theory. I. Fundamental formalism. *Ann. of Math. (2)* **150** (1999), 1029–1057.
100. H. de Snoo and H. Winkler, Canonical systems of differential equations with self-adjoint interface conditions on graphs. *Proc. Roy. Soc. Edinburgh Sect. A* **135** (2005), 297–315.
101. H. de Snoo and H. Winkler, Two-dimensional trace-normed canonical systems of differential equations and selfadjoint interface conditions. *Integral Equations Operator Theory* **51** (2005), 73–108.

102. H. Winkler, The inverse spectral problem for canonical systems. *Integral Equations Operator Theory* **22** (1995), 360–374.
103. H. Winkler, On transformations of canonical systems. In: *Operator Theory and Boundary Eigenvalue Problems (Vienna, 1993)*. Oper. Theory Adv. Appl., vol. 80. Birkhäuser, Basel, 1995, pp. 276–288.
104. H. Winkler and H. Woracek, Reparametrizations of non trace-normed Hamiltonians. In: *Spectral Theory, Mathematical System Theory, Evolution Equations, Differential and Difference equations*. Oper. Theory Adv. Appl., vol. 221. Birkhäuser/Springer Basel AG, Basel, 2012, pp. 667–690.
105. H. Winkler and H. Woracek, A growth condition for Hamiltonian systems related with Krein strings. *Acta Sci. Math. (Szeged)* **80** (2014), 31–94.
106. H. Woracek, Existence of zerofree functions N-associated to a de Branges Pontryagin space. *Monatsh. Math.* **162** (2011), 453–506.

Nevanlinna Domains and Uniform Approximation by Polyanalytic Polynomial Modules

Konstantin Fedorovskiy

1 Introduction

In this paper we consider the concept of a Nevanlinna domain and its various generalizations, modifications, and refinements (such as locally Nevanlinna domains, domains which are Nevanlinna with respect to some part of their boundaries, and g-Nevanlinna domains). All these classes of bounded simply connected domains in the complex plane appeared in connection with problems on uniform approximation of functions by polyanalytic polynomials and polyanalytic rational functions and, more generally, by elements of some special polynomial and rational modules of polyanalytic type.

For a set E in the complex plane $\mathbb{C}$ we denote by E°, $\overline{E}$ and ∂E the interior, the closure, and the boundary of E, respectively. As usual, we denote by $\widehat{\mathbb{C}} = \mathbb{C} \cup \{\infty\}$ the standard one-point compactification of $\mathbb{C}$.

Let $\mathbb{D}$ be the open unit disk $\{z \in \mathbb{C} : |z| < 1\}$, let $\mathbb{T} = \partial\mathbb{D}$ be the unit circle, and let $m(\cdot)$ stands for the (normalized) Lebesgue measure on $\mathbb{T}$.

Given an open set $U \subset \widehat{\mathbb{C}}$, let $\mathcal{O}(U)$ and $H^\infty(U)$ be the sets of all holomorphic and bounded holomorphic functions on U, respectively.

The author was partially supported by the Theoretical Physics and Mathematics Advancement Foundation “BASIS.” The results of Section 4 were obtained in frameworks of the project supported by the Russian Science Foundation, grant no. 22-11-00071 (https://rscf.ru/en/project/22-11-00071/).

K. Fedorovskiy (✉)
Faculty of Mechanics and Mathematics, Lomonosov Moscow State University, Moscow, Russia

Moscow Center for Fundamental and Applied Mathematics, Lomonosov Moscow State University, Moscow, Russia

St. Petersburg State University, St. Petersburg, Russia
e-mail: kfedorovs@yandex.ru

I. Binder et al. (eds.), *Function Spaces, Theory and Applications*, Fields Institute Communications 87, https://doi.org/10.1007/978-3-031-39270-2_6

The following definition was introduced in [10]:

Definition 1.1 A bounded simply connected domain G in $\mathbb{C}$ is said to be a Nevanlinna domain, if there exist two functions $u, v \in H^\infty(G)$ such that $v \not\equiv 0$ and the equality

$$\overline{z} = \frac{u(z)}{v(z)} \tag{1}$$

holds almost everywhere on ∂G in the sense of conformal mappings. It means that for $m(\cdot)$-almost all $\zeta \in \mathbb{T}$ the equality of angular boundary values is satisfied

$$\overline{\varphi(\zeta)} = \frac{(u \circ \varphi)(\zeta)}{(v \circ \varphi)(\zeta)}, \tag{2}$$

where φ is some conformal mapping from $\mathbb{D}$ onto G.

Note that the definition of a Nevanlinna domain does not depend on the choice of φ. Moreover, in view of the Luzin–Privalov boundary uniqueness theorem, the quotient u/v is uniquely defined in G (for a Nevanlinna domain G). If G is a Jordan domain with rectifiable boundary, then the equality (1) may be understood directly as the equality of angular boundary values almost everywhere with respect to the Lebesgue measure on ∂G. The equality (1) can be similarly understood on any rectifiable Jordan arc $\gamma \subset \partial G$ such that each point $a \in \gamma$ is not a limit point for the set $\partial G \setminus \gamma$. Note that for Jordan domains with rectifiable boundaries the concept of a Nevanlinna domain was introduced in [16] in slightly different terms.

It can be readily verified that every disk is a Nevanlinna domain, while every domain which is bounded by an ellipse which is not a circle or by a polygonal line is not Nevanlinna. Yet another interesting example of a Nevanlinna domain is Neumann's oval, that is, the domain bounded by the image of an ellipse (which is not a circle) centered at the origin under the mapping $z \mapsto 1/z$.

The main reason to introduce the concept of a Nevanlinna domain was its connections with the problem on uniform approximation by polyanalytic polynomials. Let us state this problem and present one result that shows the role of Nevanlinna domains.

A function f is said to be *polyanalytic of order n* on an open set $U \subset \mathbb{C}$, where $n \in \mathbb{N}$ is a fixed number, if it has the form

$$f(z) = \overline{z}^{n-1} f_{n-1}(z) + \cdots + \overline{z} f_1(z) + f_0(z), \tag{3}$$

where f_k, $k = 0, 1, \ldots, n-1$, are holomorphic functions in U, and $\overline{z}$ stands for the complex conjugate variable to $z = x + iy \in \mathbb{C}$, so that $\overline{z} = x - iy$. The set of all polyanalytic functions of order n in U will be denoted by $\mathcal{O}_n(U)$, so that $\mathcal{O}_1(U) = \mathcal{O}(U)$. Notice that every function $f \in \mathcal{O}_n(U)$ satisfies in U the elliptic partial differential equation $\overline{\partial}^n f = 0$, where $\overline{\partial} f = \partial f/\partial \overline{z} = \frac{1}{2}(\partial f/\partial x + i \partial f/\partial y)$ is the standard Cauchy–Riemann operator in $\mathbb{C}$. Moreover, it can be readily shown that every function f, which is continuous in U and satisfies therein the equation $\overline{\partial}^n f = 0$ in the sense of distributions, is real analytic in U and satisfies this equation in U in the classical sense; thus it has the form (3).

In what follows a polynomial and a rational function will mean a complex-valued polynomial and a rational function in the complex variable z, respectively. By polyanalytic polynomials and by polyanalytic rational functions (or order n), we mean the functions of the form (3) where $f_0, \dots, f_{n-1}$ are polynomials and rational functions, respectively. The spaces of all polyanalytic polynomials and polyanalytic rational functions (or order n) will be denoted by $\mathcal{P}_n$ and $\mathcal{R}_n$, respectively. Next, for a given set $E \subset \mathbb{C}$ we denote by $\mathcal{R}_{n,E}$ the space of all polyanalytic rational functions f of order n such that the corresponding rational functions $f_0, \dots, f_{n-1}$ from (3) have their poles outside E. We put $\mathcal{P} = \mathcal{P}_1$, $\mathcal{R} = \mathcal{R}_1$ and $\mathcal{R}_E = \mathcal{R}_{1,E}$.

For a compact set X let $C(X)$ be the space of all continuous complex-valued functions on X endowed with the uniform norm $\|f\|_X = \max_{z \in X} |f(z)|$. Let $P_n(X)$ be the space of functions that can be approximated uniformly on X by polyanalytic polynomials of order n, that is, $P_n(X)$ is the closure in $C(X)$ of the subspace $\{f|_X : f \in \mathcal{P}_n\}$. Furthermore, for a pair of compact sets X and Y such that $X \subseteq Y$, let $R_n(X, Y)$ be the closure in $C(X)$ of the subspace $\{f|_X : f \in \mathcal{R}_{n,Y}\}$, so that $R_n(X, Y)$ is the space of all functions, which can be approximated uniformly on X by polyanalytic rational functions of order n with singularities lying outside Y (which, in general, is assumed to be bigger than X). For brevity we will write $R_n(X) = R_n(X, X)$. It is clear that

$$P_n(X) \subset R_n(X, Y) \subset R_n(X) \subset A_n(X) := C(X) \cap \mathcal{O}_n(X^\circ),$$

where X° is the interior of X, and the problem we are interested in is to describe such compact sets X, for which it holds $A_n(X) = P_n(X)$.

It will be convenient for us now to defer our discussion of the history of this problem and revert to it later in Sect. 4 in a slightly more general setting.

A compact set $X \subset \mathbb{C}$ is said to be a *Carathéodory compact set*, if $\partial X = \partial \widehat{X}$, where $\widehat{X}$ is the union of X and all bounded (connected) components of $\mathbb{C} \setminus X$. The set $\widehat{X}$ is usually called the *hull* of X. By virtue of Runge's approximation theorem $\widehat{X} = \{z \in \mathbb{C} : \text{for each } h \in \mathcal{P}_n \text{ one has } |h(z)| \leq \max_{w \in X} |h(w)|\}$; thus $\widehat{X}$ is also called a *polynomial convex hull* of X. Similarly, a bounded domain G in $\mathbb{C}$ is called a *Carathéodory domain*, if $\partial G = \partial G_\infty$, where G_∞ stands for the unbounded connected component of $\widehat{\mathbb{C}} \setminus \overline{G}$. Notice that every Carathéodory domain G is simply connected and coincides with the interior of $\overline{G}$. The classes of Carathéodory sets (both domains and compact sets) appeared quite naturally in many topics of the theory of approximation by analytic functions. The following result was proved in [10, Theorem 2.2].

Theorem 1.1 *Let $n \geqslant 2$ be a fixed integer. For a Carathéodory domain $G \subset \mathbb{C}$ the following conditions are equivalent:*

(*a*) $C(\partial G) = R_n(\partial G, \overline{G})$.
(*b*) *G is not a Nevanlinna domain.*

For a Carathéodory compact set X the following conditions are equivalent:

(*a'*) $A_n(X) = P_n(X)$.
(*b'*) *Every bounded connected component of* $\mathbb{C} \setminus X$ *is not a Nevanlinna domain.*

If G in this theorem is such that the set $\mathbb{C} \setminus \overline{G}$ is connected, then we can replace (a) with the condition $C(\partial G) = P_n(\partial G)$. Indeed, it suffices to apply the Runge pole-pushing method that remains valid for polyanalytic functions.

In the proof of Theorem 4.1 in Sect. 4 we will explain some ideas and methods that underline the proof of the first part of this theorem.

It is important to notice that the approximability conditions that appeared in the problem of polyanalytic polynomial approximation are stated in terms of special analytic properties of a compact set where approximation is considered, and it distinguishes this problem significantly from the ones on uniform approximation by holomorphic and harmonic polynomials: approximability conditions in these problems are obtained in terms of topological properties of sets (we refer here the well-known classical Mergelyan and Walsh–Lebesgue theorems, respectively).

The structure of this paper is as follows. Section 2 is devoted to properties of Nevanlinna domains and their conformal mappings. In this section we show, in particular, how the properties of Nevanlinna domains are related with the properties of univalent functions belonging to model spaces (that is invariant with respect to the backward shift subspaces of the Hardy space H^2 in the unit disk) and how "complicated" (in the metrical sense) may be the structure of boundaries of Nevanlinna domains. In Sect. 3 we discuss the concept of a locally Nevanlinna domain, which was introduced in [10] and (in a different way) in [6] in connection with uniform approximation by polyanalytic polynomials on non-Carathéodory compact sets. Finally, in Sect. 4 we introduce and discuss the concept of g-Nevanlinna domain, which is a special generalization of the concept of a Nevanlinna domain related to the problem of uniform approximation of functions by elements of polyanalytic polynomial modules generated by general entire antiholomorphic functions g.

2 Nevanlinna Domains and Their Conformal Representations

In this section we explore the properties of conformal mappings from the unit disk $\mathbb{D}$ onto Nevanlinna domains and show how "complicated" (in the metrical sense) may be the boundaries of Nevanlinna domains.

The following simple proposition may be found in [10, Proposition 3.1]:

Proposition 2.1 *Let G be a bounded simply connected domain in $\mathbb{C}$ and let φ be some conformal mapping from $\mathbb{D}$ onto G. Then G is a Nevanlinna domain if and only if φ admits a (Nevanlinna type) pseudocontinuation, which means that there exist two functions $f_1, f_2 \in H^\infty(\overline{\mathbb{C}} \setminus \overline{\mathbb{D}})$ such that the equality of angular boundary values*

$$\varphi(\zeta) = \frac{f_1(\zeta)}{f_2(\zeta)}$$

holds for $m(\cdot)$-almost all $\zeta \in \mathbb{T}$.

Proof (Sketch of the proof) The proof of this proposition is straightforward and quite simple. For a function $g(\cdot)$ denote by g_* the function defined by the formula $g_*(w) = \overline{g(\overline{w})}$ for all w where this definition has sense. Now, if φ is a conformal mapping from $\mathbb{D}$ onto some Nevanlinna domain G, then one can define the functions f_1 and f_2 from Proposition 2.1 using u and v from (1) as follows: $f_1 = (u \circ \varphi \circ L)_*$ and $f_2 = (v \circ \varphi \circ L)_*$, where $L(z) = 1/z$. Conversely, if f_1 and f_2 are taken from Proposition 2.1, then the functions u and v demanded by the definition of a Nevanlinna domain can be taken as follows: $u = (f_1 \circ L)_* \circ \varphi^{-1}$ and $v = (f_2 \circ L)_* \circ \varphi^{-1}$. We omit remaining simple details. □

Let us mention two simple (and almost direct) corollaries of this proposition. The first one says that if G is a Nevanlinna domain, and if a rational function g is univalent in G and has its poles outside $\overline{G}$, then $g(G)$ is also a Nevanlinna domain. The second corollary is the following "density" property of Nevanlinna domains: any neighborhood of an arbitrary simple closed curve Γ contains an analytic Nevanlinna contour (i.e., the boundary of some Jordan Nevanlinna domain). In order to establish the latter property one needs to take some conformal mapping from the unit disk onto the domain bounded by Γ, and then approximate it uniformly on $\overline{\mathbb{D}}$ with an appropriate rate by univalent polynomials.

In what follows we need to use the essential connection between the concept of a Nevanlinna domain and the theory of model (sub)spaces. Recall that a function $\Theta \in H^\infty := H^\infty(\mathbb{D})$ is called an *inner function* if $|\Theta(\zeta)| = 1$ for $m(\cdot)$-almost all $\zeta \in \mathbb{T}$. Let us denote by H^2 the standard Hardy space (in $\mathbb{D}$). For an inner function Θ we define the space

$$K_\Theta := (\Theta H^2)^\perp = H^2 \ominus \Theta H^2.$$

In view of the Beurling theorem, the spaces $K_\Theta \subset H^2$ are exactly the invariant subspaces of the backward shift operator $f \mapsto (f(z) - f(0))/z$ in H^2. The spaces K_Θ are usually called *model spaces* (or *model subspaces*): this terminology was suggested by N. Nikolski in view of the remarkable role these spaces play in the functional model of Sz.-Nagy and Foiaş. The following result may be found in [17, Theorem 1] or [3, Theorems A and B]:

Proposition 2.2 *Let G be a bounded simply connected domain in $\mathbb{C}$ and let φ be some conformal mapping from $\mathbb{D}$ onto G. If G is a Nevanlinna domain, then there exists an inner function Θ such that $\varphi \in K_\Theta$. Reciprocally, if Θ is an inner function, then any bounded univalent function from the space K_Θ maps $\mathbb{D}$ conformally onto some Nevanlinna domain.*

Proof (Sketch of the proof) In order to prove this proposition it suffices to notice that in [14, Theorem 2.2.1] it is proved that a function $f \in H^2$ admits a (Nevanlinna type) pseudocontinuation if and only if f is not a cyclic vector for the backward shift operator $\mathcal{B}$ (let us recall that g is a cyclic vector for $\mathcal{B}$ if the linear span of elements $\{\mathcal{B}^n g : n = 0, 1, \dots\}$ is dense in H^2). Thus, if a function $f \in H^2$ belongs to some proper subspace of H^2, invariant with respect to $\mathcal{B}$, then f is not a cyclic vector for $\mathcal{B}$. It remains to recall that any proper invariant with respect to $\mathcal{B}$ subspace of H^2 has the form K_Θ for some inner function Θ. □

This proposition yields the following method for constructing Nevanlinna domains with prescribed properties: in the space K_Θ (for a special choice of inner function Θ) one needs to find a univalent function which possesses some special analytic properties. For instance, the question about possible regularity or irregularity of boundaries of Nevanlinna domains leads us to the question on whether univalent functions with the boundary behavior, corresponding to the demanded regularity of the boundary, do exist in some K_Θ.

The first important question that arose in this topic is to describe such inner functions Θ for which the corresponding space K_Θ contains bounded univalent functions. This question was recently solved in [5] and we are going to state the respective result. Recall that every inner function Θ can be expressed in the form $\Theta(z) = e^{ic}B(z)S(z)$, where c is some positive constant, while B and S are some Blaschke product and singular inner function, respectively. Recall also that a Blaschke product is a function of the form

$$B(z) = \prod_{n=1}^{\infty} \frac{\overline{a}_n}{|a_n|} \frac{z - a_n}{\overline{a}_n z - 1}, \tag{4}$$

where $(a_n)_{n=1}^{\infty}$ is some Blaschke sequence in $\mathbb{D}$ (i.e., $a_n \in \mathbb{D}$ for $n \in \mathbb{N}$ and $\sum_{n=1}^{\infty}(1 - |a_n|) < \infty$), while a singular inner function is a function of the form

$$S(z) = \exp\bigg(- \int_{\mathbb{T}} \frac{\zeta + z}{\zeta - z} \, d\mu_S(\zeta) \bigg), \tag{5}$$

where μ_S is some finite positive singular (with respect to $m(\cdot)$) measure on $\mathbb{T}$. The description of inner functions such that the corresponding model spaces contain bounded univalent functions is as follows (see [1, Theorem 1.1 and Corollary 1.2] and [5, Theorem 1]):

Theorem 2.1 *Let Θ be an inner function in $\mathbb{D}$. The space K_Θ contains bounded univalent functions if and only if one of the following two conditions is satisfied:*

(i) *Θ has a zero in $\mathbb{D}$.*
(ii) *$\Theta = S$ is a singular inner function and the measure μ_S is such that $\mu_S(E) > 0$ for some Beurling–Carleson set $E \subset \mathbb{T}$, which means that*

$$\int_{\mathbb{T}} \log \operatorname{dist}(\zeta, E)\, |d\zeta| > -\infty.$$

Beurling–Carleson sets first appeared as boundary zero sets of analytic functions in the disk which are smooth up to the boundary. It is worth to mention that property (ii) in Theorem 2.1 is also a necessary and sufficient condition for the space K_S to contain mildly smooth functions (e.g., functions belonging to the standard Dirichlet space in $\mathbb{D}$); see [15].

Theorem 2.1 was firstly proved in [1]. The obtained proof turned out to be technically complicated, and later on in [5] a new short proof was given. Let us show the main steps of this proof.

Proof (The proof of Theorem 2.1 given in [5]) First of all, let Θ have at least one zero, says a, in $\mathbb{D}$. Then the corresponding space K_Θ contains a bounded univalent function $1/(1-\overline{a}z)$. So, in the rest of the proof we are dealing with singular inner functions only. Let $\Theta = S = S_\mu$. Assume that there exists a bounded univalent function $f \in K_S$. Since f belongs to the Dirichlet space in $\mathbb{D}$, then, in view of [15, Theorem 2.1], there exists a Beurling–Carleson set E such that $\mu(E) > 0$.

Let us show that for every singular inner function $S = S_\mu$ such that $\operatorname{Supp}(\mu)$ (the support of μ) is a Carleson set, there exists a bounded univalent function $f \in K_S$. Consider an arbitrary (nonzero) function $f_0 \in K_S \cap C^1(\overline{\mathbb{D}})$, which exists in view of [15, Theorem 2.1]). Observe that

$$f_0(z) = \frac{1-S(z)}{2\pi i}\int_{\mathbb{T}} \frac{\zeta+z}{\zeta-z}\widetilde{f}(\zeta)\,d\nu(\zeta), \tag{6}$$

where ν is some appropriate Clark measure for S (see, for instance, [12]), and $\widetilde{f} \in L^2(\nu)$. Thus, f_0 can be analytically extended to $\overline{\mathbb{C}}\setminus\operatorname{Supp}(\mu)$. We have $f_0 \notin H^\infty(\overline{\mathbb{C}}\setminus\overline{\mathbb{D}})$. Indeed, otherwise the boundary values of functions $f_0|_{\mathbb{D}}$ and $f_0|_{\mathbb{C}\setminus\overline{\mathbb{D}}}$ coincide almost everywhere on $\mathbb{T}$, which is, clearly, impossible whenever $f_0 \not\equiv \text{const}$.

Fix a point a, $1 < |a| < 2$, such that $|f_0(a)| > 100(\|f_0\|_{\infty,\mathbb{T}} + \|f_0'\|_{\infty,\mathbb{T}})$. Define $f(z) = (1 - Af_0(z))/(z-a)$, where $A = 1/f_0(a)$. It is clear that f is bounded in $\mathbb{D}$. Furthermore, since f may be represented in the form of the integral from the right-hand side of (6), then $f \in K_S$. It remains to verify that the function f is univalent in $\overline{\mathbb{D}}$. Assuming by contradiction that $f(z) = f(w)$ for some pair of points $z, w \in \overline{\mathbb{D}}$, $z \neq w$, we have

$$\begin{aligned} 1 &= aA\frac{f_0(w)-f_0(z)}{w-z} + A\frac{wf_0(z)-zf_0(w)}{w-z} \\ &= aA\frac{f_0(w)-f_0(z)}{w-z} - Aw\frac{f_0(w)-f_0(z)}{w-z} + Af_0(w). \end{aligned}$$

It is clear that the moduli of all three summands in the right-hand side are less than $1/10$. The obtained contradiction completes the proof. □

Let us return to the problem on how to construct Nevanlinna domains with the prescribed properties using the abovementioned method from Propositions 2.1 and 2.2. Since in many instances questions about the regularity or irregularity of boundaries of planar domains may be reduced to the corresponding questions about the boundary regularity of conformal mappings from $\mathbb{D}$ onto the corresponding domains, we aim to find bounded univalent functions with certain properties of boundary regularity (or irregularity) in K_Θ for specially chosen inner functions Θ. It seems appropriate to consider separately two distinct cases: (i) $\Theta = B$ is a Blaschke product, and (ii) $\Theta = S$ is a singular inner function (it is clear that $K_{BS} = K_B \oplus BK_S$).

The first example of a Nevanlinna domain with nowhere analytic boundary was constructed in [19]. The respective domain was constructed as the conformal image of the unit disk under a map f of the form

$$f(z) = \sum_{n=1}^{\infty} \frac{c_n}{1 - \overline{a}_n z}, \tag{7}$$

where $(a_n)_{n\geq 1}$ is some (infinite) Blaschke sequence satisfying the Carleson condition

$$\inf_{n\in\mathbb{N}} \prod_{\substack{k=1\\k\neq n}}^{\infty} \left| \frac{a_n - a_k}{1 - a_n\overline{a}_k} \right| > 0,$$

and $(c_n)_{n\geq 1}$ is an appropriately chosen sequence of coefficients. Such Blaschke sequences are called *interpolating*, and for any interpolating Blaschke sequence $(a_n)_{n\geq 1}$ the sequence of functions

$$\left\{ \frac{\sqrt{1 - |a_n|^2}}{1 - \overline{a}_n z} \right\}$$

forms a Riesz basis in the corresponding space K_B.

In [17, Theorem 3] it was shown that for every $\alpha \in (0, 1)$ there exists a Nevanlinna domain with boundary in the class C^1 but not in the class $C^{1,\alpha}$. The construction in [17] is rather complicated and technically involved. The main idea was to use an orthonormal basis in the space K_B (namely, the Malmquist–Walsh basis) instead of the Riesz basis consisting of the corresponding Cauchy kernels. Later on it was proved in [3, Theorem 2] that for every $\alpha \in (0, 1)$ and for every closed subset $E \subseteq \mathbb{T}$ there exists an interpolating Blaschke sequence $(a_n)_{n\geq 1}$ such that the set of its limit points is equal to E, and the space K_B, where B is the corresponding Blaschke product, contains a univalent function f of the form (7) which maps $\mathbb{D}$ conformally onto a Nevanlinna domain $f(\mathbb{D})$ with boundary in the class C^1 but not in the class $C^{1,\alpha}$.

Furthermore, in [3], there is a construction of a function f of the form (7) such that f is univalent in $\mathbb{D}$ but $f' \notin H^p$ for any $p > 1$. It means that the boundary of a Nevanlinna domain $f(\mathbb{D})$ is "almost" non-rectifiable. The first example of a Jordan Nevanlinna domain with non-rectifiable boundary was constructed in [22]. The corresponding domain is also $f(\mathbb{D})$, for some function f of the form (7) univalent in the unit disk.

Next, in [23] an example of a Nevanlinna domain G such that $\dim_H(\partial G) = \log_2 3$ was constructed, where $\dim_H$ stands for the Hausdorff dimension. As before, $G = f(\mathbb{D})$ for a suitable function f of the form (7). The definition of the Hausdorff dimension may be found, for instance, in [18, Chapter 4], but we present it here for the sake of completeness. Let $D(a, r)$ stand for the open disk with center at the point $a \in \mathbb{C}$ and with radius $r > 0$. For a bounded set $E \subset \mathbb{C}$ its s-dimensional Hausdorff measure $\mathcal{H}^s(E)$, where $s > 0$, is defined as follows:

$$\mathcal{H}^s(E) = \lim_{\delta \to 0} \inf_{\{D_j\}} \sum_j r_j^s,$$

where the infimum is taken over all coverings of E by families of disks $\{D_j\}$, $D_j = D(z_j, r_j)$, of radius at most δ. By definition, the Hausdorff dimension $\dim_H(E)$ is the unique number such that $\mathcal{H}^s(E) = \infty$ for every $s < \dim_H(E)$, while $\mathcal{H}^t(E) = 0$ for every $t > \dim_H(E)$.

Finally, in [4] the following result was proved:

Theorem 2.2 *For every $\beta \in [1, 2]$ there exists a function f of the form* (7) *univalent in $\mathbb{D}$ and such that the Nevanlinna domain $G = f(\mathbb{D})$ satisfies the property $\dim_H(\partial_a G) = \beta$, where $\partial_a G$ is the accessible part of the boundary ∂G.*

In order to understand this result better, let us briefly explain the notion of the accessible part of the boundary. Given a bounded simply connected domain G, we consider the set $\partial_a G \subset \partial G$ which consists of all points of ∂G being accessible from G by some curve. According to [26, Propositions 2.14 and 2.17], the equality

$$\partial_a G = \{\varphi(\zeta) \colon \zeta \in \mathcal{F}(\varphi)\}$$

takes place, where φ is some conformal mapping from the unit disk $\mathbb{D}$ onto G and $\mathcal{F}(\varphi)$ is its Fatou set, that is, the set of all points $\zeta \in \mathbb{T}$, where the angular boundary values $\varphi(\zeta)$ exist. It can be shown that $\partial_a G$ is a Borel set (see, for instance, [11, Section 2]). It is clear that the set $\partial_a G$ depends only on the domain G but not on the choice of f.

The definition of Nevanlinna domains (see (1) and its interpretation (2)) imposes conditions only on the accessible part $\partial_a G$ of their boundaries. Exactly because of this reason it seems more accurate and adequate to study the question about the existence of Nevanlinna domains with the prescribed properties of accessible boundaries.

The results presented above show that Nevanlinna domains may have boundaries, which are "arbitrarily" complicated from the metric point of view. Notice that all aforesaid constructions of Nevanlinna domains have the form $f(\mathbb{D})$ with $f \in K_B$ for some Blaschke product B. Regarding this it is worth to note that the Paley–Wiener space $\mathcal{P}W_{[0,1]}$ (that is the Fourier image of $L^2[0, 1]$, considered as a space of functions analytic in the upper half-plane $\mathbb{C}_+$) contains bounded univalent functions.

Moreover, as it is shown in [4, Theorem 4], for every $\beta \in [1,2]$ there exists a univalent function f belonging to the space $\mathcal{P}W^{\infty}_{[0,1]}$ such that the Nevanlinna domain $G = f(\mathbb{C}_+)$ satisfies the property $\dim_H(\partial G) = \beta$. Notice, that here, in contrast to Theorem 2.2, we have the corresponding result only for ∂G, but not for $\partial_a G$.

We end this section by considering relationships between the concepts of Nevanlinna and quadrature domains. We start with the case of Jordan domains with analytic boundaries. Let Γ be a simple closed analytic curve. It is well-known (see, for instance [29, Sections 1,2]) that in this case there exist an open set U, $\Gamma \subset U$, and a function S holomorphic in U, such that

$$\Gamma = \{z \in U : \overline{z} = S(z)\}.$$

The function S is called a *Schwarz function* of Γ. Thus, the (Jordan) domain $D(\Gamma)$ bounded by Γ is a Nevanlinna domain if and only if the corresponding Schwarz function S is meromorphic in $D(\Gamma)$. As it is shown in [13, Chapter 14] it is equivalent to the fact that $D(\Gamma)$ is a conformal image of $\mathbb{D}$ under a mapping by some rational function univalent in $\mathbb{D}$ (it is clear that any such function admits a pseudocontinuation).

In fact, the concept of a Schwarz function is often considered in the following strengthened way. Let G be a bounded (not necessarily simply connected) domain possessing the following property: there exist a compact set $K \subset G$ and a function S holomorphic in $G \setminus K$, continuous up to ∂G, and such that $\overline{z} = S(z)$ on ∂G. In the latter case the aforesaid function S is called the *one-sided Schwarz function* of ∂G. According to the breakthrough result by Sakai [28, Theorem 5.2], if the boundary of some domain admits the one-sided Schwarz function, then it consists of finitely many analytic curves. The concept of a one-sided Schwarz function has deep connection with the important concept of a quadrature domain. Let us recall that a quadrature domain in the wide sense is a domain G satisfying the following property: there exists a distribution T with support $\mathrm{Supp}(T) \subset G$ such that for every holomorphic and integrable (with respect to the planar Lebesgue measure in G) function h in G we have $\iint_G h(z)\,dxdy = T(f)$. If T has finite support, then G is a quadrature domain, or quadrature domain in the classical sense. It can be shown that the boundary of any quadrature domain in the wide sense admits a one-sided Schwarz function; see [29, Section 4.2]. For instance, if G is a Jordan domain with analytic boundary and if G is, at the same time, a Nevanlinna domain, then G is a quadrature domain (the demanded distribution T in this case is the finite sum of Dirac delta functions and their derivatives supported at the poles of the Schwarz function S of $\Gamma = \partial G$ with coefficients related to the residues of S at these points).

Taking into account all these observations we can treat the concept of a Nevanlinna domain (more precisely, the function u/v from (1)) as a special generalization of the concept of a Schwarz function (both in its classical and one-sided versions). In view of Sakai's result stated above and according to Theorem 2.2, we can get far away from domains with piecewise analytic boundaries (and, therefore, from quadrature domains) if we consider Nevanlinna domains instead of domains whose boundaries admit the one-sided Schwarz function. More information about relationships between Nevanlinna and quadrature domains may be found in [7].

At the end of this section let us show how to construct a Jordan Nevanlinna domain whose Schwarz function of the boundary has the prescribed finite set of poles inside the domain under consideration. The idea of this construction was suggested by P. Paramonov in personal communication. Let $A = \{a_1, \dots, a_n\}$, $n \in \mathbb{N}$, be a finite set of points such that $a_j \in \widehat{\mathbb{C}} \setminus \overline{\mathbb{D}}$. Let f be a function holomorphic in $\widehat{\mathbb{C}} \setminus A$ such that f is univalent in $\mathbb{D}$ and injective on $\mathbb{T}$. Then the domain $f(\mathbb{D})$ is Jordan and its boundary admits a Schwarz function that has the form

$$S(z) = f_*\big(1/f^{-1}(z)\big), \quad \text{where } f_*(w) = \overline{f(\overline{w})}.$$

Moreover, the set of singularities of S consists of points $\{b_1, \dots, b_n\}$, where $b_j = f(z_j)$ as $z_j = 1/\overline{a_j}$, $j = 1, \dots, n$. Furthermore, the type of singularity of S at b_j coincides with the type of singularity of f at a_j, $j = 1, \dots, n$. For instance, if all points a_j, $j = 1, \dots, n$ are poles, then $f(\mathbb{D})$ is a Nevanlinna domain. Assume that $a_1 = \infty$, $a_j \in \mathbb{C} \setminus \overline{\mathbb{D}}$ for $j = 2, \dots, n$, and take a set of positive integers $\{m_1, \dots, m_n\}$. For $N = \sum_{j=1}^{n} m_j - n + 1$ and for sufficiently small $\varepsilon > 0$ the function

$$f(z) = z + \varepsilon \frac{z^N (z - z_2) \times \cdots \times (z - z_n)}{(z - a_2)^{m_2} \times \cdots \times (z - a_n)^{m_n}}$$

is univalent in the disk $\{|z| < r\}$ for some $r > 1$ (because the real part of its derivative is positive in this disk), and the points $a_1, \dots, a_n$ are poles of f of orders $m_1, \dots, m_n$, respectively. Thus, the domain $f(\mathbb{D})$ is the desired Jordan Nevanlinna domain, and the points $b_1, \dots, b_n$ are the poles of S of the same orders, and S has no other singularities.

At the end of this section let us mention a recent work [32], where the concept of a Nevanlinna domain is considered in connection with free boundary problems in the spirit of the aforesaid Sakai's theorem.

3 Locally Nevanlinna Domains

The concept of a locally Nevanlinna domain appeared in the problem of uniform approximation by polyanalytic polynomials on non-Carathéodory compact sets. There are two different approaches to define the local Nevanlinna property of a domain. We start with the definition given in [10, Definition 4.2] for Jordan domains with rectifiable boundaries and in [11, Definition 2] in the general case.

Definition 3.1 A bounded simply connected domain G in $\mathbb{C}$ is said to be locally Nevanlinna domain, if there exists a compact set $K \subset G$ and two functions $u, v \in H^\infty(G \setminus K)$ such that $v \not\equiv 0$ and the equality (1) holds almost everywhere on ∂G

in the sense of conformal mappings (which means that the equality (2) holds for $m(\cdot)$-almost all point $\zeta \in \mathbb{T}$, where φ is some conformal mapping from $\mathbb{D}$ onto G).

It is clear that the definition of a locally Nevanlinna domain extends the definition of a Nevanlinna one (the case when $K = \emptyset$). For a (locally) Nevanlinna domain we will put $F = u/v$. Notice that if G is a locally Nevanlinna domain, then there are many different possibilities to define the desired compact set K. We will call the pair (K, F) admissible for G, if the local Nevanlinna property is satisfied with these K and F. The following property is the direct consequence of the Luzin–Privalov boundary uniqueness theorem: if G is a locally Nevanlinna domain and if (K_1, F_1) and (K_2, F_2) are two admissible pairs for G, then $F_1 = F_2$ on $G \setminus \widehat{K_1 \cup K_2}$.

Let (K, F) be an admissible pair for some locally Nevanlinna domain G. It is always possible to assume that K is minimal, which means that F cannot be meromorphically continued from $G \setminus K$ to $G \setminus K_1$ for any proper compact subset K_1 of K. The choice of such minimal K is also not unique. Thus, in what follows we will always need to select K appropriately in order to apply the forthcoming approximation results.

In fact, the property of a domain to be a locally Nevanlinna one is closely related with one special pseudocontinuation property of a conformal mapping from the unit disk onto the domain under consideration. Indeed, if G is a locally Nevanlinna domain and f is some conformal mapping from $\mathbb{D}$ onto G, then f admits a pseudocontinuation onto some open set U such that $\overline{\mathbb{D}} \subset U$. It means that there exist two functions $f_1, f_2 \in H^\infty(U \setminus \overline{\mathbb{D}})$ such that for almost all $\zeta \in \mathbb{T}$ the equality $f(\zeta) = f_1(\zeta)/f_2(\zeta)$ is satisfied, where $f(\zeta)$ is the angular boundary value of f from $\mathbb{D}$, while $f_1(\zeta)$ and $f_2(\zeta)$ are angular boundary values of f_1 and f_2 from $U \setminus \overline{\mathbb{D}}$. The proof of this fact is exactly the same as the proof of Proposition 2.1. Notice that this pseudocontinuation property is quite weaker than the classical pseudocontinuation property obtained in Proposition 2.1 and, unfortunately, it does not yield such far reaching consequences, that the classical pseudocontinuation property yields for Nevanlinna domain.

Let us give two simple examples of locally Nevanlinna domains which are not Nevanlinna domains. The first example is a domain, bounded by an ellipse which is not a circle. Indeed, if the ellipse $\Gamma_{a,b}$ under consideration is given by the equation $x^2/a^2 + y^2/b^2 = 1$, where $a > b > 0$, then the Schwarz function $S_{a,b}$ of $\Gamma_{a,b}$ has the form $\alpha z - \beta\sqrt{z^2 - c^2}$, where $\alpha = (a^2 + b^2)/c^2$, $\beta = 2ab/c^2$ and $c = \sqrt{a^2 - b^2} > 0$. In fact, the points $\pm c$ are the foci of $\Gamma_{a,b}$. It is clear that $S_{a,b}$ is holomorphic in $G_{a,b} \setminus [-c, c]$, where $G_{a,b}$ is the domain bounded by $\Gamma_{a,b}$. The function $S_{a,b}$ cannot be represented in the form u/v with $u, v \in H^\infty(G_{a,b})$ in view of the Lusin–Privalov boundary uniqueness theorem. Thus $G_{a,b}$ is a locally Nevanlinna, but not a Nevanlinna domain. The next example is the image of $\mathbb{D}$ under the mapping $z \mapsto \exp(z)$. This domain has an analytic boundary whose Schwarz function is $\exp(1/\log z)$, where $\log z$ stands for the principal branch of the multivalued logarithm in the right-hand half-plane $\{z : \operatorname{Re} z > 0\}$. It is clear again that the domain $\exp(\mathbb{D})$ is locally Nevanlinna, but not Nevanlinna since the Schwarz function of its boundary has an essential singularity.

In order to show the role of the local Nevanlinna property in the polyanalytic polynomial approximation theory, let us state the following results which were proved in [11, Theorems 3]:

Theorem 3.1 *Let Y be a Carathéodory compact set such that $Y^\circ \neq \emptyset$, and let $X \subset Y^\circ$ be a compact set. Let $\Omega_1, \dots, \Omega_M$, $M \in \mathbb{Z}_+$, be a collection of all connected components of Y° such that $X_m := \Omega_m \cap X \neq \emptyset$, $m = 1, \dots, M$. Suppose that $P_n(X) = A_n(X)$ for some integer $n \geqslant 2$ and any connected component of Y° is not a Nevanlinna domain. Then, if one of the following two conditions is satisfied for every $m = 1, \dots, M$:*

(a) Ω_m is not a locally Nevanlinna domain, or
(b) Ω_m is a locally Nevanlinna domain, (K_m, F_m) is an admissible pair for Ω_m, and F_m cannot be meromorphically continued from $\Omega_m \setminus \widehat{K_m \cup X_m}$ to $\Omega_m \setminus \widehat{X_m}$,

then $A_n(X \cup \partial Y) = R_n(X \cup \partial Y, Y)$.

If, moreover, each bounded connected component of $\mathbb{C} \setminus Y$ is not a Nevanlinna domain, then $A_n(X \cup \partial Y) = P_n(X \cup \partial Y)$.

An interest and importance of this result and its further generalizations and refinements considered below are related to the fact that the question whether the equality $P_n(X) = A_n(X)$ holds for a given compact set X can be reduced to the question whether the equality $R_n(X \cap \overline{G}, \overline{G}) = A_n(X \cap \overline{G})$ is satisfied for certain Carathédory domains G (see [6, Theorem 3]).

In order to state next result we need to introduce one simple topological concept. Let U be a bounded open set in $\mathbb{C}$, and let $\gamma \subset \partial U$ be a closed Jordan arc. One says that γ is a *gate* for U if there exists an open connected set V such that $V \setminus \gamma = V_1 \cup V_2$, where V_1 and V_2 are nonempty open connected sets such that $V_1 \subset U$ and $V_2 \subset \mathbb{C} \setminus \overline{U}$. Using the concept of a *free arc* introduced in [27] one can say that γ is a gate for a given domain U, if and only if γ is a free arc both for U and $\mathbb{C} \setminus \overline{U}$. If U is a Jordan domain, then any closed subarc $\gamma \subset \partial U$ is a gate for U. For instance, it was shown in [11, Proposition 4] that if U is a bounded open set in $\mathbb{C}$ and μ is a measure supported on $\overline{U}$ such that μ is orthogonal to rational functions with poles located outside $\overline{U}$, and if γ is a rectifiable gate for U, then $\mu|_\gamma$ is absolutely continuous with respect to $d\zeta|_\gamma$. The following theorem was proved in [11, Theorem 4]:

Theorem 3.2 *Let D be a bounded simply connected domain in $\mathbb{C}$ and $\Gamma = \partial D$.*

(1) If D is a Nevanlinna domain, then $R_n(\Gamma, D) \neq C(\Gamma)$ for every integer $n \geqslant 1$.
(2) Let D be a locally Nevanlinna, but not a Nevanlinna domain, and suppose that there exists a rectifiable gate γ for D; then $R_2(\Gamma, D) = C(\Gamma)$.
(3) Let D and γ be as in the previous statement. Let $X \subset \overline{D}$ be a compact set such that $\Gamma \subset X$. Put $E = \overline{X \setminus \Gamma}$ and assume that:

(i) $D \setminus \widehat{E}$ is connected and γ is a gate for $D \setminus \widehat{E}$.
(ii) A pair (K, F) is an admissible pair for D, and F cannot be meromorphically continued from $D \setminus \widehat{E \cup K}$ to $D \setminus \widehat{E}$.

(*iii*) $R_n(E, \overline{D}) = A_n(E)$ *for some integer* $n \geqslant 2$.

Then $R_n(X, \overline{D}) = A_n(X)$.

Notice that in this theorem we do not require that D is a Carathéodory domain in contrast to previous results about polyanalytic approximation. The proofs of the first and the second statements of this theorem are similar to the proofs of the corresponding assertions in Theorem 4.1.

We end this section presenting yet another version of the concept of a locally Nevanlinna domain, which is slightly different to the one given above. Let E be a subset of $\mathbb{C}$. A nonempty subset $I \subset E$ is called a *border* of E (or, shortly, E-border), if every point $a \in I$ has a neighborhood V in $\mathbb{C}$ such that there is a homeomorphism h of $V \cap E$ onto a relatively open subset of the closed upper half-plane with the property $V \cap I = \{z \in V : \operatorname{Im} h(z) = 0\}$. Thus, any border is locally a Jordan arc. In what follows we also require that all borders are locally rectifiable. Let now D be a domain in $\mathbb{C}$ and I be some $\overline{D}$-border. Furthermore, let $f \in H^\infty(D)$. Then angular boundary values $f(\zeta)$ exist for almost all points $\zeta \in I$ with respect to the length on I.

Given a bounded domain $D \subset \mathbb{C}$, a relatively open set $I \subset \partial D$, and a number $\varepsilon > 0$, we define the set $D_\varepsilon(I) = \{z \in D : \operatorname{dist}(z, \partial D \setminus I) > \varepsilon\}$. If $I = \partial D$ we put $D_\varepsilon(I) = D$.

The following definition was introduced in [6, Definition 4]:

Definition 3.2 Let D be a bounded domain in $\mathbb{C}$ and I be an $\overline{D}$-border. One says that D is I-Nevanlinna, if one can find a function F which is meromorphic in D and has angular limits $F(\zeta)$ from D for almost all points $\zeta \in I$ satisfying the property $F(\zeta) = \overline{\zeta}$ almost everywhere on I, and, moreover, for every sufficiently small $\varepsilon > 0$ there are functions $u_\varepsilon, v_\varepsilon \in H^\infty(D_\varepsilon(I))$ with $v_\varepsilon \not\equiv 0$ such that $F(z) = u_\varepsilon(z)/v_\varepsilon(z)$ on $D_\varepsilon(I)$.

According to the Luzin–Privalov boundary uniqueness theorem, the function F in the definition of I-Nevanlinna domain is unique whenever it exists. Nevertheless, it is not clear whether F can be represented as a quotient of two bounded holomorphic functions on the whole domain D.

Notice that for Jordan domains D with rectifiable boundaries, one has that D is a Nevanlinna domain (in the sense of the initial definition) if and only if D is ∂D-Nevanlinna one.

The next result was obtained in [6, Theorem 4] and it points to the same direction as Theorem 3.1 and statement 3 of Theorem 3.2.

Theorem 3.3 *Let* $n \geqslant 2$, *D be a Carathéodory domain and X be a compact subset of* $\overline{D}$ *such that* $A_n(X) = R_n(X, \overline{D})$. *Assume that* $\partial D \setminus X$ *is* $\overline{D}$*-border. Let* $\{U_j\}_{j \in J}$ *be the set of all connected components of* $D \setminus X$ *such that* $E_j = \partial U_j \setminus X \neq \emptyset$. *If* U_j *is not* E_j*-Nevanlinna domain for every* $j \in J$, *then* $A_n(X \cup \partial D) = R_n(X \cup \partial D, \overline{D})$.

This result generalizes Theorem 3.1 in several situations, but to verify its conditions seems to be a bit complicated. Let us present one corollary of Theorem 3.3

that is stated in terms which can be regarded as a natural generalization of the concept of analytic independence of analytic arcs. Recall this concept. If γ is an analytic arc (i.e., an image of a segment under mapping by a function univalent in its neighborhood), then there exist a neighborhood U of γ and a function S holomorphic in U such that $\overline{z} = S(z)$ for all $z \in \gamma$; this function S is called a Schwarz function of the analytic arc γ (similarly to the Schwarz function of an analytic contour). One says that two analytic arcs γ_1 and γ_2 are analytically dependent, if the corresponding analytic elements (U_1, S_1) and (U_2, S_2) are analytic continuations of each other; otherwise γ_1 and γ_2 are analytically independent. It follows immediately from the definition of a Nevanlinna domain that if D is a bounded simply connected domain whose boundary contains two analytically independent analytic arcs both of which are gates for D (or $\overline{D}$-borders), then D is not a Nevanlinna domain.

Let now D be a bounded domain in $\mathbb{C}$ and let γ_1 and γ_2 be (open) Jordan arcs which are $\overline{D}$-borders. Let us additionally assume that for $j = 1, 2$ there are neighborhoods U_j of γ_j in $\mathbb{C}$ such that $D_j = D \cap U_j$ are γ_j-Nevanlinna domains. Moreover, let F_j, $j = 1, 2$, be the corresponding functions from Definition 3.2. One says that γ_1 and γ_2 are Nevanlinna D-independent, if there is no path in D such that the meromorphic elements (D_j, F_j) are meromorphic continuation of each other along this path; otherwise γ_1 and γ_2 are Nevanlinna D-dependent.

Applying Luzin–Privalov boundary uniqueness theorem once again, we observe that if γ_1 and γ_2 are Nevanlinna D-independent, then for every $\overline{D}$-border Υ containing $\gamma_1 \cup \gamma_2$ the domain D is not Υ-Nevanlinna. Notice that if γ_1 and γ_2 are analytically independent $\overline{D}$-borders, then they are also Nevanlinna D-independent. Thus, two sides of a polygonal domain D are Nevanlinna D-independent, if they do not belong to the same straight line.

Proposition 3.1 *Let $n \geqslant 2$ be an integer. Let D be a Carathéodory domain such that $\mathbb{C} \setminus \overline{D}$ is connected (so that the set $\overline{D}$ does not separate the plane) and ∂D contains Nevanlinna D-independent arcs γ_1 and γ_2. Let X be a compact subset of D such that $A_n(X) = P_n(X)$. Then $A_n(X \cup \partial D) = P_n(X \cup \partial D)$.*

4 g-Nevanlinna Domains

Let g_0 be an entire function, and let $g(z) = g_0(\overline{z})$ (i.e., g is an entire antiholomorphic function). In this section we are dealing with the problem on uniform approximation of functions by elements of polynomial and rational modules generated by g.

The spaces of functions $\mathcal{P}(g) = \{f_1 g + f_0 : f_0, f_1 \in \mathcal{P}\}$ and $\mathcal{R}_E(g) = \{f_1 g + f_0 : f_0, f_1 \in \mathcal{R}_E\}$ have the structure of a holomorphic module generated by g; these spaces are called polynomial and rational modules of polyanalytic type determined by g, respectively. If $g_0(z) = z$, then we are dealing with spaces of bianalytic polynomials and bianalytic rational functions with poles outside E.

Another interesting and important example appears in the case when $g_0(z) = \overline{z}^d$ for some integer $d \geqslant 2$.

Let X be a compact set in $\mathbb{C}$, and let K be a compact set such that $X \subseteq K$. Denote by $P(X, g)$ and $R_K(X, g)$ the closures in $C(X)$ of subspaces $\{f|_X : f \in \mathcal{P}(g)\}$ and $\{f|_X : f \in \mathcal{R}_K(g)\}$, respectively. In other words, $P(X, g)$ and $R_K(X, g)$ are exactly the spaces of functions that can be approximated uniformly on X by elements of polynomial and rational modules generated by g. Let $R(X, g) = R_X(X, g)$. It can be readily verified that

$$P(X, g) \subset R_K(X, g) \subset A(X, g) = \{f \in C(X) : f|_{X^\circ} \\ = gh_1 + h_0 \text{ with } h_0, h_1 \in \mathcal{O}(X^\circ)\}.$$

Notice that $P(X, \overline{z}) = P_2(X)$, $R_K(X, \overline{z}) = R_2(X, K)$, $R(X, \overline{z}) = R_2(X)$, and $A(X, \overline{z}) = A_2(X)$.

The problem we are interested in is to describe such compact sets X for which we have $P(X, g) = A(X, g)$. This problem is closely and naturally related to the question on whether $R_K(X, g) = A(X, g)$ for some appropriately chosen compact sets K, $X \subseteq K$. In connection with this problem let us mention two papers [8] and [9] where several sufficient conditions, in order that the equalities $P(X, g) = A(X, g)$ and $R(X, g) = A(X, g)$ take place, were obtained in terms of topological and metrical properties of X. Let us also mention that the uniform approximation by elements of polynomial and rational modules is studied since the last quarter of the twentieth century and we could refer to the works [25, 30, 31, 33, 35] concerning this matter.

Later on, the problem on description of compact sets X for which the equality $R_2(X) = A_2(X)$ (and, more generally, the equality $R_n(X) = A_n(X)$ and for integer $n \geqslant 2$) is satisfied was solved in [20] and [21]. The answer is as follows: this equality takes place for every compact set $X \subset \mathbb{C}$. This highly nontrivial result confirmed the conjecture stated by Verdera in [34].

The problem about description of compact sets X for which $P_n(X) = A_n(X)$ is already discussed above. For instance, Theorem 1.1 gives an approximation criterion for Carathéodory compact sets. Several sufficient approximation conditions in this problem for non-Carathéodory compact sets are presented in the previous section. In [2] the first results were obtained about uniform approximation of functions by elements of polynomial modules $\mathcal{P}(\overline{z}^d)$, $d \geqslant 2$. As in the case of polyanalytic polynomial approximation, in this case there are approximation criteria formulated in terms of new special analytic characteristics of sets in the plane, similar to the concept of a Nevanlinna domain.

Definition 4.1 Let G be a bounded simply connected domain in $\mathbb{C}$ and g be as above. The domain G is said to be *g-Nevanlinna domain*, if there are two functions $u, v \in H^\infty(G)$, $v \not\equiv 0$ such that the equality

$$g(z) = \frac{u(z)}{v(z)} \tag{8}$$

holds almost everywhere on ∂G in the sense of conformal mappings. It means that for $m(\cdot)$-almost all points $\zeta \in \mathbb{T}$ the equality of angular boundary values

$$g(\varphi(\zeta)) = \frac{u(\varphi(\zeta))}{v(\varphi(\zeta))}$$

is satisfied, where φ is some conformal mapping from $\mathbb{D}$ onto G.

Notice that for $g(z) = \overline{z}$ we have the standard definition of a Nevanlinna domain given above. In the case when $g(z) = \overline{z}^d$, $d \geqslant 2$, this definition was introduced in [2]. Notice that this definition is consistent in the sense that it does not depend on the choice of φ. Also, as in the case of the definition of a Nevanlinna domain, the quotient u/v is uniquely determined in a g-Nevanlinna domain G. Moreover, in many instances it is convenient to assume that u and v have no common zeros.

For better understanding let us present two simple examples. Let $\varphi(z) = \sqrt[d]{2-w}$ for an integer $d \geqslant 2$, where one considers the principal branch of the multivalued root function defined outside the ray $[2, +\infty)$ and possessing the property $\varphi(-2) > 0$. Then it can be readily verified that the domain $\varphi(\mathbb{D})$ is $\overline{z}^d$-Nevanlinna, but not a Nevanlinna domain. At the same time, the ellipse $G_{a,b}$ defined in the previous section is not $\overline{z}^d$-Nevanlinna domain for every integer $d \geqslant 1$.

Theorem 4.1 *Let G be a Carathéodory domain in $\mathbb{C}$ and let g be as above. The equality $C(\partial G) = R_{\overline{G}}(\partial G, g)$ is satisfied if and only if G is not a g-Nevanlinna domain.*

Proof Let φ be a conformal mapping from $\mathbb{D}$ onto G and put $\psi = \varphi^{-1}$. Let $\partial_a G$ be the accessible part of ∂G, that is, the set of all points in ∂G which are accessible from G by some curve. As it is mentioned above, by virtue of [26, Propositions 2.14 and 2.17] we have $\partial_a D = \{\varphi(\xi) : \xi \in \mathcal{F}(\varphi)\}$, where $\mathcal{F}(\varphi)$ is the Fatou set for φ. As it was shown in [11], $\partial_a G$ is a Borel set. In view of [11, Corollary. 1] the functions φ and ψ can be extended to Borel measurable functions (denoted also by φ and ψ) on $\mathbb{D} \cup \mathcal{F}(\varphi)$ and $G \cup \partial_a G$, respectively, in such a way that $\varphi(\psi(\zeta)) = \zeta$ for all $\zeta \in \partial_a G$ and $\psi(\varphi(\xi)) = \xi$ for all $\xi \in \mathcal{F}(\varphi)$.

Let ω be the measure on ∂G defined by $\omega := \varphi(d\xi)$ (see [11, Section 3]). In fact ω is a measure on $\partial_a G$ and has no atoms. Moreover, for $E \subset \partial G$ the quantity $|\omega(E)|/(2\pi)$ is the harmonic measure of E evaluated with respect to $\varphi(0)$ and G.

If $C(\partial G) \neq R_{\overline{G}}(\partial G, g)$, then there exists a nonzero measure μ on ∂G such that μ and $g\mu$ are orthogonal to $\mathcal{R}_{\overline{G}}$, which means that $\int f\,d\mu = \int f\,d(g\mu) = 0$ for every $f \in \mathcal{R}_{\overline{G}}$. In view of [11, Theorem 2] there exist two functions h_1, h_2, belonging to the Hardy space H^1 in $\mathbb{D}$, such that $h_1 \not\equiv 0$ and

$$\mu = (h_1 \circ \psi)\,\omega, \qquad g\mu = (h_2 \circ \psi)\,\omega.$$

Therefore, for $m(\cdot)$-almost all $\xi \in \mathbb{T}$ one has $g(\varphi(\xi))\, h_1(\xi) = h_2(\xi)$. Furthermore, replacing the quotient h_2/h_1 by f_2/f_1 with $f_1, f_2 \in H^\infty(\mathbb{D})$ and defining the functions u and v in D by the settings $u(z) = f_2(\psi(z))$, $v(z) = f_1(\psi(z))$, one obtains that $g(z) = u(z)/v(z)$ almost everywhere on ∂G in the sense of conformal mappings, as it is demanded.

Let us prove the inverse statement. In fact, it is possible to prove more. Namely, the following result takes place: Let G be an arbitrary bounded simply connected domain in $\mathbb{C}$ and suppose that G is g-Nevanlinna; then, $C(\partial G) \neq R_{\overline{G}}(\partial G, g)$. Following the line of arguments that were used in the proof of the first statement in Theorem 2.2 in [10], one can show that if G is a g-Nevanlinna domain and the functions $u, v \in H^\infty(G)$ are taken from the corresponding definition, then for every rational function f having at least one pole on $\{z \in G : v(z) \neq 0\}$ we have $f|_{\partial G} \notin R_{\overline{G}}(\partial G, g)$. □

In particular, Theorem 4.1 has the following corollary: if G is a Carathéodory domain in $\mathbb{C}$ such that the set $\mathbb{C} \setminus \overline{G}$ is connected and such that G is not a g-Nevanlinna domain, then $C(\partial G) = P(\partial G, g)$, that is, every continuous function on ∂G can be approximated uniformly by elements of the module $\mathcal{P}(g)$, that is, by functions of the form $gp_1 + p_0$, where p_0 and p_1 are polynomials in the complex variable.

Let us turn to the question about properties of g-Nevanlinna domains and about relationships between these classes for different g. Let G be a bounded simply connected domain in $\mathbb{C}$ and φ be some conformal mapping from $\mathbb{D}$ onto G.

First of all let us mention that G is a g-Nevanlinna domain if and only the function $g_{0,*} \circ \varphi$ admits a (Nevanlinna type) pseudocontinuation (see Proposition 2.1, and recall that $g(z) = g_0(\overline{z})$). This yields immediately that any Nevanlinna domain is $\overline{z}^d$-Nevanlinna for every integer $d \geqslant 2$ (the class of such domains was studied in [2]). At the same time, if G is a Nevanlinna domain (so that φ itself admits a pseudocontinuation), then G is not $\exp(\overline{z})$-Nevanlinna domain because the function $\exp(\varphi(z))$ does not admit a pseudocontinuation in this case. Indeed, we have $\varphi = f_1/f_2$ almost everywhere on $\mathbb{T}$, where $f_1, f_2 \in H^\infty(\widehat{\mathbb{C}} \setminus \overline{\mathbb{D}})$, and f_1/f_2 has poles in $\mathbb{C} \setminus \overline{\mathbb{D}}$. Thus, the pseudocontinuation of $\exp(\varphi(z))$ (if any) coincides with $\exp(f_1/f_2)$, but the last function has essential singularity at poles of f_1/f_2. The obtained contradiction yields the answer.

Furthermore we recall that G is a Nevanlinna domain if and only if it is an image of $\mathbb{D}$ under conformal mapping by some univalent function belonging to a certain model space K_Θ, where Θ is an inner function (see Sect. 2). Using some facts about model spaces K_Θ we are able to answer the question when a certain g-Nevanlinna domain is a Nevanlinna one for some special g.

Every nonzero function $f \in H^\infty = H^\infty(\mathbb{D})$ can be uniquely factorized in the following way $f = e^{ic}BSF$, where B and S are a Blaschke product and a singular inner function (see Sect. 2), $c \in \mathbb{R}$, and where F is an outer function that is a function of the form

$$F(z) = \exp\left(\frac{1}{2\pi}\int_0^{2\pi} \frac{e^{it}+z}{e^{it}-z}\log|f(e^{it})|\,dt\right).$$

The function $e^{ic}BS$ is called the inner factor of f.

In order to state our next result we need to recall three basic facts about model spaces K_Θ (see, e.g., [24, Lecture II]):

1. $f \in K_\Theta$ if and only if there exists a function $w \in H^2$ such that $f = \overline{z}\Theta\overline{w}$ on $\mathbb{T}$ (this equality means the respective equality of functions in L^∞).
2. Any function $f \in K_\Theta$ admits a pseudocontinuation and the ratio f_1/f_2 from Proposition 2.1 has the form

$$\frac{f_1(z)}{f_2(z)} = \frac{z^{-1}\overline{g(1/\overline{z})}}{\overline{\Theta(1/\overline{z})}}, \qquad z \in \mathbb{C}\setminus\overline{\mathbb{D}}.$$

3. For an inner function Θ let us define an operator, acting in L^2 as follows:

$$f \mapsto \widetilde{f}_\Theta := \overline{z}\Theta\overline{f}.$$

This operator is an antilinear isometric involution, and it commutes with the orthogonal projection from L^2 to K_Θ. Thus, for any function $f \in K_\Theta$, one has $\overline{z}\Theta\overline{f} \in K_\Theta$.

Since there are $\overline{z}^d$-Nevanlinna domains which are not Nevanlinna ones, it seems reasonable to consider the problem of d-root extraction in the space K_Θ in more detail. The following result was proved in [2, Theorem 4].

Theorem 4.2 *Let $k \geq 1$ be an integer, let $f \in H^\infty$, and let Θ be an inner function. Suppose that $h = f^k \in K_\Theta$. The following two assertions are equivalent:*

1. The function f admits a pseudocontinuation.
2. There exists an inner function Θ_1 such that $(zJ)^{k-1}\Theta = \Theta_1^k$, where J is the inner factor of the function $\widetilde{h}_\Theta$.

Moreover, if f admits a pseudocontinuation, then $f \in K_{I_1}$ for some inner function I_1 such that I_1 divides Θ and Θ_1 and I_1^k divides $z^{k-1}\Theta$ in the set of inner functions.

In connection with this theorem we notice that the possibility of taking k-root of the function $(zJ)^{k-1}\Theta$ essentially depends on orders of its zeros, because for any singular inner function its k-root is always well defined.

It is also interesting to analyze the mentioned above example of the domain $\varphi(\mathbb{D})$ with $\varphi(z) = \sqrt[d]{2-z}$, $d \geqslant 2$, from the point of view of Theorem 4.2 (this domain is $\overline{z}^d$-Nevanlinna domain, but not a Nevanlinna one). For $f = \varphi$ we have $h(z) = 2-z$, $z \in \mathbb{D}$. It is easy to check that $h \in K_{z^2}$, $\widetilde{h}_{z^2} = 2z-1$, $J = (2z-1)/(z-2)$, and $(zJ)^{d-1}z^2 = z^{d+1}J^{d-1}$. Thus, in order to have a pseudocontinuation property for f under consideration, one needs that $z^{k+1}J^{k-1} = \Theta_1^d$ for some inner function Θ_1, which is clearly impossible.

Let now $g_0 = P \in \mathcal{P}$. Take a function $f \in H^\infty$, suppose that $h = P(f)$ admits a pseudocontinuation (and hence belongs to some model space K_Θ), and assume additionally that f itself also admits a pseudocontinuation. Then there exist two inner functions I and Θ_1 such that $I\Theta = zJ\Theta_1^k$, where $k = \deg P$, and J is an inner factor of $\widetilde{h}_\Theta$. The proof of this fact may be obtained by an argument similar to that used in the corresponding implication in Theorem 4.2, which was given in [2].

References

1. A. Baranov, Yu. Belov, A. Borichev, K. Fedorovskiy, *Univalent functions in model spaces: revisited*, arXiv:1705.05930 [math.CV].
2. A. Baranov, J. Carmona, K. Fedorovskiy, *Density of certain polynomial modules*, J. Approx. Theory **206** (2016) 1–16.
3. A. Baranov, K. Fedorovskiy, *Boundary regularity of Nevanlinna domains and univalent functions in model subspaces*, Sb. Math. **202** (2011) 1723–1740.
4. Yu. Belov, A. Borichev, K. Fedorovskiy, *Nevanlinna domains with large boundaries*, J. Funct. Anal. **277** (2019) 2617–2643.
5. Yu. Belov, K. Fedorovskiy, *Model spaces containing univalent functions*, Russian Math. Surv. **73** (2018) 172–174.
6. A. Boivin, P. Gauthier, P. Paramonov, *On uniform approximation by n-analytic functions on closed sets in* $\mathbb{C}$, Izv. Math. **68** (2004) 447–459.
7. E. Borovik, K. Fedorovskiy, *On the relationship between Nevanlinna and quadrature domains*, Math. Notes **99** (2016) 460–464.
8. J. Carmona, *A necessary and sufficient condition for uniform approximation by certain rational modules*, Proc. Amer. Math. Soc. **86** (1982) 487–490.
9. J. Carmona, *Mergelyan's approximation theorem for rational modules*, J. Approx. Theory **44** (1985) 113–126.
10. J. Carmona, P. Paramonov, K. Fedorovskiy, *On uniform approximation by polyanalytic polynomials and the Dirichlet problem for bianalytic functions*, Sb. Math. **193** (2002) 1469–1492.
11. J. Carmona, K. Fedorovskiy, *Conformal maps and uniform approximation by polyanalytic functions*, Selected Topics in Complex Analysis, Oper. Theor. Adv. Appl. 158, Birkhäuser, Basel, 2005, pp. 109–130.
12. D. N. Clark, *One dimensional perturbations of restricted shifts*, J. Anal. Math. **25** (1972) 169–191.
13. P. Davis, *The Schwarz function and its applications*, Carus Math. Monogr. 17, Math. Ass. of America, Buffalo, NY 1974.
14. R. G. Douglas, H. S. Shapiro and A. L. Shields, *Cyclic vectors and invariant subspaces for the backward shift operator*, Annales de l'institut Fourier, **20** (1970), 37–76.
15. K. Dyakonov, D. Khavinson, *Smooth functions in star-invariant subspaces*, Recent advances in operator-related function theory, Contemp. Math. 393, Amer. Math. Soc., Providence, RI 2006, pp. 59–66.
16. K. Fedorovskiy, *On uniform approximations of functions by n-analytic polynomials on rectifiable contours in* $\mathbb{C}$, Math. Notes **59** (1996) 435–439.
17. K. Fedorovskiy, *On some properties and examples of Nevanlinna domains*, Proc. Steklov Inst. Math. **253** (2006), no. 2, 186–194.
18. P. Mattila, *Geometry of sets and measures in Euclidean spaces*, Cambridge Studies in Adv. Math. 44, Cambridge University Press, Cambridge 1995.
19. M. Mazalov, *An example of a nonconstant bianalytic function vanishing everywhere on a nowhere analytic boundary*, Math. Notes **62** (1997) 524–526.

20. M. Mazalov, *On uniform approximations by bi-analytic functions on arbitrary compact sets in* $\mathbb{C}$, Sb. Math. **195** (2004) 687–709.
21. M. Mazalov, *A criterion for uniform approximability on arbitrary compact sets for solutions of elliptic equations*, Sb. Math. **199** (2008) 13–44.
22. M. Mazalov, *An example of a nonrectifiable Nevanlinna contour*, St. Petersburg Math. J. **27** (2016) 625–630.
23. M. Mazalov, *On Nevanlinna domains with fractal boundaries*, St. Petersburg Math. J. **29** (2018) 777–791.
24. N. Nikolskiĭ, *Treatise on the shift operator*, Springer–Verlag, Berlin 1986.
25. A. G. O'Farrell, *Annihilators of rational modules*, J. Funct. Anal. **19** (1975) 373–389.
26. Ch. Pommerenke, *Boundary behaviour of conformal maps*, Springer–Verlag, Berlin 1992.
27. Ch. Pommerenke, *Conformal maps at the boundary*, Handbook of Complex analysis: Geometric Function Theory. Volume 1, 2002.
28. M. Sakai, *Regularity of a boundary having a Schwarz function*, Acta Math. **166** (1991) 263–297.
29. H. Shapiro, *The Schwarz function and its generalization to higher dimensions*, University of Arkansas Lecture Notes in the Mathematical Sciences 9, John Wiley & Sons, Inc., New York 1992.
30. T. Trent, J. L.-M. Wang, *Uniform approximation by rational modules on nowhere dense sets*, Proc. Amer. Math. Soc. **81** (1981) 62–64.
31. T. Trent, J. L.-M. Wang, *The uniform closure of rational modules*, Bull. London Math. Soc. **13** (1981) 415–420.
32. D. Vardakis, A. Volberg, *Free boundary problems in the spirit of Sakai's theorem*, Comptes Rendus Mathematique **359** (2021) 1233–1238.
33. J. Verdera, *Approximation by rational modules in Sobolev and Lipschitz norms*, J. Funct. Anal. **58** (1984) 267–290.
34. J. Verdera, *On the uniform approximation problem for the square of the Cauchy-Riemann operator*, Pacific J. Math. **159** (1993) 379–396.
35. J. L.-M. Wang, *A localization operator for rational modules*, Rocky Mountain J. Math. **19** (1989) 999–1002.

On Meromorphic Inner Functions in the Upper Half-Plane

Burak Hatinoğlu

1 Introduction

Inner functions in the upper half-plane are bounded analytic functions in the upper half-plane with unit modulus almost everywhere on the real line. An inner function is called a meromorphic inner function (MIF) if it extends meromorphically to the complex plane.

Originated from the study of complex function theory, MIFs appear in various fields such as spectral theory of differential operators [4, 27], Krein-de Branges theory of entire functions [9, 31, 43], functional model theory [33, 34], approximation theory [37], Toeplitz operators [17, 28, 39] and non-linear Fourier transform [40]. MIFs also play a critical role in the Toeplitz approach to the uncertainty principle [27, 38], which was used in the study of problems of Fourier and harmonic analysis [32, 35, 36].

In this paper we review some properties and applications of MIFs, along with their several recent results. In Sect. 2 we recall function theoretic properties of MIFs. In Sect. 3 we consider some restricted interpolation results for MIFs. In Sect. 4 we discuss several uniqueness theorems for MIFs. In Sect. 5 we consider the relation of MIFs with Toeplitz operators and their role in the Toeplitz approach to the uncertainty principle. In Sect. 6 we briefly mention the fundamental place of MIFs in the enormous literature of inverse spectral theory of differential operators. In Sect. 7 we discuss how MIFs appear in a recent theorem that proves an analog of Carleson's theorem on almost everywhere convergence of Fourier series for a version of the non-linear Fourier transform.

B. Hatinoğlu (✉)
Michigan State University, Department of Mathematics, Wells Hall East Lansing, MI, USA
e-mail: hatinogl@msu.edu

I. Binder et al. (eds.), *Function Spaces, Theory and Applications*, Fields Institute Communications 87, https://doi.org/10.1007/978-3-031-39270-2_7

2 Some Properties of MIFs

Each MIF is characterized by parameters (C, a, Ω) in the canonical (Riesz-Smirnov) factorization

$$\Theta(z) = Ce^{iaz} \prod_{n\in\mathbb{N}} c_n \frac{z - \omega_n}{z - \overline{\omega_n}}, \tag{1}$$

where C is a unimodular constant, a is a nonzero real constant, $\Omega = \{\omega_n\}_{n\in\mathbb{N}}$ is a discrete sequence (i.e. has no finite accumulation point) in $\mathbb{C}_+$ satisfying the Blaschke condition

$$\sum_{n\in\mathbb{N}} \frac{\Im\omega_n}{1 + |\omega_n|^2} < \infty,$$

and $c_n = \frac{i+\overline{\omega}_n}{i+\omega_n}$. MIFs are represented as $\Theta(x) = e^{i\phi(x)}$ on the real line, where ϕ is an increasing real analytic function.

A complex-valued function is said to be real if it maps real numbers to real numbers on its domain. A meromorphic Herglotz function m is a real meromorphic function with positive imaginary part on $\mathbb{C}_+$. It has negative imaginary part on $\mathbb{C}_-$ via the relation $m(\overline{z}) = \overline{m}(z)$. There is a one-to-one correspondence between MIFs and meromorphic Herglotz functions via equations

$$\Theta = \frac{m_\Theta - i}{m_\Theta + i}, \qquad m_\Theta = i\frac{1 + \Theta}{1 - \Theta}. \tag{2}$$

Therefore, MIFs are described by parameters (b, c, μ) via the Herglotz representation

$$m_\Theta(z) = bz + c + iS\mu_\Theta(z), \tag{3}$$

where b is a non-negative constant, c is a real constant and

$$S\mu_\Theta(z) := \frac{1}{i\pi} \int \Big(\frac{1}{t - z} - \frac{t}{1 + t^2}\Big) d\mu_\Theta(t)$$

is the Schwarz integral of the positive discrete Poisson-finite measure μ_Θ on $\mathbb{R}$, i.e.

$$\int \frac{d\mu_\Theta(t)}{1 + t^2} < \infty.$$

The number πb is considered as the point mass of μ_Θ at infinity. This extended measure μ_Θ is called the Clark (or spectral) measure of Θ. Conversely, for every positive discrete Poisson-finite measure μ and a non-negative number b, there exists an MIF Θ_μ such that the corresponding meromorphic Herglotz function m_{Θ_μ} satisfies (3).

The spectrum of the MIF Θ, denoted by $\sigma(\Theta)$, is the level set $\{\Theta = 1\}$, so μ_Θ is supported on $\sigma(\Theta)$ or $\sigma(\Theta) \cup \{\infty\}$. The point masses at $t \in \sigma(\Theta)$ are given by $\mu_\Theta(t) = 2\pi/|\Theta'(t)|$. The point mass at infinity is nonzero if and only if one of the equivalent conditions is satisfied:

1. $\Theta - 1 \in H^2(\mathbb{C}_+)$.
2. $\lim_{y\to+\infty} \Theta(iy) = 1$ and $\lim_{y\to+\infty} y^2\Theta'(iy)$ exists.
3. $\{\Im\omega_n\}_n$ is summable and $a = 0$ in the canonical factorization (2).

The spectrum $\sigma(\Theta)$ is a finite set of real numbers or a discrete sequence of real numbers, i.e. $\{\lambda_n\}_n \subset \mathbb{R}$ without any finite accumulation point.

An entire function E belongs to Hermite-Biehler class if E has no real zeros and $|E(\overline{z})| < |E(z)|$ whenever $\Im z > 0$. The identity

$$\Theta_E = \overline{E(\overline{z})}/E(z) \tag{4}$$

relates entire functions of Hermite-Biehler class with MIFs. Conversely, given any MIF, there is at least one entire function of Hermite-Biehler class satisfying (4).

All of the abovementioned properties of MIFs can be found, e.g. in [38] and references therein.

3 Interpolation with Uniformly Bounded Derivatives

A natural interpolation question is constructing a MIF from a given spectrum. If $\{\lambda_n\}_n$ is a discrete sequence of real numbers, then one can find a MIF Θ such that $\sigma(\Theta) = \{\lambda_n\}_n$. This is a well-known result and the procedure can be found, e.g. in [41].

In her 2017 paper [43], Rupam discussed the following restricted interpolation problem: Given a discrete sequence $\{\lambda_n\}$ in $\mathbb{R}$, does there exist an MIF Θ with $\sigma(\Theta) = \{\lambda_n\}_n$ such that $|\Theta'|$ is uniformly bounded on $\mathbb{R}$? Usefulness of the bounded derivative on $\mathbb{R}$ restriction is discussed in the same paper through two applications to the gap problem and Toeplitz kernels.

First, let's introduce some notations: $f(n) \asymp g(n)$ denotes the existence of constants $c_1, c_2 > 0$ such that $c_1 f(n) \le g(n) \le c_2 f(n)$ for sufficiently large n, and $f(n) \lesssim g(n)$ denotes $f(n) \le cg(n)$ for some $c > 0$ and sufficiently large n. For an increasing sequence $\{\lambda_n\} \subset \mathbb{R}$, let

$$\Delta_n := \begin{cases} \lambda_{n+1} - \lambda_n, & n > 0 \\ \lambda_n - \lambda_{n-1}, & n \le 0 \end{cases}$$

Theorem 3.1 ([43], Theorem 4) *Let $\{\lambda_n\}$ be a discrete sequence on $\mathbb{R}$ satisfying one of the following three conditions:*

1. $\Delta_{n+1} \asymp \Delta_n$ *and* $\ln|\lambda_n|/\ln\ln|\lambda_n| \lesssim \Delta_n \lesssim \ln|\lambda_n|$.
2. $\Delta_{n+1} \asymp \Delta_n$ *and* $(\ln|\lambda_n|)^2 \lesssim \Delta_n$.
3. *There exists* $d > 0$ *such that* $\{\lambda_n\}$ *can be partitioned into clusters* $\{\lambda_n^j\}_n$ *such that:*
   - *The number of points in each cluster is uniformly bounded.*
   - *For each cluster,* $a\lambda_n^j/\lambda_n^{j+1} \to 1$.
   - *Between successive clusters,* $(\lambda_{n+1}^j/a_n^k) - 1 > d > 0$.

Then, there exists an MIF Θ *with* $\sigma(\Theta) = \{\lambda_n\}_n$ *and a uniformly bounded derivative on* $\mathbb{R}$.

In the same paper Rupam also proved a partial converse result.

Theorem 3.2 ([43], Proposition 2) *Let* $\{\lambda_n\}_n$ *be a discrete sequence in* $\mathbb{R}$, *and* $D > 0$ *a constant such that given any* $N > 0$, *there exists a subset* $\{\mu_n\}_{n=1}^N$ *such that* $\{\mu_n\}_{n=1}^N \cap \{\lambda_n\}_n = \{\mu_n\}_{n=1}^N$ *for which* $\mu_2 - \mu_1 > ND$ *and* $\mu_{n+1} - \mu_n < D$ *for all* $2 \leq n \leq N-1$. *Let* Θ *be an MIF with spectrum* $\{\lambda_n\}_n$. *Then, given* $\delta > 0$, *there exists a zero* $z_n = x_n + iy_n$ *of* Θ *such that* $0 < y_n < \delta$. *Hence,* $|\Theta'|$ *is unbounded on* $\mathbb{R}$.

Some other cases for which this interpolation problem is open are when the gaps are comeasurable ($\Delta_n \asymp \Delta_{n+1}$) and

- The gaps are very small, i.e. $\Delta_n \lesssim \ln|\lambda_n|/\ln\ln\Delta_n$.
- The gaps are in between, i.e. $\ln|\lambda_n| \lesssim \Delta_n \lesssim \ln^2|\lambda_n|$.
- $\{\lambda_n\}$ has clusters and gaps, i.e. $(ln|\lambda_n|)^2 \not\lesssim \Delta_n$ and $\Delta_n \not\lesssim \ln|\lambda_n|$ [43].

4 Uniqueness Results

In the previous section we discussed that if $\{\lambda_n\}_n$ is a discrete sequence of real numbers, then one can find a MIF Θ such that $\sigma(\Theta) = \{\lambda_n\}_n$. Recalling (3), we observe that there are infinitely many such Θ, since any positive discrete Poisson-finite measure μ_Θ supported on $\sigma(\Theta)$ provides a desired MIF Θ. If we consider another spectrum, we get closer to a uniqueness result. Let $\{\mu_n\}_n$ denote the level set $\{\Theta = -1\}$, which is the same as $\sigma(-\Theta)$. This set is a discrete sequence of real numbers. Also $\{\lambda_n\}_n$ and $\{\mu_n\}_n$ are interlacing since $\Theta = e^{i\phi}$ on the real line with an increasing real analytic function ϕ. Recovery of an MIF from two spectra is a classical result proved by Lifšic and Krein.

Theorem 4.1 ([24, 26]) *Let* $\{\lambda_n\}_n$ *and* $\{\mu_n\}_n$ *be two interlacing discrete sequences on* $\mathbb{R}$. *Then, there exists an MIF* Θ *such that* $\{\lambda_n\}_n = \sigma(\Theta)$ *and* $\{\mu_n\}_n = \sigma(-\Theta)$. *Such a function is unique up to a Möbius transform, i.e. if* Φ *is another MIF, which satisfies* $\{\lambda_n\}_n = \sigma(\Phi)$ *and* $\{\mu_n\}_n = \sigma(-\Phi)$, *then*

$$\Phi = \frac{\Theta - c}{1 - c\Theta}$$

for $-1 < c < 1$.

If we consider point masses of μ_Θ (or equivalently $\Theta'(\lambda_n)$) instead of the second spectrum $\sigma(-\Theta)$, we get uniqueness of Θ up to a linear polynomial by the Herglotz representation (3). The author recently proved that with some extra conditions, Θ can be uniquely determined. Recall that for any MIF Θ, $\{\lambda_n\}_n = \sigma(\Theta)$ and $\{\mu_n\}_n = \sigma(-\Theta)$ are interlacing, so $\{(\lambda_n, \mu_n)\}_n$ is a sequence of disjoint intervals on $\mathbb{R}$. Let us denote these intervals by I_n.

Theorem 4.2 ([20], Theorem 2.3) *Let* Θ *be a MIF. If*

$$\sum_{n\in\mathbb{Z}} \frac{|I_n|}{1 + dist(0, I_n)} < \infty,$$

then $\sigma_\Theta = \{\lambda_n\}_{n\in\mathbb{Z}}$ *and* $\{\Theta'(\lambda_n)\}_{n\in\mathbb{Z}}$ *uniquely determine* Θ.

Theorem 4.3 ([20], Theorem 2.5) *Let* Θ *be a MIF. If* $\sigma_\Theta = \{\lambda_n\}_{n\in\mathbb{N}}$ *is bounded below, then* $\{\lambda_n\}_{n\in\mathbb{N}}$ *and* $\{\Theta'(\lambda_n)\}_{n\in\mathbb{N}}$ *uniquely determine* Θ.

Next, the two theorems show that the missing part of $\{\Theta'(\lambda_n)\}$ from the spectral data can be compensated by the corresponding data from $\sigma(-\Theta)$ in Theorems 4.2 and 4.3.

Theorem 4.4 ([20], Theorem 2.6) *Let* Θ *be a MIF and* $A \subseteq \mathbb{Z}$. *If*

$$\sum_{n\in\mathbb{Z}} \frac{|I_n|}{1 + dist(0, I_n)} < \infty,$$

then the spectral data consisting of

- $\{\lambda_n\}_{n\in\mathbb{Z}}$
- $\{\mu_n\}_{n\in\mathbb{Z}\setminus A}$
- $\{\Theta'(\lambda_n)\}_{n\in A}$

uniquely determine Θ.

Theorem 4.5 ([20], Theorem 2.7) *Let* Θ *be a MIF and* $A \subseteq \mathbb{N}$. *If* $\{\lambda_n\}_{n\in\mathbb{N}}$ *is bounded below, then the spectral data consisting of*

- $\{\lambda_n\}_{n\in\mathbb{N}}$
- $\{\mu_n\}_{n\in\mathbb{N}\setminus A}$
- $\{\Theta'(\lambda_n)\}_{n\in A}$

uniquely determine Θ.

5 MIFs and Toeplitz Operators

In the last 20 years, Makarov and Poltoratski extensively used MIFs in their study of Toeplitz operators and model spaces to provide original approaches to problems from the area of uncertainty principle in harmonic analysis [27, 28, 38]. In his 2018 paper [39], Poltoratski developed and systematized the function theoretic component of this Toeplitz approach by introducing a partial order on the set of inner functions induced by the action of Toeplitz operators.

The Toeplitz operator T_U with a symbol $U \in L^\infty(\mathbb{R})$ is the map

$$T_U : H^2 \to H^2, \qquad F \to P_+(UF),$$

where P_+ is the orthogonal projection from $L^2(\mathbb{R})$ onto the Hardy space $H^2(\mathbb{C}_+)$. We denote kernels of Toeplitz operators in H^2 by

$$N[U] = \ker T_U.$$

For each inner function $\theta(z)$, a model subspace of $H^2(\mathbb{C}_+)$ is given by

$$K_\theta = H^2 \ominus \theta H^2,$$

where $\ominus$ stands for the orthogonal difference. If θ is an inner function, then $N[\overline{\theta}] = K_\theta$. Now, we can state the main definitions of Toeplitz order, which were introduced in [39].

Definition 5.1 Let θ, I and J be inner functions.

1. The **(Toeplitz) dominance set** of θ, denoted by $\mathcal{D}(\theta)$, is defined as

$$\mathcal{D}(\theta) = \{I \text{ inner} \mid N[\overline{\theta} I] \neq 0\}.$$

2. I and J are **Toeplitz equivalent**, denoted by $I \overset{\mathrm{T}}{\sim} J$, if $\mathcal{D}(I) = \mathcal{D}(J)$. This is an equivalence relation, which is called **Toeplitz equivalence (TE)**.
3. We write $I \overset{\mathrm{T}}{\leq} J$ if $\mathcal{D}(I) \subset \mathcal{D}(J)$. This partial order on the set of inner functions in $\mathbb{C}_+$ is called **Toeplitz order (TO)**.
4. We say that I divides J if J/I is an inner function. An inner function $I \in \mathcal{D}(\theta)$ is a **base element** if it does not divide any other element of $\mathcal{D}(\theta)$. In other words, base elements are the maximal elements of $\mathcal{D}(\theta)$ with respect to the order by division. The set of all base elements of $\mathcal{D}(\theta)$ is denoted by $\mathcal{D}_B(\theta)$.
5. We say that $f \in N[\overline{I} J]$ is purely outer if f is outer and $\overline{I} J f = \overline{g}$, where g is outer. Then, an element I of $\mathcal{D}(\theta)$ is a **totalelement**, if $N[\overline{\theta} I]$ contains a purely outer function. The set of all total elements of $\mathcal{D}(\theta)$ is denoted by $\mathcal{D}_T(\theta)$.

Next, let us list some properties. Let θ be an inner function. The following properties are given in Proposition 1, Proposition 2 and Theorem 5 in [39].

- $\mathcal{D}_B(\theta) \subset \mathcal{D}_T(\theta) \subset \mathcal{D}(\theta)$.
- The sets $\mathcal{D}_B(\theta)$ and $\mathcal{D}_T(\theta)$ are equal if and only if θ is a Blaschke factor, i.e. of the form

$$\frac{\overline{a}}{a}\frac{z-a}{z-\overline{a}}$$

for some $a \in \mathbb{C}_+$.
- Every element of $\mathcal{D}(\theta)$ is a factor of a total element.
- Every element of $\mathcal{D}(\theta)$ is a divisor of a base element.

Let us also consider some of the main results from [39] focussing on MIFs.

Theorem 5.1 ([39], Theorem 1) *Let I and Θ be MIFs and $\Theta \in \mathcal{D}(I)$. Then, there exists an MIF J such that $\sigma(J) \subset \sigma(I)$ and $\Theta \in \mathcal{D}_T(J)$. The function J can be chosen so that the purely outer $f \in N[\overline{J}\Theta]$ is also zero-free on $\mathbb{R}$.*

Theorem 5.2 ([39], Lemma 2) *Let Θ, Φ be MIFs with the common spectrum σ. Then, $\Theta \overset{\mathrm{T}}{\sim} \Phi$ if and only if*

$$c|\Theta'| < |\Phi'| < C|\Theta'|$$

on σ for some positive constants c and C.

Theorem 5.3 ([39], Lemma 4) *Let Θ, Φ be MIFs such that*

$$c|\Theta'| < |\Phi'| < C|\Theta'|$$

on $\mathbb{R}$ for some positive constants c and C. Then, $\Theta \overset{\mathrm{T}}{\sim} \Phi$ if and only if $\arg\Theta - \arg\Phi$ *has a bounded harmonic conjugate.*

Many other results on Toeplitz order are provided in [39]. In Section 8 of the same paper, connections of Toeplitz order with some of the classical problems of harmonic analysis are also discussed. Namely, these are the Beurling-Malliavin problem, the Type problem and some problems on sampling measures. Let us review the connection between Toeplitz order and the Type problem.

Let μ be a general finite positive measure on $\mathbb{R}$. The type of μ is defined as

$$\mathcal{T}_\mu = \inf\{a \mid e^{ist}, s \in [-a, a], \text{ are complete in } L^2(\mu)\}.$$

The Type problem is to characterize $\mathcal{T}_\mu$ in terms of μ. A formula for $\mathcal{T}_\mu$ was given by Poltoratski in 2013 with the use of the Toeplitz approach [36]. More on the Toeplitz approach and the Type problem can be found in [38].

Recall that for a positive discrete Poisson-finite measure μ, we denote by Θ_μ the MIF with Clark measure μ. The same can be said for inner functions if the discreteness condition is removed, so for a positive singular Poisson-finite measure

μ, we denote by θ_μ the inner function with Clark measure μ. Then, the Type problem is given in terms of Toeplitz order.

Theorem 5.4 ([39], Theorem 12, Theorem 13) *Let μ be a positive singular Poisson-finite measure. Then,*

$$\mathcal{T}_\mu = \sup\{a \mid e^{iaz} \in \mathcal{D}(\theta_\mu)\} = \sup\{a \mid e^{iaz} \overset{\mathrm{T}}{\leq} \theta_\mu\}.$$

6 MIFs in Inverse Spectral Theory

Direct spectral problems aim to get spectral information from the given operator. In inverse spectral problems, the goal is to recover the operator from spectral information, such as the spectrum, the norming constants, the spectral measure or Weyl m-function.

Classical results of inverse spectral theory are given in terms of Schrödinger (Sturm-Liouville) equation

$$Lu = -u'' + qu = zu$$

on the interval $(0, \pi)$ with the boundary conditions

$$u(0)\cos\alpha - u'(0)\sin\alpha = 0, \qquad u(\pi)\cos\beta + u'(\pi)\sin\beta = 0,$$

and a real-valued potential $q \in L^1(0, \pi)$. The spectrum of the Schrödinger operator L corresponding to these boundary conditions defines a discrete subset of the real line, bounded from below, diverging to $+\infty$.

The first inverse spectral result on Schrödinger operators was given by Ambarzumian [1]. He considered continuous potential with Neumann boundary conditions at both endpoints ($\alpha = \beta = \pi/2$) and showed that $q \equiv 0$ if the spectrum consists of squares of integers.

Later, Borg [7] proved that an L^1-potential is uniquely recovered from two spectra, corresponding to various pairs of boundary conditions and sharing the same boundary conditions at π ($\beta_1 = \beta_2$), one of which should be Dirichlet boundary condition at 0 ($\alpha_1 = 0$). Levinson [25] extended Borg's result by removing the restriction of Dirichlet boundary condition at 0.

Furthermore, Marchenko [30] observed that Weyl m-function uniquely recovers an L^1-potential. Weyl m-function is a meromorphic Herglotz function, and the corresponding MIF is called Weyl inner function.

Another classical result is due to Hochstadt and Liebermann [22], which says that if half of an L^1-potential is known, one spectrum recovers the whole potential.

Statements of these classical results and some recent results can be found, e.g. in [18]. Classical theorems of Borg, Levinson, Marchenko, Hochstadt and Lieberman led to numerous results with different approaches such as:

- Using various spectral data (Borg-Marchenko type results) [11, 18, 29]
- Using mixture of potential and spectral data (Hochstadt-Lieberman type results) [2, 15, 16]
- Considering various smoothness classes for the potential ($q \in L^1, L^p, C^k, L^1_{loc}$) [2, 3, 15]
- Finding connections with exponential systems [4, 23, 29]
- Considering different settings (discrete, half-line, real line, canonical systems, quantum graphs) [5, 6, 10, 12–14, 19, 21, 44, 45]

Weyl m-functions and equivalently Weyl inner functions are the first and main bridge between the inverse spectral theory and the complex function theory. They also exist for canonical systems. Canonical systems are the most general class of symmetric second-order operators, so the more classical second-order operators such as Schrödinger, Jacobi, Dirac, Sturm-Liouville operators and Krein strings can be written as canonical systems [42], e.g. see Section 1.3 of [42] for rewriting the Schrödinger equation as a canonical system.

A canonical system is a differential equation of the form

$$\Omega f'(x) = -zH(x)f(x)$$

where $L > 0$, $x \in (0, L)$ and

$$\Omega := \begin{pmatrix} 0 & -1 \\ 1 & 0 \end{pmatrix}, \quad f(x) = \begin{pmatrix} f_1(x) \\ f_2(x) \end{pmatrix}.$$

We assume $H(x) \in \mathbb{R}^{2\times 2}$, $H \in L^1_{\text{loc}}(0, L)$ and $H(x) \geq 0$ a.e. on $(0, L)$. Let us consider the limit circle case, i.e. $\int_0^L \text{Tr}H(x)dx < \infty$. The self-adjoint realizations of a canonical system in the limit circle case with separated boundary conditions are exactly those described by

$$f_1(0)\sin\alpha - f_2(0)\cos\alpha = 0, \qquad f_1(L)\sin\beta - f_2(L)\cos\beta = 0, \tag{5}$$

with $\alpha, \beta \in [0, \pi)$. Such a self-adjoint realization with separated boundary conditions (5) has a discrete spectrum $\sigma_{\alpha,\beta}$.

Let $f = f(x, z)$ be a solution of $Jf' = -zHf$ that satisfies the boundary condition

$$f_1(L)\sin\beta - f_2(L)\cos\beta = 0.$$

The Weyl m-function $m_{\alpha,\beta}(z)$ is defined as

$$m_{\alpha,\beta}(z) := \frac{\cos(\alpha) f_1(0) + \sin(\alpha) f_2(0)}{-\sin(\alpha) f_1(0) + \cos(\alpha) f_2(0)}. \tag{6}$$

If we let u and v be solutions of $Ju' = -zHu$ with the initial conditions

$$u_1(0) = \cos(\alpha), \quad u_2(0) = \sin(\alpha),$$
$$v_1(0) = -\sin(\alpha), \quad v_2(0) = \cos(\alpha),$$

then the solution u satisfies the boundary condition at $x = 0$, and

$$f(x, z) = v(x, z) + m_{\alpha,\beta}(z) u(x, z) \tag{7}$$

is the unique solution of the form $v + Mu$ that satisfies the boundary condition β at $x = L$. The spectrum $\sigma_{\alpha,\beta} = \{\lambda_n\}_n$ is the set of poles of $m_{\alpha,\beta}(z)$, and the spectrum $\sigma_{\alpha+\pi/2,\beta}$ is the set of zeros of $m_{\alpha,\beta}(z)$. The norming constant $\gamma_{\alpha,\beta}^{(n)}$, for $\lambda_n \in \sigma_{\alpha,\beta}$, is defined as

$$\gamma_{\alpha,\beta}^{(n)} = \frac{1}{||u(\cdot, \lambda_n)||^2}.$$

The Weyl m-function $m_{\alpha,\beta}(z)$ is a meromorphic Herglotz function, so the corresponding MIF is defined as

$$\Theta_{\alpha,\beta} = \frac{m_{\alpha,\beta} - i}{m_{\alpha,\beta} + i}. \tag{8}$$

The Clark (spectral) measure of $\Theta_{\alpha,\beta}$ is the discrete measure supported on $\sigma_{\alpha,\beta}$ with point masses given by the corresponding norming constants, i.e.

$$\mu_{\alpha,\beta} = \sum_n \gamma_{\alpha,\beta}^{(n)} \delta_{a_n}$$

is the spectral measure. The MIF $\Theta_{\alpha,\beta}$ carries out spectral properties of the corresponding canonical system through (6), (7) and (8), so it allows us to use powerful tools from complex function theory to attack problems from the spectral theory of canonical system.

The classical theorem of Marchenko says that the Weyl m-function of a Schrödinger operator uniquely determines the operator [30]. The same result is valid for canonical systems. Note that $m_{\alpha,\beta}$ and $\Theta_{\alpha,\beta}$ are in one-to-one correspondence, so we can state Marchenko's theorem for canonical systems in terms of MIFs. We consider trace-normed canonical system, which is always possible since we are in the limit circle case, i.e. $\int_0^L \mathrm{Tr} H(x) dx < \infty$.

Theorem 6.1 ([42], Theorem 5.2) *Let $L > 0$, $\alpha_1, \alpha_2, \beta \in [0, \pi)$ and H be a trace-normed canonical system on $[0, L]$. Then, $\Theta_{\alpha,\beta}$ uniquely determines H.*

7 Carleson's Theorem for the Non-linear Fourier Transform

Carleson's theorem [8] is one of the fundamental results of classical Fourier analysis. It was proved in 1966 and says that for any $f \in L^2(\mathbb{R})$ the Fourier transform

$$\hat{f}_T(x) = \int_{-T}^{T} f(s)e^{-ixs} ds$$

converges to $\hat{f}(x) = \hat{f}_\infty(x)$ as $T \to \infty$ at almost every point $x \in \mathbb{R}$. In a recent preprint [40] Poltoratski proved an analog of Carleson's theorem in the scattering setting for a version of the non-linear Fourier transform. The scattering setting corresponds to the Dirac system with a real L^2 potential on the right half-line $\mathbb{R}_+$

$$\Omega X' = zX - QX, \tag{9}$$

where $z \in \mathbb{C}$ is a spectral parameter and

$$\Omega = \begin{pmatrix} 0 & 1 \\ -1 & 0 \end{pmatrix}, \qquad Q(t) = \begin{pmatrix} 0 & f(t) \\ f(t) & 0 \end{pmatrix}$$

for some real-valued $f \in L^2(\mathbb{R}_+)$. Note that any Dirac system is a canonical system.

The matrix function M whose columns are the Neumann and Dirichlet solutions is called the transfer matrix (or the fundamental matrix) of the system. In other words the transfer matrix M solves (9) with the initial condition that $M(0, z)$ is the 2×2 identity matrix. If

$$X(t, z) = \begin{pmatrix} u(t, z) \\ v(t, z) \end{pmatrix}$$

is a solution of (9) with a self-adjoint boundary condition at 0, then for each fixed $t \in \mathbb{R}_+$ the function $u(t, z) - iv(t, z)$ is an entire function of Hermite-Biehler class. Recall the definition of Hermite-Biehler class entire functions from Section 2. Let $E(t, z)$ and $F(t, z)$ denote the entire functions of Hermite-Biehler class corresponding to the Neumann and Dirichlet conditions of (9) at 0, respectively. The functions $E(t, z)$ and $F(t, z)$ have exponential type t ([42], Theorem 4.2, Theorem 4.19). The entries of the transfer matrix $M(t, z)$ are real entire functions and if

$$M(t,z) = \begin{pmatrix} A(t,z) & B(t,z) \\ C(t,z) & D(t,z) \end{pmatrix},$$

then $E(t,z) = A(t,z) - iC(t,z)$ and $F(t,z) = B(t,z) - iD(t,z)$. The scattering functions of the Dirac system (9) are $\mathcal{E}(t,z) = e^{itz}E(t,z)$ and $\mathcal{F}(t,z) = e^{itz}F(t,z)$. Now, let us introduce the entire functions

$$a(t,z) = \frac{\mathcal{E}(t,z) + i\mathcal{F}(t,z)}{2}, \qquad b(t,z) = \frac{\mathcal{E}(t,z) - i\mathcal{F}(t,z)}{2}. \tag{10}$$

If $h^{\#}(z)$ denotes $\overline{h(\overline{z})}$, then the matrix

$$G(t,z) = \begin{pmatrix} a^{\#}(t,z) & b^{\#}(t,z) \\ b(t,z) & a(t,z) \end{pmatrix}$$

solves the differential equation

$$\frac{\partial}{\partial t} G(t,z) = \begin{pmatrix} 0 & e^{-2izt} f(t) \\ e^{2izt} f(t) & 0 \end{pmatrix} G(t,z), \qquad G(0,z) = \begin{pmatrix} 0 & 1 \\ 1 & 0 \end{pmatrix},$$

since the scattering functions $\mathcal{E}$ and $\mathcal{F}$ satisfy

$$\frac{\partial}{\partial t}\mathcal{E}(t,z) = iz\mathcal{E}(t,z) + e^{izt}\frac{\partial}{\partial t}E(t,z) = f(t)e^{2izt}\mathcal{E}^{\#}(t,z)$$
$$\frac{\partial}{\partial t}\mathcal{F}(t,z) = iz\mathcal{F}(t,z) + e^{izt}\frac{\partial}{\partial t}F(t,z) = f(t)e^{2izt}\mathcal{F}^{\#}(t,z).$$

The version of the non-linear Fourier transform in the scattering setting is given by

$$\overset{\curlywedge}{f}(z) = \frac{\lim_{t\to\infty} b(t,z)}{\lim_{t\to\infty} a(t,z)}.$$

If f_T is the restriction of the potential function f to the interval $(0,T)$, then

$$\overset{\curlywedge}{f}_T(z) = \frac{b(T,z)}{a(T,z)}.$$

Before stating Carleson's theorem, let us list some properties of this version of the non-linear Fourier transform. Let $f \in L^2(\mathbb{R}_+)$ be real-valued. Then,

- Shifting property: If $g(t) = e^{2\pi i s t} f(t)$, then $\overset{\curlywedge}{g}(z) = \overset{\curlywedge}{f}(z - s)$.
- Time-scaling property: If $a > 0$ and $g(t) = af(at)$, then $\overset{\curlywedge}{g}(z) = \overset{\curlywedge}{f}(z/a)$.
- Non-linear Parseval's identity: $||f||^2_{L^2(\mathbb{R})} = ||\log(1 - |\overset{\curlywedge}{f}|^2)||_{L^1(\mathbb{R})}$.
- Linearization gives the Fourier transform: $\overset{\curlywedge}{\varepsilon f} = \varepsilon\hat{f} + O(\varepsilon^2)$.

- Analyticity: If f is supported on a half-line, then $\check{f}$ is analytic in a half-plane.

The main result of [40] is the following analog of Carleson's theorem:

Theorem 7.1 *For every real-valued* $f \in L^2(\mathbb{R}_+)$, $\check{f}_T(s) \to \check{f}(s)$ *as* $T \to \infty$ *for almost every* $s \in \mathbb{R}$, *where* f_T *denotes the restriction of* f *to the interval* $(0, T)$.

Let us briefly explain how MIFs play a critical role in the proof of Theorem 7.1. If H is an entire function of Hermite-Biehler class, then $\Theta_H(z) = H^\#(z)/H(z)$ is an MIF. Then, recalling the functions $E(t, z)$ and $F(t, z)$ corresponding to the Neumann and Dirichlet conditions of (9) at 0, respectively, we can introduce the MIFs $\Theta_E(z)$ and $\Theta_F(z)$ for any fixed t. These MIFs are called Dirac inner functions. Note that

$$\Theta_E(t, z) = \frac{A(t, z) + iC(t, z)}{A(t, z) - iC(t, z)}$$

and the spectra $\sigma(\Theta_E)$ and $\sigma(-\Theta_E)$ represent the Dirichlet-Dirichlet and Dirichlet-Neumann spectra of the Dirac system restricted to $(0, t)$, respectively. Moreover, Dirac inner function Θ_E satisfies the Riccati equation

$$\frac{\partial}{\partial t}\Theta(t, z) = 2iz\Theta(t, z) - f(t)(1 - \Theta^2(t, z)).$$

The same observations can be made for Θ_F. The Neumann condition at 0 for the Dirac system corresponds to the initial condition $\Theta(0, z) = 1$ for the Riccati equation, and the Dirichlet condition corresponds to $\Theta(0, z) = -1$.

The Riccati equation together with other properties of Dirac inner functions is used in [40] in the study of the behaviour of resonances of the Dirac system, which is one of the main steps of the proof of Theorem 7.1.

References

1. V. Ambarzumian. Über eine frage der eigenwerttheorie. *Zeitschrift für Physik*, 53:690–695, 1929.
2. L. Amour and T. Raoux. Inverse spectral results for Schrödinger operators on the unit interval with potentials in L^p spaces. *Inverse Problems*, 23(6):2367–2373, 2007. ISSN 0266-5611. URL https://doi.org/10.1088/0266-5611/23/6/006.
3. L. Amour, J. Faupin, and T. Raoux. Inverse spectral results for Schrödinger operators on the unit interval with partial information given on the potentials. *J. Math. Phys.*, 50(3):033505, 14, 2009. ISSN 0022-2488. URL https://doi.org/10.1063/1.3087426.
4. A. Baranov, Y. Belov, and A. Poltoratski. De Branges functions of Schroedinger equations. *Collect. Math.*, 68(2):251–263, 2017. ISSN 0010-0757. URL https://doi.org/10.1007/s13348-016-0168-0.
5. N. P. Bondarenko. A partial inverse problem for the Sturm-Liouville operator on a star-shaped graph. *Anal. Math. Phys.*, 8(1):155–168, 2018. ISSN 1664-2368. URL https://doi.org/10.1007/s13324-017-0172-x.

6. N. P. Bondarenko and C. Yang. Partial inverse problems for the Sturm-Liouville operator on a star-shaped graph with different edge lengths. *Results Math.*, 73(2):Paper No. 56, 17, 2018. ISSN 1422-6383. URL https://doi.org/10.1007/s00025-018-0817-6.
7. G. Borg. Eine Umkehrung der Sturm-Liouvilleschen Eigenwertaufgabe. Bestimmung der Differentialgleichung durch die Eigenwerte. *Acta Math.*, 78:1–96, 1946. ISSN 0001-5962. URL https://doi.org/10.1007/BF02421600.
8. L. Carleson. On convergence and growth of partial sums of Fourier series. *Acta Math.*, 116: 135–157, 1966. ISSN 0001-5962. URL https://doi.org/10.1007/BF02392815.
9. L. de Branges. *Hilbert spaces of entire functions*. Prentice Hall, 1968.
10. R. del Rio and M. Kudryavtsev. Inverse problems for Jacobi operators: I. Interior mass-spring perturbations in finite systems. *Inverse Problems*, 28(5):055007, 18, 2012. ISSN 0266-5611. URL https://doi.org/10.1088/0266-5611/28/5/055007.
11. R. del Rio, F. Gesztesy, and B. Simon. Inverse spectral analysis with partial information on the potential. III. Updating boundary conditions. *Internat. Math. Res. Notices*, (15):751–758, 1997. ISSN 1073-7928. URL https://doi.org/10.1155/S1073792897000494.
12. J. Eckhardt and G. Teschl. Uniqueness results for one-dimensional Schrödinger operators with purely discrete spectra. *Trans. Amer. Math. Soc.*, 365(7):3923–3942, 2013. ISSN 0002-9947. URL https://doi.org/10.1090/S0002-9947-2012-05821-1.
13. F. Gesztesy and B. Simon. Uniqueness theorems in inverse spectral theory for one-dimensional Schrödinger operators. *Trans. Amer. Math. Soc.*, 348(1):349–373, 1996. ISSN 0002-9947. URL https://doi.org/10.1090/S0002-9947-96-01525-5.
14. F. Gesztesy and B. Simon. m-functions and inverse spectral analysis for finite and semi-infinite Jacobi matrices. *J. Anal. Math.*, 73:267–297, 1997. ISSN 0021-7670. URL https://doi.org/10.1007/BF02788147.
15. F. Gesztesy and B. Simon. A new approach to inverse spectral theory. II. General real potentials and the connection to the spectral measure. *Ann. of Math. (2)*, 152(2):593–643, 2000. ISSN 0003-486X. URL https://doi.org/10.2307/2661393.
16. F. Gesztesy and B. Simon. Inverse spectral analysis with partial information on the potential. II. The case of discrete spectrum. *Trans. Amer. Math. Soc.*, 352(6):2765–2787, 2000. ISSN 0002-9947. URL https://doi.org/10.1090/S0002-9947-99-02544-1.
17. A. Hartmann and M. Mitkovski. Kernels of Toeplitz operators. In *Recent progress on operator theory and approximation in spaces of analytic functions*, volume 679 of *Contemp. Math.*, pages 147–177. Amer. Math. Soc., Providence, RI, 2016.
18. B. Hatinoğlu. Mixed data in inverse spectral problems for the Schrödinger operators. *J. Spectr. Theory*, 11(1):281–322, 2021. ISSN 1664-039X. URL https://doi.org/10.4171/jst/341.
19. B. Hatinoğlu. Inverse problems for Jacobi operators with mixed spectral data. *J. Difference Equ. Appl.*, 27(1):81–101, 2021. ISSN 1023-6198. URL https://doi.org/10.1080/10236198.2020.1867546.
20. B. Hatinoğlu. Uniqueness theorems for meromorphic inner functions. page arXiv:2110.12384, 2021.
21. B. Hatinoğlu, J. Eakins, W. Frendreiss, L. Lamb, S. Manage, and A. Puente. Ambarzumian-type problems for discrete Schrödinger operators. *Complex Anal. Oper. Theory*, 15(8):Paper No. 118, 13, 2021. ISSN 1661-8254. URL https://doi.org/10.1007/s11785-021-01169-5.
22. H. Hochstadt and B. Lieberman. An inverse Sturm-Liouville problem with mixed given data. *SIAM J. Appl. Math.*, 34(4):676–680, 1978. ISSN 0036-1399. URL https://doi.org/10.1137/0134054.
23. M. Horváth. Inverse spectral problems and closed exponential systems. *Ann. of Math. (2)*, 162 (2):885–918, 2005. ISSN 0003-486X. URL https://doi.org/10.4007/annals.2005.162.885.
24. M. G. Kreĭn. On the trace formula in perturbation theory. *Mat. Sbornik N.S.*, 33(75):597–626, 1953.
25. N. Levinson. The inverse Sturm-Liouville problem. *Mat. Tidsskr. B*, 1949:25–30, 1949. ISSN 0909-3540.
26. I. M. Lifšic. On a problem of the theory of perturbations connected with quantum statistics. *Uspehi Matem. Nauk (N.S.)*, 7(1(47)):171–180, 1952. ISSN 0042-1316.

27. N. Makarov and A. Poltoratski. Meromorphic inner functions, Toeplitz kernels and the uncertainty principle. In *Perspectives in analysis*, volume 27 of *Math. Phys. Stud.*, pages 185–252. Springer, Berlin, 2005. URL https://doi.org/10.1007/3-540-30434-7_10.
28. N. Makarov and A. Poltoratski. Beurling-Malliavin theory for Toeplitz kernels. *Invent. Math.*, 180(3):443–480, 2010. ISSN 0020-9910. URL https://doi.org/10.1007/s00222-010-0234-2.
29. N. Makarov and A. Poltoratski. Two-spectra theorem with uncertainty. *J. Spectr. Theory*, 9(4): 1249–1285, 2019. ISSN 1664-039X. URL https://doi.org/10.4171/jst/276.
30. V.A. Marchenko. Some questions in the theory of one-dimensional linear differential operators of the second order. *Trudy Moskov Mat. Obsc.*, 1:327–420, 1952.
31. M. Mitkovski and A. Poltoratski. Pólya sequences, Toeplitz kernels and gap theorems. *Adv. Math.*, 224(3):1057–1070, 2010. ISSN 0001-8708. URL https://doi.org/10.1016/j.aim.2009.12.014.
32. M. Mitkovski and A. Poltoratski. On the determinacy problem for measures. *Invent. Math.*, 202(3):1241–1267, 2015. ISSN 0020-9910. URL https://doi.org/10.1007/s00222-015-0588-6.
33. N. K. Nikolski. *Treatise on the shift operator*, volume 273 of *Grundlehren der Mathematischen Wissenschaften [Fundamental Principles of Mathematical Sciences]*. Springer-Verlag, Berlin, 1986. ISBN 3-540-15021-8. URL https://doi.org/10.1007/978-3-642-70151-1. Spectral function theory, With an appendix by S. V. Hruščev [S. V. Khrushchëv] and V. V. Peller, Translated from the Russian by Jaak Peetre.
34. N. K. Nikolski. *Operators, functions, and systems: an easy reading. Vol. 2*, volume 93 of *Mathematical Surveys and Monographs*. American Mathematical Society, Providence, RI, 2002. ISBN 0-8218-2876-2. Model operators and systems, Translated from the French by Andreas Hartmann and revised by the author.
35. A. Poltoratski. Spectral gaps for sets and measures. *Acta Math.*, 208(1):151–209, 2012. ISSN 0001-5962. URL https://doi.org/10.1007/s11511-012-0076-4.
36. A. Poltoratski. A problem on completeness of exponentials. *Ann. of Math.*, 178(3):983–1016, 2013. ISSN 0003-486X. URL https://doi.org/10.4007/annals.2013.178.3.4.
37. A. Poltoratski. Bernstein's problem on weighted polynomial approximation. In *Operator-related function theory and time-frequency analysis*, volume 9 of *Abel Symp.*, pages 147–171. Springer, Cham, 2015.
38. A. Poltoratski. *Toeplitz approach to problems of the uncertainty principle*, volume 121 of *CBMS Regional Conference Series in Mathematics*. Published for the Conference Board of the Mathematical Sciences, Washington, DC; by the American Mathematical Society, Providence, RI, 2015. ISBN 978-1-4704-2017-8. URL https://doi.org/10.1090/cbms/121.
39. A. Poltoratski. Toeplitz order. *J. Funct. Anal.*, 275(3):660–697, 2018. ISSN 0022-1236. URL https://doi.org/10.1016/j.jfa.2018.02.006.
40. A. Poltoratski. Pointwise convergence of the non-linear Fourier transform. page arXiv:2103.13349, 2021.
41. A. Poltoratski and D. Sarason. Aleksandrov-Clark measures. In *Recent advances in operator-related function theory*, volume 393 of *Contemp. Math.*, pages 1–14. Amer. Math. Soc., Providence, RI, 2006. URL https://doi.org/10.1090/conm/393/07366.
42. C. Remling. *Spectral theory of canonical systems*, volume 70 of *De Gruyter Studies in Mathematics*. De Gruyter, Berlin, 2018. ISBN 978-3-11-056202-6; 978-3-11-056228-6.
43. R. Rupam. Uniform boundedness of the derivatives of meromorphic inner functions on the real line. *J. Anal. Math.*, 131:189–206, 2017. ISSN 0021-7670. URL https://doi.org/10.1007/s11854-017-0006-7.
44. B. Simon. A new approach to inverse spectral theory. I. Fundamental formalism. *Ann. of Math. (2)*, 150(3):1029–1057, 1999. ISSN 0003-486X. URL https://doi.org/10.2307/121061.
45. M. Suzuki. An inverse problem for a class of canonical systems having Hamiltonians of determinant one. *J. Funct. Anal.*, 279(12):108699, 34, 2020. ISSN 0022-1236. URL https://doi.org/10.1016/j.jfa.2020.108699.

On the Norm of the Hilbert Matrix

Boban Karapetrović

1 Introduction

First introduced by Hilbert [9], the infinite matrix which has come to bear his name and whose elements are given by $(n + k + 1)^{-1}$, where $n, k \geq 0$, was initially considered in approximation theory. In the mentioned paper, Hilbert shows, among other things, some of the first inequalities on sequences induced by the elements of his matrix. A number of authors then considered the action of the Hilbert matrix on some elementary sequence spaces. The Hilbert matrix H and its action on the space ℓ^2 of all square summable sequences were first studied in [19], where Magnus described the spectrum of the Hilbert matrix. In [4, 5], Diamantopoulos and Siskakis studied for the first time the action of the Hilbert matrix on Hardy and Bergman spaces in the open unit disc of the complex plane. They obtained some results related to the questions of boundedness and exact norm of the Hilbert matrix on these particular spaces. Later, these results were improved by Dostanić, Jevtić, and Vukotić [6]. We note also that Aleman, Montes-Rodríguez, and Sarafoleanu provide a closed formula for the eigenvalues of the Hilbert matrix in a more general context [1]. For further information related to the spectrum of Hilbert matrix operator, see [20]. The Hilbert matrix has also been intensively studied on other various spaces of holomorphic functions in the unit disc of the complex plane. See [2, 11, 13, 14, 17, 18] and references therein. One of the recent more general results relates to the characterization of the boundedness of the Hilbert matrix on

The author is supported in part by Serbian Ministry of Education, Science and Technological Development, Project #174032.

B. Karapetrović (✉)
Faculty of Mathematics, University of Belgrade, Belgrade, Serbia
e-mail: bkarapetrovic@matf.bg.ac.rs

I. Binder et al. (eds.), *Function Spaces, Theory and Applications*, Fields Institute Communications 87, https://doi.org/10.1007/978-3-031-39270-2_8

generalized mixed norm spaces. The generalized mixed norm spaces represent a significant class of spaces of holomorphic functions in the open unit disc of the complex plane, which include almost all relevant spaces of such functions, like the Hardy and weighted Hardy spaces, Bergman and weighted Bergman spaces, weighted Dirichlet spaces, Hardy-Bloch spaces, classical mixed norm spaces, and some other known and important spaces.

1.1 Hilbert Matrix

The Hilbert matrix is the infinite matrix $\mathrm{H} = \left[\frac{1}{n+k+1}\right]_{n,k=0}^{\infty}$ (see Fig. 1).
It can be viewed as an operator on sequence spaces. Namely, if $\{a_n\}_{n=0}^{\infty}$ is a complex sequence, then

$$\mathrm{H} : \{a_n\}_{n=0}^{\infty} \mapsto \left\{\sum_{k=0}^{\infty} \frac{a_k}{n+k+1}\right\}_{n=0}^{\infty}.$$

In this way, we can consider the action of the Hilbert matrix on the sequence spaces ℓ^p of all p-summable sequences. Namely, let $1 \leq p \leq \infty$ and define ℓ^p to be the set of complex sequences $a = \{a_n\}_{n=0}^{\infty}$ for which

$$\|a\|_{\ell^p} = \left(\sum_{n=0}^{\infty} |a_n|^p\right)^{\frac{1}{p}} < \infty \text{ for } 1 \leq p < \infty \text{ and } \|a\|_{\ell^\infty} = \sup_{n\in\mathbb{N}_0} |a_n| < \infty.$$

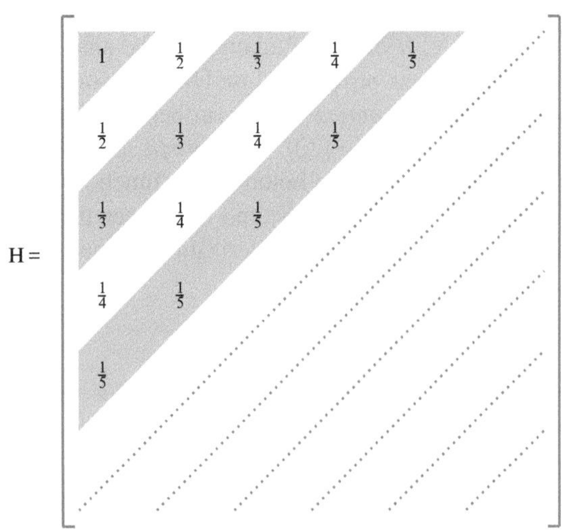

Fig. 1 All antidiagonals of the Hilbert matrix H have the same elements. Therefore, the Hilbert matrix can be viewed as a particular example of Hankel matrices in a more general settings. Namely, an infinite matrix $\left[h_{n,k}\right]_{n,k=0}^{\infty}$ is called a Hankel matrix if its entries are of the form $h_{n,k} = c_{n+k}$, where $n, k \geq 0$, for some sequence of numbers $\{c_n\}_{n=0}^{\infty}$

Using a particular form of the classical Hilbert inequality [8]

$$\left(\sum_{n=0}^{\infty}\left|\sum_{k=0}^{\infty}\frac{a_k}{n+k+1}\right|^p\right)^{\frac{1}{p}} \leq \frac{\pi}{\sin\frac{\pi}{p}}\left(\sum_{n=0}^{\infty}|a_n|^p\right)^{\frac{1}{p}}, \quad 1 < p < \infty,$$

it can be shown that the Hilbert matrix H is bounded on the sequence space ℓ^p if and only if $1 < p < \infty$ and

$$\|\mathrm{H}\|_{\ell^p \to \ell^p} = \frac{\pi}{\sin\frac{\pi}{p}}.$$

On the other hand, let $\mathrm{Hol}(\mathbb{D})$ be the space of all holomorphic functions in the open unit disc $\mathbb{D} = \{z \in \mathbb{C} : |z| < 1\}$ of the complex plane $\mathbb{C}$. The Hilbert matrix H can be also viewed as an operator on spaces of holomorphic functions in the unit disc $\mathbb{D}$ by its action on their Taylor coefficients. If

$$f(z) = \sum_{n=0}^{\infty} a_n z^n,$$

is a holomorphic function in the unit disc $\mathbb{D}$, then we set

$$\mathrm{H}f(z) = \sum_{n=0}^{\infty}\left(\sum_{k=0}^{\infty}\frac{a_k}{n+k+1}\right)z^n.$$

1.2 Function Spaces

We list some basic spaces of holomorphic functions in the unit disc, which we will discuss below in the context of the Hilbert matrix. Let $f \in \mathrm{Hol}(\mathbb{D})$, $0 < p \leq \infty$ and $0 \leq r < 1$. Then we define the integral means of order p in the following way:

$$\mathrm{M}_p(r, f) = \left(\frac{1}{2\pi}\int_0^{2\pi}\left|f\left(re^{i\theta}\right)\right|^p d\theta\right)^{\frac{1}{p}} \quad \text{for } 0 < p < \infty,$$

and

$$\mathrm{M}_\infty(r, f) = \sup_{0 \leq \theta < 2\pi}\left|f\left(re^{i\theta}\right)\right|.$$

Hardy spaces If $0 < p \leq \infty$, the *Hardy space* H^p is defined as follows:

$$\mathrm{H}^p = \left\{ f \in \mathrm{Hol}(\mathbb{D}) : \|f\|_{\mathrm{H}^p} = \sup_{0 \leq r < 1} \mathrm{M}_p(r, f) < \infty \right\}.$$

Mixed norm spaces If $0 < p, q \leq \infty$ and $0 < \alpha < \infty$, the *mixed norm space* $\mathrm{H}^{p,q,\alpha}$, is the space of all functions $f \in \mathrm{Hol}(\mathbb{D})$ for which

$$\|f\|_{\mathrm{H}^{p,q,\alpha}} = \left(\int_0^1 (1-r)^{q\alpha-1} \mathrm{M}_p^q(r, f)\, \mathrm{d}r \right)^{\frac{1}{q}} < \infty \ \text{ for } \ 0 < q < \infty,$$

and

$$\|f\|_{\mathrm{H}^{p,\infty,\alpha}} = \sup_{0 \leq r < 1} (1-r)^{\alpha}\, \mathrm{M}_p(r, f) < \infty.$$

The spaces $\mathrm{H}^{p,\infty,\alpha}$ are known as *weighted Hardy spaces*. On the other hand, if

$$f(z) = \sum_{n=0}^{\infty} a_n z^n \in \mathrm{Hol}(\mathbb{D}) \ \text{ and } \ t \in \mathbb{R},$$

we define

$$\mathrm{D}^t f(z) = \sum_{n=0}^{\infty} (n+1)^t a_n z^n, \quad z \in \mathbb{D}.$$

The *generalized mixed norm space* $\mathrm{H}_t^{p,q,\alpha}$ is the space of all functions $f \in \mathrm{Hol}(\mathbb{D})$ for which

$$\left\|\mathrm{D}^t f\right\|_{\mathrm{H}^{p,q,\alpha}} < \infty.$$

So we can write

$$\mathrm{H}_t^{p,q,\alpha} = \mathrm{D}^{-t} \mathrm{H}^{p,q,\alpha}.$$

It is a well-known fact that if $f \in \mathrm{Hol}(\mathbb{D})$, $0 < p, q \leq \infty$, $0 < \alpha, \beta < \infty$ and $t, s \in \mathbb{R}$ are such that $\alpha - t = \beta - s$, then $\left\|\mathrm{D}^t f\right\|_{\mathrm{H}^{p,q,\alpha}}$ is comparable to $\|\mathrm{D}^s f\|_{\mathrm{H}^{p,q,\beta}}$ and consequently

$$\mathrm{H}_t^{p,q,\alpha} = \mathrm{H}_s^{p,q,\beta}. \tag{1}$$

Bergman and Dirichlet spaces For $0 < p < \infty$ and $\alpha > -1$, the *weighted Bergman space* is defined as follows:

$$\mathrm{A}^p_\alpha = \left\{ f \in \mathrm{Hol}(\mathbb{D}) : \|f\|_{\mathrm{A}^p_\alpha} = \left(\frac{\alpha+1}{\pi} \int_{\mathbb{D}} |f(z)|^p \left(1-|z|^2\right)^\alpha \mathrm{dm}(z) \right)^{\frac{1}{p}} < \infty \right\},$$

where dm is the Euclidean area measure in the complex plane, that is,

$$\mathrm{dm}(z) = \mathrm{d}x\,\mathrm{d}y = r\,\mathrm{d}r\,\mathrm{d}\theta \quad \text{for} \quad z = x + \mathrm{i}y = r\mathrm{e}^{\mathrm{i}\theta}.$$

We note that if $\alpha = 0$, then $\mathrm{A}^p = \mathrm{A}^p_0$ are standard unweighted Bergman spaces. It is a well-known fact that

$$\mathrm{A}^p_\alpha = \mathrm{H}^{p,p,\frac{\alpha+1}{p}}. \tag{2}$$

The *weighted Dirichlet space* $\mathcal{D}^p_\alpha$ is the space of all functions $f \in \mathrm{Hol}(\mathbb{D})$ for which $f' \in \mathrm{A}^p_\alpha$ or equivalently $\mathrm{D}^1 f \in \mathrm{A}^p_\alpha$. Therefore,

$$\mathcal{D}^p_\alpha = \mathrm{D}^{-1}\mathrm{A}^p_\alpha = \mathrm{D}^{-1}\mathrm{H}^{p,p,\frac{\alpha+1}{p}} = \mathrm{H}_1^{p,p,\frac{\alpha+1}{p}}.$$

Besov spaces The *analytic Besov space* B^p, $0 < p < \infty$, is defined as follows:

$$\mathrm{B}^p = \mathrm{H}^{p,p,1}_{1+\frac{1}{p}}.$$

Note that for $1 < p < \infty$, we may write

$$\mathrm{B}^p = \mathrm{H}^{p,p,1}_{1+\frac{1}{p}} \overset{(1)}{=} \mathrm{H}_1^{p,p,1-\frac{1}{p}} = \mathrm{D}^{-1}\mathrm{H}^{p,p,1-\frac{1}{p}} \overset{(2)}{=} \mathrm{D}^{-1}\mathrm{A}^p_{p-2},$$

which implies that in this case B^p is the set of all functions $f \in \mathrm{Hol}(\mathbb{D})$ such that $\mathrm{D}^1 f \in \mathrm{A}^p_{p-2}$ or equivalently $f' \in \mathrm{A}^p_{p-2}$. Similarly,

$$\mathrm{B}^1 = \mathrm{H}_2^{1,1,1} = \mathrm{D}^{-2}\mathrm{H}^{1,1,1} = \mathrm{D}^{-2}\mathrm{A}^1,$$

is the set of all functions $f \in \mathrm{Hol}(\mathbb{D})$ for which

$$\mathrm{D}^2 f \in \mathrm{A}^1 \quad \text{or equivalently} \quad f'' \in \mathrm{A}^1.$$

Bloch and Hardy-Bloch spaces The *Hardy-Bloch space* $\mathcal{B}^{p,q}$, $0 < p, q \leq \infty$, is defined in the following way:

$$\mathcal{B}^{p,q} = \mathrm{H}_1^{p,q,1}.$$

As a special case, we obtain the *Bloch space* $\mathcal{B}$. Namely,

$$\mathcal{B} = \mathcal{B}^{\infty,\infty} = \mathrm{H}_1^{\infty,\infty,1}.$$

Bloch space $\mathcal{B}$ is the space of all functions $f \in \mathrm{Hol}(\mathbb{D})$ for which $\mathrm{D}^1 f \in \mathrm{H}^{\infty,\infty,1}$ or equivalently $f' \in \mathrm{H}^{\infty,\infty,1}$. This implies

$$\mathcal{B} = \left\{ f \in \mathrm{Hol}(\mathbb{D}) : \left| f'(z) \right| = \mathrm{O}\left(\frac{1}{1-|z|^2} \right) \right\}.$$

BMOA *and* VMOA The space BMOA consists of the functions $f \in \mathrm{H}^1$ whose boundary values $f(\mathrm{e}^{\mathrm{i}\theta})$ are of bounded mean oscillation on the unit circle $\mathbb{T} = \partial\mathbb{D}$, that is,

$$\sup_I \frac{1}{|I|} \int_I \left| f\left(\mathrm{e}^{\mathrm{i}\theta}\right) - f_I \right| \mathrm{d}\theta < \infty,$$

where supremum is taken over all intervals $I \subset \mathbb{T}$ and

$$f_I = \frac{1}{|I|} \int_I f\left(\mathrm{e}^{\mathrm{i}\theta}\right) \mathrm{d}\theta.$$

If

$$\lim_{|I|\to 0} \frac{1}{|I|} \int_I \left| f\left(\mathrm{e}^{\mathrm{i}\theta}\right) - f_I \right| \mathrm{d}\theta = 0,$$

then we say that $f \in \mathrm{VMOA}$. Here, as usual, $|I|$ is arc length measure of the interval $I \subset \mathbb{T}$.

For further information related to the mentioned function spaces, see the recent monograph [10].

2 Boundedness of the Hilbert Matrix

The starting point for studying the boundedness of the Hilbert matrix H on the Hardy space H^p was the paper [5] by Diamantopoulos and Siskakis. It was the first time someone considered the action of the Hilbert matrix on some spaces of holomorphic functions in the unit disc of the complex plane. They obtained a representation of the Hilbert matrix H as an average of weighted composition operators, and they proved that the Hilbert matrix H is bounded on H^p if and only if $1 < p < \infty$. Moreover, Diamantopoulos [4] obtained that the Hilbert matrix H is bounded on

Bergman space A^p if and only if $2 < p < \infty$. Lanucha, Nowak, and Pavlović [16] showed that the Hilbert matrix H is bounded on weighted Hardy spaces $H^{p,\infty,\alpha}$ if and only if $\alpha + 1/p < 1$. Continuing in the same way, Galanopoulos, Girela, Peláez, and Siskakis [7] proved that if $1 < \alpha + 2 < p$, then the Hilbert matrix H is bounded on weighted Bergman space A^p_α and if $p > 1$ and $p - 2 < \alpha \leq p - 1$, then the Hilbert matrix H is bounded on weighted Dirichlet space $\mathcal{D}^p_\alpha$. In light of the previous partial results, the following problem naturally arises: *characterize the boundedness of the Hilbert matrix* H *on weighted Bergman spaces* A^p_α *and weighted Dirichlet spaces* $\mathcal{D}^p_\alpha$. Since

$$A^p_\alpha = H^{p,p,\frac{\alpha+1}{p}} \text{ and } \mathcal{D}^p_\alpha = H_1^{p,p,\frac{\alpha+1}{p}},$$

we actually have a more general version of the previous problem: *characterize the boundedness of the Hilbert matrix* H *on generalized mixed norm spaces* $H^{p,q,\alpha}_t$. This problem was completely solved in [11] with the following surprisingly simple characterization. Namely, the Hilbert matrix H is bounded on $H^{p,q,\alpha}_t$ if and only if

$$0 < \frac{1}{p} + \alpha - t < 1.$$

Note that the above condition does not depend on the parameter q. This result allows us several interesting special cases. We list some of them below.

- H is bounded on A^p_α if and only if $1 < \alpha + 2 < p$.
- H is bounded on $\mathcal{D}^p_\alpha$ if and only if $\max\{-1, p - 2\} < \alpha < 2p - 2$.
- H is bounded on $\mathcal{B}^{p,q} = H^{p,q,1}_1$ if and only if $1 < p < \infty$.
- H is unbounded on Bloch space $\mathcal{B} = \mathcal{B}^{\infty,\infty} = H^{\infty,\infty,1}_1$.
- H is unbounded on Besov space $B^p = H^{p,p,1}_{1+\frac{1}{p}}$.

We also have the following interesting results. First, let $0 < p \leq \infty$. If $0 < q \leq 1$, then the Hilbert matrix H maps the generalized mixed norm space $H^{p,q,1}_{1+1/p}$ into the space BMOA, and if $1 < q \leq \infty$, then the Hilbert matrix H does not map the generalized mixed norm space $H^{p,q,1}_{1+1/p}$ into the Bloch space $\mathcal{B}$. This implies (after some work) the following result: the Hilbert matrix H does not map the space VMOA into Bloch space $\mathcal{B}$. See the paper [12] for more information. As already stated, the Hilbert matrix H is not bounded on Hardy space H^1. On the other hand, it can be shown [12] that the Hilbert matrix H maps continuously H^1 into the mixed norm space $H^{p,\infty,1-1/p}$, where $1 < p < \infty$, but not into the mixed norm space $H^{p,q,1-1/p}$, for any $0 < q < \infty$ (notice that $H^1 \subset H^{p,q,1-1/p} \subset H^{p,\infty,1-1/p}$).

The Hilbert matrix can also be considered on many other spaces of holomorphic function in the open unit disc. For example, the action of the Hilbert matrix on logarithmically weighted Bergman, Bloch, and Hardy-Bloch spaces is considered in [12, 13, 16].

3 Norm of the Hilbert Matrix on Weighted Bergman Spaces

As stated earlier, Diamantopoulos and Siskakis [5] proved that the Hilbert matrix H is bounded on H^p if and only if $1 < p < \infty$, and they obtained the following two integral representations of the Hilbert matrix:

$$\mathrm{H}f(z) = \int_0^1 \frac{f(t)}{1-tz}\,\mathrm{d}t, \tag{3}$$

and

$$\mathrm{H}f(z) = \int_0^1 \mathrm{T}_t f(z)\,\mathrm{d}t = \int_0^1 \frac{1}{1-(1-t)z} f\left(\frac{t}{1-(1-t)z}\right)\mathrm{d}t. \tag{4}$$

They also used the integral representation (4) to show that

$$\|\mathrm{H}\|_{\mathrm{H}^p \to \mathrm{H}^p} \le \frac{\pi}{\sin\frac{\pi}{p}} \quad \text{for } 2 \le p < \infty,$$

and

$$\|\mathrm{H}\|_{\mathrm{H}^p \to \mathrm{H}^p} \le \frac{2\pi}{\sin\frac{\pi}{p}} + \left\|\log\frac{1}{1-z}\right\|_{\mathrm{H}^p} \quad \text{for } 1 < p < 2.$$

It was conjectured that the quantity $\pi/\sin(\pi/p)$ is exactly the norm of the Hilbert matrix on Hardy space H^p in the whole range $1 < p < \infty$. This conjecture was proved by Dostanić, Jevtić, and Vukotić [6]. They used the theory of Hankel matrices and the well-known fact related to the norm of Riesz projection on Hardy spaces. Diamantopoulos [4] considered the action of the Hilbert matrix on Bergman spaces A^p and proved the following results:

- The Hilbert matrix H is bounded on A^p if and only if $2 < p < \infty$.
- If $4 \le p < \infty$, then

$$\|\mathrm{H}\|_{\mathrm{A}^p \to \mathrm{A}^p} \le \frac{\pi}{\sin\frac{2\pi}{p}}.$$

- If $2 < p < 4$, then

$$\|\mathrm{H}\|_{\mathrm{A}^p \to \mathrm{A}^p} \le \left(\frac{2^{7-p}}{9(p-2)} + 2^{4-p}\right)^{\frac{1}{p}} \frac{\pi}{\sin\frac{2\pi}{p}}.$$

Dostanić, Jevtić, and Vukotić [6] obtained

$$\|\mathrm{H}\|_{\mathrm{A}^p \to \mathrm{A}^p} = \frac{\pi}{\sin\frac{2\pi}{p}} \quad \text{for } 4 \le p < \infty,$$

and it was conjectured that the above equality also holds in the remaining case $2 < p < 4$. This conjecture was settled in [2], where a new method was introduced which uses monotonicity of integral means and reduces the conjecture about the norm to some inequalities for Beta function. See also [17]. Namely, the following inequality has been proven:

$$\left(\frac{4-p}{2}+\frac{p-2}{2}s^4\right)\mathrm{B}\left(\frac{2}{p},1-\frac{2}{p}\right)\le\int_0^1\psi_p(t)\max\left\{s^2,t^2\right\}^{\frac{p-2}{2}}\mathrm{d}t,$$

where $2 < p < 4$, $s \in [0,1]$, $\psi_p(t) = t^{\frac{2}{p}-1}(1-t)^{-\frac{2}{p}}$ and B is a classical Beta function. In the following considerations, we assume that $0 < p < \infty$ and $-1 < \alpha < \infty$. As we stated earlier, the following is known [11] to be valid:

$$\mathrm{H}\ \text{is bounded on}\ \mathrm{A}^p_\alpha\ \text{if and only if}\ 1<\alpha+2<p.$$

The preceding result opened the way to the question of the exact norm of the Hilbert matrix on weighted Bergman spaces. In [14] it was proved that

$$\|\mathrm{H}\|_{\mathrm{A}^p_\alpha\to\mathrm{A}^p_\alpha}\ge\frac{\pi}{\sin\frac{(\alpha+2)\pi}{p}}\quad\text{for } 1<\alpha+2<p, \tag{5}$$

and it was conjectured that this lower bound is actually the exact norm of the Hilbert matrix on weighted Bergman spaces.

Conjecture If $1 < \alpha + 2 < p$, then

$$\|\mathrm{H}\|_{\mathrm{A}^p_\alpha\to\mathrm{A}^p_\alpha}=\frac{\pi}{\sin\frac{(\alpha+2)\pi}{p}}.$$

This means that it is necessary to have the following upper bound:

$$\|\mathrm{H}\|_{\mathrm{A}^p_\alpha\to\mathrm{A}^p_\alpha}\le\frac{\pi}{\sin\frac{(\alpha+2)\pi}{p}}\quad\text{for } 1<\alpha+2<p,$$

to prove the mentioned conjecture. The conjecture was confirmed [14] in the case when $\alpha \ge 0$ and $p \ge 2(\alpha+2)$, which reduces the conjecture in the case $\alpha \ge 0$ to the interval $\alpha+2 < p < 2(\alpha+2)$. As already stated, when $\alpha = 0$, this was completely solved in [2] (see also [17]). Lindström, Miihkinen, and Wikman [18] confirmed the conjecture in the case $\alpha > 0$ when

$$\alpha+2+\sqrt{\alpha^2+\frac{7}{2}\alpha+3}=\alpha+2+\sqrt{(\alpha+2)^2-\frac{1}{2}(\alpha+2)}\le p<2(\alpha+2).$$

Recently, this was improved and the conjecture is confirmed [15] in the case $\alpha > 0$ when

$$\alpha + 2 + \sqrt{(\alpha + 2)^2 - \left(\sqrt{2} - \frac{1}{2}\right)(\alpha + 2)} \le p < 2(\alpha + 2).$$

3.1 Lower Bound

Below we prove the inequality (5), where the proof is taken from the paper [14]. First, we recall that Beta and Gamma functions are defined by

$$\mathrm{B}(s, t) = \int_0^1 x^{s-1}(1 - x)^{t-1}\, \mathrm{d}x = \int_0^\infty \frac{x^{s-1}}{(1 + x)^{s+t}}\, \mathrm{d}x,$$

and

$$\Gamma(t) = \int_0^\infty x^{t-1} \mathrm{e}^{-x}\, \mathrm{d}x,$$

for s, t such that $\operatorname{Re} s > 0$, $\operatorname{Re} t > 0$. The value $\mathrm{B}(s, t)$ can be expressed in terms of the Gamma function as

$$\mathrm{B}(s, t) = \frac{\Gamma(s)\Gamma(t)}{\Gamma(s + t)}.$$

Moreover, the Gamma function satisfies the functional equation

$$\Gamma(z)\Gamma(1 - z) = \frac{\pi}{\sin \pi z},$$

for nonintegral complex numbers z. As is usual, $\mathrm{F}(a, b, c; z) = {}_2\mathrm{F}_1(a, b, c; z)$, where $z \in \mathbb{D}$, denotes the hypergeometric function with parameters a, b, c, that is,

$$\mathrm{F}(a, b, c; z) = \sum_{k=0}^{\infty} \frac{\Gamma(k + a)}{\Gamma(a)} \frac{\Gamma(k + b)}{\Gamma(b)} \frac{\Gamma(c)}{\Gamma(k + c)} \frac{z^k}{k!}.$$

We will use the following integral representation of hypergeometric function:

$$\mathrm{F}(a, b, c; z) = \frac{1}{\mathrm{B}(a, c - a)} \int_0^1 \frac{t^{a-1}(1 - t)^{c-a-1}}{(1 - tz)^b}\, \mathrm{d}t, \quad \operatorname{Re} c > \operatorname{Re} a > 0. \tag{6}$$

Theorem 3.1 *If* $1 < \alpha + 2 < p$, *then*

$$\|\mathrm{H}\|_{A^p_\alpha \to A^p_\alpha} \geq \frac{\pi}{\sin \frac{(\alpha+2)\pi}{p}}.$$

Proof Let $1 < \gamma < \alpha + 2 < p$ and

$$f_\gamma(z) = \left(\frac{1}{1-z}\right)^{\frac{\gamma}{p}}, \quad \text{where } z \in \mathbb{D}.$$

An easy calculation shows that

$$\| f_\gamma \|^p_{A^p_\alpha} = \mathrm{F}\left(\frac{\gamma}{2}, \frac{\gamma}{2}, \alpha + 2; 1\right).$$

By Stirling's formula

$$\frac{\Gamma^2\left(k + \frac{\gamma}{2}\right)}{\Gamma(k + \alpha + 2)} \sim \frac{k!}{(k+1)^{\alpha+3-\gamma}}, \quad \text{when } k \to \infty.$$

Since $\alpha + 3 - \gamma > 1$, we conclude

$$\| f_\gamma \|_{A^p_\alpha} < \infty.$$

On the other hand, we have

$$\lim_{\gamma \to \alpha+2} \| f_\gamma \|_{A^p_\alpha} = \infty.$$

Using (3) and (6), we find that

$$\mathrm{H} f_\gamma(z) = \int_0^1 \frac{dt}{(1-t)^{\frac{\gamma}{p}}(1-tz)} = \mathrm{B}\left(1, 1 - \frac{\gamma}{p}\right) \mathrm{F}\left(1, 1, 2 - \frac{\gamma}{p}; z\right).$$

Thus,

$$\mathrm{H} f_\gamma(z) = \Gamma\left(\frac{\gamma}{p}\right) \Gamma\left(1 - \frac{\gamma}{p}\right) \sum_{k=0}^{\infty} \frac{\Gamma^2(k+1)}{\Gamma\left(k + 2 - \frac{\gamma}{p}\right) \Gamma\left(k + \frac{\gamma}{p}\right)} \frac{\Gamma\left(k + \frac{\gamma}{p}\right)}{\Gamma\left(\frac{\gamma}{p}\right)} \frac{z^k}{k!}.$$

Since

$$\frac{\Gamma^2(k+1)}{\Gamma\left(k + 2 - \frac{\gamma}{p}\right) \Gamma\left(k + \frac{\gamma}{p}\right)} = 1 + \mathrm{O}\left(\frac{1}{k+1}\right),$$

we obtain

$$\mathrm{H}f_\gamma(z) = \frac{\pi}{\sin\frac{\pi\gamma}{p}}\left(f_\gamma(z) + g_\gamma(z)\right),$$

where

$$\sup_{1<\gamma<\alpha+2} \|g_\gamma\|_{\mathrm{A}^p_\alpha} \le \mathrm{C}_{p,\alpha} < \infty.$$

Therefore,

$$\|\mathrm{H}\|_{\mathrm{A}^p_\alpha\to\mathrm{A}^p_\alpha} \ge \frac{\|\mathrm{H}f_\gamma\|_{\mathrm{A}^p_\alpha}}{\|f_\gamma\|_{\mathrm{A}^p_\alpha}} \ge \frac{\pi}{\sin\frac{\pi\gamma}{p}}\,\frac{\|f_\gamma\|_{\mathrm{A}^p_\alpha} - \|g_\gamma\|_{\mathrm{A}^p_\alpha}}{\|f_\gamma\|_{\mathrm{A}^p_\alpha}}.$$

Letting $\gamma \to \alpha + 2$, we get

$$\begin{aligned}\|\mathrm{H}\|_{\mathrm{A}^p_\alpha\to\mathrm{A}^p_\alpha} &\ge \lim_{\gamma\to\alpha+2}\left(\frac{\pi}{\sin\frac{\pi\gamma}{p}}\,\frac{\|f_\gamma\|_{\mathrm{A}^p_\alpha} - \|g_\gamma\|_{\mathrm{A}^p_\alpha}}{\|f_\gamma\|_{\mathrm{A}^p_\alpha}}\right)\\ &= \frac{\pi}{\sin\frac{(\alpha+2)\pi}{p}}\lim_{\gamma\to\alpha+2}\left(1 - \frac{\|g_\gamma\|_{\mathrm{A}^p_\alpha}}{\|f_\gamma\|_{\mathrm{A}^p_\alpha}}\right)\\ &= \frac{\pi}{\sin\frac{(\alpha+2)\pi}{p}},\end{aligned}$$

because

$$\lim_{\gamma\to\alpha+2}\|f_\gamma\|_{\mathrm{A}^p_\alpha} = \infty \quad\text{and}\quad \sup_{1<\gamma<\alpha+2}\|g_\gamma\|_{\mathrm{A}^p_\alpha} \le \mathrm{C}_{p,\alpha} < \infty.$$

This concludes the proof. □

3.2 Upper Bound

The main part of this subsection is based on the results of the paper [15]. Let $1 < \alpha+2 < p$. By following [14], from the continuous version of Minkowski inequality and representation (4), we have estimate

$$\|\mathrm{H}f\|_{\mathrm{A}^p_\alpha} \le \int_0^1 \|\mathrm{T}_t f\|_{\mathrm{A}^p_\alpha}\,\mathrm{d}t,$$

and

$$\|\mathrm{T}_t f\|_{\mathrm{A}^p_\alpha} = t^{\frac{\alpha+2}{p}-1}(1-t)^{-\frac{\alpha+2}{p}} \left(\frac{\alpha+1}{\pi}\int_{\mathrm{D}_t} |w|^{p-4}|f(w)|^p g_t(w)^\alpha \,\mathrm{dm}(w)\right)^{\frac{1}{p}}, \tag{7}$$

where

$$g_t(w) = \frac{2\operatorname{Re} w - t - (2-t)|w|^2}{(1-t)|w|^2}, \quad \mathrm{D}_t = \mathrm{D}(c_t, \rho_t), \quad c_t = \frac{1}{2-t} \text{ and } \rho_t = \frac{1-t}{2-t}.$$

Here, $\mathrm{D}_t = \mathrm{D}(c_t, \rho_t)$ denotes the Euclidean disc of radius ρ_t centered at the point c_t in the complex plane, that is,

$$\mathrm{D}_t = \mathrm{D}(c_t, \rho_t) = \{z \in \mathbb{C} : |z - c_t| < \rho_t\}.$$

It is easy to check that

$$g_t(w) = \frac{1}{|w|^2} \cdot \frac{\rho_t^2 - |w - c_t|^2}{\rho_t} \quad \text{for } w \in \mathrm{D}(c_t, \rho_t).$$

In the rest of the paper, we will use the following function:

$$\psi_{p,\alpha}(t) = t^{\frac{\alpha+2}{p}-1}(1-t)^{-\frac{\alpha+2}{p}},$$

where $0 < t < 1$. Therefore, we conclude

$$\|\mathrm{T}_t f\|_{\mathrm{A}^p_\alpha} = \psi_{p,\alpha}(t)\left(\frac{\alpha+1}{\pi}\int_{\mathrm{D}_t} |w|^{p-2(\alpha+2)}|f(w)|^p \left(\frac{\rho_t^2 - |w-c_t|^2}{\rho_t}\right)^\alpha \mathrm{dm}(w)\right)^{\frac{1}{p}}.$$

The previous formula is valid for all $1 < \alpha + 2 < p$ and it will be used later in this paper. On the other hand, following [14] in the special case when $\alpha > 0$, we obtain

$$g_t(w)^\alpha \le \left(\frac{1 + |w|^2 - t - (2-t)|w|^2}{(1-t)|w|^2}\right)^\alpha = \left(\frac{1 - |w|^2}{|w|^2}\right)^\alpha. \tag{8}$$

By combining (7) and (8), we get

$$\|\mathrm{T}_t f\|_{\mathrm{A}^p_\alpha} \le \psi_{p,\alpha}(t)\left(\frac{\alpha+1}{\pi}\int_{\mathrm{D}_t} |w|^{p-2(\alpha+2)}|f(w)|^p \left(1 - |w|^2\right)^\alpha \mathrm{dm}(w)\right)^{\frac{1}{p}}.$$

Let again $\alpha > 0$. We define the function Φ_α as follows:

$$\Phi_\alpha(x) = 2x^2 - (4(\alpha+2)+1)\,x + 2\sqrt{\alpha+2}\sqrt{x} + \alpha + 2,$$

where $x \in (\alpha+2, 2(\alpha+2))$. It is easy to check that Φ_α is an increasing function on interval $(\alpha+2, 2(\alpha+2))$. Also

$$\Phi_\alpha(\alpha+2) < 0 \text{ and } \Phi_\alpha(2(\alpha+2)) > 0.$$

This means that function Φ_α has a unique zero α_0 on the interval $(\alpha+2, 2(\alpha+2))$. Moreover, we get $\Phi_\alpha < 0$ on $(\alpha+2, \alpha_0)$ and $\Phi_\alpha > 0$ on $(\alpha_0, 2(\alpha+2))$. See [15] for more details. The following result is also taken from [15].

Theorem 3.2 *Let* $\alpha > 0$ *and* $\alpha_0 \le p < 2(\alpha+2)$. *Then*

$$\|\mathrm{H}\|_{\mathrm{A}^p_\alpha \to \mathrm{A}^p_\alpha} \le \frac{\pi}{\sin\frac{(\alpha+2)\pi}{p}}. \tag{9}$$

It can be checked [15] that

$$\Phi_\alpha\left(\alpha+2+\sqrt{(\alpha+2)^2-\left(\sqrt{2}-\frac{1}{2}\right)(\alpha+2)}\right) > 0.$$

The previous inequality implies

$$\alpha_0 < \alpha+2+\sqrt{(\alpha+2)^2-\left(\sqrt{2}-\frac{1}{2}\right)(\alpha+2)},$$

whence by Theorems 3.1 and 3.2, it follows:

$$\|\mathrm{H}\|_{\mathrm{A}^p_\alpha \to \mathrm{A}^p_\alpha} = \frac{\pi}{\sin\frac{(\alpha+2)\pi}{p}},$$

in the case when

$$\alpha > 0 \text{ and } \alpha+2+\sqrt{(\alpha+2)^2-\left(\sqrt{2}-\frac{1}{2}\right)(\alpha+2)} \le p < 2(\alpha+2).$$

Let also

$$\beta = \alpha+2+\sqrt{(\alpha+2)^2-(\alpha+2)}.$$

Then by straightforward calculations, we derive

$$\Phi_\alpha(\beta) = 2\underbrace{\left(\beta^2-2(\alpha+2)\beta+\alpha+2\right)}_{=0} - \left(\sqrt{\beta}-\sqrt{\alpha+2}\right)^2 < 0,$$

which implies that $\beta < \alpha_0$. In the case when $\alpha > 0$ and $\alpha + 2 < p \le \beta$, we have the following partial result [15].

Theorem 3.3 *Let* $\alpha > 0$, $\alpha+2 < p \le \alpha+2+\sqrt{(\alpha+2)^2-(\alpha+2)}$ *and suppose that the following condition holds:*

$$\int_0^1 \psi_{p,\alpha}(t)\xi_{p,\alpha}(t)\,dt \le \frac{1}{\alpha+1}\int_0^1 \psi_{p,\alpha}(t)\,dt = \frac{1}{\alpha+1}\mathrm{B}\left(\frac{\alpha+2}{p}, 1-\frac{\alpha+2}{p}\right), \tag{10}$$

where we denoted

$$\psi_{p,\alpha}(t) = t^{\frac{\alpha+2}{p}-1}(1-t)^{-\frac{\alpha+2}{p}} \quad \text{and} \quad \xi_{p,\alpha}(t) = \int_{\left(\frac{t}{2-t}\right)^2}^1 \rho^{\frac{p}{2}-(\alpha+2)}(1-\rho)^\alpha\,d\rho,$$

for $0 < t < 1$. *Then*

$$\|\mathrm{H}\|_{A^p_\alpha \to A^p_\alpha} = \frac{\pi}{\sin\frac{(\alpha+2)\pi}{p}}.$$

We note that condition (10) is not always satisfied under the given conditions even when $p/2 - (\alpha+2) + 1 > 0$, that is, $p > 2\alpha + 2$, which actually allows the convergence of the integral $\xi_{p,\alpha}(0)$. Namely, a calculation involving *Mathematica* shows that when $\alpha = 1$, then $\beta \approx 5.449$ and $\alpha_0 \approx 5.487$ and also for

$$\alpha + 2 < 2\alpha + 2 < p = 4.4 < \beta < \alpha_0,$$

we have

$$\int_0^1 \psi_{p,\alpha}(t)\xi_{p,\alpha}(t)\,dt - \frac{1}{\alpha+1}\int_0^1 \psi_{p,\alpha}(t)\,dt \approx 0.962 > 0.$$

On the other hand, if $\alpha = 1$ and $\alpha + 2 < 2\alpha + 2 < p = 5.2 < \beta < \alpha_0$, then

$$\int_0^1 \psi_{p,\alpha}(t)\xi_{p,\alpha}(t)\,dt - \frac{1}{\alpha+1}\int_0^1 \psi_{p,\alpha}(t)\,dt \approx -0.103 < 0,$$

which allows the application of Theorem 3.3 in some cases where it is not possible to apply Theorem 3.2. We also have

$$\beta = \alpha+2+\sqrt{(\alpha+2)^2-(\alpha+2)} < \alpha_0 < \alpha+2+\sqrt{(\alpha+2)^2-\left(\sqrt{2}-\frac{1}{2}\right)(\alpha+2)},$$

and since $\sqrt{2} - 1/2 \approx 0.914$ (actually $\sqrt{2} - 1/2 > 0.914$), we can write

$$\alpha+2+\sqrt{(\alpha+2)^2-(\alpha+2)} < \alpha_0 < \alpha+2+\sqrt{(\alpha+2)^2-0.914(\alpha+2)}. \quad (11)$$

From (11) we can conclude that there remains a small gap between β and α_0 to which we cannot apply Theorems 3.2 and 3.3. The following picture (Fig. 2) shows the results obtained in relation to the conjecture of Hilbert matrix norm on weighted Bergman spaces, where

$$\gamma = \alpha+2+\sqrt{(\alpha+2)^2-\frac{1}{2}(\alpha+2)}, \quad \delta = \alpha+2+\sqrt{(\alpha+2)^2-\left(\sqrt{2}-\frac{1}{2}\right)(\alpha+2)}.$$

Let $\alpha > 0$ and $\alpha+2 < p < 2(\alpha+2)$. We consider the following functions:

$$\psi_{p,\alpha}(t) = t^{\frac{\alpha+2}{p}-1}(1-t)^{-\frac{\alpha+2}{p}} \quad \text{and} \quad \xi_{p,\alpha}(t) = \int_{\left(\frac{t}{2-t}\right)^2}^{1} \rho^{\frac{p}{2}-(\alpha+2)}(1-\rho)^{\alpha}\,d\rho,$$

where $0 < t < 1$ as in the statement of Theorem 3.3. For $s \in [0,1]$, we denote

$$F_{p,\alpha}(s) = \xi_{p,\alpha}(s)\int_0^s \psi_{p,\alpha}(t)\,dt + \int_s^1 \psi_{p,\alpha}(t)\xi_{p,\alpha}(t)\,dt$$
$$-\frac{1}{\alpha+1}\left(1-\left(\frac{s}{2-s}\right)^2\right)^{\alpha+1}\int_0^1 \psi_{p,\alpha}(t)\,dt.$$

It turns out [15, Lemma 1] that

$$F_{p,\alpha}(s) \le 0 \quad \text{for all } s \in [0,1], \quad \text{in the case } \alpha_0 \le p < 2(\alpha+2). \quad (12)$$

On the other hand, if the following conditions holds:

$$\int_0^1 \psi_{p,\alpha}(t)\xi_{p,\alpha}(t)\,dt \le \frac{1}{\alpha+1}\int_0^1 \psi_{p,\alpha}(t)\,dt.$$

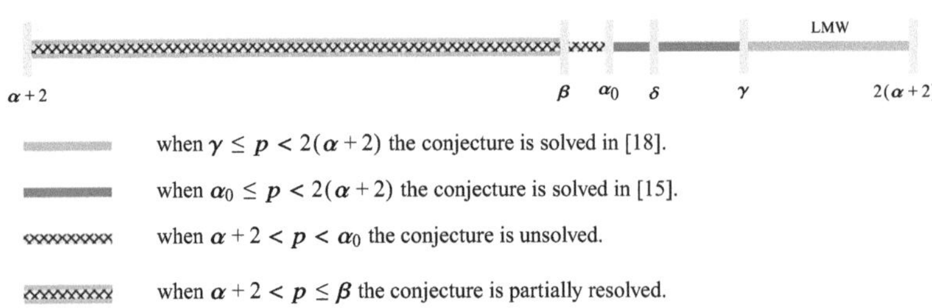

Fig. 2 We present graphically the obtained results on the interval $(\alpha+2, 2(\alpha+2))$ in the case when $\alpha > 0$

then [15, Lemma 2]

$$F_{p,\alpha}(s) \leq 0 \text{ for all } s \in [0,1], \text{ in the case } \alpha + 2 < p \leq \beta. \tag{13}$$

Let $f \in \mathrm{A}^p_\alpha$. We define

$$\varphi(r) = 2\mathrm{M}_p^p(r, f) \text{ and } \chi(r) = \varphi(r) - \varphi(0) \text{ for } 0 \leq r < 1.$$

Then φ is a nondecreasing and differentiable function on the interval (0, 1), which implies that function χ is also nondecreasing and differentiable on (0, 1). Hence,

$$\chi' \geq 0 \text{ on } (0,1) \text{ and } \chi(r) = \int_0^r \chi'(s)\,\mathrm{d}s, \tag{14}$$

for $0 \leq r < 1$. We already know that

$$\|\mathrm{H}f\|_{\mathrm{A}^p_\alpha} \leq \int_0^1 \|\mathrm{T}_t f\|_{\mathrm{A}^p_\alpha}\,\mathrm{d}t, \tag{15}$$

and

$$\|\mathrm{T}_t f\|_{\mathrm{A}^p_\alpha} \leq \psi_{p,\alpha}(t) \left(\frac{\alpha+1}{\pi} \int_{\mathrm{D}_t} |w|^{p-2(\alpha+2)} |f(w)|^p \left(1 - |w|^2\right)^\alpha \mathrm{dm}(w)\right)^{\frac{1}{p}},$$

where

$$\mathrm{D}_t = \mathrm{D}(c_t, \rho_t), \quad c_t = \frac{1}{2-t} \text{ and } \rho_t = \frac{1-t}{2-t}.$$

Also, we have

$$\mathrm{D}_t \subset \mathrm{A}(0, c_t - \rho_t, c_t + \rho_t) = \mathrm{A}\left(0, \frac{t}{2-t}, 1\right),$$

Here

$$\mathrm{A}_t = \mathrm{A}\left(0, \frac{t}{2-t}, 1\right) = \left\{z \in \mathbb{C} : \frac{t}{2-t} < |z| < 1\right\}.$$

Therefore (see Fig. 3),

$$\|\mathrm{T}_t f\|_{\mathrm{A}^p_\alpha} \leq \psi_{p,\alpha}(t) \left(\frac{\alpha+1}{\pi} \int_{\mathrm{A}_t} |w|^{p-2(\alpha+2)} |f(w)|^p \left(1 - |w|^2\right)^\alpha \mathrm{dm}(w)\right)^{\frac{1}{p}},$$

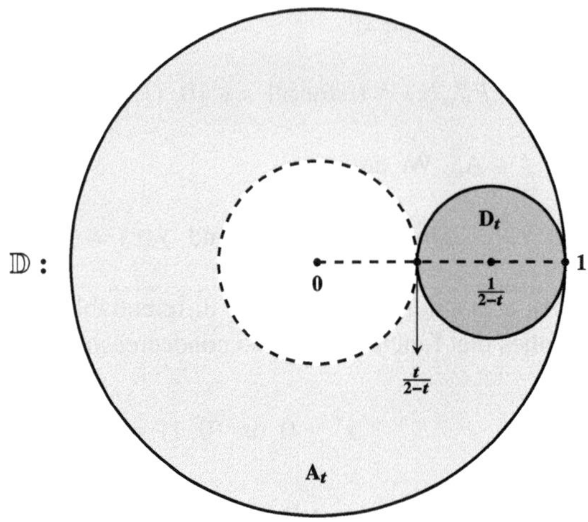

Fig. 3 Instead of the integration over the disc D_t, we switch to the integration over the annulus A_t which contains disc D_t

or equivalently

$$\|\mathrm{T}_t f\|_{\mathrm{A}^p_\alpha} \leq \psi_{p,\alpha}(t) \left((\alpha+1) \int_{\frac{t}{2-t}}^{1} r^{p-2(\alpha+2)+1} \left(1-r^2\right)^\alpha \varphi(r)\,\mathrm{d}r \right)^{\frac{1}{p}}. \tag{16}$$

Also, we have

$$\frac{\pi}{\sin \frac{(\alpha+2)\pi}{p}} \|f\|_{\mathrm{A}^p_\alpha} = \int_0^1 \psi_{p,\alpha}(t) \left((\alpha+1) \int_0^1 r \left(1-r^2\right)^\alpha \varphi(r)\,\mathrm{d}r \right)^{\frac{1}{p}} \mathrm{d}t. \tag{17}$$

From (15), (16), and (17), we can conclude that (9) holds if the following inequality is true:

$$\int_0^1 \psi_{p,\alpha}(t) \left(I_{p,\alpha}(t)^{1/p} - J_\alpha^{1/p} \right) \mathrm{d}t \leq 0, \tag{18}$$

where

$$I_{p,\alpha}(t) = \int_{\frac{t}{2-t}}^{1} r^{p-2(\alpha+2)+1} \left(1-r^2\right)^\alpha \varphi(r)\,\mathrm{d}r \quad \text{and} \quad J_\alpha = \int_0^1 r \left(1-r^2\right)^\alpha \varphi(r)\,\mathrm{d}r.$$

Using the following elementary fact

$$I_{p,\alpha}(t)^{1/p} - J_\alpha^{1/p} \leq \frac{1}{p} J_\alpha^{\frac{1}{p}-1} \left(I_{p,\alpha}(t) - J_\alpha \right),$$

we conclude that (18) holds if the following inequality is true:

$$\int_0^1 \psi_{p,\alpha}(t)\left(I_{p,\alpha}(t) - J_\alpha\right) \mathrm{d}t \leq 0,$$

or

$$\int_0^1 \psi_{p,\alpha}(t)\left(\int_{\frac{t}{2-t}}^1 r^{p-2(\alpha+2)+1}\left(1-r^2\right)^\alpha \varphi(r)\,\mathrm{d}r - \int_0^1 r\left(1-r^2\right)^\alpha \varphi(r)\,\mathrm{d}r\right) \mathrm{d}t \leq 0,$$

or equivalently

$$V_{p,\alpha} + \varphi(0) W_{p,\alpha} \leq U_{p,\alpha}, \tag{19}$$

where we denoted

$$V_{p,\alpha} = \int_0^1 \psi_{p,\alpha}(t) \int_{\frac{t}{2-t}}^1 r^{p-2(\alpha+2)+1}\left(1-r^2\right)^\alpha \chi(r)\,\mathrm{d}r\mathrm{d}t,$$

and

$$\begin{aligned} W_{p,\alpha} = &\int_0^1 \psi_{p,\alpha}(t) \int_{\frac{t}{2-t}}^1 r^{p-2(\alpha+2)+1}\left(1-r^2\right)^\alpha \mathrm{d}r\mathrm{d}t \\ &- \int_0^1 \psi_{p,\alpha}(t) \int_0^1 r\left(1-r^2\right)^\alpha \mathrm{d}r\mathrm{d}t, \end{aligned}$$

and

$$U_{p,\alpha} = \int_0^1 \psi_{p,\alpha}(t) \int_0^1 r\left(1-r^2\right)^\alpha \chi(r)\,\mathrm{d}r\mathrm{d}t.$$

After some work, by using changes of variables and Fubini theorem (see [15] for more details), it can be shown that

$$W_{p,\alpha} = \frac{1}{2} F_{p,\alpha}(0),$$

$$\begin{aligned} V_{p,\alpha} = &\int_0^1 \frac{1}{(2-u)^2} \chi'\left(\frac{u}{2-u}\right)\left(\xi_{p,\alpha}(u)\int_0^u \psi_{p,\alpha}(t)\mathrm{d}t \right. \\ &\left. + \int_u^1 \psi_{p,\alpha}(t)\xi_{p,\alpha}(t)\,\mathrm{d}t\right)\mathrm{d}u, \end{aligned}$$

and

$$U_{p,\alpha} = \int_0^1 \frac{1}{(2-u)^2}\chi'\left(\frac{u}{2-u}\right)\left(\frac{1}{\alpha+1}\left(1-\left(\frac{u}{2-u}\right)^2\right)^{\alpha+1}\right.$$
$$\left.\int_0^1 \psi_{p,\alpha}(t)\,\mathrm{d}t\right)\mathrm{d}u.$$

Therefore,

$$V_{p,\alpha} - U_{p,\alpha} = \int_0^1 \frac{1}{(2-u)^2}\chi'\left(\frac{u}{2-u}\right)F_{p,\alpha}(u)\,\mathrm{d}u.$$

To obtain (19), it is enough to prove that

$$\int_0^1 \frac{1}{(2-u)^2}\chi'\left(\frac{u}{2-u}\right)F_{p,\alpha}(u)\,\mathrm{d}u + \frac{\varphi(0)}{2}F_{p,\alpha}(0) \le 0.$$

or

$$\int_0^1 \frac{1}{(2-u)^2}\chi'\left(\frac{u}{2-u}\right)F_{p,\alpha}(u)\,\mathrm{d}u + |f(0)|^p F_{p,\alpha}(0) \le 0, \tag{20}$$

because $\varphi(0) = 2|f(0)|^p$. We are ready to prove Theorems 3.2 and 3.3.

Proof (Proof of Theorem 3.2) Let $f \in \mathrm{A}^p_\alpha$. By using (12), we have

$$F_{p,\alpha}(u) \le 0, \tag{21}$$

for all $u \in [0, 1]$. From (14) we conclude that

$$\frac{1}{(2-u)^2}\chi'\left(\frac{u}{2-u}\right) \ge 0. \tag{22}$$

Combining (21) and (22), we get

$$\int_0^1 \frac{1}{(2-u)^2}\chi'\left(\frac{u}{2-u}\right)F_{p,\alpha}(u)\,\mathrm{d}u + |f(0)|^p F_{p,\alpha}(0) \le 0,$$

which leads to inequality (20) being valid. This implies that inequality (9) is also valid, which completes the proof. □

Proof (Proof of Theorem 3.3) Let $f \in A^p_\alpha$. Then (13) implies

$$F_{p,\alpha}(u) \le 0,$$

for all $u \in [0, 1]$. Since $\chi' \geq 0$, we derive

$$\int_0^1 \frac{1}{(2-u)^2}\chi'\left(\frac{u}{2-u}\right) F_{p,\alpha}(u)\,\mathrm{d}u + |f(0)|^p F_{p,\alpha}(0) \leq 0,$$

which implies validity of the inequality (9). This finishes the proof. □

All previous considerations relating to the upper bound of the Hilbert matrix norm on weighted Bergman spaces are given in the case when $\alpha > 0$. On the other hand, it turns out that the analysis when $-1 < \alpha < 0$ is a bit more complicated. Namely, in this case, we cannot use the inequality (8), which further simplifies the estimation of the Hilbert matrix norm on weighted Bergman spaces. So in this case, we have to use the following two facts directly:

$$\|\mathrm{H}f\|_{\mathrm{A}^p_\alpha} \leq \int_0^1 \|\mathrm{T}_t f\|_{\mathrm{A}^p_\alpha}\,\mathrm{d}t,$$

and

$$\|\mathrm{T}_t f\|_{\mathrm{A}^p_\alpha} = \psi_{p,\alpha}(t)\left(\frac{\alpha+1}{\pi}\int_{\mathrm{D}_t} |w|^{p-2(\alpha+2)}|f(w)|^p\left(\frac{\rho_t^2 - |w-c_t|^2}{\rho_t}\right)^\alpha \mathrm{dm}(w)\right)^{\frac{1}{p}}.$$

If we denote

$$\eta_t(z) = \rho_t z + c_t \ \text{ for } \ z \in \mathbb{D} \ \text{ and } \ 0 < t < 1,$$

then

$$\eta_t(\mathbb{D}) = \mathrm{D}_t \subset \mathbb{D}.$$

The following result taken from [3] will be useful.

Lemma 3.1 *Let $0 < p < \infty$, $-1 < \alpha < 0$, $0 < c < 1$, and $\eta(z) = \rho z + c$ for $z \in \mathbb{D}$, where $\rho = 1 - c$. Then,*

$$\rho^{\alpha+2}\int_{\mathbb{D}} |(f\circ\eta)(z)|^p \left(1-|z|^2\right)^\alpha \mathrm{dm}(z) \leq \frac{1+3c}{(1+c)^{\alpha+1}}\int_{\mathbb{D}} |f(z)|^p\left(1-|z|^2\right)^\alpha \mathrm{dm}(z),$$

where $f \in \mathrm{Hol}(\mathbb{D})$.

Let $f \in \mathrm{A}^p_\alpha$, where $-1 < \alpha < 0$ and $\alpha + 2 < p$. We define

$$\kappa(x) = \frac{1+3x}{(1+x)^{\alpha+1}} \quad \text{for } x \in [0, 1].$$

It is easy to see that

$$\max_{x\in[0,1]} \kappa(x) = \kappa(1) = 2^{1-\alpha}.$$

Then we have the following two cases. First, let $2(\alpha+2) \le p$. Using Lemma 3.1 and the following fact

$$|w|^{p-2(\alpha+2)} \le 1 \text{ for } w \in \eta_t(\mathbb{D}) \subset \mathbb{D},$$

we obtain

$$\|\mathrm{T}_t f\|_{\mathrm{A}^p_\alpha} \le \psi_{p,\alpha}(t) \left(\frac{\alpha+1}{\pi}\right)^{\frac{1}{p}} \left(\int_{\eta_t(\mathbb{D})} |f(w)|^p \left(\frac{\rho_t^2 - |w-c_t|^2}{\rho_t}\right)^\alpha \mathrm{dm}(w)\right)^{\frac{1}{p}}$$

$$= \psi_{p,\alpha}(t) \left(\frac{\alpha+1}{\pi}\right)^{\frac{1}{p}} \left(\rho_t^{\alpha+2} \int_{\mathbb{D}} |(f\circ\eta_t)(z)|^p \left(1-|z|^2\right)^\alpha \mathrm{dm}(w)\right)^{\frac{1}{p}}$$

$$\le \psi_{p,\alpha}(t)\, \kappa(c_t)^{\frac{1}{p}} \|f\|_{\mathrm{A}^p_\alpha} \le 2^{\frac{1-\alpha}{p}} \psi_{p,\alpha}(t) \|f\|_{\mathrm{A}^p_\alpha},$$

which implies

$$\|\mathrm{H}f\|_{\mathrm{A}^p_\alpha} \le \int_0^1 \|\mathrm{T}_t f\|_{\mathrm{A}^p_\alpha} \,\mathrm{d}t \le 2^{\frac{1-\alpha}{p}} \frac{\pi}{\sin\frac{(\alpha+2)\pi}{p}} \|f\|_{\mathrm{A}^p_\alpha}.$$

Second, let $\alpha+2 < p < 2(\alpha+2)$. If $w \in \eta_t(\mathbb{D})$, then $w = \eta_t(z)$ for some $z \in \mathbb{D}$, whence it follows

$$|w| = |\rho_t z + c_t| \ge c_t - \rho_t|z| > c_t - \rho_t = \frac{t}{(2-t)},$$

that is,

$$|w|^{p-2(\alpha+2)} \le \left(\frac{2-t}{t}\right)^{2(\alpha+2)-p}.$$

Similar to the first case, we obtain the following inequality:

$$\|\mathrm{T}_t f\|_{\mathrm{A}^p_\alpha} \le 2^{\frac{1-\alpha}{p}} \left(\frac{2-t}{t}\right)^{\frac{2(\alpha+2)}{p}-1} \psi_{p,\alpha}(t) \|f\|_{\mathrm{A}^p_\alpha}.$$

On the other hand, since

$$\frac{2(\alpha+2)}{p} - 1 \in (0,1),$$

we find

$$(2-t)^{\frac{2(\alpha+2)}{p}-1} = (t+2(1-t))^{\frac{2(\alpha+2)}{p}-1} \le t^{\frac{2(\alpha+2)}{p}-1} + 2^{\frac{2(\alpha+2)}{p}-1}(1-t)^{\frac{2(\alpha+2)}{p}-1},$$

which implies

$$\left(\frac{2-t}{t}\right)^{\frac{2(\alpha+2)}{p}-1} \psi_{p,\alpha}(t) \le \psi_{p,\alpha}(t) + 2^{\frac{2(\alpha+2)}{p}-1}\psi_{p,\alpha}(1-t).$$

Then,

$$\|\mathrm{T}_t f\|_{\mathrm{A}^p_\alpha} \le 2^{\frac{1-\alpha}{p}}\left(\psi_{p,\alpha}(t) + 2^{\frac{2(\alpha+2)}{p}-1}\psi_{p,\alpha}(1-t)\right)\|f\|_{\mathrm{A}^p_\alpha},$$

which leads to

$$\|\mathrm{H}f\|_{\mathrm{A}^p_\alpha} \le \int_0^1 \|\mathrm{T}_t f\|_{\mathrm{A}^p_\alpha}\,\mathrm{d}t \le 2^{\frac{1-\alpha}{p}}\left(1 + 2^{\frac{2(\alpha+2)}{p}-1}\right)\frac{\pi}{\sin\frac{(\alpha+2)\pi}{p}}\|f\|_{\mathrm{A}^p_\alpha}.$$

In this way, we have shown that the following result is valid.

Theorem 3.4 *Let $-1 < \alpha < 0$ and $\alpha + 2 < p$.*

(i) If $2(\alpha+2) \le p$, then

$$\|\mathrm{H}\|_{\mathrm{A}^p_\alpha \to \mathrm{A}^p_\alpha} \le 2^{\frac{1-\alpha}{p}}\frac{\pi}{\sin\frac{(\alpha+2)\pi}{p}}.$$

(ii) If $\alpha + 2 < p < 2(\alpha+2)$, then

$$\|\mathrm{H}\|_{\mathrm{A}^p_\alpha \to \mathrm{A}^p_\alpha} \le 2^{\frac{1-\alpha}{p}}\left(1 + 2^{\frac{2(\alpha+2)}{p}-1}\right)\frac{\pi}{\sin\frac{(\alpha+2)\pi}{p}}.$$

On the other hand, let $\eta : \mathbb{D} \to \mathbb{D}$ be a holomorphic function, and let $0 < p < \infty$, $-1 < \alpha < 0$, and $f \in \mathrm{Hol}(\mathbb{D})$. Then we have the following well-known inequality:

$$\int_{\mathbb{D}} |(f\circ\eta)(z)|^p \left(1-|z|^2\right)^\alpha \mathrm{dm}(z) \le \left(\frac{1+|\eta(0)|}{1-|\eta(0)|}\right)^{\alpha+2} \int_{\mathbb{D}} |f(z)|^p \left(1-|z|^2\right)^\alpha \mathrm{dm}(z),$$

which can be viewed as a consequence of the Littlewood subordination principle. See [21, Theorem 11.6]. In the special case, when

$$\eta(z) = \rho z + c, \quad \text{for } z \in \mathbb{D},$$

where $0 < c < 1$ and $\rho = 1 - c$, we obtain

$$\rho^{\alpha+2} \int_{\mathbb{D}} |(f \circ \eta)(z)|^p \left(1 - |z|^2\right)^\alpha \mathrm{dm}(z) \le (1+c)^{\alpha+2}$$
$$\int_{\mathbb{D}} |f(z)|^p \left(1 - |z|^2\right)^\alpha \mathrm{dm}(z).$$

We note that this inequality, but only for $-1 < \alpha < 0$, was used in the proof of [15]. Namely, if we use the previous inequality instead of the one from Lemma 3.1 and apply the previous procedure as in the proof of the Theorem 3.4, we get the following result.

Theorem 3.5 *Let $-1 < \alpha < 0$ and $p > \alpha + 2$.*

(i) If $p \ge 2(\alpha + 2)$, then

$$\|\mathrm{H}\|_{A^p_\alpha \to A^p_\alpha} \le 2^{\frac{\alpha+2}{p}} \frac{\pi}{\sin \frac{(\alpha+2)\pi}{p}}.$$

(ii) If $\alpha + 2 < p < 2(\alpha + 2)$, then

$$\|\mathrm{H}\|_{A^p_\alpha \to A^p_\alpha} \le 2^{\frac{\alpha+2}{p}} \left(1 + 2^{\frac{2(\alpha+2)}{p} - 1}\right) \frac{\pi}{\sin \frac{(\alpha+2)\pi}{p}}.$$

See also [15, Theorem 1.3] for more details. Note that in Theorem 3.5, the constant $2^{\frac{1-\alpha}{p}}$ from Theorem 3.4, in both of its parts, is replaced by the constant $2^{\frac{\alpha+2}{p}}$. Since

$$2^{\frac{1-\alpha}{p}} < 2^{\frac{\alpha+2}{p}},$$

when $-1/2 < \alpha < 0$, Theorem 3.4 is a slight improvement of Theorem 3.5 in that case. It would be interesting to see some more precise estimates of the Hilbert matrix norm on weighted Bergman spaces in the case when $-1 < \alpha < 0$.

Acknowledgments The author thanks the reviewers for their constructive comments and recommendations that contributed to improving the quality of the work.

References

1. A. Aleman, A. Montes-Rodríguez, A. Sarafoleanu, *The eigenfunctions of the Hilbert matrix*, Constr. Approx. **36** (2012), 353–374.
2. V. Božin, B. Karapetrović, *Norm of the Hilbert matrix on Bergman spaces*, J. Funct. Anal. **274** (2018), 525–543.

3. D. Bralović, B. Karapetrović, *New upper bound for the Hilbert matrix norm on negatively indexed weighted Bergman spaces*, Bull. Malays. Math. Sci. Soc. **45** (2022), 1183–1193.
4. E. Diamantopoulos, *Hilbert matrix on Bergman spaces*, Illinois J. Math. **48** (2004), 1067–1078.
5. E. Diamantopoulos, A. G. Siskakis, *Composition operators and the Hilbert matrix*, Studia Math. **140** (2000), 191–198.
6. M. Dostanić, M. Jevtić, D. Vukotić, *Norm of the Hilbert matrix on Bergman and Hardy spaces and theorem of Nehari type*, J. Funct. Anal. **254** (2008), 2800–2815.
7. P. Galanopoulos, D. Girela, J. A. Peláez, A. G. Siskakis, *Generalized Hilbert operators*, Ann. Acad. Sci. Fenn. Math. **39** (2014), 231–258.
8. G.H. Hardy, J.E. Littlewood, G. Pólya, *Inequalities*, Cambridge University Press, Cambridge, 1959.
9. D. Hilbert, *Ein Beitrag zur Theorie des Legendre'schen Polynoms*, Acta Math. **18** (1894), 155–159.
10. M. Jevtić, D. Vukotić, M. Arsenović, *Taylor Coefficients and Coefficient Multipliers of Hardy and Bergman-Type Spaces*, RSME Springer Series 2, Springer, Cham, Switzerland (2016).
11. M. Jevtić, B. Karapetrović, *Hilbert matrix on spaces of Bergman-type*, J. Math. Anal. Appl. **453** (2017), 241–254.
12. M. Jevtić, B. Karapetrović, *Hilbert matrix operator on Besov spaces*, Publ. Math. Debrecen **90** (2017), 359–371.
13. B. Karapetrović, *Libera and Hilbert matrix operator on logarithmically weighted Bergman, Bloch and Hardy-Bloch spaces*, Czech. Math. J. **68** (2018), 559–576.
14. B. Karapetrović, *Norm of the Hilbert matrix operator on the weighted Bergman spaces*, Glasgow Math. J. **60** (2018), 513–525.
15. B. Karapetrović, *Hilbert Matrix and Its Norm on Weighted Bergman Spaces*, J. Geom. Anal. **31** (2021), 5909–5940.
16. B. Lanucha, M. Nowak, M. Pavlović, *Hilbert matrix operator on spaces of analytic functions*, Ann. Acad. Sci. Fenn. Math. **37** (2012), 161–174.
17. M. Lindström, S. Miihkinen, N. Wikman, *Norm estimates of weighted composition operators pertaining to the Hilbert matrix*, Proc. Amer. Math. Soc. **147** (2019), 2425–2435.
18. M. Lindström, S. Miihkinen, N. Wikman, *On the exact value of the norm of the Hilbert matrix operator on the weighted Bergman spaces*, Annales Fennici Mathematici **46** (2021), 201–224.
19. W. Magnus, *On the spectrum of Hilbert's matrix*, Amer. J. Math. **72** (1950), 699–704.
20. B. Silbermann, *On the Spectrum of Hilbert Matrix Operator*, Integral Equations Operator Theory **93** (21) (2021), 1–35.
21. K. Zhu, *Operator Theory in Function Spaces*, Second Edition, Mathematical Surveys and Monographs **138**, American Mathematical Society, Providence, RI (2007).

Radial Limits of Holomorphic Functions in $\mathbb{C}^n$ or the Polydisc

Paul M. Gauthier and Mohammad Shirazi

1 Introduction

Denote by $\mathbb{D}_1^1 = \mathbb{D}$ the unit disc in the complex plane $\mathbb{C}$ and by $\mathbb{T} = \partial\mathbb{D}$ the unit circle $\{z = e^{i\theta} : 0 \le \theta < 2\pi\}$. For $0 < R \le +\infty$, put $\mathbb{D}_R = \mathbb{D}_R^1 = \{z = rp : p \in \mathbb{T},\ 0 \le r < R\}$ and for $n = 2, 3, \ldots$, put

$$\mathbb{D}_R^n = \left\{z = (z_1, \ldots, z_n) \in \mathbb{C}^n : |z_j| < R,\ j = 1, \cdots, n\right\}.$$

Thus, $\mathbb{D}_1^n$ is the polydisc $\mathbb{D}^n$ and $\mathbb{D}_\infty^n$ is all of $\mathbb{C}^n$. Denote by $\mathbb{T}^n$ the n-torus,

$$\mathbb{T}^n = \left\{z \in \mathbb{C}^n : z = (e^{i\theta_1}, \ldots, e^{i\theta_n}),\ \theta_j \in [0, 2\pi),\ j = 1, \ldots, n\right\},$$

also called the *distinguished boundary* of $\mathbb{D}^n$. Moreover, denote by $[0, +\infty \cdot p)$ the radius in $\mathbb{C}^n$ passing through the point p and by $[0, R \cdot p)$ the corresponding radius in $\mathbb{D}_R^n$. For a function $u : \mathbb{D}_R^n \to (-\infty, +\infty)$ and a point $p \in \mathbb{T}^n$, denote by $u^*(p) \in [-\infty, +\infty]$ the radial limit of u on the radius $[0, R \cdot p)$ if it exists. Let $A \subset \mathbb{T}^n$ be the set of points $p \in \mathbb{T}^n$, for which the radial limit $u^*(p)$ exists. Then, we show (as noted in [24] when $n = 1$ and $A = \mathbb{T}$) that A is measurable and, if u is continuous, that $u^* : A \to [-\infty, +\infty]$ is of Baire class 0 or 1. Our main result is

First author was supported by NSERC (Canada) grant RGPIN-2016-04107.

P. M. Gauthier (✉)
Département de mathématiques et de statistique, Université de Montréal, Montréal, QC, Canada
e-mail: paul.m.gauthier@umontreal.ca

M. Shirazi
Department of Mathematics and Statistics, Georgia State University, Atlanta, GA, USA
e-mail: mshirazi@gsu.edu

I. Binder et al. (eds.), *Function Spaces, Theory and Applications*, Fields Institute Communications 87, https://doi.org/10.1007/978-3-031-39270-2_9

that this in fact yields a characterization of radial limits for holomorphic functions. Indeed, denoting by $C_R(u, p)$ the radial cluster set of u on the radius $[0, R \cdot p)$,

$$C_R(u,p) = \{w \in [-\infty,+\infty] : \textit{ there exists a sequence } 0 < r_k \nearrow R \textit{ such that } \lim_{r_k \nearrow R} u(r_k p) = w\},$$

we shall show the following.

Theorem 1.1 *For $n \in \mathbb{N}$, $0 < R \leq +\infty$, arbitrary measurable sets $A^1, A^2 \subset \mathbb{T}^n$ and arbitrary measurable functions $\psi_1 : A^1 \to [-\infty, +\infty]$, $\psi_2 : A^2 \to [-\infty, +\infty]$, there exists a function $f = u + iv$, holomorphic in $\mathbb{D}^n_R$, such that*

$$u^*(p) = \psi_1(p), \textit{ for a.e. } p \in A^1; \quad C_R(u,p) = [-\infty,+\infty], \textit{ for a.e. } p \in \mathbb{T}^n \setminus A^1;$$

$$v^*(p) = \psi_2(p), \textit{ for a.e. } p \in A^2; \quad C_R(v,p) = [-\infty,+\infty], \textit{ for a.e. } p \in \mathbb{T}^n \setminus A^2.$$

This generalizes a result of Lehto [21], who established this theorem for the unit disc $\mathbb{D}^1_1$ in $\mathbb{C}$, when both A^1 and A^2 are all of $\mathbb{T}$. Theorem 1.1 can be loosely thought of as giving a characterization of the class of *almost-everywhere* radial limit functions for holomorphic functions as simply the class of all measurable functions on measurable subsets of $\mathbb{T}^n$. In 1938, Roth [24] characterized the class of *everywhere* radial limit functions for entire functions of a single variable. The work of Roth represents the first major application of approximation theory to the boundary behaviour of holomorphic functions.

Theorem 1.1 states that we can specify the radial boundary behaviour for holomorphic functions on the polydisc at almost every point of the *distinguished boundary* $\mathbb{T}^n$ of the polydisc. From a result of Hakim and Sibony [17] (see also [20] and [7]), one can deduce a similar result for almost every point of the *entire* boundary $\partial\mathbb{D}^n$ of the polydisc. This does not imply Theorem 1.1, since the distinguished boundary is of boundary measure zero.

Denote by $\mathcal{O}(\Omega)$ the family of all holomorphic functions on a bounded domain $\Omega \subset \mathbb{C}^n$ and let $\mathcal{F}(\Omega)$ be a subfamily of $\mathcal{O}(\Omega)$. We recall that a sequence $\{z^j\}_{j=1}^\infty$ of distinct points in a domain $\Omega \subset \mathbb{C}^n$ (i.e. each z^j is an $n-$tuple) is an interpolating sequence for $\mathcal{O}(\Omega)$ if, for every sequence of complex numbers $\{w^j\}_{j=1}^\infty$, there is an $f \in \mathcal{O}(\Omega)$ such that $f(z^j) = w^j$, for $j = 1, 2, \ldots$.

It is interesting to contrast Theorem 1.1 with the fact that, since the polydisc is a domain of holomorphy, every sequence of distinct points tending to the boundary is an interpolating sequence for $\mathcal{O}(\Omega)$, and so there is an unbounded holomorphic function in the polydisc which tends to zero on a sequence having every point of the distinguished boundary as a limit point.

Boundaries are creatures for which smaller is better (perhaps not for topologists, but at least for analysts). Let us elaborate this remark. For a function f defined on a domain $\Omega \subset \mathbb{C}^n$ and a subset $E \subset \overline{\Omega}$, we denote

$$\sup_E |f| = \sup_{w \in E} \limsup_{z \to w} \{|f(z)| : z \in \Omega\}$$

and note that if $E \subset \Omega$ and f is continuous on Ω, then this agrees with the usual meaning of $\sup_E |f|$.

A closed subset $E \subset \overline{\Omega}$ is said to be a *boundary* for the family $\mathcal{F}(\Omega)$, if

$$\sup_\Omega |f| = \sup_E |f|, \quad \text{for all} \quad f \in \mathcal{F}(\Omega).$$

Since, for every bounded domain Ω, the closure $\overline{\Omega}$ is a boundary and every closed subset of a boundary is a boundary, smaller boundaries are more interesting. A *Shilov boundary* for the family $\mathcal{F}(\Omega)$ is a minimal boundary. If the family $\mathcal{F}(\Omega)$ contains a nonconstant function, then a Shilov boundary for $\mathcal{F}(\Omega)$ must be contained in the topological boundary $\partial\Omega$.

One of the most interesting Banach algebras is the *polydisc algebra* $A(\mathbb{D}^n)$, consisting of continuous functions on the closed polydisc $\overline{\mathbb{D}}^n$, which are holomorphic on the (open) polydisc $\mathbb{D}^n$. The family $A(\mathbb{D}^n)$ can be considered as a subfamily of $\mathcal{O}(\mathbb{D}^n)$, and it is well known that the distinguished boundary $\mathbb{T}^n$ of $\mathbb{D}^n$ is the (unique) Shilov boundary for the family $A(\mathbb{D}^n)$. We also have the following.

Lemma 1.1 *The distinguished boundary $\mathbb{T}^n$ of $\mathbb{D}^n$ is the unique Shilov boundary of $\mathcal{O}(\mathbb{D}^n)$. Thus,*

$$\sup_{\mathbb{D}^n} |f| = \sup_{\mathbb{T}^n} |f| \quad \textit{for all} \quad f \in \mathcal{O}(\mathbb{D}^n). \tag{1}$$

Proof Choose $f \in \mathcal{O}(\mathbb{D}^n)$ and fix a point $z = (z_1, \ldots, z_n) \in \mathbb{D}^n$. Let $\{R_j\}_{j=1}^\infty$ be a sequence increasing to 1, with $\max\{|z_1|, \ldots, |z_n|\} < R_1$. Since the Shilov boundary $R_j\mathbb{T}^n$ for each $\mathbb{D}^n_{R_j}$ is compact, we can construct a sequence of points $z^j \in R_j\mathbb{T}^n$, $j = 1, 2, \ldots,$ such that

$$|f(z)| \leq |f(z^1)| \leq \cdots \leq |f(z^j)| \leq \cdots \leq \limsup_{j \to \infty} |f(z^j)|,$$

and since $\mathbb{T}^n$ is compact, the sequence $\{z^j\}$ has a limit point $w^\circ \in \mathbb{T}^n$. Hence,

$$|f(z)| \leq \limsup_{\eta \to w^\circ} \{|f(\eta)| : \eta \in \mathbb{D}^n\} \leq \sup_{w \in \mathbb{T}^n} \limsup_{\eta \to w} \{|f(\eta)| : \eta \in \mathbb{D}^n\} = \sup_{\mathbb{T}^n} |f|.$$

Thus, $\mathbb{T}^n$ is indeed a boundary for $\mathcal{O}(\mathbb{D}^n)$. To prove uniqueness, suppose E is a Shilov boundary for $\mathcal{O}(\mathbb{D}^n)$. Then, E is a Shilov boundary for the smaller class $A(\mathbb{D}^n)$, for which the only Shilov boundary is $\mathbb{T}^n$. So $E = \mathbb{T}^n$. □

From (1) we have the striking conclusion that without even assuming that f is bounded, it follows that if $f \in \mathcal{O}(\mathbb{D}^n)$ is bounded near $\mathbb{T}^n$, then it is bounded in all of $\mathbb{D}^n$ and by the same bound.

A boundary for a family of functions is also a boundary for every subfamily; it follows that, for every intermediate family, $A(\mathbb{D}^n) \subset \mathcal{F}(\mathbb{D}^n) \subset O(\mathbb{D}^n)$, $\mathbb{T}^n$ is the unique Shilov boundary for $\mathcal{F}(\mathbb{D}^n)$. This includes all of the important function spaces on $\mathbb{D}^n$: Hardy spaces, Bergman spaces, Dirichlet spaces, etc. For these and many other reasons, the distinguished boundary $\mathbb{T}^n$ is more interesting than the topological boundary $\partial\mathbb{D}^n$ of the polydisc. In this paper, for holomorphic functions in the polydisc, we consider radial behaviour only at the distinguished boundary $\mathbb{T}^n$, not the entire boundary $\partial\mathbb{D}^n$.

2 Radial Limits for a Handful of Function Spaces

Before presenting the preparatory material towards the proof of Theorem 1.1, we recall some basic facts regarding radial limits of holomorphic and meromorphic functions of a single complex variable. Some basic results on function spaces can be found, for example, in [9] and [19] and more recent references are given in the survey [12]. For a general treatment of boundary behaviour of meromorphic functions and references to the classical literature, see [23].

Lindelöf showed that, for functions in $H^\infty(\mathbb{D})$, radial limit implies angular (non-tangential) limit, and this was extended to the class of normal meromorphic functions in $\mathbb{D}$ by Lehto and Virtanen. However, we shall be concerned only with radial limits.

A holomorphic function in $\mathbb{D}^1_R$ need not have any radial limits; however, functions in certain function classes do have radial limits. Denote, as usual, by $H^p(\mathbb{D}), 0 < p \leq +\infty$, the Hardy classes and by $\mathcal{N}(\mathbb{D})$ the Nevanlinna class in $\mathbb{D}$. Fatou showed that functions in $H^\infty(\mathbb{D})$ have radial limits a.e. on $\mathbb{T}$. Since $\mathcal{N}(\mathbb{D})$ is the quotient field of the ring $H^\infty(\mathbb{D})$, functions in $\mathcal{N}(\mathbb{D})$ also have radial limits a.e. on $\mathbb{T}$. For the Hardy classes $H^p(\mathbb{D})$, $0 < p \leq \infty$, we have the inclusions $H^\infty \subset H^p \subset \mathcal{N}$, so functions in Hardy classes also have radial boundary values almost everywhere on $\mathbb{T}$. For univalent functions in $\mathbb{D}$, there is the stronger result that they have radial limits n.e., that is, *nearly everywhere*, which means outside of a subset of $\mathbb{T}$ of capacity zero.

Theorem 1.1 gives a characterization of radial limits f^* of functions f in $O(\mathbb{D}^n)$. For $n = 1$, the unit disc, characterizations of radial limit functions are known for many function spaces $\mathcal{F}(\mathbb{D}) \subset O(\mathbb{D})$. A continuous function ψ on $\mathbb{T}$ is equal to the radial boundary values f^* of a function f in the disc algebra $A(\mathbb{D})$ if and only if the negative Fourier coefficients of ψ vanish. That is, $c_n = 0$, for $n = -1, -2, \cdots$ in the Fourier expansion

$$\psi(\theta) = \sum_{n=-\infty}^{+\infty} c_n e^{in\theta}.$$

Similarly, for $1 \leq p \leq +\infty$, a measurable function ψ on $\mathbb{T}$ is the a.e. radial limit function f^* of some function f in the Hardy space $H^p(\mathbb{D})$, if and only if

$\psi \in L^p(\mathbb{T})$ and the negative Fourier coefficients of ψ vanish. For $0 < p < +\infty$, the family of a.e. radial limit functions f^* of functions $f \in H^p(\mathbb{D})$ can also be characterized as the closure in $L^p(\mathbb{T})$ of polynomials $\sum_{n=0}^{N} a_n e^{in\theta}$, $N \in \mathbb{N}$.

3 Measure Theory

Let us begin by justifying the following fact which was mentioned in the introduction.

Lemma 3.1 *Let $u : \mathbb{D}_R^n :\to (-\infty, +\infty)$ be continuous and, for $p \in \mathbb{T}^n$, denote by $u^*(p) \in [-\infty, +\infty]$ the radial limit of u on the radius $[0, R \cdot p)$ if it exists. Let $A \subset \mathbb{T}^n$ be the set of points $p \in \mathbb{T}^n$, for which $u^*(p)$ exists. Then, A is a measurable set and the function $u^* : A \to [-\infty, +\infty]$ is of Baire class 0 or 1.*

Proof If the radial limit along some ray is infinite, then it is either $-\infty$ or $+\infty$. Indeed, suppose $u(rp) \to \infty$ and there are sequences $r_j^+ \to R$ and $r_j^- \to R$, such that $u(r_j^+ p) \to +\infty$ and $u(r_j^- p) \to -\infty$. By the intermediate value theorem, there is a sequence $r_j^\circ \to R$, such that $u(r_j^\circ p) = 0$. This contradicts the assumption that $u(rp) \to \infty$ as $r \to R$.

Let $\{r_j\}_{j=1}^\infty$ be a sequence increasing to R. For $j = 1, 2, \ldots$, we denote

$$\omega_j(u, p) = \sup \left\{ \chi\Big(u(r_j p), u(sp)\Big) : r_j < s < R \right\}$$

where χ denotes the *chordal distance*. Then, $p \in A$ if and only if $\omega_j(u, p) \to 0$, as $j \to +\infty$, so

$$A = \bigcap_{i=1}^{\infty} \bigcup_{j=1}^{\infty} A_{ij} \quad \text{with} \quad A_{ij} = \left\{ p \in \mathbb{T}^n : \omega_j(u, p) \le \frac{1}{i} \right\}.$$

Let s_k, $k = 1, 2, \ldots$, be a dense sequence in the open interval (r_j, R). Then,

$$A_{ij} = \bigcap_{k=1}^{\infty} \left\{ p : \chi\big(u(r_j p), u(s_k p)\big) \le \frac{1}{i} \right\} = \bigcap_{k=1}^{\infty} \left\{ p : \phi_{jk}(p) \in [0, +\frac{1}{i}] \right\},$$

where $\phi_{jk}(p) = \chi\big(u(r_j p), u(s_k p)\big)$. Since ϕ_{jk} is continuous, each A_{ij} is measurable and consequently A is also measurable. Finally, the function $u^* : A \to [-\infty, +\infty]$ is of Baire class 0 or 1, since it is the limit of a sequence of continuous functions. □

The following version of Lusin's theorem can be found, for example, in [18, p. 242]. By a Borel measure on a locally compact Hausdorff space, we mean a measure defined on the class of Borel sets, such that the measure of compact sets is finite.

Lemma 3.2 (Lusin) *If μ is a regular Borel measure on a locally compact Hausdorff space X, $E \subset X$ is a Borel set of finite measure and f is a (real) bounded Borel measurable function on E, then for every $\epsilon > 0$, there exists a compact set $K \subset E$ such that $\mu(E \setminus K) \leq \epsilon$ and such that the restriction $f|_K$ of f to K is continuous (on K).*

Lemma 3.3 *Let X be a separable locally compact Hausdorff space and μ a regular Borel measure on X, which does not charge points. Let $A \subset X$ be a Borel set of finite measure and ψ is a (real) bounded Borel measurable function on A. For every $\epsilon > 0$, there exists a compact nowhere dense subset $K \subset A$, such that $\psi|_K$ is continuous and $\mu(A \setminus K) < \epsilon$.*

Proof Applying Lemma 3.2, there is a compact set $Q \subset A$, such that $\mu(A \setminus Q) < \epsilon$ and $\psi|_Q$ is continuous on Q.

We show that, for an arbitrary $\eta > 0$, there exists a compact nowhere dense set $K \subset Q$ such that $\mu(Q \setminus K) < \eta$. To do so, let D be a countable dense subset of X. Since μ does not charge points, it follows that $\mu(D) = 0$. Clearly, $Q \cap D$ is dense in Q of measure zero. By the regularity of μ, there exists an open subset O of X such that $Q \cap D \subset O$ and $\mu(O) < \eta$. Let $K = Q \setminus O$, which is closed in Q (hence compact). Then,

$$\mu(Q \setminus K) = \mu(Q \setminus Q \setminus O) = \mu(Q \cap O) \leq \mu(O) \leq \eta.$$

The interior of K is empty; otherwise, it should have a point of D which is impossible. Thus, K is nowhere dense and totally disconnected.

Now, choose $\eta = \epsilon - \mu(A \setminus Q)$. Then, K is a compact nowhere dense subset of A such that $\psi|_K$ is continuous and

$$\mu(A \setminus K) = \mu(A \setminus Q) + \mu(Q \setminus K) < \epsilon.$$

4 Holomorphic Approximation

We recall some definitions and a theorem from Chacrone, Gauthier, and Nersessian [6]. Let Ω be a Riemann surface and X a closed subset of Ω. Let $A(X)$ denote the set of all continuous complex-valued functions on X which are holomorphic in the interior X°. By an *open* Riemann surface, we mean as usual a non-compact Riemann surface.

For $j = 1, \ldots, n$, let X_j be a closed subset of a Riemann surface Ω_j. Consider the closed subset $X = X_1 \times \cdots \times X_n$ of $\Omega = \Omega_1 \times \cdots \times \Omega_n$. Then, $A(X)$ is defined to be the set of all functions f, continuous on X, such that for every $p \in X$, setting

$$g_j(q) = f(p_1, \ldots, p_{j-1}, q, p_{j+1}, \ldots, p_n),$$

for $q \in X_j$, we have $g_j \in A(X_j)$, $j = 1, \ldots, n$. A set $X \subset \Omega$ is said to be a *Mergelyan* (respectively, *Carleman*) set in Ω if for every $f \in A(X)$, and every ϵ, which is a positive constant (respectively, positive continuous function on X), there exists a function g, holomorphic on Ω, such that $|f - g| < \epsilon$ on X. Of course, every Carleman set is *a fortiori* a Mergelyan set.

For a Riemann surface Ω, we denote by $\Omega^* = \Omega \cup \{*\}$ the *one-point compactification* of Ω, where $*$ denotes the ideal point at infinity. A celebrated theorem of Arakelian states that a closed subset F of a plane domain Ω is a Mergelyan set if and only if $\Omega^* \setminus X$ is connected and locally connected.

A family of subsets of a topological space Ω is said to satisfy the *long islands* condition if, for each compact set $K \subset \Omega$, there is a (larger) compact set $Q \subset \Omega$, such that each member of the family which meets K is contained in Q. The long islands condition was introduced by the senior author in [13], where it was shown that in order for a Mergelyan set F in a plane domain Ω to be a Carleman set, it was necessary that the components of F° satisfy the long islands condition in Ω. Subsequently, the condition was also shown to be sufficient in [22]. The conditions for Mergelyan and Carleman sets were shown in [15] to be necessary also for closed subsets of open Riemann surfaces. The following result of Boivin [4] completely characterizes Carleman sets on Riemann surfaces.

Theorem 4.1 *Let Ω be a non-compact Riemann surface and let F be a closed subset of Ω. Then, the following are equivalent:*

(a) F is a Carleman set.
(b) $\Omega^* \setminus F$ is connected and locally connected, and the components of F° satisfy the long islands condition.
(c) F is a Mergelyan set, and the components of F° satisfy the long islands condition.

For a survey of complex approximation on Riemann surfaces, see [1, 14, 16], and for a survey of complex approximation in several complex variables, see [11]. The Boivin-Nersesian theorem was extended to several complex variables in [6] as follows.

Theorem 4.2 *Let $\Omega = \Omega_1 \times \cdots \times \Omega_n$ be a product of open Riemann surfaces Ω_j, $j = 1, \ldots, n$, and assume $E = E_1 \times \cdots \times E_n$ is a product type closed subset of Ω. Then, E is a holomorphic Carleman set in Ω if and only if each E_j is a holomorphic Carleman set in Ω_j, $j = 1, \ldots, n$.*

See also Falcó, Gauthier, Manolaki, and Nestoridis [10, Proposition 5.3] for the case that each Ω_j is $\mathbb{C}$ and each E_j is compact.

5 Proof of Theorem 1.1 [5, 8]

The following theorem was first proved by Bagemihl and Seidel [2] (see also [23, Theorem 8]) for the unit disc (the case $R = 1$ and $n = 1$). We give a shorter proof, using Theorem 4.1, which was not available at the time [2] was written.

Lemma 5.1 *Let $U, V : \mathbb{D}^n_R \to (-\infty, +\infty)$ be continuous functions. Then, there exists a function $f = u + iv$ holomorphic in $\mathbb{D}^n_R$, such that*

$$\lim_{r\nearrow R}\big(u(rp)-U(rp)\big) = 0, \quad \textit{and} \quad \lim_{r\nearrow R}\big(v(rp)-V(rp)\big) = 0, \quad \textit{for a.e.} \quad p \in \mathbb{T}^n.$$

Proof We prove the theorem for $n = 2$, the proof for other n is a simple change of notation. By Lemma 3.3 there exists a countable collection of disjoint compact nowhere dense subsets $A_j \subset \mathbb{T}$, $j = 1, 2, \ldots$, such that setting $A = A_1 \cup A_2 \cup \cdots$ and $A^2 = A \times A$, we have $m_2(\mathbb{T}^2 \setminus A^2) = 0$ (where m_n is the Haar measure on $\mathbb{T}^n$).

Fix a sequence $0 < r_1 < \cdots < r_j < r_{j+1} < \cdots$, $r_j \nearrow R$. For the closed subsets

$$E_j = \big\{z = rp : r_j \le r < R,\ p \in A_j\big\}, \quad j = 1, 2, \ldots,$$

of $\mathbb{D}^1_R$, put $E = E_1 \cup E_2 \cup \cdots$ and $E^2 = E \times E$. By Theorem 4.1 E is a Carleman set in $\mathbb{D}^1_R$, and consequently, by Proposition 4.2, the set E^2 is a Carleman set in $\mathbb{D}^2_R$, as it is a product of two Carleman sets in $\mathbb{D}^1_R$.

The function $g = U + iV$ is a continuous complex-valued function on $\mathbb{D}^2_R$ and in particular on E^2. By the definition of a Carleman set, there is a function f holomorphic in $\mathbb{D}^2_R$, such that

$$|f(rp) - g(rp)| \to 0 \quad \text{as} \quad r \nearrow R, \quad p \in A^2.$$

Writing $f = u + iv$, we have that f satisfies the conclusion of the lemma.

We state Tietze's extension theorem for closed sets in a normal topological space (see, e.g. [25, Theorem 8.2.11]).

Theorem 5.1 (Tietze) *Fix $0 < c \le +\infty$. If E is a closed subset of the normal topological space X, then every continuous function $f : E \to (-c, +c)$ extends to a continuous function $g : X \to (-c, +c)$. Moreover, g can be chosen so that $\inf_{x\in X} g(x) = \inf_{a\in E} f(a)$ and $\sup_{x\in X} g(x) = \sup_{a\in E} f(a)$.*

This is usually stated only for $c = +\infty$, but the formulation for arbitrary c is topologically equivalent.

Lemma 5.2 *Fix* $n \in \mathbb{N}$, $0 < c \leq +\infty$ *and* $0 < R \leq +\infty$. *For an arbitrary measurable set* $A \subset \mathbb{T}^n$ *and a measurable function* $\psi : A \to [-c, +c]$, *there exists a continuous function* $\Psi : \mathbb{D}^n_R \to (-c, +c)$, *such that*

$$\lim_{r \nearrow R} \Psi(rp) = \psi(p), \quad \textit{for} \quad \textit{a.e.} \quad p \in A;$$

$$C_R(\Psi, p) = [-c, +c], \quad \textit{for} \quad \textit{a.e.} \quad p \in \mathbb{T}^n \setminus A.$$

Proof For various values of c and R, the assertions of the lemma are topologically equivalent, so we shall suppose that $c = 1$ and $R = 1$.

Invoking Lemma 3.3, construct a sequence $\widetilde{A}_1, \widetilde{A}_2, \ldots$ of disjoint compact nowhere dense subsets of A, such that, setting $\widetilde{A} = \cup_j \widetilde{A}_j$, we have $m_n(A \setminus \widetilde{A}) = 0$, and $\psi|_{\widetilde{A}_j}$ is continuous, $j = 1, 2, \ldots$. Similarly for $B = \mathbb{T}^n \setminus A$, i.e. $\widetilde{B} = \cup_j \widetilde{B}_j$, $m_n(B \setminus \widetilde{B}) = 0$. Set

$$X_k = \{z = rp : 1 - 1/k \leq r < 1, \ p \in \widetilde{A}_k\},$$

$$Y_k = \{z = rp : 1 - 1/k \leq r < 1, \ p \in \widetilde{B}_k\}.$$

Define Ψ as follows

$$\Psi(z) = \Psi(rp) := r\psi(p), \qquad \text{for } z \in \cup_k X_k,$$

$$\Psi(z) := |z| \sin\left(\frac{1}{1 - |z|}\right), \qquad \text{for } z \in \cup_k Y_k.$$

Note that, since $(\cup_k X_k) \cup (\cup_k Y_k)$ is the union of a locally finite family of disjoint closed sets, on each of which Ψ is continuous, it follows that this union is closed and Ψ is continuous on this union. Now, we extend $\Psi : (\cup_k X_k) \cup (\cup_k Y_k) \to (-1, +1)$ continuously to $\Psi : \mathbb{D}^n_1 \to (-1, +1)$, by using Theorem 5.1.

Proof (of Theorem 1.1) We first apply Lemma 5.2 (with $c = +\infty$) to each ψ_k on A^k, $k = 1, 2$, to obtain continuous functions $\Psi_k : \mathbb{D}^n_R \to (-\infty, +\infty)$ such that

$$\lim_{r \nearrow R} \Psi_k(rp) = \psi_k(p), \quad \textit{a.e.} \quad p \in A^k;$$

and

$$C_R(\Psi_k, p) = [-\infty, +\infty] \quad \textit{a.e.} \quad p \in \mathbb{T}^n \setminus A^k.$$

Now, we apply Lemma 5.1 to $\Psi_k, k = 1, 2$, to obtain a holomorphic function $f = u + iv$ in $\mathbb{D}^n_R$ such that

$$\lim_{r \nearrow R} \big(u(rp) - \Psi_1(rp)\big) = 0, \quad \textit{a.e.} \quad p \in \mathbb{T}^n, \quad \text{and}$$

$$\lim_{r\nearrow R}\big(v(rp)-\Psi_2(rp)\big)=0, \quad a.e. \quad p\in\mathbb{T}^n.$$

Combining the two above we have

$$\lim_{r\nearrow R} u(rp)=\psi_1(p), \quad a.e. \quad p\in A^1, \quad \text{and} \quad \lim_{r\nearrow R} v(rp)=\psi_2(p), \quad a.e. \quad p\in A^2.$$

Clearly,

$$C_R(u,p)=[-\infty,+\infty] \quad a.e. \quad p\in\mathbb{T}^n\setminus A^1, \quad \text{and}$$

$$C_R(v,p)=[-\infty,+\infty] \quad a.e. \quad p\in\mathbb{T}^n\setminus A^2.$$

6 Some Special Domains for Holomorphic Functions

6.1 Radial Domains

Let us recall the natural projection $\pi:\mathbb{C}^*\to\mathbb{T}$ in $\mathbb{C}^*=\mathbb{C}\setminus\{0\}$, given by $\pi(z)=z/|z|$. Then, $E'\subset\mathbb{T}$ is nowhere dense in $\mathbb{T}$ if and only if $E:=\pi^{-1}(E')$ is nowhere dense in $\mathbb{C}^*$.

By abuse of notation, we identify a point, $e^{i\theta}, 0\le\theta<2\pi$, of the unit circle $\mathbb{T}$ with the point θ of the interval $[0,2\pi)$. Let U' be a domain in $\mathbb{T}$ and $r:U'\to[0,+\infty)$, $R:U'\to(0,+\infty]$ be continuous functions, with $r<R$. In polar coordinates, (θ,ρ) for $z=\rho e^{i\theta}$, we call the domain

$$U_{r,R}=(U',r,R)=\{(\theta,\rho)\in\mathbb{T}\times\mathbb{R}^+:\theta\in U',\ r(\theta)<\rho<R(\theta)\}$$

a *radial domain.*

Theorem 6.1 *Let $U_{r,R}\subset\mathbb{C}$, be a radial domain and $F'\subset U'$ an F_σ set, which we assume to be of first category. Then, for every function φ continuous on $U_{r,R}$, there is a holomorphic function h on $U_{r,R}$ such that,*

$$\big(h-\varphi\big)(\theta,\rho)\to 0, \quad as \quad \rho\searrow r(\theta), \quad or \quad \rho\nearrow R(\theta) \quad ; \quad for\ all \quad \theta\in F'.$$

Proof We write F as a union $F'=F_1'\cup F_2'\cup\cdots$, of closed nowhere dense sets of U', as before. Let $\{r_k\}_{k=1}^\infty$ be a decreasing and $\{R_l\}_{l=1}^\infty$ be an increasing sequence of continuous functions on U' such that $r_k\searrow r$ and $R_k\nearrow R$ pointwise on U'. We set

$$F_k=\big\{(\theta,\rho):\theta\in F_k',\ \rho\in\big(r(\theta),R(\theta)\big)\setminus\big(r_k(\theta),R_k(\theta)\big)\big\}$$

and $F=\cup_{k=1}^\infty F_k$. Then, as in the proof of Theorem 1.1, the set F is a Carleman set in $U_{r,R}$. The rest of the proof is like that of Lemma 5.1 but much simpler.

6.2 Strictly Starlike Domains

We shall say that a domain $D \subset \mathbb{C}$ is *strictly starlike* with respect to the origin if for every p in $\overline{D}$, and every $0 \leq \lambda < 1$, λp is in D. We only consider strictly starlike domains with respect to the origin so we may say strictly starlike rather than strictly starlike with respect to the origin. We first show that a domain D is strictly starlike if and only if there is a unique continuous function $R : \mathbb{T} \to (0, +\infty]$, such that, in polar coordinates,

$$D = \{z = (\theta, \rho) : 0 \leq \rho < R(\theta),\ \theta \in \mathbb{T}\}.$$

The "if part" is clear as the above D is a strictly starlike domain. For the "only if part" for $\theta \in \mathbb{T}$, we let $R(\theta) = \sup\{\rho : (\theta, \rho) \in D\}$. The sup could well be infinity for some values of θ, as the domain D is not necessarily bounded.

Lemma 6.1 $R : \mathbb{T} \to (0, +\infty]$ *is continuous, where* $(0, +\infty]$ *has the usual order topology.*

Proof We first show that R is lower semicontinuous in $\mathbb{T}$. Suppose, to obtain a contradiction, that there is a point $\theta \in \mathbb{T}$, a finite value $s < R(\theta)$, and a sequence $\theta_j \to \theta$ (in $\mathbb{T}$) such that $R(\theta_j) < s$. Since D is open, it follows that no point in the segment $\{r\theta : s < r < R(\theta)\}$ lies in D, which contradicts the definition of $R(\theta)$. Note that the argument is valid even if $R(\theta) = +\infty$.

We now need to show the upper semicontinuity of R in $\mathbb{T}$. Fix $\theta \in \mathbb{T}$. If $R(\theta) = +\infty$, upper semicontinuity at θ is automatic. Suppose $R(\theta) < +\infty$ and suppose, to obtain a contradiction, that there is a sequence $\theta_j \to \theta$ and $\epsilon > 0$, such that $R(\theta_j) > R(\theta) + \epsilon$. Then, for $s = R(\theta) + \epsilon$, $s\theta_j \in D$. However, $s\theta \notin D$. Therefore, $s\theta \in \partial D$. This contradicts the definition of $R(\theta)$, as D is a strictly starlike domain.

Taking into account such a parametrization, we shall denote a strictly starlike domain D as U_R. In order to prove the next theorem, it is convenient to have the following lemma for starlike domains analogous to Lemma 5.2 for polydomains.

Lemma 6.2 *Let* U_R *be a strictly starlike domain in* $\mathbb{C}$ *and* $\psi : \mathbb{T} \to [-c, +c]$, *a measurable function, where* $0 < c \leq +\infty$. *There exists a continuous function* $\Psi : U_R \to (-c, +c)$, *such that*

$$\lim_{\rho \nearrow R(\theta)} \Psi((\theta, \rho)) = \psi(\theta), \quad a.e. \quad \theta \in \mathbb{T}.$$

Proof Again, since for various values of c the assertions of the lemma are topologically equivalent, we shall suppose that $c = 1$.

Invoking Lemma 3.3, construct a sequence $\widetilde{A}_1, \widetilde{A}_2, \ldots$ of disjoint compact nowhere dense subsets of $\mathbb{T}$, such that, setting $\widetilde{A} = \cup_j \widetilde{A}_j$, we have $m(\mathbb{T} \setminus \widetilde{A}) = 0$, and $\psi|_{\widetilde{A}_j}$ is continuous, $j = 1, 2, \ldots$.

Let $\{R_k\}_{k=1}^{\infty}$ be a sequence of positive continuous functions increasing to R on $\mathbb{T}$ and set

$$X_k = \{(\theta, \rho) \in U_R : R_k(\theta) \leq \rho < R(\theta),\ \theta \in \widetilde{A}_k\}.$$

Define Ψ as follows. Set $\Psi = \psi$ on $\widetilde{A}$ and, invoking Theorem 5.1, extend Ψ continuously from $\widetilde{A}_1$ to $X_1 \cup \widetilde{A}_1$ so that $\Psi : X_1 \to (-1, +1)$.

Invoking Theorem 5.1 recursively, extend Ψ continuously from

$$\left[\cup_{j=1}^{k-1} (X_j \cup \widetilde{A}_j)\right] \cup \widetilde{A}_k$$

to $\cup_{j=1}^{k}(X_j \cup \widetilde{A}_j)$ so that $\Psi : \cup_{j=1}^{k} X_j \to (-1, +1)$. Now, again by Theorem 5.1 we extend $\Psi : \cup_k X_k \longrightarrow (-1, +1)$ continuously to $\Psi : U_R \to (-1, +1)$.

The following result, again, is not a direct consequence of Theorem 6.1, but the proof is very similar.

Theorem 6.2 *Suppose U_R is a domain in $\mathbb{C}$, strictly starlike, and let F' be an F_σ subset of $\mathbb{T}$ which is of first category. Then, for every function φ continuous on U_R, there is a function h holomorphic on U_R such that*

$$\lim_{r \nearrow R(\theta)} \big(h - \varphi\big)((\theta, r)) = 0, \quad \textit{for all} \quad \theta \in F'.$$

Proof We write F' as a union of nowhere dense closed sets, that is, $\cup_k F'_k$, where each F'_k is a nowhere dense closed set in $\mathbb{T}$.

Let $R_k : \mathbb{T} \to (0, +\infty)$, $k = 1, 2, \ldots$, be an increasing sequence of continuous functions converging to the function R pointwise in $\mathbb{T}$. Set

$$F_k = \{(\theta, \rho) : \theta \in F'_k,\ R_k(\theta) \leq \rho < R(\theta)\},$$

and $F = F_1 \cup F_2 \cup \ldots$. Then, F is a Carleman set in U_R.

Consider on U_R the step function

$$\eta(z) = 1/k, \quad \text{for} \quad z = (\theta, r), \quad R_k(\theta) \leq r < R_{k+1}(\theta),$$

and let ϵ be a continuous function on U_R, such that $0 < \epsilon < \eta$.

By the definition of a Carleman set, there exists a holomorphic function h on U_R, such that $|h(z) - \varphi(z)| < \epsilon(z)$, for $z \in F$. Then, h satisfies the conclusion of the theorem.

6.3 Vertical Line Domains

Writing $\mathbb{C} = \mathbb{R} \times i\mathbb{R}$, let $\pi : \mathbb{C} \to \mathbb{R}$ be the associated projection; that is, $\pi(z) = x$ for $z = x + iy$. Given an interval $(a, b) \subset \mathbb{R}$, where $-\infty \leq a < b \leq +\infty$, by a *vertical line domain* $V = V_{a,b,C,D}$ in $\mathbb{C}$, with base (a, b), we mean a domain of the form

$$V_{a,b,C,D} = \{z = x + iy \in \mathbb{C} : a < x < b,\ C(x) < y < D(x)\},$$

where $C, D : (a, b) \to [-\infty, +\infty]$ are continuous functions, with $C < D$. Note that (a, b) is the projection of V in $\mathbb{R}$.

In our notation for radial domains, the slit unit disc $\mathbb{D} \setminus [0, 1)$ can be denoted by $U_{0,1} = (\mathbb{T} \setminus \{1\}, 0, 1)$.

Lemma 6.3 *For every vertical line domain $V_{a,b,C,D}$, there is a homeomorphism*

$$h : V_{a,b,C,D} \longrightarrow U_{0,1} \quad z \longmapsto \big(\theta(x), \rho(x, y)\big),$$

which maps vertical lines to radii.

Proof For $x \in (a, b)$, denote by $\mathcal{H}(a, x)$ the chordal length of the horizontal segment (a, x) and set

$$\theta(x) = \frac{\mathcal{H}(a, x)}{\mathcal{H}(a, b)} 2\pi.$$

For $z = x + iy \in V_{a,b,C,D}$, denote by $\mathcal{V}(x, y)$ the chordal length of the vertical segment from $(x, C(x))$ to (x, y) and put

$$\rho(x, y) = \frac{\mathcal{V}(x, y)}{\mathcal{V}\big(x, D(x)\big)}.$$

Then, $z \to (\theta, \rho)$ is the desired homeomorphism. Note that for a fixed x, θ is fixed and ρ increases strictly from zero to one as y increases from $C(x)$ to $D(x)$.

Theorem 6.3 *Let $V_{a,b,C,D}$ be a vertical line domain with base (a, b) and $F' \subset (a, b)$ an F_σ set, which we suppose to be of first category. Then, for every φ continuous on $V_{a,b,C,D}$, there is a holomorphic function h on $V_{a,b,C,D}$, such that, for all $z = x + iy \in V_{a,b,C,D}$, with $x \in F'$,*

$$\big(h - \varphi\big)(z) \longrightarrow 0, \quad \textit{if} \quad y \searrow C(x) \quad \textit{or} \quad y \nearrow D(x).$$

Proof We first write F' as $\cup_k F'_k$, where each F'_k is a nowhere dense closed set in (a, b). Let $\{C_k\}_{k=1}^{\infty}$ and $\{D_k\}_{k=1}^{\infty}$ be sequences of continuous functions on (a, b), with the sequence $\{C_k\}_{k=1}^{\infty}$ strictly decreasing to C, $\{D_k\}_{k=1}^{\infty}$ strictly increasing to D, and $C_1 < D_1$. For $k = 1, 2, \ldots,$ and $z = x + iy$, set

$$F_k = \{z : x \in F'_k,\ C(x) < y \le C_k(x)\} \cup \{z : x \in F'_k,\ D_k(x) \le y < D(x)\}$$

and $F = \cup_{k=1}^{\infty} F_k$.

By Lemma 6.3, there is a homeomorphism h from the vertical line domain $V_{a,b,C,D}$ onto a radial domain $U_{0,1} = (\mathbb{T} \setminus \{1\}, 0, 1)$ which maps vertical lines to radial segments. Since the conditions for a Mergelyan set are purely topological, and our previous discussion of radial domains shows that $h(F)$ is a Mergelyan set in $U_{0,1}$, it follows that F is also a Mergelyan set in $V_{a,b,C,D}$. The rest of the proof follows the pattern we have established.

6.4 Upper Domains

Let us call a domain $U \subset \mathbb{C}$, an *upper domain* if, for each $z = x + iy_\circ \in U$, every point $x + iy$, $y > y_\circ$, is also in U. If $U \subset \mathbb{C}$ is an upper domain, we denote $U' = \pi(U)$, which is an open (possibly unbounded) interval in $\mathbb{R}$. For $x \in U'$, set

$$L(x) = \inf\{y : x + iy \in U\}. \tag{2}$$

The upper domain U can be described as

$$U_L := \{z = x + iy \in U' \times i\mathbb{R} : L(x) < y\}.$$

Lemma 6.4 *If U_L is an upper domain, the function $x \mapsto L(x)$ mapping U' into $[-\infty, +\infty)$ is upper semicontinuous.*

Proof $L(x) < +\infty$ for each $x \in U_L$, since $y < +\infty$, for each $z = x + iy \in U$. To show that L is upper semicontinuous, fix $x_\circ \in U'$. We shall show that

$$\limsup_{x \to x_\circ} L(x) \leq L(x_\circ). \tag{3}$$

Fix $t > L(x_\circ)$, choose $z_\circ = x_\circ + iy_\circ \in U_L$, with $y_\circ < t$, and choose $0 < \delta < t - y_\circ$, such that

$$Q = \{x + iy \; : \; |x - x_\circ|, \; |y - y_\circ| < \delta\} \subset U_L.$$

For all $|x - x_\circ| < \delta$, there exists $z = x + iy \in Q \subset U_L$, such that $y < t$. Hence, $L(x) = \inf\{y : x + iy \in U_L\} < t$. We have shown that, for all $t > L(x_\circ)$, there is a $\delta > 0$, such that $L(x) < t$, for all $|x - x_\circ| < \delta$. This gives (3).

An upper domain U_L will be called a *strictly upper domain* if, for every $z_\circ = x_\circ + iy_\circ \in \partial U_L$, every point $x_\circ + iy$, $y > y_\circ$, is in U_L. For example, the open upper half-plane, with the vertical segment $\{0 + iy : 0 < y \leq 1\}$ deleted, is an upper domain but not a strictly upper domain.

Lemma 6.5 *If U_L is a strictly upper domain, then the function $x \mapsto L(x)$ mapping U' into $[-\infty, +\infty)$ is continuous.*

Proof In view of the previous lemma, we only need to show the lower semicontinuity. Fix $x_\circ \in U'$. If $L(x_\circ) = -\infty$, lower semicontinuity at $x_\circ$ is automatic. Suppose $L(x_\circ) > -\infty$ and suppose, to obtain a contradiction, that there is a sequence $x_j \to x_\circ$ in U' and $\epsilon > 0$, such that $L(x_j) < L(x_\circ) - \epsilon$. Then, there is a $y < L(x_\circ)$

and a sequence $x_j \to x_\circ$ with $x_j + iy \in U_L$. Therefore, $x_\circ + iy \in \partial U_L$. By strictness, $x_\circ + i\,L(x_\circ) \in U_L$, which, since U_L is open, contradicts the definition of $L(x_\circ)$.

We have the following particular case of Theorem 6.3.

Corollary 6.1 *Let $U_L \subset \mathbb{C}$ be a strictly upper domain and $F' \subset U'$ an F_σ set, which we suppose to be of first category. Then, for every φ continuous on U_L, there is a holomorphic function h on U_L, such that, for $z = x + iy \in U$, with $x \in F'$,*

$$\big(h - \varphi\big)(z) \to 0, \quad \text{if} \quad y \searrow L(x) \quad \text{or} \quad y \nearrow +\infty.$$

Proof We may write U_L as a vertical line domain $U_L = U_{a,b,C,D}$, where $C = L$, defined by (2), and $D = +\infty$, and then apply Theorem 6.3.

7 Radial Limits of Entire Functions

Theorem 1.1 for holomorphic functions in the polydisc has the following special case for entire functions of one or several complex variables.

Theorem 7.1 *For all $n = 1, 2, \cdots$ and all measurable functions $\psi_1, \psi_2 : \mathbb{T}^n \to [-\infty, +\infty]$, there exists a function $f = u + iv$, holomorphic in $\mathbb{C}^n$, such that*

$$\lim_{r \nearrow +\infty} u(rp) = \psi_1(p), \textit{ for a.e. } p \in \mathbb{T}^n; \qquad \lim_{r \nearrow +\infty} v(rp) = \psi_2(p), \textit{ for a.e. } p \in \mathbb{T}^n.$$

Since it easy to see that such $\mathbb{T}^n$ *almost-everywhere* radial limit functions ψ_1 and ψ_2 must be measurable, Theorem 7.1 gives a characterization of $\mathbb{T}^n$ almost-everywhere radial limit functions of entire holomorphic functions. We note that, for $n = 1$, Roth [24] gave a characterization of *everywhere* radial limit functions for entire functions of a single complex variable.

Acknowledgments The junior author would like to thank the senior author for introducing the problem and for very helpful discussions and warm support throughout the writing of this paper. Furthermore, the junior author is grateful to Professors Dmitry Jakobson and Jacques Hurtubise for the support provided during his postdoctorate at *McGill University*. We also wish to thank the Fields Institute, in particular, the organizers of the *Focus Program on Analytic Function Spaces and their Applications*.

References

1. Bagby, T. and Gauthier, P. M. *Harmonic approximation on closed subsets of Riemannian manifolds,* Complex potential theory (Montreal, PQ, 1993), 75–87, NATO Adv. Sci. Inst. Ser. C Math. Phys. Sci., 439, Kluwer Acad. Publ., Dordrecht, 1994.
2. Bagemihl, F. and Seidel, W. *Some boundary properties of analytic functions.* Math. Z. **61** (1954), 186–199.

3. Bagemihl, F. and Seidel, W. *Regular functions with prescribed measurable boundary values almost everywhere.* Proc. Nat. Acad. Sci. U.S.A. **41** (1955), 740–743.
4. Boivin, A. *Carleman approximation on Riemann surfaces.* Math. Ann. **275** (1986), no. 1, 57–70.
5. Boivin, A. and Gauthier, P. M. *Holomorphic and harmonic approximation on Riemann surfaces.* Approximation, complex analysis, and potential theory (Montreal, QC, 2000), 107–128, NATO Sci. Ser. II Math. Phys. Chem., **37**, Kluwer Acad. Publ., Dordrecht, 2001.
6. Chacrone, S., Gauthier, P. M., and Nersessian, A. H. *Carleman approximation on products of Riemann surfaces.* Complex Variables, Theory and Application, **37** (1998), no. 1–4, pp 97–111.
7. Charpentier, Ph.; Dupain, Y.; Mounkaila, M. *Approximation par des fonctions holomorphes à croissance controlée.* (French) [Approximation by holomorphic functions with controlled growth] Publ. Mat. **38** (1994), no. 2, 269–298.
8. Collingwood, E. F.; Lohwater, A. J. *The theory of cluster sets.* Cambridge Tracts in Mathematics and Mathematical Physics, No. **56** Cambridge University Press, Cambridge 1966. [Russian translation, Library of the Journal "Matematika"] Izdat. "Mir", Moscow, 1971.
9. Duren, P. L. *Theory of H^p spaces.* Pure and Applied Mathematics, Vol. **38** Academic Press, New York-London 1970.
10. Falcó, J. Gauthier, P. M. Manolaki, M. and Nestoridis, V. *A function algebra providing new Mergelyan type theorems in several complex variables.* Adv. Math. **381** (2021).
11. Fornæss, J. E; Forstnerič, F.; Wold, E. F. *Holomorphic approximation: the legacy of Weierstrass, Runge, Oka-Weil, and Mergelyan.* Advancements in complex analysis—from theory to practice, 133–192, Springer, Cham, [2020], 2020. arXiv: 1802.03924v [math.CV] 12 Feb. 2018.
12. Garcia, S. R.; Mashreghi, J.; Ross, W. T. *Real complex functions.* Recent progress on operator theory and approximation in spaces of analytic functions, 91–128, Contemp. Math., **679**, Amer. Math. Soc., Providence, RI, 2016.
13. Gauthier, P. *Tangential approximation by entire functions and functions holomorphic in a disc.* Izv. Akad. Nauk Armjan. SSR Ser. Mat. **4** 1969 no. 5, 319–326.
14. Gauthier, P. M. *Uniform approximation.* Complex potential theory (Montreal, PQ, 1993), 235–271, NATO Adv. Sci. Inst. Ser. C Math. Phys. Sci., **439**, Kluwer Acad. Publ., Dordrecht, 1994.
15. Gauthier, P. M. and Hengartner, W. *Uniform approximation on closed sets by functions analytic on a Riemann surface.* Approximation theory (Proc. Conf. Inst. Math., Adam Mickiewicz Univ., Poznań, 1972), pp. 63–69. Reidel, Dordrecht, 1975.
16. Gauthier, P. M. and Hengartner, W. *Approximation uniforme qualitative sur des ensembles non bornés.* Séminaire de Mathématiques Supérieures, **82.** Presses de l'Université de Montréal, Montréal, Qué., 1982.
17. Hakim, M. and Sibony, N., *Boundary properties of holomorphic functions in the ball of $\mathbb{C}^n$*, Math. Ann. **276** (1987), no. 4, 549–555.
18. Halmos, P. R. *Measure Theory.* D. Van Nostrand Company, Inc., New York, N. Y., 1950.
19. Hoffman, K. *Banach spaces of analytic functions.* Reprint of the 1962 original. Dover Publications, Inc., New York, 1988.
20. Krantz, S. G. and Min, B. *Approximation by holomorphic functions of several complex variables,* J. Korean Math. Soc. 57 (2020), No. 5, pp. 1287–1298.
21. Lehto, O. *On the first boundary value problem for functions harmonic in the unit circle.* Ann. Acad. Sci. Fenn. Ser. A. I. 1955 (1955), no. **210**, 26 pp.
22. Nersesjan, A. A. *Carleman sets.* (Russian) Izv. Akad. Nauk Armjan. SSR Ser. Mat. **6** (1971), no. 6, 465–471.
23. Noshiro, K. *Cluster Sets.* Ergebnisse der Mathematik und ihrer Grenzgebiete. N. F., Heft **28** Springer-Verlag, Berlin-Göttingen-Heidelberg 1960.
24. Roth, A. *Approximationseigenschaften und Strahlengrenzwerte meromorpher und ganzer Funktionen.* Comment. Math. Helv. **11** (1938), 77–125.
25. Singh, T. B. *Introduction to Topology,* Springer Nature Singapore Pte Ltd. 2019.

Recent Developments in the Interplay Between Function Theory and Operator Theory for Block Toeplitz, Hankel, and Model Operators

Raúl E. Curto, In Sung Hwang, and Woo Young Lee

1 Introduction

Over the last several years, the authors have studied a natural interplay between function theory and operator theory in the context of Toeplitz, Hankel, and model operators. For a long time, operator theory has had deep connections with function theory. The intensive and fruitful study of Toeplitz and Hankel operators has contributed in great measure to this synergy, including the study of the spectral properties of Toeplitz and Hankel operators, which are intrinsically determined by their symbols, i.e., by functions defined on the unit circle or the unit disk. Also, the model operator for contractions, introduced by B. Sz.-Nagy and C. Foiaş, has been a focus in the study of this field. In fact, the model operator is a truncated backward shift operator on the model space constructed from an inner function called the characteristic function of the model operator. Since the spectral theory of the model operator is naturally determined by properties of its characteristic function, one is drawn to the study of the inner functions. In particular, we pay close attention to the Beurling-Lax-Halmos Theorem, which characterizes the invariant subspaces of the shift operator acting on the vector-valued Hardy space. There are several interesting questions emerging from the Beurling-Lax-Halmos Theorem. A

R. E. Curto (✉)
Department of Mathematics, University of Iowa, Iowa, IA, USA
e-mail: raul-curto@uiowa.edu

I. S. Hwang
Department of Mathematics, Sungkyunkwan University, Suwon, Korea
e-mail: ihwang@skku.edu

W. Y. Lee
Department of Mathematics and RIM, Seoul National University, Seoul, Korea
e-mail: wylee@snu.ac.kr

I. Binder et al. (eds.), *Function Spaces, Theory and Applications*, Fields Institute Communications 87, https://doi.org/10.1007/978-3-031-39270-2_10

study of those questions invites a detailed analysis of matrix-valued functions and operator-valued functions.

In this semi-expository paper, we combine a brief survey of recent developments in function theory associated with Toeplitz, Hankel, and model operators with some new results on this subject.

For this study, we need some new notions, e.g., strong L^2-functions, complementary factors, and degree of non-cyclicity. In Sect. 2, we give a brief sketch of these notions. Section 3 is devoted to the matrix-valued function theory associated with Toeplitz, Hankel, and model operators. The first subject of Sect. 3 deals with the product of Hankel operators. It is well known that if the product of two (classical) Hankel operators is zero, then one of them must be zero. However, this is not the case for matrix-valued symbols. To get an affirmative answer for matrix-valued symbols, we introduce a notion of tensored-scalar singularity and then prove a new result under such a condition. The second subject of Sect. 3 is the spectral multiplicity of the model operator with a matrix-valued characteristic function. Here we introduce the Beurling degree of an inner matrix function and then obtain an elegant formula: the spectral multiplicity of the model operator is equal to the Beurling degree of its characteristic function. The third subject of this section is Halmos' Problem 5: Is every subnormal Toeplitz operator either normal or analytic? Abrahamse's Theorem gave a general sufficient condition for the answer to be affirmative. However, Abrahamse's Theorem may fail for Toeplitz operators with matrix-valued symbols. Despite this, one can obtain a matrix-valued version under the constraint of tensored-scalar singularity of the symbol. The last subject of Sect. 3 is an H^∞-functional calculus for compressions of the shift operator. We review this functional calculus and then extend it to an $\overline{H^\infty} + H^\infty$-functional calculus.

Section 4 is devoted to operator-valued function theory. Firstly, we review meromorphic pseudo-continuations of bounded type and give an application to C_0-contractions. Secondly, we consider a canonical decomposition of strong L^2-functions, which generalizes the Douglas-Shapiro-Shields factorization for functions of bounded type. This idea provides a description of a set F such that given a model space, i.e., a backward shift-invariant subspace, the smallest invariant subspace of the backward shift operator containing F is equal to the model space. Thirdly, we examine a question on the spectrum of the model operator. In fact, if the characteristic function of the model operator is two-sided inner, then by the operator-valued version of the Livšic-Moeller Theorem, the spectrum of the model operator is computed from the spectrum of the characteristic function. However, this is not the case for general model operators. We give a partial answer to this question for general model operators using the complementary factor of the characteristic function. Fourthly, we introduce operator-valued rational functions and give an operator-valued extension of Potapov's matrix-valued factorization theorem. Lastly, we pose some unsolved problems on hyponormality and subnormality of Toeplitz operators with operator-valued symbols.

2 Preliminaries and Basic Theory

In this section we provide the notation, basic notions, and basic results which will be used in this paper. For instance, we introduce the notions of strong L^2-functions, the Beurling-Lax-Halmos Theorem, the Douglas-Shapiro-Shields factorization, the complementary factor of an inner function, and the degree of non-cyclicity.

2.1 Basic Notions

Throughout the paper, we suppose that D and E are separable complex Hilbert spaces. We write $\mathcal{B}(D, E)$ for the set of all bounded linear operators from D to E and abbreviate $\mathcal{B}(E, E)$ as $\mathcal{B}(E)$. For $A, B \in \mathcal{B}(E)$, we let $[A, B] := AB - BA$. An operator $T \in \mathcal{B}(E)$ is said to be normal if $[T^*, T] = 0$ and hyponormal if $[T^*, T] \geq 0$. For an operator $T \in \mathcal{B}(E)$, we write $\ker T$ and $\operatorname{ran} T$ for the kernel and the range of T, respectively. For a subset $\mathcal{M} \subseteq E$, $\operatorname{cl} \mathcal{M}$ and $\mathcal{M}^\perp$ denote the closure and the orthogonal complement of $\mathcal{M}$, respectively.

If $A : D \to E$ is a linear operator whose domain is a subspace of D, then A is also a linear operator from the closure of the domain of A into E. So we will only consider those A such that the domain of A is dense in D. Such an operator A is said to be densely defined. If $A : D \to E$ is densely defined, write $\operatorname{dom} A^* = \big\{e \in E : d \mapsto \langle Ad, e\rangle$ is a bounded linear functional on $\operatorname{dom} A\big\}$. If $e \in \operatorname{dom} A^*$, then there exists a unique $f \in E$ such that $\langle Ad, e\rangle = \langle d, f\rangle$ for all $d \in \operatorname{dom} A$. Denote this unique vector f by $f \equiv A^*e$. Thus $\langle Ad, e\rangle = \langle d, A^*e\rangle$ for all $d \in \operatorname{dom} A$ and $e \in \operatorname{dom} A^*$. We call A^* the adjoint of A. It is well known from unbounded operator theory (cf. [8, 29]) that if A is densely defined, then $\ker A^* = (\operatorname{ran} A)^\perp$, so that $\ker A^*$ is closed even though $\ker A$ may not be closed.

We write $\mathbb{D}$ for the open unit disk in the complex plane $\mathbb{C}$ and $\mathbb{T}$ for the unit circle in $\mathbb{C}$. For $\phi \in L^2$, write

$$\breve{\phi}(z) := \phi(\overline{z}) \quad \text{and} \quad \widetilde{\phi}(z) := \overline{\phi(\overline{z})}.$$

For $\phi \in L^2$, write

$$\phi_+ := P_+\phi \quad \text{and} \quad \breve{\phi}_- := P_-\phi,$$

where P_+ and P_- are the orthogonal projections from L^2 onto H^2 and $L^2 \ominus H^2$, respectively. Thus, we may write $\phi = \breve{\phi}_- + \phi_+$. We recall [1, 10, 31, 44] that a meromorphic function $\phi : \mathbb{D} \to \mathbb{C}$ is said to be of *bounded type* (or in the Nevanlinna class $\mathcal{N}$) if there are functions $\psi_1, \psi_2 \in H^\infty$ such that $\phi(z) = \frac{\psi_1(z)}{\psi_2(z)}$ for almost all $z \in \mathbb{T}$. We can easily check that if $\phi \in L^2$ is of bounded type, then ϕ can be written as

$$\phi = \overline{\theta} a, \tag{1}$$

where θ is inner, $a \in H^2$, and θ and a are coprime. Write $\mathbb{D}^e := \{z : 1 < |z| \le \infty\}$. For a function $g : \mathbb{D}^e \to \mathbb{C}$, define a function $g_{\mathbb{D}} : \mathbb{D} \to \mathbb{C}$ by $g_{\mathbb{D}}(\zeta) := \overline{g(1/\overline{\zeta})}$ ($\zeta \in \mathbb{D}$). For a function $g : \mathbb{D}^e \to \mathbb{C}$, we say that g belongs to $H^p(\mathbb{D}^e)$ if $g_{\mathbb{D}} \in H^p$ ($1 \le p \le \infty$). A function $g : \mathbb{D}^e \to \mathbb{C}$ is said to be of *bounded type* if $g_{\mathbb{D}}$ is of bounded type. If $f \in H^2$, then the function $\hat{f}$ defined in $\mathbb{D}^e$ is called a *pseudo-continuation* of f if $\hat{f}$ is a function of bounded type and $\hat{f}(z) = f(z)$ for almost all $z \in \mathbb{T}$ (cf. [6, 44, 50]). Then we can easily show that $\breve{f}$ is of bounded type if and only if f has a pseudo-continuation $\hat{f}$. In this case, $\hat{f}_{\mathbb{D}}(z) = \overline{f(z)}$ for almost all $z \in \mathbb{T}$. In particular,

$$\phi \equiv \breve{\phi}_- + \phi_+ \in L^2 \text{ is of bounded type } \iff \phi_- \text{ has a pseudo-continuation.} \tag{2}$$

Let m denote the normalized Lebesgue measure on $\mathbb{T}$. For a complex Banach space X and $1 \le p \le \infty$, let

$$L^p_X \equiv L^p(\mathbb{T}, X) := \left\{ f : \mathbb{T} \to X : f \text{ is strongly measurable and } ||f||_p < \infty \right\},$$

where

$$||f||_p \equiv ||f||_{L^p_X} := \begin{cases} \left(\int_{\mathbb{T}} ||f(z)||^p_X dm(z) \right)^{\frac{1}{p}} & (1 \le p < \infty); \\ \text{ess sup}_{z \in \mathbb{T}} ||f(z)||_X & (p = \infty). \end{cases}$$

Then we can see that L^p_X forms a Banach space. For $f \in L^1_X$, the n-th Fourier coefficient of f, denoted by $\widehat{f}(n)$, is defined by

$$\widehat{f}(n) := \int_{\mathbb{T}} \overline{z}^n f(z)\, dm(z) \text{ for each } n \in \mathbb{Z}.$$

Also, $H^p_X \equiv H^p(\mathbb{T}, X)$ is defined by the set of $f \in L^p_X$ with $\widehat{f}(n) = 0$ for $n < 0$.

We write $M_{n \times m}$ for the set of $n \times m$ complex matrices and abbreviate $M_{n \times n}$ to M_n. We observe that $L^2_{\mathbb{C}^n} = L^2 \otimes \mathbb{C}^n$ and $H^2_{\mathbb{C}^n} = H^2 \otimes \mathbb{C}^n$. For a matrix-valued function $\Phi \equiv (\varphi_{ij}) \in L^\infty_{M_n}$, we say that Φ is of *bounded type* if each entry φ_{ij} is of bounded type, and we say that Φ is *rational* if each entry φ_{ij} is a rational function. Let $\Phi \equiv (\varphi_{ij}) \in L^\infty_{M_n}$ be such that Φ^* is of bounded type. Then each $\overline{\varphi}_{ij}$ is of bounded type. Thus in view of (1), we may write $\varphi_{ij} = \theta_{ij} \overline{b}_{ij}$, where θ_{ij} is inner and θ_{ij} and b_{ij} are coprime; in other words, there does not exist a nonconstant common inner divisor of θ_{ij} and b_{ij}. Thus if θ is the least common multiple of $\{\theta_{ij} : i, j = 1, 2, \cdots, n\}$, then we may write

$$\Phi = (\varphi_{ij}) = (\theta_{ij} \overline{b}_{ij}) = (\theta \overline{a}_{ij}) \equiv \theta A^* \quad (\text{where } A \equiv (a_{ji}) \in H^2_{M_n}). \tag{3}$$

In particular, $A(\alpha)$ is nonzero whenever $\theta(\alpha) = 0$ and $|\alpha| < 1$.

2.2 Strong L^2-Functions

We often need to consider operator-valued functions defined on the unit circle constructed by arranging the vectors in a given set $F \subseteq H^2_s(\mathcal{B}(D, E))$ as their column vectors. Using this viewpoint, we consider operator-valued functions whose "column" vectors are L^p-functions. Note that (bounded linear) operators between separable Hilbert spaces may be represented as infinite matrices, so that column vectors of operators are well justified. This viewpoint leads us to define (operator-valued) strong L^p-functions.

For $1 \le p < \infty$, we define the class $L^p_s(\mathcal{B}(D, E)) \equiv L^p_s(\mathbb{T}, \mathcal{B}(D, E))$ as the set of all (WOT) measurable $\mathcal{B}(D, E)$-valued functions Φ on $\mathbb{T}$ such that $\Phi(\cdot)x \in L^p_E$ for each $x \in D$. A function $\Phi \in L^p_s(\mathcal{B}(D, E))$ is called a *strong L^p-function*. Thus if $\Phi \in L^1_s(\mathcal{B}(D, E))$ and $x \in D$, then $\Phi(\cdot)x \in L^1_E$, so that the n-th Fourier coefficient $\widehat{\Phi(\cdot)x}(n)$ of $\Phi(\cdot)x$ is given by $\widehat{\Phi(\cdot)x}(n) = \int_{\mathbb{T}} \overline{z}^n \Phi(z)x \, dm(z)$. Now define the n-th Fourier coefficient of $\Phi \in L^1_s(\mathcal{B}(D, E))$, denoted by $\widehat{\Phi}(n)$, by

$$\widehat{\Phi}(n)x := \widehat{\Phi(\cdot)x}(n) \quad (n \in \mathbb{Z}, \ x \in D).$$

Define

$$H^p_s(\mathcal{B}(D, E)) \equiv H^p_s(\mathbb{T}, \mathcal{B}(D, E)) := \left\{\Phi \in L^p_s(\mathcal{B}(D, E)) : \ \widehat{\Phi}(n) = 0 \text{ for } n < 0\right\},$$

or equivalently, $H^p_s(\mathcal{B}(D, E))$ is the set of all (WOT) measurable functions Φ on $\mathbb{T}$ such that $\Phi(\cdot)x \in H^p_E$ for each $x \in D$. The terminology of a "strong H^2-function" is reserved for the operator-valued functions on the unit disk $\mathbb{D}$, following N.K. Nikolskii [44]: A function $\Phi : \mathbb{D} \to \mathcal{B}(D, E)$ is called a *strong H^2-function* if $\Phi(\cdot)x \in H^2(\mathbb{D}, E)$ for each $x \in D$. Let $L^\infty(\mathcal{B}(D, E))$ be the space of all bounded (WOT) measurable $\mathcal{B}(D, E)$-valued functions on $\mathbb{T}$ and let

$$H^\infty(\mathcal{B}(D, E)) := \left\{\Phi \in L^\infty(\mathcal{B}(D, E)) : \ \widehat{\Phi}(n) = 0 \text{ for } n < 0\right\}.$$

We can show that (cf. [16, Appendix A]):

(a) If $\dim D < \infty$, then $L^2_s(\mathcal{B}(D, E)) = L^2_{\mathcal{B}(D,E)}$ and $H^2_s(\mathcal{B}(D, E)) = H^2_{\mathcal{B}(D,E)}$
(b) $L^\infty_{\mathcal{B}(D,E)} \subseteq L^\infty(\mathcal{B}(D, E)) \subseteq L^p_s(\mathcal{B}(D, E))$

A function $\Delta \in H^\infty(\mathcal{B}(D, E))$ is called an *inner* function with values in $\mathcal{B}(D, E)$ if $\Delta(z)$ is an isometric operator from D into E for almost all $z \in \mathbb{T}$, i.e., $\Delta^*\Delta = I_D$ a.e. on $\mathbb{T}$. Also, Δ is called a *two-sided inner* function if $\Delta\Delta^* = I_E$ a.e. on $\mathbb{T}$ and $\Delta^*\Delta = I_D$ a.e. on $\mathbb{T}$. If Δ is an inner function with values in $\mathcal{B}(D, E)$, we may assume that D is a subspace of E, and if further Δ is two-sided inner, then we may assume that $D = E$. We write $\mathcal{P}_D$ for the set of all polynomials with values in D, i.e., $p(z) = \sum_{k=0}^n \widehat{p}(k)z^k$, where $\widehat{p}(k) \in D$. If $F \in H^2_s(B(D, E))$, then the function Fp belongs to H^2_E for all $p \in \mathcal{P}_D$. A function $F \in H^2_s(B(D, E))$ is called *outer* if cl $F\mathcal{P}_D = H^2_E$. We then have an analogue of the scalar factorization theorem:

Inner-Outer Factorization for H^2_s-functions If $F \in H^2_s(B(D, E))$, then F can be expressed in the form $F = F^i F^e$, where F^e is an outer function with values in $\mathcal{B}(D, E')$ and F^i is an inner function with values in $\mathcal{B}(E', E)$ for some subspace E' of E.

The proof of the above inner-outer factorization for H^2_s-functions is identical to the proof for strong H^2-function (cf. [44, Corollary I.9]).

For a function $\Phi : \mathbb{T} \to \mathcal{B}(D, E)$, write

$$\breve{\Phi}(z) := \Phi(\overline{z}), \quad \widetilde{\Phi} := \breve{\Phi}^*.$$

We call $\breve{\Phi}$ the *flip* of Φ. For $\Phi \in L^2_s(\mathcal{B}(D, E))$, we denote by $\breve{\Phi}_- \equiv \mathbb{P}_-\Phi$ and $\Phi_+ \equiv \mathbb{P}_+\Phi$ the functions

$$((\mathbb{P}_-\Phi)(\cdot))x := P_-(\Phi(\cdot)x) \quad \text{a.e. on } \mathbb{T} \quad (x \in D);$$
$$((\mathbb{P}_+\Phi)(\cdot))x := P_+(\Phi(\cdot)x) \quad \text{a.e. on } \mathbb{T} \quad (x \in D),$$

where P_+ and P_- are the orthogonal projections from L^2_E onto H^2_E and $L^2_E \ominus H^2_E$, respectively. Then we may write $\Phi \equiv \breve{\Phi}_- + \Phi_+$. Note that if $\Phi \in L^2_s(\mathcal{B}(D, E))$, then $\Phi_+, \ \Phi_- \in H^2_s(\mathcal{B}(D, E))$.

For a function $\Phi \in H^2_s(\mathcal{B}(D, E))$, we say that an inner function Δ with values in $\mathcal{B}(D', E)$ is a *left inner divisor* of Φ if $\Phi = \Delta A$ for $A \in H^2_s(\mathcal{B}(D, D'))$. For $\Phi \in H^2_s(\mathcal{B}(D_1, E))$ and $\Psi \in H^2_s(\mathcal{B}(D_2, E))$, we say that Φ and Ψ are *left coprime* if the only common left inner divisor of both Φ and Ψ is a unitary operator. Also, we say that Φ and Ψ are *right coprime* if $\widetilde{\Phi}$ and $\widetilde{\Psi}$ are left coprime. The determination of left or right coprime-ness seems to be a somewhat delicate problem. For matrix-valued functions, left and right coprime-ness was developed in [12–15, 25].

If θ is an inner function, write $I_\theta := \theta I$ (where I is the identity operator). We also write left-g.c.d.$(\cdot)$ and left-l.c.m.$(\cdot)$ for the greatest common left inner divisor and the least common left inner multiple, respectively. By contrast with scalar-valued functions, in (3), I_θ and A need not be (right) coprime. If $\Omega =$ left-g.c.d. $\{I_\theta, A\}$ in the representation (3), that is, $\Phi = \theta A^*$, then $I_\theta = \Omega\Omega_\ell$ and $A = \Omega A_\ell$ for some inner matrix Ω_ℓ (where $\Omega_\ell \in H^2_{M_n}$ because det$(I_\theta) \neq 0$) and some $A_l \in H^2_{M_n}$. Therefore if $\Phi^* \in L^\infty_{M_n}$ is of bounded type, then we can write

$$\Phi = A_\ell{}^*\Omega_\ell, \quad \text{where } A_\ell \text{ and } \Omega_\ell \text{ are left coprime.} \tag{4}$$

In this case, $A_\ell^*\Omega_\ell$ is called the *left coprime factorization* of Φ. Similarly, we can write

$$\Phi = \Omega_r A_r^*, \quad \text{where } A_r \text{ and } \Omega_r \text{ are right coprime.} \tag{5}$$

In this case, $\Omega_r A_r^*$ is called the *right coprime factorization* of Φ. We also say that Ω_ℓ and Ω_r are called the *inner parts* of those factorizations.

For an inner function $\Delta \in H^\infty(\mathcal{B}(D, E))$, $\mathcal{H}(\Delta)$ denotes the orthogonal complement of the subspace ΔH_D^2 in H_E^2, i.e.,

$$\mathcal{H}(\Delta) := H_E^2 \ominus \Delta H_D^2.$$

The space $\mathcal{H}(\Delta)$ is often called a *model space* or a *de Branges-Rovnyak space* (cf. [19, 49, 51]). The name "model space" comes from the model theory of Sz.-Nagy and Foiaş contractions (see Sect. 2.3).

2.3 The Douglas-Shapiro-Shields Factorization

We first review the definition of (vectorial) Toeplitz operators and (vectorial) Hankel operators, and for that we will use [7, 20, 21, 44, 45, 47] for general references. For $\Phi \in L_s^2(\mathcal{B}(D, E))$, the Hankel operator $H_\Phi : H_D^2 \to H_E^2$ is a densely defined operator given by

$$H_\Phi p := J P_-(\Phi p) \quad (p \in \mathcal{P}_D),$$

where J denotes the unitary operator from L_E^2 to L_E^2 given by $(Jg)(z) := \overline{z} g(\overline{z})$ for $g \in L_E^2$. Also a Toeplitz operator $T_\Phi : H_D^2 \to H_E^2$ is a densely defined operator defined by

$$T_\Phi p := P_+(\Phi p) \quad (p \in \mathcal{P}_D).$$

The following basic properties can be easily derived: For $\Phi \in L^\infty(\mathcal{B}(D, E))$ and $\Psi \in L^\infty(\mathcal{B}(D', D))$

$$T_\Phi^* = T_{\Phi^*}, \; H_\Phi^* = H_{\widetilde{\Phi}}; \tag{6}$$

$$T_{\Phi\Psi} - T_\Phi T_\Psi = H_{\Phi^*}^* H_\Psi. \tag{7}$$

The *shift* operator on H_E^2 is defined by

$$(S_E f)(z) := z f(z) \quad \text{for each } f \in H_E^2.$$

The following theorem characterizes the invariant subspaces for the shift operator.

The Beurling-Lax-Halmos Theorem [5, 25, 33, 41, 47] A subspace M of H_E^2 is invariant for the shift operator S_E on H_E^2 if and only if

$$M = \Delta H_{E'}^2,$$

where E' is a subspace of E and Δ is an inner function with values in $\mathcal{B}(E', E)$. Furthermore, Δ is unique up to a unitary constant right factor, i.e., if $M = \Theta H^2_{E''}$, where Θ is an inner function with values in $\mathcal{B}(E'', E)$, then $\Delta = \Theta V$, where V is a unitary operator from E' onto E''.

As customarily done, we say that two inner functions $A, B \in H^\infty(\mathcal{B}(D, E))$ are *equal* if they are equal up to a unitary constant right factor.

By the Beurling-Lax-Halmos Theorem, $\mathcal{H}(\Delta)$ is an invariant subspace for the backward shift operator S_E^*. Thus the truncated backward shift $S_E^*|_{\mathcal{H}(\Delta)}$ is well-defined. We here recall the Model Theorem [44, 51]: If $T \in \mathcal{B}(\mathcal{H})$ for a Hilbert space $\mathcal{H}$ is a contraction (i.e., $||T|| \leq 1$) satisfying

$$\lim_{n\to\infty} T^n x = 0 \quad \text{for each } x \in \mathcal{H}, \tag{8}$$

then T is unitarily equivalent to a truncated backward shift $S_E^*|_{\mathcal{H}(\Delta)}$ for some inner function Δ with values in $\mathcal{B}(E', E)$. In this case, $S_E^*|_{\mathcal{H}(\Delta)}$ is called the *model operator* of T and Δ is called the *characteristic function* of T. We often write $T \in C_{0\cdot}$ for a contraction $T \in \mathcal{B}(\mathcal{H})$ satisfying the condition (8).

On the other hand, if $\Phi \in L^\infty(\mathcal{B}(D, E))$, then $H_{\Phi^*} S_E = S_E^* H_{\Phi^*}$, which implies that the kernel of H_{Φ^*} is an invariant subspace of the shift operator S_E on H^2_E. Thus, by the Beurling-Lax-Halmos Theorem, $\ker H_{\Phi^*} = \Delta H^2_{E'}$ for some inner function Δ with values in $\mathcal{B}(E', E)$. We note that Δ need not be two-sided inner.

Theorem 2.1 ([16, Lemma 2.4]) *If $\Phi \in L^\infty(\mathcal{B}(D, E))$ and Δ is a two-sided inner function with values in $\mathcal{B}(E)$, then the following are equivalent:*

(a) $\ker H_{\Phi^*} = \Delta H^2_E$.
(b) $\Phi = \Delta A^*$, *where $A \in H^\infty(\mathcal{B}(E, D))$ is such that Δ and A are right coprime.*

The factorization in Theorem 2.1(b) is called the *Douglas-Shapiro-Shields* (briefly, DSS) *factorization* of $\Phi \in L^\infty(\mathcal{B}(D, E))$ (see [23, 24, 27]; in particular, [27] contains many important applications of the DSS factorization to linear system theory). Consequently, Theorem 2.1 may be rephrased as follows: If $\Phi \in L^\infty(\mathcal{B}(D, E))$, then the following are equivalent:

(a) Φ admits a DSS factorization.
(b) $\ker H_{\Phi^*} = \Delta H^2_E$ for some two-sided inner function $\Delta \in H^\infty(\mathcal{B}(E))$.

It is known (cf. [14, 24, 31]) that if $\Phi \in L^\infty(\mathcal{B}(D, E))$ admits a DSS factorization $\Phi = \Delta A^*$, then Δ can be obtained from the equation

$$\ker H_{\Phi^*} = \Delta H^2_E; \tag{9}$$

in this case, Δ and A are right coprime. The DSS factorization satisfying (9) is called *canonical*. Consequently, each function that admits a DSS factorization can be arranged in a canonical form. It is also known that if Φ is a matrix-valued function, then (cf. [16, 31])

$$\Phi^* \text{ is of bounded type} \iff \Phi \text{ admits a (canonical) DSS factorization.} \tag{10}$$

If the condition "Δ is two-sided" is dropped in a DSS factorization $\Phi = \Delta A^*$, what can we say about a DSS factorization? More concretely, we may ask: If $\Phi \in L^\infty(\mathcal{B}(E', E))$ is expressed as $\Phi = \Delta A^*$, where $\Delta \in H^\infty(\mathcal{B}(D, E))$ is inner and $A \in H^\infty(\mathcal{B}(D, E'))$, does it follow that Δ can be obtained from the equation $\ker H_{\Phi^*} = \Delta H^2_E$?

An answer to this question is affirmative.

Theorem 2.2 ([32]) *If $\Phi \in L^\infty(\mathcal{B}(E', E))$ is expressed as $\Phi = \Delta A^*$, where $\Delta \in H^\infty(\mathcal{B}(D, E))$ is inner and $A \in H^\infty(\mathcal{B}(D, E'))$, then we can write $\Phi = \Delta_A B_0^*$, where $B_0 \in H^\infty(\mathcal{B}(E_0, E'))$ and $\Delta_A \in H^\infty(\mathcal{B}(E_0, E))$ is an inner function which comes from the equation* $\ker H_{\Phi^*} = \Delta_A H^2_{E_0}$ *for some Hilbert space E_0. Moreover, Δ_A and B_0 are right coprime.*

The expression $\Phi = \Delta A^*$ in Theorem 2.2 is called a *pseudo-DSS factorization*, and the expression $\Phi = \Delta_A B_0^*$ is called a *canonical* pseudo-DSS factorization. Thus Theorem 2.2 says that if a function $\Phi \in L^\infty(\mathcal{B}(E', E))$ admits a pseudo-DSS factorization, then we can always arrange the pseudo-DSS factorization of Φ in a canonical form.

2.4 Complementary Factors

We now consider the following question: If Δ is an inner function with values in $\mathcal{B}(D, E)$, what is the kernel of H_{Δ^*} ? We are tempted to guess that the answer is ΔH^2_E, which is wrong in general. We examine an answer to the above question.

For $\Phi \in L^\infty(\mathcal{B}(D, E))$, we symbolically define the kernel of Φ by

$$\ker \Phi := \{f \in H^2_D : \Phi(z) f(z) = 0 \text{ for almost all } z \in \mathbb{T}\}.$$

Note that the kernel of Φ consists of functions in H^2_D, but not in L^2_D, such that $\Phi f = 0$ a.e. on $\mathbb{T}$. Since $\ker \Phi$ is an invariant subspace for S_D, it follows from the Beurling-Lax-Halmos Theorem that $\ker \Phi = \Omega H^2_{D'}$, for some inner function $\Omega \in H^\infty(D', D)$.

We recall a notion from classical Banach space theory, regarding a vector as an operator acting on the scalars. This notion is important as motivation for the study of strong L^2-functions. Let E be a separable complex Hilbert space. For a function $f : \mathbb{T} \to E$, define $[f] : \mathbb{T} \to \mathcal{B}(\mathbb{C}, E)$ by

$$[f](z)\alpha := \alpha f(z) \quad (\alpha \in \mathbb{C}). \tag{11}$$

Let Δ be an inner function with values in $\mathcal{B}(D, E)$. If $g \in \ker \Delta^*$, then $g \in H^2_E$, so that we can see that $[g] \in H^2_s(\mathcal{B}(\mathbb{C}, E))$. Write

$$[g] = [g]^i [g]^e \quad \text{(inner-outer factorization)},$$

where $[g]^e$ is an outer function with values in $\mathcal{B}(\mathbb{C}, E')$ and $[g]^i$ is an inner function with values in $\mathcal{B}(E', E)$ for some subspace E' of E. If $g \neq 0$, then $[g]^e$ is a nonzero outer function, so that $E' = \mathbb{C}$. Thus, $[g]^i \in H^\infty(\mathcal{B}(\mathbb{C}, E))$. If instead $g = 0$, then $E' = \{0\}$. Therefore, in this case, $[g]^i \in H^\infty(\mathcal{B}(\{0\}, E))$.

The following theorem provides a description of the kernels of H_{Δ^*} and Δ^* [16, Lemma 2.7].

Theorem 2.3 *Let Δ be an inner function with values in $\mathcal{B}(D, E)$. Then* $\ker \Delta^* = \Delta_c H^2_{D'}$, *where*

$$\Delta_c := \textit{left-g.c.d.}\left\{ [g]^i : g \in \ker \Delta^* \right\},$$

which is an inner function with values in $B(D', E)$. Moreover, $[\Delta, \Delta_c]$ is an inner function with values in $\mathcal{B}(D \oplus D', E)$ and

$$\ker H_{\Delta^*} = [\Delta, \Delta_c] H^2_{D \oplus D'} \equiv \Delta H^2_D \bigoplus \Delta_c H^2_{D'}. \tag{12}$$

The inner function Δ_c is called the complementary factor of the inner function Δ.

Example 2.4 If $\Delta = \begin{bmatrix} z \\ 0 \end{bmatrix}$, then Δ is inner. A straightforward calculation shows that $\Delta_c = \begin{bmatrix} 0 \\ 1 \end{bmatrix}$. Indeed,

$$\ker H_{\Delta^*} = \ker \left[H_{\bar{z}} \ 0 \right] = zH^2 \oplus H^2 = \begin{bmatrix} z & 0 \\ 0 & 1 \end{bmatrix} H^2_{\mathbb{C}^2} = \left[\Delta \ \Delta_c \right] H^2_{\mathbb{C}^2}.$$

On the other hand, it is known [1, Lemma 4] that if $\phi \in L^\infty$, then

$$\phi \text{ is of bounded type} \iff \ker H_\phi \neq \{0\}. \tag{13}$$

The following corollary shows that (13) still holds for L^2-functions [16, Corollaries 2.18 and 2.19].

Corollary 2.5 *If $\phi \in L^2$, then ϕ is of bounded type if and only if* $\ker H^*_\phi \neq \{0\}$. *Moreover, if Δ is an inner matrix function, then the following are equivalent:*

(a) Δ^* is of bounded type.
(b) $\breve{\Delta}$ is of bounded type.
(c) $[\Delta, \Delta_c]$ is two-sided inner.

R.G. Douglas and J.W. Helton [22] have considered a problem from engineering circuit theory called Darlington synthesis which mathematically translates to the following: given a contractive analytic operator-valued function S on the unit disk, can one embed S into a two-sided 2×2 inner matrix function $\Theta = \left[\begin{smallmatrix} S & A \\ B & C \end{smallmatrix}\right]$? The special case where $S = \Delta$ is inner and the second block-row is vacuous amounts to our problem of finding Ω so that $[\Delta, \Omega]$ is two-sided inner. Thus, Corollary 2.5 can be obtained from [22, Theorem].

2.5 Degree of Non-cyclicity

By the Beurling-Lax-Halmos Theorem, every invariant subspace for the backward shift operator S_E^* on H_E^2 is a model space $\mathcal{H}(\Delta)$ for some inner function Δ with values in $\mathcal{B}(D, E)$. For a subset F of H_E^2, let E_F^* denote the smallest S_E^*-invariant subspace containing F, i.e.,

$$E_F^* := \bigvee \{S_E^{*n} F : n \geq 0\}.$$

Then by the Beurling-Lax-Halmos Theorem, $E_F^* = \mathcal{H}(\Delta)$ for an inner function Δ with values in $\mathcal{B}(D, E)$. In general, if $\dim E = 1$, then every S_E^*-invariant subspace M admits a cyclic vector, i.e., $M = E_f^*$ for some $f \in H^2$. However, if $\dim E \geq 2$, then this is not such a case. For example, if $M = \mathcal{H}(\Delta)$ with $\Delta = \left[\begin{smallmatrix} z & 0 \\ 0 & z \end{smallmatrix}\right]$, then M does not admit a cyclic vector, i.e., $M \neq E_f^*$ for any vector $f \in H_{\mathbb{C}^2}^2$.

We thus introduce the following:

Definition 2.6 Let $F \subseteq H_E^2$. The *degree of non-cyclicity* of F, denoted by $\mathrm{nc}(F)$, is defined by the number

$$\mathrm{nc}(F) := \sup_{\zeta \in \mathbb{D}} \dim\{g(\zeta) : g \in H_E^2 \ominus E_F^*\}.$$

We will often refer to $\mathrm{nc}(F)$ as the nc-*number* of F. Since E_F^* is an invariant subspace for S_E^*, it follows from the Beurling-Lax-Halmos Theorem that $E_F^* = \mathcal{H}(\Delta)$ for some inner function Δ with values in $\mathcal{B}(D, E)$. Thus

$$\mathrm{nc}(F) = \sup_{\zeta \in \mathbb{D}} \dim \{g(\zeta) : g \in \Delta H_D^2\} = \dim D.$$

In particular, $\mathrm{nc}(F) \leq \dim E$. We note that $\mathrm{nc}(F)$ may take the value ∞. So it is customary to make the following conventions: (i) if n is a natural number, then $n + \infty = \infty$; (ii) $\infty + \infty = \infty$.

If $\dim E = r < \infty$, then $\mathrm{nc}(F) \leq r$ for every subset $F \subseteq H_E^2$. If $F \subseteq H_E^2$ and $\dim E = r < \infty$, then the *degree of cyclicity* of F, denoted by $\mathrm{dc}(F)$, is defined by the number (cf. [52]): $\mathrm{dc}(F) := r - \mathrm{nc}(F)$. In particular, if $E_F^* = \mathcal{H}(\Delta)$, then Δ is two-sided inner if and only if $\mathrm{dc}(F) = 0$.

To understand E_F^* for a subset $F \subseteq H_E^2$, we need to consider the kernels of the adjoints of unbounded Hankel operators with strong L^2-symbols involved with

F. Thus we deal with unbounded Hankel operators H_Φ with strong L^2-symbols Φ. However, the adjoint of an unbounded Hankel operator need not be a Hankel operator. But if Φ is an L^∞-function, then $H_{\Phi^*} = H^*_{\check{\Phi}}$. Thus for a bounded symbol Φ, we may use the notations H_{Φ^*} and $H^*_{\check{\Phi}}$ interchangeably. By contrast, for a strong L^2-function Φ, H_{Φ^*} may not be equal to $H^*_{\check{\Phi}}$ even when Φ^* is a strong L^2-function. In particular, the kernel of an unbounded Hankel operator H_{Φ^*} is likely to be trivial because it is defined on the dense subset of polynomials. From this viewpoint, to avoid potential technical issues in our arguments, we deal with the operator $H^*_{\check{\Phi}}$ in place of H_{Φ^*}. In spite of this, and since the kernel of the adjoint of an unbounded operator is always closed, we can show that via the Beurling-Lax-Halmos Theorem, the kernel of $H^*_{\check{\Phi}}$, with strong L^2-symbol Φ, is still of the form $\Delta H^2_{E'}$ (see [16, Corollary 2.6]).

The next question arises naturally from the Beurling-Lax-Halmos Theorem.

Question 2.7 Recall that, by the Beurling-Lax-Halmos Theorem, $\ker H^*_{\check{\Phi}} = \Theta H^2_{E'}$ for some inner function Θ. Which property of Φ determines the dimension of the space E'? In particular, if Φ is an $n \times m$ matrix-valued L^2-function and $\dim E' = r$, which property of Φ determines the number r?

If $\Phi \in H^2_s(\mathcal{B}(D, E))$ and $\{d_k\}_{k\geq 1}$ is an orthonormal basis for D, write

$$\phi_k := \Phi d_k \in H^2_E \cong H^2_s(\mathcal{B}(\mathbb{C}, E)).$$

We then define

$$\{\Phi\} := \{\phi_k\}_{k\geq 1} \subseteq H^2_E.$$

Hence, $\{\Phi\}$ may be regarded as the set of "column" vectors ϕ_k (in H^2_E) of Φ, in which case we may think of Φ as an infinite matrix-valued function. Then it was shown in [16, Lemma 2.9] that for $\Phi \in H^2_s(\mathcal{B}(D, E))$,

$$E^*_{\{\Phi\}} = \text{cl ran}\, H_{\bar{z}\check{\Phi}}. \tag{14}$$

By definition, $\{\Phi\}$ depends on the orthonormal basis of D. However, (14) shows that $E^*_{\{\Phi\}}$ is independent of a particular choice of the orthonormal basis of D because the right-hand side of (14) is independent of the orthonormal basis of D.

The following theorem gives an answer to Question 2.7 (cf. [16, Theorem 2.13]).

Theorem 2.8 *Let Φ be a strong L^2-function with values in $\mathcal{B}(D, E)$. In view of the Beurling-Lax-Halmos Theorem, we may write $E^*_{\{\Phi_+\}} = \mathcal{H}(\Delta)$ and $\ker H^*_{\check{\Phi}} = \Theta H^2_{E'}$ for some inner functions Δ and Θ with values in $\mathcal{B}(E'', E)$ and $\mathcal{B}(E', E)$, respectively. Then $\Delta = \Theta\Delta_1$ for some two-sided inner function Δ_1 with values in $\mathcal{B}(E'', E')$. Hence, in particular,*

$$\ker H^*_{\check{\Phi}} = \Theta H^2_{E'} \iff \text{nc}\{\Phi_+\} = \dim E'. \tag{15}$$

Example 2.9 If

$$\Phi := \begin{bmatrix} z & 0 \\ 0 & z \\ 0 & 0 \\ a & 0 \end{bmatrix} \quad \text{(with } a \in H^\infty \text{ and } \bar{a} \text{ is not of bounded type)},$$

then a straightforward calculation shows that

$$\ker H^*_{\breve{\Phi}} = \ker \begin{bmatrix} H_{\bar{z}} & 0 & 0 & H_{\bar{a}} \\ 0 & H_{\bar{z}} & 0 & 0 \end{bmatrix} = zH^2 \oplus zH^2 \oplus H^2 \oplus \{0\} = \begin{bmatrix} z & 0 & 0 \\ 0 & z & 0 \\ 0 & 0 & 1 \\ 0 & 0 & 0 \end{bmatrix} H^2_{\mathbb{C}^3},$$

which implies, by (15), $\text{nc}\{\Phi\} = 3$.

On the other hand, if Φ is an $n \times m$ matrix L^2-function, then Theorem 2.8 gives

$$\ker H^*_{\Phi} = \Delta H^2_{\mathbb{C}^n} \text{ for some two-sided inner matrix function } \Delta \iff \text{dc}\,\{\widetilde{\Phi_-}\} = 0. \tag{16}$$

Also, by an argument of [44, p.47],

$$\text{dc}\,\{\widetilde{\Phi_-}\} = 0 \iff \Phi \text{ is of bounded type.} \tag{17}$$

Consequently, by (16) and (17), we can see that if $\Phi \in L^2_{M_{n\times m}}$, then

$$\Phi \text{ is of bounded type} \iff \ker H^*_{\Phi} = \Delta H^2_{\mathbb{C}^n} \text{ for some two-sided inner matrix function } \Delta. \tag{18}$$

3 Matrix-Valued Function Theory

In this section, we consider matrix-valued function theory associated with Toeplitz and Hankel operators. We focus on the notions of tensored-scalar singularity, Beurling degree of inner matrix functions, Abrahamse's Theorem for matrix-valued symbols, and the H^∞-functional calculus for the compressions of the shift.

3.1 Tensored-Scalar Singularities

We ask a question: How does one define a singularity for matrix functions? Conventionally, the singularity (or the existence of a pole) of matrix L^∞-functions is defined by a singularity (or a pole) of some entry of the matrix function (cf. [3, 4]). However we propose another notion of singularity which is more useful for the study of Hankel and Toeplitz operators. To do so, we recall that if the product of two Hankel operators is zero, then one of the operators must be zero. However, this is not such the case for Hankel operators with matrix-valued symbols. For example, if we take

$$\Psi = \begin{bmatrix} 1 & 0 & \overline{z} \\ 0 & 1 & 0 \end{bmatrix} \quad \text{and} \quad \Phi = \begin{bmatrix} \overline{z} & 0 \\ 0 & \overline{z} \\ 0 & 0 \end{bmatrix}, \tag{19}$$

then

$$H_\Psi H_\Phi = \begin{bmatrix} 0 & 0 & H_{\overline{z}} \\ 0 & 0 & 0 \end{bmatrix} \begin{bmatrix} H_{\overline{z}} & 0 \\ 0 & H_{\overline{z}} \\ 0 & 0 \end{bmatrix} = 0,$$

but $H_\Phi \neq 0$ and $H_\Psi \neq 0$. In this section, we examine the following question: Under what conditions does it follow that

$$H_\Psi H_\Phi = 0 \implies H_\Psi = 0 \ \text{ or } \ H_\Phi = 0\,? \tag{20}$$

To examine this question, in [15] the notion of "tensored-scalar singularity" was introduced for square matrix functions of bounded type. In [15, Theorem 5.4], it was shown that the answer to question (20) is affirmative for bounded type matrix functions such that Φ or Ψ has a tensored-scalar singularity. In this subsection we extend the above result for general matrix-valued L^∞-functions having a tensored-scalar singularity. To see this, we introduce the notion of tensored-scalar singularity for general L^∞-matrix functions.

Definition 3.1 For $\Phi \in L^\infty_{n\times m}$, we say that Φ has a *tensored-scalar singularity (with respect to* θ*)* if there exists a nonconstant inner function θ such that

$$\{0\} \neq \ker H_\Phi \subseteq \theta H^2_{\mathbb{C}^m}\,.$$

Note 3.1 Every bounded type function $\varphi \in L^2$ has, trivially, a tensored-scalar singularity; for, if φ is of bounded type, then by Beurling's Theorem, $\ker H_\varphi = \theta H^2$ for an inner function $\theta \in H^\infty$. Also we can easily see that if $\Phi \in L^\infty_{M_n}$ is of bounded type and $\Phi = A\Theta^*$ (right coprime), then (cf. [15, Lemma 5.2])

$$\Phi \text{ has a tensored-scalar singularity} \Longleftrightarrow \Theta \text{ has a diagonal-constant inner divisor.} \tag{21}$$

From this viewpoint, if $\Phi \in L^\infty_{M_n}$ has the coprime factorization $\Phi_- = \theta B^*$ (coprime), then, clearly, Φ has a tensored-scalar singularity.

Example 3.2 In Definition 3.1, we cannot conclude that $m \le n$. To see this, let $h(z) := e^{\frac{1}{z-3}}$. Then $h \in H^\infty$ and $\overline{h}$ is not of bounded type. Let

$$f(z) := \frac{h(z)}{\sqrt{2}||h||_\infty}.$$

Clearly, $\overline{f}$ is not of bounded type. Let $h_1(z) := \sqrt{1-|f(z)|^2}$. Then $h_1 \in L^\infty$ and $|h_1| \ge \frac{1}{\sqrt{2}}$. Thus there exists an outer function g such that $|h_1| = |g|$ a.e. on $\mathbb{T}$ (see [20, Corollary 6.25]). Let

$$\Delta := \begin{bmatrix} zf \\ zg \end{bmatrix}.$$

Then

$$\ker H_{\Delta^*} = \begin{bmatrix} zf \\ zg \end{bmatrix} \subseteq zH^2_{\mathbb{C}^2},$$

which implies Δ^* has a tensored-scalar singularity with respect to z.

Remark 3.3 Let $\Phi \in L^2_{n\times m}$. Then we may ask whether Φ has a tensored-scalar singularity if and only if $\widetilde{\Phi}$ has a tensored-scalar singularity. The answer is negative. For example, let

$$\Phi := \begin{bmatrix} \overline{z} & 0 \\ 0 & \overline{z} \\ 0 & 0 \end{bmatrix}.$$

Then a straightforward calculation shows that Φ has a tensored-scalar singularity with respect to z, but $\widetilde{\Phi}$ has no tensored-scalar singularity.

We now have the following:

Theorem 3.4 *Let $\Phi \in L^\infty_{n\times m}$ and $\Psi \in L^\infty_{r\times n}$. If $\widetilde{\Phi}$ or Ψ has a tensored-scalar singularity, then*

$$H_\Psi H_\Phi = 0 \Longrightarrow H_\Phi = 0 \text{ or } H_\Psi = 0.$$

Proof Suppose that $\widetilde{\Phi}$ has a tensored-scalar singularity with respect to δ. Then by the Beurling-Lax-Halmos Theorem, there exists an inner matrix function $\Delta \in H^\infty_{n\times p}$

such that $\ker H_\Phi^* = \Delta H^2_{\mathbb{C}^p} \subseteq \delta H^2_{\mathbb{C}^n}$. If $H_\Psi H_\Phi = 0$, then we have

$$\mathcal{H}(I_\delta) \subseteq \text{cl ran}\, H_\Phi \subseteq \ker H_\Psi \equiv \Omega H^2_{\mathbb{C}^q}, \tag{22}$$

where $\Omega \equiv (\omega_{ij})$ is an $n \times q$ inner function. Since δ is not constant, $\mathcal{H}(\delta)$ has at least an outer function g that is invertible in H^∞ (cf. [13, Lemma 3.4]). Put

$$e_1 = \begin{bmatrix} g \\ 0 \\ \vdots \\ 0 \end{bmatrix}, e_2 = \begin{bmatrix} 0 \\ g \\ \vdots \\ 0 \end{bmatrix}, \cdots, e_n = \begin{bmatrix} 0 \\ 0 \\ \vdots \\ g \end{bmatrix}.$$

Then for each $j = 1, 2, \cdots, n$, $e_j \in \mathcal{H}(I_\delta)$ and hence by (22), $e_j \in \Omega H^2_{\mathbb{C}^q}$. Then

$$q = \max\{\text{rank}\, \Omega(\zeta) : \zeta \in \mathbb{D}\} \geq n,$$

which implies $n = q$. For all $j = 1, 2, \cdots, n$, it follows from (22) that $\Omega^* e_j \in H^2_{\mathbb{C}^n}$, which implies that

$$\overline{\omega}_{ij}\, g \in H^2 \quad \text{for each } i, j = 1, 2, \cdots, n,$$

so that $\overline{\omega}_{ij} \in \frac{1}{g} H^2 \subseteq H^2$ for each i, j. Therefore each ω_{ij} is constant and hence, Ω is a unitary matrix. Thus, again by (22), $\ker H_\Psi = H^2_{\mathbb{C}^q}$. Therefore $H_\Psi = 0$. Suppose that Ψ has a tensored-scalar singularity and $H_\Psi H_\Phi = 0$. Then $H_{\widetilde{\Phi}} H_{\widetilde{\Psi}} = 0$, so that $H_{\widetilde{\Phi}} = 0$ by what we proved just above, and hence $H_\Phi = 0$. This completes the proof. □

In Theorem 3.4, the condition "$\widetilde{\Phi}$ or Ψ has a tensored-scalar singularity" cannot be replaced by the condition "Φ or Ψ has a tensored-scalar singularity." For example, if Φ, Ψ are given in (19), then by Remark 3.3, Φ has a tensored-scalar singularity, but $\widetilde{\Phi}$ has no tensored-scalar singularity. Note that, in such a case, $H_\Psi H_\Phi = 0$, but $H_\Phi \neq 0$ and $H_\Psi \neq 0$.

On the other hand, we recall [31, Theorem 3.3] that if $\Phi \in L^\infty_{M_n}$, then

$$H^*_{\Phi^*} H_{\Phi^*} \geq H^*_\Phi H_\Phi \iff \exists\, K \in H^\infty_{M_n} \text{ such that } ||K||_\infty \leq 1 \text{ and } \Phi - K\Phi^* \in H^\infty_{M_n}. \tag{23}$$

The following theorem extends [15, Theorem 5.5] for general matrix-valued L^∞-functions.

Theorem 3.5 *Let $\Phi, \Psi \in L^\infty_{M_n}$. If Φ or Ψ has a tensored-scalar singularity, then*

$$H^*_\Phi H_\Phi = H^*_\Psi H_\Psi \iff \Phi - U\Psi \in H^\infty_{M_n}$$

for some unitary matrix $U \in M_n$.

Proof Suppose $H_\Phi^* H_\Phi = H_\Psi^* H_\Psi$. Then $\ker H_\Psi = \ker H_\Phi$, which implies that Φ has a tensored-scalar singularity if and only if Ψ does. Thus, without loss of generality, we may assume that Φ has a tensored-scalar singularity with respect to θ. Then $\ker H_\Phi \subseteq \theta H^2_{\mathbb{C}^n}$, so that

$$\mathcal{H}(I_\theta) \subseteq \text{cl ran } H_{\widetilde{\Phi}}. \tag{24}$$

By our assumption together with (23), there exist $U, U' \in H^\infty_{M_n}$ with $||U||_\infty \leq 1$ and $||U'||_\infty \leq 1$ such that

$$\Phi_-^* - U\Psi_-^* \in H^2_{M_n} \quad \text{and} \quad \Psi_-^* - U'\Phi_-^* \in H^2_{M_n}, \tag{25}$$

which implies $H_{\Phi_-^*} - H_{UU'\Phi_-^*} = 0$, and hence $(I - T^*_{\widetilde{UU'}})H_\Phi = 0$. By the same argument as in [15, Proof of Theorem 5.5], we can see that $UU' = I_n$. Therefore U is a unitary constant and by (25), $\Phi - U\Psi \in H^\infty_{M_n}$. The converse is clear. □

If $\Phi \in L^\infty_{M_n}$, then by (7),

$$[T_\Phi^*, T_\Phi] = H^*_{\Phi^*} H_{\Phi^*} - H_\Phi^* H_\Phi - T_{\Phi^*\Phi - \Phi\Phi^*}.$$

Thus if Φ is normal, i.e., $\Phi^*\Phi = \Phi\Phi^*$ a.e. on $\mathbb{T}$, then it follows at once that

$$T_\Phi \text{ is normal} \iff H^*_{\Phi^*} H_{\Phi^*} = H_\Phi^* H_\Phi. \tag{26}$$

Therefore if $\Phi \in L^\infty_{M_n}$ is normal and has a tensored-scalar singularity, then by Theorem 3.5 and (26), T_Φ is normal if and only if $\Phi - U\Phi^* \in H^\infty_{M_n}$ for some unitary matrix $U \in M_n$.

3.2 The Beurling Degree of an Inner Matrix Function

In this subsection we consider the following question: *If Δ is an $n \times r$ inner matrix function, does there exist a function $\Phi \in L^2_{M_{n\times m}}$ satisfying the equation*

$$\ker H^*_{\breve{\Phi}} = \Delta H^2_{\mathbb{C}^r}\ ? \tag{27}$$

To formulate an answer to question (27), we consider whether there exists an inner function Ω satisfying $\ker H_{\Omega^*} = \Delta H^2_{\mathbb{C}^r}$ when Δ is an $n \times r$ inner matrix function. In fact, the answer to this question is negative. Indeed, if $\ker H_{\Omega^*} = \Delta H^2_{\mathbb{C}^r}$ for some inner matrix function $\Omega \in H^\infty_{M_{n\times m}}$, then by Theorem 2.3, we have $[\Omega, \Omega_c] = \Delta$, and hence $\Delta_c = 0$. Conversely, if $\Delta_c = 0$, then again by Theorem 2.3, we should have $\ker H_{\Delta^*} = \Delta H^2_{\mathbb{C}^r}$. Consequently, $\ker H_{\Omega^*} = \Delta H^2_{\mathbb{C}^r}$ for some inner function Ω

if and only if $\Delta_c = 0$. Thus if $\Delta := \begin{bmatrix} 1 \\ 0 \end{bmatrix}$, then there exists no inner function Ω such that $\ker H_{\Omega^*} = \Delta H^2$. On the other hand, we note that the solution Φ is not unique although there exists an inner function Φ satisfying Equation (27). For example, if $\Delta := \text{diag}\,(z, 1, 1)$, then the following Φs are such solutions:

$$\Phi = \begin{bmatrix} z \\ 0 \\ 0 \end{bmatrix}, \quad \begin{bmatrix} z & 0 \\ 0 & 1 \\ 0 & 0 \end{bmatrix}, \quad \Delta.$$

The following theorem gives an affirmative answer to question (27): indeed, we can always find an analytic solution $\Phi \in H^\infty_{M_{n\times m}}$ [16, Corollary 4.2].

Theorem 3.6 *For a given $n \times r$ inner matrix function Δ, there exists at least a solution $\Phi \in H^\infty_{M_{n\times m}}$ (with $m \leq r + 1$) of the equation $\ker H^*_{\breve{\Phi}} = \Delta H^2_{\mathbb{C}^r}$.*

Remark 3.7 In view of Theorem 3.6, it is reasonable to ask whether such a solution $\Phi \in L^2_{M_{n\times m}}$ of the equation $\ker H^*_{\breve{\Phi}} = \Delta H^2_{\mathbb{C}^r}$ (Δ an $n \times r$ inner matrix function) exists for each $m = 1, 2, \cdots$ even though it exists for some m. However, the answer is negative in general, i.e., a solution exists for some m, but may not exist for another $m_0 < m$. To see this, let

$$\Delta := \begin{bmatrix} z & 0 & 0 \\ 0 & z & 0 \\ 0 & 0 & 1 \\ 0 & 0 & 0 \end{bmatrix} \in H^\infty_{M_{4\times 3}}. \tag{28}$$

Then Δ is inner. A straightforward calculation shows that there exists no solution Φ, in $L^2_{M_{4\times 1}}$ (the case $m = 1$), of the equation $\ker H^*_{\breve{\Phi}} = \Delta H^2_{\mathbb{C}^3}$. By contrast, if $m = 2$, then we can find a solution $\Phi \in L^2_{M_{4\times 2}}$. Indeed, let

$$\Phi := \begin{bmatrix} z & 0 \\ 0 & z \\ 0 & 0 \\ a & 0 \end{bmatrix},$$

where $a \in H^\infty$ is such that $\overline{a}$ is not of bounded type. Then $\ker H_{\Phi^*} = \Delta H^2_{\mathbb{C}^3}$. Thus we obtain a solution for $m = 2$ although there exists no solution for $m = 1$.

Let Δ be an $n \times r$ inner matrix function. In view of Remark 3.7, we may ask how to determine a possible value m for which there exists a solution $\Phi \in L^2_{M_{n\times m}}$ of the equation $\ker H^*_{\breve{\Phi}} = \Delta H^2_{\mathbb{C}^r}$. In fact, if we have a solution $\Phi \in L^2_{M_{n\times m}}$ of the equation $\ker H^*_{\breve{\Phi}} = \Delta H^2_{\mathbb{C}^r}$, then a solution $\Psi \in L^2_{M_{n\times q}}$ also exists if $q \geq m$: indeed, if $\mathbf{0}$ denotes the zero operator in $M_{n\times(q-m)}$ and $\Psi := [\Phi, \mathbf{0}]$, then it follows that

$\ker H^*_{\breve{\Phi}} = \ker H^*_{\breve{\Psi}}$. Thus we would like to ask what is the infimum of the set of positive integers m such that there exists a solution $\Phi \in L^2_{M_{n\times m}}$ of the equation $\ker H^*_{\breve{\Phi}} = \Delta H^2_{\mathbb{C}^r}$. To answer this question, we introduce a notion of the "Beurling degree" for an inner function.

Definition 3.8 If Δ is an $n \times r$ inner matrix function, then the *Beurling degree* of Δ, denoted by $\deg_B(\Delta)$, is defined by

$$\deg_B(\Delta) = \inf\{m : \ker H^*_{\breve{\Phi}} = \Delta H^2_{\mathbb{C}^r} \text{ for some } \Phi \in L^2_{M_{n\times m}}\}. \tag{29}$$

We recall that the *spectral multiplicity* for $T \in \mathcal{B}(E)$ is defined by the number μ_T:

$$\mu_T := \inf \dim \bigvee\{f : f \in F\},$$

where $F \subseteq E$, the infimum being taken over all generating subsets F, i.e., subsets such that $\bigvee\{T^n F : n \geq 0\} = E$. It was shown in [16, Theorem 4.6] that if Δ is the characteristic function of the model operator T with values in $M_{n\times r}$, then the spectral multiplicity of the model operator is equal to the Beurling degree of Δ.

Theorem 3.9 (The Beurling degree and the spectral multiplicity) *Given an $n \times r$ inner matrix function Δ, let $T := S^*_E|_{\mathcal{H}(\Delta)}$. Then*

$$\mu_T = \deg_B(\Delta). \tag{30}$$

Example 3.10 If Δ is given by (28), then by the preceding argument and (29), $\deg_B(\Delta) = 2$. Thus if $T := S^*_{\mathbb{C}^4}|_{\mathcal{H}(\Delta)}$, then by (30), $\mu_T = 2$. □

For an inner matrix function $\Delta \in H^\infty_{M_n}$, write

$$\mathbf{S}_\Delta := P_{\mathcal{H}(\Delta)} S_{\mathbb{C}^n}|_{\mathcal{H}(\Delta)}.$$

It is known (cf. [44, Appendix]) that if $\Delta \in H^\infty_{M_n}$ is square-inner, then the spectral multiplicity of $\mathbf{S}_\Delta$ can be computed from its characteristic function Δ by using the Jordan model theory for C_0-contractions due to Sz.-Nagy and Foiaş (cf. [51]) and the Moore-Nordgren theory of quasi-equivalence (cf. [42, 46]). To see this, for an inner matrix function $\Delta \in H^\infty_{M_n}$, write

$$\delta_k := \text{g.c.d.}\{\text{all inner parts of the minors of order } n-k \text{ of } \Delta\} \tag{31}$$

for $k = 0, 1, 2, \cdots, n$. If $T := \mathbf{S}_\Delta$, then

$$\mu_T = \max\{k : \delta_{k-1} \neq \delta_k \text{ up to constant}\}. \tag{32}$$

It is also known [44, Lemma on the Function Θ^t, p.75] that $S^*_{\mathbb{C}^n}|_{\mathcal{H}(\Delta)}$ is unitarily equivalent to $\mathbf{S}_{\widetilde{\Delta}}$. Moreover, if Δ is an inner matrix function in $H^\infty_{M_n}$ and δ_k is given in (31), then we can easily check that for $k = 0, 1, 2, \cdots, n$,

$$\widetilde{\delta_k} = \text{g.c.d.}\big\{\text{all inner parts of the minors of order } n-k \text{ of } \widetilde{\Delta}\big\}. \tag{33}$$

Therefore, by (32) and Theorem 3.9 together with this argument, we can get the following:

Corollary 3.11 *If Δ is an inner matrix function in $H^\infty_{M_n}$, then*

$$deg_B(\Delta) = \max\big\{k : \delta_{k-1} \neq \delta_k \textit{ up to constant}\big\}. \tag{34}$$

We would like to pose the following:

3.1 Give a direct (matricial) proof for (34) without using the formula (32), which relies upon the Jordan model theory of Sz.-Nagy and Foiaş.

In general, we wish to ask:

3.2 If Δ is an $n \times r$ inner matrix function, how do we describe $\deg_B(\Delta)$ in terms of entries of Δ (or matrices involving Δ)?

3.3 Abrahamse's Theorem for Matrix-Valued Symbols

In 1970, P.R. Halmos addressed the following problem, listed as Problem 5 in his lecture "Ten problems in Hilbert space" [34, 35]:

Is every subnormal Toeplitz operator either normal or analytic?

Any analytic Toeplitz operator T_φ ($\varphi \in H^\infty$) is easily seen to be subnormal: indeed, $T_\varphi h = P(\varphi h) = \varphi h = M_\varphi h$ for $h \in H^2$, where M_φ is the normal operator of multiplication by φ on L^2. The question is natural because normal and analytic Toeplitz operators are fairly well understood and they are both subnormal. In 1984, Halmos' Problem 5 was answered in the negative by C. Cowen and J. Long [11]. However, Cowen and Long's construction does not provide an intrinsic connection between subnormality and the theory of Toeplitz operators. Until now researchers have been unable to characterize subnormal Toeplitz operators in terms of their symbols. Thus, we may reformulate Halmos' Problem 5: *Which Toeplitz operators are subnormal?* The most interesting partial answer to Halmos' Problem 5 was given by M.B. Abrahamse [1], who gave a general sufficient condition for the answer to Halmos' Problem 5 to be affirmative.

Abrahamse's Theorem **([1])** *Let $\varphi \in L^\infty$ be such that φ or $\overline{\varphi}$ is of bounded type. If T_φ is hyponormal and* $\ker[T_\varphi^*, T_\varphi]$ *is invariant under T_φ, then T_φ is normal or analytic.*

Consequently, if $\varphi \in L^\infty$ is such that φ or $\overline{\varphi}$ is of bounded type, then every subnormal Toeplitz operator must be either normal or analytic, since $\ker[S^*, S]$ is invariant under S for every subnormal operator S. We say that a Toeplitz operator T_Φ with matrix-valued symbol Φ is *analytic* if $\Phi \in H^\infty_{M_n}$. Evidently, any analytic Toeplitz operator with a normal symbol is subnormal because the multiplication operator M_Φ is a normal extension of T_Φ. As a first inquiry in the above reformulation of Halmos' Problem 5, the following question can be raised: *Is Abrahamse's Theorem valid for Toeplitz operators with matrix-valued symbols*? In general, a straightforward matrix-valued version of Abrahamse's Theorem is doomed to fail: for instance, if $\Phi := \begin{bmatrix} z+\overline{z} & 0 \\ 0 & z \end{bmatrix}$, then, clearly, both Φ and Φ^* are of bounded type and $T_\Phi = \begin{bmatrix} U+U^* & 0 \\ 0 & U \end{bmatrix}$ (where U is the shift on H^2) is subnormal, but neither normal nor analytic.

We here extend the above result to the case of bounded type symbols: we shall say that T_Φ has a *bounded type symbol* if both Φ and Φ^* are of bounded type.

It was shown in [13] that if T_Φ has a bounded type symbol with

$$\Phi_- = \theta B^* \quad \text{(coprime)} \tag{35}$$

and if T_Φ is hyponormal and $\ker[T_\varphi^*, T_\varphi]$ is invariant under T_φ, then T_Φ is normal or analytic. However, the condition (35) forces the inner part of the right coprime factorization (5) of Φ_- to be diagonal-constant. Also, it was shown in [12] that if Φ is a matrix-valued *rational* function, then the condition (35) can be weakened to the condition that the inner part of the right coprime factorization (5) of Φ_- has a nonconstant diagonal-constant inner divisor. We note that in view of (21), those conditions of [13] and [12] are special cases of the condition of "having a tensored-scalar singularity." Indeed, it was shown in [15, Theorem 7.3] that for bounded type symbols $\Phi \in L^\infty_{M_n}$, if Φ has a tensored-scalar singularity, then we get a full-fledged matrix-valued version of Abrahamse's Theorem.

Theorem 3.12 (Abrahamse's Theorem for matrix-valued symbols) *Let $\Phi \in L^\infty_{M_n}$ be such that Φ and Φ^* are of bounded type. Assume Φ has a tensored-scalar singularity. If T_Φ is hyponormal and* $\ker[T_\Phi^*, T_\Phi]$ *is invariant under T_Φ, then T_Φ is normal. Hence, in particular, if T_Φ is subnormal, then T_Φ is normal.*

Remark 3.13

(a) We note that the assumption "Φ has a tensored-scalar singularity" is essential in Theorem 3.12. As we saw before, if $\Phi := \begin{bmatrix} \overline{z}+z & 0 \\ 0 & z \end{bmatrix}$, then T_Φ is neither normal nor analytic. But since $\ker H_\Phi = \ker H_{\begin{bmatrix} \overline{z} & 0 \\ 0 & 0 \end{bmatrix}} = \begin{bmatrix} z & 0 \\ 0 & 1 \end{bmatrix} H^2_{\mathbb{C}^n}$, it follows that $\Theta \equiv \begin{bmatrix} z & 0 \\ 0 & 1 \end{bmatrix}$ does not have any nonconstant diagonal-constant inner divisor, so that Φ does not have a tensored-scalar singularity.

(b) If $n = 1$, then $\Theta \equiv \theta \in H^\infty$ is vacuously diagonal-constant, so that Theorem 3.12 reduces to the original Abrahamse's Theorem.

3.4 H^∞-Functional Calculus for the Compressions of the Shift

It is well known that the functional calculus for polynomials of compressions of the shift results in the Hermite-Fejér matrix via the classical Hermite-Fejér interpolation problem. We now extend the polynomial calculus to an H^∞-functional calculus (so-called the Sz.-Nagy-Foiaş functional calculus) via the triangularization theorem and then extend it further to an $\overline{H^\infty} + H^\infty$-functional calculus for compressions of the shift operator. Let

$$\mathbf{S}_\theta := P_{\mathcal{H}(\theta)} S|_{\mathcal{H}(\theta)}$$

be the compression of the shift operator S to $\mathcal{H}(\theta)$. We briefly review a functional calculus for polynomials of $\mathbf{S}_\theta$ (cf. [25]).

Let θ be a finite Blaschke product of degree d. Let W be the unitary operator from $\bigoplus_1^d \mathbb{C}^n$ onto $\mathcal{H}(I_\theta)$. It is known [25, Theorem X.1.5] that $\mathbf{S}_\theta$ is unitarily equivalent to the lower triangular matrix M on $\mathbb{C}^d$. Now let $P(z) \in H^\infty_{M_n}$ be a matrix polynomial of degree k. Then the matrix $P(M)$ on $\mathbb{C}^{n\times d}$ is defined by

$$P(M) := \sum_{i=0}^{k} P_i \otimes M^i, \quad \text{where } P(z) = \sum_{i=0}^{k} P_i z^i. \tag{36}$$

If the matrix $P(M)$ is the *Hermite-Fejér matrix* determined by (36) and if $(T_P)_\Theta := P_{\mathcal{H}(\Theta)} T_P|_{\mathcal{H}(\Theta)}$ is the compression of T_P to $\mathcal{H}(\Theta)$ (where $\Theta := I_\theta$ for an inner function θ), then it is known [25, Theorem X.5.6] that

$$W^*(T_P)_\Theta W = P(M), \tag{37}$$

which says that $P(M)$ is a matrix representation for $(T_P)_\Theta$.

We now extend the representation (37) to the case of matrix H^∞-functions. We refer to [2] and [44] for details on this representation. For an explicit criterion, we need to introduce the triangularization theorem concretely. There are three cases to consider.

Case 1: Let B be a Blaschke product and let $\Lambda := \{\lambda_n : n \geq 1\}$ be the sequence of zeros of B counted with their multiplicities. Write

$$\beta_1 := 1, \quad \beta_k := \prod_{n=1}^{k-1} \frac{\lambda_n - z}{1 - \overline{\lambda}_n z} \cdot \frac{|\lambda_n|}{\lambda_n} \qquad (k \geq 2),$$

and let

$$\delta_j := \frac{d_j}{1-\overline{\lambda}_j z}\beta_j \qquad (j \ge 1),$$

where $d_j := (1-|\lambda_j|^2)^{\frac{1}{2}}$. Let μ_B be the measure on $\mathbb{N}$ given by $\mu_B(\{n\}) := \frac{1}{2}d_n^2$, $(n \in \mathbb{N})$. Then the map $V_B : L^2(\mu_B) \to \mathcal{H}(B)$ defined by

$$V_B(c) := \frac{1}{\sqrt{2}}\sum_{n\ge 1} c(n) d_n \delta_n, \quad c \equiv \{c(n)\}_{n\ge 1}, \tag{38}$$

is unitary.
Case 2: Let s be a singular inner function with continuous representing measure $\mu \equiv \mu_s$. Let μ_λ be the projection of μ onto the arc $\{\zeta : \zeta \in \mathbb{T},\ 0 < \arg\zeta \le \arg\lambda\}$ and let

$$s_\lambda(\zeta) := \exp\Big(-\int_{\mathbb{T}} \frac{t+\zeta}{t-\zeta} d\mu_\lambda(t)\Big) \ \ (\zeta \in \mathbb{D}).$$

Then the map $V_s : L^2(\mu) \to \mathcal{H}(s)$ defined by

$$(V_s c)(\zeta) = \sqrt{2}\int_{\mathbb{T}} c(\lambda) s_\lambda(\zeta) \frac{\lambda d\mu(\lambda)}{\lambda - \zeta} \quad (\zeta \in \mathbb{D}) \tag{39}$$

is unitary.
Case 3: Let Δ be a singular inner function with pure point representing measure $\mu \equiv \mu_\Delta$. We enumerate the set $\{t \in \mathbb{T} : \mu(\{t\}) > 0\}$ as a sequence $\{t_k\}_{k\in\mathbb{N}}$. Write $\mu_k := \mu(\{t_k\})$, $k \ge 1$. Further, let μ_Δ be a measure on $\mathbb{R}_+ = [0,\infty)$ such that $d\mu_\Delta(\lambda) = \mu_{[\lambda]+1} d\lambda$ and define a function Δ_λ on the unit disk $\mathbb{D}$ by the formula

$$\Delta_0 := 1, \quad \Delta_\lambda(\zeta) := \exp\left\{-\sum_{k=1}^{[\lambda]} \mu_k \frac{t_k+\zeta}{t_k-\zeta} - (\lambda - [\lambda])\mu_{[\lambda]+1}\frac{t_{[\lambda]+1}+\zeta}{t_{[\lambda]+1}-\zeta}\right\},$$

where $[\lambda]$ is the integer part of λ $(\lambda \in \mathbb{R}_+)$. Then the map $V_\Delta : L^2(\mu_\Delta) \to \mathcal{H}(\Delta)$ defined by

$$(V_\Delta c)(\zeta) := \sqrt{2}\int_{\mathbb{R}_+} c(\lambda)\Delta_\lambda(\zeta)(1-\overline{t}_{[\lambda]+1}\zeta)^{-1} d\mu_\Delta(\lambda) \ (\zeta \in \mathbb{D}) \tag{40}$$

is unitary.

Collecting together the above three cases, we get the following:

Triangularization Theorem ([44, p.123]) *Let θ be an inner function with the canonical factorization $\theta = B \cdot s \cdot \Delta$, where B is a Blaschke product and*

s and Δ are singular inner functions with representing measures μ_s and μ_Δ, respectively, with μ_s continuous and μ_Δ a pure point measure. Then the map $V : L^2(\mu_B) \times L^2(\mu_s) \times L^2(\mu_\Delta) \to \mathcal{H}(\theta) \equiv \mathcal{B}(B) \oplus B\mathcal{H}(s) \oplus Bs\mathcal{H}(\Delta)$ defined by

$$V := \begin{bmatrix} V_B & 0 & 0 \\ 0 & BV_s & 0 \\ 0 & 0 & BsV_\Delta \end{bmatrix} \tag{41}$$

is unitary, where $V_B, \mu_B, V_S, \mu_S, V_\Delta, \mu_\Delta$ are defined in (38), (39), and (40) and $M := V^*\mathbf{S}_\theta V$ is a lower-triangular operator.

Now we note that every compression of the shift operator is completely non-unitary. Therefore M is an absolutely continuous contraction. Thus if $\Phi \in H^\infty_{M_n}$, then we can define $\Phi(M)$ as a H^∞-functional calculus (the Sz.-Nagy-Foiaş functional calculus). The following theorem was proved in [15, Theorem 6.3].

Theorem 3.14 *Let $\Phi \in H^\infty_{M_n}$ and let $\theta \in H^\infty$ be an inner function. If we write*

$$M := V^*\mathbf{S}_\theta V \quad \text{and} \quad \mathcal{V} := V \otimes I_n, \tag{42}$$

where $V : L^2(\mu_B) \times L^2(\mu_s) \times L^2(\mu_\Delta) \to \mathcal{H}(\theta)$ is unitary as in (41), then

$$\mathcal{V}^*(T_\Phi)_\Theta \mathcal{V} = \Phi(M), \tag{43}$$

where $\Theta := I_\theta$ and $(T_\Phi)_\Theta := P_{\mathcal{H}(\Theta)} T_\Phi|_{\mathcal{H}(\Theta)}$

Remark 3.15 $\Phi(M)$ is called a *matrix representation* for $(T_\Phi)_\Theta$.

An application of the functional calculus In [31], it was shown that (i) if $\Phi \in L^\infty_{M_n}$ is such that T_Φ is hyponormal, i.e., $[T^*_\Phi, T_\Phi] \geq 0$, then Φ is normal, i.e., $\Phi^*\Phi = \Phi\Phi^*$ a.e. on $\mathbb{T}$, and that (ii) if $\Phi \in L^\infty_{M_n}$ is normal and

$$C(\Phi) := \{K \in H^\infty_{M_n} : \Phi - K\Phi^* \in H^\infty_{M_n}\},$$

then

$$T_\Phi \text{ is hyponormal} \iff \exists\, K \in C(\Phi) \text{ with } ||K||_\infty \leq 1.$$

In [14, Theorem 3.3], it was shown from (43) that if $\Phi := \Phi^*_- + \Phi_+ \in L^\infty_{M_n}$ is normal of the form

$$\Phi_+ = \theta_0\theta_1 A^* \quad \text{and} \quad \Phi_- = \theta_1 B^* \;\; (A, B \in H^\infty_{M_n}) \tag{44}$$

and if $(T_A)_{\theta_0\theta_1} := P_{\mathcal{H}(\theta_0\theta_1)} T_A|_{\mathcal{H}(\theta_0\theta_1)}$ has dense range, then

$$K(M) \text{ is contractive} \iff T_\Phi \text{ is hyponormal}, \tag{45}$$

where $K \in \mathcal{C}(\Phi)$ and M is given by (42) with $\theta := \theta_0\theta_1$. Note that the form (44) is a necessary condition for T_Φ to be hyponormal when Φ and Φ^* are of bounded type.

We consider a revealing example that illustrates (45).

Example 3.16 Let θ be an inner function and consider the matrix-valued function

$$\Phi := \begin{bmatrix} \overline{z} & \overline{z\theta} + z\theta \\ \overline{z\theta} + z\theta & \overline{z} \end{bmatrix}.$$

We now use the equivalence (45) to determine the hyponormality of T_Φ. Under the above notation, we have

$$\theta_0 = 1, \;\; \theta_1 = z\theta, \;\; A = \begin{bmatrix} 0 & 1 \\ 1 & 0 \end{bmatrix}, \;\; B = \begin{bmatrix} \theta & 1 \\ 1 & \theta \end{bmatrix}.$$

If we put $K(z) := \left[\begin{smallmatrix} 1 & \theta \\ \theta & 1 \end{smallmatrix}\right]$, then a straightforward calculation shows that $K \in \mathcal{C}(\Phi)$. We can also see that

$$(T_A)_{\theta_0\theta_1} = \begin{bmatrix} 0 & 1 \\ 1 & 0 \end{bmatrix} \quad \text{is invertible.}$$

But since $K(M) = \left[\begin{smallmatrix} 1 & \theta(M) \\ \theta(M) & 1 \end{smallmatrix}\right]$, it follows that

$$I - K(M)^*K(M) = -\begin{bmatrix} \theta(M)^*\theta(M) & \theta(M) + \theta(M)^* \\ \theta(M) + \theta(M)^* & \theta(M)^*\theta(M) \end{bmatrix},$$

which is not positive (simply by looking at the upper-left entry). It therefore follows from (45) that T_Φ is not hyponormal. □

We next extend the representation (43) to $\overline{H^\infty_{M_n}} + H^\infty_{M_n}$ (where $\overline{H^\infty_{M_n}}$ denotes the set of $n \times n$ matrix functions whose entries belong to $\overline{H^\infty} := \{g : \overline{g} \in H^\infty\}$). Let $Q \in \overline{H^\infty_{M_n}} + H^\infty_{M_n}$ be of the form $Q = Q_-^* + Q_+$. If $\Theta := I_\theta$ for an inner function θ, then we define

$$(T_Q)_\Theta := P_{\mathcal{H}(\Theta)} T_Q|_{\mathcal{H}(\Theta)}.$$

Then

$$(T_Q)_\Theta = (T_{Q_-^*})_\Theta + (T_{Q_+})_\Theta = (T_{Q_-})_\Theta^* + (T_{Q_+})_\Theta.$$

If $M := V^*\mathbf{S}_\theta V$, where $V : L \equiv L^2(\mu_B) \times L^2(\mu_s) \times L^2(\mu_\Delta) \to \mathcal{H}(\theta)$ is unitary as in (41), we also define $Q(M)$ by

$$Q(M) := (Q_-(M))^* + Q_+(M), \tag{46}$$

where $Q_\pm(M)$ is defined by the Sz.-Nagy-Foiaş functional calculus.

We then have [15, Lemma 6.8] the following:

Theorem 3.17 *Let $Q \in \overline{H^\infty_{M_n}} + H^\infty_{M_n}$ and $\Theta := I_\theta$ for an inner function θ. Then $\mathcal{V}^*(T_Q)_\Theta \mathcal{V} = Q(M)$ and $Q(M)^* = Q^*(M)$, where $\mathcal{V}$ and M are given by* (42).

Remark 3.18 We are tempted to guess that $Q(M)^*Q(M) = (Q^*Q)(M)$. But this is not the case. To see this, let $\theta = z^3$ and let $Q(z) := \begin{bmatrix} z^2 & 0 \\ 0 & z^2 \end{bmatrix}$. Then we have

$$M = \begin{bmatrix} 0&0&0 \\ 1&0&0 \\ 0&1&0 \end{bmatrix}$$

and

$$Q(M) \equiv \begin{bmatrix} 1&0 \\ 0&1 \end{bmatrix} \otimes M^2 = \begin{bmatrix} 0&0&0&0&0&0 \\ 0&0&0&0&0&0 \\ 0&0&0&0&0&0 \\ 0&0&0&0&0&0 \\ 1&0&0&0&0&0 \\ 0&1&0&0&0&0 \end{bmatrix}.$$

Thus

$$Q(M)^*Q(M) = \begin{bmatrix} 1&0&0&0&0&0 \\ 0&1&0&0&0&0 \\ 0&0&0&0&0&0 \\ 0&0&0&0&0&0 \\ 0&0&0&0&0&0 \\ 0&0&0&0&0&0 \end{bmatrix} \quad \text{and} \quad (Q^*Q)(M) = \begin{bmatrix} 1&0&0&0&0&0 \\ 0&1&0&0&0&0 \\ 0&0&1&0&0&0 \\ 0&0&0&1&0&0 \\ 0&0&0&0&1&0 \\ 0&0&0&0&0&1 \end{bmatrix},$$

which gives $Q^*(M)Q(M) \neq (Q^*Q)(M)$.

4 Operator-Valued Function Theory

In this section we consider operator-valued function theory related with Hankel, Toeplitz, and model operators. We focus on meromorphic pseudo-continuation of bounded type, a canonical decomposition of strong L^2-functions, spectra of model operators, an operator-valued version of Potapov's matrix-valued factorization theorem, and hyponormality and subnormality of Toeplitz operators with operator-valued symbols.

4.1 Meromorphic Pseudo-continuations of Bounded Type

We introduce the notion of "bounded type" for strong L^2-functions. Recall that a matrix-valued function of bounded type was defined by a matrix whose entries are of bounded type. But this definition is not appropriate for operator-valued functions, in particular strong L^2-functions, even though the terminology of matrix "entry" can be properly interpreted. Thus we need a new idea for a suitable definition of "bounded type" strong L^2-function, which is equivalent to the condition that each entry is of bounded type when the function is matrix-valued. Our motivation stems from the equivalence in (18) for the case of matrix-valued functions.

Definition 4.1 A function $\Phi \in L^2_s(\mathcal{B}(D, E))$ with values in $\mathcal{B}(D, E)$ is said to be of *bounded type* if $\ker H^*_{\Phi} = \Theta H^2_E$ for some two-sided inner function Θ with values in $\mathcal{B}(E)$.

It is known that if Δ is an inner function with values in $\mathcal{B}(D, E)$, then (cf. [16, Corollary 2.25])

$$\breve{\Delta} \text{ is of bounded type } \iff [\Delta, \Delta_c] \text{ is two-sided inner.} \tag{47}$$

On the other hand, in general, if a strong L^2-function Φ is of bounded type, then we cannot guarantee that each entry $\phi_{ij} \equiv \langle \Phi d_j,\ e_i \rangle$ is of bounded type, where $\{d_j\}$ and $\{e_i\}$ are orthonormal bases of D and E, respectively. We may ask:

Question 4.2 Under what conditions can we guarantee that each entry of Φ is of bounded type?

To examine this question, given a function

$$\Psi : \mathbb{D}^e \equiv \{z : 1 < |z| \le \infty\} \to \mathcal{B}(D, E),$$

we define $\Psi_{\mathbb{D}} : \mathbb{D} \to \mathcal{B}(E, D)$ by $\Psi_{\mathbb{D}}(\zeta) := \Psi^*(1/\overline{\zeta})$ for $\zeta \in \mathbb{D}$. If $\Psi_{\mathbb{D}}$ is a strong H^2-function, inner, and two-sided inner with values in $\mathcal{B}(E, D)$, then we shall say that Ψ is a strong H^2-function, inner, and two-sided inner in $\mathbb{D}^e$ with values in $\mathcal{B}(D, E)$, respectively. A $\mathcal{B}(D, E)$-valued function Ψ is said to be *meromorphic of bounded type* in $\mathbb{D}^e$ if it can be represented by $\Psi = \frac{G}{\theta}$, where G is a strong H^2-function in $\mathbb{D}^e$, with values in $\mathcal{B}(D, E)$, and θ is a scalar inner function in $\mathbb{D}^e$ (cf. [26, 27]). A function $\Phi \in L^2_s(\mathcal{B}(D, E))$ is said to have a *meromorphic pseudo-continuation* $\hat{\Phi}$ *of bounded type* in $\mathbb{D}^e$ if $\hat{\Phi}$ is meromorphic of bounded type in $\mathbb{D}^e$

and Φ is the nontangential SOT limit of $\hat{\Phi}$, that is, for all $x \in D$,

$$\Phi(z)x = \hat{\Phi}(z)x := \lim_{rz \to z} \hat{\Phi}(rz)x \quad \text{for almost all } z \in \mathbb{T}.$$

Note that for almost all $z \in \mathbb{T}$,

$$\Phi(z)x = \lim_{rz \to z} \hat{\Phi}(rz)x = \lim_{rz \to z} \hat{\Phi}^*_{\mathbb{D}}(r^{-1}z)x = \hat{\Phi}^*_{\mathbb{D}}(z)x \quad (x \in D).$$

The following proposition was proved in [26] under the more restrictive setting of $H^\infty(\mathcal{B}(D, E))$.

Proposition 4.3 ([16, Lemma 2.28 and Corollary 2.29]) *Let $\Phi \in L^\infty(\mathcal{B}(D, E)) \cup L^2_{\mathcal{B}(D,E)}$. Then the following are equivalent:*

(a) Φ has a meromorphic pseudo-continuation of bounded type in $\mathbb{D}^e$.
(b) $\theta H^2_E \subseteq \ker H_{\Phi^*}$ for some scalar inner function θ.

The following proposition gives an answer (for $L^2_{\mathcal{B}(D,E)}$) to Question 4.2 (cf. [16, Proposition 2.30]).

Proposition 4.4 *Let D and E be separable complex Hilbert spaces and let $\{d_j\}$ and $\{e_i\}$ be orthonormal bases of D and E, respectively. If $\Phi \in L^2_{\mathcal{B}(D,E)}$ has a meromorphic pseudo-continuation of bounded type in $\mathbb{D}^e$, then $\breve{\phi}_{ij}(z) \equiv \langle \breve{\Phi}(z)d_j, e_i \rangle_E$ is of bounded type for each i, j.*

Remark 4.5 For a function $\Phi \in L^2_s(\mathcal{B}(D, E))$, we can show (cf. [16, Lemma 2.27]) that if Φ has a meromorphic pseudo-continuation of bounded type in $\mathbb{D}^e$, then $\breve{\Phi}$ is of bounded type and that the converse is also true when Φ is matrix-valued. However, the converse is not true in general. To see this, let $\{\alpha_n\}$ be a sequence of distinct points in $\mathbb{D}$ such that $\sum_{n=1}^\infty (1 - |\alpha_n|) = \infty$ and put $\Delta := \text{diag}(b_{\alpha_n})$, where $b_{\alpha_n}(z) := \frac{z-\alpha_n}{1-\overline{\alpha}_n z}$. Then Δ is two-sided inner, and hence $\breve{\Delta}$ is of bounded type. On the other hand, by Lemma 2.3, $\ker H_{\Delta^*} = \Delta H^2_{\ell^2}$. Thus if Δ had a meromorphic pseudo-continuation of bounded type in $\mathbb{D}^e$, then by Proposition 4.3, we would have $\theta H^2_{\ell^2} \subseteq \Delta H^2_{\ell^2}$ for a scalar inner function θ, so that we should have $\theta(\alpha_n) = 0$ for each $n = 1, 2, \cdots$, and hence $\theta = 0$, a contradiction. Therefore, Δ cannot have a meromorphic pseudo-continuation of bounded type in $\mathbb{D}^e$.

Remark 4.6 By contrast to the matrix-valued case, it may happen that an L^∞-function Φ is not of bounded type in the sense of Definition 4.1 even though each entry ϕ_{ij} of Φ is of bounded type. To see this, let $\{\alpha_j\}$ be a sequence of distinct points in $(0, 1)$ satisfying $\sum_{j=1}^\infty (1-\alpha_j) < \infty$. For each $j \in \mathbb{Z}_+$, choose a sequence $\{\alpha_{ij}\}$ of distinct points on the circle $C_j := \{z \in \mathbb{C} : |z| = \alpha_j\}$. Let

$$B_{ij} := \frac{\overline{b}_{\alpha_{ij}}}{(i+j)!} \quad (i, j \in \mathbb{Z}_+),$$

where $b_\alpha(z) := \frac{z-\alpha}{1-\overline{\alpha}z}$, and let

$$\Phi := [B_{ij}] = \begin{bmatrix} \frac{\overline{b}_{\alpha_{11}}}{2!} & \frac{\overline{b}_{\alpha_{12}}}{3!} & \frac{\overline{b}_{\alpha_{13}}}{4!} & \cdots \\ \frac{\overline{b}_{\alpha_{21}}}{3!} & \frac{\overline{b}_{\alpha_{22}}}{4!} & \frac{\overline{b}_{\alpha_{23}}}{5!} & \cdots \\ \frac{\overline{b}_{\alpha_{31}}}{4!} & \frac{\overline{b}_{\alpha_{32}}}{5!} & \frac{\overline{b}_{\alpha_{33}}}{6!} & \cdots \\ \vdots & \vdots & \vdots & \end{bmatrix}.$$

Observe that

$$\sum_{i,j} |B_{ij}(z)|^2 = \sum_i \frac{i}{((1+i)!)^2} \leq \sum_i \frac{1}{(1+i)^2} < \infty,$$

which implies that $\Phi \in L^\infty(\mathcal{B}(\ell^2))$. For a function $f \in H^2_{\ell^2}$, we write $f = (f_1, f_2, f_3, \cdots)^t$ $(f_n \in H^2)$. Thus if $f = (f_1, f_2, f_3, \cdots)^t \in \ker H_\Phi$, then $\sum_j \frac{\overline{b}_{\alpha_{ij}}}{(i+j)!} f_j \in H^2$ for each $i \in \mathbb{Z}_+$, which forces that $f_j(\alpha_{ij}) = 0$ for each i, j. Thus $f_j = 0$ for each j (by the Identity Theorem). Therefore we can conclude that $\ker H^*_{\widetilde{\Phi}} = \{0\}$, so that $\widetilde{\Phi}$ is not of bounded type. But we note that every entry of $\widetilde{\Phi}$ is of bounded type.

We next consider an application to C_0-contractions.

The class C_{00} denotes the set of all contractions $T \in \mathcal{B}(\mathcal{H})$ such that $\lim_{n\to\infty} T^n x = 0$ and $\lim_{n\to\infty} T^{*n} x = 0$ for each $x \in \mathcal{H}$. It is known [44, p.43] that if T is a C_0.-contraction with characteristic function Δ (i.e., $T \cong S_E^*|_{\mathcal{H}(\Delta)}$), then

$$T \in C_{00} \iff \Delta \text{ is } \textit{two-sided} \text{ inner}. \tag{48}$$

A contraction $T \in \mathcal{B}(\mathcal{H})$ is called a completely non-unitary (c.n.u.) if there exists no nontrivial reducing subspace on which T is unitary. The class C_0 is the set of all c.n.u. contractions T such that there exists a nonzero function $\varphi \in H^\infty$ annihilating T, i.e., $\varphi(T) = 0$, where $\varphi(T)$ is given by the Sz.-Nagy-Foiaş functional calculus. We can easily check that $C_0 \subseteq C_{00}$. Moreover, it is well known [44, p.73] that if $T := P_{\mathcal{H}(\Delta)} S_E|_{\mathcal{H}(\Delta)} \in C_{00}$ and $\varphi \in H^\infty$, then

$$\varphi(T) = 0 \iff \exists\, G \in H^\infty(\mathcal{B}(E)) \text{ such that } G\Delta = \Delta G = \varphi I_E. \tag{49}$$

The theory of spectral multiplicity for operators of class C_0 has been well developed (see [44, Appendix 1], [51]). If $T \in C_0$, then there exists an inner function m_T such that $m_T(T) = 0$ and

$$\varphi \in H^\infty,\ \varphi(T) = 0 \implies \varphi/m_T \in H^\infty.$$

The function m_T is called the *minimal annihilator* of the operator T.

In view of (48), we may ask: What is a condition on the characteristic function Δ of T for a $C_0.$-contraction T to belong to the class C_0? An answer to this question was given in [16, Proposition 2.34].

Theorem 4.7 *Let $T := S_E^*|_{\mathcal{H}(\Delta)}$ for an inner function Δ with values in $\mathcal{B}(D, E)$. Then $T \in C_0$ if and only if Δ is two-sided inner and has a meromorphic pseudo-continuation of bounded type in $\mathbb{D}^e$. Hence, in particular, if Δ is an inner matrix function, then $T \in C_0$ if and only if $T \in C_{00}$.*

4.2 A Canonical Decomposition of Strong L^2-Functions

There exists a matrix-valued L^2-function that does not admit a DSS factorization. To see this, assume that θ_1 and θ_2 are coprime inner functions. Consider

$$\Phi := \begin{bmatrix} \theta_1 & 0 & 0 \\ 0 & \theta_2 & 0 \\ 0 & 0 & a \end{bmatrix} \equiv [\phi_1, \phi_2, \phi_3] \in H^\infty_{M_3},$$

where $a \in H^\infty$ is such that $\overline{a}$ is not of bounded type. Then a direct calculation shows that

$$\ker H_{\Phi^*} = \begin{bmatrix} \theta_1 & 0 \\ 0 & \theta_2 \\ 0 & 0 \end{bmatrix} H^2_{\mathbb{C}^2} \equiv \Delta H^2_{\mathbb{C}^2}.$$

Since Δ is not two-sided inner, it follows from Lemma 2.1 that Φ does not admit a DSS factorization. For a decomposition of Φ, suppose that $\Phi = \Omega A^*$, where $\Omega, A \in H^2_{M_{3\times k}}$ $(k = 1, 2)$, Ω is an inner function, and Ω and A are right coprime. We then have $\Phi^*\Omega = A \in H^2_{M_{3\times k}}$. But since $\overline{a}$ is not of bounded type, it follows that the third row vector of Ω is zero. Thus we must have $a = 0$, a contradiction. Therefore we could not get any decomposition of the form $\Phi = \Omega A^*$ with a $3 \times k$ inner matrix function Ω for each $k = 1, 2, 3$. To get to another idea, we note that $\ker \Delta^* = [0\ 0\ 1]^t H^2 \equiv \Delta_c H^2$. Then by a direct manipulation, we can get

$$\Phi = \begin{bmatrix} \theta_1 & 0 & 0 \\ 0 & \theta_2 & 0 \\ 0 & 0 & a \end{bmatrix} = \begin{bmatrix} \theta_1 & 0 \\ 0 & \theta_2 \\ 0 & 0 \end{bmatrix} \begin{bmatrix} 1 & 0 \\ 0 & 1 \\ 0 & 0 \end{bmatrix}^* + \begin{bmatrix} 0 \\ 0 \\ 1 \end{bmatrix} \begin{bmatrix} 0 & 0 & a \end{bmatrix} \equiv \Delta A^* + \Delta_c C \tag{50}$$

where Δ and A are right coprime because $\widetilde{\Delta} H^2_{\mathbb{C}^3} \bigvee \widetilde{A} H^2_{\mathbb{C}^3} = H^2_{\mathbb{C}^2}$. The example (50) seems to signal that the decomposition of matrix-valued H^2-functions Φ satisfying $\ker H^*_{\breve{\Phi}} = \Delta H^2_{\mathbb{C}^n}$ may be affected by the kernel of Δ^* and, in turn, the complementary factor Δ_c of Δ. The following theorem gives a canonical decomposition of strong L^2-functions which realizes the idea inside the above example [16, Theorem 3.1].

Theorem 4.8 (A canonical decomposition of strong L^2-functions) *If Φ is a strong L^2-function with values in $\mathcal{B}(D, E)$, then Φ can be expressed in the form*

$$\Phi = \Delta A^* + B, \tag{51}$$

where:

(i) *Δ is an inner function with values in $\mathcal{B}(E', E)$, $\widetilde{A} \in H_s^2(\mathcal{B}(D, E'))$, and $B \in L_s^2(\mathcal{B}(D, E))$*
(ii) *Δ and A are right coprime*
(iii) *$\Delta^* B = 0$*
(iv) *$\text{nc}\{\Phi_+\} \le \dim E'$*

In particular, if $\dim E' < \infty$ (for instance, if $\dim E < \infty$), then the expression (51) *is unique (up to a unitary constant right factor).*

If $\breve{\Phi}$ is of bounded type, then Δ should be two-sided, and hence $B = 0$ in (51). Thus if $\breve{\Phi}$ is of bounded type, the decomposition (51) reduces to the DSS factorization. Since the decomposition in (51) is unique (except when $\dim E' = \infty$), we call (51) the *canonical* decomposition of Φ. On the other hand, we can show (cf. [16, Proof of Theorem 4.16]) that the inner function Δ in (51) associated with a strong L^2-function Φ can be obtained from the equation $\ker H^*_{\breve{\Phi}} = \Delta H^2_{E'}$ which is guaranteed by the Beurling-Lax-Halmos Theorem. In this case, the expression (51) will be called the *BLH-canonical decomposition* of Φ, from the viewpoint that Δ comes from the Beurling-Lax-Halmos Theorem. However, if $\dim E' = \infty$ (even though $\dim D < \infty$), then it is possible to get another inner function Θ of a canonical decomposition (51) for the same function: in this case, $\ker H^*_{\breve{\Phi}} \neq \Theta H^2_{E''}$. Indeed, we can show that the canonical decomposition (51) is not unique in general. Indeed, if $\dim E' = \infty$ (even though $\dim D < \infty$), the canonical decomposition (51) may not be unique even if $\breve{\Phi}$ is of bounded type. To see this, let Φ be an inner function with values in $\mathcal{B}(\mathbb{C}^2, \ell^2)$ defined by

$$\Phi := \begin{bmatrix} \theta_1 & 0 \\ 0 & 0 \\ 0 & \theta_2 \\ 0 & 0 \\ 0 & 0 \\ 0 & 0 \\ \vdots & \vdots \end{bmatrix},$$

where θ_1 and θ_2 are scalar inner functions. Then

$$\ker H^*_{\breve{\Phi}} = \ker H_{\Phi^*} = \text{diag}(\theta_1, 1, \theta_2, 1, 1, 1, \cdots)H^2_{\ell^2} \equiv \Theta H^2_{\ell^2},$$

which implies that $\breve{\Phi}$ is of bounded type since Θ is two-sided inner (see Definition 4.1). Let

$$A := \Phi^*\Theta = \begin{bmatrix} 1\,0\,0\,0\,0\,0\,0\,\cdots \\ 0\,0\,1\,0\,0\,0\,0\,\cdots \end{bmatrix} \quad \text{and} \quad B := 0.$$

Then $\widetilde{A}$ belongs to $H^2_s(\mathcal{B}(\mathbb{C}^2, \ell^2))$ and $\widetilde{\Theta}H^2_{\ell^2} \bigvee \widetilde{A}H^2_{\mathbb{C}^2} = H^2_{\ell^2}$, which implies that Θ and A are right coprime. Clearly, $\Theta^* B = 0$ and $\text{nc}\{\Phi_+\} \leq \dim \ell^2 = \infty$. Therefore, $\Phi = \Theta A^*$ is the BLH-canonical decomposition of Φ. On the other hand, to get another canonical decomposition of Φ, let

$$\Delta := \begin{bmatrix} \theta_1 & 0 & 0\,0\,0\,0\,0\,0 & \cdots \\ 0 & 0 & 0\,0\,0\,0\,0\,0 & \cdots \\ 0 & \theta_2 & 0\,0\,0\,0\,0 & \cdots \\ 0 & 0 & 1\,0\,0\,0\,0 & \cdots \\ 0 & 0 & 0\,1\,0\,0\,0 & \cdots \\ 0 & 0 & 0\,0\,1\,0\,0 & \cdots \\ \vdots & \vdots & \vdots\,\vdots\,\vdots\,\vdots\,\vdots & \ddots \end{bmatrix}.$$

Then Δ is an inner function. If we define

$$A_1 := \begin{bmatrix} 1\,0\,0\,0\,0\,0\,0\,\cdots \\ 0\,1\,0\,0\,0\,0\,0\,\cdots \end{bmatrix} \quad \text{and} \quad B := 0,$$

then $\widetilde{A}_1$ belongs to $H^2_s(B(\mathbb{C}^2, \ell^2))$ such that Δ and A_1 are right coprime, $\Delta^* B = 0$, and $\text{nc}\{\Phi_+\} \leq \dim \ell^2 = \infty$. Therefore $\Phi = \Delta A_1^*$ is also a canonical decomposition of Φ. In this case, $\ker H^*_{\breve{\Phi}} \neq \Delta H^2_{\ell^2}$. Therefore, the canonical decomposition of Φ is not unique.

We are also interested in the following problem: *Given an inner function* $\Delta \in \mathcal{B}(E', E)$, *with* $\dim E' < \infty$, *describe the set* F *in* H^2_E *such that* $\mathcal{H}(\Delta) = E^*_F$. As a corollary of Theorem 4.8, we get an answer to the problem [16, Corollary 4.9]:

Corollary 4.9 *Suppose* Δ *is an inner function with values in* $\mathcal{B}(E', E)$, *with* $\dim E' < \infty$. *If* $\Phi = \Delta A^* + B$ *is a canonical decomposition of* Φ *in* $L^2_s(\mathcal{B}(D, E))$, *we define a function* F *by* $F(z) := \overline{z}\big(\Phi_+(z) - \hat{\Phi}(0)\big)$. *We then have* $E^*_{\{F\}} = \mathcal{H}(\Delta)$.

4.3 Spectra of Model Operators

If θ is a scalar inner function, then the spectrum, $\sigma(\theta)$, of θ is defined by the complement (in $\mathrm{cl}\,\mathbb{D}$) of the set of all points $\lambda \in \mathrm{cl}\,\mathbb{D}$, such that the function $\frac{1}{\theta}$ can be continued analytically into a (full) neighborhood of λ. We recall that if θ is a scalar inner function, then we may write

$$\theta(\zeta) = B(\zeta)\exp\left(-\int_{\mathbb{T}} \frac{z+\zeta}{z-\zeta} d\mu(z)\right),$$

where B is a Blaschke product and μ is a singular measure on $\mathbb{T}$. It is known that the spectrum $\sigma(\theta)$ of θ is given as

$$\sigma(\theta) = \mathrm{cl}\,\theta^{-1}(0) \bigcup \mathrm{supp}\,\mu \qquad (cf.[44, \text{p.63}]). \tag{52}$$

It is also known that if $\mathbf{S} := P_{\mathcal{H}(\Delta)} S_E|_{\mathcal{H}(\Delta)} \in C_0$, then (cf. [44, p.72])

$$\sigma(\mathbf{S}) = \sigma(m_{\mathbf{S}}), \tag{53}$$

where $m_{\mathbf{S}}$ is the minimal annihilator of $\mathbf{S}$.

On the other hand, if Δ is a two-sided inner function with values in $\mathcal{B}(E)$, then the *spectrum*, denoted by $\sigma(\Delta)$, of Δ is defined as the complement (in $\mathrm{cl}\,\mathbb{D}$) of the set of values $\lambda \in \mathrm{cl}\,\mathbb{D}$ for which $\Delta(\zeta)^{-1}$ exists in a certain relative neighborhood of λ and admits an analytic continuation to a (full) neighborhood of λ. By the operator-valued version of the Livšic-Moeller Theorem [44, p.75], we know that if Δ is two-sided inner and $T := S_E^*|_{\mathcal{H}(\Delta)}$, then

$$\sigma(T) = \sigma(\widetilde{\Delta}). \tag{54}$$

Now, we may ask, what is the spectrum of the model operator $S_E^*|_{\mathcal{H}(\Delta)}$ (for a one-sided inner function Δ)? Here is a partial answer.

Proposition 4.10 *Let $T := S_E^*|_{\mathcal{H}(\Delta)}$ for an inner function Δ with values in $\mathcal{B}(D, E)$. If $\breve{\Delta}$ is of bounded type, then*

$$\sigma(T) \subseteq \sigma(\widetilde{[\Delta, \Delta_c]}) \bigcup \overline{\sigma(S_E|_{\Delta_c H^2_{D'}})}, \tag{55}$$

where Δ_c is the complementary factor of Δ, with values in $\mathcal{B}(D', E)$. Moreover, if Δ has a meromorphic pseudo-continuation of bounded type in $\mathbb{D}^e$, then

$$\sigma(\widetilde{[\Delta, \Delta_c]}) \subseteq \sigma(T). \tag{56}$$

Proof Suppose first that $\breve{\Delta}$ is of bounded type. Then by (47), $[\Delta, \Delta_c]$ is two-sided inner. Observe that

$$[\Delta, \Delta_c]H^2_{D\oplus D'} = \Delta H^2_D \oplus \Delta_c H^2_{D'},$$

and hence

$$\mathcal{H}(\Delta) = \mathcal{H}([\Delta, \Delta_c]) \oplus \Delta_c H^2_{D'}.$$

Thus we may write

$$T = \begin{bmatrix} T_1 & * \\ 0 & T_2 \end{bmatrix} : \begin{bmatrix} \mathcal{H}([\Delta, \Delta_c]) \\ \Delta_c H^2_{D'} \end{bmatrix} \to \begin{bmatrix} \mathcal{H}([\Delta, \Delta_c]) \\ \Delta_c H^2_{D'} \end{bmatrix}. \tag{57}$$

Note that $T_1 = S_E^*|_{\mathcal{H}([\Delta,\Delta_c])}$. Thus by (54), $\sigma(T_1) = \sigma(\widetilde{[\Delta, \Delta_c]})$. Also, since $T_2 = P_{\Delta_c H^2_{D'}} S_E^*|_{\Delta_c H^2_{D'}}$, it follows that $\sigma(T_2) = \overline{\sigma(T_2^*)} = \overline{\sigma(S_E|_{\Delta_c H^2_{D'}})}$. Since $\sigma(T) \subseteq \sigma(T_1) \cup \sigma(T_2)$, the assertion (55) follows at once.

Suppose next that Δ has a meromorphic pseudo-continuation of bounded type in $\mathbb{D}^e$. Then $\breve{\Delta}$ is of bounded type, so that again by (47), $[\Delta, \Delta_c]$ is two-sided inner. Observe, by (12),

$$\ker H_{\Delta^*} = [\Delta, \Delta_c]H^2_{D\oplus D'} = \ker H_{[\Delta,\Delta_c]^*}.$$

Thus by Proposition 4.3, $[\Delta, \Delta_c]$ has a meromorphic pseudo-continuation of bounded type in $\mathbb{D}^e$. Then by Theorem 4.7, $T_1 \equiv S_E^*|_{\mathcal{H}([\Delta,\Delta_c])}$ belongs to the class C_0. By the Model Theorem, T_1 is equivalent to $D \equiv P_{\mathcal{H}(\widetilde{[\Delta,\Delta_c]})} S_E|_{\mathcal{H}(\widetilde{[\Delta,\Delta_c]})}$ (cf. [44, p.75]). Thus by (53),

$$\sigma(T_1) = \sigma(m_D) = \operatorname{cl} m_D^{-1}(0) \bigcup \operatorname{supp} \mu_D$$

(μ_D is a singular measure corresponding to m_D), which has no interior points. Thus $\sigma(T_1) \cap \sigma(T_2)$ has no interior points. Thus we have $\sigma(T) = \sigma(T_1) \cup \sigma(T_2)$ because in the Banach space setting, the passage from $\sigma\left(\begin{smallmatrix} A & C \\ 0 & B \end{smallmatrix}\right)$ to $\sigma(A) \cup \sigma(B)$ consists of filling in certain holes in $\sigma\left(\begin{smallmatrix} A & C \\ 0 & B \end{smallmatrix}\right)$, occurring in $\sigma(A) \cap \sigma(B)$ (cf. [36]). Therefore, (55) holds with equality in place of inclusion, which gives (56). This completes the proof. □

We would like to pose the following:

Problem 4.11 If $T := S_E^*|_{\mathcal{H}(\Delta)}$ for an inner function Δ with values in $\mathcal{B}(D, E)$ such that $\breve{\Delta}$ is of bounded type, describe the spectrum of T in terms of Δ.

4.4 An Extension of Potapov's Factorization Theorem

In 1955, V.P. Potapov [48] showed that every rational inner $n \times n$ matrix-valued function can be written as a finite Blaschke-Potapov product. We extend this result to the case of operator-valued functions.

Note that if V is a unitary operator in $B(E)$, then for every $\Phi \in H^\infty(\mathcal{B}(E))$,

$$\Phi = V(V^*\Phi) = (\Phi V^*)V,$$

which implies that V is an inner divisor of Φ. If M is a nonzero closed subspace of a Hilbert space E, then a function of the form

$$b_\alpha P_M + (I_E - P_M) \quad (\text{where } b_\alpha(z) := \frac{z-\alpha}{1-\overline{\alpha}z})$$

is called a *(operator-valued) Blaschke-Potapov factor*, where P_M is the orthogonal projection of E onto M. A function D is called a *(operator-valued) finite Blaschke-Potapov product* if D is of the form

$$D = V \prod_{m=1}^{M} \Big(b_m P_m + (I - P_m) \Big),$$

where V is a unitary operator, b_m is a Blaschke factor, and P_m is a nonzero orthogonal projection in E for each $m = 1, \cdots, M$. In particular, a scalar-valued function D reduces to a finite Blaschke product $D = \nu \prod_{m=1}^{M} b_m$, where $\nu = e^{i\omega}$. It is known (cf. [48]) that an $n \times n$ matrix function D is rational and inner if and only if it can be represented as a finite Blaschke-Potapov product.

To proceed, we consider the following question: What is a left inner divisor of zI_n? For this question, we may guess that each left inner divisor of zI_n is a Blaschke-Potapov factor. More specifically, we wonder if a left inner divisor of $\left[\begin{smallmatrix} z & 0 \\ 0 & z \end{smallmatrix}\right] \equiv zI_2$ should be of the following form up to a unitary constant right factor (also up to unitary equivalence):

$$\begin{bmatrix} 1 & 0 \\ 0 & 1 \end{bmatrix}, \quad \begin{bmatrix} 1 & 0 \\ 0 & z \end{bmatrix}, \quad \begin{bmatrix} z & 0 \\ 0 & 1 \end{bmatrix} \quad \text{or} \quad \begin{bmatrix} z & 0 \\ 0 & z \end{bmatrix}.$$

For example, $A \equiv \frac{1}{\sqrt{2}} \left[\begin{smallmatrix} 1 & -z \\ 1 & z \end{smallmatrix}\right]$ is a left inner divisor of $\left[\begin{smallmatrix} z & 0 \\ 0 & z \end{smallmatrix}\right] \equiv zI_2$: indeed,

$$A \cdot \frac{1}{\sqrt{2}} \begin{bmatrix} z & z \\ -1 & 1 \end{bmatrix} = \begin{bmatrix} z & 0 \\ 0 & z \end{bmatrix}.$$

In this case, if we take a unitary operator $V := \frac{1}{\sqrt{2}} \left[\begin{smallmatrix} 1 & 1 \\ -1 & 1 \end{smallmatrix}\right]$, then

$$\begin{bmatrix} 1 & 0 \\ 0 & z \end{bmatrix} = V \cdot \frac{1}{\sqrt{2}} \begin{bmatrix} 1 & -z \\ 1 & z \end{bmatrix} = \left[V \cdot \frac{1}{\sqrt{2}} \begin{bmatrix} 1 & -z \\ 1 & z \end{bmatrix} \cdot V^* \right] \cdot V.$$

In fact, it was shown in [13, Lemma 2.5] that

$$\text{every left inner divisor of } zI_n \in H^\infty_{M_n} \text{ is a Blaschke-Potapov factor.} \tag{58}$$

This fact is useful for the study of coprime-ness of functions (cf. [14]). In [15, p.23], the authors asked:

Question 4.12 Is the statement in (58) still true for operator-valued functions?

We will call a function of the form zI_E the operator-valued coordinate function. This allows us to rephrase Question 4.12 as follows: Is every left inner divisor of the operator-valued coordinate function a Blaschke-Potapov factor? For Question 4.12, we wish to study a more general case, and we first observe that if A is an inner divisor of $z^N I_E$, then there exists a function $\Omega \in H^\infty(\mathcal{B}(E))$ such that $A\Omega = \Omega A = z^N I_E$ a.e. on $\mathbb{T}$, so that $A^* z^N I_E = \Omega \in H^\infty(\mathcal{B}(E))$, which implies that A is a polynomial of degree at most N.

In [17, Theorem 2.1], inner divisors of $z^N I_E$ were characterized:

Theorem 4.13 *Let A be a polynomial of degree N. Then A is an inner divisor of $z^N I_E$ if and only if A is a finite Blaschke-Potapov product of the form*

$$A(z) = V \prod_{m=1}^{N} \Big(zP_m + (I - P_m) \Big),$$

where P_m is the orthogonal projection from E onto $\bigoplus_{n=m}^{N} \operatorname{ran} \widehat{A}(n)^*$, and

$$V := \operatorname{diag}\Big(\widehat{A}(0)|_{\operatorname{ran} \widehat{A}(0)^*}, \widehat{A}(1)|_{\operatorname{ran} \widehat{A}(1)^*}, \cdots, \widehat{A}(N)|_{\operatorname{ran} \widehat{A}(N)^*} \Big).$$

The following corollary gives an affirmative answer to Question 4.12.

Corollary 4.14 *If $\Delta \in H^\infty(\mathcal{B}(E))$ is a left inner divisor of zI_E, then Δ is a Blaschke-Potapov factor.*

For $\Phi \in H^\infty(\mathcal{B}(E))$ and $\alpha \in \mathbb{D}$, write

$$\Phi_\alpha := \Phi \circ b_\alpha.$$

Then we can easily check the following:

(a) $\Phi_\alpha \in H^\infty(\mathcal{B}(E))$.
(b) If Δ is an inner function with values in $\mathcal{B}(E)$, then so is Δ_α.

Then the following corollary follows from Theorem 4.13 at once.

Corollary 4.15 *Let $A \in H^\infty(\mathcal{B}(E))$. Then A is an inner divisor of $b_\alpha^N I_E$ if and only if A is a finite Blaschke-Potapov product of the form*

$$A = V \prod_{m=1}^{N} \Big(b_\alpha P_m + (I - P_m) \Big),$$

where P_m and V are defined as in Theorem 4.13.

We now introduce the notion of operator-valued "rational" function. Recall that a matrix-valued function is rational if its entries are rational functions. But this definition is not appropriate for operator-valued functions, in particular H^∞-functions, even though the terminology of matrix "entry" may be properly interpreted. Thus, the new idea should capture and encapsulate a definition of operator-valued rational function which is equivalent to the condition that each entry is rational when the function is matrix-valued. In the sequel, we give a formal definition of operator-valued rational function.

To do so, we recall Definition 4.1 – the definition of bounded type: A function $\Phi \in L^2_s(B(D, E))$ is said to be of bounded type if $\ker H^*_\Phi = \Theta H^2_E$ for some two-sided inner function Θ with values in $\mathcal{B}(E)$.

We now introduce the following:

Definition 4.16 A function $\Phi \in H^\infty(\mathcal{B}(D, E))$ is said to be *rational* if

$$\theta H^2_E \subseteq \ker H_{\Phi^*} \tag{59}$$

for some finite Blaschke product θ.

Observe that if $\Phi \in H^\infty(\mathcal{B}(D, E))$, then

$$\Phi \text{ is rational} \Longrightarrow \breve{\Phi} \text{ is of bounded type.} \tag{60}$$

To see this, suppose Φ is rational. By definition and the Beurling-Lax-Halmos Theorem, there exist a finite Blaschke product θ and an inner function $\Delta \in H^\infty(\mathcal{B}(E', E))$ such that

$$\theta H^2_E \subseteq \ker H_{\Phi^*} = \Delta H^2_{E'},$$

which implies that Δ is a left inner divisor of θI_E. Thus Δ is two-sided inner, so that $\breve{\Phi}$ is of bounded type, which proves (60).

Also, if $\Phi \equiv (\phi_{ij}) \in H^\infty_{M_{m\times n}}$ is a rational function in the sense of Definition 4.16, then each entry ϕ_{ij} is rational. To see this suppose a matrix-valued function Φ satisfies the condition (59). Put $A := \theta \Phi^*$. Then $A \in H^\infty_{M_{n\times m}}$ and $\Phi = \theta A^*$. Thus ϕ_{ij} can be written as $\phi_{ij} = \theta \overline{a_{ij}}$ ($a_{ij} \in H^\infty$). Via Kronecker's Lemma [44, p.183], we can see that

$$\phi_{ij} \text{ is rational} \iff \phi_{ij} = \theta\overline{a_{ij}} \text{ with a finite Blaschke product } \theta,$$

which says that each ϕ_{ij} is rational.

In [17, Corollary 3.7], Potapov's matrix-valued factorization theorem was extended to operator-valued functions.

Theorem 4.17 *A two-sided inner function $\Phi \in H^\infty(\mathcal{B}(E))$ is rational if and only if it can be represented as a finite Blaschke-Potapov product.*

We now present a key difference between matrix-valued functions and operator-valued functions.

If Φ and Ψ are not left coprime, then there exists a common left inner divisor Δ of both Φ and Ψ such that Δ is not a unitary operator. However, this is not the case for right coprime-ness. As we know, right coprime-ness is defined via left coprime-ness. As a result, there exist examples of two functions with no common right inner divisor and which are not right coprime. Example 4.18 below shows such an instance. To see this, we first recall that if $\theta \in H^\infty$ is an inner function, then θ can be written as $\theta = cbs$, where c is a constant of modulus 1, b is the Blaschke product formed from the zeros of θ in $\mathbb{D}$, and s is a singular inner function given by

$$s(\zeta) = \exp\left(-\int_{-\pi}^{\pi} \frac{e^{it}+\zeta}{e^{it}-\zeta} d\mu(t)\right) \quad (\zeta \in \mathbb{D}), \tag{61}$$

where μ is a finite positive Borel measure on $\mathbb{T}$ which is singular with respect to Lebesgue measure. If s is the singular inner function given by (61), then it is well known that

$$\lim_{r\to 1_-} s(rz) = 0 \quad \text{for } \mu \text{ almost all } z \in \mathbb{T}. \tag{62}$$

We then have the following:

Example 4.18 Let f and g be defined as in Example 3.2. Put

$$\Delta := \begin{bmatrix} f \\ g \end{bmatrix} \quad (f, g \in H^\infty).$$

Then Δ is an inner function. Note that

$$\Delta(z) \neq 0 \quad \text{for all } z \in \mathbb{C}. \tag{63}$$

Put $\Phi := \widetilde{\Delta}$. Clearly, Δ and Φ are not right coprime. We claim that Δ and Φ do not have a common right inner divisor. Assume to the contrary that Φ has a right inner divisor Θ that is not a unitary matrix, i.e.,

$$\Phi = \Psi\Theta \quad (\Theta \text{ is inner and not a unitary matrix}).$$

Then $\Psi \in H^\infty_{M_{1\times 2}}$ and $\Theta \in H^\infty_{M_2}$, and hence Θ is two-sided inner. Thus it follows that Θ^* is of bounded type. Thus we can write

$$\Theta^* = \begin{bmatrix} \overline{\theta_1} b_1 & \overline{\theta_2} b_2 \\ \overline{\theta_3} b_3 & \overline{\theta_4} b_4 \end{bmatrix}, \quad \text{i.e.,} \quad \Theta = \begin{bmatrix} \theta_1 \overline{b_1} & \theta_3 \overline{b_3} \\ \theta_2 \overline{b_2} & \theta_4 \overline{b_4} \end{bmatrix},$$

where θ_i is inner and $b_i \in H^\infty$. Let θ be the least common multiple of $\{\theta_i : 1 \le i \le 4\}$. Then we can write

$$\Theta = \theta \begin{bmatrix} \overline{a_1} & \overline{a_3} \\ \overline{a_2} & \overline{a_4} \end{bmatrix} = \theta A^* \quad (\text{where } A \in H^\infty_{M_2}).$$

Thus $\Delta = \widetilde{\Theta}\widetilde{\Psi} = \widetilde{\theta}\breve{A}\widetilde{\Psi}$. If θ has a zero in $\mathbb{D}$, then so is Δ, which contradicts (63). If instead θ is a singular inner function of the form

$$\theta(\zeta) = \exp\left(- \int_{-\pi}^{\pi} \frac{e^{it} + \zeta}{e^{it} - \zeta} d\mu(t) \right),$$

then by (62), $\lim_{r \to 1_-} \Delta(rz) = 0$ for μ almost all $z \in \mathbb{T}$, which is a contradiction. □

4.5 Hyponormality and Subnormality of Toeplitz Operators

In the literature, many authors have considered the special cases of the following (scalar-valued or operator-valued) interpolation problem (cf. [9, 14, 15, 25, 28, 30, 31, 37–40, 43, 53]).

Problem 4.19 For $\Phi \in L^\infty(\mathcal{B}(E))$, when does there exist a function $K \in H^\infty(\mathcal{B}(E))$ with $||K||_\infty \le 1$ satisfying

$$\Phi - K\Phi^* \in H^\infty(\mathcal{B}(E))\,? \tag{64}$$

For notational convenience, we write, for $\Phi \in L^\infty(\mathcal{B}(E))$,

$$C(\Phi) := \Big\{ K \in H^\infty(\mathcal{B}(E)) : \ \Phi - K\Phi^* \in H^\infty(\mathcal{B}(E)) \Big\}.$$

We then have the following:

Theorem 4.20 ([16, Theorem C.19]) *Let $\Phi \equiv \breve{\Phi}_- + \Phi_+ \in L^\infty(\mathcal{B}(E))$. If $C(\Phi)$ is non-empty, then*

$$\ker H^*_{\breve{\Phi}_+} \subseteq \ker H^*_{\Phi^*_-}. \tag{65}$$

In particular, $\mathrm{nc}\{\Phi_+\} \le \mathrm{nc}\{\widetilde{\Phi_-}\}$.

If Φ is a matrix-valued rational function, then using Kronecker's Lemma we can show that the condition (65) is sufficient for $C(\Phi) \neq \emptyset$ [14, Proposition 3.9]. Moreover, in this case, the solution $K \in C(\Phi)$ is given by a polynomial via the classical Hermite-Fejér interpolation problem. However we were unable to determine whether the condition (65) is sufficient for the existence of a solution $K \in C(\Phi)$ when $\Phi \in L^\infty_{M_n}$ is such that Φ and Φ^* are of bounded type.

Problem 4.21 Let $\Phi \in L^\infty_{M_n}$ be such that Φ and Φ^* are of bounded type. If $\ker H_{\Phi_+^*} \subseteq \ker H_{\Phi_-^*}$, does there exist a solution $K \in H^\infty_{M_n}$ satisfying $\Phi - K\Phi^* \in H^\infty_{M_n}$?

For $\Phi \in L^\infty(\mathcal{B}(E))$, write

$$\mathcal{E}(\Phi) := \Big\{K \in H^\infty(\mathcal{B}(E)) : \Phi - K\Phi^* \in H^\infty(\mathcal{B}(E)) \text{ and } ||K||_\infty \le 1\Big\},$$

i.e., $\mathcal{E}(\Phi) = \{K \in C(\Phi) : ||K||_\infty \le 1\}$. If $\dim E = 1$ and $\Phi \equiv \varphi$ is a scalar-valued function, then an elegant theorem of C. Cowen (cf. [9, 18, 43]) says that $\mathcal{E}(\varphi)$ is non-empty if and only if T_φ is hyponormal, i.e., the self-commutator $[T_\varphi^*, T_\varphi]$ is positive semi-definite. Cowen's Theorem is to recast the operator-theoretic problem of hyponormality into the problem of finding a solution of an interpolation problem. In [31], it was shown that Cowen's theorem still holds for a Toeplitz operator T_Φ with a matrix-valued symbol $\Phi \in L^\infty_{M_n}$.

As we saw in Sect. 3.3, Halmos' Problem 5 has been partially answered in the affirmative by many authors. Despite considerable efforts, to date researchers have been unable to characterize subnormal Toeplitz operators in terms of their symbols. For cases of matrix-valued symbols, the subnormality of Toeplitz operators was studied in [12–15]. Also Theorem 3.12 shows that if the matrix-valued symbol Φ satisfies a general condition and T_Φ is subnormal, then it is either normal or analytic. Also in [12], it was conjectured that every subnormal Toeplitz operator with matrix-valued rational symbol is unitarily equivalent to a direct sum of a normal operator and a Toeplitz operator with analytic symbol. In fact, if an $n \times n$ matrix-valued function Φ is analytic, then the normal extension of T_Φ is the multiplication operator M_Φ, so clearly T_Φ is subnormal. However, this is not the case for the operator-valued symbols. We have such an example (see Example 4.23). On the other hand, if Φ is matrix-valued and T_Φ is subnormal (even hyponormal), then Φ should be normal, i.e., $\Phi^*\Phi = \Phi\Phi^*$ a.e. on $\mathbb{T}$ (cf. [31]). However this may also fail for operator-valued symbols.

Example 4.22 Let $S := T_z$ on H^2 and $\Phi(z) = Sz^n \in H^\infty(\mathcal{B}(H^2))$ $(n \ge 0)$. Then

$$T_\Phi^* T_\Phi = T_{S^*S} = I_{H^2(\mathcal{B}(H^2))},$$

so that T_Φ is quasinormal and hence subnormal. However,

$$\Phi(z)\Phi^*(z) = SS^* \neq S^*S = \Phi^*(z)\Phi(z) \quad \text{for all } z \in \mathbb{T},$$

which implies that Φ is not normal. Here we don't need to expect that the multiplication operator $M_\Phi : L^2(\mathcal{B}(H^2)) \to L^2(\mathcal{B}(H^2))$ is a normal extension of T_Φ. Indeed, it is easy to show that M_Φ is not normal, and hence M_Φ can never be a normal extension of T_Φ. What is a normal extension of T_Φ ? Let $B := M_z$ on L^2 and $\Psi(z) := Bz^n \in H^\infty(\mathcal{B}(L^2))$. Then a straightforward calculation shows that the multiplication operator $M_\Psi : L^2(\mathcal{B}(L^2)) \to L^2(\mathcal{B}(L^2)))$ is a normal extension of T_Φ.

The following simple example shows that analytic Toeplitz operators with operator-valued symbols need not be even hyponormal.

Example 4.23 Let $\Phi(z) = S^* \in H^\infty(\mathcal{B}(H^2))$ and e_0 be the constant function $\mathbf{1} \in H^2(\mathbb{T})$. If $f(z) = e_0 z$, then

$$\langle (T_{\Phi^*}T_\Phi - T_\Phi T_{\Phi^*})f,\ f\rangle = \langle -e_0 z,\ e_0 z\rangle = -1 < 0,$$

which implies that T_Φ is not hyponormal and hence not subnormal even though Φ is analytic.

We would like to pose the following:

Question 4.24 Which analytic Toeplitz operators with operator-valued symbols are subnormal?

For a sufficient condition, one may be tempted to conjecture that if $\Phi \in H^\infty(\mathcal{B}(H^2))$ and if $\Phi(z)$ is subnormal for almost all $z \in \mathbb{T}$, then T_Φ is subnormal. We have not been able to decide whether this is true.

Acknowledgments The authors are grateful to the referee for a careful reading of the manuscript and for many valuable suggestions that helped improved the presentation. The work of the second named author was supported by NRF (Korea) grant No. 2019R1A2C1005182. The work of the third named author was supported by NRF (Korea) grant No. 2021R1A2C1005428.

References

1. M.B. Abrahamse, *Subnormal Toeplitz operators and functions of bounded type*, Duke Math. J. **43** (1976), 597–604.
2. P.R. Ahern and D.N. Clark, *On functions orthogonal to invariant subspaces*, Acta Math. **124** (1970), 191–204.
3. J.A. Ball, I. Gohberg, and L. Rodman, *Interpolation of Rational Matrix Functions*, Oper. Th. Adv. Appl. vol. 45, Birkhäuser, Basel, 1990.
4. J.A. Ball and M.W. Raney, *Discrete-time dichotomous well-posed linear systems and generalized Schur-Nevanlinna-Pick interpolation*, Complex Anal. Oper. Theory **1(1)** (2007), 1–54.
5. A. Beurling, *On two problems concerning linear transformations in Hilbert space*, Acta Math. **81** (1949), 239–255.

6. S. Bochner and H.F. Bohnenblust, *Analytic functions with almost periodic coefficients*, Ann. Math. **35** (1934), 152–161.
7. A. Böttcher and B. Silbermann, *Analysis of Toeplitz Operators*, Springer, Berlin-Heidelberg, 2006.
8. J.B. Conway, *A Course in Functional Analysis*, Springer, New York, 1990.
9. C. Cowen, *Hyponormality of Toeplitz operators*, Proc. Amer. Math. Soc. **103** (1988), 809–812.
10. C. Cowen, *Hyponormal and subnormal Toeplitz operators*, Surveys of Some Recent Results in Operator Theory, I (J.B. Conway and B.B. Morrel, eds.), Pitman Research Notes in Mathematics, Vol **171** (1988), 155–167.
11. C. Cowen and J. Long, *Some subnormal Toeplitz operators*, J. Reine Angew. Math. **351** (1984), 216–220.
12. R.E. Curto, I.S. Hwang, D. Kang and W.Y. Lee, *Subnormal and quasinormal Toeplitz operators with matrix-valued rational symbols*, Adv. Math. **255** (2014), 561–585.
13. R.E. Curto, I.S. Hwang and W.Y. Lee, *Which subnormal Toeplitz operators are either normal or analytic ?*, J. Funct. Anal. **263(8)** (2012), 2333–2354.
14. R.E. Curto, I.S. Hwang and W.Y. Lee, *Hyponormality and subnormality of block Toeplitz operators*, Adv. Math. **230** (2012), 2094–2151.
15. R.E. Curto, I.S. Hwang and W.Y. Lee, *Matrix functions of bounded type: An interplay between function theory and operator theory*, Mem. Amer. Math. Soc. **260** (2019), no. 1253, vi+100.
16. R.E. Curto, I.S. Hwang and W.Y. Lee, *The Beurling-Lax-Halmos Theorem for infinite multiplicity*, J. Funct. Anal. **280(6)** (2021), 108884.
17. R.E. Curto, I.S. Hwang and W.Y. Lee, *Operator-valued rational functions*, J. Funct. Anal. **283(9)** (2022), 109640.
18. R.E. Curto and W.Y. Lee, *Joint hyponormality of Toeplitz pairs*, Mem. Amer. Math. Soc. **150** (2001), no. 712, x+65.
19. L. de Branges and J. Rovnyak, *Square Summable Power Series*, Holt, Rinehart and Winston, New York-Toronto, 1966.
20. R.G. Douglas, *Banach Algebra Techniques in Operator Theory*, Academic Press, New York, 1972.
21. R.G. Douglas, *Banach Algebra Techniques in the Theory of Toeplitz Operators*, CBMS vol. 15, Amer. Math. Soc., Providence, 1973.
22. R.G. Douglas and J.W. Helton, *Inner dilations of analytic matrix functions and Darlington synthesis*, Acta Sci. Math. (Szeged) **34** (1973), 61–67.
23. R.G. Douglas, H. Shapiro, and A. Shields, *Cyclic vectors and invariant subspaces for the backward shift operator*, Ann. Inst. Fourier(Grenoble) **20** (1970), 37–76.
24. A. Frazho and W. Bhosri, *An Operator Perspective on Signals and Systems*, Oper. Th. Adv. Appl. vol. 204, Birkhäuser, Basel, 2010.
25. C. Foiaş and A. Frazho, *The Commutant Lifting Approach to Interpolation Problems*, Oper. Th. Adv. Appl. vol. 44, Birkhäuser, Boston, 1993.
26. P. Fuhrmann, *On Hankel operator ranges, meromorphic pseudo-continuation and factorization of operator-valued analytic functions*, J. London Math. Soc.(2), **13** (1975), 323–327.
27. P. Fuhrmann, *Linear Systems and Operators in Hilbert Spaces*, McGraw-Hill, New York, 1981.
28. J. Garnett, *Bounded Analytic Functions*, Academic Press, New York, 1981.
29. S. Goldberg, *Unbounded Linear Operators*, Dover, New York, 2006.
30. C. Gu, *A generalization of Cowen's characterization of hyponormal Toeplitz operators*, J. Funct. Anal. **124** (1994), 135–148.
31. C. Gu, J. Hendricks and D. Rutherford, *Hyponormality of block Toeplitz operators*, Pacific J. Math. **223** (2006), 95–111.
32. C. Gu, I.S. Hwang, and W.Y. Lee, *On Douglas-Shapiro-Shields factorizations*, Filomat **34:4** (2020) 1053–1060.
33. P.R. Halmos, *Shifts on Hilbert spaces*, J. Reine Angew. Math. **208** (1961), 102–112.
34. P.R. Halmos, *Ten problems in Hilbert space*, Bull. Amer. Math. Soc. **76** (1970), 887–933.
35. P.R. Halmos, *Ten years in Hilbert space*, Integral Equations Operator Theory **2** (1979), 529–564.

36. J.K. Han, H.Y. Lee, and W.Y. Lee, *Invertible completions of* 2×2 *triangular operator matrices*. Proc. Amer. Math. Soc. **128**(2000), 119–123.
37. I.S. Hwang, I.H. Kim, and W.Y. Lee, *Hyponormality of Toeplitz operators with polynomial symbols*, Math. Ann. **313(2)** (1999), 247–261.
38. I.S. Hwang, I.H. Kim and W.Y. Lee, *Hyponormality of Toeplitz operators with polynomial symbols: An extremal case*, Math. Nachr. **231** (2001), 25–38.
39. I.S. Hwang and W.Y. Lee, *Hyponormality of trigonometric Toeplitz operators*, Trans. Amer. Math. Soc. **354** (2002), 2461–2474.
40. I.S. Hwang and W.Y. Lee, *Hyponormality of Toeplitz operators with rational symbols*, Math. Ann. **335** (2006), 405–414.
41. P.D. Lax, *Translation invariant subspaces*, Acta Math. **101** (1959), 163–178.
42. B. Moore III and E. A. Nordgren, *On quasi-equivalence and quasi-similarity*, Acta. Sci. Math. (Szeged) **34** (1973), 311–316.
43. T. Nakazi and K. Takahashi, *Hyponormal Toeplitz operators and extremal problems of Hardy spaces*, Trans. Amer. Math. Soc. **338** (1993), 753–767.
44. N.K. Nikolskii, *Treatise on the Shift Operator*, Springer, New York, 1986.
45. N.K. Nikolskii, *Operators, Functions, and Systems: An Easy Reading Volume I: Hardy, Hankel, and Toeplitz*, Mathematical Surveys and Monographs, vol. 92, Amer. Math. Soc., Providence, 2002.
46. E. A. Nordgren, *On quasi-equivalence of matrices over* H^∞, Acta. Sci. Math. (Szeged) **34** (1973), 301–310.
47. V.V. Peller, *Hankel Operators and Their Applications*, Springer, New York, 2003.
48. V.P. Potapov, *On the multiplicative structure of J-nonexpansive matrix functions*, Tr. Mosk. Mat. Obs. (1955), 125–236 (in Russian); English trasl. in: Amer. Math. Soc. Transl. (2) 15(1960), 131–243.
49. D. Sarason, *Sub-Hardy Hilbert Spaces in the Unit Disk*, John Wiley & Sons, Inc., New York, 1994.
50. H.S. Shapiro, *Generalized analytic continuation*, In: Symposia on Theoretical Physics and Mathematics, vol. 8, pp. 151–163, Plenum Press, 1968.
51. B. Sz.-Nagy, C. Foiaş, H. Bercovici, and L. Kérchy, *Harmonic Analysis of Operators on Hilbert Space*, Springer, New York, 2010.
52. V.I. Vasyunin and N.K. Nikolskii, *Classification of* H^2*-functions according to the degree of their cyclicity*, Math. USSR Izvestiya **23(2)** (1984), 225–242.
53. K. Zhu, *Hyponormal Toeplitz operators with polynomial symbols*, Integral Equations Operator Theory **21** (1996), 376–381.

Sarason's Ha-plitz Product Problem

Kehe Zhu

1 Introduction

Let $\mathbb{C}$ be the complex plane, $\mathbb{D}$ be the open unit disk, and dA be the area measure normalized so that the unit disk has area 1. Let $H(\mathbb{D})$ and $H(\mathbb{C})$ be the space of all analytic functions on $\mathbb{D}$ and $\mathbb{C}$, respectively. We will be concerned with three classical Hilbert spaces of analytic functions: the Hardy space H^2, the Bergman space A^2, and the Fock space F^2.

Recall that $H^2(\mathbb{D})$ is the space of functions $f \in H(\mathbb{D})$ such that

$$\|f\|_{H^2}^2 = \sup_{0<r<1} \frac{1}{2\pi} \int_0^{2\pi} |f(re^{it})|^2 \, dt < \infty.$$

Every function $f \in H^2$ has radial limits

$$f(e^{it}) = \lim_{r \to 1^-} f(re^{it})$$

for almost all $t \in [0, 2\pi]$. This allows us to identify $H^2(\mathbb{D})$ with a closed subspace of $L^2(\mathbb{T})$ for the unit circle $\mathbb{T}$, which will still be denoted by H^2. The Bergman space A^2 of the unit disk is simply defined by

$$A^2 = L^2(\mathbb{D}, dA) \cap H(\mathbb{D}),$$

Research partially supported by NNSF of China (Grant no. 12271328).

K. Zhu (✉)
Department of Mathematics and Statistics, SUNY at Albany, Albany, NY, USA
e-mail: kzhu@albany.edu

I. Binder et al. (eds.), *Function Spaces, Theory and Applications*, Fields Institute Communications 87, https://doi.org/10.1007/978-3-031-39270-2_11

which is a closed subspace of $L^2(\mathbb{D}, dA)$. Similarly, the Fock space F^2 is a closed subspace of $L^2(\mathbb{C}, d\lambda)$, namely,

$$F^2 = L^2(\mathbb{C}, d\lambda) \cap H(\mathbb{C}),$$

where

$$d\lambda(z) = e^{-|z|^2} \, dA(z)$$

is the Gaussian measure on the complex plane. See [17] and [18] for more information about Hardy, Bergman, and Fock spaces.

Each of the spaces $H = H^2, A^2, F^2$ is contained in an L^2 space. We will use $P : L^2 \to H$ to denote the orthogonal projection. For a reasonably good function φ (on $\mathbb{T}$, or $\mathbb{D}$, or $\mathbb{C}$), the Toeplitz operator T_φ on H is defined by $T_\varphi(f) = P(\varphi f)$. Similarly, the Hankel operator $H_\varphi : H \to L^2 \ominus H$ is defined by $H_\varphi(f) = (I - P)(\varphi f)$, where I is the identity operator on L^2. When $\varphi \in L^\infty$, it is easy to see that T_φ and H_φ are both bounded with $\|T_\varphi\| \le \|\varphi\|_\infty$ and $\|H_\varphi\| \le \|\varphi\|_\infty$. We will be mostly interested in situations where φ is unbounded but the operators T_φ and H_φ are densely defined.

In [11], "Linear and Complex Analysis Problem Book 3" – Springer Lecture Notes in Mathematics 1573, Don Sarason raised the following question about Toeplitz operators on the classical Hardy and Bergman spaces of the unit disk: for which analytic symbols f and g is the product $T_f T_{\overline{g}}$ bounded? Sarason also formulated a conjecture in the form of a necessary and sufficient condition in terms of the Berezin transform and provided a proof for one direction of the conjecture.

Sarason's problem and conjecture also make sense for the Fock space, in exactly the same form as in the Hardy and Bergman space cases. The purpose of this article is to give a survey on research about Sarason's problem, including some of its natural extensions, over the past three decades or so. Here is a summary of progress so far:

- Sarason's original conjecture has been disproved by counterexamples for both the Hardy and Bergman spaces.
- Sarason's original question remains more or less open for both the Hardy and Bergman spaces, although necessary and sufficient conditions have been formulated in terms of two-weight conditions in harmonic analysis.
- The Fock space version of Sarason's conjecture has been proved in the affirmative, and we now have explicit descriptions of entire functions f and g such that $T_f T_{\overline{g}}$ is bounded on the Fock space.
- There has been substantial progress in recent years about the natural extension of Sarason's problem and conjecture to products of Hankel operators as well as mixed products of Hankel and Toeplitz operators on the Fock space.
- There are sharp contrast and surprising differences between the Hardy/Bergman space settings and the Fock space setting.

The paper is based on a lecture for the "Focus Program on Analytic Function Spaces and their Applications" at the Fields Institute. I thank the organizers for their invitation to speak at this wonderful program.

2 The Berezin Transform

Suppose Ω is a domain in the complex plane (including $\mathbb{C}$ itself) and H is a Hilbert space of analytic functions on Ω. If for each $a \in \Omega$ the point evaluation $f \mapsto f(a)$ is a bounded linear functional on H, then by Riesz representation there exists a unique function $K_a \in H$ such that $f(a) = \langle f, K_a \rangle$ for all $f \in H$. Such a space H is called a reproducing kernel Hilbert space, and the associated function $K(z, w) = K_w(z)$ on $\Omega \times \Omega$ is called the reproducing kernel of H.

If H has the additional property that for each $a \in \Omega$ there exists some function $f \in H$ such that $f(a) \neq 0$, then $K_a \neq 0$ and so

$$K(a, a) = \langle K_a, K_a \rangle = \|K_a\|^2 > 0.$$

In this case, the unit vectors $k_a = K_a/\sqrt{K(a, a)}$ are called the normalized reproducing kernels. Given any linear operator T on H such that the domain of T contains all normalized reproducing kernels, we can then define a function $\widetilde{T}$ on Ω by

$$\widetilde{T}(z) = \langle T k_z, k_z \rangle, \qquad z \in \Omega.$$

The function $\widetilde{T}$ is called the Berezin symbol (or sometimes the Berezin transform) of T. In particular, if T is bounded, then by the Cauchy-Schwarz inequality, we have $\widetilde{T} \in L^\infty(\Omega)$ with $\|\widetilde{T}\|_\infty \leq \|T\|$.

It is well known that H^2, A^2, and F^2 are all reproducing kernel Hilbert spaces, whose reproducing kernels $K(z, w)$ are given by

$$\frac{1}{1 - z\overline{w}}, \qquad \frac{1}{(1 - z\overline{w})^2}, \qquad e^{z\overline{w}},$$

respectively. If $T = T_\varphi$ is a Toeplitz operator on H^2, A^2, or F^2 with bounded symbol φ, then its Berezin transform is given by

$$\widetilde{T}(z) = \frac{1}{2\pi} \int_0^{2\pi} \frac{1 - |z|^2}{|1 - \overline{z}e^{it}|^2} \varphi(e^{it})\, dt,$$

$$\widetilde{T}(z) = \int_{\mathbb{D}} \frac{(1 - |z|^2)^2}{|1 - \overline{z}w|^4} \varphi(w)\, dA(w),$$

$$\widetilde{T}(z) = \int_{\mathbb{C}} e^{-|z-w|^2} \varphi(w)\, dA(w),$$

respectively. It is clear that the integrals above converge for many more functions than the bounded ones. We will use the notation $\widetilde{\varphi}(z)$ for these integrals, whenever they converge, and call them the Berezin transform of φ. In particular, in the case of the unit circle $\mathbb{T}$ and the unit disk $\mathbb{D}$, the Berezin transform is well defined for every function in L^1. The Berezin transform of a function $\varphi \in L^1(\mathbb{T})$ is nothing but the classical Poisson transform (or harmonic extension) of φ.

The Berezin transform has played a key role in function-theoretic operator theory over the past few decades. It is especially useful in the study of Hankel and Toeplitz operators on the Bergman and Fock space. For example, in each of these two settings, it is known that:

(a) If $\varphi \geq 0$, then T_φ is bounded if and only if $\widetilde{\varphi}$ is bounded.
(b) If φ is real-valued, then H_φ is bounded if and only if $\widetilde{|\varphi|^2} - |\widetilde{\varphi}|^2$ is bounded.

In the case of the Hardy space, part (b) above is also true, and part (a) remains valid even without the extra assumption $\varphi \geq 0$. See [17, 18].

If f and g are analytic symbol functions such that the Toeplitz product $T_f T_{\overline{g}}$ is densely defined on the space $H = H^2$, A^2, or F^2, then its Berezin transform is given by

$$\begin{aligned}\widetilde{T_f T_{\overline{g}}}(z) &= \langle T_f T_{\overline{g}} k_z, k_z \rangle = \langle T_{\overline{g}} k_z, T_{\overline{f}} k_z \rangle \\ &= \langle \overline{g(z)} k_z, \overline{f(z)} k_z \rangle = f(z)\overline{g(z)}.\end{aligned}$$

It follows from the Cauchy-Schwarz inequality that

$$\|f g\|_\infty \leq \|T_f T_{\overline{g}}\|. \tag{1}$$

Thus, a necessary condition for the boundedness of $T_f T_{\overline{g}}$ on H is that the function fg be bounded.

3 Toeplitz Products on H^2 and A^2

The following is Sarason's original Toeplitz product problem. See [11].

3.1 Characterize functions $f, g \in H^2$ such that the product $T_f T_{\overline{g}}$ is bounded on H^2.

Sarason also commented in [11] that the problem is interesting for the Bergman space. Every function in H^2 can be factored into the product of an inner function and an outer function. Since every inner function induces an isometric Toeplitz operator on H^2, Sarason's problem for the Hardy space reduces to the case where both f

and g are outer functions. That is how Problem 3.1 was formulated in [11]. There is clearly no such reduction in the cases of A^2.

The motivation for Sarason's problem was his study of exposed points in H^1; see [10]. A necessary condition (weaker than the boundedness of fg) for the boundedness of $T_f T_{\overline{g}}$ on H^2 was given in [11]. The condition is in terms of the Berezin transform and it leads to a natural conjecture. The following is a discussion of this necessary condition for both H^2 and A^2.

For any $a \in \mathbb{D}$ let φ_a be the Mobius map defined by

$$\varphi_a(z) = \frac{a - z}{1 - \overline{a}z}, \qquad z \in \mathbb{D}.$$

In the case of H^2 and A^2, the operator U_a defined by $U_a f = f \circ \varphi_a k_a$ is self-adjoint and unitary. Also, it is well known and easy to verify that $U_a T_g U_a = T_{g \circ \varphi_a}$.

On the Hardy space, T_z is the forward shift and $T_{\overline{z}}$ is the backward shift. Therefore,

$$T_z T_{\overline{z}} f = f - f(0), \qquad (I - T_z T_{\overline{z}}) f = f(0).$$

It follows that

$$\begin{aligned}(I - T_{\varphi_a} T_{\overline{\varphi_a}}) f &= U_a (I - T_z T_{\overline{z}}) U_a f \\ &= U_a [f(a) k_a(0)] = f(a) k_a(0) k_a \\ &= \frac{1 - |a|^2}{1 - \overline{a}z} f(a) = (k_a \otimes k_a) f,\end{aligned}$$

so $I - T_{\varphi_a} T_{\overline{\varphi_a}} = k_a \otimes k_a$. Since any two Toeplitz operators with analytic symbols commute, we obtain

$$T_{\varphi_a} T_f T_{\overline{g}} T_{\overline{\varphi_a}} = T_f T_{\varphi_a} T_{\overline{\varphi_a}} T_{\overline{g}} = T_f (I - k_a \otimes k_a) T_{\overline{g}}.$$

It is easy to check that

$$T_f (k_a \otimes k_a) T_{\overline{g}} = (f k_a) \otimes (g k_a).$$

Thus,

$$(f k_a) \otimes (g k_a) = T_f T_{\overline{g}} - T_{\varphi_a} T_f T_{\overline{g}} T_{\overline{\varphi_a}}.$$

If $T_f T_{\overline{g}}$ is bounded on H^2, then the identity above together with $\|T_{\varphi_a}\| = \|T_{\overline{\varphi_a}}\| \le 1$ yields

$$\|f k_a\| \|g k_a\| = \|(f k_a) \otimes (g k_a)\| \le 2 \|T_f T_{\overline{g}}\|.$$

By a change of variables, we have

$$\|fk_a\|^2 = \widetilde{|f|^2}(a), \qquad \|gk_a\|^2 = \widetilde{|g|^2}(a).$$

This proves that the boundedness of $T_f T_{\overline{g}}$ on H^2 implies that the function $\widetilde{|f|^2}\widetilde{|g|^2}$ is bounded on $\mathbb{D}$, with

$$\sup_{z\in\mathbb{D}} \widetilde{|f|^2}(z)\widetilde{|g|^2}(z) \le 4\|T_f T_{\overline{g}}\|^2. \tag{2}$$

For the Bergman space A^2, a direct computation shows that

$$(I - 2T_z T_{\overline{z}} + T_{z^2} T_{\overline{z}^2}) f = f(0).$$

Replacing f by $U_a f$, we obtain

$$(I - 2T_z T_{\overline{z}} + T_{z^2} T_{\overline{z}^2}) U_a f = f(a) k_a(0).$$

Applying U_a to both sides and using the identity $U_a T_\varphi U_a = T_{\varphi\circ\varphi_a}$, we arrive at

$$I - 2T_{\varphi_a} T_{\overline{\varphi_a}} + T_{\varphi_a^2} T_{\overline{\varphi_a}^2} = k_a \otimes k_a.$$

Just like the Hardy space case, we then have

$$\begin{aligned}(fk_a) \otimes (gk_a) &= T_f (k_a \otimes k_a) T_{\overline{g}} \\ &= T_f T_{\overline{g}} - 2T_{\varphi_a} T_f T_{\overline{g}} T_{\overline{\varphi_a}} + T_{\varphi_a^2} T_f T_{\overline{g}} T_{\overline{\varphi_a}^2}\end{aligned}$$

for analytic symbols f and g. If $T_f T_{\overline{g}}$ is bounded, then

$$\|fk_a\|\|gk_a\| = \|(fk_a) \otimes (gk_a)\| \le 4\|T_f T_{\overline{g}}\|.$$

By a change of variables again, we have $\|fk_a\|^2 = \widetilde{|f|^2}(a)$. Therefore, the boundedness of the Toeplitz product $T_f T_{\overline{g}}$ on A^2 implies that the function $\widetilde{|f|^2}\widetilde{|g|^2}$ is bounded on $\mathbb{D}$. More precisely,

$$\sup_{z\in\mathbb{D}} \widetilde{|f|^2}(z)\widetilde{|g|^2}(z) \le 16\|T_f T_{\overline{g}}\|^2. \tag{3}$$

The following is referred to as Sarason's conjecture in the literature.

Conjecture 1 Suppose f and g are functions in H^2 (or A^2). Then, the Toeplitz product $T_f T_{\overline{g}}$ is bounded on H^2 (or A^2) if and only if the product function $\widetilde{|f|^2}\,\widetilde{|g|^2}$ is bounded on $\mathbb{D}$.

Note that we are using the same notation for the Berezin transform in the Hardy and Bergman space settings, although they are clearly different integral transforms.

Sarason's Toeplitz product problem turns out to be highly nontrivial. It has a close and natural connection to a classical problem in harmonic analysis. More precisely, the following formulation of Sarason's problem can be traced back to [5].

Proposition 3.1 *Given functions f and g in A^2, the Toeplitz product $T_f T_{\overline{g}}$ is bounded on A^2 if and only if the Bergman projection maps $L^2(\mathbb{D}, |g|^{-2}dA)$ boundedly into $L^2(\mathrm{d}, |f|^2 dA)$. Similarly, for functions f and g in H^2, the Toeplitz product $T_f T_{\overline{g}}$ is bounded on H^2 if and only if the Riesz-Szego projection (or the Hilbert transform) maps $L^2(\mathbb{T}, |g|^{-2}dt)$ boundedly into $L^2(\mathbb{T}, |f|^2 dt)$.*

More generally, if ω_1 and ω_2 are two weight functions on the unit circle, then the so-called two-weight boundedness problem for the Riesz-Szego projection (or, equivalently, the Hilbert transform or the Hardy-Littlewood maximal operator)

$$P : L^2(\mathbb{D}, \omega_1 dt) \to L^2(\mathbb{D}, \omega_2 dt) \tag{4}$$

is a classical problem in harmonic analysis. When $\omega_1 = \omega_2 =: \omega$, it is well known that the boundedness of P in (4) is characterized by the Muckenhoupt A_2-condition

$$\sup_{z \in \mathbb{D}} \widetilde{\omega}(z)\, \widetilde{\omega^{-1}}(z) < \infty,$$

where $\widetilde{\omega}$ is the Hardy space Berezin transform (viz., the Poisson transform) of ω. It was an open problem in harmonic analysis for some time whether the two-weight problem in (4) can be characterized by the condition

$$\sup_{z \in \mathbb{D}} \widetilde{\omega_2}(z)\, \widetilde{\omega_1^{-1}}(z) < \infty, \tag{5}$$

which is a natural extension of the Muckenhoupt A_2-condition. It is clear that Sarason's Toeplitz product conjecture for the Hardy space is a special case of this two-weight problem: $\omega_1 = 1/|g|^2$ and $\omega_2 = |f|^2$.

In contrast to the one-weight problem, Nazarov [8] proved in 1997 by a counterexample that the condition in (5) cannot characterize the boundedness of the Riesz-Szego projection for the two-weight problem. Furthermore, Nazarov's counterexample also disproved Sarason's conjecture for Toeplitz products on H^2.

In the case of the Bergman space, the general two-weight problem concerns the following boundedness of the Bergman projection:

$$P : L^2(\mathbb{D}, \omega_1 dA) \to L^2(\mathbb{D}, \omega_2 dA), \tag{6}$$

where ω_1 and ω_2 are weight functions on $\mathbb{D}$. When $\omega_1 = \omega_2 =: \omega$, it is known that the boundedness of P above is characterized by the Bekolle-Bonami condition

$$\sup_{z\in\mathbb{D}} \widetilde{\omega}(z)\, \widetilde{\omega^{-1}}(z) < \infty,$$

where $\widetilde{\omega}$ is the Bergman space Berezin transform of ω. It was then natural to ask whether the two-weight problem in (6) can be characterized by the condition

$$\sup_{z\in\mathbb{D}} \widetilde{\omega_2}(z)\, \widetilde{\omega_1^{-1}}(z) < \infty. \tag{7}$$

Once again, Sarason's Toeplitz product conjecture for the Bergman space is a special case of this: $\omega_1 = 1/|g|^2$ and $\omega_2 = |f|^2$.

Similar to the Hardy space case, a counterexample was produced by Aleman, Pott, and Reguera in [1] that shows that condition (7) does not characterize the boundedness of P in the two-weight problem (6) and Sarason's Toeplitz product conjecture for the Bergman space fails as well. On the other hand, the following solution to Sarason's Toeplitz product problem was obtained in [1].

Theorem 3.2 *For f and g in A^2, the following conditions are equivalent:*

(a) The Toeplitz product $T_f T_{\overline{g}}$ is bounded on A^2.
(b) There exists a positive constant C such that

$$\|P^+(|g|^2\chi_Q)\|_{L^2(\mathbb{D},|f|^2dA)} \leq C\|\chi_Q\|_{L^2(\mathbb{D},|g|^2dA)}$$

and

$$\|P^+(|f|^2\chi_Q)\|_{L^2(\mathbb{D},|g|^2dA)} \leq C\|\chi_Q\|_{L^2(\mathbb{D},|f|^2dA)}$$

for all Carleson squares Q in $\mathbb{D}$, where χ_Q is the characteristic function for Q and

$$P^+h(z) = \int_{\mathrm{d}} \frac{h(z)\, dA(z)}{|1 - z\overline{w}|^2}.$$

Although the conditions in (b) gave a characterization for the boundedness of the Toeplitz product $T_f T_{\overline{g}}$ on A^2, they are likely to be of limited use. The same is true for Sarason's Toeplitz product problem for the Hardy space: a similar necessary and sufficient condition in terms of a two-weight problem for the Hilbert transform from harmonic analysis is not likely to be of practical use. Thus, we consider Sarason's Toeplitz product problem for H^2 and A^2 essentially still open. On the other hand, it may be the nature of the problem that there are simply no "reasonably elementary conditions" for its solution.

We will see in the next section that the situation for the Fock space is completely different.

4 Toeplitz Products on F^2

From the perspective of operator theory, Sarason's Toeplitz product problem and the associated Sarason conjecture make perfect sense for the Fock space.

4.1 Characterize functions f and g in F^2 such that the Toeplitz product $T_f T_{\overline{g}}$ is bounded on F^2. Furthermore, is it true that $T_f T_{\overline{g}}$ is bounded on F^2 if and only if the product function $\widetilde{|f|^2}\,\widetilde{|g|^2}$ is bounded on $\mathbb{C}$?

Here, $\widetilde{h}$ means the Berezin transform of h in the Fock space setting.

Recall from the previous section that Sarason's conjecture has been disproved for both the Hardy and Bergman space. It turns out that the Fock space version of Sarason's conjecture is true. In fact, more can be said here. See [4] for the following result.

Theorem 4.1 *Let $f, g \in F^2$. Then, the densely defined operator $T_f T_{\overline{g}}$ on F^2 is bounded if and only if there exist complex constants a, b, and c such that*

$$f(z) = ae^{cz}, \qquad g(z) = be^{-cz}. \tag{8}$$

If f and g are given by (8), then a direct calculation shows that

$$T_f T_{\overline{g}} = a\overline{b}e^{|c|^2/2}W_c, \tag{9}$$

where W_c is the Weyl operator on F^2 defined by

$$W_c h(z) = h(z-c)k_c(z).$$

Each W_c is a unitary operator on F^2. These operators are the Fock space version of the operators U_a defined in the previous section.

Thus, we see that, for the Fock space, if the operator $T_f T_{\overline{g}}$ is bounded, it must be a constant times a unitary operator. Consequently, operators of the form $T_f T_{\overline{g}}$, where $f, g \in F^2$, cannot be compact unless it is zero. Another immediate consequence of Theorem 4.1 is that fg must be constant whenever $T_f T_{\overline{g}}$ is bounded on F^2. In the case of Hardy and Bergman spaces, we can only conclude that fg is bounded on $\mathbb{D}$. The difference is easily explained by Liouville's theorem.

Corollary 4.2 *The Fock space version of Sarason's conjecture holds, that is, for f and g in F^2, the Toeplitz product $T_f T_{\overline{g}}$ is bounded on F^2 if and only if the product function $\widetilde{|f|^2}\,\widetilde{|g|^2}$ is bounded on $\mathbb{C}$.*

Proof Suppose $f(z) = e^{cz}$ and $g(z) = e^{-cz}$, where c is a complex constant. A direct computation yields

$$\widetilde{|f|^2}(z) = e^{|c|^2+\overline{c}z+c\overline{z}}, \qquad \widetilde{|g|^2}(z) = e^{|c|^2-\overline{c}z-c\overline{z}}. \tag{10}$$

This together with Theorem 4.1 shows that the boundedness of the operator $T_f T_{\overline{g}}$ implies the boundedness of the function $\widetilde{|f|^2}\,\widetilde{|g|^2}$.

On the other hand, if $\widetilde{|f|^2}\,\widetilde{|g|^2}$ is bounded, then fg must be bounded, because for $f \in F^2$ we always have $|f|^2 \le \widetilde{|f|^2}$ (by Cauchy-Schwarz and the fact that $f(z) = \widetilde{f}(z) = \langle fk_z, k_z\rangle$ for $f \in F^2$). The proof of Theorem 4.1 in [4] then shows that f and g must be of the form $f(z) = ae^{cz}$ and $g(z) = be^{-cz}$. By Theorem 4.1, $T_f T_{\overline{g}}$ is bounded. □

Recall from the previous section that, for Hardy and Bergman spaces, there exists a positive constant C (e.g., we can take $C = 4$ for H^2 and $C = 16$ for A^2) such that

$$\sup_{z\in\mathbb{D}} \left[\widetilde{|f|^2}(z)\,\widetilde{|g|^2}(z)\right] \le C\|T_f T_{\overline{g}}\|^2, \tag{11}$$

where the constant is independent of f and g. Surprisingly, no such constant exists for F^2, although Sarason's conjecture holds for the Fock space! In fact, for $f(z) = e^{cz}$ and $g(z) = e^{-cz}$, it follows from (9) and (10) that

$$\|T_f T_{\overline{g}}\|^2 = e^{|c|^2}, \qquad \widetilde{|f|^2}\,\widetilde{|g|^2} = e^{2|c|^2}.$$

Thus, the Fock space analog of (11) cannot possibly hold as $|c| \to \infty$.

5 Products of Hankel Operators

It is well known and easy to check that

$$H^*_{\overline{f}} H_{\overline{g}} = T_{f\overline{g}} - T_f T_{\overline{g}} \tag{12}$$

on H (the Hardy space, Bergman space, or Fock space). Note that when f and g are analytic, we always have $H_f = H_g = 0$.

Recall from Sect. 2 that a necessary condition for $T_f T_{\overline{g}}$ to be bounded is that fg (and $f\overline{g}$) be bounded. When $f\overline{g}$ is bounded, the identity in (12) shows that the boundedness of $T_f T_{\overline{g}}$ on H is equivalent to the boundedness of $H^*_{\overline{f}} H_{\overline{g}}$ on H. Thus, the following is a natural companion to Sarason's Toeplitz product problem.

5.1 For $H = H^2$, A^2, or F^2, characterize functions $f, g \in H$ such that the Hankel product $H^*_{\overline{f}} H_{\overline{g}}$ is bounded on H.

In the cases of Hankel operators on H^2 and A^2, using ideas similar to those in Sect. 3, it can be shown that there exists an absolute constant C such that

$$\|H_{\overline{f}} k_z\|\,\|H_{\overline{g}} k_z\| \le C\|H^*_{\overline{f}} H_{\overline{g}}\|. \tag{13}$$

See [12] for details. For an analytic symbol f, we have

$$P(\overline{f}k_z) = T_{\overline{f}}k_z = \overline{f(z)}k_z.$$

It follows that

$$\|H_{\overline{f}}k_z\|^2 = \|\overline{f}k_z - P(\overline{f}k_z)\|^2 = \|fk_z\|^2 - |f(z)|^2 = \widetilde{|f|^2}(z) - |\widetilde{f}(z)|^2.$$

This proves the following.

Proposition 5.1 *Suppose $f, g \in H$ (the Hardy space or the Bergman space). Then, the boundedness of the operator $H^*_{\overline{f}}H_{\overline{g}}$ implies the boundedness of the following function on $\mathbb{D}$:*

$$D(f,g) =: \left[\widetilde{|f|^2} - |\widetilde{f}|^2\right]\left[\widetilde{|g|^2} - |\widetilde{g}|^2\right] \tag{14}$$
$$= \left[\widetilde{|f|^2} - |f|^2\right]\left[\widetilde{|g|^2} - |g|^2\right].$$

Although the proof for inequality (13) in [12] does not work for the Fock space, it is still tempting to make the following conjecture. Recall from the previous section that the inequality $\widetilde{|f|^2}(z)\widetilde{|g|^2}(z) \le C\|T_f T_{\overline{g}}\|$ is invalid for the Fock space, but eventually it turned out that the boundedness of $T_f T_{\overline{g}}$ on F^2 still implies the boundedness of $\widetilde{|f|^2}\,\widetilde{|g|^2}$ on $\mathbb{C}$.

Conjecture 2 Suppose f and g are functions in $H = H^2$, A^2, or F^2. Then, the Hankel product $H^*_{\overline{f}}H_{\overline{g}}$ is bounded on H if and only if the function $D(f,g)$ in (14) is bounded.

The converse of Proposition 5.1 (viz., the "if part" of the conjecture above) for H^2 and A^2 is still an open question. Some partial results in this direction can be found in [12]. As for the Fock space, unlike the Toeplitz product problem, the conjecture above is actually false. But we will see that the situation for F^2 is opposite to the situations for H^2 and A^2. More specifically, the Fock space version of Proposition 5.1 is false, while the converse of the Fock space version of Proposition 5.1 is true. The following theorem can be found in [6].

Theorem 5.2 *Suppose f and g are functions in F^2 and $D(f,g)$ is the function defined in (14). Then, the following conditions are equivalent:*

(a) $D(f,g)$ is bounded on $\mathbb{C}$.
(b) $D(f,g)$ is constant on $\mathbb{C}$.
(c) At least one of f and g is constant, or both f and g are linear polynomials, or f and g are of the form

$$f(z) = A + e^{az+b}, \qquad g(z) = B + e^{-az+c}, \tag{15}$$

where A, B, a, b, c are constants.

As a consequence of the theorem, we easily obtain the following converse of the Fock space version of Proposition 5.1.

Corollary 5.3 *Suppose f and g are functions in F^2. If $D(f, g)$ is bounded on $\mathbb{C}$, then the Hankel product $H_{\overline{f}}^* H_{\overline{g}}$ is bounded on F^2.*

Proof We just need to check the boundedness of $T =: H_{\overline{f}}^* H_{\overline{g}}$ for each of the three cases in part (c) of Theorem 5.2.

If one of f and g is constant, then $T = 0$. If both f and g are linear polynomials, then $H_{\overline{f}}$ and $H_{\overline{g}}$ are both bounded (see, e.g., [18]), so T is bounded.

If f and g are given by (15), then it is clear that

$$H_{\overline{f}}^* H_{\overline{g}} = e^{b+c} H_{\overline{\phi}}^* H_{\overline{\psi}},$$

where $\phi(z) = e^{az}$ and $\psi(z) = e^{-az}$. Recall that

$$H_{\overline{\phi}}^* H_{\overline{\psi}} = T_{\phi\overline{\psi}} - T_\phi T_{\overline{\psi}}.$$

The operator $T_{\phi\overline{\psi}}$ is bounded since $\phi\overline{\psi}$ is bounded. The operator $T_\phi T_{\overline{\psi}}$ is bounded by Theorem 4.1. This shows that T is bounded and completes the proof. □

The next example shows that the converse of Corollary 5.3 is false, which disproves Conjecture 2 for the Fock space and implies that there is no Fock space analog of (13).

Example 5.4 Suppose $f(z) = e^{2\pi i z}$ and $g(z) = e^z$. Then, $H_{\overline{f}}^* H_{\overline{g}} = 0$ but $D(f, g)$ is unbounded on $\mathbb{C}$.

Proof For any $a, b \in \mathbb{C}$, it follows from the reproducing property of the kernel function $K(z, w)$ in F^2 that

$$\begin{aligned}
\langle H_{\overline{f}}^* H_{\overline{g}} K_a, K_b \rangle &= \langle H_{\overline{g}} K_a, H_{\overline{f}} K_b \rangle \\
&= \left\langle (\overline{g} - \overline{g(a)}) K_a, (\overline{f} - \overline{f(b)}) K_b \right\rangle \\
&= \int_{\mathbb{C}} \left(e^{2\pi i z} - e^{2\pi i b} \right) \left(e^{\overline{z}} - e^{\overline{a}} \right) e^{\overline{a} z + b\overline{z}} \, d\lambda(z) \\
&= e^{(2\pi i + \overline{a})(b+1)} - e^{(2\pi i + \overline{a})b + \overline{a}} - e^{2\pi i b + \overline{a}(b+1)} + e^{2\pi i b + \overline{a}(b+1)} \\
&= 0.
\end{aligned}$$

Since a and b are arbitrary and the set of all finite linear combinations of kernel functions is dense in F^2, we conclude that $H_{\overline{f}}^* H_{\overline{g}} = 0$.

On the other hand, a direct calculation shows that for any function of the form $\varphi(z) = e^{cz}$, we have

$$\widetilde{|\varphi|^2}(z) = e^{|c|^2}|\varphi(z)|^2.$$

It follows that for our functions f and g,

$$\begin{aligned} D(f,g)(z) &= \left[\widetilde{|f|^2}(z) - |f(z)|^2\right]\left[\widetilde{|g|^2}(z) - |g(z)|^2\right] \\ &= (e^{4\pi^2} - 1)(e - 1)|e^{(2\pi i+1)z}|^2, \end{aligned}$$

which is obviously unbounded on $\mathbb{C}$. □

The example above is a special case of the following result which was proved in [6].

Theorem 5.5 *Let f and g be functions in F^2. Then, the following conditions are equivalent:*

(a) $H_{\overline{f}}^* H_{\overline{g}} = 0$.
(b) The Berezin transform of $H_{\overline{f}}^* H_{\overline{g}}$ is identically zero.
(c) f and g satisfy $\widetilde{f\overline{g}} = f\overline{g}$.
(d) f and g satisfy

$$f(z)\overline{g(w)} = e^{-z\overline{w}} \int_{\mathbb{C}} f(\zeta)\overline{g(\zeta)} e^{z\overline{\zeta} + \overline{w}\zeta}\, d\lambda(\zeta), \qquad z, w \in \mathbb{C}.$$

Once again, we have disproved Sarason's conjecture for Hankel products on the Fock space, but we have not obtained a characterization for $f, g \in F^2$ such that $H_{\overline{f}}^* H_{\overline{g}}$ is bounded on F^2. Therefore, the Hankel product *problem* remains open for all three spaces H^2, A^2, and F^2.

6 Mixed Products of Ha-plitz Operators

Another natural companion to Sarason's original Toeplitz product problem is the following.

6.1 Let f and g be functions in H, where $H = H^2$ or A^2 or F^2. When is the mixed-product $H_{\overline{f}} T_{\overline{g}}$ bounded on H?

Just like Toeplitz and Hankel products, one direction is relatively easy here for both the Hardy and Bergman spaces. See [12] for the following result.

Proposition 6.1 *Suppose f and g are functions in H, where $H = H^2$ or A^2. Then, the boundedness of the mixed product $H_{\overline{f}}T_{\overline{g}}$ implies that the function*

$$M(f, g)(z) =: \left[\widetilde{|f|^2}(z) - |\widetilde{f}(z)|^2\right] \widetilde{|g|^2}(z) \tag{16}$$
$$= \left[\widetilde{|f|^2}(z) - |f(z)|^2\right] \widetilde{|g|^2}(z)$$

is bounded on $\mathbb{D}$. Furthermore, there exists an absolute constant $C > 0$ such that

$$\sup_{z\in\mathbb{D}} M(f, g)(z) \leq C\|H_{\overline{f}}T_{\overline{g}}\|^2. \tag{17}$$

It is then natural to wonder if the converse of the above is also true, which is still an open question. We formally make the following conjecture, including the Fock space version.

Conjecture 3 Suppose $f, g \in H$, where H is H^2 or A^2 or F^2. Then, the mixed-product $H_{\overline{f}}T_{\overline{g}}$ is bounded on H if and only if the function $M(f, g)$ defined in (16) is bounded.

We will see that, like the situation for Toeplitz products, the Fock space version of (17) is false, but the conjecture above holds for F^2. The following result is proved in [7].

Theorem 6.2 *Suppose $f, g \in F^2$. Then, the following conditions are equivalent.*

(a) The operator $H_{\overline{f}}T_{\overline{g}}$ is bounded on F^2.
(b) The function $M(f, g)$ is bounded on $\mathbb{C}$.
(c) f is constant, or g is identically zero, or f is a linear polynomial and g is a nonzero constant, or

$$f(z) = A + e^{az+b}, \qquad g(z) = e^{-az+c} \tag{18}$$

for some constants A, a, b, c.

Note that when $M(f, g)$ is bounded, it must be constant. This can be verified using the specific forms of f and g listed in part (c) above. In fact, $M(f, g) = 0$ when f is constant or g is identically zero. If f is a linear polynomial, say $f(z) = Cz + D$, it is easy to check that

$$\widetilde{|f|^2}(z) - |f(z)|^2 = |C|^2 \int_{\mathbb{C}} |w|^2\, d\lambda(w)$$

is constant. Therefore, if f is a linear polynomial and g is a nonzero constant, the function $M(f, g)$ is constant. If f and g are given by (18), a direct computation (see

Example 5.4) shows that

$$\widetilde{|f|^2}(z) - |f(z)|^2 = e^{|a|^2}|e^{az+b}|^2, \qquad \widetilde{|g|^2}(z) = e^{|a|^2}|e^{-az+c}|^2,$$

so

$$M(f, g)(z) = e^{|a|^2}|e^{b+c}|^2$$

is constant.

Note also that the boundedness of the function $M(f, g)$ is equivalent to the boundedness of the function $\left[\widetilde{|f|^2} - |f|^2\right]|g|^2$, although $|g|^2 \le \widetilde{|g|^2}$ in general. See [7] for details.

7 Further Remarks

As was mentioned earlier, for both the Hardy and Bergman spaces, the boundedness of the function $\widetilde{|f|^2}\,\widetilde{|g|^2}$ does not imply the boundedness of the operator $T_f T_{\overline{g}}$. However, a slightly weaker version of this implication is true. More specifically, if for some positive number ε (greater than a certain cutoff constant) the function $\widetilde{|f|^{2+\varepsilon}}\,\widetilde{|g|^{2+\varepsilon}}$ is bounded, then $T_f T_{\overline{g}}$ is bounded. A similar result also holds for Hankel products and mixed products on H^2 and A^2. See [12].

The boundedness of the products $T_f T_{\overline{g}}$, $H^*_{\overline{f}} H_{\overline{g}}$, $H_{\overline{f}} T_{\overline{g}}$ on H^2 and A^2 is much different than their Fock space counterparts. What is surprising is that, for Sarason type conjectures, the easy direction for H^2 and A^2 becomes difficult for the Fock space, and vice versa. One advantage in the Fock space setting is Liouville's theorem, which often helps us simplify certain arguments in this setting.

There has been additional work in recent years about analogs of Sarason's problems and conjectures for Toeplitz operators on weighted Fock spaces. Weighted Fock spaces provide a convenient setting, because Liouville's theorem is still available. What is lacking (which makes the problem difficult in general) is the natural action of the Heisenberg group for more general weighted Fock spaces. See [2, 3, 16].

The classical Liouville's theorem is enough for dealing with Toeplitz products $T_f T_{\overline{g}}$ on F^2. However, to tackle Hankel products $H^*_{\overline{f}} H_{\overline{g}}$ and mixed products $H_{\overline{f}} T_{\overline{g}}$, the following vector version of Liouville's theorem is needed.

Theorem 7.1 *Suppose X is a complex Banach space and $f : \mathbb{C} \to X$ is an entire function, that is, for any $F \in X^*$ the function $z \mapsto F(f(z))$ is analytic on the whole complex plane. If there exists a positive constant C such that $\|f(z)\| \le C$ for all $z \in \mathbb{C}$, then $f(z) = f(0)$ for all $z \in \mathbb{C}$. Furthermore, if*

$$\lim_{z\to\infty} \|f(z)\| = 0,$$

then $f = 0$.

When studying products of Hankel and Toeplitz operators, there are several other natural choices for the form of the product. For example, we can also consider Hankel products of the form $H_{\overline{f}} H_{\overline{g}}^*$. There does not seem to be any work in the literature about such products.

There has also been work about Sarason's problems and conjectures for Toeplitz and Hankel operators on weighted Bergman spaces of the unit ball and unit polydisk. See [9, 13–15].

Finally, we mention that Sarason also raised the following question in [11] in addition to the boundedness of $T_f T_{\overline{g}}$, namely, if the operator $T_f T_{\overline{g}}$ is bounded on H^2 or A^2, does it belong to the Toeplitz algebra (the C^*-algebra generated by all bounded Toeplitz operators)? This is certainly an interesting problem, but there does not seem to be any work done about it yet.

References

1. A. Aleman, S. Pott, and C. Reguera, Sarason's conjecture on the Bergman space, *Intern. Math. Research Notices* **14** (2017), 4320–4349.
2. H. Bommier-Hato, E. H. Youssfi, K. Zhu, Sarason's Toeplitz product problem for a class of Fock spaces, *Bull. Sci. Math.* **141** (2017), 408–442.
3. J. Chen, X. Wang, J. Xia, and G. Cao, Sarason's Toeplitz product problem on the Fock-Sobolev space, *Acta Math. Sin.* (Engl. Ser.) **34** (2018), 288–296.
4. H. R. Cho, J.-D. Park, K. Zhu, Products of Toeplitz operators on the Fock space, *Proc. Amer. Math. Soc.* **142** (2014), 2483–2489.
5. D. Cruz-Uribe, The invertibility of the product of unbounded Toeplitz operators, *Integral Equations and Operator Theory* **20** (1994), 231–237.
6. P. Ma, F. Yan, D. Zheng, and K. Zhu, Products of Hankel operators on the Fock space, *J. Funct. Anal.* **277** (2019), 2644–2663.
7. P. Ma, F. Yan, D. Zheng, and K. Zhu, Mixed products of Toeplitz and Hankel operators on the Fock space, *J. Operator Theory* **84** (2020), 35–47.
8. F. Nazarov, A counterexample to Sarason's conjecture, preprint, 1997.
9. S. Pott and E. Strouse, Products of Toeplitz operators on the Bergman spaces A^2_α, *Algebra i Analiz* **18** (2006), 144–161; reprinted in *St. Petersburg Math. J.* **18** (2007), 105–118.
10. D. Sarason, Exposed points in H^1, *Oper. Theory, Adv. Appl.* **48** (1990), 333–347.
11. D. Sarason, Products of Toeplitz operators, in: V. P. Khavin, N. K. Nikolski (Eds.), Linear and Complex Analysis Problem Book 3, Part I, Lecture Notes in Math. **1573**, Springer, Berlin, 1994, 318–319.
12. K. Stroethoff and D. Zheng, Products of Hankel and Toeplitz operators on the Bergman space, *J. Funct. Anal.* **169** (1999), 289–313.
13. K. Stroethoff and D. Zheng, Bounded Toeplitz products on the Bergman space of the polydisk, *J. Math. Anal. Appl.* **278** (2003), 125–135.
14. K. Stroethoff and D. Zheng, Bounded Toeplitz products on Bergman spaces of the unit ball, *J. Math. Anal. Appl.* **325** (2007), 114–129.
15. K. Stroethoff and D. Zheng, Bounded Toeplitz products on weighted Bergman spaces, *J. Operator Theory* **59** (2008), 277–308.

16. X. Wang, J. Chen, and J. Xia, Sarason's conjecture of Toeplitz operators on Fock-Sobolev type spaces, *J. Funct. Spaces* (2018), Art. ID 1571365, 7 pages.
17. K. Zhu, *Operator Theory in Function Spaces*, 2nd edition, American Mathematical Society, Providence, RI, 2007.
18. K. Zhu, *Analysis on Fock Spaces*, GTM **263**, Springer Verlag, New York, 2012.

Sub-Hardy-Hilbert Spaces in the Non-commutative Unit Row Ball

Michael T. Jury and Robert T. W. Martin

1 Introduction

The purpose of this paper is to consider some factorization problems for non-commutative analogues of Toeplitz operators, in terms of various interconnections between these operators and (non-commutative versions of) the theory of de Branges–Rovnyak spaces. To explain what we mean, we will first recall some well-known results from the classical theory of Toeplitz operators.

A bounded Toeplitz operator, T, is the compression of multiplication by an L^∞ function, f, on the complex unit circle, $\partial\mathbb{D}$, to the Hardy space, H^2, of L^2 functions which extend analytically to the complex unit disk. That is, $T = T_f := P_{H^2} M_f|_{H^2}$ and $f \in L^\infty(\partial\mathbb{D})$ is called the *symbol* of T. Here, H^2 can also be defined as the Hilbert space of analytic functions in the complex unit disk, $\mathbb{D}$, with square-summable Taylor series coefficients at 0, and H^2 embeds isometrically into L^2 of the circle by taking non-tangential limits almost everywhere with respect to normalized Lebesgue measure, m. A theorem of Brown and Halmos [1, Theorem 6] characterizes the bounded Toeplitz operators as the bounded linear operators, $T \in \mathscr{L}(H^2)$, obeying the simple algebraic condition:

Supported by NSF grant DMS-1900364
Supported by NSERC grant 2020-05683

M. T. Jury (✉)
Department of Mathematics, University of Florida, Gainesville, FL, USA
e-mail: mjury@ad.ufl.edu

R. T. W. Martin
Department of Mathematics, University of Manitoba, Winnipeg, MB, Canada
e-mail: Robert.Martin@umanitoba.ca

I. Binder et al. (eds.), *Function Spaces, Theory and Applications*, Fields Institute Communications 87, https://doi.org/10.1007/978-3-031-39270-2_12

$$S^*TS = T; \qquad S := M_z,$$

where S, the isometry of multiplication by the independent variable, z, is called the *shift* on H^2. We employ the notation $h(S) = T_h$, clearly $p(S) = T_p$ for any complex polynomial $p \in \mathbb{C}[z]$. We say that $T_f \geq 0$ is *factorable* if we can write $T_f = T_{\overline{h}} T_h = h(S)^* h(S)$, for some outer $h \in H^\infty$, where H^∞ is the unital Banach algebra of uniformly bounded holomorphic functions in the disk. Here, we say an element $h \in H^\infty$ is *outer* if the analytic Toeplitz operator $h(S) = T_h$ has dense range and we say h is *inner* if $h(S)$ is an isometry.

A celebrated theorem of G. Szegö (later strengthened by Kolmogoroff and Kreĭn) implies that $T_f \geq 0$ is factorable if and only if its symbol, $f \geq 0$, is log-integrable,

$$\int_{\partial \mathbb{D}} \log f > -\infty,$$

[2]; see [3, Chapter IV]. As a particular corollary, suppose $T = T_f$ has a *factorable minorant*: that is, there is an H^∞ function g so that $T_f \geq T_g^* T_g$. In symbols, this means that $f \geq |g|^2$ on the unit circle. Since $\int \log |g|^2 \, dm > -\infty$, the same holds for f, and thus there in fact exists an (outer) $a \in H^\infty$ so that $T_f = T_a^* T_a$. Thus,

Theorem 1 $T = T_f \geq 0$ *is factorable if and only if it has a non-trivial factorable minorant.*

Factorizability of positive semi-definite Toeplitz operators is intimately connected to the theory of extreme points of the closed unit ball of H^∞, as well as to the factorization of functions in the Smirnov class. Indeed, it is known that a contractive analytic function, $b \in [H^\infty]_1$, is an extreme point if and only if $1 - |b|$ is not log-integrable on the circle and this is in turn equivalent to b being *column-extreme* (CE) in the sense that if $a \in H^\infty$ is any multiplier and $c := \binom{b}{a}$ is contractive as a multiplier from one to two copies of H^2, then $a \equiv 0$. Note that c will be contractive if and only if $I - b(S)^* b(S) \geq a(S)^* a(S)$, so that b is not CE if and only if the Toeplitz operator $T = I - b(S)^* b(S)$ has a non trivial factorable minorant. Hence there are three possibilities: (i) $I - b(S)^* b(S)$ vanishes identically, in which case b is inner, (ii) $I - b(S)^* b(S) \neq 0$ has no non-trivial factorable minorant, in which case b is an extreme point (or equivalently CE), or, (iii) b is not an extreme point and $I - b(S)^* b(S)$ is factorable. Given a non-extreme $b \in [H^\infty]_1$, the existence of an outer a so that the column $c = \binom{b}{a}$ is inner is in fact equivalent to the statement that a positive semi-definite Toeplitz operator is factorable if and only if it has factorable minorant. Namely, if $T = T_f \geq 0$ has a factorable minorant and we set $\mu = f \cdot m$ as before, then $\mu = \mu_b$ is the *Aleksandrov–Clark measure* of a contractive $b \in [H^\infty]_1$. Here, recall that there is (essentially) a bijection between positive, finite and regular Borel measures on the circle and contractive analytic functions in the disk [4–6]. Since μ_b is absolutely continuous, Fatou's theorem implies that b is not inner and that

$$T = (I - b(S)^*)^{-1} (I - b(S)^* b(S)) (I - b(S))^{-1}.$$

From this formula, it is clear that since T has a factorable minorant, so does the numerator $I - b(S)^*b(S)$, so that b cannot be an extreme point. Moreover, T will be factorable if and only if $I - b(S)^*b(S)$ is factorable, i.e. if and only if there exists an outer a so that $c = \binom{b}{a}$ is inner. Assuming that $b \in [H^\infty]_1$ is non-CE so that $I - b(S)^*b(S)$ is factorable by Theorem 1, there is then an outer function, $a \in [H^\infty]_1$, so that $a(0) > 0$ and $a(S)^*a(S) = I - b(S)^*b(S)$. This outer function, a, is uniquely determined by b and we call a the *Sarason outer function of b* after D. Sarason who performed a detailed analysis and computed the Taylor coefficients of a in [7, 8].

The Smirnov class, $\mathscr{N}^+$, consists of the analytic functions in the disk obtained as ratios of H^∞ functions with outer denominators. By a theorem of D. Sarason, any $f \in \mathscr{N}^+$ factors uniquely as $f = \frac{b}{a}$, where $b, a \in [H^\infty]_1$ are contractive, a is outer and the column $c := \binom{a}{b}$ is inner [9, Proposition 3.1]. This is essentially a consequence of the general form of Beurling's theorem – given any $f = \frac{b_0}{a_0} \in \mathscr{N}^+$, multiplication by f is densely defined on $\operatorname{Ran} a_0(S)$ as a_0 is outer and it is easily seen to be closable. Applying the Beurling theorem to the closed $S \otimes I_2$-invariant graph, $\operatorname{Gr} f(S) \subseteq H^2 \otimes \mathbb{C}^2$, where $f(S)$ is the closure of multiplication by f, gives a two-component column–inner $c = \binom{a}{b}$ so that $\operatorname{Gr} f(S) = \operatorname{Ran} c(S)$ and a is necessarily outer since $T_f = f(S)$ is densely defined. One can further show that, equipping the domains of $f(S)$ and $f(S)^*$ with the graph norm yields

$$\operatorname{Dom} f(S) = \mathscr{M}(a) \quad \text{and} \quad \operatorname{Dom} f(S)^* = \mathscr{H}(b),$$

where $\mathscr{M}(a) = \mathscr{M}(a(S))$ is the operator-range space of $a(S)$ and $\mathscr{H}(b) := \mathscr{M}\sqrt{I - b(S)b(S)^*}$ is the *de Branges–Rovnyak space* of b, the complementary space of $b(S)$. Here, if $A : \mathcal{H} \to \mathcal{J}$ is a linear operator, the operator-range space, $\mathscr{M}(A)$, is a Hilbert space equal to $\operatorname{Ran} A$ as a vector space, equipped with the norm that makes A a co-isometry onto its range. Both $\mathscr{M}(a)$ and $\mathscr{H}(b)$ are contractively contained in H^2. In summary, given any non-extreme contractive $b \in H^\infty$, there is a unique outer $a \in [H^\infty]_1$ so that $c = \binom{a}{b}$ is inner and $\operatorname{Ran} c(S) = \operatorname{Gr} f(S)$, for $f = b/a \in \mathscr{N}^+$, and equipping the domains of $f(S)$ and $f(S)^*$ with the graph norm yields $\operatorname{Dom} f(S) = \mathscr{M}(a)$ and $\operatorname{Dom} f(S)^* = \mathscr{H}(b)$. In particular, the de Branges–Rovnyak space of any non-extreme $b \in [H^\infty]_1$ is the domain of the adjoint of some (generally unbounded) closed Smirnov multiplier and vice versa.

It turns out that a surprising amount of the above discussion can be carried over to the non-commutative realm, where we consider the so-called 'multi-Toeplitz' operators, first considered by Popescu. We now briefly describe the non-commutative framework.

The full Fock space, $\mathbb{H}^2_d$, is a natural multi-variable and non-commutative (NC) generalization of H^2. The Fock space can be defined as the Hilbert space of all square-summable power series in several non-commuting variables, $\mathfrak{z} := (\mathfrak{z}_1, \cdots, \mathfrak{z}_d)$, for some fixed $d \in \mathbb{N} \cup \{+\infty\}$. Namely, let $\mathbb{F}^d$ denote the *free monoid*, the set of all finite *words* in the d letters $1, ..., d$. Any word, $\omega \in \mathbb{F}^d$, has the form $\omega = i_1 \cdots i_n$, $i_k \in \{1, \cdots, d\}$ and the unit of $\mathbb{F}^d$ is the *empty word*, $\emptyset$, containing no letters. Here, $|\omega| = n$ is called the length of the word ω and $|\emptyset| := 0$. Employing the notation $\mathfrak{z}^{\omega} := \mathfrak{z}_{i_1}\mathfrak{z}_{i_2}\cdots\mathfrak{z}_{i_n}$ and $1 := \mathfrak{z}^{\emptyset}$, any $h \in \mathbb{H}^2_d$ is a formal power series:

$$h(\mathfrak{z}) = \sum_{\omega\in\mathbb{F}^d} \hat{h}_{\omega}\mathfrak{z}^{\omega}; \qquad \hat{h}_{\omega} \in \mathbb{C}, \qquad \sum_{\omega\in\mathbb{F}^d} |\hat{h}_{\omega}|^2 = \sum_{k=0}^{\infty}\sum_{|\omega|=k} |\hat{h}_{\omega}|^2 < +\infty.$$

As first shown by Popescu, these formal power series converge absolutely in operator norm when evaluated at any strict contraction $Z = (Z_1, \cdots, Z_d) : \mathcal{H}\otimes\mathbb{C}^d \to \mathcal{H}$, $Z_k \in \mathscr{L}(\mathcal{H})$ from d copies of a Hilbert space $\mathcal{H}$ into itself, so that elements of $\mathbb{H}^2_d$ can be viewed as analytic, non-commutative functions on the open *NC unit row ball* of all strict *row contractions* on a fixed, separable Hilbert space [10]. Moreover, as in the single-variable setting, the unital Banach algebra, $\mathbb{H}^{\infty}_d$, of all uniformly bounded NC functions in the NC unit row ball, can be identified, completely isometrically, with the *left multiplier algebra* of $\mathbb{H}^2_d$, the algebra of all NC functions in the unit row ball which left multiply $\mathbb{H}^2_d$ into itself [10, 11]. In particular, left multiplication by any of the d non-commuting variables defines a d-tuple of isometries, $L_k := M^L_{\mathfrak{z}_k}$, with pairwise orthogonal ranges which we call the *left free shifts*. It follows that the row d-tuple $L = (L_1, \cdots, L_d) : \mathbb{H}^2_d \otimes \mathbb{C}^d \to \mathbb{H}^2_d$ is an isometry from several copies of $\mathbb{H}^2_d$ into itself. That is, the *left free shift*, L, is a *row isometry* and this plays the role of the shift in this non-commutative and multi-variable Hardy space theory. It follows that we can view Fock space, $\mathbb{H}^2_d$, as the *free* or *NC Hardy space* and $\mathbb{H}^{\infty}_d$ as the *NC Hardy algebra*.

Toeplitz operators in this NC setting were first defined and studied by Popescu [12–14]. A bounded operator $T \in \mathscr{L}(\mathbb{H}^2_d)$ will be called *left Toeplitz* if it obeys the left NC Brown–Halmos Toeplitz condition:

$$L_j^* T L_k = \delta_{j,k} T.$$

Popescu also initiated the study of factorization of positive left Toeplitz operators in [12–14]. Namely, in addition to the left free shifts, $L_k := M^L_{\mathfrak{z}_k}$, one can also define the right free shifts, $R_k := M^R_{\mathfrak{z}_k}$, on $\mathbb{H}^2_d$. As before, these are isometries with pairwise orthogonal ranges and the weakly closed, unital operator algebra they generate can be identified with the *right multiplier algebra* of Fock space. Here, a formal power series, $f(\mathfrak{z})$, is said to be a *right multiplier* if right multiplication by $f(\mathfrak{z})$ maps $\mathbb{H}^2_d$ into itself. A positive left Toeplitz operator, $T > 0$, is then said to be *factorable* if there is a right multiplier, $f(\mathfrak{z})$, so that $T = (M^R_f)^* M^R_f$. There is an exact version of the inner–outer factorization in this NC setting and any right (or left) multiplier, f, factors as $\Theta \cdot h$, where Θ is an *inner*, i.e. isometric right multiplier and h is *outer*, i.e. M^R_h has dense range [13, Theorem 4.2] [15, Corollary 2.2]. Hence if $T = (M^R_f)^* M^R_f$ then $T = (M^R_h)^* M^R_h$, and we can assume that $f = h$ is outer. Here, observe that right multipliers and left multipliers commute, so that any operator of the form $T = (M^R_f)^* M^R_f$ is left Toeplitz and positive. By applying his NC Wold decomposition for row isometries and using a similar operator–theoretic proof as in that of a corresponding weakened form of Theorem 1, Popescu established the following factorization results in Theorem 2 [14]. In the statement below, the *entropy* of a positive semi-definite left Toeplitz operator, T, is defined as the Szegö-type quantity,

$$\mathrm{e}(T) := \log \inf_{p \in \mathbb{C}\{\mathfrak{z}\}_0} \|\sqrt{T}(1-p)\|^2_{\mathbb{H}^2} \in \mathbb{R} \cup \{-\infty\},$$

where $\mathbb{C}\{\mathfrak{z}\} = \mathbb{C}\{\mathfrak{z}_1, \cdots, \mathfrak{z}_d\}$ denotes the ring of free or non-commutative complex polynomials in several NC variables and $\mathbb{C}\{\mathfrak{z}\}_0$ denotes the free polynomials which vanish at $0 := (0, \cdots, 0) \in \mathbb{B}^d_1$. Equivalently, $\mathbb{C}\{\mathfrak{z}\}_0$ is the ring of all complex free polynomials with vanishing constant term, $\hat{p}_\emptyset = 0$.

Theorem 2 (Popescu) *A positive left Toeplitz operator, $T \in \mathscr{L}(\mathbb{H}^2_d)$, has a non-zero factorable minorant if and only if it has finite entropy $\mathrm{e}(T) > -\infty$. If a positive left Toeplitz operator is bounded below, $T \geq \epsilon I > 0$, $\epsilon > 0$, then it is factorable.*

If $T \geq T_0 \geq 0$ are positive semi-definite left Toeplitz operators and $T_0 = (M^R_h)^*(M^R_h)$ is a factorable left Toeplitz operator bounded above by T, where h is outer, we say that T_0 is a *factorable Toeplitz minorant* or, more simply, a *factorable minorant* of T. If T_0 is maximal, it is called the *maximal factorable minorant* and h is the *maximal outer factor*. A natural question, posed and left unresolved in [14], is whether Theorem 1 holds in this setting:

Question 1 Let $T \geq 0$ be a positive semi-definite left Toeplitz operator on the Fock space. If T has a non-trivial factorable minorant, is T factorable?

While we do not have an answer to this question at this time, we will provide a detailed analysis and our results, Theorem 5.1 and Corollary 5.3, will show that this question is equivalent to several open questions on (column) extreme points of $[\mathbb{H}^\infty_d]_1$, free de Branges–Rovnyak spaces, purity of free Gleason solutions and free Smirnov multipliers of $\mathbb{H}^2_d$.

One of the central results of the paper is the following theorem, where we explicitly construct the maximal outer factor, a of $I - (M^R_b)^* M^R_b$, which we call the *free* or *NC Sarason function of* b, where b is a non-column-extreme contractive left multiplier of the Fock space and we generalize several results of D. Sarason on the relationship between b and a; see Sect. 3 and Corollary 3.3.

Theorem *Given a non-CE $b \in [\mathbb{H}^\infty_d]_1$, let $\mathscr{H}^{\mathrm{t}}(b) := \mathscr{M}\sqrt{I - b(R)b(R)^*}$ be the right free de Branges–Rovnyak space of b, let $X := L^*|_{\mathscr{H}^{\mathrm{t}}(b)}$ and set $\boldsymbol{b} := L^* b^{\mathrm{t}} \in \mathscr{H}^{\mathrm{t}}(b) \otimes \mathbb{C}^d$. Define a formal power series, $a(\mathfrak{z}) := \sum_{\omega \in \mathbb{F}^d} \hat{a}_\omega \mathfrak{z}^\omega$, by*

$$a(0) = \hat{a}_\emptyset := 1 - |b(0)|^2 - \|\boldsymbol{b}\|^2_{\mathscr{H}^{\mathrm{t}}(b)} > 0$$

and

$$\hat{a}_\omega := -a(0) \left\langle b^{\mathrm{t}}, X^\omega b^{\mathrm{t}} \right\rangle_{\mathscr{H}^{\mathrm{t}}(b)}, \quad \omega \neq \emptyset.$$

Then a belongs to $[\mathbb{H}^\infty_d]_1$*, a is outer, and* $a(R)^*a(R)$ *is the* ***maximal*** *factorable left Toeplitz minorant of* $I - b(R)^*b(R)$ *so that* $c := \binom{b}{a}$ *is CE and*

$$a(0)^2 = \inf_{p \in \mathbb{C}\{\mathfrak{z}\}_0} \left\langle 1 - p, (I - b(R)^*b(R))(1 - p) \right\rangle_{\mathbb{H}^2}.$$

In the above statement, $b^{\mathrm{t}} \in \mathbb{H}^2_d$ is the NC function obtained by reversing the order of products in all NC monomials in the NC Taylor series of b and $b(R) = M^R_{b^{\mathrm{t}}}$. The above formulas for the Taylor series coefficients of a recover Sarason's formulas when $d = 1$ [7]. As described above, Theorem 5.1 then connects factorizability of positive semi-definite left Toeplitz operators, graphs of free Smirnov multipliers and doubly free shift-invariant de Branges–Rovnyak spaces to the condition that $c := \binom{b}{a}$ is inner:

Theorem 5.1 Given a non-column-extreme left multiplier, $b \in [\mathbb{H}^\infty_d]_1$, with outer Sarason function, $a \in [\mathbb{H}^\infty_d]_1$, let $\binom{B}{A} D := CD$ be the inner–outer factorization of $c := \binom{b}{a}$. The following are equivalent:

(i) $c = \binom{b}{a}$ is inner so that $D \equiv 1$.
(ii) If $X := L^*|_{\mathscr{H}^{\mathrm{t}}(b)}$, then X^* is *pure*, i.e. $\lim \sum_{|\omega|=n} \|X^\omega h\|_{\mathscr{H}^{\mathrm{t}}(b)} = 0$ for any $h \in \mathscr{H}^{\mathrm{t}}(b)$.
(iii) X^* obeys the *weak purity condition*, $\sum_{|\omega|=n} \|X^\omega b^{\mathrm{t}}\|^2_{\mathscr{H}^{\mathrm{t}}(b)} \to 0$.
(iv) The free polynomials are dense in $\mathscr{H}^{\mathrm{t}}(b)$.
(v) The right free de Branges–Rovnyak space, $\mathscr{H}^{\mathrm{t}}(b)$, is the domain of the adjoint of the closed, densely defined right Smirnov multiplier, $H(R) = b(R)a(R)^{-1}$.

As a consequence of this theorem, Corollary 5.3 shows that Question 1 has a positive answer if and only if the above equivalent conditions hold for every non-column-extreme $b \in [\mathbb{H}^\infty_d]_1$:

Corollary 5.3 The following statements are equivalent:

(i) The equivalent conditions of Theorem 5.1 hold for every non-column-extreme $b \in [\mathbb{H}^\infty_d]_1$.
(ii) A positive semi-definite left Toeplitz operator, $T \in \mathscr{L}(\mathbb{H}^2_d)$, is factorable if and only if it has a factorable minorant.

While the above theorem, Theorem 5.1, is admittedly formidable and technical in appearance, it turns out that in many cases of interest its conditions are in fact checkable. In particular, it turns out to give very satisfactory results in the case of NC rational symbols $\mathfrak{b}$. In Sect. 6 we apply our results to the special case of contractive, non-commutative rational multipliers of the Fock space to establish the following dichotomy:

Theorem 6.3 If $\mathfrak{b}$ is a contractive, non-commutative rational multiplier of the Fock space, $\mathfrak{b}$ is either *inner* or $\mathfrak{b}$ is not column-extreme. If $\mathfrak{b}$ is not column-extreme, then its NC outer Sarason function, $\mathfrak{a} \in [\mathbb{H}^\infty_d]_1$, is also NC rational and the column $\mathfrak{c} := \binom{\mathfrak{b}}{\mathfrak{a}}$ is inner.

This fact certainly holds in one variable as a consequence of the Fejér–Riesz theorem and we further show that a Fejér–Riesz theorem holds for NC rational multipliers of the Fock space; see Theorem 6.5.

Theorem (NC rational Fejér–Riesz) *If* $\mathfrak{r}$ *is an NC rational left multiplier of Fock space and* $T := \mathrm{Re}\,\mathfrak{r}(R) \geq 0$ *is a positive semi-definite left Toeplitz operator, then* T *is factorable,* $T = \mathfrak{h}(R)^*\mathfrak{h}(R)$*, where* $\mathfrak{h}$ *is an outer NC rational multiplier.*

In fact, we obtain a more detailed conclusion, which gives some 'degree' control over $\mathfrak{h}$ (where 'degree' is defined in a suitable sense, namely, as the size of a so-called minimal realization), and we also obtain some control over the domain of regularity of $\mathfrak{h}$. We prove the theorem by converting the problem to an instance of computing a Sarason function a, and read off consequences from the de Branges–Rovnyak realization. As a consequence, we obtain an exact analogue of a classical corollary to Fatou's theorem for contractive NC rational multipliers in Corollary 6.6: A contractive NC rational left multiplier $\mathfrak{b} \in [\mathbb{H}_d^\infty]_1$ is inner if and only if its NC Clark measure is singular with respect to NC Lebesgue measure, m. Here, NC measures are defined as positive linear functionals on the *free disk system*, $\mathscr{A}_d := (\mathbb{A}_d + \mathbb{A}_d^*)^{-\|\cdot\|}$, where $\mathbb{A}_d := \mathrm{Alg}\{I, L_1, \cdots, L_d\}^{-\|\cdot\|}$ is the *free disk algebra*. When $d = 1$, NC measures can be identified with positive linear functionals on the C^*-algebra of continuous functions on the unit circle and hence with positive, finite and regular Borel measures by the Riesz–Markov theorem. Theorem 6.7 then provides an explicit formula for the non-commutative Radon–Nikodym derivative of the absolutely continuous part of $\mu_{\mathfrak{b}}$ with respect to a canonical NC Lebesgue measure.

2 Background and Notation

2.1 *Multipliers of Fock Space*

Left multiplications by the d independent NC variables, $\mathfrak{z} = (\mathfrak{z}_1, \cdots, \mathfrak{z}_d)$, define isometries on the Fock space with pairwise orthogonal ranges:

$$L_k := M^L_{\mathfrak{z}_k}, \qquad L_j^* L_k = \delta_{j,k} I.$$

It follows that the row d-tuple $L := (L_1, \cdots, L_d) : \mathbb{H}^2_d \otimes \mathbb{C}^d \to \mathbb{H}^2_d$ is an isometry from several copies of $\mathbb{H}^2_d$ into itself. Such an isometry is called a *row isometry*; we call L the *left free shift* and its components left free shifts. Similarly, one can define the right free shifts $R_k := M^R_{\mathfrak{z}_k}$ and the row isometric right free shift, R. The letter reversal map $\mathrm{t} : \mathbb{F}^d \to \mathbb{F}^d$, which reverses the order of letters in any word $\omega \in \mathbb{F}^d$, defines an involution on the free monoid,

$$\omega = i_1 \cdots i_n \mapsto \omega^{\mathrm{t}} := i_n \cdots i_1.$$

The free monomials $\{e_\omega := \mathfrak{z}^\omega |\ \omega \in \mathbb{F}^d\}$ define a standard orthonormal basis of $\mathbb{H}^2_d \simeq \ell^2(\mathbb{F}^d)$ and the letter reversal map gives rise to a unitary involution of the Fock space, U_{t}, defined by $U_{\mathrm{t}}\mathfrak{z}^\omega = \mathfrak{z}^{\omega^{\mathrm{t}}}$. Here, $e_\emptyset = \mathfrak{z}^\emptyset =: 1$ is called the vacuum vector of the Fock space. It is straightforward to verify that $U_{\mathrm{t}} L_k U_{\mathrm{t}} = R_k$, so that the left shifts are isomorphic to the right shifts.

The NC Hardy algebra, $\mathbb{H}^\infty_d$, of uniformly bounded NC functions can be identified, completely isometrically, with the unital Banach algebra of left multipliers of the NC Hardy space, $\mathbb{H}^2_d$. That is, given any NC function, $F \in \mathbb{H}^\infty_d$ and $h \in \mathbb{H}^2_d$, the left multiplication operator $M^L_F : \mathbb{H}^2_d \to \mathbb{H}^2_d$, defined by

$$h(Z) \mapsto F(Z) \cdot h(Z),$$

is bounded and $\|M^L_F\| = \|F\|_\infty$, where $\|\cdot\|_\infty$ denotes the supremum norm over $\mathbb{B}^d_{\mathbb{N}}$ [11, Theorem 3.1], [10, Theorem 3.1]. For any free polynomial, $p \in \mathbb{C}\{\mathfrak{z}\}$, one can check that $p(L) = M^L_p$ and so we employ the notation $F(L) := M^L_F$. Similarly, if $p \in \mathbb{C}\{\mathfrak{z}\}$, then $M^R_{p^{\mathrm{t}}} = p(R) = U_{\mathrm{t}} p(L) U_{\mathrm{t}}$, where if h is a formal power series $h(\mathfrak{z}) = \sum \hat{h}_\omega \mathfrak{z}^\omega$,

$$h^{\mathrm{t}}(\mathfrak{z}) := \sum \hat{h}_\omega \mathfrak{z}^{\omega^{\mathrm{t}}} = \sum \hat{h}_{\omega^{\mathrm{t}}} \mathfrak{z}^\omega.$$

In particular, if $h \in \mathbb{H}^2_d$, $h^{\mathrm{t}} = U_{\mathrm{t}} h$. The left and right multiplier algebras of $\mathbb{H}^2_d$ are unitarily equivalent via the unitary letter reversal involution U_{t} and can be identified with the left and right analytic Toeplitz algebras, $\mathscr{L}^\infty_d := \mathrm{Alg}\{I, L_1, \cdots, L_d\}^{-WOT}$ and $\mathscr{R}^\infty_d = \{I, R_1, \cdots, R_d\}^{-WOT} = U_{\mathrm{t}} \mathscr{L}^\infty_d U_{\mathrm{t}}$. Here, WOT denotes the weak operator topology. Since $p(R) = M^R_{p^{\mathrm{t}}}$ for any $p \in \mathbb{C}\{\mathfrak{z}\}$, we will write $G(R) = M^R_{G^{\mathrm{t}}}$ for any $G \in \mathbb{H}^\infty_d$. Namely, G belongs to the left multiplier algebra, $\mathbb{H}^\infty_d$, if and only if G^{t} belongs to the right multiplier algebra, $\mathbb{H}^{\infty;\mathrm{t}}_d := \mathrm{t} \circ \mathbb{H}^\infty_d$. (If $F = F(R) = M^R_{F^{\mathrm{t}}}$ for $F \in \mathbb{H}^\infty_d$, one can show that $\|F(R)\|$ is equal to the supremum norm of $F^{\mathrm{t}}(Z)$ over an *NC unit column ball.*) We will use the following terminology: A left or right multiplier is *inner* if it is isometric and *outer* if it has dense range.

2.2 Non-commutative Reproducing Kernel Hilbert Spaces

As in classical Hardy space theory, the Fock space is a (non-commutative) reproducing kernel Hilbert space, in the sense that for any $Z \in \mathbb{B}^d_n$ and vectors $y, v \in \mathbb{C}^n$, the *matrix-entry point evaluation*, $\ell_{Z,y,v} : \mathbb{H}^2_d \to \mathbb{C}$,

$$h \mapsto y^* h(Z) v,$$

is a bounded linear functional. Equivalently, $h \mapsto h(Z)$ is bounded as a linear map from $\mathbb{H}^2_d$ into the Hilbert space $\mathbb{C}^{n\times n}$ equipped with the Hilbert–Schmidt inner product. By the Riesz lemma, $\ell_{Z,y,v}$ is implemented by inner products against vectors $K\{Z, y, v\} \in \mathbb{H}^2_d$ which we call *NC Szegö kernel vectors.*

More generally, a Hilbert space, $\mathcal{H}$, of free non-commutative functions in $\mathbb{B}^d_{\mathbb{N}}$ is a *non-commutative reproducing kernel Hilbert space* (NC-RKHS), if the linear point evaluation functional, $\ell_{Z,y,v}$, is bounded on $\mathcal{H}$ for any $Z \in \mathbb{B}^d_n$, $y, v \in \mathbb{C}^n$ and $n \in \mathbb{N}$ [16]. (NC-RKHS can, of course, be defined on general NC sets. However, all NC-RKHS in this paper are Hilbert spaces of NC functions in $\mathbb{B}^d_{\mathbb{N}}$ so we omit the general definition.) As before, the Riesz lemma implies that these functionals are implemented by taking the inner products against *point evaluation* or *NC kernel vectors* $k\{Z, y, v\} \in \mathcal{H}$. Given any such NC-RKHS, $Z \in \mathbb{B}^d_n$ and $W \in \mathbb{B}^d_m$, one can define a completely bounded linear map on $n \times m$ complex matrices by $k(Z, W)[\cdot] : \mathbb{C}^{n\times m} \to \mathbb{C}^{n\times m}$,

$$y^* k(Z, W)[vu^*]x := \langle k\{Z, y, v\}, k\{W, x, u\}\rangle_{\mathcal{H}},$$

and this map is completely positive if $Z = W$ [16]. Following [16], we call $k(Z, W)[\cdot]$ the *completely positive non-commutative (CPNC) reproducing kernel* of $\mathcal{H}$ and we write $\mathcal{H} = \mathcal{H}_{nc}(k)$. One can check that adjoints of left and right multipliers of an NC-RKHS have a familiar action on NC kernel vectors:

$$(M^L_F)^* k\{Z, y, v\} = k\{Z, F(Z)^* y, v\} \quad \text{and} \quad (M^R_G)^* k\{Z, y, v\} = k\{Z, y, G(Z)v\}.$$

All NC-RKHS in this paper will be Hilbert spaces of free NC functions in the unit row ball $\mathbb{B}^d_{\mathbb{N}}$,

$$\mathbb{B}^d_{\mathbb{N}} = \bigsqcup \mathbb{B}^d_n, \quad \mathbb{B}^d_n := \left\{ Z \in \mathbb{C}^{n\times n} \otimes \mathbb{C}^{1\times d} \middle| \; ZZ^* = Z_1 Z_1^* + \cdots + Z_d Z_d^* < I_n \right\}.$$

In the case of the Fock space, $\mathbb{H}^2_d = \mathcal{H}_{nc}(K)$, where K is the *NC Szegö kernel*: Given $Z \in \mathbb{B}^d_n$, $W \in \mathbb{B}^d_m$ and $P \in \mathbb{C}^{n\times m}$,

$$K(Z, W) := \left(\mathrm{id}_{n,m}[\cdot] - \mathrm{Ad}_{Z,W^*}[\cdot]\right)^{-1} \circ P = \sum_{j=0}^{\infty} \mathrm{Ad}^{(j)}_{Z,W^*}[P] = \sum_{\omega \in \mathbb{F}^d} Z^{\omega} P W^{*\omega},$$

$$\mathrm{Ad}_{Z,W^*}[P] := Z_1 P W_1^* + \cdots + Z_d P W_d^*.$$

3 Column-Extreme Multipliers and the Sarason Outer Function

Recall that an element $b \in [\mathbb{H}_d^\infty]_1$ is said to be *column-extreme (CE)* if there is no non-zero $a \in \mathbb{H}_d^\infty$ so that the two-component column:

$$\begin{pmatrix} b(L) \\ a(L) \end{pmatrix} : \mathbb{H}_d^2 \to \mathbb{H}_d^2 \otimes \mathbb{C}^2$$

is a contractive left multiplier from one to two copies of Fock space [17]. Any CE multiplier of Fock space is necessarily an extreme point of the closed convex set of contractive left multipliers [17, Corollary 6.8]. Classically, a multiplier of H^2 is an extreme point if and only if it is CE and this also holds in the commutative multi-variable setting of Drury–Arveson space as a consequence of the column–row property of M.P. Hartz [18, Theorem 1.2, Theorem 1.6]. However, the column–row property does not hold in the NC setting of Fock space and so whether or not every extreme point of $[\mathbb{H}_d^\infty]_1$ is CE remains an open problem [19].

This concept of column-extreme extends readily to the case of operator-valued left multipliers between vector-valued Fock spaces [17]. Let $\mathcal{H}, \mathcal{J}$ be separable Hilbert spaces. We define $\mathbb{H}_d^\infty \otimes \mathscr{L}(\mathcal{H}, \mathcal{J})$ as the closure of this algebraic tensor product in the weak operator topology (WOT) of $\mathscr{L}\left(\mathbb{H}_d^2 \otimes \mathcal{H}, \mathbb{H}_d^2 \otimes \mathcal{J}\right)$. Elements of $\mathbb{H}_d^\infty \otimes \mathscr{L}(\mathcal{H}, \mathcal{J})$ can be viewed as operator-valued multipliers from $\mathbb{H}_d^2 \otimes \mathcal{H}$ into $\mathbb{H}_d^2 \otimes \mathcal{J}$. A contractive left multiplier $b \in [\mathbb{H}_d^\infty \otimes \mathscr{L}(\mathcal{H}, \mathcal{J})]_1$ is said to be CE if and only if $a \in [\mathbb{H}_d^\infty \otimes \mathscr{L}(\mathcal{H}, \mathcal{J})]_1$ and $c := \left(\begin{smallmatrix} b \\ a \end{smallmatrix}\right) \in [\mathbb{H}_d^\infty \otimes \mathscr{L}(\mathcal{H}, \mathcal{J}) \otimes \mathbb{C}^2]_1$ implies that $a \equiv 0$.

Recall that any contractive $b \in [H^\infty]_1$ has a *de Branges–Rovnyak realization*:

$$b(z) = D + C(I - zA)^{-1} zB, \qquad z \in \mathbb{D},$$

where:

$$A := S^*|_{\mathscr{H}(b)}, \quad B := S^* b, \quad C := (k_0^b)^* \quad \text{and} \quad D := b(0),$$

and $\mathscr{H}(b)$ is the de Branges–Rovnyak space of b [20, Theorem 1.1, Theorem 1.2, Theorem 1.3]. This space is a Hilbert space of analytic functions in $\mathbb{D}$ which is contractively contained in H^2, and it is a RKHS with reproducing kernel:

$$k^b(z, w) := \frac{1 - b(z)b(w)^*}{1 - zw^*}.$$

In particular, b is inner if and only if $\mathscr{H}(b)$ is contained isometrically in H^2 and in this case $\mathscr{H}(b) = (bH^2)^\perp$. The contractive multiplier, b, is said to be the *transfer function* of the (observable, co-isometric) *de Branges–Rovnyak colligation*, U_b, defined by:

$$U_b := \begin{pmatrix} A & B \\ C & D \end{pmatrix} : \begin{pmatrix} \mathscr{H}(b) \\ \mathbb{C} \end{pmatrix} \to \begin{pmatrix} \mathscr{H}(b) \\ \mathbb{C} \end{pmatrix};$$

see [20].

These de Branges–Rovnyak realizations were extended to the NC multivariable setting of contractive multipliers between vector-valued Fock spaces in [20]. Given $b \in [\mathbb{H}^\infty_d \otimes \mathscr{L}(\mathcal{H}, \mathcal{J})]_1$, we define the *right free de Branges–Rovnyak space* of b, $\mathscr{H}^{\mathrm{t}}(b)$, as the complementary space of $b(R)$, $\mathscr{H}^{\mathrm{t}}(b) := \mathscr{M}\left(\sqrt{I - b(R)b(R)^*}\right)$. We will sometimes denote the norm and inner product of $\mathscr{H}^{\mathrm{t}}(b)$ by $\| \cdot \|_b := \| \cdot \|_{\mathscr{H}^{\mathrm{t}}(b)}$ and $\langle \cdot, \cdot \rangle_b := \langle \cdot, \cdot \rangle_{\mathscr{H}^{\mathrm{t}}(b)}$, respectively. We will also employ the notations $\mathscr{M}^{\mathrm{t}}(b) := \mathscr{M}(b(R))$, $\mathscr{M}^{\mathrm{t}}(b^*) := \mathscr{M}(b(R)^*)$ and $\mathscr{H}^{\mathrm{t}}(b^*) := \mathscr{M}\sqrt{I - b(R)^* b(R)}$, where recall that if $A \in \mathscr{L}(\mathcal{H}, \mathcal{J})$ is a bounded linear operator, $\mathscr{M}(A)$ is the operator-range space of A. The right free de Branges–Rovnyak space, $\mathscr{H}^{\mathrm{t}}(b)$, is a $\mathcal{J}$–valued NC-RKHS on $\mathbb{B}^d_{\mathbb{N}}$ with CPNC kernel

$$K^b(Z, W)[P \otimes I_{\mathcal{H}}] := K(Z, W)[P] \otimes I_{\mathcal{H}} - K(Z, W) \otimes \mathrm{id}_{\mathcal{H}}[b^{\mathrm{t}}(Z)(P \otimes I_{\mathcal{H}}) b^{\mathrm{t}}(W)^*],$$

for $Z \in \mathbb{B}^d_n$, $W \in \mathbb{B}^d_m$ and $P \in \mathbb{C}^{n \times m}$. Set $X := L^* \otimes I_{\mathcal{J}}|_{\mathscr{H}^{\mathrm{t}}(b)}$ and $\boldsymbol{b} := L^* \otimes I_{\mathcal{J}} b(R) 1 \otimes I_{\mathcal{H}} \in \mathscr{L}\left(\mathcal{H}, \mathscr{H}^{\mathrm{t}}(b) \otimes \mathbb{C}^d\right)$. Namely,

$$\boldsymbol{b} h := L^* \otimes I_{\mathcal{J}} b(R) 1 \otimes h \in \mathscr{H}^{\mathrm{t}}(b) \otimes \mathbb{C}^d,$$

for any $h \in \mathcal{H}$. Here, in parallel with the classical theory, any right free de Branges–Rovnyak space, $\mathscr{H}^{\mathrm{t}}(b)$ for $b \in [\mathbb{H}^\infty_d]_1$ is contractively contained in $\mathbb{H}^2_d$, is left shift co-invariant, and while b^{t} generally does not belong to $\mathscr{H}^{\mathrm{t}}(b)$, its backward left shifts always do, $L^* b^{\mathrm{t}} \in \mathscr{H}^{\mathrm{t}}(b) \otimes \mathbb{C}^d$ [20, Proposition 4.2]. Analogous statements hold for operator-valued b.

Any such $b \in [\mathbb{H}^\infty_d \otimes \mathscr{L}(\mathcal{H}, \mathcal{J})]_1$ is realized as the transfer function of the co-isometric right *free de Branges–Rovnyak colligation* U_b, defined by

$$U_b := \begin{pmatrix} A & B \\ C & D \end{pmatrix} : \begin{pmatrix} \mathscr{H}^{\mathrm{t}}(b) \\ \mathcal{H} \end{pmatrix} \to \begin{pmatrix} \mathscr{H}^{\mathrm{t}}(b) \otimes \mathbb{C}^d \\ \mathcal{J} \end{pmatrix},$$

where $A := X := L^* \otimes I_{\mathcal{J}}|_{\mathscr{H}^{\mathrm{t}}(b)}$,

$$B := L^* \otimes I_{\mathcal{J}} b(R)\, 1 \otimes I_{\mathcal{H}} = \boldsymbol{b}, \quad C := (K^b_0)^* \quad \text{and} \quad D := b(0);$$

see [20]. For any $b \in [\mathbb{H}^\infty_d \otimes \mathscr{L}(\mathcal{H}, \mathcal{J})]_1$, recall that

$$X^* X \leq I - K^b_0 (K^b_0)^* \quad \text{and} \quad \boldsymbol{b}^* \boldsymbol{b} \leq I_{\mathcal{H}} - b(0)^* b(0),$$

[20, Proposition 4.2]. Further recall that b is column-extreme if and only if equality holds in either (and hence both) of these formulas [17, Theorem 6.4]. That is, b is CE if and only if

$$X^*X = I - K_0^b(K_0^b)^*,$$

or equivalently,

$$\boldsymbol{b}^*\boldsymbol{b} = I_{\mathcal{H}} - b(0)^*b(0).$$

Given $X := L^* \otimes I_{\mathcal{J}}|_{\mathscr{H}^{\mathrm{t}}(b)}$ for some $b \in [\mathbb{H}_d^\infty \otimes \mathscr{L}(\mathcal{H}, \mathcal{J})]_1$, let

$$X^{(n)} := (X \otimes I_d \otimes I_{n-1}) \cdots (X \otimes I_d \otimes I_d)(X \otimes I_d)X,$$

so that for any $h \in \mathscr{H}^{\mathrm{t}}(b)$,

$$\|X^{(n)}h\|_b^2 = \sum_{|\alpha|=n} \|X^\alpha h\|_b^2 = \left\langle h, \mathrm{Ad}_{X^*,X}^{(n)}(I)h \right\rangle_b, \tag{1}$$

where recall that $\|\cdot\|_b$ and $\langle \cdot, \cdot \rangle_b$ denote the norm and inner product in $\mathscr{H}^{\mathrm{t}}(b)$. Here, given a row d-tuple of operators, $T = (T_1, \cdots, T_d) : \mathcal{K} \otimes \mathbb{C}^d \to \mathcal{K}$, the completely positive linear map of *adjunction by T and T^**, $\mathrm{Ad}_{T,T^*} : \mathscr{L}(\mathcal{K}) \to \mathscr{L}(\mathcal{K})$, is defined as

$$\mathrm{Ad}_{T,T^*}(A) := T_1AT_1^* + \cdots + T_dAT_d^*. \tag{2}$$

The row d-tuple, T, is said to be *pure* if $\mathrm{Ad}_{T,T^*}^{(n)}(I) \overset{SOT}{\to} 0$, where SOT denotes the strong operator topology.

Theorem 3.1 *Let $b \in [\mathbb{H}_d^\infty \otimes \mathscr{L}(\mathcal{H}, \mathcal{J})]_1$ be column-extreme (CE). Then b is inner if and only if X^* obeys the weak purity condition, $\|X^{(n)}\boldsymbol{b}h\| \to 0$ for any $h \in \mathcal{H}$. In particular, X^* is pure if and only if b is inner.*

By [20, Theorem 4.6], $b \in [\mathbb{H}_d^\infty \otimes \mathscr{L}(\mathcal{H}, \mathcal{J})]_1$ is inner if and only if X^* is pure (or *strongly stable* in the language of [20]). The above theorem is a slight weakening of this assumption in the case of a CE multiplier – instead of checking that $\mathrm{Ad}_{X^*,X}^{(n)}(I)x$ converges to 0 for any $x \in \mathscr{H}^{\mathrm{t}}(b)$, it suffices to show that $\|X^{(n)}\boldsymbol{b}h\|^2 = \left\langle \boldsymbol{b}h, \mathrm{Ad}_{X^*,X}^{(n)}(I)\boldsymbol{b}h \right\rangle \to 0$ for any $h \in \mathcal{H}$. (Here, recall that WOT convergence of a sequence of self-adjoint operators to 0 implies SOT convergence.) In particular, if $b \in [\mathbb{H}_d^\infty]_1$ is scalar, it suffices to check that $\|X^{(n)}\boldsymbol{b}\| \to 0$.

Proof If b is inner then $\mathscr{H}^{\mathrm{t}}(b) = \mathrm{Ran}\, b(R)^\perp \subseteq \mathbb{H}_d^2 \otimes \mathcal{J}$, so that $\mathscr{H}^{\mathrm{t}}(b)$ is a closed, co-invariant subspace for $L \otimes I_{\mathcal{J}}$. It is then easily checked that X^* is pure since $L \otimes I_{\mathcal{J}}$ is.

Conversely, if b is CE then we have that $X^*X = I - K_0^b(K_0^b)^*$. Suppose now that X is pure and consider $\|X^{(n)}\boldsymbol{b}h\|_b^2$ where $\boldsymbol{b}h = L^* \otimes I_{\mathcal{J}}\, b(R)1 \otimes h \in \mathscr{H}^{\mathrm{t}}(b) \otimes \mathbb{C}^d$ and $\|h\|_{\mathcal{H}} = 1$. Hence $\|X^{(n)}\boldsymbol{b}h\|_b^2 \to 0$ as $n \uparrow +\infty$. Then,

$$\begin{aligned}
\|X^{(n)}\boldsymbol{b}h\|^2 &= \left\langle X^{(n-1)}\boldsymbol{b}h,\, X^*X \otimes I_d \otimes I_{n-1}X^{(n-1)}\boldsymbol{b}h\right\rangle_b \\
&= \|X^{(n-1)}\boldsymbol{b}h\|_b^2 - \left\langle X^{(n-1)}\boldsymbol{b}h,\, K_0^b(K_0^b)^* \otimes I_d \otimes I_{n-1}X^{(n-1)}\boldsymbol{b}h\right\rangle_b \\
&= \|X^{(n-1)}\boldsymbol{b}h\|_b^2 - \left\langle L^{*(n)}b(R)1 \otimes h,\, K_0^b(K_0^b)^* \otimes I_d \otimes I_{n-1}L^{*(n)} \right. \\
&\quad \left. b(R)1 \otimes h\right\rangle_b \\
&= \|X^{(n-1)}\boldsymbol{b}h\|_b^2 - \sum_{|\alpha|=n} \|\hat{b}_\alpha h\|_{\mathcal{J}}^2 \\
\cdots &= \|\boldsymbol{b}h\|_b^2 - \sum_{|\alpha|=1}^{n} \|\hat{b}_\alpha h\|_{\mathcal{J}}^2 \\
&= 1 - \sum_{|\alpha|\le n} \|\hat{b}_\alpha h\|_{\mathcal{J}}^2 \\
&\to 1 - \sum_{\alpha} \|\hat{b}_\alpha h\|_{\mathcal{J}}^2 = 1 - \|b(R)1 \otimes h\|_{\mathbb{H}^2 \otimes \mathcal{J}}^2.
\end{aligned}$$

In the above we used that if b is CE, then $\|\boldsymbol{b}h\|_b^2 = 1 - \|b(0)h\|_{\mathcal{J}}^2$. Since X^* is pure, it follows that $\|b(R)1 \otimes h\|_{\mathbb{H}^2 \otimes \mathcal{J}}^2 = 1$. Since we also have that $b \in [\mathbb{H}_d^\infty \otimes \mathscr{L}(\mathcal{H}, \mathcal{J})]_1$, an observation of Davidson–Pitts implies that b is inner [15, Proposition 2.4]. Indeed, if $b \in [\mathbb{H}_d^\infty \otimes \mathscr{L}(\mathcal{H}, \mathcal{J})]_1$ is a contractive multiplier so that $\|b(R)1 \otimes h\|_{\mathbb{H}^2 \otimes \mathcal{J}} = 1$ for any unit norm $h \in \mathcal{H}$, then for any $\omega \in \mathbb{F}^d$,

$$\begin{aligned}
\|b(R)L^\omega 1 \otimes h\|_{\mathbb{H}^2 \otimes \mathcal{J}}^2 &= \|L^\omega \otimes I_{\mathcal{J}}\, b(R)1 \otimes h\|_{\mathbb{H}^2 \otimes \mathcal{J}}^2 \\
&= \|b(R)1 \otimes h\|^2 = 1 = \|L^\omega 1 \otimes h\|_{\mathbb{H}^2 \otimes \mathcal{H}}^2.
\end{aligned}$$

Hence,

$$\begin{aligned}
1 &= \langle L^\omega 1 \otimes h, b(R)^* b(R) L^\omega 1 \otimes h\rangle_{\mathbb{H}^2 \otimes \mathcal{H}} \\
&\le \|L^\omega 1 \otimes h\| \|b(R)^* b(R) L^\omega 1 \otimes h\| \\
&\le \|b(R)^* b(R)\| \|L^\omega 1 \otimes h\| \le 1.
\end{aligned}$$

Since equality holds in the Cauchy–Schwarz inequality, we must have that $b(R)^* b(R) L^\omega 1 \otimes h = \zeta L^\omega 1 \otimes h$, for some $\zeta \in \mathbb{C}$. Since $b(R)^* b(R)$ is positive semi-definite, $\zeta \ge 0$ and since $\|b(R)L^\omega 1 \otimes h\| = 1$, $\zeta = 1$. It follows that for any free polynomial, $p \in \mathbb{C}\{\mathfrak{z}\}$ and unit norm $h \in \mathcal{H}$,

$$b(R)^* b(R) p(L) 1 \otimes h = b(R)^* b(R) \sum_{|\omega|\le N} \hat{p}_\omega L^\omega 1 \otimes h = p(L) 1 \otimes h.$$

By the density of the free polynomials in Fock space, $b(R)^*b(R) = I$ and b is inner. □

Proposition 3.2 *$b \in [\mathbb{H}^\infty_d \otimes \mathscr{L}(\mathcal{H}, \mathcal{J})]_1$ is column-extreme if and only if its canonical de Branges–Rovnyak colligation is both isometric and co-isometric.*

Proof For simplicity, assume that $b \in [\mathbb{H}^\infty_d]_1$. Proof of the general case is analogous. The de Branges–Rovnyak colligation of b is always co-isometric [20]. Here, the colligation is:

$$U_b := \begin{pmatrix} A & B \\ C & D \end{pmatrix} : \begin{pmatrix} \mathscr{H}^{\mathrm{t}}(b) \\ \mathbb{C} \end{pmatrix} \to \begin{pmatrix} \mathscr{H}^{\mathrm{t}}(b) \otimes \mathbb{C}^d \\ \mathbb{C} \end{pmatrix},$$

$$A := X := L^*|_{\mathscr{H}^{\mathrm{t}}(b)}, \ B := \boldsymbol{b} := L^* b^{\mathrm{t}} \in \mathscr{H}^{\mathrm{t}}(b) \otimes \mathbb{C}^d, \ C := (K_0^b)^*, \ D := b(0).$$

As we know, b is CE if and only if $X^*X = I - K_0^b(K_0^b)^*$, or equivalently, $\boldsymbol{b}^*\boldsymbol{b} = 1 - |b(0)|^2$. So now, we calculate:

$$U_b^* U_b = \begin{pmatrix} \overbrace{X^*X + K_0^b(K_0^b)^*}^{=I} & X^*\boldsymbol{b} + K_0^b b(0) \\ * & \underbrace{\boldsymbol{b}^*\boldsymbol{b} + |b(0)|^2}_{=1} \end{pmatrix}.$$

It remains to show the off-diagonal components vanish. Indeed,

$$\begin{aligned}
(K_Z^b)^* \left(X^*\boldsymbol{b} + K_0^b b(0) \right) &= (X K_Z^b)^* \boldsymbol{b} + (K_Z^b)^* K_0^b b(0) \\
&= \left(K_Z^b Z^* - \boldsymbol{b} b^{\mathrm{t}}(Z)^* \right)^* \boldsymbol{b} + (I - b^{\mathrm{t}}(Z) b(0)^*) b(0) \\
&= Z\boldsymbol{b}(Z) - b^{\mathrm{t}}(Z)\boldsymbol{b}^*\boldsymbol{b} + b(0) - b^{\mathrm{t}}(Z)|b(0)|^2 \\
&= b^{\mathrm{t}}(Z) - b(0) - b^{\mathrm{t}}(Z) + b^{\mathrm{t}}(Z)|b(0)|^2 + b(0) \\
&\quad - b^{\mathrm{t}}(Z)|b(0)|^2 \\
&= 0.
\end{aligned}$$

□

3.1 The Sarason Function

In one variable, as discussed in the introduction, $b \in [H^\infty]_1$ is not an extreme point if and only if $1 - |b|^2$ is log-integrable on the circle and in this case there is a unique outer function, $a \in H^\infty$, so that $a(0) > 0$ and

$$|a(\zeta)|^2 = 1 - |b(\zeta)|^2; \qquad \zeta \ a.e. \ \partial\mathbb{D}.$$

This implies, in turn, that $c := \binom{b}{a}$ is inner. If $b \in [H^\infty]_1$ is non-extreme, D. Sarason showed that this unique outer a is realized as the transfer function of the colligation:

$$U_a := \begin{pmatrix} A & B \\ C_a & D_a \end{pmatrix} : \begin{pmatrix} \mathscr{H}(b) \\ \mathbb{C} \end{pmatrix} \to \begin{pmatrix} \mathscr{H}(b) \\ \mathbb{C} \end{pmatrix},$$

where

$$U_b = \begin{pmatrix} A & B \\ C & D \end{pmatrix} : \begin{pmatrix} \mathscr{H}(b) \\ \mathbb{C} \end{pmatrix} \to \begin{pmatrix} \mathscr{H}(b) \\ \mathbb{C} \end{pmatrix},$$

is the de Branges–Rovnyak realization of b,

$$C_a := -a(0) \langle b, \cdot \rangle_{\mathscr{H}(b)}, \quad D_a := a(0) > 0 \quad \text{and}$$
$$a(0)^2 := 1 - \|S^* b\|^2_{\mathscr{H}(b)} - |b(0)|^2 > 0.$$

The column multiplier, $c = \binom{b}{a}$, is then the transfer function of the colligation:

$$U_c := \begin{pmatrix} A & B \\ C & D \\ C_a & D_a \end{pmatrix} : \begin{pmatrix} \mathscr{H}(b) \\ \mathbb{C} \end{pmatrix} \to \begin{pmatrix} \mathscr{H}(b) \\ \mathbb{C} \\ \mathbb{C} \end{pmatrix}.$$

(Sarason didn't state his results in the language of realization theory, but he computes the Taylor series coefficients of a and these are easily seen to coincide with the formula given by the above realization [7].)

Given a non-CE $b \in [\mathbb{H}^\infty_d]_1$ (or more generally any non-CE $b \in [\mathbb{H}^\infty_d \otimes \mathscr{L}(\mathcal{H}, \mathcal{J})]_1$) we can construct an $a \in [\mathbb{H}^\infty_d]_1$ so that the column $c = \binom{b}{a} \in [\mathbb{H}^\infty_d \otimes \mathbb{C}^2]_1$ is contractive by checking that the analogous colligation:

$$U_c := \begin{pmatrix} - & \boxed{U_b} & - \\ -a(0) \left\langle b^{\mathrm{t}}, \cdot \right\rangle_{\mathscr{H}^{\mathrm{t}}(b)} & & a(0) \end{pmatrix} : \begin{pmatrix} \mathscr{H}^{\mathrm{t}}(b) \\ \mathbb{C} \end{pmatrix} \to \begin{pmatrix} \mathscr{H}^{\mathrm{t}}(b) \otimes \mathbb{C}^d \\ \mathbb{C}^2 \end{pmatrix},$$

is contractive. Here, as above, we define $a(0) > 0$ and

$$a(0)^2 := 1 - |b(0)|^2 - \boldsymbol{b}^* \boldsymbol{b} > 0.$$

In fact, in [17], we check that U_c is both isometric and co-isometric and it follows that c is column-extreme by Proposition 3.2. Further recall that $b^{\mathrm{t}} \in \mathscr{H}^{\mathrm{t}}(b)$ if and only if b is non-CE [17, Theorem 6.4]. In particular, it immediately follows that

$$a(Z) = a(0)\left(I - \sum_{\omega \neq \emptyset} Z^{\omega} \langle b^{t}, X^{\omega} b^{t} \rangle_b \right),$$

so that the non-vacuum coefficients of a are given by the Sarason formulas:

$$\hat{a}_{\omega} = -a(0) \langle b^{t}, X^{\omega} b^{t} \rangle_b ; \qquad \omega \neq \emptyset.$$

To simplify notation, we will assume for the remainder of this section that $b \in [\mathbb{H}_d^{\infty}]_1$ or $c = \binom{b}{a} \in [\mathbb{H}_d^{\infty} \otimes \mathbb{C}^2]_1$, where a is the Sarason outer function of b. The proofs for general $b \in [\mathbb{H}_d^{\infty} \otimes \mathscr{L}(\mathcal{H}, \mathcal{J})]_1$ are virtually identical.

Corollary 3.3 *Let $b \in [\mathbb{H}_d^{\infty}]_1$ be non-CE and let $a \in [\mathbb{H}_d^{\infty}]_1$ be its unique Sarason function. Then a is the maximal outer factor of $T = I - b(R)^*b(R)$. In particular,*

$$a(0)^2 = \inf_{\substack{p \in \mathbb{C}\{\mathfrak{z}\};\\ p(0)=0}} \langle 1 - p, (I - b(R)^*b(R))(1 - p) \rangle_{\mathbb{H}^2}.$$

If $c = \binom{b}{a}$, then c is CE and $\|\boldsymbol{c}\|_c^2 = \|\boldsymbol{b}\|_b^2$, where $\boldsymbol{c} = L^* \otimes I_2 c^{t}$, $\boldsymbol{b} = L^* b^{t}$.

Proof Consider $T = I - b(R)^*b(R)$ and let $T_0 = A(R)^*A(R) \leq T$ be the maximal factorable minorant with $A \in [\mathbb{H}_d^{\infty}]_1$ outer. By [14, Theorem 1.3],

$$|A(0)|^2 = \inf_{p \in \mathbb{C}\{\mathfrak{z}\}_0} \langle 1 - p, (I - b(R)^*b(R))(1 - p) \rangle_{\mathbb{H}^2},$$

and A is maximal in the sense that if $D \in \mathbb{H}_d^{\infty} \otimes \mathcal{H}$ is any column outer so that $D(R)^*D(R) \leq T$, then $D(R)^*D(R) \leq A(R)^*A(R)$. We can and do assume that $A(0) > 0$. Note that $D(R)^*D(R) \leq T$ would imply that the column $C := \binom{b}{D}$ is contractive. By applying Douglas factorization, there is a contraction, F, so that $D(R)^* = A(R)^*F^*$, or equivalently, $FA(R) = D(R)$. Since A is outer, F commutes with the left shifts so that $F = F(R) \in [\mathbb{H}_d^{\infty}]_1$ [15, Theorem 1.2]. In fact, F is also outer since D is. It follows that the column $\binom{b}{A}$ is CE. If not, then there is some non-zero D so that

$$\begin{pmatrix} b \\ A \\ D \end{pmatrix}$$

is contractive and we can assume without loss in generality that A, D are outer. The previous argument then implies that $A(R)^*A(R) + D(R)^*D(R) \leq A(R)^*A(R)$ since A is maximal and hence $D \equiv 0$.

To prove that $A = a$, where A is the Popescu maximal outer factor of $I - b^*b$ and a is the outer Sarason function of b, consider the column: $C := \binom{b}{A}$. This is contractive and CE, so that

$$\|\boldsymbol{C}\|_C^2 = \boldsymbol{C}^*\boldsymbol{C} = 1 - |b(0)|^2 - A(0)^2$$
$$= \|\boldsymbol{b}\|_b^2 + a(0)^2 - A(0)^2,$$

where recall that $a(0) > 0$ by definition and we can assume that $A(0) > 0$ as well. It follows that

$$\|\boldsymbol{C}\|_C^2 - \|\boldsymbol{b}\|_b^2 = a(0)^2 - A(0)^2.$$

The CPNC kernel for $\mathscr{H}^{\mathrm{t}}(C)$ is:

$$K^C(Z,W) := K(Z,W)[\cdot] - K(Z,W)\left[\begin{pmatrix} b^{\mathrm{t}}(Z)(\cdot)b^{\mathrm{t}}(W)^* & * \\ * & A^{\mathrm{t}}(Z)(\cdot)A^{\mathrm{t}}(W)^* \end{pmatrix}\right] \geq 0.$$

It follows that the norm of $\|\boldsymbol{C}\|_C^2$ is the infimum over all $t > 0$ so that

$$\boldsymbol{C}(Z)(\cdot)\boldsymbol{C}(W)^* = \begin{pmatrix} \boldsymbol{b}(Z) \\ \boldsymbol{A}(Z) \end{pmatrix}[\cdot]\,(\,\boldsymbol{b}(W)^*,\ \boldsymbol{A}(W)^*\,) \leq t^2 K^C(Z,W) \otimes I_d\,[\cdot],$$

as CPNC kernels [17, Lemma 6.6]. Comparing (1, 1) entries, we see that

$$\|\boldsymbol{b}\|_b^2 \leq \|\boldsymbol{C}\|_C^2.$$

In particular,

$$0 \leq \|\boldsymbol{C}\|_C^2 - \|\boldsymbol{b}\|_b^2 = a(0)^2 - A(0)^2,$$

and $a(0) \geq A(0)$.

If a was not outer, then $a = \Theta \cdot F$ for some non-trivial inner Θ and outer F and then $a(0)^2 < |F(0)|^2$ and $F(R)^*F(R) = a(R)^*a(R) \leq I - b(R)^*b(R) =: T$. However, A is the maximal outer factor of T, so that $F(R)^*F(R) \leq A(R)^*A(R)$ and $F(R) = h(R)A(R)$ for some contractive and non-trivial outer h [21, Theorem 3.1]. In particular $a(0)^2 < |F(0)|^2 < A(0)^2$, contradicting that $a(0)^2 \geq A(0)^2$. This proves that $A = a$ and it follows that if $c = \binom{b}{a}$, then $c = C$ is column-extreme and

$$\|\boldsymbol{c}\|_c^2 = \left\| \begin{pmatrix} L^*b^{\mathrm{t}} \\ L^*a^{\mathrm{t}} \end{pmatrix} \right\|_c^2 = \|\boldsymbol{b}\|_b^2.$$

□

Corollary 3.4 *Let $b = \theta \cdot f$ be the inner–outer factorization of a non-CE $b \in [\mathbb{H}_d^\infty]_1$ where $f(0) > 0$. If a is the Sarason outer function of b, then f is the Sarason outer function of a.*

Corollary 3.5 *Suppose that $b \in [\mathbb{H}_d^\infty]_1$ is non-CE and let $a \in [\mathbb{H}_d^\infty]_1$ be its Sarason function. We have that $c = \binom{b}{a}$ is inner if and only if $\|X^{(n)}\boldsymbol{b}\|_b^2 = \|X^{(n)}\boldsymbol{c}\|_c^2 \to 0$. In particular, c is inner if and only if $X_c^* = \left(L^* \otimes I_2|_{\mathscr{H}^{\mathrm{t}}(c)}\right)^*$ is pure, or equivalently if and only if X_b^* is pure.*

Lemma 3.1 *If $b \in [\mathbb{H}_d^\infty]_1$ is not CE, then X obeys the rank-two defect condition:*

$$X^*X = I - K_0^b(K_0^b)^* - a(0)^2 \left\langle b^{\mathrm{t}}, \cdot \right\rangle_b b^{\mathrm{t}},$$

where a is the free Sarason outer function of b.

This is an analogue of [7, Equation (7)].

Proof This is readily established using the formula of [17, Proposition 5.6], the fact that $b^{\mathrm{t}} \in \mathscr{H}^{\mathrm{t}}(b)$ by [17, Theorem 6.4] since b is non-CE and that $a(0)^2 = 1 - |b(0)|^2 - \|\boldsymbol{b}\|_b^2$ by definition of the Sarason function. Namely,

$$\begin{aligned}
&\|XK^b\{Z, y, v\}\|^2_{\mathscr{H}^{\mathrm{t}}(b)} \\
&= \|K^b\{Z, Z^*y, v\} - \boldsymbol{b}v^*b^{\mathrm{t}}(Z)^*y\|^2 \\
&= y^*K^b(Z, Z)[vv^*]y - 2\mathrm{Re}\left\langle K^b\{Z, Z^*y, v\}, \boldsymbol{b}v^*b^{\mathrm{t}}(Z)^*y\right\rangle + \boldsymbol{b}^*\boldsymbol{b}\left|v^*b^{\mathrm{t}}(Z)^*y\right|^2 \\
&= \|K^b\{Z, y, v\}\|^2 - y^*\left(vv^* - b^{\mathrm{t}}(Z)vv^*b^{\mathrm{t}}(Z)^*\right)y - 2\left|v^*b^{\mathrm{t}}(Z)^*y\right|^2 \\
&\quad + 2\mathrm{Re}\, b(0)y^*vv^*b^{\mathrm{t}}(Z)^*y + \left(1 - |b(0)|^2 - a(0)^2\right)\left|v^*b^{\mathrm{t}}(Z)^*y\right|^2 \\
&= \|K^b\{Z, y, v\}\|^2 - yvv^*y + 2\mathrm{Re}\, b(0)y^*vv^*b^{\mathrm{t}}(Z)^*y \\
&\quad - \left(|b(0)|^2 + a(0)^2\right)\left|v^*b^{\mathrm{t}}(Z)^*y\right|^2.
\end{aligned}$$

On the other hand, applying the formula from the lemma statement,

$$\begin{aligned}
&\left\langle K^b\{Z, y, v\}, \left(I - K_0^b(K_0^b)^* - a(0)^2\left\langle b^{\mathrm{t}}, \cdot\right\rangle_{\mathscr{H}^{\mathrm{t}}(b)} b^{\mathrm{t}}\right) K^b\{Z, y, v\}\right\rangle \\
&= \|K^b\{Z, y, v\}\|^2 - \left|\left\langle K^b\{Z, y, v\}, K_0^b\right\rangle\right|^2 - a(0)^2\left|y^*b^{\mathrm{t}}(Z)v\right|^2,
\end{aligned}$$

yields the same expression. The claim then follows by polarization. □

Proof (of Corollary 3.5) If c is inner then $\mathscr{H}^{\mathrm{t}}(c) = \mathrm{Ran}\, c(R)^\perp \subseteq \mathbb{H}_d^2 \otimes \mathbb{C}^2$, so that $\mathscr{H}^{\mathrm{t}}(c)$ is a closed, co-invariant subspace for $L \otimes I_2$. It is then easily checked that X^* is pure since L is. Conversely if $\|X^{(n)}\boldsymbol{b}\|_b^2 \to 0$, then c is inner by Theorem 3.1.

To see that $\|X^{(n)}\boldsymbol{b}\|_b^2 = \|X^{(n)}\boldsymbol{c}\|_c^2$ for all $n \geq 0$, we apply the rank<Dmathdollar>-two defect condition of Lemma 3.1 repeatedly: For all $n \in \mathbb{N}$,

$$
\begin{aligned}
\|X^{(n)}\boldsymbol{b}\|_b^2 &= \left\langle b^{\mathrm{t}}, \mathrm{Ad}_{X^*,X}^{(n+1)}(I)b^{\mathrm{t}}\right\rangle_b \\
&= \left\langle b^{\mathrm{t}}, \mathrm{Ad}_{X^*,X}^{(n)}(I)b^{\mathrm{t}}\right\rangle_b - \left|\left\langle K_0^b, X^{(n)}b^{\mathrm{t}}\right\rangle_b\right|^2 - a(0)^2\left|\left\langle b^{\mathrm{t}}, X^{(n)}b^{\mathrm{t}}\right\rangle_b\right|^2 \\
&= \left\langle b^{\mathrm{t}}, \mathrm{Ad}_{X^*,X}^{(n)}(I)b^{\mathrm{t}}\right\rangle_b - \sum_{|\omega|=n}|\hat{b}_\omega|^2 - \sum_{|\omega|=n}|\hat{a}_\omega|^2 \\
\dots &= \|Xb^{\mathrm{t}}\|_b^2 - \sum_{0<|\omega|\leq n}|\hat{b}_\omega|^2 - \sum_{0<|\omega|\leq n}|\hat{a}_\omega|^2 \\
&= \|\boldsymbol{c}\|_c^2 - \sum_{0<|\omega|\leq n}\hat{c}_\omega^*\hat{c}_\omega; \qquad \hat{c}_\omega \in \mathbb{C}^2 \\
&= 1 - |c(0)|^2 - \sum_{0<|\omega|\leq n}\hat{c}_\omega^*\hat{c}_\omega \\
&= 1 - \sum_{|\omega|\leq n}\hat{c}_\omega^*\hat{c}_\omega = \|X^{(n)}\boldsymbol{c}\|_c^2.
\end{aligned}
$$

In the above we used that $\|\boldsymbol{c}\|_c = \|\boldsymbol{b}\|_b$ as proven in Corollary 3.3. □

4 Free de Branges–Rovnyak Space and Smirnov Graph Analysis

If $c = \binom{b}{a} \in [\mathbb{H}_d^\infty \otimes \mathbb{C}^2]_1$ is inner and a is outer, we will say that (a, b) is a *Smirnov column–inner pair*. This terminology is motivated by the results of [22]. Namely, the *left free Smirnov class*, $\mathscr{N}_d^+$, is the set of all free NC functions, $h \in \mathscr{O}(\mathbb{B}_{\mathbb{N}}^d)$, defined as 'fractions' of bounded left multipliers with outer denominators. That is, $h \in \mathscr{N}_d^+$ if there are $a, b \in \mathbb{H}_d^\infty$ with a outer so that $h(Z) = b(Z)a(Z)^{-1}$. In [22] we showed that a closed and densely defined linear operator $A \in \mathscr{L}(\mathbb{H}_d^2)$ is *affiliated to* $\mathscr{L}_d^\infty$ in the sense that Dom A is R-invariant and $R_k Ax = AR_k x$ for all $x \in$ Dom A if and only if $A = M_h^L = h(L)$ is an unbounded, closed and densely defined left multiplier with symbol $h \in \mathscr{N}_d^+$. Here, recall that the commutant of $\mathscr{L}_d^\infty$ is $\mathscr{R}_d^\infty$ [15, Theorem 1.2]. Moreover, the graph, $\mathrm{Gr}\, h(L) \subseteq \mathbb{H}_d^2 \otimes \mathbb{C}^2$, is then $R \otimes I_2$-invariant and

$$
\mathrm{Gr}\, h(L) = \mathrm{Ran}\, c(L), \quad c(L) : \mathbb{H}_d^2 \to \mathbb{H}_d^2 \otimes \mathbb{C}^2 \quad c(L) =: \begin{pmatrix} a(L) \\ b(L) \end{pmatrix},
$$

where $c(L)$ is inner so that $\mathrm{Ran}\, a(L) = \mathrm{Dom}\, h(L)$, $a, b \in [\mathbb{H}_d^\infty]_1$ and $a(L)$ is outer since $h(L)$ is densely defined [22, Corollary 4.26, Corollary 4.27]. Hence (a, b) is a Smirnov column–inner pair and $h(L) = b(L)a(L)^{-1}$. Smirnov column–inner pairs which represent a given $h \in \mathscr{N}_d^+$ are unique [22, Corollary 5.2]. Equivalently, $h \in \mathscr{N}_d^+$ if and only if $h(R) := M^R_{h^t}$, $h(R) = b(R)a(R)^{-1}$, is a densely defined and closed operator affiliated to $\mathscr{R}_d^\infty$ and we write $h(R) \sim \mathscr{R}_d^\infty$ in this case.

Assume that $b \in [\mathbb{H}_d^\infty]_1$ is non-CE and let a be its unique Sarason outer function so that the column $c := \begin{pmatrix} a \\ b \end{pmatrix}$ is CE. Consider the NC inner–outer factorization of c, $c = C \cdot D$ where C is inner and D is outer. Here,

$$\operatorname{Ran} C(R) = \operatorname{Ran} c(R)^{-\|\cdot\|_{\mathbb{H}^2}},$$

so that $\operatorname{Ran} C(R)$ is a closed, $L \otimes I_2$-invariant subspace of $\mathbb{H}_d^2 \otimes \mathbb{C}^2$ and it has the $L \otimes I_2$-cyclic vector $x := \begin{pmatrix} b^t \\ a^t \end{pmatrix} = c^t$. Since $\operatorname{Ran} C(R)$ has a cyclic vector, it follows that the wandering space of this $L \otimes I_2$-invariant subspace is also one-dimensional so that

$$C(R) = \begin{pmatrix} A(R) \\ B(R) \end{pmatrix},$$

for some Smirnov column–inner pair $A, B \in [\mathbb{H}_d^\infty]_1$. Hence $D(R) \in [\mathbb{H}_d^\infty]_1$ is outer, and since $a(R) = A(R)D(R)$ is outer, $A(R)$ must also be outer. Moreover, since $C(R)$ is inner,

$$I - B(R)^* B(R) = A(R)^* A(R),$$

so that if we set $A(0) > 0$, A is the Sarason outer function of B by Corollary 3.3. The outer multiplier D must be CE:

$$\begin{aligned} I - b(R)^* b(R) &= I - D(R)^* B(R)^* B(R) D(R) \\ &= I - D(R)^* D(R) + D(R)^* A(R)^* A(R) D(R) \\ &= I - D(R)^* D(R) + a(R)^* a(R). \end{aligned}$$

Since $a(R)$ is the maximal outer factor of $I - b(R)^* b(R)$, $I - D(R)^* D(R)$ cannot have any non-trivial factorable minorant, i.e. D is column-extreme. If we set $H(R) := B(R)A(R)^{-1}$, then this is a closed and densely defined right Smirnov multiplier of the Fock space, with $\operatorname{Gr} H(R) = \operatorname{Ran} C(R)$ [22]. Since C is inner, basic facts about operator-range spaces and complementary spaces imply that

$$\mathscr{H}^t(c) = \mathscr{H}^t(C) \oplus C(R)\mathscr{H}^t(D),$$

and that $C(R)$ is an isometric right multiplier of $\mathscr{H}^t(D)$ into $\mathscr{H}^t(c)$ [23, Theorem 18.8].

Theorem 4.1 *Given $b = BD$ as above,*

$$\mathscr{H}^t(b) = \mathscr{H}^t(B) \oplus B(R)\mathscr{H}^t(D),$$

and $B(R)$ is an isometric right multiplier of $\mathscr{H}^{t}(D)$ into $\mathscr{H}^{t}(b)$. If we define $E_1 : \mathscr{H}^{t}(c) \to \mathscr{H}^{t}(b)$ by projection onto the first co-ordinate, $E_1 \left(\begin{smallmatrix} f \\ g \end{smallmatrix}\right) = f$, then E_1 is an onto isometry.

Proposition 4.2 *Let $b_n \in [\mathbb{H}^\infty_d]_1$ be a sequence (or net) of contractive non-CE multipliers which converge $SOT - *$ to a contractive, non-CE $b \in [\mathbb{H}^\infty_d]_1$. Then the corresponding Sarason outer multipliers $a_n(R)$ converge to $a(R)$ in the weak operator topology. If $c := \left(\begin{smallmatrix} b \\ a \end{smallmatrix}\right)$ is inner then $a_n(R)$ converges to $a(R)$ SOT$-*$.*

This lemma applies, for example, to the free polynomial sequence b_n of nth Cesàro sums of b, or to the nets $b(rR)$ or $rb(R)$ for $0 < r < 1$, taking the limit as $r \uparrow 1$.

Proof Recall that

$$a_n(0)^2 = 1 - |b_n(0)|^2 - \boldsymbol{b}_n^* \boldsymbol{b}_n,$$

$a_n(0) > 0$ and $\boldsymbol{b}_n = L^* b_n^{t}$. Since b is non-CE, its Sarason function, a, also obeys $a(0) > 0$ and $a(0)^2 = 1 - |b(0)|^2 - \boldsymbol{b}^* \boldsymbol{b} > 0$. Since $b_n(R) \overset{SOT-*}{\to} b(R)$, it follows that $a_n(0) \to a(0) > 0$.

Since every a_n is a contractive multiplier, we can assume that (by possibly passing to a subsequence) $a_n(R) \overset{WOT}{\to} \widetilde{a}(R)$, where $\widetilde{a} \in [\mathbb{H}^\infty_d]_1$ and so that

$$\widetilde{a}(0) = a(0) > 0.$$

Hence,

$$c_n(R) = \begin{pmatrix} b_n(R) \\ a_n(R) \end{pmatrix} \overset{WOT}{\to} \begin{pmatrix} b(R) \\ \widetilde{a}(R) \end{pmatrix},$$

so that $\left(\begin{smallmatrix} b \\ \widetilde{a} \end{smallmatrix}\right) \in [\mathbb{H}^\infty_d \otimes \mathbb{C}^2]_1$. Since $\widetilde{a}(0) = a(0) > 0$, $\widetilde{a} \neq 0$, and since $\widetilde{a}$ is not CE, it has an inner–outer factorization $\widetilde{a} = \Theta \hat{a}$ where $\Theta(R)$ is inner and $\hat{a}(R)$ is outer. We can and do further assume that $\hat{a}(0) > 0$. Now observe that

$$\begin{pmatrix} b(R) \\ \hat{a}(R) \end{pmatrix} = \begin{pmatrix} I & 0 \\ 0 & \Theta(R)^* \end{pmatrix} \begin{pmatrix} b(R) \\ \widetilde{a}(R) \end{pmatrix} \in [\mathbb{H}^\infty_d \otimes \mathbb{C}^2]_1,$$

so that it follows that $\left(\begin{smallmatrix} b \\ \hat{a} \end{smallmatrix}\right)$ is a contractive multiplier, where now $\hat{a}$ is outer. However, the Sarason function, $a(R)$, is the maximal outer factor of $I - b(R)^* b(R)$ by Corollary 3.3. This means that since $\hat{a}(R)^* \hat{a}(R) \leq I - b(R)^* b(R)$, it must be that

$$\hat{a}(R)^* \hat{a}(R) \leq a(R)^* a(R).$$

By Douglas factorization, there is a contraction C^* so that

$$\hat{a}(R)^* = a(R)^* C^* \quad \Rightarrow \quad \hat{a}(R) = C a(R).$$

Since both $\hat{a}, a$ are outer, it follows that C commutes with the left shifts, so that $C = C(R) \in [\mathscr{R}_d^\infty]_1$ is a contractive right multiplier [15, Theorem 1.2]. This would imply that

$$\hat{a}(0) = C(0)a(0) = C(0)\widetilde{a}(0) = C(0)\Theta(0)\hat{a}(0),$$

so that

$$C(0)\Theta(0) = 1.$$

However, $C(R)\Theta(R)$ is a contractive right multiplier, so that the NC maximum modulus principle implies that $C(R)\Theta(R) \equiv I$ [11, Lemma 6.11]. Since $\Theta(R)$ is inner, it is not invertible unless $\Theta(R) \equiv I$. It follows that $\Theta(R) = I = C(R)$ and we conclude that $\hat{a}(R) = \widetilde{a}(R) = a(R)$. We can conclude that the entire sequence $a_n(R) \overset{WOT}{\rightarrow} a(R)$, since any WOT-convergent subsequence must converge to $a(R)$ by the above argument.

Now suppose that $c = \binom{b}{a}$ is inner. Let $h_n(R) := b_n(R)a_n(R)^{-1}$ and $\tau_n := h_n(R)^*h_n(R)$. Then for any $z \in \mathbb{C} \setminus [0, +\infty)$,

$$\begin{aligned}(\tau_n - zI)^{-1} &= \left(a_n(R)^{-*}(b_n(R)^*b_n(R) - za_n(R)^*a_n(R))a_n(R)^{-1}\right)^{-1} \\ &= a_n(R)\left((1+z)b_n(R)^*b_n(R) - zI\right)^{-1}a_n(R)^*.\end{aligned}$$

Since $a_n(R) \overset{WOT}{\rightarrow} a(R)$ we have that $a_n(R)^* \overset{SOT}{\rightarrow} a(R)^*$ and $b_n(R) \overset{SOT-*}{\rightarrow} b(R)$ by [24, Lemma 9], so that the above converges in the weak operator topology to

$$\begin{aligned}&a(R)\left((1+z)b(R)^*b(R) - zI\right)^{-1}a(R)^* \\ &= a(R)\left((1+z)b(R)^*b(R) - zb(R)^*b(R) - za(R)^*a(R)\right)^{-1}a(R)^* \\ &= (\tau - zI)^{-1}\end{aligned}$$

where $\tau = h(R)^*h(R)$ and $h(R) = b(R)a(R)^{-1}$. In the above we used the assumption that $a(R)^*a(R)+b(R)^*b(R) = I$. Hence τ_n converges to τ in the weak resolvent sense [25]. A simple argument using the resolvent formula then shows that $(\tau_n - zI)^{-1}$ converges in the strong operator topology to $(\tau - zI)^{-1}$ for any $z \in \mathbb{C} \setminus [0, +\infty)$. In particular, taking $z = -1$ gives

$$(\tau + I)^{-1} = a(R)\left((1+z)b(R)^*b(R) - zI\right)^{-1}a(R)^* = a(R)a(R)^*,$$

so that $a_n(R)a_n(R)^* \overset{SOT}{\rightarrow} a(R)a(R)^*$ and it follows easily from this that $a_n(R) \overset{SOT}{\rightarrow} a(R)$. Recall that for any net of right (or left) multipliers, WOT convergence implies SOT convergence of $a_n(R)^*$; hence, $a_n(R) \overset{SOT-*}{\rightarrow} a(R)$ by [24, Lemma 9]. □

Consider the (generally) unbounded multiplier $H(R) := b(R)a(R)^{-1}$ with dense domain $\mathrm{Dom}\, H(R) := \mathrm{Ran}\, a(R)$. By [22, Corollary 3.9], $H(R)$ is closeable and both the linear span of the NC Szegö kernel vectors $K\{Z, y, v\}$ and the free polynomials are cores for $H(R)^*$ [22, Corollary 3.13]. Here, recall that a dense linear subspace, $\mathscr{D} \subseteq \mathrm{Dom}\, T$ in the domain of a closed, densely defined linear operator T is called a *core* for T if $\mathrm{Gr}\, T = \{x \oplus Tx \mid x \in \mathscr{D}\}^{-\|\cdot\|}$. That is, T is the closure of its restriction to $\mathscr{D}$.

Given any $b \in [\mathbb{H}_d^\infty]_1$, for each $0 \leq r < 1$, let a_r be the (unique, invertible) outer multiplier defined by

$$I - r^2 b(R)^* b(R) = a_r(R)^* a_r(R).$$

Lemma 4.1 *Let $b \in [\mathbb{H}_d^\infty]_1$ be a contractive multiplier. Then $f \in \mathscr{H}^{\mathrm{t}}(b)$ if and only if*

$$\sup_{0<r<1} \|a_r^{-*}(R)b(R)^* f\|_{\mathbb{H}^2} < \infty,$$

in which case

$$\begin{aligned}\|f\|^2_{\mathscr{H}^{\mathrm{t}}(b)} &= \|f\|^2_{\mathbb{H}^2} + \lim_{r\uparrow 1} \|a_r^{-*}(R)b(R)^* f\|_{\mathbb{H}^2} \\ &= \|f\|^2_{\mathbb{H}^2} + \|H(R)^* f\|^2_{\mathbb{H}^2} = \|f\|^2_{\mathrm{Gr}(H(R)^*)}.\end{aligned}$$

Proof This follows from basic facts on operator-range spaces and complementary spaces; see [23, Theorem 16.17, Theorem 16.18]. □

The following can be pieced together from results in [23]; in particular see [23, Theorem 18.8].

Theorem 4.3 *Let $a = bc$ where $a, b, c \in [\mathbb{H}_d^\infty]_1$. Then, $\mathscr{H}^{\mathrm{t}}(b^*) \cap \mathscr{H}^{\mathrm{t}}(c) = \{0\}$ if and only if $\mathscr{H}^{\mathrm{t}}(b) \cap b(R)\mathscr{H}^{\mathrm{t}}(c) = \{0\}$ and in this case:*

1. *$\mathscr{H}^{\mathrm{t}}(a) = \mathscr{H}^{\mathrm{t}}(b) \oplus b(R)\mathscr{H}^{\mathrm{t}}(c)$.*
2. *$\mathscr{H}^{\mathrm{t}}(b)$ embeds isometrically into $\mathscr{H}^{\mathrm{t}}(a)$.*
3. *$M^R_{b^{\mathrm{t}}}$ is an isometric multiplier of $\mathscr{H}^{\mathrm{t}}(c)$ into $\mathscr{H}^{\mathrm{t}}(a)$.*

Proof We prove the first equivalent statement; the rest of the proof can be found in [23, Theorem 18.8]. If $x \in \mathscr{H}^{\mathrm{t}}(b^*) \cap \mathscr{H}^{\mathrm{t}}(c)$ then $b(R)x \in \mathscr{H}^{\mathrm{t}}(b) \cap b(R)\mathscr{H}^{\mathrm{t}}(c)$. Conversely, if $b(R)x \in \mathscr{H}^{\mathrm{t}}(b) \cap b(R)\mathscr{H}^{\mathrm{t}}(c)$ for some $x \in \mathscr{H}^{\mathrm{t}}(c)$ then $b(R)x \in \mathscr{H}^{\mathrm{t}}(b) \cap \mathscr{M}^{\mathrm{t}}(b) = b(R)\mathscr{H}^{\mathrm{t}}(b^*)$ by [23, Lemma 16.20]. Hence, $b(R)x = b(R)y$ for some $y \in \mathscr{H}^{\mathrm{t}}(b^*)$ and $x = y \in \mathscr{H}^{\mathrm{t}}(c) \cap \mathscr{H}^{\mathrm{t}}(b^*)$. □

Proof (of Theorem 4.1) Consider any NC Szegö kernel vector $K\{Z, y, v\}$. Then,

$$\sup_{0<r<1} \|a_r^{-*}(R)b(R)^* K\{Z, y, v\}\|_{\mathbb{H}^2} = \sup \|K\{Z, y, a_r^{\mathrm{t}}(Z)^{-1} b^{\mathrm{t}}(Z)v\}\| < +\infty,$$

since $a_r(R) \overset{WOT}{\rightarrow} a(R)$ by Proposition 4.2 and WOT convergence implies pointwise convergence. We conclude, by Lemma 4.1, that the linear span of the NC Szegö kernels is contained in $\mathscr{H}^{\mathrm{t}}(b)$. (Since b is non-CE, this also follows from the results of [17].) Again by Lemma 4.1,

$$\begin{aligned} \|K\{Z, y, v\}\|_b^2 &= \|K\{Z, y, v\}\|_{\mathbb{H}^2}^2 + \lim_{r\uparrow 1} \|K\{Z, y, a_r^{\mathrm{t}}(Z)^{-1} b^{\mathrm{t}}(Z)v\}\|_{\mathbb{H}^2}^2 \\ &= \|K\{Z, y, v\}\|_{\mathbb{H}^2}^2 + \|K\{Z, y, H^{\mathrm{t}}(Z)v\}\|_{\mathbb{H}^2}^2 \\ &= \|K\{Z, y, v\}\|_B^2. \end{aligned}$$

Recall that $C = \begin{pmatrix} A \\ B \end{pmatrix}$ is inner, $\operatorname{Ran} C(R) = \operatorname{Gr} H(R)$ and the NC Szegö kernels are a core for $H(R)^*$. By [22, Theorem 4.17], $\mathscr{H}^{\mathrm{t}}(B) = \operatorname{Dom} H(R)^*$ if we equip $\operatorname{Dom} H(R)^*$ with the graph norm. It follows that the NC Szegö kernels (or free polynomials) are dense in $\mathscr{H}^{\mathrm{t}}(B)$ and that $\mathscr{H}^{\mathrm{t}}(B)$ embeds isometrically into $\mathscr{H}^{\mathrm{t}}(b)$.

The fact that $\operatorname{Ran} b(R) \subseteq \operatorname{Ran} B(R)$ and $b(R)b(R)^* = B(R)D(R)D(R)^*B(R)^* \leq B(R)B(R)^*$ implies that $\mathscr{H}^{\mathrm{t}}(B)$ is contractively contained in $\mathscr{H}^{\mathrm{t}}(b)$. Moreover, by [23, Theorem 16.23],

$$\mathscr{H}^{\mathrm{t}}(b) = \mathscr{H}^{\mathrm{t}}(B) + B(R)\mathscr{H}^{\mathrm{t}}(D),$$

$M^R_{B^{\mathrm{t}}}$ defines a contraction from $\mathscr{H}^{\mathrm{t}}(D)$ into $\mathscr{H}^{\mathrm{t}}(b)$ and if $x \in \mathscr{H}^{\mathrm{t}}(B)$, $y \in \mathscr{H}^{\mathrm{t}}(D)$ then,

$$\|x + B(R)y\|_b^2 \leq \|x\|_B^2 + \|y\|_D^2.$$

Using that $\mathscr{H}^{\mathrm{t}}(B)$ is isometrically contained in $\mathscr{H}^{\mathrm{t}}(b)$,

$$\|x + B(R)y\|_b^2 = \|x\|_B^2 + 2\mathrm{Re}\,\langle x, B(R)y\rangle_b + \|B(R)y\|_b^2.$$

Subtracting these two formulas gives:

$$0 \leq \|y\|_D^2 - \|B(R)y\|_b^2 - 2\mathrm{Re}\,\langle x, B(R)y\rangle_b\,.$$

Here,

$$0 \leq \|y\|_D^2 - \|B(R)y\|_b^2,$$

since $B(R)y = M^R_{B^{\mathrm{t}}}y$ and right multiplication by B^{t} defines a contraction from $\mathscr{H}^{\mathrm{t}}(D)$ into $\mathscr{H}^{\mathrm{t}}(b)$. Hence,

$$2\mathrm{Re}\,\langle x, B(R)y\rangle_b \leq \|y\|_D^2 - \|B(R)y\|_b^2.$$

Given any such x, y, choose $\lambda \in \partial\mathbb{D}$ so that $\langle \lambda x, B(R)y\rangle_b = |\langle x, B(R)y\rangle_b|$, then for any $r > 0$,

$$2r|\langle x, B(R)y\rangle_b| = 2\mathrm{Re}\,\langle r\lambda x, B(R)y\rangle_b \leq \|y\|_D^2 - \|B(R)y\|_b^2,$$

and it follows that $\langle x, B(R)y\rangle_b = 0$. That is,

$$\mathscr{H}^{\mathrm{t}}(b) = \mathscr{H}^{\mathrm{t}}(B) \oplus B(R)\mathscr{H}^{\mathrm{t}}(D).$$

By reproducing kernel theory, this is equivalent to the condition that $\mathscr{H}^{\mathrm{t}}(B) \cap B(R)\mathscr{H}^{\mathrm{t}}(D) = 0$, so that the remainder of the claim follows from the previous theorem. □

Corollary 4.4 *With a, b, A, B and D as above,*

$$\mathscr{H}^{\mathrm{t}}(D) \cap \mathscr{M}^{\mathrm{t}}(A^*) = \{0\}.$$

Proof By RKHS theory, if K, k are CPNC kernels on the same NC set, $\mathcal{H}_{nc}(K + k) = \mathcal{H}_{nc}(K) \oplus \mathcal{H}_{nc}(k)$ if and only if

$$\mathcal{H}_{nc}(K) \cap \mathcal{H}_{nc}(k) = \{0\},$$

[26, Section 6], [27, Theorem 4.4]. By Theorem 4.1,

$$\mathscr{H}^{\mathrm{t}}(b) = \mathscr{H}^{\mathrm{t}}(B) \oplus B(R)\mathscr{H}^{\mathrm{t}}(D).$$

Adding together the CPNC kernels, K^B, of $\mathscr{H}^{\mathrm{t}}(B)$ and k of $B(R)\mathscr{H}^{\mathrm{t}}(D)$ gives:

$$\begin{aligned} &K^B(Z, W)[\cdot] + k(Z, W)[\cdot] \\ &= K(Z, W)\left[(\cdot) - B^{\mathrm{t}}(Z)(\cdot)B^{\mathrm{t}}(W)^*\right] \\ &\quad + K(Z, W)\left[B^{\mathrm{t}}(Z)(\cdot)B^{\mathrm{t}}(W)^* - D^{\mathrm{t}}(Z)B^{\mathrm{t}}(Z)(\cdot)B^{\mathrm{t}}(W)^*D^{\mathrm{t}}(W)^*\right] \\ &= K(Z, W)[\cdot] - K(Z, W)[b^{\mathrm{t}}(Z)(\cdot)b^{\mathrm{t}}(W)^*] = K^b(Z, W)[\cdot]. \end{aligned}$$

Hence, $\mathscr{H}^{\mathrm{t}}(b) = \mathcal{H}_{nc}(K^B + k)$ and the orthogonal decomposition of Theorem 4.1 then implies that

$$\mathscr{H}^{\mathrm{t}}(B) \cap B(R)\mathscr{H}^{\mathrm{t}}(D) = \{0\}.$$

By the theory of operator-range spaces [23, Section 16.6],

$$\mathscr{H}^{\mathrm{t}}(B) \cap \mathscr{M}^{\mathrm{t}}(B) = B(R)\mathscr{H}^{\mathrm{t}}(B^*).$$

Hence,

$$\{0\} = \mathscr{H}^{\mathrm{t}}(B) \cap B(R)\mathscr{H}^{\mathrm{t}}(D) \subseteq \mathscr{H}^{\mathrm{t}}(B) \cap \mathscr{M}^{\mathrm{t}}(B) = B(R)\mathscr{H}^{\mathrm{t}}(B^*).$$

Since $B(R)$ is injective,

$$\mathscr{H}^{\mathrm{t}}(D) \cap \mathscr{H}^{\mathrm{t}}(B^*) = \{0\}.$$

Finally, the column $\left(\begin{smallmatrix} A \\ B \end{smallmatrix}\right)$ is inner so that

$$\mathscr{H}^{\mathrm{t}}(B^*) = \mathrm{Ran}\,\sqrt{I - B(R)^*B(R)} = \mathrm{Ran}\,\sqrt{A(R)^*A(R)} = \mathscr{M}^{\mathrm{t}}(A^*).$$

□

If $b \in [\mathbb{H}^\infty_d]_1$ is non-CE, then by [17, Corollary 6.15], the right free de Branges–Rovnyak space $\mathscr{H}^{\mathrm{t}}(b)$ is both L-co-invariant and L-invariant and, following [8], we then say that $\mathscr{H}^{\mathrm{t}}(b)$ is *doubly free shift invariant*. In this case, let $Y := L|_{\mathscr{H}^{\mathrm{t}}(b)}$.

Lemma 4.2 *Let $b \in [\mathbb{H}^\infty_d]_1$ be non-CE. The restriction, Y, of L to $\mathscr{H}^{\mathrm{t}}(b)$ obeys the formula:*

$$Y = X^* + \langle \boldsymbol{b}, \cdot \rangle_b\, b^{\mathrm{t}}.$$

Proof Given any $\boldsymbol{h} \in \mathscr{H}^{\mathrm{t}}(b) \otimes \mathbb{C}^d$, we calculate $X^*\boldsymbol{h}$:

$$\begin{aligned}
\left\langle K^b\{Z, y, v\}, X^*\boldsymbol{h}\right\rangle_b &= \left\langle XK^b\{Z, y, v\}, \boldsymbol{h}\right\rangle_b \\
&= \left\langle L^*\left(K\{Z, y, v\} - K\{Z, y, b^{\mathrm{t}}(Z)v\}b^{\mathrm{t}}\right), \boldsymbol{h}\right\rangle_b \\
&= \left\langle \left(K\{Z, Z^*y, v\} - K\{Z, Z^*y, b^{\mathrm{t}}(Z)v\}b^{\mathrm{t}} \right.\right. \\
&\quad \left.\left. - K\{Z, y, b^{\mathrm{t}}(Z)v\}(0)\boldsymbol{b}\right), \boldsymbol{h}\right\rangle_b \\
&= \left\langle K^b\{Z, Z^*y, v\} - \overline{y^*b^{\mathrm{t}}(Z)v}\boldsymbol{b}, \boldsymbol{h}\right\rangle_b \\
&= y^*Z\boldsymbol{h}^{\mathrm{t}}(Z)v - y^*b^{\mathrm{t}}(Z)v\, \langle \boldsymbol{b}, \boldsymbol{h}\rangle_b\,.
\end{aligned}$$

This proves that

$$X^* = Y - \langle \boldsymbol{b}, \cdot \rangle_b\, b^{\mathrm{t}},$$

and the formula for Y follows. □

The next two lemmas provide an analogue of a formula of Sarason for the norm of b^{t} in $\mathscr{H}^{\mathrm{t}}(b)$ when b is non-CE [7, Lemma 2].

Lemma 4.3 *Let $A, B \in [\mathbb{H}^\infty_d]_1$ be a Smirnov column–inner pair and $C = \left(\begin{smallmatrix} A \\ B \end{smallmatrix}\right)$ be inner. Then,*

$$\|B^{\mathrm{t}}\|_B^2 = \frac{1}{A(0)^2} - 1.$$

Proof This follows from [22, Theorem 4.17]: $\operatorname{Ran} C(R)$ is the graph, $\operatorname{Gr} H(R)$ of the right Smirnov multiplier $H(R) := B(R)A(R)^{-1}$ and $\mathscr{H}^{\mathrm{t}}(B) = \operatorname{Dom} H(R)^*$. For any $x \in \mathscr{H}^{\mathrm{t}}(B) = \operatorname{Dom} H(R)^*$,

$$\|x\|_B^2 = \|x\|_{\mathbb{H}^2}^2 + \|H(R)^* x\|_{\mathbb{H}^2}^2.$$

In particular,

$$\begin{aligned}
\|B^{\mathrm{t}}\|_B^2 &= \|B^{\mathrm{t}}\|_{\mathbb{H}^2}^2 + \|A(R)^{-*}B(R)^*B(R)1\|_{\mathbb{H}^2}^2 \\
&= \|B^{\mathrm{t}}\|_{\mathbb{H}^2}^2 + \left\langle A(R)^{-*}B(R)^*B(R)1, A(R)^{-*}(I - A(R)^*A(R))1\right\rangle_{\mathbb{H}^2} \\
&= \|B^{\mathrm{t}}\|_{\mathbb{H}^2}^2 + \left\langle A(R)^{-*}B(R)^*B(R)1, A(R)^{-*}1\right\rangle_{\mathbb{H}^2} \\
&\quad - \left\langle A(R)^{-*}B(R)^*B(R)1, A(R)1\right\rangle_{\mathbb{H}^2} \\
&= \|B^{\mathrm{t}}\|_{\mathbb{H}^2}^2 + \frac{1}{A(0)} \left\langle A(R)^{-*}(I - A(R)^*A(R)1, 1\right\rangle_{\mathbb{H}^2} - \|B^{\mathrm{t}}\|_{\mathbb{H}^2}^2 \\
&= \frac{1}{A(0)^2} - \frac{1}{A(0)} \langle 1, A(R)1\rangle_{\mathbb{H}^2} \\
&= \frac{1}{A(0)^2} - 1.
\end{aligned}$$

□

The conclusion of the previous lemma holds more generally with a different argument:

Lemma 4.4 *If $b \in [\mathbb{H}_d^\infty]_1$ is non-CE and $a \in [\mathbb{H}_d^\infty]_1$ is its outer Sarason function, then*

$$\|b^{\mathrm{t}}\|_b^2 = \frac{1}{a(0)^2} - 1.$$

Proof For $\omega \neq \emptyset$, recall that the Taylor coefficients of a are given by the formula:

$$\hat{a}_\omega = -a(0)\left\langle b^{\mathrm{t}}, X^\omega b^{\mathrm{t}}\right\rangle_b.$$

Note that we can write $b^{\mathrm{t}} = YXb^{\mathrm{t}} + b(0)1$ where $Y = L|_{\mathscr{H}^{\mathrm{t}}(b)}$. Substituting this into the above formula yields:

$$\hat{a}_\omega = -a(0)\left(\left\langle YXb^{\mathrm{t}}, X^\omega b^{\mathrm{t}}\right\rangle_b + \overline{b(0)}\left\langle 1, X^\omega b^{\mathrm{t}}\right\rangle_b\right).$$

Then, applying Lemma 4.2 and Lemma 3.1,

$$\begin{aligned}\langle 1, X^{\omega} b^{\mathrm{t}} \rangle_b &= \left\langle K_0^b + \overline{b}(0) b^{\mathrm{t}}, X^{\omega} b^{\mathrm{t}} \right\rangle_b \\ &= \hat{b}_{\omega} - \frac{b(0)}{a(0)} \hat{a}_{\omega}.\end{aligned}$$

Now calculate,

$$\begin{aligned}-a(0) \langle Y X b^{\mathrm{t}}, X^{\omega} b^{\mathrm{t}} \rangle_b &= -a(0) \left(\langle X^* X b^{\mathrm{t}}, X^{\omega} b^{\mathrm{t}} \rangle_b + \boldsymbol{b}^* \boldsymbol{b} \langle b^{\mathrm{t}}, X^{\omega} b^{\mathrm{t}} \rangle_b \right) \\ &= -a(0) \langle b^{\mathrm{t}}, X^{\omega} b^{\mathrm{t}} \rangle_b + a(0) \overline{b(0)} b_{\omega} \\ &\quad + a(0)^3 \| b^{\mathrm{t}} \|_b^2 \langle b^{\mathrm{t}}, X^{\omega} b^{\mathrm{t}} \rangle_b \qquad (3) \\ &\quad - a(0)(1 - |b(0)|^2 - a(0)^2) \langle b^{\mathrm{t}}, X^{\omega} b^{\mathrm{t}} \rangle_b \\ &= \hat{a}_{\omega} + a(0) \overline{b(0)} \hat{b}_{\omega} - a(0)^2 \| b^{\mathrm{t}} \|_b^2 \hat{a}_{\omega} \\ &\quad + (1 - |b(0)|^2 - a(0)^2) \hat{a}_{\omega}.\end{aligned}$$

Putting this all together gives:

$$\begin{aligned}\hat{a}_{\omega} &= \hat{a}_{\omega} + a(0) \overline{b(0)} \hat{b}_{\omega} - a(0)^2 \| b^{\mathrm{t}} \|_b^2 \hat{a}_{\omega} + (1 - |b(0)|^2 - a(0)^2) \hat{a}_{\omega} \\ &\quad - a(0) \overline{b}(0) \left(\hat{b}_{\omega} - \frac{b(0)}{a(0)} \hat{a}_{\omega} \right) \\ &= 2\hat{a}_{\omega} - a(0)^2 \| b^{\mathrm{t}} \|_b^2 \hat{a}_{\omega} - (|b(0)|^2 + a(0)^2) \hat{a}_{\omega} + |b(0)|^2 \hat{a}_{\omega}.\end{aligned}$$

Solving for $\| b^{\mathrm{t}} \|_b^2$ gives:

$$a(0)^2 \| b^{\mathrm{t}} \|_b^2 \hat{a}_{\omega} = \hat{a}_{\omega} - a(0)^2 \hat{a}_{\omega},$$

or equivalently,

$$\| b^{\mathrm{t}} \|_b^2 = \frac{1}{a(0)^2} - 1.$$

This is of course provided that $\hat{a}_{\omega} \neq 0$ for at least one $\omega \neq \emptyset$. However, if this were true $a = a(0)1$ which can only happen trivially. □

Proposition 4.5 *The vector b^{t} has the direct sum decomposition*

$$b^{\mathrm{t}} = \frac{1}{D(0)} B^{\mathrm{t}} \oplus \frac{-1}{D(0)} B(R) K_0^D,$$

with respect to $\mathscr{H}^{\mathrm{t}}(b) = \mathscr{H}^{\mathrm{t}}(B) \oplus B(R) \mathscr{H}^{\mathrm{t}}(D)$.

Proof As before we write $b^{\mathrm{t}} = YXb^{\mathrm{t}} + b(0)1$. Since B is non-CE, we have that $1 \in \mathscr{H}^{\mathrm{t}}(B)$. To compute the projection of YXb^{t} onto $\mathscr{H}^{\mathrm{t}}(B)$, calculate $\langle K^B\{Z, y, v\}, YXb^{\mathrm{t}}\rangle_b$. First,

$$\begin{aligned}
Y^*K^B\{Z, y, v\} &= XK^B\{Z, y, v\} + \left\langle b^{\mathrm{t}}, K^B\{Z, y, v\}\right\rangle_b \boldsymbol{b} \\
&= L^*\left(K\{Z, y, v\} - K\{Z, y, B^{\mathrm{t}}(Z)v\}B^{\mathrm{t}}\right) + \left\langle b^{\mathrm{t}}, K^B\{Z, y, v\}\right\rangle_b \boldsymbol{b} \\
&= K\{Z, Z^*y, v\} - K\{Z, Z^*y, B^{\mathrm{t}}Zv\}B^{\mathrm{t}} - K\{Z, y, B^{\mathrm{t}}(Z)v\}(0)\boldsymbol{B} \\
&\quad + \left\langle b^{\mathrm{t}}, K^B\{Z, y, v\}\right\rangle_b \boldsymbol{b} \\
&= K^B\{Z, Z^*y, v\} - \overline{y^*B^{\mathrm{t}}(Z)v}\boldsymbol{B} + \left\langle b^{\mathrm{t}}, K^B\{Z, y, v\}\right\rangle_b \boldsymbol{b}.
\end{aligned}$$

Putting this all together yields:

$$\begin{aligned}
\left\langle K^B\{Z, y, v\}, YXb^{\mathrm{t}}\right\rangle_b &= \left\langle Y^*K^B\{Z, y, v\}, \boldsymbol{b}\right\rangle_b \\
&= \left\langle Y^*K^B\{Z, y, v\}, B(R)\boldsymbol{D} \oplus D(0)\boldsymbol{B}\right\rangle_b \\
&= D(0)\left\langle K^B\{Z, Z^*y, v\} - \overline{y^*B^{\mathrm{t}}(Z)v}\boldsymbol{B}, \boldsymbol{B}\right\rangle_B \\
&\quad + \left\langle K^B\{Z, y, v\}, b^{\mathrm{t}}\right\rangle_b \boldsymbol{b}^*\boldsymbol{b} \\
&= D(0)y^*(B^t(Z) - B(0))v \\
&\quad - D(0)y^*B^{\mathrm{t}}(Z)v(1 - |B(0)|^2 - A(0)^2) \\
&\quad + \left\langle K^B\{Z, y, v\}, b^{\mathrm{t}}\right\rangle (1 - |b(0)|^2 - a(0)^2) \\
&= -b(0)y^*v + D(0)(|B(0)|^2 + A(0)^2)y^*B^{\mathrm{t}}(Z)v \\
&\quad + \left\langle K^B\{Z, y, v\}, b^{\mathrm{t}}\right\rangle (1 - |b(0)|^2 - a(0)^2).
\end{aligned}$$

Finally,

$$\begin{aligned}
\left\langle K^B\{Z, y, v\}, b^{\mathrm{t}}\right\rangle_b &= -b(0)y^*v + D(0)(|B(0)|^2 + A(0)^2)y^*B^{\mathrm{t}}(Z)v \\
&\quad + \left\langle K^B\{Z, y, v\}, b^{\mathrm{t}}\right\rangle (1 - |b(0)|^2 - a(0)^2) + y^*vb(0).
\end{aligned}$$

Solving for $\langle K^B\{Z, y, v\}, b^{\mathrm{t}}\rangle$ gives:

$$D(0)^2(|B(0)|^2 + A(0)^2)\left\langle K^B\{Z, y, v\}, b^{\mathrm{t}}\right\rangle = D(0)(|B(0)|^2 + A(0)^2)y^*B^{\mathrm{t}}(Z)v.$$

Equivalently,

$$Pb^{\mathrm{t}} = \frac{1}{D(0)} B^{\mathrm{t}},$$

where $P := P_{\mathscr{H}^{\mathrm{t}}(B)}$ denotes the orthogonal projection of $\mathscr{H}^{\mathrm{t}}(b)$ onto the subspace $\mathscr{H}^{\mathrm{t}}(B)$.

Let $Q := P_{B(R)\mathscr{H}^{\mathrm{t}}(D)} : \mathscr{H}^{\mathrm{t}}(b) \to B(R)\mathscr{H}^{\mathrm{t}}(D)$ be the orthogonal projection so that $Q = I - P$. It follows that

$$Qb^{\mathrm{t}} = b^{\mathrm{t}} - \frac{1}{D(0)} B^{\mathrm{t}}.$$

We can write:

$$\begin{aligned} Qb^{\mathrm{t}} &= D^{\mathrm{t}} B^{\mathrm{t}} - \frac{1}{D(0)} B^{\mathrm{t}} \\ &= -\frac{1}{D(0)} \big(B^{\mathrm{t}} - D^{\mathrm{t}} B^{\mathrm{t}} D(0) \big) \\ &= -\frac{1}{D(0)} B(R) K_0^D, \end{aligned}$$

so that the direct sum decomposition of b^{t} with respect to $\mathscr{H}^{\mathrm{t}}(b) = \mathscr{H}^{\mathrm{t}}(B) \oplus B(R)\mathscr{H}^{\mathrm{t}}(D)$ is:

$$b^{\mathrm{t}} = \frac{1}{D(0)} B^{\mathrm{t}} \oplus \frac{-1}{D(0)} B(R) K_0^D.$$

□

Remark 4.1 Under the above assumptions, $a = AD$, $b = BD$, A and B are a Smirnov column–inner pair, A and a are the Sarason outer functions of B and b, $D \in [\mathbb{H}_d^{\infty}]_1$ is outer and CE and $\mathscr{H}^{\mathrm{t}}(D) \cap \mathscr{M}^{\mathrm{t}}(A^*) = \{0\}$. If it is true that $c = \binom{b}{a}$ is always inner, whenever $a, b \in [\mathbb{H}_d^{\infty}]_1$ and a is the NC Sarason outer function of b, then $c = CD$ where $C := \binom{B}{A}$ is also inner. However, if left multiplication by $c = CD$ and C are both isometries, this implies that $D(L)$ is also an isometry, i.e. inner so that D is both inner and outer which means that $D(L)$ is an isometry with dense range, i.e. a unitary. By Davidson–Pitts [15, Corollary 1.5], the left multiplier algebra of $\mathbb{H}_d^2$ contains no non-constant normal elements. Hence, $D = \zeta$ for some $\zeta \in \partial\mathbb{D}$ and since we can assume $D(0) > 0$, $D = 1$ and $c = C$.

In order to show that $c = \binom{b}{a}$ is always inner, one might hope to show that the fact that $a = AD$ is the Sarason function of $b = BD$ where A is the Sarason function of B can only hold if $D = 1$ is constant. However, these assumptions do not seem to place any such restrictions on D. Indeed, since $a(R) = A(R)D(R)$, if

$$A^{\mathrm{t}} := \sum_{\beta} \hat{A}^{\mathrm{t}}_{\beta} L^{\beta} 1, \quad \text{and} \quad D^{\mathrm{t}} = \sum_{\alpha} \hat{D}^{\mathrm{t}}_{\alpha} L^{\alpha} 1,$$

then

$$a^{\mathrm{t}} = D^{\mathrm{t}} A^{\mathrm{t}} = \sum_{\gamma} \hat{a}^{\mathrm{t}}_{\gamma} L^{\gamma} 1,$$

where

$$\hat{a}^{\mathrm{t}}_{\gamma} = \sum_{\alpha \cdot \beta = \gamma} \hat{D}^{\mathrm{t}}_{\alpha} \hat{A}^{\mathrm{t}}_{\beta}.$$

Using that $\hat{a}^{\mathrm{t}}_{\gamma} = \hat{a}_{\gamma^{\mathrm{t}}}$ we obtain

$$\begin{aligned}\hat{a}_{\gamma} &= \sum_{\alpha \cdot \beta = \gamma^{\mathrm{t}}} \hat{D}_{\alpha^{\mathrm{t}}} \hat{A}_{\beta^{\mathrm{t}}} \\ &= \sum_{\lambda \cdot \sigma = \gamma} \hat{D}_{\sigma} \hat{A}_{\lambda}.\end{aligned}$$

On the other hand, using the Sarason formula for the non-vacuum Taylor coefficients of a,

$$\begin{aligned}\hat{a}_{\gamma} &= -a(0) \left\langle b^{\mathrm{t}}, X^{\gamma} b^{\mathrm{t}} \right\rangle_b \\ &= -a(0) \left\langle \frac{1}{D(0)} B^{\mathrm{t}} \oplus \frac{-1}{D(0)} B(R) K^D_0, \sum_{\substack{\lambda \cdot \sigma = \gamma \\ \lambda \neq \emptyset}} \hat{D}_{\sigma} X^{\lambda} B^{\mathrm{t}} \oplus B(R) X^{\gamma} D^{\mathrm{t}} \right\rangle_b \\ &= -A(0) \sum_{\substack{\lambda \cdot \sigma = \gamma \\ \lambda \neq \emptyset}} \hat{D}_{\sigma} \left\langle B^{\mathrm{t}}, X^{\lambda} B^{\mathrm{t}} \right\rangle_B + A(0) \left\langle K^D_0, X^{\gamma} D^{\mathrm{t}} \right\rangle_D \\ &= \sum_{\substack{\lambda \cdot \sigma = \gamma \\ \lambda \neq \emptyset}} \hat{D}_{\sigma} \hat{A}_{\lambda} + A(0) \hat{D}_{\gamma},\end{aligned}$$

which is the same as the previous formula for any $\gamma \neq \emptyset$. In the case of $\hat{a}_{\emptyset} = a(0)$,

$$\begin{aligned}a(0)^2 &= \frac{1}{1 + \|b^{\mathrm{t}}\|^2_b}, && \text{by Lemma 4.4,} \\ &= \frac{1}{1 + \left(\frac{\|B^{\mathrm{t}}\|^2_B + 1 - D(0)^2}{D(0)^2} \right)}, && \text{by Proposition 4.5,} \\ &= D(0)^2 \frac{1}{1 + \|B^{\mathrm{t}}\|^2_B} = D(0)^2 A(0)^2, && \text{by Lemma 4.4 again.}\end{aligned}$$

Remark 4.2 Given a, b, c, A, B, C and D as above, one can readily check that

$$D(0)^2 = \frac{1 + \|B^t\|_B^2}{1 + \|b^t\|_b^2} = \frac{1 - \|\boldsymbol{b}\|_b^2}{1 - \|\boldsymbol{B}\|_B^2},$$

where recall that $B = bD$.

Also by Corollary 3.4, if $b \in [\mathbb{H}_d^\infty]_1$ is non-CE and outer with Sarason outer function a, then b is the Sarason outer function of a, so that by Theorem 4.1, we have that

$$\begin{aligned} \mathscr{H}^t(c) &= \mathscr{H}^t(C) \oplus C(R)\mathscr{H}^t(D), \\ \mathscr{H}^t(b) &= \mathscr{H}^t(B) \oplus B(R)\mathscr{H}^t(D), \qquad \text{and} \\ \mathscr{H}^t(a) &= \mathscr{H}^t(A) \oplus A(R)\mathscr{H}^t(D). \end{aligned}$$

Moreover, one can verify, in this case, that for any $x \in \mathscr{H}^t(B)$ and $y \in \mathscr{H}^t(D)$, the linear map

$$x \oplus B(R)y \overset{U}{\mapsto} H(R)^* x \oplus -A(R)y; \qquad H(R) := B(R)A(R)^{-1},$$

is an onto isometry that intertwines $X^{(b)} = L^*|_{\mathscr{H}^t(b)}$ with $X^{(a)} = L^*|_{\mathscr{H}^t(a)}$.

5 Smirnov Column–Inner Pairs and Toeplitz Factorization

We are now sufficiently prepared to prove one of our main results which shows how the open Question 1 on factorizability of positive semi-definite left Toeplitz operators with factorable minorants is equivalent to several corresponding questions regarding de Branges–Rovnyak spaces for non-CE b, the relationship between a non-CE b and its outer NC Sarason function and the graphs of left Smirnov multipliers and their adjoints:

Theorem 5.1 *Given a non-column-extreme left multiplier $b \in [\mathbb{H}_d^\infty]_1$ with outer Sarason function $a \in [\mathbb{H}_d^\infty]_1$, let $\binom{B}{A} D = CD$ be the inner–outer factorization of $c = \binom{b}{a}$. The following are equivalent:*

(i) *c is inner so that $D \equiv 1$. That is, a, b are a column–inner Smirnov pair.*
(ii) *If $X := L^*|_{\mathscr{H}^t(b)}$, then X^* is pure.*
(iii) *X^* obeys the weak purity condition $\|X^{(n)} b^t\|_b^2 \to 0$.*
(iv) *The NC Szegö kernels or the free polynomials are dense in $\mathscr{H}^t(b)$.*
(v) *$\mathscr{H}^t(b)$ is the domain of the adjoint of the closed, densely defined Smirnov multiplier $H(R) = b(R)a(R)^{-1}$.*

Proof Equivalence of (i), (ii) and (iii) was proven in Corollary 3.5.

(i)⇔(iv): Since b is non-CE, $\mathscr{H}^{\mathrm{t}}(b)$ always contains $\mathbb{C}\{\mathfrak{z}\}$ by [17, Corollary 6.14]. Moreover, $b^{\mathrm{t}} \in \mathscr{H}^{\mathrm{t}}(b)$ by [17, Theorem 4.2] and $K_0^b = 1 - b^{\mathrm{t}}\overline{b(0)}$ so that $1 \in \mathscr{H}^{\mathrm{t}}(b)$ as well. Alternatively, Theorem 4.1 implies that $\mathscr{H}^{\mathrm{t}}(b) = \mathscr{H}^{\mathrm{t}}(B) \oplus B(R)\mathscr{H}^{\mathrm{t}}(D)$ and $\mathscr{H}^{\mathrm{t}}(B) = \mathrm{Dom}\, H(R)^*$, equipped with the graph norm, where $H(R) = B(R)A(R)^{-1}$ is a closed right Smirnov multiplier by [22, Theorem 4.17]. By [22, Corollary 3.13], the free polynomials and the NC Szegö kernels are both cores for $H(R)^*$, hence dense in $\mathscr{H}^{\mathrm{t}}(B) \subseteq \mathscr{H}^{\mathrm{t}}(b)$. It follows that the free polynomials/NC Szegö kernels are dense in $\mathscr{H}^{\mathrm{t}}(b)$ if and only if $\mathscr{H}^{\mathrm{t}}(B) = \mathscr{H}^{\mathrm{t}}(b)$ in which case $D \equiv 1$ and $B = b$. This also establishes (iv)⇔(v). □

In one variable, each of the above conditions is true. For example, here is an argument which proves that $c = C$ is inner using Theorem 4.1 and Corollary 4.4.

Proposition 5.2 *If $d = 1$ then $c = C$ is inner.*

Proof If $x \in \mathscr{H}(D)$ is any non-zero vector then

$$A(S)^*x \in \mathscr{H}(D) \cap \mathscr{M}(A^*),$$

since $\mathscr{H}(D)$ is S^*-invariant. Corollary 4.4 then implies that $A(S)^*x = 0$ so that $x = 0$ since A is outer. Hence, $\mathscr{H}(D) = \{0\}$, $D = 1$, and $c = C$ is inner. □

When $d > 1$ the above does not work since $A(R)^*x$ need not belong to $\mathscr{H}^{\mathrm{t}}(D)$ as $\mathscr{H}^{\mathrm{t}}(D)$ is L-co-invariant but generally not R-co-invariant.

Corollary 5.3 *The following two statements are equivalent:*

(i) *The equivalent conditions of Theorem 5.1 hold for every non-column-extreme $b \in [\mathbb{H}^\infty_d]_1$.*
(ii) *A positive semi-definite left Toeplitz operator, $T \in \mathscr{L}(\mathbb{H}^2_d)$, is factorable if and only if it has a non-trivial, positive semi-definite and factorable left Toeplitz minorant.*

Before proceeding with the proof, it will be convenient to recall the concept of a *positive NC measure*. A positive NC measure, μ, is any positive linear functional on the *free disk system*, $\mathscr{A}_d$,

$$\mathscr{A}_d := \left(\mathbb{A}_d + \mathbb{A}_d^*\right)^{-\|\cdot\|}, \quad \text{where} \quad \mathbb{A}_d := \mathrm{Alg}\{I, L_1, \cdots, L_d\}^{-\|\cdot\|},$$

is the *free disk algebra*; see [17, 24, 27, 28] for details. We denote the set of all positive NC measures by $(\mathscr{A}_d)^\dagger_+$. When $d = 1$, $\mathscr{A}_d \simeq \mathscr{C}(\partial\mathbb{D})$ can be identified with the commutative C^*-algebra of continuous functions on the complex unit circle, $\partial\mathbb{D}$, and then the Riesz–Markov theorem identifies any positive NC measure $\mu \in (\mathscr{A}_1^\dagger)_+$ with a positive, finite and regular Borel measure on $\partial\mathbb{D}$. The appropriate analogue of normalized Lebesgue measure is the so-called vacuum state, m,

$$m(L^\omega) := \langle 1, L^\omega 1\rangle_{\mathbb{H}^2} = \delta_{\omega,\emptyset}.$$

Here, if $d = 1$,

$$m(S^k) = \left\langle 1, S^k 1 \right\rangle_{H^2} = \int_{\partial\mathbb{D}} \zeta^k \check{m}(d\zeta) = \delta_{k,0},$$

is the unique positive linear functional corresponding to normalized Lebesgue measure, $\check{m}$, on the circle. If $T \geq 0$ is any positive semi-definite left Toeplitz operator, observe that the linear functional

$$\mu_T(L^\omega) := \langle 1, TL^\omega 1\rangle_{\mathbb{H}^2},$$

extends to a positive NC measure on the free disk system, in which case T can be thought of as the *NC Radon–Nikodym derivative* of μ_T with respect to m.

Proof (i) $\Leftrightarrow$ (ii): To connect Toeplitz factorization to the theory we have developed for non-CE multipliers, define the positive NC measure $\mu_T \in (\mathscr{A}_d)^\dagger_+$ by

$$\mu_T(L^\omega) := \langle 1, TL^\omega 1\rangle_{\mathbb{H}^2}.$$

This is an absolutely continuous NC measure and T is its NC Radon–Nikodym derivative. By [24], since T is bounded, there is a $b_T \in [\mathbb{H}^\infty_d]_1$ so that

$$T = (I - b_T(R)^*)^{-1}(I - b_T(R)^* b_T(R))(I - b_T(R))^{-1} \geq 0.$$

In more detail, since T is bounded, $\mu_T \leq \|T\| m$, where recall that $m(L^\omega) := \langle 1, L^\omega 1\rangle_{\mathbb{H}^2}$ is NC Lebesgue measure [27]. Hence, μ_T is absolutely continuous in the sense of the NC Lebesgue decomposition of [24, 27]. By the NC Fatou theorem of [24], if $\mu_T = \mu_{b_T}$ is the NC Clark measure of $b_T \in [\mathbb{H}^\infty_d]_1$,

$$\mu_T(p(L)^* q(L)) = \langle p, Tq\rangle_{\mathbb{H}^2} = \left\langle \widetilde{T}^{\frac{1}{2}} p, \widetilde{T}^{\frac{1}{2}} q \right\rangle_{\mathbb{H}^2},$$

where $\widetilde{T}$ is the closed, positive semi-definite left Toeplitz operator obtained as the strong resolvent limit of the bounded and positive semi-definite left Toeplitz operators

$$T_r := (I - b_T(rR)^*)^{-1}(I - b_T(rR)^* b_T(rR))(I - b_T(rR))^{-1},$$

as $r \uparrow 1$. The above equation shows that T and $\widetilde{T}$ define the same bounded and positive semi-definite quadratic forms, so that $T = \widetilde{T}$ and $\widetilde{T}$ is bounded. Strong resolvent convergence of a bounded net of positive semi-definite operators to a bounded and positive semi-definite operator is equivalent to convergence in the strong operator topology [25, Theorem VIII.20]. Hence $T_r \overset{SOT}{\rightarrow} T$. Finally, $b_T(rR) \overset{SOT-*}{\rightarrow} b_T(R)$ and the formula for T follows.

Since we can assume that $T \neq 0$, $b_T(R)$ is not inner and it is clear that T has a factorable minorant/is factorable if and only if the numerator $I - b_T(R)^* b_T(R)$ has factorable minorant/is factorable. In particular, if such a T with a factorable minorant is always factorable, this is equivalent to saying that any non-CE $b \in [\mathbb{H}^\infty_d]_1$ is such that $c = \binom{b}{a}$, where a is the Sarason function of b, is always inner.

□

6 Application to Rational Multipliers

In this section we will establish a Fejér–Riesz theorem for NC rational multipliers of Fock space. The classical Fejer-Riesz theorem states that if a trigonometric polynomial

$$p(e^{i\theta}) = \sum_{n=-N}^{N} \hat{p}_n e^{in\theta}$$

is nonnegative for all $\theta \in [0, 2\pi]$, then there is a single analytic polynomial $q(z) = \sum_{n=0}^{N} \hat{q}_n z^n$, of the same degree as p, so that

$$p(e^{i\theta}) = |q(e^{i\theta})|^2$$

for all $\theta \in [0, 2\pi]$. As a trivial consequence, there is a similar factorization in the rational case: Namely, if $\mathfrak{r}(z) = \frac{p_1(z)}{p_2(z)}$ is a rational function, regular in $|z| \leq 1$, and

$$\mathfrak{r}(e^{i\theta}) + \overline{\mathfrak{r}(e^{i\theta})} \geq 0$$

for all θ, then there is another rational function $\mathfrak{q}$, regular in $|z| \leq 1$ and with degree at most the degree of $\mathfrak{r}$, such that

$$\mathfrak{r}(e^{i\theta}) + \overline{\mathfrak{r}(e^{i\theta})} = |\mathfrak{q}(e^{i\theta})|^2 \tag{4}$$

for all θ. Indeed, suppose deg $p_1, p_2 \leq N$. Since, by hypothesis,

$$\frac{p_1(e^{i\theta})}{p_2(e^{i\theta})} + \overline{\frac{p_1(e^{i\theta})}{p_2(e^{i\theta})}} \geq 0,$$

we can clear the denominators to obtain a trigonometric polynomial of degree at most N, which is positive on the unit circle, and hence factors:

$$\overline{p_2(e^{i\theta})} p_1(e^{i\theta}) + \overline{p_1(e^{i\theta})} p_2(e^{i\theta}) =: |\widetilde{p}(e^{i\theta})|^2,$$

where $\widetilde{p} \in \mathbb{C}[e^{i\theta}]$ is an analytic polynomial of degree at most N. Dividing by $|p_2|^2$ again, and setting $\mathfrak{q} = \widetilde{p}/p_2$, we arrive at (4). We see immediately from the calculation that $\mathfrak{q}$ is regular everywhere that $\mathfrak{r}$ is, and the degree of $\mathfrak{q}$ (i.e. the maximum of the degrees of numerator and denominator) does not exceed the degree of p.

Relatedly, if now $\mathfrak{b}$ is a rational function regular in $|z| \leq 1$ and bounded by 1 there, we can consider the function

$$1 - |\mathfrak{b}(e^{i\theta})|^2 \geq 0$$

for $\theta \in [0, 2\pi]$. If $\mathfrak{b}$ is a rational inner function (a finite Blaschke product) then this expression is identically 0. Otherwise, we can apply the same argument as above to obtain a rational function $\mathfrak{a}$ (in fact an outer function, i.e. having no zeroes in $|z| < 1$), of degree not exceeding the degree of $\mathfrak{b}$, and regular everywhere that $\mathfrak{b}$ is, so that

$$1 - |\mathfrak{b}(e^{i\theta})|^2 = |\mathfrak{a}(e^{i\theta})|^2 \tag{5}$$

for all θ. The function $\mathfrak{a}$ is sometimes called the *Pythagorean mate* of $\mathfrak{b}$. Thus, we observe a dichotomy: Either $\mathfrak{b}$ is inner, or $1 - \mathfrak{b}^*\mathfrak{b}$ 'factors completely' as $\mathfrak{a}^*\mathfrak{a}$. In the latter case, it follows from Szegö's theorem that $\log(1 - |\mathfrak{b}|^2)$ is integrable on the unit circle, and hence $\mathfrak{b}$ is not an extreme point of the unit ball of $H^\infty(\mathbb{D})$. (This dichotomy breaks down with the rationality assumption; there exists non-inner b so that $1 - |b|^2$ is not log-integrable, and hence $1 - b^*b$ is nonzero, but does not factor. Recall that such a b is necessarily a non-inner extreme point of $[H^\infty]_1$.)

The Fejér–Riesz theorem has been extended to Fock space by G. Popescu [12, Theorem 1.6]:

Theorem (NC Fejér–Riesz. Popescu) *Let $T := \mathrm{Re}\, p(R) \geq 0$ be a positive semi-definite left Toeplitz operator on $\mathbb{H}^2_d$, where $p \in \mathbb{C}\{\mathfrak{z}\}$ and deg $p = N$. Then there is an outer free polynomial $q \in \mathbb{C}\{\mathfrak{z}\}$ with deg $q = N$ so that $q(R)^*q(R) = T$.*

However, since NC rational functions are not, in general, expressible as quotients of polynomials, the simple arguments just given in the one-variable case do not generalize to the NC setting. Nonetheless, it turns out there are NC versions of (4) and (5) available, though the proof will require completely different methods. In fact, (4) can be more or less reduced to (5), by means of the Cayley transform, and (5) can be proven directly by constructing an appropriate *realization* of a from a suitable realization of b as in Sect. 3. In particular, the realizations we work with arise from the de Branges–Rovnyak style functional model, and our various claims can be established by analysis of this model.

In particular, we replace 'degree' with 'size of a minimal Fornasini–Marchesini realization' (an analogue of the MacMillan degree) and are able to prove an NC version of (5) with sharp degree bounds and an NC version of (4) with a reasonable degree bound (but we do not know if it is sharp). Likewise in the NC version of (5) we are able to prove that $\mathfrak{a}$ is regular everywhere that $\mathfrak{b}$ is, while in the NC version of (4) we settle for a somewhat weaker claim (that the $\mathfrak{q}$ in the right-hand side will at least be regular across the boundary of the row ball); we do not know if it can in fact be chosen to be regular everywhere that $\mathfrak{r}$ is.

6.1 Preliminaries on NC Rational Functions

Recall that a complex NC rational expression is any valid combination of the several NC variables $\mathfrak{z}_1, \cdots, \mathfrak{z}_d$, the complex scalars, $\mathbb{C}$, the operations $+, \cdot, ^{-1}$ and parentheses (,) with domain

$$\mathrm{Dom}\,\mathrm{r} = \bigsqcup_{n=1}^{\infty} \mathrm{Dom}_n\,\mathrm{r},$$

$$\mathrm{Dom}_n\,\mathrm{r} := \bigsqcup_{n=1}^{\infty} \left\{ X = (X_1, \cdots, X_d) \in \mathbb{C}^{n\times n} \otimes \mathbb{C}^{1\times d} \middle|\ \mathrm{r}(X) \text{ is defined.} \right\}.$$

We will use the notation $\mathbb{C}_n^d := \mathbb{C}^{n\times n} \otimes \mathbb{C}^{1\times d}$ for a row d-tuple of complex $n \times n$ matrices. Such an NC rational expression is said to be *valid* if its domain is not empty. An NC rational function, $\mathfrak{r}$, is then the equivalence class of valid NC rational expressions with respect to the relation $\mathrm{r}_1 \equiv \mathrm{r}_2$ if $\mathrm{r}_1(X) = \mathrm{r}_2(X)$ for all $X \in \mathrm{Dom}\,\mathrm{r}_1 \bigcap \mathrm{Dom}\,\mathrm{r}_2$ (this intersection is always non-empty by [29, Footnote, page 52]), with domain $\mathrm{Dom}\,\mathfrak{r} := \cup_{\mathrm{r}\in\mathfrak{r}} \mathrm{Dom}\,\mathrm{r}$, and we write $\mathfrak{r}(Z) := \mathrm{r}(Z)$ if $Z \in \mathrm{Dom}\,\mathrm{r}$ and $\mathrm{r} \in \mathfrak{r}$. Any NC rational function in $d-$variables, $\mathfrak{r}$, which is regular at 0 has a unique (up to joint similarity) *minimal Fornasini–Marchesini (FM) realization.* Namely, there is a quadruple (A, B, C, D) with $A \in \mathbb{C}_n^d$, $B = \begin{pmatrix} B_1 \\ \vdots \\ B_d \end{pmatrix} \in \mathbb{C}^n \otimes \mathbb{C}^d$, $C \in \mathbb{C}^{1\times n}$ and $D \in \mathbb{C}$ so that for any $X \in \mathbb{C}_m^d \cap \mathrm{Dom}\,\mathfrak{r}$,

$$\mathfrak{r}(X) = DI_m + C \otimes I_m L_A(X)^{-1} B \otimes X; \qquad L_A(X) := I_n \otimes I_m - \sum A_j \otimes X_j,$$

where $B \otimes X := B_1 \otimes X_1 + \cdots + B_d \otimes X_d$ and this realization is minimal in the sense that n is as small as possible. We will also sometimes write $A \otimes X$ in place of $\sum A_j \otimes X_j$. Here, $L_A(\cdot)$ is called a (monic, affine) *linear pencil.* A realization is minimal if and only if it is both *controllable*:

$$\bigvee_{\substack{\omega\in\mathbb{F}^d \\ 1\leq j\leq d}} A^\omega b_j = \mathbb{C}^n,$$

and *observable*

$$\bigvee A^{*\omega}C^* = \mathbb{C}^n;$$

see, e.g. [30, Subsection 3.1.2]. Minimal realizations are unique up to joint similarity [31, Theorem 2.1]. If (A, B, C, D) is a minimal FM realization of $\mathfrak{r}$ and $A \in \mathbb{C}_n^d$, we will say that the *minimal FM realization size* of $\mathfrak{r}$ is n and write $\dim_{FM} \mathfrak{r} = n$. We will also have occasion to use *descriptor realizations*. Any NC rational function, $\mathfrak{r}$, which is *regular at* 0, i.e. so that $0 = (0, \cdots, 0) \in \mathrm{Dom}\,\mathfrak{r}$, has a minimal descriptor realization $(A, b, c) \in \mathbb{C}_m^d \times \mathbb{C}^m \times \mathbb{C}^m$: For any $Z \in \mathrm{dom}_n\,\mathfrak{r} := \mathrm{Dom}\mathfrak{r} \cap \mathbb{C}_n^d$,

$$\mathfrak{r}(Z) = b^* \otimes I_n L_A(Z)^{-1} c \otimes I_n;$$

see [32]. Minimality again means that n is as small as possible, and this is equivalent to c being A-cyclic (the realization is controllable) and b being A^*-cyclic (the realization is observable), and again, minimal descriptor realizations are unique up to joint similarity. We will sometimes write $\dim \mathfrak{r} = m$ for the *minimal descriptor realization size* of $\mathfrak{r}$. The following lemma shows how minimal FM realizations can be constructed from minimal descriptor realizations.

Lemma 6.1 *Let* $(A, b, c) \in \mathbb{C}_n^d \times \mathbb{C}^n \times \mathbb{C}^n$ *be a descriptor realization of an NC rational function* $\mathfrak{r}$ *with* $0 \in \mathrm{Dom}\,\mathfrak{r}$. *Let* $\mathscr{M}_0 := \bigvee_{\omega \neq \emptyset} A^\omega c \subseteq \mathbb{C}^n$ *with projector* Q_0 *and define* $(A^{(0)}, B^{(0)}, C^{(0)}, D^{(0)})$ *by*

$$A_j^{(0)} := A_j|_{\mathscr{M}_0}, \quad B_j^{(0)} := A_j c, \quad C^{(0)} := (Q_0 b)^* \quad \text{and} \quad D^{(0)} = \mathfrak{r}(0).$$

This is an FM realization of $\mathfrak{r}$ *and it is minimal if* (A, b, c) *is minimal. In particular,* $\dim \mathfrak{r} - 1 \leq \dim_{FM} \mathfrak{r} \leq \dim \mathfrak{r}$.

Proof It is easily checked that $(A^{(0)}, B^{(0)}, C^{(0)}, D^{(0)})$ is a realization of $\mathfrak{r}$ and this realization will be both controllable and observable, hence minimal, if (A, b, c) is. Finally, $\dim \mathfrak{r} - 1 \leq \dim_{FM} \mathfrak{r} := \dim \mathscr{M}_0 \leq \dim \mathfrak{r}$. □

As proven in [33, Theorem A], given an NC rational function, $\mathfrak{r}$, regular at 0, the following are equivalent:

(i) $\mathfrak{r} \in \mathbb{H}_d^2$.
(ii) $\mathfrak{r} \in \mathbb{H}_d^\infty$.
(iii) $\mathfrak{r} = K\{Z, y, v\}$ is an NC Szegö kernel for some $Z, y, v \in \mathbb{B}_n^d \times \mathbb{C}^n \times \mathbb{C}^n$.
(iv) $r\mathbb{B}_{\mathbb{N}}^d \subseteq \mathrm{Dom}\,\mathfrak{r}$ for some $r > 1$.

Given an NC rational $\mathfrak{b} \in [\mathbb{H}_d^\infty]_1$, its minimal FM realization can be constructed by cutting down its de Branges–Rovnyak realization. Namely, let

$$\mathscr{M}_0(\mathfrak{b}) := \bigvee_{\omega \neq \emptyset} L^{*\omega} \mathfrak{b}^{\mathrm{t}} \subseteq \mathscr{H}^{\mathrm{t}}(\mathfrak{b}),$$

with orthogonal projection P_0. It is easy to check that this space is finite-dimensional since $\mathfrak{b}^{\mathrm{t}} = K\{Z, y, v\} \in \mathbb{H}^2_d$ is also NC rational, hence an NC Szegö kernel and $L^{*\omega} K\{Z, y, v\} = K\{Z, Z^{*\omega} y, v\}$; see [34, Lemma 2]. Then defining

$$A := L^*|_{\mathscr{M}_0(\mathfrak{b})}, \quad B := L^* \mathfrak{b}^{\mathrm{t}}, \quad C := (P_0 K_0^{\mathfrak{b}})^* \quad \text{and} \quad D := \mathfrak{b}(0),$$

yields a minimal FM realization of $\mathfrak{b}$:

Lemma 6.2 *Let $\mathfrak{b} \in [\mathbb{H}^\infty_d]_1$ be a contractive and NC rational multiplier of Fock space. The above finite FM realization (A, B, C, D) of $\mathfrak{b}$ obtained by compression of the de Branges–Rovnyak realization of $\mathfrak{b}$ to $\mathscr{M}_0(\mathfrak{b})$ is minimal.*

Proof It is easily checked that (A, B, C, D) is a finite-dimensional realization of $\mathfrak{b}$. This realization is controllable by construction since

$$\bigvee_{\substack{\omega \in \mathbb{F}^d \\ 1 \leq j \leq d}} A^\omega B_j = \bigvee_{\omega \neq \emptyset} L^{*\omega} \mathfrak{b}^{\mathrm{t}} = \mathscr{M}_0(\mathfrak{b}).$$

Suppose that there is an $h \in \mathscr{M}_0(\mathfrak{b})$ so that h is orthogonal to $\bigvee_{\omega \in \mathbb{F}^d} A^{*\omega} C^*$. Then, for any $\omega \in \mathbb{F}^d$,

$$\begin{aligned} 0 = \left\langle A^{*\omega^{\mathrm{t}}} C^*, h \right\rangle &= \left\langle P_0 K_0^{\mathfrak{b}}, A^\omega h \right\rangle \\ &= \left\langle K_0^{\mathfrak{b}}, L^{*\omega} h^{\mathrm{t}} \right\rangle_{\mathfrak{b}} = \hat{h}_\omega. \end{aligned}$$

It follows that all Taylor coefficients of $h \in \mathscr{M}_0(\mathfrak{b}) \subseteq \mathbb{H}^2_d$ vanish so that $h \equiv 0$. This proves that the realization is also observable, hence minimal. □

Proposition 6.2 *If $\mathfrak{b} \in [\mathbb{H}^\infty_d]_1$ is non-CE and NC rational, then its Sarason function $\mathfrak{a} \in [\mathbb{H}^\infty_d]_1$ is also NC rational. The minimal FM realization size of $\mathfrak{a}$ is at most that of $\mathfrak{b}$.*

Proof (of Proposition 6.2) As above, let P_0 denote the orthogonal projection of $\mathscr{H}^{\mathrm{t}}(\mathfrak{b})$ onto $\mathscr{M}_0(\mathfrak{b})$, and let (A, B, C, D) be the minimal and finite FM realization of $\mathfrak{b}$ obtained by compression of the de Branges–Rovnyak realization of $\mathfrak{b}$ to $\mathscr{M}_0(\mathfrak{b})$. This compression also yields a finite-dimensional FM colligation and hence realization for $\mathfrak{c} = \begin{pmatrix} \mathfrak{b} \\ \mathfrak{a} \end{pmatrix}$:

$$U_{\mathfrak{c}} = \begin{pmatrix} A & B \\ C & D \\ -\mathfrak{a}(0) \langle P_0 \mathfrak{b}^{\mathrm{t}}, \cdot \rangle_{\mathfrak{b}} & \mathfrak{a}(0) \end{pmatrix} : \begin{pmatrix} \mathscr{M}_0(\mathfrak{b}) \\ \mathbb{C} \end{pmatrix} \to \begin{pmatrix} \mathscr{M}_0(\mathfrak{b}) \otimes \mathbb{C}^d \\ \mathbb{C}^2 \end{pmatrix}.$$

In particular,

$$U_{\mathfrak{a}} := \begin{pmatrix} A & B \\ -\mathfrak{a}(0)\left\langle P_0 \mathfrak{b}^{\mathrm{t}}, \cdot \right\rangle_{\mathfrak{b}} & \mathfrak{a}(0) \end{pmatrix} : \begin{pmatrix} \mathscr{M}_0(\mathfrak{b}) \\ \mathbb{C} \end{pmatrix} \to \begin{pmatrix} \mathscr{M}_0(\mathfrak{b}) \otimes \mathbb{C}^d \\ \mathbb{C} \end{pmatrix}$$

is a finite-dimensional and contractive FM colligation for $\mathfrak{a}$, so that $\mathfrak{a}$ is NC rational and $\dim_{FM} \mathfrak{a} \leq \dim \mathscr{M}_0(\mathfrak{b}) = \dim_{FM} \mathfrak{b}$. □

Theorem 6.3 *If* $\mathfrak{b} \in [\mathbb{H}_d^\infty]_1$ *is NC rational, then either* $\mathfrak{b}$ *is inner or it is not column-extreme. If* $\mathfrak{b}$ *is non-CE and* $\mathfrak{a}$ *is its NC rational Sarason function, then* $\mathfrak{c} := \binom{\mathfrak{b}}{\mathfrak{a}}$ *is inner, so that* $T := I - \mathfrak{b}(R)^*\mathfrak{b}(R) = \mathfrak{a}(R)^*\mathfrak{a}(R)$ *is factorable.*

Proof Suppose that $\mathfrak{b} \in [\mathbb{H}_d^\infty]_1$ is NC rational. Consider the right free de Branges–Rovnyak space $\mathscr{H}^{\mathrm{t}}(\mathfrak{b})$ of $\mathfrak{b}$. We define the minimal de Branges–Rovnyak FM realization (A, B, C, D) as above. Recall that since $\mathfrak{b} \in \mathbb{H}_d^2$ is NC rational, it also has a *minimal descriptor realization*, $(A', b', c') \in \mathbb{C}_n^d \times \mathbb{C}^n \times \mathbb{C}^n$. By [33, Theorem A], A' is jointly similar to a strict row contraction. By Lemma 6.1, one can construct a minimal FM realization, $(A^{(0)}, B^{(0)}, C^{(0)}, D^{(0)})$, from the minimal descriptor realization (A', b', c') by defining

$$\mathscr{M}_0 := \bigvee_{\omega \neq \emptyset} A'^{w} c,$$

with projector Q_0 and

$$A^{(0)} := A'|_{\mathscr{M}_0}, \quad B^{(0)} := A'c', \quad C^{(0)} := (Q_0 b')^* \quad \text{and} \quad D^{(0)} = \mathfrak{b}(0).$$

Since $A^{(0)}$ is the restriction of A' to an invariant subspace and A' is pure, we obtain that $A^{(0)}$ is also pure. Namely,

$$\|\mathrm{Ad}_{(A^{(0)})^{(n)}}(I)\|^2 = \|\mathrm{Ad}_{A'}^{(n)}(Q_0)\|^2 \leq \|\mathrm{Ad}_{A'}^{(n)}(I)\|^2 \to 0,$$

since A' is pure. Since minimal FM realizations are unique up to joint similarity, A is jointly similar to $A^{(0)}$ and is hence pure and similar to a strict row contraction. Hence, A^* is jointly similar to a strict column contraction. By [34, Lemma 1], A^* is then jointly similar to a strict row contraction and A^* defines a pure row d-tuple, $(A_1^*, \cdots, A_d^*)$. Since $A^* \in \mathbb{C}_m^d$ is pure, we necessarily have that $\mathfrak{b}$ is inner if $\mathfrak{b}$ is CE and $\mathfrak{c}$ is inner if $\mathfrak{b}$ is non-CE by Theorem 3.1. □

Remark 6.1 Setting $d = 1$, it follows that any contractive rational multiplier of Hardy space is either not an extreme point or it is inner. Indeed, if $\mathfrak{b}$ is a contractive rational multiplier of H^2 which is not inner, then $I - \mathfrak{b}(S)^*\mathfrak{b}(S) > 0$ is factorable by the Fejér–Riesz theorem, so that there is a contractive, outer and NC rational multiplier, $\mathfrak{a} \in H^\infty$, so that the two-component column $\binom{\mathfrak{b}}{\mathfrak{a}}$ is inner. That is, $\mathfrak{b}$ is non-CE. Further recall that a contractive multiplier of H^2 is column-extreme if and only if it is an extreme point. Hence, the assertion of Theorem 6.3 that any contractive NC rational multiplier of Fock space is either inner or non-CE is an exact analogue of classical fact.

Remark 6.2 If $b \in \mathbb{C}\{\mathfrak{z}\}$ is also a free polynomial, then the A from a minimal FM realization of b is jointly nilpotent and it follows that the Sarason function $a \in \mathbb{C}\{\mathfrak{z}\}$ is also a free polynomial. Also note that if $p \in [\mathbb{H}^\infty_d]_1$, then either p is inner or p is non-CE by the NC Fejér–Riesz theorem of Popescu [12, Theorem 1.6].

The above results also hold if $b = K\{Z, y, v\}$ is an NC Szegö kernel at an infinite point $Z \in \mathbb{B}^d_\infty$. In this case, since Z is a strict row contraction, the Sarason function a is also an NC Szegö kernel at a potentially infinite point and so is the outer factor D, of $c = \binom{b}{a}$.

6.2 NC Rational Fejér–Riesz

Proposition 6.4 *Let* $\mathfrak{r} \in \mathbb{H}^\infty_d$ *be an NC rational function. Then* $\mathfrak{r}(R)^*\mathfrak{r}(R) = \operatorname{Re}\mathfrak{H}(R) \geq 0$ *where* $\mathfrak{H} \in \mathbb{H}^\infty_d$ *is NC rational and Herglotz and* $\dim_{FM}\mathfrak{H} \leq \dim_{FM}\mathfrak{r} + 1$.

In the above statement, recall that an NC function $H \in \mathscr{O}(\mathbb{B}^d_\mathbb{N})$ is called a *left free Herglotz function*, if $\operatorname{Re} H(Z) \geq 0$ for all $Z \in \mathbb{B}^d_\mathbb{N}$. An NC function $G \in \mathscr{O}(B^d_\mathbb{N})$ is then called a *right free Herglotz function* if G^{t} is a left free Herglotz function. Since, in the above statement, $\operatorname{Re} H(R) \geq 0$, it follows also that $\operatorname{Re} H(L) = U_{\mathrm{t}}\operatorname{Re} H(R)U_{\mathrm{t}} \geq 0$, and taking inner products against NC Szegö kernels shows that H is a left NC Herglotz function and H^{t} is right Herglotz.

Proof Since $\mathfrak{r}$ is NC rational, $\mathfrak{r}^{\mathrm{t}} \in \mathbb{H}^2_d$ is also NC rational by [34, Lemma 2] and $\mathfrak{r}, \mathfrak{r}^{\mathrm{t}}$ have the same size minimal descriptor realizations. Hence, $\mathfrak{r}^{\mathrm{t}} \in \mathbb{H}^\infty_d$ is an NC kernel, $\mathfrak{r}^{\mathrm{t}} = K\{Z, y, v\}$ and $\mathfrak{r}(R)^*\mathfrak{r}^{\mathrm{t}} = K\{Z, y, \mathfrak{r}^{\mathrm{t}}(Z)v\} =: \mathfrak{s}^{\mathrm{t}} \in \mathbb{H}^\infty_d$ is NC rational, and the size of the minimal descriptor realization of $\mathfrak{s}$ is at most that of $\mathfrak{r}$. In more detail, if $\mathfrak{r}$ has minimal descriptor realization (A, b, c), then the minimal descriptor realization of $\mathfrak{r}^{\mathrm{t}}$ is $(A^{\mathrm{t}}, \overline{c}, \overline{b})$, where $A^{\mathrm{t}} := (A^{\mathrm{t}}_1, \cdots, A^{\mathrm{t}}_d)$ denotes component-wise matrix transpose and $\overline{c}$ denotes entry-wise complex conjugation [34, Lemma 2]. Moreover, if $\mathfrak{r} \in \mathbb{H}^2_d$ has minimal descriptor realization (A, b, c), then by [33, Theorem A], we can assume without loss in generality that $A \in \mathbb{B}^d_\mathbb{N}$ is a strict row contraction and in this case, $\mathfrak{r}$ is equal to the NC Szegö kernel vector $K\{\overline{A}, \overline{b}, \overline{c}\}$. As before $\overline{A} = (\overline{A}_1, \cdots, \overline{A}_d)$ and $\overline{A}_j$ denotes entry-wise complex conjugation. Putting this all together, if (A, b, c) is a minimal descriptor realization of $\mathfrak{r}$, then $\mathfrak{s}$ has a descriptor realization $(A^{\mathrm{t}}, \overline{c}, \overline{\mathfrak{r}^{\mathrm{t}}}(A^{\mathrm{t}})\overline{b})$, where $\overline{\mathfrak{r}^{\mathrm{t}}}$ denotes the power series obtained by complex conjugation of the Taylor coefficients of $\mathfrak{r}^{\mathrm{t}}$. Applying Lemma 6.1 to construct FM realizations from these descriptor realizations for $\mathfrak{r}$ and $\mathfrak{s}$ shows that the minimal FM realization size of $\mathfrak{s}$ obeys $\dim_{FM}\mathfrak{s} \leq \dim_{FM}\mathfrak{r} + 1$.

In particular, $\mathfrak{s}(R)$ is a bounded right multiplier. Define the bounded left Toeplitz operator,

$$T := \mathfrak{r}(R)^*\mathfrak{r}(R) - \mathfrak{s}(R)^*.$$

Since T is a bounded left Toeplitz operator, it is completely determined by its 'Fourier coefficients'

$$\hat{T}_\omega := \left\langle L^\omega 1, T1 \right\rangle_{\mathbb{H}^2} \quad \text{and} \quad \left\langle 1, TL^\omega 1 \right\rangle_{\mathbb{H}^2} =: \hat{T}_{-\omega},$$

and T is the strong operator topology limit of its partial Cesàro sums [15, Lemma 1.1]. Each of these partial sums is a left Toeplitz operator of the form $p(R)+q(R)^*$, for $p, q \in \mathbb{C}\{\mathfrak{z}\}$. We claim that $T \in \mathfrak{R}_d^\infty$ is an analytic left Toeplitz operator, i.e. that $T = y(R)$ is a bounded right multiplier, $y \in \mathbb{H}_d^\infty$. To prove this, it suffices to show that the 'negative Fourier coefficients' of T vanish. For any $\omega \in \mathbb{F}^d$,

$$\begin{aligned}
\hat{T}_{-\omega} &= \left\langle 1, TL^\omega 1 \right\rangle_{\mathbb{H}^2} \\
&= \left\langle T^* 1, L^\omega 1 \right\rangle = \left\langle \mathfrak{r}(R)^* \mathfrak{r}^{\mathrm{t}} - \mathfrak{s}^{\mathrm{t}}, L^\omega 1 \right\rangle \\
&= \left\langle \mathfrak{s}^{\mathrm{t}} - \mathfrak{s}^{\mathrm{t}}, L^\omega 1 \right\rangle = 0,
\end{aligned}$$

and it follows that the partial Cesàro sums of T have the form

$$\sigma_n(T) = p_n(R), \qquad p_n \in \mathbb{C}\{\mathfrak{z}\}.$$

Since the partial Cesàro sum map, $T \mapsto \sigma_n(T)$, is a completely contractive linear map on $\mathscr{L}(\mathbb{H}_d^2)$, it follows that $p_n(R)$ is a uniformly bounded sequence of free polynomial right multipliers that converge to T in the strong operator topology so that $T \in \mathfrak{R}_d^\infty$, $T = y(R)$ for some $y \in \mathbb{H}_d^\infty$ [15, Lemma 1.1]. Also note that $\hat{y}_\emptyset = 0$. Hence, for any $\omega \neq \emptyset$,

$$\begin{aligned}
\hat{y}_\omega &= \left\langle L^{\omega^{\mathrm{t}}} 1, y(R)1 \right\rangle_{\mathbb{H}^2} \\
&= \left\langle L^{\omega^{\mathrm{t}}} 1, \mathfrak{s}^{\mathrm{t}} \right\rangle_{\mathbb{H}^2} - \left\langle L^{\omega^{\mathrm{t}}} 1, 1 \right\rangle \overline{\mathfrak{s}(0)} \\
&= \hat{\mathfrak{s}}_\omega,
\end{aligned}$$

so that $y(R) = \mathfrak{s}(R) - cI$ with $\overline{\mathfrak{s}(0)} = c$. In particular, $y = \mathfrak{y}$ is NC rational. Moreover,

$$0 \le \mathfrak{r}(R)^*\mathfrak{r}(R) = \mathfrak{y}(R) + \mathfrak{s}(R)^*$$

is positive so that $c \ge 0$ and we conclude that

$$\mathfrak{r}(R)^*\mathfrak{r}(R) = \mathfrak{y}(R) + \mathfrak{y}(R)^* + cI =: \mathrm{Re}\, \mathfrak{H}(R),$$

with $\mathfrak{H} := 2\mathfrak{y} + c \in \mathbb{H}_d^\infty$, a bounded and NC rational Herglotz multiplier. □

Lemma 6.3 *Let* $T := \mathrm{Re}\, \mathfrak{H}(R) \geq 0$ *be a positive semi-definite and NC rational left Toeplitz operator. If we define the positive NC measure* $\mu_T(L^\alpha) := \langle 1, TL^\alpha 1\rangle_{\mathbb{H}^2}$, *then* μ_T *is the* NC Clark measure *of a contractive, NC rational* $\mathfrak{b}_T \in [\mathbb{H}^\infty_d]_1$ *where* $\mathfrak{b}_T = \frac{\mathfrak{H}-1}{\mathfrak{H}+1}$.

Recall that if $\mu \in (\mathscr{A}_d)^\dagger_+$ is any positive NC measure, then one can define its left *free Herglotz–Riesz transform*,

$$H_\mu(Z) := \mathrm{id}_n \otimes \mu \circ (I + ZL^*)(I - ZL^*)^{-1}; \qquad Z \in \mathbb{B}^d_n,$$

where $ZL^* := Z_1 \otimes L_1^* + \cdots + Z_d \otimes L_d^*$ [17, 28]. The NC function, $H_\mu \in \mathscr{O}(\mathbb{B}^d_{\mathbb{N}})$, is an *NC Herglotz function*, i.e. for any $Z \in \mathbb{B}^d_n$, $\mathrm{Re}\, H_\mu(Z) \geq 0$. As in the classical, $d = 1$ setting, the inverse Cayley transform of H_μ,

$$b_\mu(Z) := (H_\mu(Z) - I_n)(H_\mu(Z) + I_n)^{-1}; \qquad Z \in \mathbb{B}^d_n,$$

is then a contractive left multiplier of Fock space so that $\mu = \mu_{b_\mu}$ is the *NC Clark measure* of b_μ [17, 28].

Proof The (left) NC Herglotz–Riesz transform of μ_T can be expanded as a power series:

$$\begin{aligned}
\mathfrak{H}_T(Z) &= 2\sum_\alpha Z^{\alpha^t} \left\langle L^\alpha 1, T1\right\rangle_{\mathbb{H}^2} - \langle 1, T1\rangle_{\mathbb{H}^2} \\
&= \sum_\alpha Z^{\alpha^t} \left\langle L^\alpha 1, (\mathfrak{H}(R) + \mathfrak{H}(R)^*)1\right\rangle_{\mathbb{H}^2} - \langle 1, \mathrm{Re}\, \mathfrak{H}(R)1\rangle_{\mathbb{H}^2} \\
&= \mathfrak{H}(Z) + I_n\overline{\mathfrak{H}(0)} - I_n \mathrm{Re}\, \mathfrak{H}(0) \\
&= \mathfrak{H}(Z) - iI_n \mathrm{Im}\, \mathfrak{H}(0);
\end{aligned}$$

see [17, 28]. Since $\mathfrak{H}_T$ and $\mathfrak{H}$ are NC Herglotz functions which differ by an imaginary constant, μ_T is the NC Clark measure of both $\mathfrak{b}_T$ and $\mathfrak{b}$, where $\mathfrak{b}_T$ is the inverse Cayley transform of $\mathfrak{H}_T$ and $\mathfrak{b}$ is the inverse Cayley transform of $\mathfrak{H}$ [34]. □

Theorem 6.5 (NC rational Fejér–Riesz) *Any NC rational, positive semi-definite left Toeplitz operator,* $T = \mathrm{Re}\, \mathfrak{H}(R) \geq 0$, $\mathfrak{H} \in \mathbb{H}^\infty_d$, *is factorable. If* $\mathfrak{b} = (\mathfrak{H} - 1)(\mathfrak{H} + 1)^{-1} \in [\mathbb{H}^\infty_d]_1$ *is the NC rational and contractive inverse Cayley transform of* $\mathfrak{H}$, *then the factorization is*

$$T = \mathfrak{D}(R)^*\mathfrak{D}(R), \quad \textit{where} \quad \mathfrak{D}(R) = \mathfrak{a}(R)(I - \mathfrak{b}(R))^{-1} \in \mathbb{H}^\infty_d,$$

$\mathfrak{a} \in [\mathbb{H}^\infty_d]_1$ *is the NC rational Sarason outer function of* $\mathfrak{b}$, $\dim_{FM} \mathfrak{a} \leq \dim_{FM} \mathfrak{b} = \dim_{FM} \mathfrak{H}$ *and* $\dim_{FM} \mathfrak{D} \leq 2\dim_{FM} \mathfrak{H}$. *The NC domain of* $\mathfrak{D}$ *contains a row ball,* $r\mathbb{B}^d_{\mathbb{N}}$ *of radius* $r > 1$ *and* $\mathrm{Dom}\, \mathfrak{D} \supseteq \mathrm{Dom}\, \mathfrak{a} \cap \mathrm{Dom}\, \mathfrak{H}$.

Proof Let $T = \text{Re}\,\mathfrak{H}(R) \geq 0$ be a bounded, positive and NC rational left Toeplitz operator. That is, $\mathfrak{H} \in \mathbb{H}_d^\infty$ is bounded, NC Herglotz and NC rational. Since we assume that T is bounded, the NC Fatou theorem implies, as before in the proof of Theorem 5.1, that

$$T = (I - \mathfrak{b}(R)^*)^{-1}(I - \mathfrak{b}(R)^*\mathfrak{b}(R))(I - \mathfrak{b}(R))^{-1},$$

where $\mathfrak{b} = (\mathfrak{H} - I)(\mathfrak{H} + I)^{-1} \in [\mathbb{H}_d^\infty]_1$ is the NC rational inverse Cayley transform of $\mathfrak{H}$ [24]. We can assume that $T \neq I$, as then it is trivially factorable. If $\mathfrak{b}$ is inner then observe that $T \equiv 0$, which is again trivially factorable. By Theorem 6.3, if $\mathfrak{b}$ is not inner, it cannot be CE. Since $\mathfrak{b}$ is non-CE, Theorem 6.3 implies that $I - \mathfrak{b}(R)^*\mathfrak{b}(R) = \mathfrak{a}(R)^*\mathfrak{a}(R)$ is factorable, where $\mathfrak{a} \in [\mathbb{H}_d^\infty]_1$ is the NC rational Sarason outer function of $\mathfrak{b}$. Hence,

$$T = (I - \mathfrak{b}(R)^*)^{-1}\mathfrak{a}(R)^*\mathfrak{a}(R)(I - \mathfrak{b}(R))^{-1} =: \mathfrak{D}(R)^*\mathfrak{D}(R).$$

Since we assume that T is bounded, so is $\mathfrak{D}$ so that $\mathfrak{D} \in \mathbb{H}_d^\infty$. In particular, since $\mathfrak{D} \in \mathbb{H}_d^\infty$ is NC rational, Dom $\mathfrak{D}$ contains a row ball, $r\mathbb{B}_{\mathbb{N}}^d$ of radius $r > 1$, and by its definition, Dom $\mathfrak{D} \supseteq$ Dom $(1 - \mathfrak{b})^{-1} \cap$ Dom $\mathfrak{a}$.

Let (A, B, C, D) be a minimal FM realization of $\mathfrak{H}$. Since $\mathfrak{H} = 2(1 - \mathfrak{b})^{-1} - 1$, the minimal FM realization of $(1 - \mathfrak{b})^{-1}$ is (A, B, C', D') with $C' = \frac{1}{2}C$ and $D' = (1 - \mathfrak{b}(0))^{-1}$. The minimal FM realization of $1 - \mathfrak{b}$ is then given by the 'flip' realization (A^X, B^X, C^X, D^X) where

$$A_j^X := A_j - \frac{1}{2D'}B_jC, \quad B^X = \frac{1}{D'}B, \quad C^X = -\frac{1}{2D'}C \quad \text{and} \quad D^X = 1 - \mathfrak{b}(0);$$

see [35, Section 5.2]. Finally, by the construction of the minimal FM realization of the NC Sarason function, $\mathfrak{a}$ of $\mathfrak{b}$ in Proposition 6.2, $\mathfrak{a}$ has the finite FM realization, $(A^X, \frac{1}{D'}B, \widetilde{C}, \widetilde{D})$. A finite and generally non-minimal FM realization for $\mathfrak{D} = \mathfrak{a}(1-\mathfrak{b})^{-1}$ is then given by $(\hat{A}, \hat{B}, \hat{C}, \hat{D})$ where

$$\hat{A}_j = \begin{pmatrix} A_j^X & \frac{1}{2D'}B_jC \\ & A_j \end{pmatrix}.$$

In particular, the minimal FM realization size of $\mathfrak{D}$ is at most twice that of $\mathfrak{H}$. □

In [24, 27], we developed the Lebesgue decomposition of any positive NC measure, $\mu \in (\mathscr{A}_d)_+^\dagger$ with respect to NC Lebesgue measure, $m \in (\mathscr{A}_d)_+^\dagger$. Namely, $\mu = \mu_{ac} + \mu_s$, where μ_{ac} is a absolutely continuous with respect to m and μ_s is singular with respect to m.

Corollary 6.6 *Let $\mathfrak{b} \in [\mathbb{H}_d^\infty]_1$ be a contractive NC rational left multiplier. Then the NC Clark measure $\mu_\mathfrak{b} \in (\mathscr{A}_d)_+^\dagger$ is singular if and only if $\mathfrak{b}$ is inner.*

Proof In [24, Corollary 6.29], we proved that if $b \in [\mathbb{H}_d^\infty]_1$ is inner then μ_b is singular with respect to NC Lebesgue measure, m. Following the proof, we observed that if $T := 2I - b(R) - b(R)^* > 0$ is a factorable left Toeplitz operator then the converse also holds and we further proved that $T \geq T_0 = (I - b(R)^*)(I - b(R)) > 0$ so that T has a factorable left Toeplitz minorant. By the NC rational Fejér–Riesz Theorem, if $b = \mathfrak{b}$ is NC rational, T is factorable and the claim follows. □

Theorem 6.7 *Let* $\mathfrak{b} \in [\mathbb{H}_d^\infty]_1$ *be a non-CE and contractive NC rational left multiplier. Then* $\mu_{ac} := \mu_{\mathfrak{b};ac}$ *is of type*$-L$ *and*

$$\mu_{ac}(L^\omega) = \langle \mathfrak{h}, L^\omega \mathfrak{h} \rangle_{\mathbb{H}^2} = \langle 1, \mathfrak{h}^t(R)^* \mathfrak{h}^t(R) L^\omega 1 \rangle_{\mathbb{H}^2},$$

where $\mathfrak{h}^t = \mathfrak{a}(1 - \mathfrak{b})^{-1} \in \mathbb{H}_d^\infty$, *with* $\mathfrak{a}$ *the outer Sarason function of* $\mathfrak{b}$. *The function* $\mathfrak{h} := \mathfrak{h}^t(R)1 \in \mathbb{H}_d^2$ *is NC rational and hence in* $\mathbb{H}_d^\infty$ *and* $T = \mathfrak{h}^t(R)^* \mathfrak{h}^t(R)$ *is a bounded left Toeplitz operator.*

In the above statement, recall that a Gelfand–Naimark–Segal (GNS) construction applied to any $\mu \in (\mathscr{A}_d)_+^\dagger$ produces a GNS–Hilbert space, $\mathbb{H}_d^2(\mu)$, and a GNS–row isometry, Π_μ, acting on $\mathbb{H}_d^2(\mu)$ [17, 27]. A positive NC measure, $\mu \in (\mathscr{A}_d)_+^\dagger$, is said to be of *type*$-L$, if Π_μ is jointly unitarily equivalent to L [27].

Proof By the NC Fatou theorem, if

$$T_r := (I - \mathfrak{b}(rR)^*)^{-1}(I - \mathfrak{b}_r(R)^* \mathfrak{b}_r(R))(I - \mathfrak{b}(rR))^{-1},$$

then $(I + T_r)^{-1} \overset{SOT}{\rightarrow} (I + T)^{-1}$ where

$$\mu_{ac}(L^\omega) = \left\langle \sqrt{T}1, \sqrt{T} L^\omega 1 \right\rangle_{\mathbb{H}^2},$$

and T is a closed, positive semi-definite and densely defined left Toeplitz operator with the free polynomials as a core; see [24]. Here,

$$I + T_r = (I - \mathfrak{b}(rR)^*)^{-1} \left(2I - \mathfrak{b}(rR) - \mathfrak{b}(rR)^*\right) (I - \mathfrak{b}(R))^{-1}.$$

By the NC rational Fejér–Riesz theorem,

$$I + T_r = (I - \mathfrak{b}(rR)^*)^{-1} \mathfrak{d}_r(R)^* \mathfrak{d}_r(R) (I - \mathfrak{b}(rR))^{-1},$$

where $\|\mathfrak{d}_r(R)\|^2 \leq 2$, $\mathfrak{d} \in \mathbb{H}_d^\infty$ is NC rational and $\mathfrak{d}_r(R)^* \mathfrak{d}_r(R) = 2I - \mathfrak{b}(rR) - \mathfrak{b}(rR)^*$. Hence,

$$(I + T_r)^{-1} = (I - \mathfrak{b}(rR)) \mathfrak{d}_r(R)^{-1} \mathfrak{d}_r(R)^{-*} (I - \mathfrak{b}(rR)^*).$$

Since $(I + T_r)^{-1} \overset{SOT}{\rightarrow} (I + T)^{-1}$, taking inner products against NC kernels shows that $\mathfrak{d}_r^t(Z)^{-1}$ and hence $\mathfrak{d}_r^t(Z)$ converges pointwise to some $\widetilde{\mathfrak{d}}^t(Z)$ in the unit row ball. This and uniform boundedness of the net $\mathfrak{d}_r(R)$ imply WOT convergence of $\mathfrak{d}_r(R)$ to some $\widetilde{\mathfrak{d}}(R)$, $\widetilde{\mathfrak{d}} \in \mathbb{H}^\infty_d$. In particular, it follows that

$$(I+T)^{-1} = (I - \mathfrak{b}(R))\widetilde{\mathfrak{d}}(R)^{-1}\widetilde{\mathfrak{d}}(R)^{-*}(I - \mathfrak{b}(R)^*). \tag{6}$$

On the other hand, setting $\tau_r = 2I - \mathfrak{b}(rR) - \mathfrak{b}(rR)^*$ for $0 < r \leq 1$, consider the NC measure

$$\mu_r(L^\omega) := \left\langle 1, \tau_r L^\omega 1 \right\rangle_{\mathbb{H}^2}.$$

This has left free Herglotz–Riesz transform:

$$\begin{aligned} \mathfrak{H}_r(Z) &:= 2\sum_\omega Z^\omega \left\langle L^{\omega^{\mathrm{t}}} 1, \tau_r 1 \right\rangle_{\mathbb{H}^2} - I_n \left\langle 1, \tau_r 1 \right\rangle_{\mathbb{H}^2} \\ &= 4I - 2\overline{\mathfrak{b}(0)} I_n - 2\mathfrak{b}(rZ) - 2I + (\mathfrak{b}(0) + \overline{\mathfrak{b}(0)}) I_n \\ &= (2 + 2i \operatorname{Im} \mathfrak{b}(0)) I_n - 2\mathfrak{b}(rZ), \end{aligned}$$

so that $\mathfrak{H}_r(Z) = \mathfrak{H}(rZ)$ is NC rational. This has inverse Cayley transform

$$\mathfrak{B}(rZ) = (\mathfrak{H}(rZ) - I_n)(\mathfrak{H}(rZ) + I_n)^{-1}, \qquad r \in (0, 1],$$

for some NC rational $\mathfrak{B} \in [\mathbb{H}^\infty_d]_1$. Since $\mathfrak{B}(rR) \overset{SOT-*}{\rightarrow} \mathfrak{B}(R)$ by [24, Lemma 4], and $\mathfrak{C} := \begin{pmatrix} \mathfrak{B} \\ \mathfrak{A} \end{pmatrix}$ is inner since $\mathfrak{B}$ is NC rational, Proposition 4.2 implies that if $\mathfrak{A}_r$ is the NC rational Sarason outer function of $\mathfrak{B}(rZ)$ for $r \in (0, 1]$ $\mathfrak{A}_r(R) \overset{SOT-*}{\rightarrow} \mathfrak{A}(R)$. In particular it follows that

$$\mathfrak{d}_r^{\mathrm{t}}(Z) = (I_n - \mathfrak{B}^{\mathrm{t}}(rZ))^{-1}\mathfrak{A}_r^{\mathrm{t}}(Z) \to \mathfrak{d}^{\mathrm{t}}(Z) = (I_n - \mathfrak{B}^{\mathrm{t}}(Z))^{-1}\mathfrak{A}^{\mathrm{t}}(Z).$$

This pointwise convergence in the row ball and uniform norm boundedness imply that $\mathfrak{d}_r(R) \overset{WOT}{\rightarrow} \mathfrak{d}(R)$. Hence,

$$\begin{aligned} (I+T)^{-1} &= (I - \mathfrak{b}(R))\widetilde{\mathfrak{d}}(R)^{-1}\widetilde{\mathfrak{d}}(R)^{-*}(I - \mathfrak{b}(R)^*) \\ &= SOT - \lim (I - \mathfrak{b}(rR))\mathfrak{d}_r(R)^{-1}\mathfrak{d}_r(R)^{-*}(I - \mathfrak{b}(rR)^*). \end{aligned}$$

Taking inner products against NC Szegö kernels and using that $\mathfrak{d}_r^{\mathrm{t}}(Z)$ converges pointwise to $\mathfrak{d}^{\mathrm{t}}(Z)$ where

$$\mathfrak{d}(R)^*\mathfrak{d}(R) = \tau = 2I - \mathfrak{b}(R) - \mathfrak{b}(R)^*,$$

shows that $\mathfrak{d} = \widetilde{\mathfrak{d}}$. In conclusion,

$$I + T = (I - \mathfrak{b}(R)^*)^{-1} \underbrace{\tau}_{=\mathfrak{d}(R)^*\mathfrak{d}(R)} (I - \mathfrak{b}(R))^{-1},$$

so that

$$\begin{aligned} T &= (I - \mathfrak{b}(R)^*)^{-1}(\mathfrak{d}(R)^*\mathfrak{d}(R) - (I - \mathfrak{b}(R)^*)(I - \mathfrak{b}(R)))(I - \mathfrak{b}(R))^{-1} \\ &= (I - \mathfrak{b}(R)^*)^{-1}(I - \mathfrak{b}(R)^*\mathfrak{b}(R))(I - \mathfrak{b}(R))^{-1} \\ &= (I - \mathfrak{b}(R)^*)^{-1}\mathfrak{a}(R)^*\mathfrak{a}(R)(I - \mathfrak{b}(R))^{-1}. \end{aligned}$$

By Proposition 6.2, $\mathfrak{a} \in [\mathbb{H}^\infty_d]_1$ is outer and NC rational. By [34, Theorem 1, Theorem 2], $\mu_{ac} := \mu_{\mathfrak{b};ac}$ is purely of type-L. Hence, $\mu_{ac}(L^\omega) = m_{\mathfrak{h}}(L^\omega) = \langle \mathfrak{h}, L^\omega \mathfrak{h} \rangle_{\mathbb{H}^2}$ is a positive symmetric vector functional for some $\mathfrak{h} \in \mathbb{H}^2_d$ [27, Corollary 6.23]. By [22], $\mathfrak{h}^{\mathrm{t}}(R) := M^R_{\mathfrak{h}}$ is a closed, densely defined right multiplier with the free polynomials as a core, so that $\mathfrak{h}^{\mathrm{t}}(R)^*\mathfrak{h}^{\mathrm{t}}(R) \geq 0$ is a closed positive semi-definite and self-adjoint operator. By the uniqueness of the Riesz representation of closed, densely defined and positive semi-definite sesquilinear forms [36, Chapter VI, Theorem 2.1, Theorem 2.23], it follows that

$$\mathfrak{h}^{\mathrm{t}}(R)^*\mathfrak{h}^{\mathrm{t}}(R) = T = (I - \mathfrak{b}(R)^*)^{-1}\mathfrak{a}(R)^*\mathfrak{a}(R)(I - \mathfrak{b}(R))^{-1}.$$

The unbounded Douglas factorization lemma [37, Theorem 2] then implies that there are contractions C, D so that

$$C\mathfrak{h}^{\mathrm{t}}(R) = \mathfrak{a}(R)(I - \mathfrak{b}(R))^{-1} \quad \text{and} \quad D\mathfrak{a}(R)(I - \mathfrak{b}(R))^{-1} = \mathfrak{h}^{\mathrm{t}}(R).$$

Since $\mathfrak{a}(R)(I - \mathfrak{b}(R))^{-1}$ has dense range, we conclude that C, D commute with the left shifts and $CD = DC = I$. Since both C, D are contractive, it follows that C, D must both be isometries and $D = C^*$ so that C is unitary. By [15, Theorem 1.2], $C = C(R)$ and $D = D(R) \in \mathscr{R}^\infty_d$ are unitary right multipliers, so that $C(R) = \zeta I$, $\zeta \in \partial\mathbb{D}$ is constant by [15, Corollary 1.5]. In conclusion we can assume, without loss in generality that

$$\mathfrak{h}^{\mathrm{t}}(R) = \mathfrak{a}(R)(I - \mathfrak{b}(R))^{-1}.$$

Hence, $\mathfrak{h} = \mathfrak{a}(R)(I - \mathfrak{b}(R))^{-1}1 \in \mathbb{H}^2_d$ is NC rational so that $\mathfrak{h}^{\mathrm{t}} \in \mathbb{H}^\infty_d$ by [33, Theorem A]. In particular, $T = \mathfrak{h}^{\mathrm{t}}(R)^*\mathfrak{h}^{\mathrm{t}}(R)$ is a bounded left Toeplitz operator. □

Suppose that $\mathfrak{H} \sim \mathbb{H}^\infty_d$ is an NC rational Smirnov multiplier of Fock space. That is, $\mathfrak{H} \in \mathscr{O}(\mathbb{B}^d_{\mathbb{N}})$ is a locally bounded NC function and left multiplication by $\mathfrak{H}$ is a densely defined and closed operator on its maximal domain in $\mathbb{H}^2_d$. Equivalently, right multiplication by $\mathfrak{H}^{\mathrm{t}}$, $M^R_{\mathfrak{H}^{\mathrm{t}}} = \mathfrak{H}(R)$, is densely defined and closed on its maximal domain. The above NC Radon–Nikodym formula allows us to extend our NC rational Fejér–Riesz theorem to the case where $\mathrm{Re}\,\mathfrak{H}(R) \geq 0$ and $\mathfrak{H} \sim \mathbb{H}^\infty_d$ belongs to the Smirnov class. Here, given any $H \sim \mathbb{H}^\infty_d$, we say that $\mathrm{Re}\,H(R) \geq 0$ if this holds in the *quadratic form sense*:

$$q_{\mathrm{Re}\,H}(p,q) := \frac{1}{2}\langle H(R)^* p, q\rangle_{\mathbb{H}^2} + \frac{1}{2}\langle p, H(R)^* q\rangle_{\mathbb{H}^2} \geq 0,$$

for any $p, q \in \mathbb{C}\{\mathfrak{z}\}$. That is, H is positive in this quadratic form sense if and only if H is an NC Herglotz function. Suppose then that $\mathfrak{H}$ is an NC rational Herglotz function. This also defines a positive NC measure on the free disk system by $\mu_{\mathfrak{H}}(L^\omega) := q_{\mathrm{Re}\,\mathfrak{H}}(1, L^\omega 1)$. In fact, all positive NC measures are obtained in this way [17].

Corollary 6.8 *Let* $\mathfrak{H} \sim \mathbb{H}^\infty_d$ *be an NC Herglotz and NC rational Smirnov multiplier so that* $\mathrm{Re}\,\mathfrak{H} \geq 0$ *in the quadratic form sense, and suppose that* $q_{\mathrm{Re}\,\mathfrak{H}}$ *is a closeable form, i.e. that* $\mu_{\mathfrak{H}} \in (\mathscr{A}_d)^\dagger_+$ *is an absolutely continuous NC measure. Then,* $\mathrm{Re}\,\mathfrak{H}$ *is factorable in the sense that*

$$q_{\mathrm{Re}\,\mathfrak{H}}(p,p') = \langle p, Tp'\rangle_{\mathbb{H}^2}\,; \qquad p, p' \in \mathbb{C}\{\mathfrak{z}\},$$

where

$$T = (I - \mathfrak{b}(R)^*)^{-1}\mathfrak{a}(R)^*\mathfrak{a}(R)(I - \mathfrak{b}(R))^{-1}, \qquad \mathfrak{b} = (\mathfrak{H} - I)(\mathfrak{H} + I)^{-1} \in [\mathbb{H}^\infty_d]_1,$$

$\mathfrak{b}$ *is NC rational and non-CE, and* $\mathfrak{a}$ *is the NC rational outer Sarason function of* $\mathfrak{b}$*. Moreover,* T *and* $\mathfrak{a}(I - \mathfrak{b})^{-1} \in \mathbb{H}^\infty_d$ *are bounded.*

Proof Since we assume that $\mathfrak{H}$ is NC rational and Herglotz, its Cayley transform $\mathfrak{b} \in [\mathbb{H}^\infty_d]_1$ is NC rational and contractive. Since we assume that $\mu_{\mathfrak{H}}$ is absolutely continuous, $\mathfrak{b}$ must be non-CE, so that it has a non-zero, NC rational Sarason outer function $\mathfrak{a} \in [\mathbb{H}^\infty_d]_1$ so that $\binom{\mathfrak{b}}{\mathfrak{a}}$ is inner and $\mu_{\mathfrak{H}} = \mu_{\mathfrak{b}}$ is the NC Clark measure of $\mathfrak{b}$. The claim now follows from the previous theorem. □

References

1. Paul R. Halmos and Arlen Brown. Algebraic properties of Toeplitz operators. *J. Reine Angew. Math.*, 213:89–102, 1963.
2. Gabor Szegö. Beiträge zur theorie der Toeplitzschen formen. *Math. Zeit.*, 6:167–202, 1920.
3. Kenneth Hoffman. *Banach spaces of analytic functions*. Prentice–Hall, Inc., 1962.
4. Douglas N. Clark. One dimensional perturbations of restricted shifts. *J. Anal. Math.*, 25:169–191, 1972.
5. Alexei B. Aleksandrov. On the existence of nontangential boundary values of pseudocontinuable functions. (Russian). *Zapiski Nauchnykh Seminarov POMI*, 222:5–17, 1995.

6. Alexei B. Aleksandrov. Multiplicity of boundary values of inner functions. (Russian). *Izv. Akad. Nauk Arm. SSR*, 22:490–503, 1987.
7. Donald Sarason. Shift-invariant spaces from the Brangesian point of view. In *The Bieberbach conjecture: Proceedings of the symposium on the occasion of the proof*, pages 153–166, 1986.
8. Donald Sarason. Doubly shift–invariant spaces in H^2. *J. Operator Theory*, pages 75–97, 1986.
9. Donald Sarason. Unbounded Toeplitz operators. *Integral Equations Operator Theory*, 61:281–298, 2008.
10. Gelu F. Popescu. Free holomorphic functions on the unit ball of $B(\mathcal{H})^n$. *J. Funct. Anal.*, 241:268–333, 2006.
11. Guy Salomon, Orr M. Shalit, and Eli Shamovich. Algebras of noncommutative functions on subvarieties of the noncommutative ball: the bounded and completely bounded isomorphism problem. *J. Funct. Anal.*, 278, 2020.
12. Gelu F. Popescu. Multi-analytic operators on Fock spaces. *Math. Ann.*, 303:31–46, 1995.
13. Gelu F. Popescu. Multi–analytic operators and some factorization theorems. *Indiana Univ. Math. J.*, 38:693–710, 1989.
14. Gelu F. Popescu. *Entropy and multivariable interpolation*. American Mathematical Society, 2006.
15. Kenneth R. Davidson and David R. Pitts. Invariant subspaces and hyper-reflexivity for free semigroup algebras. *Proc. Lond. Math. Soc.*, 78:401–430, 1999.
16. Joseph A. Ball, Gregory Marx, and Victor Vinnikov. Noncommutative reproducing kernel Hilbert spaces. *J. Funct. Anal.*, 271:1844–1920, 2016.
17. Michael T. Jury and Robert T. W. Martin. Column-extreme multipliers of the free Hardy space. *J. Lond. Math. Soc.*, 101:457–489, 2020.
18. Michael P. Hartz. Every complete Pick space satisfies the column-row property. arXiv:2005.09614, 2020.
19. Méric L. Augat, Michael T. Jury, and James E. Pascoe. Effective noncommutative Nevanlinna–Pick interpolation in the row ball, and applications. *J. Math. Anal. Appl.*, 492:124457, 2020.
20. Joseph A. Ball, Vladimir Bolotnikov, and Quanlei Fang. Schur–class multipliers on the Fock space: de Branges–Rovnyak reproducing kernel spaces and transfer–function realizations. In *Operator Theory, Structured Matrices, and Dilations, Theta Series Adv. Math., Tiberiu Constantinescu Memorial Volume*, volume 7, pages 101–130. Theta, Bucharest, 2007.
21. Robert T.W. Martin and Eli Shamovich. A de Branges–Beurling theorem for the full Fock space. *Journal of Mathematical Analysis and Applications*, 496:124765, 2021.
22. Michael T. Jury and Robert T. W. Martin. Operators affiliated to the free shift on the free Hardy space. *J. Funct. Anal.*, 277:108285, 2019.
23. Emmanuel Fricain and Javad Mashreghi. *The theory of $\mathscr{H}(b)$ spaces*, volume 2. Cambridge University Press, 2016.
24. Michael T. Jury and Robert T. W. Martin. Fatou's theorem for non-commutative measures. *Adv. Math.*, 400:108293, 2022.
25. Michael Reed and Barry Simon. *Methods of Modern Mathematical Physics vol.* 1, *Functional Analysis*. Academic Press, San Diego, CA, 1980.
26. Nachman Aronszajn. Theory of reproducing kernels. *Trans. Amer. Math. Soc.*, 68:337–404, 1950.
27. Michael T. Jury and Robert T. W. Martin. Lebesgue decomposition of non-commutative measures. *Int. Math. Res. Not.*, 2022:2968–3030, 2022.
28. Michael T. Jury and Robert T. W. Martin. Non-commutative Clark measures for the free and abelian Toeplitz algebras. *J. Math. Anal. Appl.*, 456:1062–1100, 2017.
29. Dmitry S. Kaliuzhnyi-Verbovetskyi and Victor Vinnikov. Noncommutative rational functions, their difference–differential calculus and realizations. *Multidimens. Syst. Signal Process.*, 23:49–77, 2012.
30. John William Helton, Tobias Mai, and Roland Speicher. Applications of realizations (a.k.a. linearizations) to free probability. *J. Funct. Anal.*, 274:1–79, 2018.

31. Joseph A. Ball, Gilbert Groenewald, and Tanit Malakorn. Structured noncommutative multidimensional linear systems. *SIAM J. Control Optim.*, 44:1474–1528, 2005.
32. Jurij Volčič. On domains of noncommutative rational functions. *Linear Algebra Appl.*, 516:69–81, 2017.
33. Michael T. Jury, Robert T. W. Martin, and Eli Shamovich. Non-commutative rational functions in the full Fock space. *Trans. Amer. Math. Soc.*, 374:6727–6749, 2021.
34. Michael T. Jury, Robert T. W. Martin, and Eli Shamovich. Non-commutative rational Clark measures. Canad. J. Math., In press.
35. John William Helton, Igor Klep, and Jurij Volčič. Geometry of free loci and factorization of noncommutative polynomials. *Adv. Math.*, 331:589–626, 2018.
36. Tosio Kato. *Perturbation theory for linear operators*. Springer, 2013.
37. Ronald G. Douglas. On majorization, factorization, and range inclusion of operators on Hilbert space. *Proc. Amer. Math. Soc.*, 17:413–415, 1966.

The Relationship of the Gaussian Curvature with the Curvature of a Cowen-Douglas Operator

Soumitra Ghara and Gadadhar Misra

1 Introduction

Let X be an arbitrary set and let $K : X \times X \to \mathcal{M}_n(\mathbb{C})$, $n \geq 1$, be a function. We say that K is a non-negative definite kernel (resp. positive definite kernel) if for any subset $\{x_1, \ldots, x_p\}$ of X, the $np \times np$ matrix $\left(\!\!\left(K(x_i, x_j) \right)\!\!\right)_{i,j=1}^{p}$ is non-negative definite (resp. positive definite). A Hilbert space $\mathcal{H}$ consisting of functions on X is said to be a reproducing kernel Hilbert space with reproducing kernel K if:

(i) For each $x \in X$ and $\eta \in \mathbb{C}^n$, $K(\cdot, x)\eta \in \mathcal{H}$.
(ii) For each $f \in \mathcal{H}$ and $x \in X$, $\langle f, K(\cdot, x)\eta\rangle_{\mathcal{H}} = \langle f(x), \eta\rangle_{\mathbb{C}^n}$.

The kernel K of a reproducing kernel Hilbert space $\mathcal{H}$ is non-negative definite. Conversely, corresponding to each non-negative definite kernel K, there exists a unique reproducing kernel Hilbert space $(\mathcal{H}, K)$ whose reproducing kernel is K (see [2, 16]). For $K : X \times X \to \mathcal{M}_n(\mathbb{C})$, we write $K(x, y) \succeq 0$, $(x, y) \in X \times X$, whenever K is non-negative definite. Analogously, we write $K(x, y) \preceq 0$, $(x, y) \in$

Support for the work of S. Ghara was provided by SPM Fellowship of the CSIR and a Fields-Laval post-doctoral fellowship. Support for the work of G. Misra was provided in the form of a MATRICS grant and the J C Bose National Fellowship, Science and Engineering Research Board. Some of the results in this paper are from the PhD thesis of the first named author submitted to the Indian Institute of Science.

S. Ghara (✉)
Department of Mathematics and Statistics, Indian Institute of Technology Kanpur, Kanpur, India

G. Misra
Statistics and Mathematics Unit, Indian Statistical Institute, Bangalore, India

Department of Mathematics, Indian Institute of Technology, Gandhinagar, India
e-mail: gm@isibang.ac.in

I. Binder et al. (eds.), *Function Spaces, Theory and Applications*, Fields Institute Communications 87, https://doi.org/10.1007/978-3-031-39270-2_13

$X \times X$ if $-K$ is non-negative definite. For $K_1, K_2 : X \times X \to \mathcal{M}_n(\mathbb{C})$, we write $K_1(x, y) \succeq K_2(x, y)$ if $K_1(x, y) - K_2(x, y) \succeq 0$ and we write $K_1(x, y) \preceq K_2(x, y)$ if $K_1(x, y) - K_2(x, y) \preceq 0$, $(x, y) \in X \times X$. For any domain Ω in $\mathbb{C}^m$, $m \geq 1$, a function $K : \Omega \times \Omega \to \mathcal{M}_n(\mathbb{C})$ is said to be sesqui-analytic if it is holomorphic in first m-variables and anti-holomorphic in the second set of m-variables. In this paper, we will deal with non-negative definite kernels which are sesqui-analytic.

We now discuss an important class of operators introduced by Cowen and Douglas (see [4, 7]). Let $\boldsymbol{T} := (T_1, \ldots, T_m)$ be a m-tuple of commuting bounded linear operators on a separable Hilbert space $\mathcal{H}$. Let $D_{\boldsymbol{T}} : \mathcal{H} \to \mathcal{H} \oplus \cdots \oplus \mathcal{H}$ be the operator defined by $D_{\boldsymbol{T}}(x) = (T_1 x, \ldots, T_m x)$, $x \in \mathcal{H}$.

Definition 1.1 (Cowen-Douglas class operator) Let $\Omega \subset \mathbb{C}^m$ be a bounded domain. A commuting m-tuple $\boldsymbol{T}$ on $\mathcal{H}$ is said to be in the Cowen-Douglas class $B_n(\Omega)$ if $\boldsymbol{T}$ satisfies the following requirements:

(i) $\dim \ker D_{\boldsymbol{T}-w} = n, \quad w \in \Omega$.
(ii) $\operatorname{ran} D_{\boldsymbol{T}-w}$ is closed for all $w \in \Omega$.
(iii) $\bigvee\{\ker D_{\boldsymbol{T}-w} : w \in \Omega\} = \mathcal{H}$.

If $\boldsymbol{T} \in B_n(\Omega)$, then for each $w \in \Omega$, there exist functions $\gamma_1, \ldots, \gamma_n$ holomorphic in a neighbourhood $\Omega_0 \subseteq \Omega$ containing w such that $\ker D_{\boldsymbol{T}-w'} = \bigvee\{\gamma_1(w'), \ldots, \gamma_n(w')\}$ for all $w' \in \Omega_0$ (cf. [5]). Consequently, every $\boldsymbol{T} \in B_n(\Omega)$ corresponds to a rank n holomorphic Hermitian vector bundle $E_{\boldsymbol{T}}$ defined by

$$E_{\boldsymbol{T}} = \{(w, x) \in \Omega \times \mathcal{H} : x \in \ker D_{\boldsymbol{T}-w}\}$$

and $\pi(w, x) = w$, $(w, x) \in E_{\boldsymbol{T}}$. For a bounded domain Ω in $\mathbb{C}^m$, let $\Omega^* = \{z : \bar{z} \in \Omega\}$. It is known that if T is an operator in $B_n(\Omega^*)$, then for each $w \in \Omega$, T is unitarily equivalent to the adjoint of the multiplication tuple $M = (M_1, \ldots, M_m)$ on some reproducing kernel Hilbert space $(\mathcal{H}, K) \subseteq \operatorname{Hol}(\Omega_0, \mathbb{C}^n)$ for some open subset $\Omega_0 \subseteq \Omega$ containing w. If $T \in B_1(\Omega^*)$, the curvature matrix $\mathcal{K}_T(\bar{w})$ at a fixed but arbitrary point $\bar{w} \in \Omega^*$ is defined by

$$\mathcal{K}_T(\bar{w}) = -\Big(\!\Big(\partial_i \bar{\partial}_j \log \|\gamma(\bar{w})\|^2 \Big)\!\Big)_{i,j=1}^{m},$$

where γ is a holomorphic frame of E_T defined on some open subset $\Omega_0^* \subseteq \Omega^*$ containing $\bar{w}$. Here, ∂_i and $\bar{\partial}_j$ denote $\frac{\partial}{\partial w_i}$ and $\frac{\partial}{\partial \bar{w}_j}$, respectively. If T is realized as the adjoint of the multiplication tuple M on some reproducing kernel Hilbert space $(\mathcal{H}, K) \subseteq \operatorname{Hol}(\Omega_0)$, where $w \in \Omega_0$, the curvature $\mathcal{K}_T(\bar{w})$ is then equal to

$$-\Big(\!\Big(\partial_i \bar{\partial}_j \log K(w, w) \Big)\!\Big)_{i,j=1}^{m}.$$

Let $\Omega \subset \mathbb{C}$ be open and $\rho : \Omega \to \mathbb{R}_+$ be a C^2-smooth function. The Gaussian curvature of the metric ρ is given by the formula

$$\mathcal{G}_\rho(z) = -\frac{(\partial\bar{\partial}\log\rho)(z)}{\rho(z)^2}, z \in \Omega. \tag{1}$$

If $K : \Omega \times \Omega \to \mathbb{C}$ is a non-negative definite kernel with $K(z, z) > 0$, then the function $\frac{1}{K}$ defines a metric on Ω and its Gaussian curvature is given by the formula

$$\mathcal{G}_{K^{-1}}(z) = K(z, z)^2\Big(\partial\bar{\partial}\log K\Big)(z, z),\ \ z \in \Omega.$$

Since $\mathcal{G}_{K^{-1}}(z)$ can also be written as $K(z, z)\partial\bar{\partial}K(z, z) - \partial K(z, z)\bar{\partial}K(z, z)$, it follows that $\mathcal{G}_{K^{-1}}(z)$ can be extended to a sesqui-analytic function $\mathcal{G}_{K^{-1}}(z, w)$ on $\Omega \times \Omega$. It is therefore natural to extend the definition of the Gaussian curvature to an open subset $\Omega \subset \mathbb{C}^m$. Thus, for any non-negative definite kernel K on Ω, we define

$$\mathcal{G}_{K^{-1}}(z, w) := \Big(\!\Big(K(z, w)\partial_i\bar{\partial}_j K(z, w) - \partial_i K(z, w)\bar{\partial}_j K(z, w)\Big)\!\Big)_{i,j=1}^m,\ \ z, w \in \Omega, \tag{2}$$

where, with a slight abuse of notation, we let the symbols ∂_i and $\bar{\partial}_j$ also stand $\frac{\partial}{\partial z_i}$ and $\frac{\partial}{\partial \bar{w}_j}$, respectively.

Proposition 1.1 *[14, Proposition 2.3] Let $\Omega \subset \mathbb{C}^m$ be a domain and $K : \Omega \times \Omega \to \mathbb{C}$ be a sesqui-analytic function. Let α, β be two positive real numbers. Suppose that K^α and K^β, defined on $\Omega \times \Omega$, are non-negative definite for some $\alpha, \beta > 0$. Then the function $\mathbb{K}^{(\alpha,\beta)} : \Omega \times \Omega \to \mathcal{M}_m(\mathbb{C})$ defined by*

$$\mathbb{K}^{(\alpha,\beta)}(z, w) := K^{\alpha+\beta}(z, w)\Big(\!\Big(\big(\partial_i\bar{\partial}_j \log K\big)(z, w) \Big)\!\Big)_{i,j=1}^m,\ \ z, w \in \Omega,$$

is a non-negative definite kernel on $\Omega \times \Omega$ taking values in $\mathcal{M}_m(\mathbb{C})$.

We obtain the following corollary, saying that $\mathcal{G}_{K^{-1}}(z, w)$ is a non-negative definite kernel whenever K is non-negative definite, by setting $\alpha = 1 = \beta$.

Corollary 1.1 *Let Ω be a domain in $\mathbb{C}^m$. Suppose that $K : \Omega \times \Omega \to \mathbb{C}$ is a sesqui-analytic non-negative definite kernel. Then $\mathcal{G}_{K^{-1}}$ is also a non-negative definite kernel on Ω taking values in $\mathcal{M}_m(\mathbb{C})$.*

The introduction of the Gaussian curvature has many advantages and Corollary 1.1 serves as a handy tool for many proofs. This is already apparent from [3]; many more examples are given in Sect. 2 of this paper. We have attempted to strengthen the curvature inequality in the hope of obtaining a criterion for contractivity of operators in $B_1(\mathbb{D})$. We haven't succeeded in doing this yet, but several partial answers that we have obtained indicate that one of these inequalities may do the job. In Sect. 2, we establish a monotonicity property of the Gaussian curvature. We conclude Sect. 2 by showing that the partial derivatives from $(\mathcal{H}, K)$

to $(\mathcal{H}, \mathcal{G}_{K^{-1}})$ are bounded. In the third section, we discuss the decomposition of the tensor product of two Hilbert modules, say $\mathcal{M}_1 \subset \text{Hol}(\Omega)$ and $\mathcal{M}_2 \subset \text{Hol}(\Omega)$. For the basic notions of Hilbert modules, submodules, and quotient modules, we refer the reader to [13]. The tensor product $\mathcal{M}_1 \otimes \mathcal{M}_2$ consists of holomorphic functions on $\Omega \times \Omega$. We consider the nested set of submodules $\mathcal{M}_1 \otimes \mathcal{M}_2 \supset \mathcal{A}_0 \supset \mathcal{A}_1 \supset \cdots \supset \mathcal{A}_k \supset \cdots$, where $\mathcal{A}_k$ is the submodule of functions in $\mathcal{M}_1 \otimes \mathcal{M}_2$ vanishing on the diagonal subset Δ of $\Omega \times \Omega$ along with their derivatives to order k. Setting $\mathcal{S}_k := \mathcal{A}_{k-1} \ominus \mathcal{A}_k$, we have the direct sum decomposition

$$\mathcal{M}_1 \otimes \mathcal{M}_2 = \bigoplus_{k=1}^{\infty} \mathcal{S}_k,$$

which one may think of as the Clebsch-Gordan decomposition for Hilbert modules. We also have the short exact sequence of Hilbert modules:

$$0 \longrightarrow \mathcal{A}_0 \xrightarrow{\ i\ } \mathcal{M}_1 \otimes \mathcal{M}_2 \xrightarrow{\ \pi\ } \mathcal{S}_0 \longrightarrow 0.$$

It is important to be able to find invariants for $\mathcal{S}_0$ from the inclusion $\mathcal{A}_0 \subset \mathcal{M}_1 \otimes \mathcal{M}_2$. In Sect. 3, in a large class of examples, we find such an invariant; see Theorem 3.2 and the remark following it. In this section, we also show that the Gaussian curvature can be obtained as a limit that we believe is revealing; see Corollary 3.1 and the discussion preceding it.

2 Remarks on Curvature Inequality

In this section, we will discuss the curvature inequality for a contractive operator $T : \mathcal{H} \to \mathcal{H}$ in the Cowen-Douglas class $B_1(\mathbb{D})$ taking into account Corollary 1.1. First, since the operator $T \in B_1(\mathbb{D})$, it follows that the map $\gamma_T : \mathbb{D} \to Gr(\mathcal{H}, 1)$, $\gamma_T(w) = \ker(T - w)$, $w \in \mathbb{D}$, is holomorphic. Here, $Gr(\mathcal{H}, 1)$ is the Grassmannian of $\mathcal{H}$ consisting of the *one*-dimensional subspaces. Second, any operator T in $B_1(\mathbb{D})$ is unitarily equivalent to the adjoint M^* of the operator M of multiplication by the coordinate function z on some reproducing kernel Hilbert space $(\mathcal{H}, K) \subseteq \text{Hol}(\mathbb{D})$. In particular, any contraction T in $B_1(\mathbb{D})$, modulo unitary equivalence, is of this form. Also, $M^*K(\cdot, w) = \overline{w}K(\cdot, w)$, therefore we can take the map $\gamma_T(\overline{w}) = \mathbb{C}[K(\cdot, w)]$, and with a slight abuse of notation, we shall write $\gamma_T(\overline{w}) = K(\cdot, w)$. It is then easy to verify that $(M^* - \overline{w}I)\bar{\partial}K(\cdot, w) = K(\cdot, w)$. Consequently, setting $\mathcal{N}(w)$ to be the *two*-dimensional space spanned by $\{K(\cdot, w), \bar{\partial}K(\cdot, w)\}$, we have that $(M - wI)^*_{|\mathcal{N}(w)} = \begin{pmatrix} 0 & 1 \\ 0 & 0 \end{pmatrix}$. However if we represent $(M - wI)^*_{|\mathcal{N}(w)}$ with respect to the orthonormal basis $e_1(w), e_2(w)$ obtained by applying the Gram-Schmidt process to the pair of vectors $K(\cdot, w), \bar{\partial}K(\cdot, w)$, then we have the representation:

$$N_T(w) := (M - wI)^*_{|\mathcal{N}(w)} = \begin{pmatrix} 0 & (-\mathcal{K}_T(\overline{w}))^{-\frac{1}{2}} \\ 0 & 0 \end{pmatrix}, \quad w \in \mathbb{D}.$$

The contractivity of the operator M, or, equivalently, that of M^*, implies that the local operators $N_T(w) + \overline{w}I$, $w \in \mathbb{D}$, must be contractive. Since a 2×2 matrix of the form $\begin{pmatrix} w & \lambda \\ 0 & w \end{pmatrix}$ is contractive if and only if $|\lambda| \le 1 - |w|^2$, we obtain the curvature inequality of [15] reproduced in the form of a proposition below.

Proposition 2.1 *If T is contraction in $B_1(\mathbb{D})$, then the curvature of T is bounded above by the curvature of the backward shift operator S^*.*

Without loss of generality, we may assume that the operator T has been realized as the adjoint of the multiplication operator M on some Hilbert space of holomorphic functions $(\mathcal{H}, K)$. Note that $-\mathcal{K}_T(w) = \partial\bar{\partial} \log K(w, w)$ and the curvature $\mathcal{K}_{S^*}(w)$ of the backward shift operator S^* is $-\partial\bar{\partial} \log \mathbb{S}_{\mathbb{D}}(z, z)$, where $\mathbb{S}_{\mathbb{D}}(z, w) = \frac{1}{1-z\overline{w}}$ is the Szegö kernel of the unit disc. In other words, for a contractive operator M^* in $B_1(\mathbb{D})$, the curvature inequality takes the form (see [3])

$$-\partial\bar{\partial} \log K(z, z) \le -\partial\bar{\partial} \log \mathbb{S}_{\mathbb{D}}(z, z) = -\tfrac{1}{(1-|z|^2)^2}, \quad z \in \mathbb{D}. \tag{3}$$

From the discussion preceding Proposition 2.1, it is clear that the curvature inequality of a contractive operator in $B_1(\mathbb{D})$ is nothing but the contractivity of its restriction to the *two*-dimensional subspaces $\mathcal{N}(w)$, $w \in \mathbb{D}$. So, it is clear that the curvature inequality, in general, is not enough to ensure contractivity. We reproduce an example from [3] illustrating this phenomenon.

Let $K_0(z, w) = \frac{8+8z\bar{w}-(z\bar{w})^2}{1-z\bar{w}}$, $z, w \in \mathbb{D}$. Note that $K_0(z, w)$ can be written in the form $8 + 16z\bar{w} + 15\frac{(z\bar{w})^2}{1-z\bar{w}}$; therefore it defines a non-negative definite kernel on the unit disc. It is not hard to see that, in this case

$$\mathcal{K}_{M^*}(w) - \mathcal{K}_{S^*}(w) = -\frac{8(8 - 4|w|^2 - |w|^4)}{(8 + 8|w|^2 - |w|^4)^2} \leqslant 0, \quad w \in \mathbb{D}.$$

Recall that for any reproducing kernel Hilbert space $(\mathcal{H}, K)$, the operator M^* on $(\mathcal{H}, K)$ is a contraction if and only if the function $G(z, w) := (1 - z\bar{w})K(z, w)$ is non-negative definite on $\mathbb{D} \times \mathbb{D}$ (see [1, Corollary 2.37]). Since $(1 - z\bar{w})K_0(z, w) = 8 + 8z\bar{w} - (z\bar{w})^2$ which is not a non-negative definite kernel on the unit disc, it follows that the operator M^* on $(\mathcal{H}, K_0)$ is not a contraction.

Since the curvature is a complete unitary invariant in the class $B_1(\mathbb{D})$, one attempts to strengthen the curvature inequality in the hope of finding a criterion for contractivity in terms of the curvature. One such possibility is discussed in the paper [3] replacing the point-wise inequality of (3) by requiring that $0 \preceq \partial\bar{\partial} \log K(z, w) - \partial\bar{\partial} \log \mathbb{S}_{\mathbb{D}}(z, w)$, that is,

$$\Big(\!\!\Big(\partial\bar{\partial}\log K(w_i, w_j) - \partial\bar{\partial}\log \mathbb{S}_{\mathbb{D}}(w_i, w_j)\Big)\!\!\Big)_{i,j=1}^{n}$$

is non-negative definite for all finite subsets $\{w_1, \ldots, w_n\}$ of $\mathbb{D}$ and $n \in \mathbb{N}$. Here, we discuss a different strengthening of the curvature inequality (3).

Proposition 2.2 *Let $T \in B_1(\mathbb{D})$ be a contraction. Assume that T is unitarily equivalent to the operator M^* on $(\mathcal{H}, K)$ for some non-negative definite kernel K on the unit disc. Then the following inequality holds:*

$$K^2(z, w) \preceq \mathbb{S}_{\mathbb{D}}^{-2}(z, w)\mathcal{G}_{K^{-1}}(z, w), \quad (z, w) \in \mathbb{D} \times \mathbb{D}, \tag{4}$$

that is, the matrix

$$\Big(\!\!\Big(\mathbb{S}_{\mathbb{D}}^{-2}(w_i, w_j)\mathcal{G}_{K^{-1}}(w_i, w_j) - K^2(w_i, w_j)\Big)\!\!\Big)_{i,j=1}^{n}$$

is non-negative definite for every subset $\{w_1, \ldots, w_n\}$ of $\mathbb{D}$ and $n \in \mathbb{N}$.

Proof Setting $G(z, w) = (1 - z\bar{w})K(z, w)$, we see that

$$\begin{aligned} &- G(z, w)^2\partial\bar{\partial}\log G(z, w) \\ &\quad = (1 - z\bar{w})^2K^2(z, w)\big(- \partial\bar{\partial}\log K(z, w) + \partial\bar{\partial}\log \mathbb{S}_{\mathbb{D}}(z, w)\big), \; z, w \in \mathbb{D}. \end{aligned}$$

Therefore, since $G(z, w)$ is non-negative definite on $\mathbb{D} \times \mathbb{D}$, applying Corollary 1.1 for $G(z, w)$, we obtain that

$$(1 - z\bar{w})^2K(z, w)^2\big(- \partial\bar{\partial}\log K(z, w) + \partial\bar{\partial}\log \mathbb{S}_{\mathbb{D}}(z, w)\big) \preceq 0.$$

Since $\mathbb{S}_{\mathbb{D}}(z, w)^{-2}\partial\bar{\partial}\log \mathbb{S}_{\mathbb{D}}(z, w) = 1$, the proof is complete.

In particular, evaluating (4) at a fixed but arbitrary point, the inequality (3) is evident. However, for any contraction T in $B_1(\mathbb{D})$ (realized as M^* on $(\mathcal{H}, K)$), the inequality (4) gives a much stronger (curvature) inequality as shown in the computation given below. Conversely, whether it is strong enough to force contractivity of the operator M^* is not clear. For a different approach, see [18].

In order to show that the inequality (4) is stronger than the inequality (3), it suffices to prove the kernel K_0 does not satisfy (4). Setting $G_0(z, w) = (1 - z\bar{w})K_0(z, w)$, we get $G_0(z, w) = 8 + 8z\bar{w} - (z\bar{w})^2$, $z, w \in \mathbb{D}$. Thus

$$\begin{aligned} &G_0(z, w)^2\partial\bar{\partial}\log G_0(z, w) \\ &\quad = G_0(z, w)\partial\bar{\partial}G_0(z, w) - \partial G_0(z, w)\bar{\partial}G_0(z, w) \\ &\quad = (8 + 8z\bar{w} - (z\bar{w})^2)(8 - 4z\bar{w}) - (8z - 2z^2\bar{w})(8\bar{w} - 2z\bar{w}^2) \\ &\quad = 64 - 32z\bar{w} - 8(z\bar{w})^2, \end{aligned}$$

which is clearly not a non-negative definite kernel. Hence the operator M^* on $(\mathcal{H}, K_0)$ does not satisfy inequality (4).

Remark 2.1 We now have the following remarks.

(i) Under the assumptions of Proposition 2.2, it follows from [16, Theorem 5.1] that the Hilbert space $(\mathcal{H}, K^2)$ is contained in the Hilbert space $(\mathcal{H}, \mathbb{S}_{\mathbb{D}}^{-2}\mathcal{G}_{K^{-1}})$ and the inclusion map from $(\mathcal{H}, K^2)$ to $(\mathcal{H}, \mathbb{S}_{\mathbb{D}}^{-2}\mathcal{G}_{K^{-1}})$ is contractive.
(ii) Recall that unitary equivalence class of the operator M acting on a reproducing kernel Hilbert space $(\mathcal{H}, K)$ is determined by the kernel K modulo pre- and post-multiplication by a nonvanishing holomorphic function and its conjugate; see [7, Theorem 3.7] and the remark following it. The Gaussian curvature $\mathcal{G}_{K^{-1}}$ of a non-negative definite kernel K clearly depends on the choice of the kernel K, and therefore it is not a function of the unitary equivalence class of the operator M. However, we note that the validity of the inequality (4) depends only on the unitary equivalence class of the operator M.

Let Ω be a finitely connected bounded planar domain and $\mathrm{Rat}(\Omega^*)$ be the ring of rational functions with poles off $\overline{\Omega^*}$. Let T be an operator in $B_1(\Omega^*)$ with $\sigma(T) = \overline{\Omega^*}$. Suppose that the homomorphism $q_T : \mathrm{Rat}(\Omega^*) \to B(\mathcal{H})$ given by

$$q_T(f) = f(T), \quad f \in \mathrm{Rat}(\Omega^*),$$

is contractive, that is, $\|f(T)\| \le \|f\|_{\Omega^*,\infty}$, $f \in \mathrm{Rat}(\Omega^*)$. As before, we think of T as the adjoint M^* of the multiplication operator M on some reproducing kernel Hilbert space $(\mathcal{H}, K) \subset \mathrm{Hol}(\Omega)$. Setting $G_f(z, w) = (1 - f(z)\overline{f(w)})K(z, w)$ and using the contractivity of $f(M^*)$, $\|f\|_{\infty,\Omega} \le 1$, we have that $G_f \succeq 0$. Applying Corollary 1.1, we conclude that

$$\begin{aligned} 0 \preceq & G_f(z, w)^2 \partial\bar{\partial} \log G_f(z, w) \\ = & G_f(z, w)^2\Big(- \tfrac{f'(z)\overline{f'(w)}}{(1-f(z)\overline{f(w)})^2} + \partial\bar{\partial} \log K(z, w)\Big) \\ = & - K(z, w)^2 f'(z)\overline{f'(w)} + (1 - f(z)\overline{f(w)})^2 K(z, w)^2 \partial\bar{\partial} \log K(z, w) \end{aligned}$$

for any rational function f with poles off $\overline{\Omega}$ and $|f(z)| \le 1$, $z \in \Omega$. Also, if f' is a nonvanishing function on Ω, then the pull-back of the metric induced by the Szegö kernel is the metric $f^*(\mathbb{S}_{\mathbb{D}})(z, z) = \frac{|f'(z)|}{1-|f(z)|^2}$, $z \in \Omega$. Thus, if f' is not zero on Ω, then the curvature inequality takes the form

$$K(z, w)^2 \preceq f^*(\mathbb{S}_{\mathbb{D}})(z, w)^{-2}\mathcal{G}_{K^{-1}}(z, w), \; z, w \in \Omega,$$

where $f^*(\mathbb{S}_{\mathbb{D}})(z, w)^2$ is the kernel $\frac{f'(z)\overline{f'(w)}}{(1-f(z)\overline{f(w)})^2}$. As in the case of the disc, in particular, evaluating this inequality at a fixed but arbitrary point $z \in \Omega$, we have

$$\partial\bar{\partial}\log K(z,z) \geq \sup\left\{\frac{|f'(z)|^2}{(1-|f(z)|^2)^2} : f \in \mathrm{Rat}(\Omega), \|f\|_{\Omega,\infty} \leq 1\right\} = \mathbb{S}_\Omega(z,z)^2,$$

where $\mathbb{S}_\Omega$ is the Szegö kernel of the domain Ω. This is the curvature inequality for contractive homomorphisms (see [15, Corollary 1.2']) and also [17]).

We now show that an analogue of Proposition 2.2 is also valid for spherical contractions in $B_1(\mathbb{B}^m)$, where $\mathbb{B}^m$ is the m-dimensional unit ball in $\mathbb{C}^m$. Recall that a commuting m-tuple $T = (T_1, \ldots, T_m)$ of operators on $\mathcal{H}$ is said to be a row contraction if $\sum_{i=1}^m T_i T_i^* \leq I$. Let $K : \mathbb{B}^m \times \mathbb{B}^m \to \mathbb{C}$ be a sesqui-analytic positive definite kernel. Assume that the commuting m-tuple $M = (M_1, \ldots, M_m)$ of multiplication by the coordinate functions on $(\mathcal{H}, K)$ is in $B_1(\mathbb{B}^m)$. We let $B_m(z,w) := \frac{1}{1-\langle z,w\rangle}$, $z, w \in \mathbb{B}^m$, be the reproducing kernel of the Drury-Arveson space. By [11, Corollary 2], M is a row contraction if and only if $B_m^{-1}(z,w)K(z,w)$ is non-negative definite on $\mathbb{B}^m$. Thus, if M on $(\mathcal{H}, K)$ is a row contraction in $B_1(\mathbb{B}^m)$, applying Corollary 1.1 for $B_m^{-1}(z,w)K(z,w)$, we obtain the following inequality:

$$K^2(z,w)B_m^{-2}(z,w)\Big(\!\Big(\partial_i\bar{\partial}_j \log B_m(z,w)\Big)\!\Big)_{i,j=1}^m \preceq B_m^{-2}(z,w)\mathcal{G}_{K^{-1}}(z,w). \tag{5}$$

As before, evaluating at a fixed but arbitrary point z in $\mathbb{B}^m$, we obtain [3, Corollary 2.3].

We now prove that the Gaussian curvature $\mathcal{G}_{K^{-1}}$ is monotone.

Proposition 2.3 *Let $\Omega \subset \mathbb{C}^m$ be a domain. Suppose that K_1 and K_2 are two scalar valued positive definite kernels on Ω satisfying $K_1(z,w) \succeq K_2(z,w)$, $z, w \in \Omega$. Then*

$$\mathcal{G}_{K_1^{-1}}(z,w) \succeq \mathcal{G}_{K_2^{-1}}(z,w), \ (z,w) \in \Omega \times \Omega.$$

Proof Set $K_3 = K_1 - K_2$. By hypothesis, K_3 is non-negative definite on Ω. For $1 \leq i, j \leq m$, a straightforward computation shows that

$$\begin{aligned} K_1^2\partial_i\bar{\partial}_j \log K_1 = K_2^2\partial_i\bar{\partial}_j \log K_2 + K_3^2\partial_i\bar{\partial}_j \log K_3 \\ + K_2\partial_i\bar{\partial}_j K_3 + K_3\partial_i\bar{\partial}_j K_2 - \partial_i K_2\bar{\partial}_j K_3 - \partial_i K_3\bar{\partial}_j K_2. \end{aligned} \tag{6}$$

Now set $\gamma_i(w) = K_2(\cdot,w) \otimes \bar{\partial}_i K_3(\cdot,w) - \bar{\partial}_i K_2(\cdot,w) \otimes K_3(\cdot,w)$, $1 \leq i \leq m$, $w \in \Omega$. For $1 \leq i, j \leq m$ and $z, w \in \Omega$, then we have

$$\begin{aligned} &\big\langle \gamma_j(w), \gamma_i(z)\big\rangle \\ &= \big(K_2\partial_i\bar{\partial}_j K_3\big)(z,w) + \big(K_3\partial_i\bar{\partial}_j K_2\big)(z,w) \\ &- \big(\partial_i K_2\bar{\partial}_j K_3\big)(z,w) - \big(\partial_i K_3\bar{\partial}_j K_2\big)(z,w). \end{aligned} \tag{7}$$

Combining (6) and (7), we obtain

$$\begin{aligned}\big(\big(K_1^2\partial_i\bar{\partial}_j \log K_1\big)(z,w)\big)_{i,j=1}^m \\ &= \big(\big(K_2^2\partial_i\bar{\partial}_j \log K_2\big)(z,w)\big)_{i,j=1}^m + \big(\big(K_3^2\partial_i\bar{\partial}_j \log K_3\big)(z,w)\big)_{i,j=1}^m \\ &\quad + \big(\langle\gamma_j(w),\gamma_i(z)\rangle\big)_{i,j=1}^m.\end{aligned}$$

Note that $(z,w)\mapsto \big(\langle\gamma_j(w),\gamma_i(z)\rangle\big)_{i,j=1}^m$ is a non-negative definite kernel on Ω (see [14, Lemma 2.1]). The proof is now complete since the sum of two non-negative definite kernels remains non-negative definite.

As a consequence of Proposition 2.3, we obtain the following inequality for row contractions involving the Gaussian curvature.

Corollary 2.1 *Let $K:\mathbb{B}^m\times\mathbb{B}^m\to\mathbb{C}$ be a sesqui-analytic positive definite kernel. Assume that K is normalized at the origin, that is, $K(z,0)=1,\ z\in\mathbb{B}^m$. Suppose that the commuting tuple M of multiplication by the coordinate functions is a row contraction on $(\mathcal{H},K)$. Then*

$$\mathcal{G}_{K^{-1}}(z,w)\succeq\mathcal{G}_{B_m^{-1}}(z,w),\ \ (z,w)\in\mathbb{B}^m\times\mathbb{B}^m. \tag{8}$$

Proof Since the tuple M on $(\mathcal{H},K)$ is a row contraction, $\tilde{K}(z,w):=B_m^{-1}(z,w)K(z,w)$ defines a non-negative definite kernel on $\mathbb{B}^m$. The kernel $\tilde{K}$ is normalized at 0 since K is normalized at 0. Thus $1=\tilde{K}(\cdot,0)\in(\mathcal{H},\tilde{K})$ and

$$\|1\|^2_{(\mathcal{H},\tilde{K})}=\langle\tilde{K}(\cdot,0),\tilde{K}(\cdot,0)\rangle_{(\mathcal{H},\tilde{K})}=\tilde{K}(0,0)=1.$$

Hence it follows from [16, Theorem 3.11] that $\tilde{K}\succeq 1$. Since the product of two non-negative definite kernels remains non-negative definite, multiplying both sides with B_m, we get $K\succeq B_m$. The proof is now complete by applying Proposition 2.3.

Remark 2.2 We point out that Corollary 2.1 can also be derived from (5). In particular, in case $m=1$, Corollary 2.1 is a consequence of Proposition 2.2. But since the kernel K_0 satisfies the inequality $K_0(z,w)\succeq\mathbb{S}_{\mathbb{D}}(z,w)$, $(z,w)\in\mathbb{D}\times\mathbb{D}$, it follows from Theorem 2.1 that $\mathcal{G}_{K_0^{-1}}(z,w)\succeq\mathcal{G}_{\mathbb{S}_{\mathbb{D}}^{-1}}(z,w)$, $(z,w)\in\mathbb{D}\times\mathbb{D}$. Therefore, the inequality (8) is weaker than the inequality (4) in case $m=1$.

After establishing a lower bound for the Gaussian curvature of a non-negative definite kernel, we show that the partial derivatives are bounded from $(\mathcal{H},K)$ to $(\mathcal{H},\mathcal{G}_{K^{-1}})$. We recall from [3, Lemma 3.1] that $\Big(\Big(\partial_i\bar{\partial}_jK(z,w)\Big)\Big)_{i,j=1}^m$ is a non-negative definite kernel whenever K is non-negative definite.

Theorem 2.1 *Let $\Omega\subset\mathbb{C}^m$ be a domain. Let $K:\Omega\times\Omega\to\mathbb{C}$ be a non-negative definite kernel. Suppose that the Hilbert space $(\mathcal{H},K)$ contains the constant function 1. Then*

$$\Big(\!\Big(\partial_i\bar{\partial}_j K(z,w)\Big)\!\Big)_{i,j=1}^{m} \preceq c\, \mathcal{G}_{K^{-1}}(z,w), \quad z,w\in\Omega,$$

where $c = \|1\|^2_{(\mathcal{H},K)}$.

Proof Set $c = \|1\|^2_{(\mathcal{H},K)}$. Choose an orthonormal basis $\{e_n(z)\}_{n\geq 0}$ of $(\mathcal{H},K)$ with $e_0(z) = \frac{1}{\sqrt{c}}$. Then

$$K(z,w) - \frac{1}{c} = \sum_{i=1}^{\infty} e_i(z)\overline{e_i(w)},\ z,w\in\Omega.$$

Hence $K(z,w) - \frac{1}{c}$ is non-negative definite on $\Omega\times\Omega$, or equivalently $cK-1$ is non-negative definite on $\Omega\times\Omega$. Therefore, by Corollary 1.1, it follows that

$$(((cK-1)^2\partial_i\bar{\partial}_j\log(cK-1)))_{i,j=1}^{m}$$

is non-negative definite on $\Omega\times\Omega$. Note that, for $z,w\in\Omega$, we have

$$\begin{aligned}
&\big((cK-1)^2\partial_i\bar{\partial}_j\log(cK-1)\big)(z,w)\\
&=(cK-1)(z,w)\big(\partial_i\bar{\partial}_j(cK-1)\big)(z,w)\\
&\qquad-\big(\partial_i(cK-1)\big)(z,w)\big(\bar{\partial}_j(cK-1)\big)(z,w)\\
&=c^2K(z,w)\partial_i\bar{\partial}_jK(z,w) - c\partial_i\bar{\partial}_jK(z,w) - c^2\partial_iK(z,w)\bar{\partial}_jK(z,w)\\
&=c^2K^2\partial_i\bar{\partial}_j\log K(z,w) - c\partial_i\bar{\partial}_jK(z,w).
\end{aligned}$$

Hence we conclude that

$$\Big(\!\Big(\partial_i\bar{\partial}_j K(z,w)\Big)\!\Big)_{i,j=1}^{m} \preceq c\, \mathcal{G}_{K^{-1}}(z,w),\ z,w\in\Omega.$$

Corollary 2.2 *Let* $\Omega\subset\mathbb{C}^m$ *be a domain. Let* $K:\Omega\times\Omega\to\mathbb{C}$ *be a non-negative definite kernel. Then the linear operator* $\boldsymbol{\partial}:(\mathcal{H},K)\to(\mathcal{H},\mathcal{G}_{K^{-1}})$*, where* $\boldsymbol{\partial} f = (\partial_1 f,\ldots,\partial_m f)^{\mathrm{tr}}$*,* $f\in(\mathcal{H},K)$*, is bounded with* $\|\boldsymbol{\partial}\|\leq\|1\|_{(\mathcal{H},K)}$*. Moreover if* K *is normalized at the point* $w_0\in\Omega$*, that is,* $K(\cdot,w_0)$ *is the constant function 1, then the linear operator* $\boldsymbol{\partial}:(\mathcal{H},K)\to(\mathcal{H},\mathcal{G}_{K^{-1}})$ *is contractive.*

Proof To prove the first assertion of the corollary, note that the map $\boldsymbol{\partial}$ is unitary from $\ker\boldsymbol{\partial}^{\perp}$ to $(\mathcal{H},(\partial_i\bar{\partial}_jK)_{i,j=1}^{m})$ and therefore is contractive from $(\mathcal{H},K)$ to $(\mathcal{H},(\partial_i\bar{\partial}_jK)_{i,j=1}^{m})$. To complete the proof, it is therefore enough to show that $(\mathcal{H},(\partial_i\bar{\partial}_jK)_{i,j=1}^{m})$ is contained in $(\mathcal{H},\mathcal{G}_{K^{-1}})$ and the inclusion map is bounded by

$\|1\|_{(\mathcal{H},K)}$. This follows from Theorem 2.1 using [16, Theorem 6.25]. For the second assertion, note that $\|1\|^2_{(\mathcal{H},K)} = \langle K(\cdot, w_0), K(\cdot, w_0)\rangle_{(\mathcal{H},K)} = K(w_0, w_0) = 1$ by hypothesis and use Theorem 2.1 to complete the proof.

3 A limit Computation

Let Ω be a bounded domain in $\mathbb{C}^m$. Let $\boldsymbol{M}^* \in B_1(\Omega^*)$ be the adjoint of the m-tuple $\boldsymbol{M}$ of multiplication by the coordinate functions on a reproducing kernel Hilbert space $(\mathcal{H}, K)$ consisting of holomorphic functions on $\Omega \subset \mathbb{C}^m$. Let $\mathcal{A}(\Omega)$ be the function algebra of all those functions holomorphic in some open neighbourhood of the compact set $\overline{\Omega}$ equipped with the supremum norm on $\overline{\Omega}$. The map $\mathbf{m}_f : h \mapsto f \cdot h$, $f \in \mathcal{A}(\Omega)$, $h \in (\mathcal{H}, K)$, where $(f \cdot h)(z) = f(z)h(z)$, defines a module multiplication for $(\mathcal{H}, K)$ over the algebra $\mathcal{A}(\Omega)$. We let $\mathcal{M} := (\mathcal{H}, K)$ denote this Hilbert module. Let $\mathcal{M}_0 \subseteq \mathcal{M}$ be a submodule. We now have a short exact sequence of Hilbert modules

$$0 \longrightarrow \mathcal{M}_0 \stackrel{i}{\longrightarrow} \mathcal{M} \stackrel{\pi}{\longrightarrow} \mathcal{Q} \longrightarrow 0$$

where i is the inclusion map and π is the quotient map. The problem of finding invariants for $\mathcal{Q}$ given the inclusion $\mathcal{M}_0 \subset \mathcal{M}$ has been studied in several papers (cf. [10, 12]). A variant of this problem occurs by replacing the inclusion map with some other module map, for instance, one might set $\mathcal{M}_0 = \varphi\mathcal{M}$ for some $\varphi \in \mathcal{A}(\Omega)$; see [13, p. 94] for the case of the Hardy module over the disc and the Beurling phenomenon. Here we are going to consider the case of submodules $\mathcal{M}_0$ consisting of the maximal set of functions in $\mathcal{M}$ vanishing on some fixed subset $\mathcal{Z}$ of Ω. A description of the specific examples we consider here follows.

Let K_1 and K_2 be two scalar valued non-negative definite kernels on Ω. Assume that both the kernels are sesqui-analytic. It is well known that $(\mathcal{H}, K_1) \otimes (\mathcal{H}, K_2)$ is the reproducing kernel Hilbert space determined by the non-negative definite kernel $K_1 \otimes K_2$, where $K_1 \otimes K_2 : (\Omega \times \Omega) \times (\Omega \times \Omega) \to \mathbb{C}$ is given by

$$(K_1 \otimes K_2)(z, \zeta; w, \rho) = K_1(z, w)K_2(\zeta, \rho), \quad z, \zeta, w, \rho \in \Omega.$$

We assume that the operator M_{z_i} of multiplication by the coordinate function z_i is bounded on $(\mathcal{H}, K_1)$ as well as on $(\mathcal{H}, K_2)$ for $i = 1, \ldots, m$. Then $(\mathcal{H}, K_1) \otimes (\mathcal{H}, K_2)$ may be realized as a Hilbert module over the polynomial ring $\mathbb{C}[z_1, \ldots, z_{2m}]$ with the module action defined by

$$\mathbf{m}_p(h) = ph, \ h \in (\mathcal{H}, K_1) \otimes (\mathcal{H}, K_2), \ p \in \mathbb{C}[z_1, \ldots, z_{2m}].$$

The Hilbert space $(\mathcal{H}, K_1) \otimes (\mathcal{H}, K_2)$ admits a natural direct sum decomposition as follows:

For a non-negative integer k, let $\mathcal{A}_k$ be the subspace of $(\mathcal{H}, K_1) \otimes (\mathcal{H}, K_2)$ defined by

$$\mathcal{A}_k := \left\{ f \in (\mathcal{H}, K_1) \otimes (\mathcal{H}, K_2) : \left(\left(\tfrac{\partial}{\partial \zeta}\right)^{\boldsymbol{i}} f(z, \zeta) \right)_{|\Delta} = 0, \ |\boldsymbol{i}| \leq k \right\}, \tag{9}$$

where $\boldsymbol{i} = (i_1, \ldots, i_m) \in \mathbb{Z}_+^m$, $|\boldsymbol{i}| = i_1 + \cdots + i_m$, $\left(\frac{\partial}{\partial \zeta}\right)^{\boldsymbol{i}} = \frac{\partial^{|\boldsymbol{i}|}}{\partial \zeta_1^{i_1} \cdots \partial \zeta_m^{i_m}}$, and $\left(\left(\frac{\partial}{\partial \zeta}\right)^{\boldsymbol{i}} f(z, \zeta)\right)_{|\Delta}$ is the restriction of $\left(\frac{\partial}{\partial \zeta}\right)^{\boldsymbol{i}} f(z, \zeta)$ to the diagonal set $\Delta := \{(z, z) : z \in \Omega\}$. It is easily verified that each of the subspaces $\mathcal{A}_k$ is closed and invariant under multiplication by any polynomial in $\mathbb{C}[z_1, \ldots, z_{2m}]$ and therefore they are submodules of $(\mathcal{H}, K_1) \otimes (\mathcal{H}, K_2)$. Setting $\mathcal{S}_0 = \mathcal{A}_0^{\perp}$, $\mathcal{S}_k := \mathcal{A}_{k-1} \ominus \mathcal{A}_k$, $k = 1, 2, \ldots$, we obtain a direct sum decomposition of the Hilbert space $(\mathcal{H}, K_1) \otimes (\mathcal{H}, K_2)$ as follows:

$$(\mathcal{H}, K_1) \otimes (\mathcal{H}, K_2) = \bigoplus_{k=0}^{\infty} \mathcal{S}_k.$$

Define a linear map $\mathcal{R}_1 : (\mathcal{H}, K^{\alpha}) \otimes (\mathcal{H}, K^{\beta}) \to \operatorname{Hol}(\Omega, \mathbb{C}^m)$ by setting

$$\mathcal{R}_1(f) = \frac{1}{\sqrt{\alpha\beta(\alpha + \beta)}} \begin{pmatrix} (\beta \partial_1 f - \alpha \partial_{m+1} f)_{|\Delta} \\ \vdots \\ (\beta \partial_m f - \alpha \partial_{2m} f)_{|\Delta} \end{pmatrix} \tag{10}$$

for $f \in (\mathcal{H}, K^{\alpha}) \otimes (\mathcal{H}, K^{\beta})$. Let $\iota : \Omega \to \Omega \times \Omega$ be the map $\iota(z) = (z, z)$, $z \in \Omega$. Any Hilbert module $\mathcal{M}$ over the polynomial ring $\mathbb{C}[z_1, \ldots, z_m]$ may be thought of as a module $\iota_{\star}\mathcal{M}$ over the ring $\mathbb{C}[z_1, \ldots, z_{2m}]$ by re-defining the multiplication: $\mathbf{m}_p(h) = (p \circ \iota)h$, $h \in \mathcal{M}$ and $p \in \mathbb{C}[z_1, \ldots, z_{2m}]$. The module $\iota_{\star}\mathcal{M}$ over $\mathbb{C}[z_1, \ldots, z_{2m}]$ is defined to be the pushforward of the module $\mathcal{M}$ over $\mathbb{C}[z_1, \ldots, z_m]$ under the inclusion map ι.

Recall that for $\alpha, \beta > 0$ and a sesqui-analytic function $K : \Omega \times \Omega \to \mathbb{C}$, the function $\mathbb{K}^{(\alpha,\beta)} : \Omega \times \Omega \to \mathcal{M}_m(\mathbb{C})$ is defined by

$$\mathbb{K}^{(\alpha,\beta)}(z, w) := K^{\alpha+\beta}(z, w) \Big(\big((\partial_i \bar{\partial}_j \log K)(z, w) \big) \Big)_{i,j=1}^{m}, \quad z, w \in \Omega.$$

Theorem 3.1 *[14, Theorem 3.5.] Suppose $K : \Omega \times \Omega \to \mathbb{C}$ is a sesqui-analytic function such that the functions K^{α} and K^{β}, defined on $\Omega \times \Omega$, are non-negative definite for some $\alpha, \beta > 0$. Then the following hold:*

(1) $\ker \mathcal{R}_1 = \mathcal{S}_1^{\perp}$ and $\mathcal{R}_1$ maps $\mathcal{S}_1$ isometrically onto $\big(\mathcal{H}, \mathbb{K}^{(\alpha,\beta)}\big)$.
(2) Suppose that the operator M_i of multiplication by the coordinate function z_i is bounded on both $(\mathcal{H}, K^{\alpha})$ and $(\mathcal{H}, K^{\beta})$ for $i = 1, 2, \ldots, m$. Then the

Hilbert module $\mathcal{S}_1$ is isomorphic to the pushforward module $\iota_\star\big(\mathcal{H}, \mathbb{K}^{(\alpha,\beta)}\big)$ via the module map $\mathcal{R}_{1|\mathcal{S}_1}$.

We consider the example of the Hardy space. Let $K_1(z, w) = K_2(z, w) = \frac{1}{1-z\bar{w}}$, $z, w \in \mathbb{D}$, be the Szegö kernel of the unit disc $\mathbb{D}$. In this case $(\mathcal{H}, K_1) \otimes (\mathcal{H}, K_2)$ is the Hardy space on the bidisc $\mathbb{D}^2$, and it is often denoted by $H^2(\mathbb{D}^2)$. Now, we can compute the kernel functions for $\mathcal{S}_0$ and $\mathcal{A}_0$ in this example as follows (see [9]): The vectors $\left\{\frac{e_k}{\sqrt{k+1}}\right\}_{k\geq 0}$ form an orthonormal basis of $\mathcal{S}_0$, where e_k is given by

$$e_k(z_1, z_2) = \sum_{j=0}^{k} z_1^j z_2^{k-j}, \; z_1, z_2 \in \mathbb{D}.$$

Therefore the reproducing kernel $K_{\mathcal{S}_0}$ of $\mathcal{S}_0$ is given by

$$K_{\mathcal{S}_0}(\boldsymbol{z}, \boldsymbol{w}) = \sum_{k\geq 0} \frac{e_k(\boldsymbol{z})\overline{e_k(\boldsymbol{w})}}{k+1} \; \boldsymbol{z} = (z_1, z_2), \boldsymbol{w} = (w_1, w_2) \in \mathbb{D}^2.$$

A closed expression for $K_{\mathcal{S}_0}$ is easily obtained:

$$K_{\mathcal{S}_0}(\boldsymbol{z}, \boldsymbol{z}) = \frac{1}{|z_1 - z_2|^2} \log \frac{|1 - z_1\bar{z}_2|^2}{(1 - |z_1|^2)(1 - |z_2|^2)}, \; \boldsymbol{z} = (z_1, z_2) \in \mathbb{D}^2.$$

Therefore it follows that

$$\begin{aligned}
&K_{\mathcal{A}_0}(\boldsymbol{z}, \boldsymbol{z})\\
&= \frac{1}{(1 - |z_1|^2)(1 - |z_2|^2)} - K_{\mathcal{S}_0}(\boldsymbol{z}, \boldsymbol{z})\\
&= \frac{1}{(1 - |z_1|^2)(1 - |z_2|^2)} - \frac{1}{|z_1 - z_2|^2} \log \frac{|1 - z_1\bar{z}_2|^2}{(1 - |z_1|^2)(1 - |z_2|^2)}\\
&= \frac{1}{(1 - |z_1|^2)(1 - |z_2|^2)} - \frac{1}{|z_1 - z_2|^2}\Big(\frac{|z_1 - z_2|^2}{(1 - |z_1|^2)(1 - |z_2|^2)}\\
&\quad - \frac{1}{2}\frac{|z_1 - z_2|^4}{(1 - |z_1|^2)^2(1 - |z_2|^2)^2} + \cdots\Big).
\end{aligned}$$

Thus, we see that

$$\lim_{z_2\to z_1} \frac{K_{\mathcal{A}_0}(\boldsymbol{z}, \boldsymbol{z})}{|z_1 - z_2|^2} = \frac{1}{2}\frac{1}{(1 - |z_1|^2)^4}, \; (z_1, z_2) \in \mathbb{D}^2, \tag{11}$$

and hence, this limit coincides (after taking the pushforward under the map ι) with the metric of the module $\mathcal{S}_1$, up to a scalar multiple; see Theorem 3.1(2).

Consider the short exact sequence $0 \to \mathcal{A}_0 \to H^2(\mathbb{D}^2) \to \mathcal{S}_0 \to 0$. It is known that the quotient module $\mathcal{S}_0$ is the pushforward of the Bergman module on the disc. Also, by Theorem 3.1(2), the module $\mathcal{S}_1$ is the pushforward of $(\mathcal{H}, \frac{1}{(1-z\overline{w})^4})$. It then follows that

$$\mathcal{K}_{\mathcal{S}_1}(\mathbf{z}) = \mathcal{K}_{\mathcal{S}_0}(\mathbf{z}) - \frac{2}{(1-|z|^2)^2}, \ \mathbf{z} = (z, z), z \in \mathbb{D}.$$

Thus the restriction of $\mathcal{K}_{\mathcal{A}_0}$ to the zero set Δ might serve as an invariant for the inclusion $\mathcal{M}_0 \subset \mathcal{M}$. This possibility is explored below in a class of examples.

Let $\Omega \subset \mathbb{C}^m$ be a bounded domain and $K : \Omega \times \Omega \to \mathbb{C}$ be a sesqui-analytic function such that the functions K^α and K^β are non-negative definite on $\Omega \times \Omega$ for some $\alpha, \beta > 0$. For a non-negative integer p, let $K_{\mathcal{A}_p}$ be the reproducing kernel of $\mathcal{A}_p$, where $\mathcal{A}_p$ is defined in (9).

To prove the main result of this section, we need the following two lemmas. One way to prove both of the lemmas is to make the change of variables

$$u_1(z, \zeta) = \tfrac{1}{2}(z_1 - \zeta_1), \ldots, u_m(z, \zeta) = \tfrac{1}{2}(z_m - \zeta_m);$$
$$v_1(z, \zeta) = \tfrac{1}{2}(z_1 + \zeta_1), \ldots, v_m(z, \zeta) = \tfrac{1}{2}(z_m + \zeta_m).$$

We give the details for the proof of the first lemma. The proof for the second one follows by similar arguments.

Lemma 3.1 *Let $\Omega \subset \mathbb{C}^m$ be a domain and let Δ be the diagonal set $\{(z, z) : z \in \Omega\}$. Suppose that $f : \Omega \times \Omega \to \mathbb{C}$ is a holomorphic function satisfying $f_{|\Delta} = 0$. Then for each $z_0 \in \Omega$, there exists a neighbourhood $\Omega_0 \subset \Omega$ (independent of f) of z_0 and holomorphic functions $f_1, f_2, \ldots, f_m$ on $\Omega_0 \times \Omega_0$ such that*

$$f(z, \zeta) = \sum_{i=1}^{m} (z_i - \zeta_i) f_i(z, \zeta), \ z = (z_1, \ldots, z_m), \zeta = (\zeta_1, \ldots, \zeta_m) \in \Omega_0.$$

Proof Note that the image of the diagonal set $\Delta \subseteq \Omega \times \Omega$ under the map $\varphi : \Omega \times \Omega \to \mathbb{C}^{2m}$, where

$$\varphi(z, \zeta) := (u_1(z, \zeta), \ldots, u_m(z, \zeta), v_1(z, \zeta), \ldots, v_m(z, \zeta)),$$

is the set $\{(0, w) : w \in \Omega\}$. Therefore we may choose a neighbourhood of $(0, z_0)$ which is a polydisc contained in $\widehat{\Omega} := \varphi(\Omega \times \Omega)$. Suppose f is a holomorphic function on $\Omega \times \Omega$ vanishing on the set Δ. Setting $g := f \circ \varphi^{-1}$ on $\hat{\Omega}$, we see that g is a holomorphic function on $\hat{\Omega}$ vanishing on the set $\{(0, w) : w \in \Omega\}$. Therefore g has a power series representation around $(0, z_0)$ of the form $\sum_{i,j \in \mathbb{Z}_+^m} a_{ij} u^i (v - z_0)^j$, where $\sum_{j \in \mathbb{Z}_+^m} a_{0j}(v - z_0)^j = 0$ on the chosen polydisc. Hence $a_{0j} = 0$ for all $j \in \mathbb{Z}_+^m$, and the power series of g is of the form $\sum_{\ell=1}^m u_\ell g_\ell(u, v)$, where

$$g_\ell(u, v) = \sum_{ij} a_{ij} u^{i - e_\ell} (v - z_0)^j, \ 1 \le \ell \le m.$$

Here the sum is over all multi-indices $i = (i_1, \dots, i_m)$ satisfying $i_1 = 0, \dots, i_{\ell-1} = 0$, $i_\ell \ge 1$ while j remains arbitrary. Pulling this expression back to $\Omega \times \Omega$ under the bi-holomorphic map φ, we obtain the expansion of f in a neighbourhood of (z_0, z_0) as prescribed in the Lemma 3.1.

Lemma 3.2 *Suppose that $f : \Omega \times \Omega \to \mathbb{C}$ is a holomorphic function satisfying $f_{|\Delta} = 0$ and $\left(\left(\frac{\partial}{\partial \zeta_j}\right) f(z, \zeta)\right)_{|\Delta} = 0$, $j = 1, \dots, m$. Then for each $z_0 \in \Omega$, there exists a neighbourhood $\Omega_0 \subset \Omega$ (independent of f) of z_0 and holomorphic functions f_{ij}, $1 \le i \le j \le m$, on $\Omega_0 \times \Omega_0$ such that*

$$f(z, \zeta) = \sum_{1 \le i \le j \le m} (z_i - \zeta_i)(z_j - \zeta_j) f_{ij}(z, \zeta), \ z, \zeta \in \Omega_0.$$

Theorem 3.2 *Let $\Omega \subset \mathbb{C}^m$ be a bounded domain and $K : \Omega \times \Omega \to \mathbb{C}$ be a sesqui-analytic function such that the functions K^α and K^β are non-negative definite on $\Omega \times \Omega$ for some $\alpha, \beta > 0$. For z in Ω and $1 \le i, j \le m$, we have*

$$\lim_{\substack{\zeta_i \to z_i \\ \zeta_j \to z_j}} \left(\frac{K_{\mathcal{A}_0}(z, \zeta; z, \zeta)}{(z_i - \zeta_i)(\bar{z}_j - \bar{\zeta}_j)} \Bigg|_{\zeta_l = z_l, l \ne i, j} \right) = \frac{\alpha\beta}{(\alpha+\beta)} K(z, z)^{\alpha+\beta} \partial_i \bar{\partial}_j \log K(z, z),$$

where $K_{\mathcal{A}_0}$ is the reproducing kernel of the subspace $\mathcal{A}_0$ and $\frac{K_{\mathcal{A}_0}(z,\zeta;z,\zeta)}{(z_i - \zeta_i)(\bar{z}_j - \bar{\zeta}_j)}\Big|_{\zeta_l = z_l, l \ne i, j}$ is the restriction of the function $\frac{K_{\mathcal{A}_0}(z,\zeta;z,\zeta)}{(z_i - \zeta_i)(\bar{z}_j - \bar{\zeta}_j)}$ to the set $\{(z, \zeta) \in \Omega \times \Omega : z_l = \zeta_l, l = 1, \dots, m, l \ne i, j\}$.

Proof Let $K_{\mathcal{A}_0 \ominus \mathcal{A}_1}(z, \zeta; w, \nu)$ be the reproducing kernels of $\mathcal{A}_0 \ominus \mathcal{A}_1$. Fix a point z_0 in Ω. Choose a neighbourhood Ω_0 of z_0 in Ω such that the conclusions of Lemmas 3.1 and 3.2 are valid. Now we restrict the kernels K^α and K^β to $\Omega_0 \times \Omega_0$.

Let f be an arbitrary function in $\mathcal{A}_1$. Then, by definition, f satisfies the hypothesis of Lemma 3.2, and therefore, it follows that

$$\lim_{\zeta_i \to z_i} \left(\frac{f(z, \zeta)}{(z_i - \zeta_i)} \Bigg|_{z_l = \zeta_l, l \ne i} \right) = 0, \ \ i = 1, \dots, m. \tag{12}$$

Let $\{h_n\}_{n \in \mathbb{Z}_+}$ be an orthonormal basis of $\mathcal{A}_1$. Since the series $\sum_{n=0}^\infty h_n(z, \zeta) \overline{h_n(z, \zeta)}$ converges uniformly to $K_{\mathcal{A}_1}(z, \zeta; z, \zeta)$ on the compact subsets of $\Omega_0 \times \Omega_0$, using (12) we see that

$$\lim_{\substack{\zeta_i\to z_i\\ \zeta_j\to z_j}}\left(\frac{K_{\mathcal{A}_1}(z,\zeta;z,\zeta)}{(z_i-\zeta_i)(\bar z_j-\bar\zeta_j)}\Big|_{\zeta_l=z_l,l\neq i,j}\right)=\sum_{n=0}^{\infty}\lim_{\zeta_i\to z_i}\left(\frac{h_n(z,\zeta)}{(z_i-\zeta_i)}\Big|_{z_l=\zeta_l,l\neq i}\right)\overline{\lim_{\zeta_j\to z_j}\left(\frac{h_n(z,\zeta)}{(z_j-\zeta_j)}\Big|_{z_l=\zeta_l,l\neq j}\right)}=0.$$

Since $K_{\mathcal{A}_0}=K_{\mathcal{A}_0\ominus\mathcal{A}_1}+K_{\mathcal{A}_1}$, the above equality leads to

$$\lim_{\substack{\zeta_i\to z_i\\ \zeta_j\to z_j}}\left(\frac{K_{\mathcal{A}_0}(z,\zeta;z,\zeta)}{(z_i-\zeta_i)(\bar z_j-\bar\zeta_j)}\Big|_{\zeta_l=z_l,l\neq i,j}\right)=\lim_{\substack{\zeta_i\to z_i\\ \zeta_j\to z_j}}\left(\frac{K_{\mathcal{A}_0\ominus\mathcal{A}_1}(z,\zeta;z,\zeta)}{(z_i-\zeta_i)(\bar z_j-\bar\zeta_j)}\Big|_{\zeta_l=z_l,l\neq i,j}\right).$$

Now let $\{e_n\}_{n\in\mathbb{Z}_+}$ be an orthonormal basis of $\mathcal{A}_0\ominus\mathcal{A}_1$. Since each $e_n\in\mathcal{A}_0$, by Lemma 3.1, there exist holomorphic functions $e_{n,i}$, $1\le i\le m$, on $\Omega_0\times\Omega_0$ such that

$$e_n(z,\zeta)=\sum_{i=1}^{m}(z_i-\zeta_i)e_{n,i}(z,\zeta),\ \ z,\zeta\in\Omega_0.$$

Thus, for $1\le i\le m$, we have

$$\lim_{\zeta_i\to z_i}\left(\frac{e_n(z,\zeta)}{(z_i-\zeta_i)}\Big|_{z_l=\zeta_l,l\neq i}\right)=e_{n,i}(z,z),\ \ z\in\Omega_0. \tag{13}$$

Since the series $\sum_{n=0}^{\infty}e_n(z,\zeta)\overline{e_n(z,\zeta)}$ converges to $K_{\mathcal{A}_0\ominus\mathcal{A}_1}$ uniformly on compact subsets of $\Omega_0\times\Omega_0$, using (13), we see that

$$\lim_{\substack{\zeta_i\to z_i\\ \zeta_j\to z_j}}\left(\frac{K_{\mathcal{A}_0\ominus\mathcal{A}_1}(z,\zeta;z,\zeta)}{(z_i-\zeta_i)(\bar z_j-\bar\zeta_j)}\Big|_{\zeta_l=z_l,l\neq i,j}\right)=\sum_{n=0}^{\infty}e_{n,i}(z,z)\overline{e_{n,j}(z,z)},\ \ z\in\Omega_0. \tag{14}$$

Recall that by Theorem 3.1, the map $\mathcal{R}_1:\mathcal{A}_0\ominus\mathcal{A}_1\to(\mathcal{H},\mathbb{K}^{(\alpha,\beta)})$ given by

$$\mathcal{R}_1f=\frac{1}{\sqrt{\alpha\beta(\alpha+\beta)}}\begin{pmatrix}(\beta\partial_1f-\alpha\partial_{m+1}f)_{|\Delta}\\ \vdots\\ (\beta\partial_mf-\alpha\partial_{2m}f)_{|\Delta}\end{pmatrix},\ \ f\in\mathcal{A}_0\ominus\mathcal{A}_1$$

is unitary. Hence $\{\mathcal{R}_1(e_n)\}_n$ is an orthonormal basis for $(\mathcal{H},\mathbb{K}^{(\alpha,\beta)})$ and consequently,

$$\sum_{n=0}^{\infty} \mathcal{R}_1(e_n)(z)\mathcal{R}_1(e_n)(w)^* = \mathbb{K}^{(\alpha,\beta)}(z,w), \ z, w \in \Omega_0. \tag{15}$$

A direct computation shows that

$$\big((\beta\partial_i - \alpha\partial_{m+i})e_n(z,\zeta)\big)_{|\Delta} = (\alpha+\beta)e_{n,i}(z,\zeta)_{|\Delta}, \ 1 \le i \le m, \ n \ge 0.$$

Therefore $\mathcal{R}_1(e_n)(z) = \sqrt{\frac{\alpha+\beta}{\alpha\beta}} \begin{pmatrix} e_{n,1}(z,z) \\ \vdots \\ e_{n,m}(z,z) \end{pmatrix}$. Thus, using (15), we obtain

$$\sum_{n=0}^{\infty} \begin{pmatrix} e_{n,1}(z,z) \\ \vdots \\ e_{n,m}(z,z) \end{pmatrix} \begin{pmatrix} e_{n,1}(z,z) \\ \vdots \\ e_{n,m}(z,z) \end{pmatrix}^* = \frac{\alpha\beta}{(\alpha+\beta)} \mathbb{K}^{(\alpha,\beta)}(z,z), \ \ z \in \Omega_0.$$

Now the proof is complete using (14).

Remark 3.1 Let $\mathcal{M}$ be Hilbert modules over the polynomial ring $\mathbb{C}[z]$ and $\mathcal{M}_0 \subseteq \mathcal{M}$ be a submodule. Let $\mathcal{Q}$ be the quotient module, i.e.

$$0 \longrightarrow \mathcal{M}_0 \longrightarrow \mathcal{M} \longrightarrow \mathcal{Q} \longrightarrow 0 \tag{16}$$

is a short exact sequence of Hilbert modules. Finding an invariant for the equivalence class of $\mathcal{Q}$ from that of the pair $\mathcal{M}_0, \mathcal{M}$ is one of the main problems of Sz.-Nagy-Foias model theory. An invariant is an object associated to the pair $\mathcal{M}_0, \mathcal{M}$ that depends only on the quotient $\mathcal{M}/\mathcal{M}_0$. Such an invariant is also said to be an invariant of the short exact sequence (16). Indeed, taking $\mathcal{M} = H^2(\mathbb{D})$ and $\mathcal{S}_i$, $i = 1, 2$, to be any two submodules of $H^2(\mathbb{D})$, then we know that while $\mathcal{S}_1$ is always equivalent to $\mathcal{S}_2$, the quotient modules $\mathcal{Q}_i := H^2(\mathbb{D})/\mathcal{S}_i$ are not necessarily equivalent; see the discussion after the theorem on page 65 of [6]. Thus, finding invariants for quotient modules is an interesting problem.

Let $H(z) = \big(\langle s_i(z), s_j(z)\rangle\big)_{i,j=1}^n$, $z \in \Omega$, be the Hermitian metric of a holomorphic (trivial) vector bundle E defined on Ω relative to the holomorphic frame $\{s_1, \ldots, s_n\}$. The curvature $\mathbf{K}_H$ of the vector bundle E is the (1, 1) form

$$\sum_{i,j=1}^{n} \bar{\partial}_j\big(H^{-1}\partial_i H\big)d\bar{z}_j \wedge dz_i.$$

The trace of the curvature $\mathbf{K}_H$ is obtained by replacing each of the coefficients $\bar{\partial}_j\big(H^{-1}\partial_i H\big)$ by their trace. Recall that the determinant bundle det E is a line bundle determined by the holomorphic frame $s_1 \wedge \cdots \wedge s_n$ and the Hermitian metric:

$H(z) := \det H(z)$. The trace of the curvature of the vector bundle E and the curvature of the determinant bundle $\det E$ is equal, i.e. $\text{trace}(\mathbf{K}_H) = \mathbf{K}_{\det H}$; see [8, Equation (4.6)].

Now, from Theorem 3.2, we see that the Hermitian structure for the Hilbert module $\mathcal{S}_1$ is $\mathbb{K}^{(\alpha,\beta)}$. Also, we have the following equality:

$$\text{trace}\big(\mathbf{K}_{\mathbb{K}^{(\alpha,\beta)}}\big) = m\mathbf{K}_{K^{\alpha+\beta}} + \mathbf{K}_{\det(\mathcal{K}_K)}.$$

Thus, in these examples, we see that $\text{trace}\big(\mathbf{K}_{\mathbb{K}^{(\alpha,\beta)}}\big)$ is a function of $\alpha+\beta$. Consequently, if $\alpha+\beta = \alpha'+\beta'$, then we have

$$\begin{array}{ccccccccc} 0 & \longrightarrow & \mathcal{A}_0 & \xrightarrow{i} & (\mathcal{H}, K^{\alpha}) \otimes (\mathcal{H}, K^{\beta}) & \xrightarrow{\pi} & (\mathcal{H}, K^{\alpha+\beta}) & \longrightarrow & 0 \\ & & & & & & \| & & \\ 0 & \longrightarrow & \mathcal{A}_0 & \xrightarrow{i} & (\mathcal{H}, K^{\alpha'}) \otimes (\mathcal{H}, K^{\beta'}) & \xrightarrow{\pi} & (\mathcal{H}, K^{\alpha'+\beta'}) & \longrightarrow & 0, \end{array}$$

and $\text{trace}\big(\mathbf{K}_{\mathbb{K}^{(\alpha,\beta)}}\big) = \text{trace}\big(\mathbf{K}_{\mathbb{K}^{(\alpha',\beta')}}\big)$. Replacing the equality of the quotient modules $(\mathcal{H}, K^{\alpha+\beta})$ and $(\mathcal{H}, K^{\alpha'+\beta'})$ by an isomorphism does not change the conclusion. In general, replace K^{α} by K_1; K^{β} by K_2; $K^{\alpha'}$ by K_1'; and $K^{\beta'}$ by K_2', and assume that $K_1'(w,w)K_2'(w,w) = \varphi(w)K_1(w,w)K_2(w,w)\overline{\varphi(w)}$ for some nonvanishing holomorphic function defined on an open subset $U \subset \Omega$. This means that the quotient modules $\mathcal{S}_0$ and $\mathcal{S}_0'$ are equivalent. A straightforward computation then shows that $\text{trace}\big(\mathbf{K}_{K_{12}}\big) = \text{trace}\big(\mathbf{K}_{K_{12}'}\big)$, where $K_{12} = \mathcal{G}_{(K_1K_2)^{-1}}$ and, similarly, $K_{12}' = \mathcal{G}_{(K_1'K_2')^{-1}}$. Hence $\text{trace}\big(\mathbf{K}_{K_{12}}\big)$ is an invariant of the short exact sequences of the form

$$0 \longrightarrow \mathcal{A}_0 \xrightarrow{i} (\mathcal{H}, K_1) \otimes (\mathcal{H}, K_2) \xrightarrow{\pi} (\mathcal{H}, K_1K_2) \longrightarrow 0.$$

We expect this to be the case in much greater generality.

The following corollary is immediate by choosing $\alpha = 1 = \beta$ in Theorem 3.2. It also gives an alternative for computing the Gaussian curvature defined in (1) whenever the metric is of the form $K(z,z)^{-1}$ for some positive definite kernel K defined on $\Omega \times \Omega$, where $\Omega \subset \mathbb{C}$ is a bounded domain. Indeed, the assumption that $\boldsymbol{T}$ is in $B_1(\Omega)$ is not necessary to arrive at the formula in the corollary below.

Corollary 3.1 *Let $\boldsymbol{T}$ be a commuting m-tuple in the Cowen-Douglas class $B_1(\Omega)$ realized as the adjoint of the m-tuple $\boldsymbol{M}$ of multiplication operators by coordinate functions on a reproducing kernel Hilbert space $(\mathcal{H}, K) \subseteq \text{Hol}(\Omega_0)$, for some open subset Ω_0 of Ω. Then the curvature $\mathcal{K}_{\boldsymbol{T}}(z) = \big(\mathcal{K}_{\boldsymbol{T}}(z)_{i,j}\big)_{i,j=1}^{m}$ is given by the formula*

$$\mathcal{K}_{\boldsymbol{T}}(z)_{i,j} = \frac{2}{K(z,z)^2} \lim_{\substack{\zeta_i \to z_i \\ \zeta_j \to z_j}} \left(\frac{K_{\mathcal{A}_0}(z,\zeta;z,\zeta)}{(z_i-\zeta_i)(\bar{z}_j-\bar{\zeta}_j)}\bigg|_{\zeta_l = z_l, l \neq i,j} \right), \; z \in \Omega,\; 1 \le i,j \le m.$$

Acknowledgments We thank the reviewers of this paper for several very useful suggestions and for spotting many typographical errors in the original draft.

References

1. J. Agler and J. E. McCarthy, Pick interpolation and Hilbert function spaces, Graduate Studies in Mathematics, vol. 44, American Mathematical Society, Providence, RI, 2002.
2. N. Aronszajn, *Theory of reproducing kernels*, Trans. Amer. Math. Soc. **68** (1950), 337–404.
3. S. Biswas, D. K. Keshari, and G. Misra, *Infinitely divisible metrics and curvature inequalities for operators in the Cowen-Douglas class*, J. Lond. Math. Soc. (2) **88** (2013), 941–956.
4. M. J. Cowen and R. G. Douglas, *Complex geometry and operator theory*, Acta Math. **141** (1978), no. 3–4, 187–261.
5. ——, *Operators possessing an open set of eigenvalues*, Functions, series, operators, Vol. I, II (Budapest, 1980), Colloq. Math. Soc. János Bolyai, vol. 35, North-Holland, Amsterdam, (1983), 323–341.
6. —— *On moduli of invariant subspaces*, In: Apostol, C., Douglas, R.G., Sz.-Nagy, B., Voiculescu, D., Arsene, G. (eds) Invariant Subspaces and Other Topics. Operator Theory: Advances and Applications, vol 6. Birkhäuser, Basel.
7. R. E. Curto and N. Salinas, *Generalized bergman kernels and the cowen-douglas theory*, American Journal of Mathematics **106** (1984), 447–488.
8. J.-P. Demailly, *Complex Analytic and Differential Geometry*, Universite de Grenoble, 2007. http://www-fourier.ujf-grenoble.fr/demailly/manuscripts/agbook.pdf
9. R. G. Douglas and G. Misra, *Some calculations for Hilbert modules*, J. Orissa Math. Soc., **12-15** (1993–96), 75–85.
10. ——, *On quotient modules*, In: Recent advances in operator theory and related topics (Szeged, 1999), 203–209, Oper. Theory Adv. Appl., 127, Birkhäuser, Basel, 2001.
11. R. G. Douglas, G. Misra and J. Sarkar, *Contractive Hilbert modules and their dilations*, Israel J. Math. **187** (2012), 141–165
12. R. G. Douglas, G. Misra and C. Varughese, *On quotient modules - the case of arbitrary multiplicity*, J. Funct. Anal. 174 (2000), 364–398.
13. R. G. Douglas and V. I. Paulsen, *Hilbert modules over function algebras*, Longman Sc & Tech, 1989.
14. S. Ghara and G. Misra, *Decomposition of the tensor product of two Hilbert modules*, Operator theory, operator algebras and their interactions with geometry and topology – Ronald G. Douglas memorial volume, Birkhäuser/Springer, Cham (2020), 221–265
15. ——, *Curvature inequalities and extremal properties of bundle shifts*, J. Operator Theory **11** (1984), 305–317.
16. V. I. Paulsen and M. Raghupathi, *An introduction to the theory of reproducing kernel Hilbert spaces*, Cambridge Studies in Advanced Mathematics **152**, Cambridge University Press, Cambridge, (2016).
17. Md. R. Reza, *Curvature inequalities for operators in the Cowen-Douglas class of a planar domain*, Indiana Univ. Math. J. **67** (2018), 1255–1279.
18. K. Wang and G. Zhang, *Curvature inequalities for operators of the Cowen-Douglas class*, Israel J. Math. **222** (2017), 279–296.

Weighted Polynomial Approximation on the Cubes of the Nonzero Integers

Paul Koosis

Dédié à ma très chère compagne Suzanne Gervais, cinéaste et artiste peintre

1 Introduction

Throughout this article, the set $\{n^3 : n \in \mathbb{Z}, n \neq 0\}$ is designated by $\mathbb{K}$ (the symbol $\mathbb{C}$ being already in use). Omission of $n = 0$ is made mainly for simplification of the computations.

Given a function $W(n^3) \geq 1$ on $\mathbb{K}$ such that $n^b / W(n^3) \to 0$ for $n \to \pm\infty$ and each $b \geq 0$, one asks whether the ratios $P(n^3)/W(n^3)$ are dense or not in $\mathscr{C}_\circ(\mathbb{K})$, the set of sequences $\{c_{n^3}; n^3 \in \mathbb{K}\}$ such that $c_{n^3} \to 0$ for $n \to \pm\infty$.

Duality will be used here to study this problem, so it is simpler to pose the equivalent question of whether the $P(n^3)/W(n^3)$ *are not* $\|\ \|_\infty$ *dense* in $\mathscr{C}_\circ(\mathbb{K})$, and that is how we shall proceed.

The problem is mentioned at the end of [1], where its difficulty was observed. But it was taken up previously by Borichev and Sodin in [2], and Sodin has returned to it, together with his coworkers, since then. There are interesting examples towards the end of [2].

What will be obtained here consists of two conditions for *non-density* for *even* weights $W(n^3)$, one necessary and another sufficient but slightly stronger, involving the existence of a certain smooth version of W. (The preparation involves no loss of generality.) The *logarithms* of these weights have a very special form, each expressible as a *sum of two terms*, reminiscent of the Fefferman-Stein representations of functions in BMO but different from the latter.

P. Koosis (✉)
Outremont, QC, Canada

I. Binder et al. (eds.), *Function Spaces, Theory and Applications*, Fields Institute Communications 87, https://doi.org/10.1007/978-3-031-39270-2_14

Because of the sparsity of $\mathbb{K}$ in $\mathbb{R}$, few analytical tools are available for beginning work on the problem treated here. The theorems of Mergelian and of de Branges are about all we have for that. For the first of these, see [3], pp 147–157, and de Branges' proof of his theorem is also reproduced there, on pp 184–203. A new alternative approach to the latter result is given in [4].

Before proceeding to the mathematical exposition of this paper, a possible alternative approach to the problem considered in it should be mentioned.

Instead of studying weighted polynomial approximation on $\mathbb{K}$, one can do so on

$$\mathbb{K}_\delta = \bigcup_{n^3 \in \mathbb{K}} [n^3 - \delta, n^3 + \delta],$$

where $0 < \delta < 1$. Then more analytical tools become available and Benedicks gave, in [7], a criterion for non-density of the polynomials, verifying its sufficiency. This work is described on pp 443–445 of [3], and a complete discussion, including a proof of the necessity of Benedicks' criterion, is given by Borichev and Sodin in [8].

Now, for the same problem on

$$\bigcup_{n \in \mathbb{Z}: n \neq 0} [n - \delta, n + \delta], 0 < \delta < \frac{1}{2},$$

(treated in [3], pp 386–434), a corresponding criterion has long been known [3], p 433. What is remarkable is that *here* one can pass from the *sufficiency* result for $\delta > 0$ to one for the limiting case with $\delta = 0$, whose treatment had originally been very difficult – see [3], p 445–526. The procedure for passing to the case $\delta = 0$, based partly on Pedersen's work in [6], is due to Pedersen and the author and described at the end of [9].

Can that procedure be *adapted* so as to pass from the sets $\mathbb{K}_\delta$ to $\mathbb{K}$? If that is feasible, one would have a sufficient condition for non-density different from the one obtained in what follows.

2 Initial Simplifications of the Problem

Our questions involving general weights $W(n^3)$ on $\mathbb{K}$ can always be reduced to one about other weights thereon having a particular kind of behaviour.

If $W(n^3)$ is any weight such as specified in the introduction, we define, for $n^3 \in \mathbb{K}$, its *Mergelian minorant*, $\underline{W}(n^3)$, by the formula

$$\underline{W}(n^3) = \frac{1}{|n^3|} \sup \left\{ |P(n^3)|;\ P \text{ a polynomial, } |P(j^3)| \leq |j^3| W(j^3) \text{ for all } j^3 \in \mathbb{K} \right\}.$$

It is clear that $\underline{W}(n^3) \leq W(n^3)$ for $n^3 \in \mathbb{K}$. But we have the following:

Theorem 2.1 *The ratios $P(n^3)/W(n^3)$ with polynomials P and $n^3 \in \mathbb{K}$ are $\| \|_\infty$ dense in $\mathscr{C}_\circ(\mathbb{K})$ if and only if the $P(n^3)/\underline{W}(n^3)$ are.*

Proof By Mergelian's theorem, the $P(n^3)/W(n^3)$ are $\| \|_\infty$ dense in $\mathscr{C}_\circ(\mathbb{K})$ if and only if

$$\sup\{|P(i)|;\ P \text{a polynomial},\ |P(j^3)| \le |j^3|W(j^3) \text{ for all } j^3 \in \mathbb{K}\}$$

is *infinite*. But $|P(j^3)| \le |j^3|W(j^3)$ for all $j^3 \in \mathbb{K}$ if and only if that is so when $W(j^3)$ is replaced by $\underline{W}(j^3)$, according to the very definition of the latter. □

Treatment of our problem is simpler when $W(n^3) = W(-n^3)$. And this *symmetry* of the weight is also relevant to the study of the general case.

For the next result, we write $W_1(n^3) = \min\big(W(n^3), W(-n^3)\big)$ and $W_2(n^3) = \max\big(W(n^3), W(-n^3)\big)$. Then W_1 and W_2 are both even, and $W_1(n^3) \le W(n^3) \le W_2(n^3)$.

Theorem 2.2 *If the $P(n^3)/W_1(n^3)$ formed from polynomials P are dense in $\mathscr{C}_\circ(\mathbb{K})$, so are the $P(n^3)/W(n^3)$. And if the latter are dense in $\mathscr{C}_\circ(\mathbb{K})$, so are the $P(n^3)/W_2(n^3)$.*

Proof Is the same for both statements; consider the first one. If the $P(n^3)/W(n^3)$ are *not* dense, there is a nonzero $s(n^3)$ in $\ell_1(\mathbb{K})$ such that $\Sigma_{n^3\in\mathbb{K}} s(n^3)\frac{P(n^3)}{W(n^3)} = 0$ for all polynomials P. But then $\Sigma_{n^3\in\mathbb{K}} \frac{W_1(n^3)}{W(n^3)} s(n^3) \cdot \frac{P(n^3)}{W_1(n^3)} = 0$ for all such P, and $\frac{W_1(n^3)}{W(n^3)} s(n^3)$ belongs to $\ell_1(\mathbb{K})$ and is not identically zero. □

Because of this result, *even* weights play a determining role in the present subject. *Therefore, a lot of the work in this paper will henceforth be devoted to their study.*

The following result will be used in the next section.

Theorem 2.3 *If $W(n^3)$ is even and the ratios $P(n^3/W(n^3)$ formed from the odd polynomials P are $\| \|_\infty$ dense in the set of odd sequences in $\mathscr{C}_o(\mathbb{K})$, such ratios formed from all polynomials P are $\| \|_\infty$ dense in $\mathscr{C}_o(\mathbb{K})$. And conversely.*

Proof The converse is *clear* because if, for any *odd* sequence $\{a_{n^3}\}$ in $\mathscr{C}_o(\mathbb{K})$, there are polynomials P_ν with $P_\nu(n^3)/W(n^3)$ tending uniformly to a_{n^3} for $\nu \to \infty$, then the same is true with the $P_\nu(n^3)$ replaced by the *odd* polynomials $\frac{1}{2}(P_\nu(n^3) - P_\nu(-n^3))$.

It therefore suffices to assume the initial statement in the hypothesis and to deduce that, given any *even* sequence $\{a_{n^3}\}$ in $\mathscr{C}_\circ(\mathbb{K})$, there are *even* polynomials $Q_\nu(x)$ with $Q_\nu(n^3)/W(n^3) \underset{\nu}{\to} a_{n^3}$ uniformly on $\mathbb{K}$.

Taking, then, the even sequence $\{a_{n^3}\}$, we put, for $\varepsilon > 0$,

$$b_{n^3}^{(\varepsilon)} = \frac{1}{1+\varepsilon|n^3|} a_{n^3};$$

without loss of generality $|a_{n^3}| \le 1$ for $n^3 \in \mathbb{K}$. Then

$$n^3 b_{n^3}^{(\varepsilon)} = \frac{n^3}{1+\varepsilon|n^3|} a_{n^3}$$

is *odd* and belongs to $\mathscr{C}_\circ(\mathbb{K})$. We have

$$|a_{n^3} - b_{n^3}^{(\varepsilon)}| = \frac{\varepsilon|n^3|}{1+\varepsilon|n|^3}|a_{n^3}| = \frac{\varepsilon}{|n|^{-3}+\varepsilon}|a_{n^3}|.$$

If $|n|^{-3} > \sqrt{\varepsilon}$, this last is $< \sqrt{\varepsilon}|a_{n^3}| \le \sqrt{\varepsilon}$. And if $|n|^{-3} \le \sqrt{\varepsilon}$, $|n^3| \ge \frac{1}{\sqrt{\varepsilon}}$ making $|a_{n^3}| < \eta(\varepsilon)$, a quantity tending to 0 with ε since $\{a_{n^3}\} \varepsilon\, \mathscr{C}_\circ(\mathbb{K})$. Therefore $||a_{n^3} - b_{n^3}^{(\varepsilon)}||_\infty$ tends to 0 as $\varepsilon \to 0$.

Fixing $\varepsilon > 0$ sufficiently small, we can now get P, an *odd* polynomial, with

$$\left|\frac{P(n^3)}{W(n^3)} - n^3 b_{n^3}^{(\varepsilon)}\right| < \eta(\varepsilon)$$

for $n^3 \in \mathbb{K}$, and this surely makes

$$\left|\frac{Q(n^3)}{W(n^3)} - b_{n^3}^{(\varepsilon)}\right| \le \eta(\varepsilon)$$

with the *even* polynomial $Q(x) = P(x)/x$. Thence, by a previous relation,

$$\left|\frac{Q(n^3)}{W(n^3)} - a_{n^3}\right| < 2\eta(\varepsilon),$$

and we are done. □

3 Use of de Branges' Theorem

Considering, now, $W(n^3)$ to be *even* and assuming the ratios $P(n^3)/W(n^3)$ formed from polynomials P to be non-dense in $\mathscr{C}_\circ(\mathbb{K})$, $\|\ \|_\infty$, we proceed to deduce and apply a special form of de Branges' theorem. This we do by a change of variable.

According to the last theorem in Sect. 2, we may as well assume that the ratios formed using *odd* polynomials P are *not* $\|\ \|_\infty$ dense in the subspace of odd sequences belonging to $\mathscr{C}_\circ(\mathbb{K})$, and we do that.

Assuming, then, that there is some *odd* sequence $\{a_{n^3}\}$ in $\mathscr{C}_\circ(\mathbb{K})$ such that

$$\sup_{n^3 \in \mathbb{K}} \left|\frac{Q(n^3)}{W(n^3)} - a_{n^3}\right| \ge \text{ some } \eta > 0$$

for any *odd* polynomial $Q(x)$, we note that any such $Q(x)$ has the form $xR(x^2)$ with some polynomial R (and conversely). Thus, the ratios $n^3R(n^6)/W(n^3)$ cannot be used to uniformly approximate a_{n^3} for $n^3 \in \mathbb{K}$. But that is *equivalent* to the property that the expressions

$$\frac{|n^3|}{W(n^3)}R(n^6) = \frac{n^3}{W(n^3)}R(n^6)\ \mathrm{sgn}(n)$$

cannot be used to approximate the *even* sequence $a_{n^3}\ \mathrm{sgn}(n)$ on $\mathbb{K}$.

We can thus take $Q = \{n^6; n \in \mathbb{N}\}$, *a subset* of $(0, \infty)$, and put $\Omega(n^6) = W(n^3)/|n^3|$ and find that the ratios $R(n^6)/\Omega(n^6)$ formed with *polynomials* R are *not* $\|\ \|_\infty$ dense in $\mathscr{C}_\circ(\mathbb{Q})$.

The standard form of de Branges' theorem now says that there is an infinite subset S of $\mathbb{Q}$ ($\subset \mathbb{N}$!) such that for the entire function

$$\Phi(z) = \prod_{n^6 \in S}\left(1 - \frac{z}{n^6}\right)$$

we have both

$$\sum_{n^6 \in S} \frac{\Omega(n^6)}{|\Phi'(n^6)|} < \infty \text{ and } \sum_{n^6 \in S} \frac{R(n^6)}{\Phi'(n^6)} = 0$$

for all polynomials R.

Let now $\Lambda = \{n^3 \in \mathbb{K}; n^6 \in S\}$ and note that $\Lambda = -\Lambda$. Denote $\Lambda \cap \mathbb{N}$ by Λ_+, and put

$$F_\Lambda(z) = \prod_{n^3 \in \Lambda_+}\left(1 - \frac{z^2}{n^6}\right);$$

$F_\Lambda(z)$ is an *even* entire function and its zeros are at the points of Λ.

In terms of $\Phi(z)$, we have $F_\Lambda(z) = \Phi(z^2)$, so for $n^3 \in \Lambda$, $F'_\Lambda(n^3) = 2n^3\Phi'(n^6)$. Since $\Omega(n^6) = W(n^3)/|n^3|$, the *first* of the above relations reads

$$\sum_{n^3 \in \Lambda} \frac{W(n^3)}{|F'_\Lambda(n^3)|} < \infty,$$

and the second becomes

$$\sum_{n^3 \in \Lambda} \frac{n^3 P(n^3)}{F'_\Lambda(n^3)} = \sum_{n^6 \in S} \frac{R(n^6)}{\Phi'(n^6)} = 0,$$

where $P(x)$ is any *even* polynomial (equal) to the corresponding $R(x^2)$ and hence $xP(x)$ any *odd* one. Thus, $\sum_{n^3\in\Lambda} \frac{Q(n^3)}{F'_\Lambda(n^3)} = 0$ for all *odd* polynomials $Q(x)$. But, $F_\Lambda(z)$ being even, $F'_\Lambda(n^3)$ is *itself odd*, so the relation holds also with *even* polynomials Q, since $\Lambda = -\Lambda$. Finally, then, the relation holds for all polynomial Q, and we have the following:

Theorem 3.1 *If $W(n^3)$ is even and the ratios $\frac{P(n^3)}{W(n^3)}$ formed from polynomials P are not $\| \|_\infty$ dense in $\mathscr{C}_\circ(\mathbb{K})$, there is an infinite subset Λ_+ of $\mathbb{K}\cap\mathbb{N}$ such that, for the entire function $F_\Lambda(z) = \prod_{n^3\in\Lambda_+} (1 - \frac{z^2}{n^6})$, we have, with $\Lambda = \Lambda_+ \cup (-\Lambda_+)$,*

$$\sum_{n^3\in\Lambda} \frac{W(n^3)}{|F'_\Lambda(n^3)|} < \infty \text{ and } \sum_{n^3\in\Lambda} \frac{P(n^3)}{F'_\Lambda(n^3)} = 0$$

for all polynomials P.

4 Formation of a Majorant for $\log \underline{W}(n^3)$

Lemma 4.1 *With $F_\Lambda(z)$ the entire function figuring in the preceding theorem, we have*

$$\frac{1}{F_\Lambda(z)} = \sum_{n^3\in\Lambda} \frac{1}{(z-n^3)F'_\Lambda(n^3)}.$$

Proof By the last theorem, we surely have

$$\sum_{n^3\in\Lambda} \frac{1}{|F'_\Lambda(n^3)|} < \infty$$

(since $W(n^3) \geq 1$), and the entire function $F_\Lambda(z)$, of order $\frac{1}{3}$ and hence surely of (minimal) exponential type, has all its zeros in $\Lambda \subseteq \mathbb{K} \subseteq \mathbb{R}$. All the conditions of the theorem on p. 203 of [3] are thus verified or easily seen to hold. The result follows. □

Theorem 4.1 *If P is any polynomial,*

$$P(z) = F_\Lambda(z) \sum_{n^3\in\Lambda} \frac{P(n^3)}{(z-n^3)F'_\Lambda(n^3)}.$$

Proof For *fixed* $z_\circ$, $\frac{P(z)-P(z_\circ)}{z_\circ - z} = Q(z)$ *is* a polynomial in z, so

$$\sum_{n^3 \in \Lambda} \frac{Q(n^3)}{F'_\Lambda(n^3)} = 0$$

by the theorem of the preceding section. Thence,

$$\sum_{n^3 \in \Lambda} \frac{P(n^3)}{(z_\circ - n^3)F'_\Lambda(n^3)} = \sum_{n^3 \in \Lambda} \frac{P(z_\circ)}{(z_\circ - n^3)F'_\Lambda(n^3)},$$

and the result follows from the lemma. □

If now P is any polynomial with $|P(n^3)| \le |n^3|W(n^3)$ (sic!) for $n^3 \in \mathbb{K}$, the last theorem enables us to write

$$|P(z)| \le |F_\Lambda(z)| \sum_{n^3 \in \Lambda} \frac{|P(n^3)|}{|n^3|W(n^3)} \cdot \frac{|n^3|}{|n^3 - z|} \cdot \frac{W(n^3)}{|F'_\Lambda(n^3)|} \le C|F_\Lambda(z)| \sup_{n^3 \in \Lambda} \frac{|n^3|}{|n^3 - z|},$$

where C is a constant depending on $W(n^3)$ and on Λ.

For $z = m^3 \in \mathbb{K} \sim \Lambda$ (assuming that Λ is not *equal* to $\mathbb{K}$!), the ratio $|n^3|/|n^3 - m^3|$ is as large as possible when $n^3 \in \Lambda$ is as close as possible to m^3, from one side or the other. The *worst* possible choices for n^3 would thus be $(m-1)^3$ and $(m+1)^3$, and for both of these, we get a value of $|n^3|/|n^3 - m^3|$ asymptotic to $|m|/3$ for large $|m|$. Therefore, for $m^3 \in \mathbb{K} \sim \Lambda$, and $P(x)$ a polynomial with $|P(n^3)| \le |n^3|W(n^3)$ for $n^3 \in \Lambda$, we have, for $m^3 \in \mathbb{K} \sim \Lambda$,

$$|P(m^3)| \le C|m|\,|F_\Lambda(m^3)|.$$

Dividing by $|m^3|$ and taking the supremum for all the P just specified, we find that, for $m^3 \in \mathbb{K} \sim \Lambda$,

$$\underline{W}(m^3) \le C|F_\Lambda(m^3)|/m^2.$$

For $m^3 \in \Lambda$, all we know is that

$$\underline{W}(m^3) = W(m^3) \le C|F'_\Lambda(m^3)|.$$

We henceforth call $F_\Lambda(z)$, formed as in Sect. 3, a *de Branges function* for our problem, and we can state the following:

Theorem 4.2 *If, for $W(n^3)$, $n^3 \in \mathbb{K}$, even, we have non-density of the $P(n^3)/W(n^3)$ in $\mathscr{C}_\circ(\mathbb{K})$, $\| \ \|_\infty$ for polynomials P, the relations*

$$\underline{W}(m^3) \le C\frac{|F_\Lambda(m^3)|}{m^2}, \quad m^3 \in \mathbb{K} \sim \Lambda$$

$$\underline{W}(m^3) \le C|F'_\Lambda(m^3)|, \quad m^3 \in \Lambda$$

hold for the Mergelian minorant $\underline{W}(m^3)$ *of* $W(m^3)$*. Here, C is a constant depending on* $W(m^3)$ *and* Λ.

Remark 4.1 Of course we may (wlog) *take* $\underline{W}(m^3) = W(m^3)$ according to the theorem at the beginning of this paper. But we *want* the estimates in terms of the entire function $F_\Lambda(z)$!

From now on we will mainly be concerned with *logarithms* of $|F_\Lambda(z)|$ and $|F'_\Lambda(z)|$ where $F_\Lambda(z)$ is one of the de Branges functions figuring in the last theorem. In order to make the analysis more tractable, we proceed to *smooth out* the distributions of their zeros.

The simplest and most fundamental example of a de Branges function is obtained by taking $\Lambda = \mathbb{K}$, thus getting the function

$$C_\circ(z) = \prod_1^\infty \left(1 - \frac{z^2}{n^6}\right).$$

(That this is a possible de Branges function was observed in [2].)

From the Euler product formula and the relation

$$(u^6 - 1) = (u^2 - 1)(u^2 - \omega^2)(u^2 - \overline{\omega}^2)$$

with

$$\omega = e^{\pi i/3} = \frac{1}{2} + \frac{\sqrt{3}}{2}i,$$

we get

$$C_\circ(\xi^3) = \frac{\sin \pi\xi}{\pi\xi} \cdot \frac{\sin \pi\omega\xi}{\pi\omega\xi} \cdot \frac{\sin \pi\overline{\omega}\xi}{\pi\overline{\omega}\xi}.$$

Thence, for $\xi^3 = x > 0$,

$$C_\circ(x) = \frac{1}{\pi^3 x} \sin(\pi x^{1/3}) \sin(\pi\omega x^{1/3}) \sin(\pi\overline{\omega} x^{1/3}).$$

Therefore, at $x = n^3$, $n \in \mathbb{N}$,

$$C'_\circ(n^3) = (-1)^n \cdot \frac{\pi}{3n^2} \cdot \frac{\sin(\pi\omega n)\sin(\pi\overline{\omega}n)}{\pi^3 n^3}.$$

Now for $\xi > 0$ large, we have

$$\sin \pi \overline{\omega} \xi = \overline{\sin \pi \omega \xi} \sim \frac{1}{2i} e^{\sqrt{3}\pi \xi/2},$$

making $\sin(\omega\xi)\sin(\overline{\omega}\xi) \sim e^{\sqrt{3}\pi\xi}/4$, whence

$$C_\circ'(n^3) \sim (-1)^n \frac{e^{\sqrt{3}\pi n}}{12\pi^3 n^5}$$

for $n \in \mathbb{N}$ *large*. A similar relation holds for $-n \in \mathbb{N}$ ($C_\circ'(\xi)$ is *odd*!), and we can write

$$\log |C_\circ'(n^3)| = \sqrt{3}\pi |n| - 5 \log |n| + O(1)$$

for $|n| \in \mathbb{N}$ large.

Considering that

$$\log |C_\circ(x)| = \sum_1^\infty \log \left| 1 - \frac{x^2}{n^6} \right|$$

and that $t^{1/3} = n$ for $t = n^3$, we proceed to look at the *integral*

$$V_\circ(x) = \int_0^\infty \log \left| 1 - \frac{x^2}{t^2} \right| d(t^{1/3})$$

which will play an important role in the following work.

Let us compute $V_\circ(x)$. Integrating by parts, we get

$$\int_0^\infty \log \left| 1 - \frac{x^2}{t^2} \right| d(t^{1/3}) = x \cdot ⨍_0^\infty \frac{2x}{x^2 - t^2} \cdot \frac{t^{1/3}}{t} dt.$$

On the right we have, as a Cauchy principal value, πx times the *Hilbert transform* of the *even* function $t^{-2/3}$. Now the function $f(z) = e^{\pi i/3} z^{-2/3}$, with $\arg z$ taken between 0 and π, is *analytic* in $\Im z > 0$, and for its values on $\mathbb{R}$,

$$f(x) = \frac{1}{2}|x|^{-2/3} + \frac{\sqrt{3}}{2} i |x|^{-2/3} \operatorname{sgn}(x).$$

Therefore, the expression on the right is equal to

$$2\pi x \cdot \frac{\sqrt{3}}{2} |x|^{-2/3} \operatorname{sgn}(x) = \sqrt{3}\pi |x|^{1/3},$$

and

$$V_\circ(x) = \sqrt{3}\pi |x|^{1/3}, \quad x \in \mathbb{R}.$$

This formula will be used often in what follows, where $V_\circ(x)$ will serve as a kind of standard to which other functions will be compared.

From $V_\circ(x)$ we get another function, $V(x)$, closely related to $\log|C'_\circ(x)|$. Referring to a relation involving $\log|C'_\circ(n^3)|$ deduced above, we see that we would like to have

$$V(n^3) = V_\circ(n^3) - 5\log|n| + O(1),$$

and that this would follow with

$$V(x) = \int_p^\infty \log\left|1 - \frac{x^2}{t^2}\right| d(t^{1/3})$$

on making

$$\int_0^p \log\left|1 - \frac{x^2}{t^2}\right| d(t^{1/3}) = \frac{5}{3}\log|x| + O(1).$$

For large | |, the last condition boils down to

$$\log|x|^2 \cdot \int_0^p d(t^{1/3}) = \frac{5}{3}\log|x|,$$

whence $p = (\frac{5}{6})^3$. And we have

$$V(x) = \int_{(5/6)^3}^\infty \log\left|1 - \frac{x^2}{t^2}\right| d(t^{1/3}).$$

This function $V(x)$, with obvious harmonic extensions to the upper and lower half planes yielding a function *subharmonic* in $\mathbb{C}$, takes values very close (within $O(1)$) to $\log|C'_\circ(n^3)|$ for $x = n^3 \in \mathbb{K}$.

We come now to the de Branges function $F_\Lambda(z)$ with Λ *properly* contained in $\mathbb{K}$; for these a *smoothing* of the corresponding zero distributions is carried out.

We may, in the first place, always assume that $1 \notin \Lambda$. For if 1 *is* in Λ, we let $n_1^3 > 1$ be the first positive member of $\mathbb{K}$ *not* in Λ. Then, if $F_\Lambda(z)$ is associated with the weight $W(n^3)$,

$$\frac{F_\Lambda(z)}{1 - z^2}\left(1 - \frac{z^2}{n_1^6}\right)$$

is evidently *another* de Branges function associated with $W(n^3)$. *We henceforth assume that this replacement has been made.*

That being the case, we denote the successive members of Λ_+ by $n_1^3 > 1, n_2^3 > n_1^3$, and so forth, with $n_k \in \mathbb{N}$. In terms of these, we form the *increasing* function $\nu(t)$ on $[0, \infty]$ by putting $\nu'(t) = 0$ on $[0, 1)$ and on each of the nonempty intervals $[n_k^3, (n_{k+1} - 1)^3]$, and

$$\nu'(t) = \frac{1}{3t^{2/3}}, \quad (n_k - 1)^3 \leq t < n_k^3,$$

$k = 1, 2, \cdots$. We then take

$$w(z) = \int_0^\infty \log\left|1 - \frac{z^2}{t^2}\right| d\nu(t)$$

and proceed to obtain relations between $w(m^3)$ and $\log|F_\Lambda(m^3)|$ for $m^3 \in \mathbb{K} \sim \Lambda$ and between $w(m^3)$ and $\log|F'_\Lambda(m^3)|$ for $m^3 \in \Lambda$.

It is convenient to write $n_\circ = 1$, remembering always that $n_\circ^3 \notin \Lambda$.

Consider first the case where $m^3 = 1 = n_0^3$. Then, with $w_\circ(1) = \int_1^\infty \log|1 - \frac{1}{t^2}|d\nu(t)$ and $\log|F_\Lambda(1)| = \sum_{l=1}^{\infty} \log|1 - \frac{1}{n_l^6}|$, we note that $\nu(t)$ increases by 1 on each interval $[n_k^3, n_{k+1}^3]$, $k \geq 0$ (sic!) (here $n_1^3 \geq 8$). But then, for $l \geq 0$,

$$\log\left|1 - \frac{1}{n_l^6}\right| \leq \int_{n_l^3}^{n_{l+1}^3} \log\left|1 - \frac{1}{t^2}\right| d\nu(t)$$

since $\log\left|1 - \frac{1}{t^2}\right| \geq \log\left|1 - \frac{1}{n_l^6}\right|$ in the interval $[n_l^3, n_{l+1}^3]$. This already makes

$$\log|F_\Lambda(1)| \leq \int_1^\infty \log\left|1 - \frac{1}{t^2}\right| d\nu(t).$$

If now $m^3 \in \mathbb{K} \sim \Lambda$ is > 1, it lies in some interval $[n_k^3, n_{k+1}^3]$, with $k \geq 0$. For $l < k$ (if there *are* such l), we have, as above,

$$\log\left|1 - \frac{m^6}{n_{l+1}^6}\right| \leq \int_{n_l^3}^{n_{l+1}^3} \log\left|1 - \frac{m^6}{t^2}\right| d\nu(t),$$

and if $l \geq k + 1$,

$$\log\left|1 - \frac{m^6}{n_l^6}\right| \leq \int_{n_l^3}^{n_{l+1}^3} \log\left|1 - \frac{m^6}{t^2}\right| d\nu(t).$$

Therefore,

$$\log|F_\Lambda(m^3)| = \sum_{l\geq 1}\log\left|1-\frac{m^6}{n_l^6}\right| \leq \left(\int_{n_0^3}^{n_k^3}+\int_{n_{k+1}^3}^{\infty}\right)\log\left|1-\frac{m^6}{t^2}\right|d\nu(t).$$

Since $n_\circ = 1$, the expression on the right is equal to

$$w(m^3) - \int_{n_k^3}^{n_{k+1}^3}\log\left|1-\frac{m^6}{t^2}\right|\,d\nu(t).$$

We need an approximation to the last integral, keeping in mind that

$$n_k^3 < m^3 < n_{k+1}^3$$

Here, we make all the ratios m^6/t^2, $n_k^3 < t < n_{k+1}^3$, *closer* to 1 by moving n_k^3 up to $(m-1)^3$ (if it is not already equal to that) and n_{k+1}^3 down to $(m+1)^3$. At the same time, we replace $d\nu(t)$ on $[n_k^3, n_{k+1}^3]$ by $\frac{1}{2}d(t^{1/3})$ on $[(m-1)^3, (m+1)^3]$. With the same reasoning as above, we then find that

$$\int_{n_k^3}^{n_{k+1}^3}\log\left|1-\frac{m^6}{t^2}\right|d\nu(t) \geq \frac{1}{2}\int_{(m-1)^3}^{(m+1)^3}\log\left|1-\frac{m^6}{s^2}\right|d(s^{1/3})$$

The right-hand expression is equal to

$$\begin{aligned}
&\frac{1}{2}\int_{m-1}^{m+1}\log\left|1+\frac{m^3}{\tau^3}\right|d\tau + \frac{1}{2}\int_{m-1}^{m+1}\log\left|1-\frac{m^3}{\tau^3}\right|d\tau\\
&= O(1) + \frac{1}{2}\int_{m-1}^{m+1}\log\left|1+\frac{m}{\tau}+\frac{m^2}{\tau^2}\right|d\tau + \frac{1}{2}\int_{m-1}^{m+1}\log\left|1-\frac{m}{\tau}\right|d\tau\\
&= O(1) + \frac{1}{2}\int_{m-1}^{m+1}\log|\tau-m|d\tau + \frac{1}{2}\int_{m-1}^{m+1}\log\frac{1}{\tau}d\tau\\
&= O(1) + \log\frac{1}{m}.
\end{aligned}$$

Referring to the first of the above relations, involving $w(n^3)$, we find that

$$\log|F_\Lambda(m^3)| = \sum_{l\geq 1}\log|1-\frac{m^6}{n_l^6}| \leq w(m^3) + \log m + O(1).$$

Thence, we *already have*

$$\log(F_\Lambda(m^3)/m^2) \leq w(m^3) - \log m + O(1)$$

for large $m \in \mathbb{K} \sim \Lambda$, and thus, by the theorem appearing near the beginning of this section (the functions involved being *even*), $\log \underline{W}(m^3) \le w(m^3) + O(1)$ for $m^3 \in \mathbb{K} \sim \Lambda$ with $|m|$ large.

Turn we now to $\log |F'_\Lambda(m^3)|$ for $m^3 \in \Lambda$. For such m we have, when $n_k^3 < m^3 = n_{k+1}^3 < n_{k+2}^3$,

$$F'_\Lambda(m^3) = -\frac{2}{m^3} \prod_{\substack{1 \le l \le k, \\ l > k+2}} \left(1 - \frac{m^6}{n_l^6}\right).$$

Comparison of $\log |F'_\Lambda(m^3)|$ with $\int_1^\infty \log |1 - \frac{m^6}{t^2}| d\nu(t)$ can now be made just as was done previously for $m^3 \in \mathbb{K} \sim \Lambda$, save that *now* $\nu(t)$ increases by 2 on the "omitted" interval $[n_k^3, n_{k+2}^3]$, instead of only by 1 (as *before* on $[n_k^3, n_{k+1}^3]$). The result is that

$$\log |F'_\Lambda(m^3)| \le \log \frac{2}{|m^3|} + w_\circ(m^3) + 2\log|m| + O(1) = w(m^3) - \log|m^3| + O(1)$$

for $m^3 \in \Lambda$ with $|m|$ large. The theorem just referred to now shows that *again* $\log \underline{W}(m^3) \le w(m^3) + O(1)$, *here* for $m^3 \in \Lambda$ with $|m|$ large.

We now include, among the functions $w(z)$ just constructed,

$$V(z) = \int_{(\frac{5}{6})^3}^\infty \log \left|1 - \frac{z^2}{t^2}\right| d(t^{1/3})$$

obtained earlier in this section and corresponding to $F_\mathbb{K}(z) = C_\circ(z)$. And we can state the following:

Theorem 4.3 *Given that the ratios $P(n^3)/W(n^3)$ formed from polynomials P are not $\| \|_\infty$ dense in $\mathscr{C}_\circ(\mathbb{K})$, we have one of the functions $w(z)$ specified above with*

$$\log \underline{W}(n^3) \le w(n^3) + C, \quad n^3 \in \mathbb{K},$$

C being a constant.

And we shall henceforth refer to $w(n^3)$ as a *majorant* for $\log \underline{W}(n^3)$, or simply to $w(z)$ as a *majorant*.

5 Examination of the Majorant

The majorant $w(z)$ specified in the preceding section has some important properties. Leaving aside for now the simple special case $w(z) = V(z)$, we look at

$$w(z) = \int_0^\infty \log\left|1 - \frac{z^2}{t^2}\right| d\nu(t)$$

with $\nu(t)$ the increasing function described in Sect. 4.

Because $\nu(t) = 0(t^{1/3})$ for large t, with $v(0) = 0$, the right side of the preceding relation may be integrated by parts twice successively, leading to the formula

$$w(x) = x \fint_0^\infty \frac{2x}{x^2 - t^2} \frac{\nu(t)}{t} dt = -x \int_0^\infty \log\left|\frac{x+t}{x-t}\right| d(\frac{\nu(t)}{t}),$$

the first integral being a Cauchy principal value. Thence, the

Lemma 5.1

$$w(x) = \mathrm{sgn}(x) \int_0^\infty \frac{2}{\tau^{\frac{2}{3}} + \tau^{\frac{4}{3}} + \tau^2} \frac{\nu(x\tau)}{\tau^{1/3}} d\tau - x^{1/3} \int_0^\infty \log\left|\frac{x+t}{x-t}\right| d\left(\frac{\nu(t)}{t^{1/3}}\right),$$

$x^{1/3}$ being taken as odd.

Proof Refer to the above formula and write, for $x > 0$,

$$\begin{aligned} w(x) &= \fint_0^\infty \frac{2x^2}{x^2 - t^2} \cdot \frac{\nu(t)}{x^{2/3} t^{1/3}} dt + \int_0^\infty \frac{2x^2}{x^2 - t^2} \left(\frac{1}{t^{2/3}} - \frac{1}{x^{2/3}}\right) \frac{\nu(t)}{t^{1/3}} dt \\ &= x^{1/3} \fint_0^\infty \frac{2x}{x^2 - t^2} \cdot \frac{\nu(t)}{t^{1/3}} dt + \int_0^\infty \frac{2x^2 (x^{2/3} - t^{2/3})}{x^{2/3} t^{2/3} (x^2 - t^2)} \cdot \frac{\nu(t)}{t^{1/3}} dt \\ &= -x^{1/3} \int_0^\infty \log\left|\frac{x+t}{x-t}\right| d\left(\frac{\nu(t)}{t^{1/3}}\right) + \int_0^\infty \frac{2x^{4/3}}{t^{2/3}\left(x^{4/3} + x^{2/3} t^{2/3} + t^{4/3}\right)} \cdot \frac{\nu(t)}{t^{1/3}} dt. \end{aligned}$$

Writing $t = x\tau$ in the second of the last two integrals, we find it equal to

$$\int_0^\infty \frac{2}{\tau^{2/3} + \tau^{4/3} + \tau^2} \cdot \frac{\nu(x\tau)}{\tau^{1/3}} d\tau,$$

and the lemma is proved for $x > 0$.

For $x < 0$ we note that $w(x)$ is *even* but $\nu(t)$ is normally taken as *odd*. So sgn$x$ must be placed in front of the *last* integral, but the other term in the lemma's statement is even as it stands. □

In Sect. 4, we found, for the special function

$$w(x) = V_\circ(x) = \int_0^\infty \log|1 - \frac{x^2}{t^2}| d(t^{1/3}),$$

that $V_\circ(x) = \sqrt{3}\pi\,|x|^{1/3}$. From the lemma we thus have

$$\int_0^\infty \frac{2d\tau}{\tau^{2/3}+\tau^{4/3}+\tau^2} = \sqrt{3}\pi.$$

Therefore if, in the formula for $w(x)$, $\nu(t)/t^{1/3}$ is always $\leq$ some $\alpha \leq 1$, the *first* right-hand term in the relation furnished by the lemma is always $\leq \sqrt{3}\pi\,|x|^{1/3}$, and it is increasing for $x \geq 0$. With $\alpha = 1$, the upper limit for that term is practically attained in the extreme case where we have (Sect. 3) $\Lambda = \mathbb{K}$ and $V(x)$ (Sect. 4) standing in place of $w(x)$. A naive guess would be that, in the general case of non-density of the $P(n^3)/W(n^3)$ in $C_\circ(\mathbb{K})$ for polynomials P, the majorant $w(n^3)$ (if not essentially *equal* to $V(n^3)$) would always be $\leq \sqrt{3}\pi\,|n^3|$.

But that is not so, because, when $\Lambda \subseteq \mathbb{K}$ has large gaps, $F_\Lambda(x)/|x|^{1/3}$ (see Sect. 3) and then $w(x)/|x|^{1/3}$ can become *significantly larger* than $\sqrt{3}\pi$ in some places although always remaining $\leq 2\pi$. (For a *general* function $\int_0^\infty \log|1-\frac{x^2}{t^2}|d\mu(t)$ formed from a $\mu(t)$ increasing with $\mu'(t) \leq 1/3t^{2/3}$, a best upper bound for the overshoot is not hard to compute.) This phenomenon is a principal cause of the difficulty in the present work.

One now sees the role of the *second term*, $-x^{1/3}\int_0^\infty \log|\frac{x+t}{x-t}|d\left(\frac{\nu(t)}{t^{1/3}}\right)$, in the formula for $w(x)$ furnished by the lemma. It is to account for the *deviation* of $w(x)$ from the *regularly increasing* function, $\leq \sqrt{3}\pi\,|x|^{1/3}$, given by the *first term* in that formula.

We proceed to examine that second term; it involves a factor of $x^{1/3}$ and then the integral

$$\int_0^\infty \log\left|\frac{x+t}{x-t}\right| d\left(\frac{\nu(t)}{t^{1/3}}\right),$$

the value, for $z = x$, of a *Green potential*,

$$G(z) = \int_0^\infty \log\left|\frac{z+t}{z-t}\right| d\left(\frac{\nu(t)}{t^{1/3}}\right).$$

The first thing to observe is that $G(x)$, $x \in \mathbb{R}$, is an *odd* function, *bounded above and below*. Indeed, we have $\left|\frac{d}{dt}\left(\frac{\nu(t)}{t^{1/3}}\right)\right| \leq c/|t|$ for $t \in \mathbb{R}$, with a constant c. So we have

$$-c\int_0^\infty \log\left|\frac{x+t}{x-t}\right|\frac{dt}{t} \leq G(x) \leq c\int_0^\infty \log\left|\frac{x+t}{x-t}\right|\frac{dt}{t},$$

with $\int_0^\infty \log\left|\frac{x+t}{x-t}\right|\frac{dt}{t} = \int_0^\infty \log\left|\frac{1+s}{1-s}\right|\frac{ds}{s}$, a finite quantity, for $x > 0$.

But the function $G(z)$ has a deeper property. Being clearly *harmonic* in the upper half plane, it turns out to have *a finite Dirichlet integral* there. That property will be crucial for the remainder of this paper, and since the establishment of it is somewhat long, that is carried out in the next section.

6 Estimate of a Dirichlet Integral

A. Borichev and M. Sodin have observed that my earlier different (and somewhat simplistic) treatment of this section's material was wrong because of a stupid mistake in computation. I thank Prof. Sodin for having informed me of that and hope that what follows has now been put to right.

With

$$G(z) = \int_0^\infty \log\left|\frac{z+t}{z-t}\right| \frac{d}{dt}\left(\frac{\nu(t)}{t^{1/3}}\right) dt$$

we proceed to show that the Dirichlet integral

$$\int_0^\infty \int_{-\infty}^\infty |\underset{\rightarrow}{\nabla} G(z)|^2 dx\, dy$$

is finite. It is better to start by taking $A > 0$ *large* and working with

$$\nu_A(t) = \begin{cases} \nu(t), & 0 < t \le A, \\ \dfrac{\nu(A)}{A^{1/3}} t^{1/3}, & t > A, \end{cases}$$

taking

$$G_A(z) = \int_0^\infty \log\left|\frac{z+t}{z-t}\right| \frac{d}{dt}\left(\frac{\nu_A(t)}{t^{1/3}}\right) dt.$$

Here the integration actually runs from $\alpha > 0$ to A, $[0, \alpha]$ being the interval on which $\nu(t) = 0$. Usually $\alpha = 1$ but in one special case (see Sect. 4) $\alpha = (5/6)^3$. And we find an upper bound on

$$\int_0^\infty \int_{-\infty}^\infty |\underset{\rightarrow}{\nabla} G_A(z)|^2 dx\, dy$$

which is *independent* of A. The desired result for $G(z)$ will then follow by making $A \to \infty$.

We have

$$\frac{\partial G_A(z)}{\partial y} = \frac{1}{2}\frac{\partial}{\partial y}\int_0^\infty \log\left|\frac{(x+t)^2+y^2}{(x-t)^2+y^2}\right| \frac{d}{dt}\left(\frac{\nu_A(t)}{t^{1/3}}\right) dt =$$

$$= \int_\alpha^A \left(\frac{y}{(x+t)^2+y^2} - \frac{y}{(x-t)^2+y^2}\right)\frac{d}{dt}\left(\frac{\nu_A(t)}{t^{1/3}}\right) dt,$$

because, $\nu_A(t)/t^{1/3}$ being *even*, $\frac{d}{dt}(\nu_A(t)/t^{1/3})$ is *odd*.

Here $\frac{d}{dt}(\nu_A(t)/t^{1/3})$ vanishes for $|t| < \alpha$ and is $\mathrm{O}(1/|t|)$ elsewhere, so it belongs to $L_2(\mathbb{R})$, making $\int_{-\infty}^\infty \left(\frac{\partial G_A(z)}{\partial x}\right)^2 dx$ *finite*. But then

$$\int_{-\infty}^\infty \left(\frac{\partial G_A(z)}{\partial x}\right)^2 dx = \int_{-\infty}^\infty \left(\frac{\partial G_A(z)}{\partial y}\right)^2 dx,$$

$\partial G_A(z)/\partial x$ being, for each $y > 0$, the *harmonic conjugate* (Hilbert transform) of $\partial G_A(z)/\partial y$. Therefore,

$$\int_0^\infty \int_{-\infty}^\infty |(\underset{\rightarrow}{\nabla} G_A)(z)|^2 dx\, dy = 2\int_0^\infty \int_{-\infty}^\infty \left(\frac{\partial G_A(z)}{\partial y}\right)^2 dx\, dy,$$

and we proceed to estimate the right-hand integral.

From the *second* of the above formulas for $\frac{\partial G_A(z)}{\partial y}$, we have

$$\left(\frac{\partial G_A(z)}{\partial y}\right)^2 = \int_\alpha^A \int_\alpha^A \left(\frac{y}{(x+s)^2+y^2}\cdot\frac{y}{(x+t)^2+y^2}\right.$$
$$+ \frac{y}{(x-s)^2+y^2}\cdot\frac{y}{(x+t)^2+y^2} - \frac{y}{(x+s)^2+y^2}\cdot\frac{y}{(x-t)^2+y^2}$$
$$\left. - \frac{y}{(x-s)^2+y^2}\cdot\frac{y}{(x-t)^2+y^2}\right)\frac{d}{ds}\left(\frac{\nu_A(s)}{s^{1/3}}\right)\frac{d}{dt}\left(\frac{\nu_A(t)}{t^{1/3}}\right).$$

We now do $\frac{1}{\pi}\int_{-\infty}^\infty (\quad)dx$ on *each* of the four products

$$\frac{y}{(x \pm s)^2+y^2}\cdot\frac{y}{(x \pm t)^2+y^2}.$$

By the version of the Poisson kernel for the upper half plane, that will yield successively

$$\frac{2y}{(t-s)^2+4y^2},\quad \frac{2y}{(t+s)^2+4y^2},\quad \frac{2y}{(-s-t)^2+4y^2},\quad \text{and}\quad \frac{2y}{(s-t)^2+4y^2},$$

which, used with the preceding formula, gives

$$\int_{-\infty}^{\infty}\left(\frac{\partial G_A(z)}{\partial y}\right)^2 dx = \pi\int_{\alpha}^{A}\int_{\alpha}^{A}\left(\frac{4y}{(s-t)^2+4y^2}-\frac{4y}{(s+t)^2+4y^2}\right) \times \frac{d}{ds}\left(\frac{\nu_A(s)}{s^{1/3}}\right)\frac{d}{dt}\left(\frac{\nu_A(t)}{t^{1/3}}\right)ds\,dt,$$

remembering that $\frac{d}{ds}\left(\nu_A(s)/s^{1/3}\right)$ and $\frac{d}{dt}\left(\nu_A(t)/t^{1/3}\right)$ are odd.

Finally, we integrate the right side of this last relation with respect to y, from 0 to a large quantity R, using the formula

$$\int_0^R \frac{4y}{(s\pm t)^2+4y^2}dy = \frac{1}{2}\log\left(\frac{(s\pm t)^2+4R^2}{(s\pm t)^2}\right).$$

This will yield

$$\frac{1}{2}\log\left(\frac{(s-t)^2+R^2}{(s+t)^2+R^2}\right)+\frac{1}{2}\log\frac{(s+t)^2}{(s-t)^2}$$

inside the first parentheses of that right side and then, since the integrations against $\frac{d}{ds}\left(\nu_A(s)/s^{1/3}\right)$ and $\frac{d}{dt}\left(\nu_A(t)/t^{1/3}\right)$ run from $\alpha > 0$ to A, we can make $R \to \infty$, getting

$$\pi\int_{\alpha}^{A}\int_{\alpha}^{A}\log\left|\frac{s+t}{s-t}\right|\frac{d}{ds}\left(\frac{\nu_A(s)}{s^{1/3}}\right)\frac{d}{dt}\left(\frac{\nu_A(t)}{t^{1/3}}\right)ds\,dt.$$

This, then, is the value of

$$\int_0^{\infty}\int_{-\infty}^{\infty}\left(\frac{\partial G_A(z)}{\partial y}\right)^2 dx\,dy,$$

and, as observed above,

$$\int_0^{\infty}\int_{-\infty}^{\infty}\left|(\underset{\rightarrow}{\nabla} G_A)(z)\right|^2 dx\,dy,$$

is *twice* that.

So we have the following:

Lemma 6.1 *The integral*

$$\int_0^{\infty}\int_{-\infty}^{\infty}\left|(\underset{\rightarrow}{\nabla} G_A)(z)\right|^2 dx\,dy$$

is equal to

$$2\pi \int_\alpha^\infty \int_\alpha^\infty \log\left|\frac{s+t}{s-t}\right| \frac{d}{ds}\left(\frac{\nu_A(s)}{s^{1/3}}\right) \frac{d}{dt}\left(\frac{\nu_A(t)}{t^{1/3}}\right) ds\, dt.$$

Our object now is to find an upper bound independent of A on the right-hand quantity.

We have

$$\frac{d}{ds}\left(\frac{\nu_A(s)}{s^{1/3}}\right) = \frac{s^{2/3}\nu_A'(s) - 3\nu_A(s)/s^{1/3}}{s}$$

(and the analogous relation for $\frac{d}{dt}\left(\nu_A(t)/t^{1/3}\right)$), where *both quantities* in the numerator are ≥ 0 and *bounded* by numbers *independent* of A. Henceforth we write

$$s^{2/3}\nu_A'(s) = \varphi(s) \quad \text{and} \quad \frac{3\nu_A(s)}{s^{1/3}} = \psi(s)$$

(with analogous relations for the variable t) and seek to estimate

$$\int_\alpha^\infty \left(\int_\alpha^\infty \log\left|\frac{s+t}{s-t}\right| \frac{\varphi(s)}{s} ds\right) \frac{d}{dt}\left(\frac{\nu_A(t)}{t^{1/3}}\right) dt;$$

here $\varphi(s)$ depends *also* on A but is ≥ 0 and has a finite upper bound independent of A. The upper limit on the inner integral *really is* ∞, because, although

$$\frac{d}{ds}\left(\frac{\nu_A(s)}{s^{1/3}}\right) = \frac{\varphi(s)}{s} - \frac{\psi(s)}{s}$$

vanishes for $s > A$, the *individual terms* on the right *do not.*

We have

$$\int_\alpha^\infty \log\left|\frac{s+t}{s-t}\right| \frac{\varphi(s)}{s} ds = \bar{\varphi} \int_\alpha^\infty \log\left|\frac{s+t}{s-t}\right| \frac{ds}{s},$$

where $\bar{\varphi}$, a mean value of $\varphi(s)$, depends on A and t but is ≥ 0 and *bounded above.* Putting $s = ut$ in the right-hand integral, it becomes

$$\bar{\varphi} \int_{\alpha/t}^\infty \log\left|\frac{1+u}{1-u}\right| \frac{du}{u}.$$

For $t > 0$ *fairly small* (making α/t *large*), the last expression, equal to

$$\bar{\varphi} \int_{\alpha/t}^\infty \log\left|\frac{1+\frac{1}{u}}{1-\frac{1}{u}}\right| \frac{du}{u},$$

is

$$\bar{\varphi}\int_{\alpha/t}^{\infty}\left(\frac{2}{u^2}+O\left(\frac{1}{u^3}\right)\right)du = \Omega t + O(t^2),$$

where Ω, having $\bar{\varphi}$ as a factor, depends on A but is positive and bounded above.

The right side of this last relation is now integrated against $\frac{d}{dt}\left(\nu_A(t)/t^{1/3}\right)$, from $t = 0$ to, say, $t = 2$, taking now $A > 2$. That the derivative is of both signs doesn't matter at this point, and we get

$$\int_0^2 (\Omega t + O(t^2))\frac{O(1)}{t}dt \le K_1,$$

with a numerical constant K_1 chosen large enough to ensure the relation's holding for all values > 2 of A.

We must now look at

$$\bar{\varphi}\int_{\alpha/t}^{\infty}\log\left|\frac{1+u}{1-u}\right|\frac{du}{u}$$

for $t \ge 2$. This we can write as

$$\bar{\varphi}\int_0^{\infty}\log\left|\frac{1+u}{1-u}\right|\frac{du}{u} - \bar{\varphi}\int_0^{\alpha/t}\log\left|\frac{1+u}{1-u}\right|\frac{du}{u}.$$

With

$$C = \int_0^{\infty}\log\left|\frac{1+u}{1-u}\right|\frac{du}{u},$$

this is, α/t being $\le \alpha/2$, of the form

$$\bar{\varphi}C - \bar{\varphi}\left(\frac{2\alpha}{t}+O\left(\frac{1}{t^2}\right)\right),$$

where the $O(1/t^2)$ term does not involve any parameters. This is to be integrated from 2 to ∞ against $\frac{d}{dt}\left(\nu_A(t)/t^{1/3}\right)$, where $A > 2$. (Later on, A will be made to tend to ∞.) The derivative vanishes unless $\alpha < t < A$ and here $t \ge 2 > \alpha$. So, for A large, the integral takes the form

$$\bar{\varphi}C\int_2^A \frac{d}{dt}\left(\frac{\nu_A(t)}{t^{1/3}}\right)dt - \bar{\varphi}\int_2^A\left(\frac{2\alpha}{t}+O\Big(\frac{1}{t^2}\Big)\right)\frac{d}{dt}\left(\frac{\nu_A(t)}{t^{1/3}}\right)dt.$$

Here

$$\int_2^A \frac{d}{dt}\left(\frac{\nu_A(t)}{t^{1/3}}\right)dt = \frac{\nu(A)}{A^{1/3}} - \frac{\nu(2)}{2^{1/3}}$$

is bounded above by 1 and below by -1. And the *second* of the integrals in the above expression is of the form

$$\int_2^A \left(\frac{2\alpha}{t} + \frac{O(1)}{t^2} \right) \frac{O(1)}{t} dt,$$

where the $O(1)$ terms depend on A (and the second one also on t) but are bounded above (and *below*) by quantities independent of A (and t). The value of the last integral, equal to

$$\int_2^A \left(\frac{2\alpha O(1)}{t^2} + \frac{O(1)}{t^3} \right) dt,$$

is thus *finite*, depending on A but bounded (*above and below*) by numbers independent of A. The same is then true of

$$\bar{\varphi} C \int_2^A \frac{d}{dt} \left(\frac{\nu_A(t)}{t^{1/3}} \right) dt - \bar{\varphi} \int_2^A \left(\frac{2\alpha}{t} + O\!\left(\frac{1}{t^2}\right) \right) \frac{d}{dt} \left(\frac{\nu_A(t)}{t^{1/3}} \right) dt.$$

Combined with the former estimate of

$$\int_0^2 (\Omega t + O(t^2)) \frac{d}{dt} \left(\frac{\nu_A(t)}{t^{1/3}} \right) dt,$$

this yields the

Lemma 6.2 *The double integral*

$$\int_\alpha^\infty \int_\alpha^\infty \log \left| \frac{s+t}{s-t} \right| \frac{d}{ds} \left(\frac{\nu_A(s)}{s^{1/3}} \right) \frac{d}{dt} \left(\frac{\nu_A(t)}{t^{1/3}} \right) ds\, dt$$

is bounded above by a quantity independent of A.

From this and the preceding lemma, we now have the

Theorem 6.1

$$\int_0^\infty \int_{-\infty}^\infty |\underset{\rightarrow}{\nabla} G_A(z)|^2 dx\, dy \le K,$$

a constant independent of A, for all $A > \alpha$.

In particular, for any $H > 0$ and *all* A,

$$\int_0^H \int_{-H}^H |(\underset{\rightarrow}{\nabla} G_A)(z)|^2 dx\, dy \le K.$$

Making now $A \to \infty$, whereby $|(\underset{\to}{\nabla} G_A)(z)| \to |(\underset{\to}{\nabla} G)(z)|$, we have, by Fatou's lemma, $\int_0^H \int_{-H}^H |(\underset{\to}{\nabla} G)(z)|^2 dx\, dy \le K$, this for all H. So finally, we have the following:

Theorem 6.2 *The Dirichlet integral*

$$\int_0^\infty \int_{-\infty}^\infty |(\underset{\to}{\nabla} G)(z)|^2 dx\, dy$$

is finite.

7 Further Examination of the Majorant

The majorant

$$w(x) = -x^{1/3} \int_0^\infty \log\left|\frac{x+t}{x-t}\right| d\left(\frac{\nu(t)}{t^{1/3}}\right) + \operatorname{sgn}(x) \int_0^\infty \frac{2}{\tau^{2/3} + \tau^{4/3} + \tau^2} \cdot \frac{\nu(x\tau)}{\tau^{1/3}} d\tau$$

furnished by the lemma of Sect. 5 has a general description in terms of its properties. As already observed, the *second* term on the right is a regularly increasing function. It is in every case actually $< \sqrt{3}\pi |x|^{1/3}$ but *never equal to that* for $x > 0$, since

$$\sqrt{3}\pi |x|^{1/3} = V_\circ(x) = \int_0^\infty \log\left|1 - \frac{x^2}{t^2}\right| \frac{dt}{3t^{2/3}},$$

whilst

$$\nu'(t) \le \frac{1}{3t^{2/3}}$$

for $t > 0$ is not always *equal* to $1/3t^{2/3}$ there but is sometimes zero. This also makes the *derivative* of that second term with respect to x *less*, for $x > 0$, than $\sqrt{3}\pi/3x^{2/3}$.

As remarked previously, the *first* term indicates how much $w(x)$ *deviates* from the very regular second one. For

$$G(x) = \int_0^\infty \log\left|\frac{x+t}{x-t}\right| d\left(\frac{\nu(t)}{t^{1/3}}\right)$$

we now know that $G(x)$ is bounded above and below. It is continuous. Indeed, since $\nu(t)$ vanishes in any event for $0 < t < (5/6)^3$ and $\left|\frac{d}{dt}\left(\nu(t)/t^{1/3}\right)\right| \le \frac{1}{t}$, the integrals

$$I_\epsilon(x) = \int_{(1-\epsilon)x}^{(1+\epsilon)x} \log\left|\frac{x+t}{x-t}\right| d\left(\frac{\nu(t)}{t^{1/3}}\right)$$

tend uniformly to 0 as $\epsilon \to 0$ on any interval $0 < a \le x \le b$, whereas the *differences* $G(x) - I_\epsilon(x)$ are obviously continuous.

It is especially important that the harmonic extension of $G(x)$ to the upper half plane has *a finite Dirichlet integral.*

We now write

$$J(x) = \operatorname{sgn}(x) \int_0^\infty \frac{2}{\tau^{2/3} + \tau^{4/3} + \tau^2} \cdot \frac{\nu(\tau x)}{\tau^{1/3}} d\tau$$

and see that, save for the special case where $w(x) = V(x)$ (see Sect. 4), the majorant $w(x)$ can be expressed as $-x^{1/3}G(x) + J(x)$, where $G(x)$ and $J(x)$ have the special properties described above.

But there remains that special case. Again the lemma of Sect. 5 applies, and we have

$$V(x) = -\,x^{1/3} \int_0^\infty \log\left|\frac{x+t}{x-t}\right| \frac{d}{dt}\left(\frac{(t^{1/3}-\alpha)^+}{t^{1/3}}\right) dt$$
$$+ \int_0^\infty \frac{2}{\tau^{2/3} + \tau^{4/3} + \tau^2} \cdot \frac{\left((x\tau)^{1/3} - \alpha\right)^+}{\tau^{1/3}} d\tau,$$

where $\alpha = 5/6$. Here the *second term* is well of the form $J(x)$ and has the properties described above for that function.

The *first term* is equal to

$$-\,x^{1/3} \cdot \int_\alpha^\infty \log\left|\frac{x+t}{x-t}\right| \frac{\alpha}{3t^{4/3}} dt.$$

With $t = x\tau$ this becomes

$$-\int_{\alpha/x}^\infty \log\left|\frac{1+\tau}{1-\tau}\right| \frac{\alpha d\tau}{3\tau^{4/3}}.$$

This integral is always absolutely convergent, and for $x \to \infty$ it tends to

$$-\int_0^\infty \log\left|\frac{1+\tau}{1-\tau}\right| \frac{\alpha d\tau}{\tau^{4/3}},$$

a finite negative quantity. For $x \to 0$ it behaves like

$$-\int_{\alpha/x}^\infty \frac{2\alpha d\tau}{3\tau^{7/3}} = C^{te} \cdot x^{4/3},$$

which tends to 0. So the first term is *negative* and $O(1)$. *Here* we have $w(x) = O(1) + J(x)$ with $J(x)$ one of the functions described previously.

Now what we *require* of a majorant $w(x)$ is that $w(n^3) + C$ be $\geq \log \underline{W}(n^3)$ for the Mergelian minorant $\underline{W}(n^3)$. So we have no need to keep the $O(1)$ term in the last formula for $w(x)$, especially since it is negative, and may as well drop it. We have thus covered all of the cases of incompleteness and have the following:

Theorem 7.1 *If the ratios $P(n^3)/W(n^3)$ with polynomials P are not dense in $\mathscr{C}_\circ(\mathbb{K})$, the Mergelian minorant $\underline{W}(n^3)$ of $W(n^3)$ is subject to the inequality*

$$\log \underline{W}(n^3) \leq C^{te} - nG(n^3) + J(n^3),$$

where $G(x)$ and $J(x)$ have the following properties:

i) $G(x)$ is odd, bounded above and below, and continuous.
ii) The harmonic (Poisson integral) extension of $G(x)$ to the upper half plane has a finite Dirichlet integral.
iii) $J(x)$ is even and, for $x > 0$, increasing and $< \sqrt{3}\pi x^{1/3}$.
iv) For $x > 0$, $J'(x) < \sqrt{3}\pi/3x^{2/3}$.
v) The harmonic extension $\mathrm{v}(z) + J(z)$ of $-x^{1/3}G(x) + J(x)$ to $\Im z > 0$ is an increasing function of $y = \Im z$ for each z there and

$$\frac{\mathrm{v}(iy) + J(iy)}{\log y} \longrightarrow \infty$$

as $y \to \infty$ (refer to Sect. 5).

8 Beginning of the Discussion of Sufficiency

It will turn out that the necessary conditions for incompleteness furnished by the theorem of the preceding section also become *sufficient* when point iii of that theorem's statement is somewhat strengthened. Seeing this will involve the use of more Dirichlet integrals and it is best to first consider how that comes about.

Let us look more closely at the odd function $G(x)$ having the properties stated in the above-mentioned theorem.

Given a finite set E of bounded closed intervals on $\mathbb{R}$ with $E = -E$ and $0 \notin E$, we take the harmonic measure $d\omega_D(x, 0)$ of E for the domain $D = \mathbb{C} \sim E$, as seen from $0 \in D$. And we will need the integral $\int_E G(x) \cdot x^{1/3} d\omega_D(x, 0)$. As we shall see, this can be expressed in terms of $(\underset{\rightarrow}{\nabla} G)(z)$, where $G(z)$ is the harmonic extension of $G(x)$ to $\Im z > 0$.

We henceforth write, for $\Im z \geq 0$,

$$U(z) = \frac{1}{2\pi} \int_E \log \left| \frac{z+t}{z-t} \right| \cdot t^{1/3} d\omega_D(t, 0).$$

Here, as always, $t^{1/3}$ is taken as odd. The above integral involving $G(x)$ can be expressed in terms of G and U. We have indeed

$$\frac{\partial}{\partial y}\log\left|\frac{z+t}{z-t}\right| = \frac{1}{2}\frac{\partial}{\partial y}\log\left(\frac{(x+t)^2+y^2}{(x-t)^2+y^2}\right) = \frac{y}{(x+t)^2+y^2} - \frac{y}{(x-t)^2+y^2},$$

so for $y > 0$,

$$-\frac{\partial U(z)}{\partial y} = \frac{1}{2\pi}\int_E \left(\frac{y}{(x-t)^2+y^2} - \frac{y}{(x+t)^2+y^2}\right)\cdot t^{1/3} d\omega_D(t,0).$$

Here $d\omega_D(t,0)/dt$ is *even* in t since $E = -E$, but $t^{1/3}$ and the part of the integrand in () are both odd in t, and the last relation becomes

$$-\frac{\partial U(z)}{\partial y} = \frac{1}{\pi}\int_E \frac{y}{(x-t)^2+y^2}\cdot t^{1/3} d\omega_D(t).$$

From this we see that as $y \downarrow 0$, $-\partial U(x+iy)/\partial y$ tends to $x^{1/3} d\omega_D(x,0)/dx$ save at the endpoints of the intervals in E. And especially in the following sense: given any function $f(x)$ continuous on an interval $(-A, A) \supset E$, we have

$$-\int_{-A}^{A} f(x)\frac{\partial U(x+iy)}{\partial y}dx \to \int_E f(x)\cdot x^{1/3}\frac{d\omega_D(x,0)}{dx}dx$$

as $y \downarrow 0$.

This property will be used with the function $G(x)$ and shows that when $(-A, A) \supset E$,

$$-\int_{-A}^{A} G(x)\frac{\partial U(x+iy)}{\partial y}dx \to \int_E G(x)\cdot x^{1/3} d\omega_D(x,0)$$

as $y \downarrow 0$. And we have the following:

Lemma 8.1

$$-\int_{-A}^{A} G(x+iy)\frac{\partial U(x+iy)}{\partial y}dx$$

tends to

$$\int_E G(x)\cdot x^{1/3} d\omega_D(x,0)$$

as $y \downarrow 0$ provided that $E \subset (-A, A)$.

Proof By property $i)$ stated in the theorem of Sect. 7, $G(x)$ is continuous and bounded, so it follows by Poisson's formula for the harmonic extension of $G(x)$ to $\Im z > 0$ that G(x+iy) tends uniformly to $G(x)$ for $-A \le x \le A$ when $y \downarrow 0$. At the same time,

$$\int_{-\infty}^{\infty} \left| \frac{\partial U(x+iy)}{\partial y} \right| dx \le \frac{1}{\pi} \int_{-\infty}^{\infty} \int_E \frac{y|t|^{1/3}}{(x-t)^2+y^2} \cdot \frac{d\omega_D(t,0)}{dt} dt\, dx$$

$$= \int_E |t|^{1/3} d\omega_D(t,0) \le p^{1/3},$$

where p is such that $E \subseteq [-p, p]$. The result follows. □

From now on, p (taken as $\frac{1}{2}$ the diameter of E) will be kept *fixed*, but $A > p$ will be allowed to increase. With $h > 0$ and $B > h$ large, take now the rectangle Q shown here:

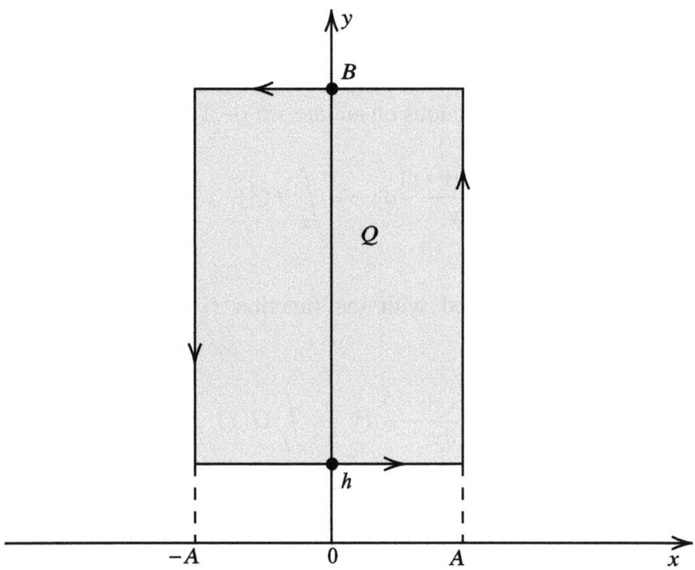

Here the vertical sides are at $x = A$ and $x = -A$, and as always $A > p$. And we consider the line integral

$$S = \oint_{\partial Q} \left(G(z) \frac{\partial U(z)}{\partial x} dy - G(z) \frac{\partial U(z)}{\partial y} dx \right)$$

taken in the usual counterclockwise direction.

Note first of all that the contribution to S from the *lower* horizontal side of Q tends, by the above lemma, to

$$\int_E G(x)\cdot x^{1/3}\frac{d\omega_D(x,0)}{dx}dx$$

as $h \downarrow 0$. And it turns out that this is the *main* contribution.

For the contribution of the *top* side to S we have the following:

Lemma 8.2 *For $A > p$ (see above) fixed and $B \to \infty$,*

$$\left|-\int_{-A}^{A} G(x+iB)U_y(x+iB)\,dx\right| \le \text{const.}\,\frac{p^{1/3}A^2}{B^2}.$$

Proof $G(z)$ is the harmonic extension of $G(x)$ to the upper half plane and one of G's properties is that $|G(x)| \le$ const. for $x \in \mathbb{R}$. Also $G(x)$ is *odd*. So for $x > 0$, we have

$$|G(x+iB)| \le \frac{\text{const.}}{\pi}\int_0^\infty\left(\frac{B}{(x-t)^2+B^2}-\frac{B}{(x+t)^2+B^2}\right)dt,$$

and the same, mutatis mutandis, for $x < 0$.

For $0 \le x \le A$, the right side of this relation is $\le (\text{const.}/\pi)/(\varphi-\psi)$, where φ and ψ are the two angles shown in the following diagram:

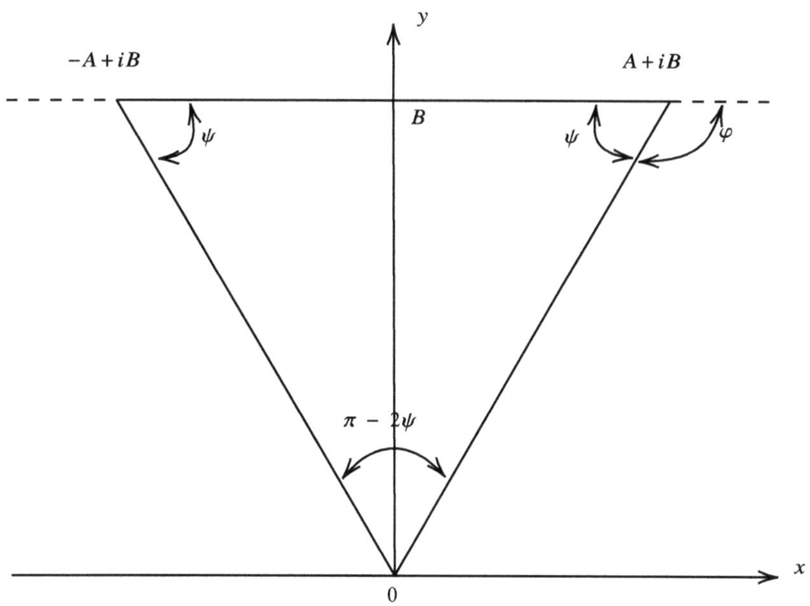

We have $(\varphi - \psi)/\pi = (\pi - 2\psi)/\pi$ and for $A > p$ *fixed* and B *large*, this is $\leq \text{const.}A/B$. Again, for $y = B$,

$$\left|\frac{\partial U(z)}{\partial y}\right| \leq \frac{1}{\pi}\int_E \frac{1}{B}|t|^{1/3} d\omega_D(t,0) \leq \frac{p^{1/3}}{\pi B}.$$

The lemma follows immediately on substituting these two estimates into the integral it concerns. □

There remain the contributions to S from the two *vertical* sides of Q. Dealing with *these* involves somewhat more work; fortunately, rather crude estimates are sufficient.

Consider the line integral along the *right side*, equal to $\int_0^B G(A+iy)\, U_x(A+iy)\, dy$. For $B > A$, which we always assume here, we see from the diagram used in proving the preceding lemma that $|G(A+iy)| < \text{const.}\, A/y$ for $y > A$, but for $0 < y \leq A$ the properties of G give only $|G(A+iy)| \leq \text{const.}$ So we have to break up the preceding integral as $\int_0^A + \int_A^B$. Let us look first at the second term. Leaving aside a multiplicative constant, it is in absolute value $\leq \int_A^B \frac{A}{y}|U_x(A+iy)|\, dy$.

Here,

$$U_x(x+iy) = \frac{1}{2\pi}\int_E \frac{\partial}{\partial x}\log\left|\frac{x+t+iy}{x-t+iy}\right| \cdot t^{1/3}\, d\omega_D(t,0)$$

and we have, by a calculation like one made in proving the first lemma of this section, that

$$\frac{\partial}{\partial x}\log\left|\frac{x+t+iy}{x-t+iy}\right| = \frac{x+t}{(x+t)^2+y^2} - \frac{x-t}{(x-t)^2+y^2}.$$

This is

$$\frac{(x+t)(x-t)^2 + y^2(x+t) - (x-t)(x+t)^2 - y^2(x-t)}{(x^2+t^2+y^2+2xt)(x^2+t^2+y^2-2xt)}$$

$$= \frac{2y^2t + (x^2-t^2)\big((x-t)-(x+t)\big)}{(x^2+y^2+t^2)^2 - 4x^2t^2}$$

$$= 2t \cdot \frac{y^2 - (x^2-t^2)}{(x^2+y^2+t^2)^2 - 4x^2t^2}.$$

We want this and its *absolute value* for $x = A$. The latter is

$$\leq 2|t|\frac{A^2+y^2+t^2}{(A^2+y^2+t^2)^2 - 4A^2t^2}.$$

Here we recall that in doing the integration involving t, that t ranges over $E \subseteq [-p, p]$. And we are going to *assume henceforth that* A is *several times larger* than p; later on it will be made to tend to ∞. Hence the preceding expression will be

$$\leq 2p \frac{A^2 + y^2 + p^2}{(A^2 + y^2)^2 - 4A^2 p^2}$$

which in turn comes out to be

$$\text{const.} \left(\frac{p}{A^2 + y^2} + \frac{p^3}{(A^2 + y^2)^2} \right),$$

where the constant is small if P/A is small.

Integration of this over E against $|t^{1/3}| d\omega_D(t, 0)$ yields

$$\text{const.} \left(\frac{p^{4/3}}{A^2 + y^2} + \frac{p^{10/3}}{(A^2 + y^2)^2} \right).$$

Thence, the integral

$$\int_A^B |G(A + iy)| \, |U_x(A + iy)| \, dy$$

is seen to be

$$\leq \int_A^\infty \frac{A}{y} \left(\frac{p^{4/3}}{A^2 + y^2} + \frac{p^{10/3}}{(A^2 + y^2)^2} \right) dy$$

for *all* $B > A$. Putting $y = A\eta$ in the last expression, it becomes

$$\frac{1}{A} \int_1^\infty \frac{p^{4/3}}{\eta(1 + \eta^2)} \, d\eta + \frac{1}{A^3} \int_1^\infty \frac{p^{10/3}}{\eta(1 + \eta^2)^2} \, d\eta = C_1 \frac{p^{4/3}}{A} + C_2 \frac{p^{10/3}}{A^3}$$

where C_1 and C_2 are constants.

Now we can estimate

$$\int_0^A |G(A + iy)| \, |U_x(A + iy)| \, dy.$$

Using the estimate obtained above for $U_x(A + iy)$ and $|G(A + iy)| \leq$ const. for $0 \leq y \leq A$, we obtain the integral

$$\int_0^A \left(\frac{p^{4/3}}{A^2 + y^2} + \frac{p^{10/3}}{(A^2 + y^2)^2} \right) dy$$

which, with the substitution made before, comes out to

$$\frac{1}{A}\int_0^1 \frac{p^{4/3}}{1+\eta^2}\,d\eta + \frac{1}{A^3}\int_0^1 \frac{p^{10/3}}{(1+\eta^2)^2}\,d\eta = C_3\frac{p^{4/3}}{A} + C_4\frac{p^{10/3}}{A^3},$$

the same as for the previous integral but with different constants.

And we have proved the following:

Lemma 8.3 *For the contribution of the right vertical side of Q to the line integral S around ∂Q, we have, assuming always that p, the demi-diameter of E, is fixed and A is at least several times larger than p, the estimate*

$$\int_h^B |G(A+iy)|\,|U_x(A+iy)|\,dy \le \text{const.}\left(\frac{p^{4/3}}{A} + \frac{p^{10/3}}{A^3}\right)$$

for all $h \in (0, A)$ and $B > A$.

Remark 8.1 Because $G(x+iy)$, like $G(x)$, is odd in x and $U_x(x+iy)$ even in x ($U(x+iy)$ being odd in x), we have exactly the same estimate for the line integral along the left vertical side of ∂Q.

Putting now together the results obtained by the preceding three lemmas and the discussion preceding them, we have the following:

Theorem 8.1 *For E (and hence p) fixed, A several times larger than p, B several times larger than A, and $h > 0$ tending to 0, we have*

$$\oint_{\partial Q} \big(G(z)U_x(z)\,dy - G(z)U_y(z)\,dx\big)$$

$$= \int_E G(x)\cdot x^{1/3}\,d\omega_D(x,0) + o(1) + O\left(\frac{p^{1/3}A^2}{B^2}\right) + O\left(\frac{p^{4/3}}{A}\right) + O\left(\frac{p^{10/3}}{A^3}\right).$$

Here, the $o(1)$ tends to 0 with h when A is fixed, and A is taken as fixed when $B \to \infty$.

Remark 8.2 One should refer to the figure preceding the second lemma in this section.

Here is where Dirichlet integrals come in. By Green's theorem, we have

$$\oint_{\partial Q} (G(z)U_x(z)\,dy - G(z)U_y(z)\,dx)$$

$$= \iint_Q \left(\frac{\partial}{\partial x}(G(z)U_x(z)) + \frac{\partial}{\partial y}(G(z)U_y(z))\right) dx\,dy$$

$$= \iint_Q \left(\frac{\partial G(z)}{\partial x}\frac{\partial U(z)}{\partial x} + G(z)\frac{\partial^2 U(z)}{\partial x^2} + \frac{\partial G(z)}{\partial y}\frac{\partial U(z)}{\partial y} + G(z)\frac{\partial^2 U(z)}{\partial y^2}\right) dx\,dy.$$

But $U(z)$ is, by the formula for it, *harmonic* in $\Im z > 0$, so

$$\frac{\partial^2 U(z)}{\widehat{\partial x}^2} + \frac{\partial^2 U(z)}{\widehat{\partial y}^2} = 0$$

there, and hence in Q. And, for the above theorem, we have the following:

Corollary 8.1 *Under the conditions of the preceding theorem,*

$$\int_E G(x) \cdot x^{1/3} d\omega_D(x, 0) = \iint_Q (\underset{\rightarrow}{\nabla} G(z)) \cdot (\underset{\rightarrow}{\nabla} U(z))\, dx\, dy$$
$$+ o(1) + O\left(\frac{p^{1/3} A^2}{B^2}\right) + O\left(\frac{p^{4/3}}{A}\right) + O\left(\frac{p^{10/3}}{A^3}\right).$$

Now, *provided* that

$$\iint_{\Im z>0} |(\underset{\rightarrow}{\nabla} U)(z)|^2\, dx\, dy$$

is finite, we can go through the following steps:

First, make $h \downarrow 0$ and get

$$\int_E G(x) \cdot x^{1/3} d\omega_D(x, 0) = \iint_{Q_0} (\underset{\rightarrow}{\nabla} G(z)) \cdot (\underset{\rightarrow}{\nabla} U(z))\, dx\, dy$$
$$+ O\left(\frac{p^{1/3} A^2}{B^2}\right) + O\left(\frac{p^{4/3}}{A}\right) + O\left(\frac{p^{10/3}}{A^3}\right).$$

where $Q_0 = \{(x, y) : -A < x < A \;\&\; 0 < y < B\}$. This is because

$$\iint_{\Im z>0} |(\underset{\rightarrow}{\nabla} G)(z)|^2\, dx\, dy < \infty,$$

and by Schwarz' inequality.

Then, make $B \to \infty$, yielding

$$\int_E G(x) \cdot x^{1/3} d\omega_D(x, 0) = \int_0^\infty \int_{-A}^{A} (\underset{\rightarrow}{\nabla} G(z)) \cdot (\underset{\rightarrow}{\nabla} U(z))\, dx\, dy$$
$$+ O\left(\frac{p^{4/3}}{A}\right) + O\left(\frac{p^{10/3}}{A^3}\right).$$

And finally, make $A \to \infty$ to get

$$\int_E G(x) \cdot x^{1/3} d\omega_D(x, 0) = \int_0^\infty \int_{-\infty}^{\infty} (\underset{\rightarrow}{\nabla} G(z)) \cdot (\underset{\rightarrow}{\nabla} U(z))\, dx\, dy.$$

The integral on the left involving harmonic measure has been expressed as a Dirichlet inner product.

Our next step must now be to verify that the Dirichlet integral $\iint_{\Im z>0} |(\underset{\to}{\nabla} U(z))|^2 dx\, dy$ is finite. And so we proceed.

9 A Parameter and Its Use

Given the collection $E = -E$ on $\mathbb{R}$ with $0 \notin E$, we continue to write $\mathcal{D} = \mathbb{C} \sim E$ and to denote by $d\omega_{\mathcal{D}}(t, 0)$ the harmonic measure on E, *in* $\mathcal{D}$ and evaluated at 0. The expression $t^{1/3} d\omega_{\mathcal{D}}(t, 0)$ plays such an important role in the present work that we need a parameter for it.

We henceforth assume that E, a finite collection of intervals, may contain two *infinite* ones, $[A, \infty)$ and $(-\infty, -A]$, where $A > 0$. And we then define an *odd* function $\Omega_{\mathcal{D}}(x)$, $x \in \mathbb{R}$, by putting

$$\Omega_{\mathcal{D}}(x) = \begin{cases} \omega_{\mathcal{D}}([x, \infty), 0)\,, & x > 0 \\ -\omega_{\mathcal{D}}([|x|, \infty), 0)\,, & x < 0 \end{cases}$$

and note that

$$\int_E |x|^{1/3} d\omega_{\mathcal{D}}(x, 0) = -2 \int_0^\infty x^{1/3} d\Omega_{\mathcal{D}}(x).$$

If E includes the interval $[A, \infty)$, it is known that $\Omega_{\mathcal{D}}(x)$ is $O(1/x)$ for large x, so integration by parts of the right-hand integral converts it to

$$\frac{2}{3} \int_0^\infty \frac{\Omega_{\mathcal{D}}(x)}{x^{2/3}}\, dx,$$

always finite, although no finite upper bound can be set for it concerning different domains $\mathcal{D}$. From now on, we write

$$\lambda_{\mathcal{D}} = \int_0^\infty \frac{\Omega_{\mathcal{D}}(x)}{x^{2/3}}\, dx.$$

This is our parameter, and it will appear frequently in what follows.

It can, in the first place, be expressed in terms of *Green's function*, $G_{\mathcal{D}}(\cdot, \cdot)$, for $\mathcal{D}$. (*Do not confound* $G_{\mathcal{D}}(\cdot, \cdot)$ with G, the function appearing frequently in the preceding section, and which shall do so again.) We have

$$G_{\mathcal{D}}(z, 0) = \log \frac{1}{|z|} + \int_E \log|z - t|\, d\omega_{\mathcal{D}}(t, 0) = \int_0^\infty \log\left|1 - \frac{t^2}{z^2}\right| d\omega_{\mathcal{D}}(t, 0)$$

thanks to the symmetry of $\mathscr{D}$. This is

$$-\int_0^\infty \log\left|1-\frac{t^2}{z^2}\right| d\Omega_{\mathscr{D}}(t)$$

which, after the usual integration by parts, becomes

$$-\fint_0^\infty \frac{2t}{x^2-t^2}\Omega_{\mathscr{D}}(t)\,dt = -\fint_{-\infty}^\infty \frac{\Omega_{\mathscr{D}}(t)}{x-t}\,dt,$$

$\Omega_{\mathscr{D}}(t)$ being odd. This exhibits $G_{\mathscr{D}}(x,0)$ as $-\pi$ times $\tilde{\Omega}_{\mathscr{D}}(x)$, the *Hilbert transform* of $\Omega_{\mathscr{D}}(x)$.

We have, however,

$$\lambda_{\mathscr{D}} = \frac{1}{2}\fint_{-\infty}^\infty \Omega_{\mathscr{D}}(x)\frac{\operatorname{sgn}(x)}{|x|^{2/3}}\,dx,$$

and we have seen in Sect. 4 that $\operatorname{sgn}(x)/|x|^{2/3}$ is the Hilbert transform of $1/\sqrt{3}|x|^{2/3}$. And this last function is the sum of *two*: *one* (its restriction to $(-\infty,-1]\cup[1,\infty)$) in $L_p(\mathbb{R})$ for $p>\frac{3}{2}$ and the *other* (its restriction to $[-1,1]$) in $L_p(\mathbb{R})$ for $p<\frac{3}{2}$. *At the same time*, $\Omega_{\mathscr{D}}(x)$ is in $L_q(\mathbb{R})$ for all $q>1$. This justifies use of the duality relation for Hilbert transforms, and we have

$$\int_{-\infty}^\infty \Omega_{\mathscr{D}}(x)\frac{\operatorname{sgn}(x)}{|x|^{2/3}}\,dx = -\int_{-\infty}^\infty \tilde{\Omega}_{\mathscr{D}}(x)\frac{dx}{\sqrt{3}|x|^{2/3}} = \frac{1}{\sqrt{3}\pi}\int_{-\infty}^\infty \frac{G_{\mathscr{D}}(x,0)}{|x|^{2/3}}\,dx.$$

Thus,

$$\lambda_{\mathscr{D}} = \frac{1}{2\sqrt{3}\pi}\int_{-\infty}^\infty \frac{G_{\mathscr{D}}(x,0)}{|x|^{2/3}}\,dx.$$

We proceed to deduce some relations involving the parameter $\lambda_{\mathscr{D}}$. In the first place, the fact that $\Omega_{\mathscr{D}}(x)$ decreases for $x>0$ and the definition of $\lambda_{\mathscr{D}}$ make

$$3x^{1/3}\Omega_{\mathscr{D}}(x) \le \int_0^x \frac{\Omega_{\mathscr{D}}(t)}{t^{2/3}}\,dt \le \lambda_{\mathscr{D}},$$

yielding the (weak) estimate

$$\Omega_{\mathscr{D}}(x) \le \frac{\lambda_{\mathscr{D}}}{3x^{1/3}}$$

for $x>0$.

Again, from the formula

$$G_{\mathscr{D}}(x,0) = -\fint_0^\infty \frac{2}{x^2-t^2}\cdot t\Omega_{\mathscr{D}}(t)\,dt$$

we get, by (still another!) integration by parts,

$$G_{\mathscr{D}}(x,0) = \frac{1}{x}\int_0^\infty \log\left|\frac{x+t}{x-t}\right| d(t\Omega_{\mathscr{D}}(t)).$$

From this and the preceding, we now have the following:

Lemma 9.1

$$\int_0^\infty \int_0^\infty \log\left|\frac{x+t}{x-t}\right| d(t,\Omega_{\mathscr{D}}(t))\frac{d(x\Omega_{\mathscr{D}}(x))}{x^{4/3}} \;\le\; \frac{\pi}{\sqrt{3}}\lambda_{\mathscr{D}}^2.$$

Proof The iterated integral in question is $\int_0^\infty G_{\mathscr{D}}(x,0)d(x\Omega_{\mathscr{D}}(x))/x^{1/3}$, but $G_{\mathscr{D}}(x,0)=0$ on $E=\partial\mathscr{D}$ and $\Omega'_{\mathscr{D}}(x)$ vanishes on $\mathbb{R}\sim\partial\mathscr{D}$. The last expression is thus just

$$\int_0^\infty G_{\mathscr{D}}(x,0)\frac{\Omega_{\mathscr{D}}(x)}{x^{1/3}}\,dx$$

which, $G_{\mathscr{D}}(,)$ being ≥ 0, is in term $\le (\lambda_{\mathscr{D}}/3)\int_0^\infty G_{\mathscr{D}}(x)dx/x^{2/3}$ by the above estimate for $\Omega_{\mathscr{D}}(x)$.

Again, $G_{\mathscr{D}}(x,0)$ is even, so we have finally the upper bound

$$\frac{\lambda_{\mathscr{D}}}{6}\int_{-\infty}^\infty \frac{G_{\mathscr{D}}(x,0)}{|x|^{2/3}}\,dx = \frac{\pi}{\sqrt{3}}\lambda_{\mathscr{D}}^2$$

on the integral in the statement. □

To proceed with further work, we shall need to know that the double integral figuring in the lemma is *absolutely convergent*, and that will also be required of certain symmetrized versions of it. These are of the form

$$\int_0^\infty \int_0^\infty \log\left|\frac{x+t}{x-t}\right| d\rho(t)\,d\rho(x),$$

where the measure $d\rho$ may be *signed*; it is known that the absolute convergence of such expressions makes them *positive*.

Since $G_{\mathscr{D}}(x,0) = -\pi\tilde{\Omega}_{\mathscr{D}}(x)$, the inversion formula for the Hilbert transform yields

$$\Omega_{\mathscr{D}}(x) = \frac{1}{\pi^2}\fint_{-\infty}^\infty \frac{G_{\mathscr{D}}(t,0)}{x-t}\,dt.$$

If now E is *bounded*, we have $\Omega_{\mathscr{D}}(x) = 0$ for all x sufficiently large in absolute value. *Otherwise*, E includes two *infinite* intervals $(-\infty, -A]$, $[A, \infty)$ and then $G_{\mathscr{D}}(t, 0)$ vanishes *outside* the bounded set $\mathbb{R} \sim E$. In that event we get from the preceding relation the convergent power series

$$\Omega_{\mathscr{D}}(x) = \sum_{k=0}^{\infty} \frac{a_{2k+1}}{x^{2k+1}}$$

valid for $|x|$ large. Here the even powers are missing because $G_{\mathscr{D}}(t, 0)$, like E, is even and

$$a_{2k+1} = \frac{1}{\pi^2} \int_{\mathbb{R}\sim E} t^{2k} G_{\mathscr{D}}(t, 0)\, dt.$$

The quantities a_{2k+1} are all ≥ 0 since $G_{\mathscr{D}}(t, 0) \geq 0$.

Lemma 9.2 *The integrals*

$$\int_0^{\infty} \int_0^{\infty} \log \left| \frac{x+t}{x-t} \right| \cdot d(t\Omega_{\mathscr{D}}(t)) \cdot \frac{d(x\Omega_{\mathscr{D}}(x))}{x^{4/3}},$$

$$\int_0^{\infty} \int_0^{\infty} \log \left| \frac{x+t}{x-t} \right| \cdot \frac{d(t\Omega_{\mathscr{D}}(t))}{t^{2/3}} \cdot \frac{d(x\Omega_{\mathscr{D}}(x))}{x^{2/3}},$$

and

$$\int_0^{\infty} \int_0^{\infty} \log \left| \frac{x+t}{x-t} \right| \cdot t^{1/3} d\Omega_{\mathscr{D}}(t) \cdot x^{1/3} d\Omega_{\mathscr{D}}(x)$$

are absolutely convergent.

Proof Consider the *first* of the above integrals. We write

$$\int_0^{\infty} \log \left| \frac{x+t}{x-t} \right| \cdot |d(t\Omega_{\mathscr{D}}(t))|$$

as

$$\int_0^{R} \log \left| \frac{x+t}{x-t} \right| \cdot |d(t\Omega_{\mathscr{D}}(t))| + \int_R^{\infty} \log \left| \frac{x+t}{x-t} \right| \cdot |d(t\Omega_{\mathscr{D}}(t))|$$

where R is taken large enough to make the above-mentioned power series available for $\Omega_{\mathscr{D}}(t)$ when $t \geq R$. (If the infinite interval $[A, \infty)$ does not occur among those of E, the second integral will *vanish* for $R > A$.) From that series (when it is needed), we get

$$d(t\Omega_{\mathscr{D}}(t)) = -\left(\frac{2a_3}{t^3} + \frac{4a_5}{t^5} + \cdots\right) dt,$$

where the coefficients are all ≥ 0. Thence, for large enough R,

$$\int_R^\infty \log\left|\frac{x+t}{x-t}\right| \cdot |d(t\Omega_{\mathscr{D}}(t))| \leq \text{const.} \int_R^\infty \log\left|\frac{x+t}{x-t}\right| \cdot \frac{dt}{t^3}$$
$$= \frac{\text{const.}}{x^2} \int_{R/x}^\infty \log\left|\frac{1+\tau}{1-\tau}\right| \cdot \frac{d\tau}{\tau^3}.$$

When $x \to 0$ (whereby $R/x \to \infty$!), the expression on the right is

$$\leq \frac{\text{const.}}{x^2} \int_{R/x}^\infty \frac{d\tau}{\tau^4} = \frac{\text{const.}}{x^2} \left(\frac{x}{R}\right)^3 = O(x),$$

and when $x \to \infty$, it behaves like

$$\frac{\text{const.}}{x^2} \int_{R/x}^1 \frac{d\tau}{\tau^2} = O\left(\frac{1}{x}\right).$$

The power series expansion of $\Omega_D(x)$ shows that

$$\frac{|d(x\Omega_{\mathscr{D}}(x))|}{x^{4/3}} \leq \frac{\text{const.}}{x^{13/3}}\, dx$$

for large x, so convergence of

$$\int_R^\infty \int_R^\infty \log\left|\frac{x+t}{x-t}\right| \cdot |d(t\Omega_{\mathscr{D}}(t))| \cdot \frac{|d(x\Omega_{\mathscr{D}}(x))|}{x^{4/3}}$$

is manifest.

When $x > 0$ is near 0, $d(x\Omega_{\mathscr{D}}(x)) = \frac{1}{2}dx$ (sic!), and the convergence of

$$\int_0^\eta \int_R^\infty \log\left|\frac{x+t}{x-t}\right| \cdot |d(t\Omega_{\mathscr{D}}(t))| \cdot \frac{|d(x\Omega_{\mathscr{D}}(x))|}{x^{4/3}}$$

for small $\eta > 0$ thus depends on that of $\int_0^\eta O(x)dx/x^{4/3}$ (see above), which, however, *does hold.*

Concerning

$$\int_\eta^R \int_R^\infty \log\left|\frac{x+t}{x-t}\right| \cdot |d(t\Omega_{\mathscr{D}}(t))| \cdot \frac{|d(x\Omega_{\mathscr{D}}(x))|}{x^{4/3}},$$

the only obstacle to its convergence would be caused by the behaviour of $\Omega'_{\mathscr{D}}(t)$ near the endpoints of the finite number of intervals belonging to E. If $a > 0$ is one of these, $|\Omega'_{\mathscr{D}}(t)|$ is like const.$/\sqrt{|t-a|}$ for t near a. But *there*,

$$\int_{a-k}^{a+k} \log\left|\frac{x+t}{x-t}\right| |d(t\Omega_{\mathscr{D}}(t))|$$

is bounded for small $k > 0$, even for x near a, or equal to it. And again,

$$\int_{a-k}^{a+k} |d(x\Omega_{\mathscr{D}}(x))|$$

is finite.

We thus see that the preceding double integral *is* finite and thus finally that

$$\int_0^R \int_R^\infty \log\left|\frac{x+t}{x-t}\right| \cdot |d(t\Omega_{\mathscr{D}}(t))| \cdot \frac{|d(x\Omega_{\mathscr{D}}(x))|}{x^{4/3}}$$

does converge.

Looking at $\int_0^R \log\left|\frac{x+t}{x-t}\right| |d(t\Omega_{\mathscr{D}}(t))|$, we see that it is $O(1/x)$ for $x \to \infty$, so its integral against $|d(x\Omega_{\mathscr{D}}(x))|/x^{4/3}$ is certainly convergent at ∞.

For $t > 0$ near 0, $|d(t\Omega_{\mathscr{D}}(t))| = \frac{1}{2}dt$, so for small $x > 0$ the preceding integral behaves like

$$\begin{aligned}\frac{1}{2}\int_0^1 \log\left|\frac{x+t}{x-t}\right| dt + O(x) &= \frac{x}{2}\int_0^2 \log\left|\frac{1+\tau}{1-\tau}\right| d\tau \\ &\quad + \frac{x}{2}\int_2^{1/x} \log\left|\frac{1+\frac{1}{\tau}}{1-\frac{1}{\tau}}\right| d\tau + O(x) \\ &\le Cx + C'x\int_2^{1/x}\frac{d\tau}{\tau} = C'x + Cx\log\frac{1}{x}.\end{aligned}$$

Here, $\int_0^1 x\log(1/x)\cdot x^{-4/3}\,dx$ is *still* finite, so

$$\int_0^\infty \int_0^R \log\left|\frac{x+t}{x-t}\right| \cdot |d(t\Omega_{\mathscr{D}}(t))| \cdot \frac{|d(x\Omega_{\mathscr{D}}(x))|}{x^{4/3}}$$

is *also* finite, and absolute convergence of the *first* double integral in the lemma's statement is confirmed.

We turn, therefore, to the *second* double integral figuring therein. From the power series expansion of $\Omega_{\mathscr{D}}(x)$ for large x, we have, for $x \ge R$,

$$\frac{|d(x\Omega_{\mathscr{D}}(x)|}{x^{2/3}} \le \frac{\text{const.}}{x^{11/3}}\,dx$$

(with, of course, a similar relation when x is replaced by $t \geq R$). This estimate is valid when E does not include two infinite intervals; there it holds with const. = 0.

Near 0, where $\Omega_{\mathscr{D}}(x) = \frac{1}{2}$, we get

$$\frac{|d(x\Omega_{\mathscr{D}}(x))|}{x^{2/3}} = \frac{dx}{2x^{2/3}}.$$

Using the *first* of these relations (with t in place of x), we find without trouble that for $x \to 0$,

$$\int_R^\infty \log\left|\frac{x+t}{x-t}\right| \cdot \frac{|d(t\Omega_{\mathscr{D}}(t))|}{t^{2/3}} \leq \frac{\text{const.}}{x^{8/3}} \int_{R/x}^\infty \log\left|\frac{1+\tau}{1-\tau}\right| \frac{d\tau}{\tau^{11/3}} = O(x),$$

while, for $x \to \infty$, the integral is

$$O\left(\frac{1}{x^{8/3}}\left(\frac{x}{R}\right)^{5/3}\right) = O\left(\frac{1}{x}\right).$$

Integration of $\int_R^\infty$ against $|d(x\Omega_{\mathscr{D}}(x))|/x^{2/3}$ from 0 to ∞ thus leads to a finite result in view of the above relations.

There remains

$$\int_0^R \log\left|\frac{x+t}{x-t}\right| \frac{|d(t\Omega_{\mathscr{D}}(t))|}{t^{2/3}}.$$

As in the preceding case, this is $O(1/x)$ for $x \to \infty$, and for $x \to 0$ we use the *second* of the previous relations (again with t in place of x) to obtain the estimate

$$\begin{aligned}\text{const.} \int_0^1 \log\left|\frac{x+t}{x-t}\right| \frac{dt}{t^{2/3}} + O(x) &= \text{const.}\, x^{1/3} \int_0^{1/x} \log\left|\frac{1+\tau}{1-\tau}\right| \frac{d\tau}{\tau^{2/3}} + O(x)\\ &= O(x^{1/3})\end{aligned}$$

for the integral $\int_0^R$. Integration of $\int_0^R$ against $|d(x\Omega_{\mathscr{D}}(x))|/x^{2/3}$ (from 0 to ∞) is thus *again alright* (both at 0 and at ∞), and we see that

$$\int_0^\infty \int_0^R \log\left|\frac{x+t}{x-t}\right| \cdot \frac{|d(t\Omega_{\mathscr{D}}(t))|}{t^{2/3}} \cdot \frac{|d(x\Omega_{\mathscr{D}}(x))|}{x^{2/3}}$$

is finite. Absolute convergence of our *second* double integral is verified.

Absolute convergence of the *third* double integral can be deduced from that of the second one by using a property of the bilinear form

$$\langle d\rho, d\sigma\rangle = \int_0^\infty \int_0^\infty \log\left|\frac{x+t}{x-t}\right| d\rho(t)\, d\sigma(x).$$

It is *positive definite*, at least when absolute convergence holds. That may be seen by observing that then, for the function

$$F(z) = \int_0^\infty \log\left|\frac{z+t}{z-t}\right| d\rho(t),$$

we have

$$\int_0^\infty F(x)\, d\rho(x) = \frac{1}{2\pi} \iint\limits_{\Re z>0} |(\underset{\rightarrow}{\nabla} F(z))|^2\, dx\, dy.$$

This (well-known) relation may in turn be deduced by following a procedure used near the end of Sect. 8, replacing both G and U by the function F.

From positive definiteness of the form $\langle d\rho, d\sigma\rangle$, we see that it is a (pre-Hilbert space) *inner* product, making $\sqrt{\langle d\rho, d\rho\rangle} = \|d\rho\|$ *a norm* and that Schwarz' inequality holds.

In our present situation, we observe that

$$\frac{d(t\Omega_{\mathscr{D}}(t))}{t^{2/3}} = \frac{\Omega_{\mathscr{D}}(t)}{t^{2/3}} dt + t^{1/3} d\Omega_{\mathscr{D}}(t),$$

making

$$t^{1/3}|d\Omega_{\mathscr{D}}(t)| \le \frac{|d(t\Omega_{\mathscr{D}}(t))|}{t^{2/3}} dt + \frac{\Omega_{\mathscr{D}}(t)}{t^{2/3}} dt.$$

So here we put $d\mu(t) = t^{1/3}|d\Omega_{\mathscr{D}}(t)|$,

$$d\rho(t) = \frac{d(t\Omega_{\mathscr{D}}(t))|}{t^{2/3}}, \quad d\sigma(t) = \frac{\Omega_{\mathscr{D}}(t)dt}{t^{2/3}},$$

and use the relation

$$|d\mu| \le |d\rho| + |d\sigma|$$

to find that

$$\begin{aligned}
&\int_0^\infty \int_0^\infty \log\left|\frac{x+t}{x-t}\right| \cdot t^{1/3}|d\Omega_{\mathscr{D}}(t)| \cdot x^{1/3}|d\Omega_{\mathscr{D}}(x)| \\
&\le 2\int_0^\infty \int_0^\infty \log\left|\frac{x+t}{x-t}\right| \cdot \frac{|d(t\Omega_{\mathscr{D}}(t))|}{t^{2/3}} \cdot \frac{|d(x\Omega_{\mathscr{D}}(x))|}{x^{2/3}} \\
&\quad + 2\int_0^\infty \int_0^\infty \log\left|\frac{x+t}{x-t}\right| \cdot \frac{\Omega_{\mathscr{D}}(t)}{t^{2/3}} \cdot \frac{\Omega_{\mathscr{D}}(x)}{x^{2/3}}\, dt\, dx.
\end{aligned}$$

Knowing now already that the *first* integral on the *right* is finite, we will have the same for the integral on the left if that is so for the *second* one on the right.

But this is true. We have, since $\Omega_{\mathscr{D}}(t) \le \lambda_{\mathscr{D}}/3t^{1/3}$ by an earlier estimate in this section,

$$\int_0^\infty \log\left|\frac{x+t}{x-t}\right| \cdot \frac{\Omega_{\mathscr{D}}(t)}{t^{2/3}}\,dt \le \int_0^\infty \log\left|\frac{x+t}{x-t}\right| \cdot \frac{\lambda_{\mathscr{D}}}{3t}\,dt$$
$$= \frac{\lambda_{\mathscr{D}}}{3}\int_0^\infty \log\left|\frac{1+\tau}{1-\tau}\right| \frac{d\tau}{\tau} = C\frac{\lambda_{\mathscr{D}}}{3}.$$

And then

$$\int_0^\infty C\frac{\lambda_{\mathscr{D}}}{3} \cdot \frac{\Omega_{\mathscr{D}}(x)}{x^{2/3}}\,dx = \frac{C\lambda_{\mathscr{D}}^2}{3}$$

by the definition of $\lambda_{\mathscr{D}}$. Hence

$$\int_0^\infty\int_0^\infty \log\left|\frac{x+t}{x-t}\right| \cdot t^{1/3}|d\Omega_{\mathscr{D}}(t)| \cdot x^{1/3}|d\Omega_{\mathscr{D}}(x)|$$

is finite, and the lemma is proved. □

The *first* of the preceding two lemmas has furnished an *upper bound*, involving $\lambda_{\mathscr{D}}^2$, on the *first integral* figuring in the *second* lemma's statement. The *other two* integrals in that statement admit a similar bound.

Theorem 9.1 *There is a numerical constant* $\alpha > 0$ *such that*

$$\int_0^\infty\int_0^\infty \log\left|\frac{x+t}{x-t}\right| \cdot d(t\Omega_{\mathscr{D}}(t)) \cdot \frac{d(x\Omega_{\mathscr{D}}(x))}{x^{4/3}}$$
$$\ge \alpha\int_0^\infty\int_0^\infty \log\left|\frac{x+t}{x-t}\right| \cdot \frac{d(t\Omega_{\mathscr{D}}(t))}{t^{2/3}} \cdot \frac{d(x\Omega_{\mathscr{D}}(x))}{x^{2/3}}.$$

Proof By the use of Mellin transforms. (See [5]; consideration of Mellin transforms occurs throughout that book.)

According to the preceding lemma, both integrals in question are absolutely convergent, so the order of integrations in the *first* one can be changed. *Doing* that, and then interchanging t and x, one obtains the expression

$$\int_0^\infty\int_0^\infty \log\left|\frac{x+t}{x-t}\right| \cdot \frac{d(t\Omega_{\mathscr{D}}(t))}{t^{4/3}} \cdot d(x\Omega_{\mathscr{D}}(x)),$$

so

$$\int_0^\infty\int_0^\infty \log\left|\frac{x+t}{x-t}\right| \cdot d(t\Omega_{\mathscr{D}}(t)) \cdot \frac{d(x\Omega_{\mathscr{D}}(x))}{x^{4/3}}$$

$$= \frac{1}{2}\int_0^\infty \int_0^\infty \left(\left(\frac{x}{t}\right)^{2/3} + \left(\frac{t}{x}\right)^{2/3}\right) \log\left|\frac{x+t}{x-t}\right| \cdot \frac{d(t\Omega_{\mathscr{D}}(t))}{t^{2/3}} \cdot \frac{d(x\Omega_{\mathscr{D}}(x))}{x^{2/3}},$$

and we have to show that this is

$$\geq \alpha \int_0^\infty \int_0^\infty \log\left|\frac{x+t}{x-t}\right| \cdot \frac{d(t\Omega_{\mathscr{D}}(t))}{t^{2/3}} \cdot \frac{d(x\Omega_{\mathscr{D}}(x))}{x^{2/3}}$$

for a suitable numerical $\alpha > 0$. The integrals involved here are absolutely convergent and the measures figuring in them absolutely continuous; with an appropriate choice of a function Φ defined on $(0, \infty)$, they can be cast in the respective forms

$$\frac{1}{2}\int_0^\infty \int_0^\infty \left(\left(\frac{x}{t}\right)^{2/3} + \left(\frac{t}{x}\right)^{2/3}\right) \log\left|\frac{x+t}{x-t}\right| \frac{\Phi(t)\Phi(x)}{tx}\, dt\, dx$$

and

$$\int_0^\infty \int_0^\infty \log\left|\frac{x+t}{x-t}\right| \frac{\Phi(t)\Phi(x)}{tx}\, dt\, dx.$$

We have to show that the first of these dominates a positive multiple of the *second.*

Both integrals are of (multiplicative) *convolution type*, being of the form

$$\int_0^\infty \int_0^\infty P\left(\frac{x}{t}\right) \frac{\Phi(t)}{t}\frac{\Phi(x)}{x}\, dt\, dx.$$

Since they are *absolutely convergent*, it suffices to verify the domination in question for functions $\Phi(t)$ of *compact support* in $(0, \infty)$, even though that is not the case for our *present* function Φ. For this it suffices to compare the *Mellin transforms* of the two multiplicative convolution "kernels" involved in them, viz.,

$$\log\left|\frac{1+\tau}{1-\tau}\right| \quad \text{and} \quad \frac{\tau^{2/3}+\tau^{-2/3}}{2} \log\left|\frac{1+\tau}{1-\tau}\right|.$$

The *first* of these has the Mellin transform

$$K(\lambda) = \int_0^\infty \log\left|\frac{1+\tau}{1-\tau}\right| \cdot \frac{\tau^{i\lambda}}{\tau} d\tau = \frac{\pi}{\lambda} \tanh\frac{\pi\lambda}{2};$$

that can be seen by noting that

$$K(\lambda) = 2\Re \int_0^1 \log\left|\frac{1+\tau}{1-\tau}\right| \cdot \frac{\tau^{i\lambda}}{\tau} d\tau$$

and working out the expression on the right by integrating $\log\left|\frac{1+z}{1-z}\right| z^{i\lambda-1}\,dz$ around the semicircle with diameter $[-1, 1]$ lying in the closed upper half plane.

We still have, however,

$$L(\lambda) = \frac{1}{2}\int_0^\infty (\tau^{2/3} + \tau^{-2/3}) \log\left|\frac{1+\tau}{1-\tau}\right| \frac{\tau^{i\lambda}}{\tau}\,d\tau$$
$$= \Re K\left(\lambda + \frac{2i}{3}\right) = \Re \frac{\pi \tanh\left(\frac{\pi\lambda}{2} + \frac{\pi i}{3}\right)}{\lambda + \frac{2i}{3}}.$$

It is now claimed that $L(\lambda)/K(\lambda) \geq \alpha$ on R, where $\alpha > 0$.

Let us compute

$$\frac{L(\lambda)}{K(\lambda)} = \frac{\lambda}{\tanh\frac{\pi\lambda}{2}} \cdot \Re \frac{\tanh\left(\frac{\pi\lambda}{2} + \frac{\pi i}{3}\right)}{\lambda + \frac{2i}{3}}$$
$$= \frac{\lambda}{\tanh\frac{\pi\lambda}{2}} \cdot \Re\left(\frac{\tanh\frac{\pi\lambda}{2} + i\tan\frac{\pi}{3}}{\left(\lambda + \frac{2i}{3}\right)\left(1 + i\tan\frac{\pi}{3}\tanh\frac{\pi\lambda}{2}\right)}\right) =$$
$$= \frac{\lambda}{\tanh\frac{\pi\lambda}{2}} \cdot \Re\left(\frac{\tanh\frac{\pi\lambda}{2} + \sqrt{3}i}{\left(\lambda + \frac{2i}{3}\right)\left(1 + \sqrt{3}i\tanh\frac{\pi\lambda}{2}\right)}\right)$$
$$= \Re\left(\frac{1 + \sqrt{3}i\coth\frac{\pi\lambda}{2}}{1 + \sqrt{3}i\tanh\frac{\pi\lambda}{2}} \cdot \frac{\lambda}{\lambda + \frac{2i}{3}}\right)$$
$$= \Re\left(\frac{\left(4 + 3i\left(\coth\frac{\pi\lambda}{2} - \tanh\frac{\pi\lambda}{2}\right)\right)\left(\lambda^2 - \frac{2}{3}i\lambda\right)}{\left(1 + 3\tanh^2\frac{\pi\lambda}{2}\right)\left(\lambda^2 + \frac{4}{9}\right)}\right)$$
$$= \frac{4\lambda^2 + \frac{2}{\sqrt{3}}\coth\frac{\pi\lambda}{2}\left(1 - \tanh^2\frac{\pi\lambda}{2}\right)}{\left(1 + 3\tanh^2\frac{\pi\lambda}{2}\right)\lambda^2 + \frac{4}{3}\tanh^2\frac{\pi\lambda}{2} + \frac{4}{9}}$$
$$= \frac{4\lambda^2 + \frac{2\lambda}{\sqrt{3}\tanh(\pi\lambda/2)}\operatorname{sech}^2\frac{\pi\lambda}{2}}{\left(1 + 3\tanh^2\frac{\pi\lambda}{2}\right)\lambda^2 + \frac{4}{3}\tanh^2\frac{\pi\lambda}{2} + \frac{4}{9}}.$$

This expression is > 0 for $\lambda \in \mathbb{R}$ and *even*. When $\lambda \to 0$, it tends to $(4/\sqrt{3}\pi)/(4/9) = 3\sqrt{3}/\pi$ and when $\lambda \to \infty$, it tends to 1. Therefore $L(\lambda)/K(\lambda)$ has a *strictly positive lower bound*, say, α, for $\lambda \in \mathbb{R}$, and that implies that

$$\frac{1}{2}\int_0^\infty \int_0^\infty \left(\left(\frac{x}{t}\right)^{2/3} + \left(\frac{t}{x}\right)^{2/3}\right) \log\left|\frac{x+t}{x-t}\right| \cdot \frac{\Phi(t)}{t} \cdot \frac{\Phi(x)}{x}\,dt\,dx$$
$$\geq \alpha \int_0^\infty \int_0^\infty \log\left|\frac{x+t}{x-t}\right| \cdot \frac{\Phi(t)}{t} \cdot \frac{\Phi(x)}{x}\,dt\,dx,$$

first for continuous Φ of compact support in $(0, \infty)$ and *then* for the functions Φ involved in our present situation. The theorem is thus proved. □

Referring to the first lemma of this section, we get the following:

Corollary 9.1

$$\int_0^\infty \int_0^\infty \log\left|\frac{x+t}{x-t}\right| \cdot \frac{d(t\Omega_{\mathscr{D}}(t))}{t^{2/3}} \cdot \frac{d(x\Omega_{\mathscr{D}}(x))}{x^{2/3}} \le C\lambda_{\mathscr{D}}^2$$

where C *is a numerical constant.*

From this we get, as at the end of the preceding lemma's proof, but with $d\Omega_{\mathscr{D}}(t)$ in place of $|d\Omega_{\mathscr{D}}(t)|$ and $d(t\Omega_{\mathscr{D}}(t))$ instead of $|d(t\Omega_{\mathscr{D}}(t))|$, the following:

Theorem 9.2

$$\int_0^\infty \int_0^\infty \log\left|\frac{x+t}{x-t}\right| \cdot t^{1/3} d\Omega_{\mathscr{D}}(t) \cdot x^{1/3} d\Omega_{\mathscr{D}}(x) \le C'\lambda_{\mathscr{D}}^2,$$

where C' *is a numerical constant.*

Now we recall, from the beginning of Sect. 8, the formula

$$U(z) = \frac{1}{2\pi}\int_E \log\left|\frac{z+t}{z-t}\right| \cdot t^{1/3}\, d\omega_{\mathscr{D}}(t,0),$$

where $\Im z > 0$. In our present notation, the expression on the right is equal to

$$-\frac{1}{\pi}\int_0^\infty \log\left|\frac{z+t}{z-t}\right| \cdot t^{1/3}\, d\Omega_{\mathscr{D}}(t)$$

because $E = -E, 0 \notin E$, and $\log|(z+t)(z-t)|$, like $t^{1/3}$, is, for each z, an *odd* function of t.

For the above function $U(z)$, it is known (cf the work with Green's theorem near the end of Sect. 8) that

$$\iint_{\Im z>0} |(\underset{\rightarrow}{\nabla} U)(z)|^2\, dx\, dy = \frac{1}{2\pi}\int_E\int_E \log\left|\frac{z+t}{z-t}\right| \cdot t^{1/3}\, d\omega_{\mathscr{D}}(t,0) \cdot x^{1/3} d\omega_{\mathscr{D}}(x,0).$$

This result is important for several reasons. First of all it shows that the relation

$$\int_E G(x) \cdot x^{1/3}\, d\omega_{\mathscr{D}}(x,0) = \int_0^\infty \int_{-\infty}^\infty (\underset{\rightarrow}{\nabla} G(z)) \cdot (\underset{\rightarrow}{\nabla} U(z))\, dx\, dy,$$

appearing at the end of Sect. 8, *is valid*, finiteness of G's Dirichlet integral being one of that function's properties. Again, applying Schwarz' inequality to the *right side* of the preceding relation, we find, referring to the last theorem, that

$$\left|\int_E G(x)\cdot x^{1/3}\,d\omega_{\mathscr{D}}(x,0)\right| \le \text{const.}\lambda_{\mathscr{D}}\sqrt{\iint\limits_{\Im z>0} |(\underset{\rightarrow}{\nabla} G)(z)|^2\,dx\,dy},$$

and this relation will be important for us later on.

For the work at the end of Sect. 8, it was *assumed*, however, that the set E is *bounded*. It is now time to lift that restriction, extending the above observations to the case where two infinite intervals, $[A,\infty)$ and $(-\infty,-A]$, are present in the set E used to form

$$U(z) = \frac{1}{2\pi}\int_E \log\left|\frac{z+t}{z-t}\right|\cdot t^{1/3}\,d\omega_{\mathscr{D}}(t,0).$$

(Here, as always, $\mathscr{D}$ denotes $\mathbb{C}\sim E$.)

To do this, we first replace $[A,\infty)$ and $(-\infty,-A]$ by $[A,N]$ and $[-N,A]$, respectively, where $N > A$, obtaining, instead of E, a set E_N. We then write $\mathscr{D}_N = \mathbb{C}\sim E_N$ and put

$$U_N(z) = \frac{1}{2\pi}\int_{E_N} \log\left|\frac{z+t}{z-t}\right|\cdot t^{1/3}\,d\omega_{\mathscr{D}_N}(t,0).$$

Corresponding to the parameter

$$\lambda_{\mathscr{D}} = \frac{3}{2}\int_E |x|^{1/3} d\omega_{\mathscr{D}}(x,0)$$

(see the beginning of this section), we now write

$$\lambda_N = \frac{3}{2}\int_{E_N} |x|^{1/3} d\omega_{\mathscr{D}_N}(x,0)$$

These formulas show that the λ_N *increase* with N and tend to $\lambda_{\mathscr{D}}$.

For each N, we *do have*

$$\int_{E_N} G(x)\cdot x^{1/3}\,d\omega_{\mathscr{D}_N}(x,0) = \iint\limits_{\Im z>0} \big((\underset{\rightarrow}{\nabla} G)(z)\big)\cdot\big((\underset{\rightarrow}{\nabla} U_N)(z)\big)\,dx\,dy.$$

And we fix R so large as to make

$$\int_{|z|>R,\,\Im z>0} |(\underset{\rightarrow}{\nabla} G)(z)|^2\,dx\,dy < \epsilon^2.$$

Observe that, as $N\to\infty$,

$$\iint\limits_{\substack{|z|\leq R \\ |\Im z|>0}} \big((\underset{\rightarrow}{\nabla} G)(z)\big)\cdot\big((\underset{\rightarrow}{\nabla} U_N)(z)\big)\,dx\,dy \longrightarrow \iint\limits_{\substack{|z|\leq R \\ |\Im z|>0}} \big((\underset{\rightarrow}{\nabla} G)(z)\big)\cdot\big((\underset{\rightarrow}{\nabla} U)(z)\big)\,dx\,dy$$

whilst

$$\left|\iint\limits_{\substack{|z|>R \\ \Im z>0}} \big((\underset{\rightarrow}{\nabla} G)(z)\big)\cdot\big((\underset{\rightarrow}{\nabla} U_N)(z)\big)\,dx\,dy\right| \leq \epsilon \sqrt{\iint\limits_{\Im z>0} \big|\underset{\rightarrow}{\nabla} U_N(z)\big|^2\,dx\,dy}$$

$$= \text{const.}\,\epsilon\,\lambda_N \leq \text{const.}\,\epsilon\,\lambda_{\mathscr{D}}.$$

Again,

$$\left|\iint\limits_{\substack{|z|>R \\ \Im z>0}} \big((\underset{\rightarrow}{\nabla} G)(z)\big)\cdot\big((\underset{\rightarrow}{\nabla} U)(z)\big)\,dx\,dy\right| \leq \text{const.}\,\epsilon\,\lambda_{\mathscr{D}}.$$

Now, $G(x)$ being bounded, we have

$$\int_{E_N} G(x)\cdot x^{1/3}\,d\omega_{\mathscr{D}_N}(x,0) \underset{N}{\rightarrow} \int_E G(x)\cdot x^{1/3}\,d\omega_{\mathscr{D}}(x,0)$$

so finally

$$\int_E G(x)\cdot x^{1/3}\,d\omega_{\mathscr{D}}(x,0) = \iint\limits_{\substack{|z|<R \\ \Im z>0}} \big((\underset{\rightarrow}{\nabla} G)(z)\big)\cdot\big((\underset{\rightarrow}{\nabla} U_N)(z)\big)\,dx\,dy + \text{const.}\,\epsilon\,\lambda_{\mathscr{D}}.$$

Since the Dirichlet integrals of G and U are both finite, this last relation yields the desired result on making $R \to \infty$, whereby $\epsilon \to 0$.

10 Partition of the Function *G*

We proceed to show that the function $G(x)$, with its properties (see the end of Sect. 7), can, given any $\epsilon > 0$, be expressed as $G_1(x)+G_2(x)$, where $|x|^{1/3}|G_1(x)|$ is *bounded* and the harmonic extension of $G_2(x)$ (also bounded) to $\Im z > 0$ has a *Dirichlet integral* $< \epsilon^2$.

To this end we take, for each large $A > 0$, the function

$$\Phi_A(x) = \begin{cases} 1, & -A \le x \le A \\ A^{1/3}/|x|^{1/3}, & |x| > A, \end{cases}$$

and also put

$$\Psi_A(x) = \begin{cases} 0, & |x| \le A \\ 1 - \Phi_A(x), & |x| > A. \end{cases}$$

Of course we have $\Phi_A(x) + \Psi_A(x) = 1$.

In order to estimate the Dirichlet integral corresponding to Ψ_A's harmonic extension to the upper half plane, we put $\breve{\Psi}_A(z) = 0$ for $|z| < A$ and $\Im z > 0$. But then, for $|z| > A$ we take

$$\breve{\Psi}_A(z) = 1 - \frac{A^{1/3}}{|z|^{1/3}},$$

thus making $\breve{\Psi}_A(z)$ a function of $r = |z|$. And we denote that function by $\psi(r)$.

We have, for $\Im z > 0$, $|\underset{\rightarrow}{\nabla}\breve{\Psi}_A(z)| = |\psi'(r)|$, making

$$\iint_{\Im z>0} |\underset{\rightarrow}{\nabla}\breve{\Psi}_A(z)|^2\,dx\,dy = \int_0^{\pi}\int_0^{\infty} (\psi'(r))^2 r\,dr\,d\theta$$

$$= \pi A^{2/3}\int_A^{\infty}\left(\frac{dr^{-1/3}}{dr}\right)^2 r\,dr = \pi A^{2/3}\int_A^{\infty}\frac{1}{9}\left(\frac{1}{r^{4/3}}\right)^2 r\,dr$$

$$= \frac{\pi A^{2/3}}{9}\int_A^{\infty}\frac{dr}{r^{5/3}} = \frac{\pi A^{2/3}}{9}\cdot\frac{3}{2}A^{-2/3} = \frac{\pi}{6}.$$

This quantity is independent of A.

From this if follows, since $\breve{\Psi}_A(z)$, $G(x)$ (and $G(z)$) are *bounded*, that with

$$\underset{\rightarrow}{\nabla}(\breve{\Psi}_A(z)G(z)) = G(z)\underset{\rightarrow}{\nabla}\breve{\Psi}_A(z) + \breve{\Psi}_A(z)\underset{\rightarrow}{\nabla}G(z),$$

we have

$$\iint_{\Im z>0} \left|\underset{\rightarrow}{\nabla}(\breve{\Psi}_A(z)G(z))\right|^2 dx\,dy$$

$$\le 2\iint_{\Im z>0}(G(z))^2\left|\underset{\rightarrow}{\nabla}\breve{\Psi}_A(z)\right|^2 dx\,dy + 2\iint_{\Im z>0}(\breve{\Psi}_A(z))^2\left|\underset{\rightarrow}{\nabla}G(z)\right|^2 dx\,dy$$

$$\le \frac{\pi}{6}\cdot C + 2\iint_{\Im z>0}\left|\underset{\rightarrow}{\nabla}G(z)\right|^2 dx\,dy \le K,$$

a quantity independent of A.

Here $\breve{\Psi}_A(z)G(z)$ is *not* the *harmonic* extension of $\Psi_A(x)G(x)$ to $\Im z > 0$. But *Dirichlet's principle* states that the *actual* harmonic extension of $\Psi_A(x)G(x)$ to $\Im z > 0$ must have a *smaller* Dirichlet integral than the quantity just estimated, i.e. *surely* of value $< K$. (True, the *derivatives* of $\breve{\Psi}_A(z)G(z)$ have a discontinuity along the semicircle $|z| = A$ in the upper half plane, but, insofar as Dirichlet's principle is concerned, *that is allowed*. See [11], p.447, and Lemma C.1., p.449.)

The product $\Psi_A(x)G(x)$ *vanishes* for $-A \le x \le A$. This, together with the result just found, implies that when $A \to \infty$, the quantities $|\underset{\rightarrow}{\nabla}(\Psi_A G)(z)|$ (where $(\Psi_A G)(z)$ designates the *harmonic* extension of $\Psi_A(x)G(x)$ to $\Im z > 0$) *tend weakly* to 0 in $L_2(\Im z > 0)$. And as we have just seen,

$$\iint_{\Im z>0} |\underset{\rightarrow}{\nabla}(\Psi_A G)(z)|^2 dx\, dy \le K$$

for all A.

From a known result, it thence follows that *a sequence of convex linear combinations of* the $\underset{\rightarrow}{\nabla}(\Psi_A G)(z)$ *tends to* 0 *in the norm* of $L_2(\Im z > 0)$. Those linear combinations are of the form

$$\sum_{k=1}^{N} \lambda_k \underset{\rightarrow}{\nabla}(\Psi_{A_k} G)(z)$$

where $\sum_{k=1}^{N} \lambda_k = 1$ with the $\lambda_k > 0$; here N, like the A_k, is arbitrary.

We *take* one of these, chosen to have $L_2(\Im z > 0)$ norm $< \epsilon^2$, and note then that for the Ψ_{A_k} and corresponding functions Φ_{A_k} (see the beginning of this section), we have

$$\sum_{k=1}^{N} \lambda_k \Phi_{A_k}(x)G(x) + \sum_{k=1}^{N} \lambda_k \Psi_{A_k}(x)G(x) = G(x).$$

And we denote the *first* sum on the left by $G_1(x)$ and the *second* by $G_2(x)$. Here $|x|^{1/3}|G_1(x)|$ is *bounded* (look again at the definition of the $\Phi_A(x)$), and we have the following:

Theorem 10.1 *Given* $\epsilon > 0$ *we have*

$$G(x) = G_1(x) + G_2(x)$$

where $|x^{1/3}G_1(x)|$ *is bounded and*

$$\iint_{\Im z>0} |\underset{\rightarrow}{\nabla} G_2(z)|^2 dx\, dy < \epsilon^2$$

for the harmonic extension $G_2(z)$ *of* $G_2(x)$ *to* $\Im z > 0$.

11 Change in Notation and a Least Superharmonic Majorant

In what follows, the sum

$$-x^{1/3}G(x) + J(x)$$

will occur frequently and the minus sign may bring cause for confusion. *It is therefore better to write* $\hat{G}(z)$ for $-G(z)$. This change in sign *will not affect* the general properties of $G(z)$ noted above, including the result of the preceding section. If a majorant of $\log \underline{W}(n^3)$ having the form

$$x^{1/3}\hat{G}(x) + J(x)$$

is forthcoming, where $\hat{G}(x)$, bounded, continuous, and odd, with its harmonic extensions to $\Im z > 0$ and $\Im z < 0$ has a finite Dirichlet integral and $J(x)$, even and increasing for $x \geq 0$, is $\leq \alpha \cdot \sqrt{3}\pi |x|^{1/3}$ for some $\alpha < 1$, we aim now to show that the $P(n^3)/W(n^3)$ formed from polynomials P are *not dense* in $\mathscr{C}_\circ(\mathbb{K})$.

This will depend on the fact that, with $\mathrm{v}(z)$ the harmonic extension of $\mathrm{v}(x) = x^{1/3}\hat{G}(x)$ to $\Im z > 0$ and $\mathrm{v}(\bar{z}) = \mathrm{v}(z)$, with the same for $J(x)$ and $J(z)$, and with $Q(z)$ the harmonic extension of $|x|^{1/3}$ to $\Im z > 0$ and $\Im z < 0$, the function

$$\mu(v(z) + J(z)) - \sqrt{3}\pi\, Q(z)$$

has, for an appropriate $\mu > 1$, a *finite least superharmonic majorant.* (For purposes of the present discussion, the function identically equal to ∞ is *also* called *superharmonic*.)

This approach has been taken before in connection with a different problem and is fully set out on pp 365–396 of [10].

Here are some of the observations made there about the least superharmonic majorant:

i) It is either *finite everywhere* or *identically infinite*.
ii) If *finite*, it is harmonic wherever the function it majorizes is harmonic.
iii) If finite, the least superharmonic majorant of a *continuous* function is larger than that function precisely on an *open set*, and is then *harmonic* on that open set.
iv) The least superharmonic majorant of an even function $F(x)$, continuous on $\mathbb{R}$ and having harmonic (Poisson) extensions $F(z)$ to the upper and lower half planes, *is finite* provided that for some $M < \infty$ we have

$$\int_E F(x)d\omega_{\mathscr{D}}(x, 0) \leq M$$

for *all* finite sets $E = -E$, $0 \notin E$, of disjoint closed intervals on $\mathbb{R}$. (As usual, $\mathscr{D} = \mathbb{C} \sim E$.)

Proof of this, although straightforward, is somewhat fussy, so it is deferred to an *addendum*, Sect. 14.

We are going to verify that the relation considered in iv holds for $F(x)$ equal to the function mentioned above for $z = x$. For this, we will use the result of the preceding section together with the one obtained at the end of Sect. 9.

Remembering that $J(x)$ is $< \sqrt{3}\pi\alpha|x|^{1/3}$ with $\alpha < 1$, we now take a number $\mu > 1$ so close to 1 that $\mu\alpha$ is still less than 1.

Theorem 11.1 *Under the above conditions, the function*

$$\mu x^{1/3}\hat{G}(x) + \mu J(x) - \sqrt{3}\pi|x|^{1/3},$$

with its harmonic extensions to the upper and lower half planes, has a finite least superharmonic majorant.

Proof Following the preceding remark iv, we show that there is a finite quantity M such that

$$\int_E \left(\mu x^{1/3}\hat{G}(x) + \mu J(x) - \sqrt{3}\pi|x|^{1/3}\right) d\omega_{\mathscr{D}}(x, 0) \leq M$$

for all sets E of intervals of the form specified there.

Using the theorem at the end of the preceding section, we write

$$\hat{G}(x) = \hat{G}_1(x) + \hat{G}_2(x),$$

where $|x^{1/3}\hat{G}_1(x)| \leq C_\delta$ and $\hat{G}_2(x)$, together with its harmonic extensions to $\Im z > 0$ and $\Im z < 0$, has a Dirichlet integral

$$\iint\limits_{\Im z>0} |\underset{\rightarrow}{\nabla} G_2(z)|^2 dx\, dy$$

of magnitude δ^2, δ being *small in relation to* $\sqrt{3}\pi(1 - \mu\alpha)$.

Since $\int_E dw_{\mathscr{D}}(x, 0) = 1$, we have

$$\int_E |x^{1/3}\hat{G}_1(x)| dw_{\mathscr{D}}(x, 0) \leq C_\delta.$$

As for the similar integral with $\hat{G}_2$, we have, from Sect. 9, the relation

$$\left|\int_E \hat{G}_2(x) \cdot x^{1/3} dw_{\mathscr{D}}(x, 0)\right| \leq \text{const.}\, \lambda_{\mathscr{D}} \sqrt{\iint\limits_{\Im z>0} |\underset{\rightarrow}{\nabla} \hat{G}_2(z)|^2 dx\, dy},$$

which, *here*, is const.$\delta\lambda_{\mathscr{D}}$.

Again,

$$\int_E \mu J(x)\,d\omega_{\mathscr{D}}(x,0) < \int_E \sqrt{3}\pi\,\mu\alpha|x|^{1/3} d\omega_{\mathscr{D}}(x,0)$$

whilst, from near the beginning of Sect. 9,

$$\int_E \sqrt{3}\pi\,|x|^{1/3} d\omega_{\mathscr{D}}(x,0) = \frac{2\pi}{\sqrt{3}}\lambda_{\mathscr{D}}.$$

It follows from all of this that

$$\int_E \left(\mu x^{1/3}\hat{G}(x) + \mu J(x) - \sqrt{3}\pi\,|x|^{1/3}\right) d\omega_{\mathscr{D}}(x,0) \le C_\delta + \text{const.}\delta\lambda_{\mathscr{D}}$$
$$+ \frac{2\pi}{\sqrt{3}}(\mu\alpha - 1)\lambda_{\mathscr{D}}.$$

Here $\mu\alpha - 1 < 0$ and $\delta > 0$ can be chosen as small as we like. Then the last expression has a finite upper bound for all values of $\lambda_{\mathscr{D}}$ (which is always > 0). The theorem is proved. □

12 Use of the Least Superharmonic Majorant

Knowing now that the function $\mu(x^{1/3}\hat{G}(x)+J(x))-\sqrt{3}\pi\,|x|^{1/3}$, with its harmonic extensions to $\Im z > 0$ and $\Im z < 0$, has a finite superharmonic majorant, we denote the latter by $-S(z)$ (sic!) and observe first of all that $S(0)$ is *finite* (that was how we proved that $-S(z)$ exists!). The reader is reminded once more that we are now denoting by $\hat{G}(z)$ the function that was denoted by $-G(z)$ in Sects. 5–8.

We wrote $-S(z)$ above because then the function $S(z)$, of main interest here, is *subharmonic*, and we have

$$S(z) \le V_\circ(z) - \mu(v(z) + J(z)),$$

$V_\circ(z)$ being $\sqrt{3}\pi\,|x|^{1/3}$ together with its above-mentioned harmonic extensions and v(z) the same for $x^{1/3}\hat{G}(x)$. It will be very important here that μ, close to 1, is actually > 1.

The function

$$\sqrt{3}\pi\,|x|^{1/3} - \mu(x^{1/3}\hat{G}(x) + J(x))$$

is visibly $O(|x|^{1/3})$ for $-\infty < x < \infty$, and from this Poisson's formula shows that the function's harmonic extensions to $\Im z > 0$ and $\Im z < 0$ are $O(|z|^{1/3})$. This and the preceding relation then make

$$S(z) \le C|z|^{1/3}$$

for $|z| \geq 1$, say.

From remarks i–iii near the beginning of Sect. 11, we see that $S(z)$ is *harmonic* for $\Im z > 0$ and for $\Im z < 0$. Hence all of its Riesz mass must be distributed on $\mathbb{R}$. We denote the resulting (*positive*!) measure on $\mathbb{R}$ by σ; then Jensen's formula says that

$$S(0) + \int_0^R \frac{\sigma(-r, r)}{r} dr = \frac{1}{2\pi} \int_{-\pi}^{\pi} S(Re^{i\theta})\, d\theta$$

According to what has just been said, the *right side* of this relation is $\leq$ const.$R^{1/3}$, at least for $R \geq 1$. And since $\sigma(-r, r)$ is an increasing function of r, the integral on the *left* is $\geq \sigma\left(-\frac{R}{e}, \frac{R}{e}\right)$.

Finally, we note that $\sigma(0, r)$ must be *equal* to $\sigma(-r, 0)$ since, for each y, $S(x+iy)$ is an *even* function of x (like $\mu(\mathrm{v}(\mathrm{x}+\mathrm{iy}) + \mathrm{J}(\mathrm{x}+\mathrm{iy}) - \mathrm{V}_\circ(\mathrm{x}+\mathrm{iy}))$. And we denote, as is customary, $\sigma([0, t]) = \sigma([-t, 0])$ by $\sigma(t)$, making $\sigma(t)$ an *even* function of t, *increasing* for $t > 0$. (σ can have no point mass at 0 because $S(0)$ is finite!) Then the above relations make

$$S(0) + \sigma(t) \leq \text{const.}\, t^{1/3},$$

at least for $t > 1$, say.

All this is enough to ensure that

$$S(z) = \text{const.} + \int_{-1}^{1} \log|z - t| d\sigma(t) + \int_1^{\infty} \log\left|1 - \frac{z^2}{t^2}\right| d\sigma(t).$$

Here, since $S(0)$ is finite, we have

$$\int_{-1}^{1} \log|t| d\sigma(t) > -\infty,$$

and so we can write

$$S(z) = S(0) + \int_0^{\infty} \log\left|1 - \frac{z^2}{t^2}\right| d\sigma(t).$$

From properties ii and iii of a least superharmonic majorant enumerated in Sect. 11, it follows that the *closed support* of σ, on $\mathbb{R}$, is precisely the set thereon where

$$-S(x) = \mu(x^{1/3}\hat{G}(x) + J(x)) - \sqrt{3}\pi|x|^{1/3}$$

And thence we have the following:

Lemma 12.1

$$\frac{\partial S(x+i0)}{\partial y^+} = 0 \quad \text{on } \mathbb{R},$$

save on the closed set $E \subseteq \mathbb{R}$ where

$$S(x) = \sqrt{3}\pi|x|^{1/3} - \mu(x^{1/3}\hat{G}(x) + J(x)).$$

There,

$$0 \le \frac{\partial S(x+i0)}{\partial y^+} \le \frac{\pi}{3|x|^{2/3}}.$$

Proof For $x \notin E$, hence outside the support of σ, the above formula shows that

$$\frac{\partial S(x+i0)}{\partial y^+} = 0.$$

For $x \in E$, we have

$$S(x) = \sqrt{3}\pi|x|^{1/3} - \mu(x^{1/3}\hat{G}(x) + J(x)),$$

whereas, for $z = x + iy$, $y > 0$,

$$S(x) < V_\circ(z) - \mu\big(v(x+iy) + J(x+iy)\big).$$

where

$$V_\circ(z) = \int_0^\infty \log\left|1 - \frac{z^2}{t^2}\right| \frac{dt}{3t^{2/3}}.$$

From this last formula, we get

$$\frac{\partial V_\circ(x+i0)}{\partial y^+} = \frac{\pi}{3|x|^{2/3}},$$

and again

$$v(x+iy) + J(x+iy)$$

increases with y (property v) noted at the end of Sect. 7). Thence,

$$\frac{\partial S(x+i0)}{\partial y^+} \le \frac{\pi}{3|x|^{2/3}}$$

and the left side is ≥ 0 by the above formula for $S(z)$ preceding the lemma. □

Corollary 12.1 *For* $x \in \mathbb{R}$ *and* $y > 0$,

$$S(x+iy) - S(x) \leq V_\circ(x+iy) - \sqrt{3}\pi|x|^{1/3}.$$

Proof

$$\frac{\partial S(x+iy)}{\partial y} \quad \text{and} \quad \frac{\partial V_\circ(x+iy)}{\partial y}$$

are both harmonic functions of z for $\Im z > 0$, given by the Poisson extension of their values on $\mathbb{R}$ to the upper half plane. The function $-\mathrm{v}(\mathrm{x}+\mathrm{iy})$ decreases when $y \geq 0$ increases (end of Sect. 7, remark ν). □

Lemma 12.2 *In terms of polar coordinates, we have, in the upper half plane,*

$$V_\circ(re^{i\theta}) = 2\pi r^{1/3}\cos\left(\frac{\theta}{3} - \frac{\pi}{6}\right), \quad 0 \leq \theta \leq \pi.$$

Proof $2\pi\Re\big(e^{-\pi i/6}(re^{i\theta})^{1/3}\big)$ is a harmonic function of $z = re^{i\theta}$ in the upper half plane, $(re^{i\theta})^{1/3}$ being analytic there, and it takes the value $\sqrt{3}\pi r^{1/3}$ for $\theta = 0$ and for $\theta = \pi$. □

We use this lemma together with the preceding corollary to estimate $S(n^3+i) - S(n^3)$. Without loss of generality, take $n > 0$. Then, with $r = |n^3+i|$ and $\tan\theta = 1/n^3$, we have

$$r^{1/3} = (n^6+i)^{1/6} = n\left(1+\frac{1}{n^6}\right)^{1/6} < n + \frac{1}{6n^5},$$

and

$$\sin\frac{\theta}{3} \;<\; \tan\frac{\theta}{3} \;<\; \frac{1}{3n^2},$$

whilst $\cos(\theta/3) < 1$. All this makes

$$\begin{aligned} V_\circ(n^3+i) - V_\circ(n^3) &= 2\pi(n^6+1)^{1/6}\cos\left(\frac{\theta}{3}-\frac{\pi}{6}\right) - \sqrt{3}\pi n \\ &< 2\pi\left(n+\frac{1}{6n^5}\right)\left(\frac{\sqrt{3}}{2}+\frac{1}{6n^2}\right) - \sqrt{3}\pi n \\ &= \frac{\pi}{3n} + \frac{\sqrt{3}\pi}{6n^5} + \frac{\pi}{18n^7}. \end{aligned}$$

Thus, with the preceding corollary, we have the following:

Theorem 12.1 *For* $n^3 \in \mathbb{K}$,

$$S(n^3 + i) - S(n^3) \ < \ \frac{\pi}{3n} + O\left(\frac{1}{n^5}\right).$$

The function we shall actually use is constructed from $S(z)$, equal, as we recall, to

$$S(0) + \int_0^\infty \log\left|1 - \frac{z^2}{t^2}\right| d\sigma(t).$$

We put

$$\Sigma(z) = S(0) + \int_0^\infty \log\left|1 - \frac{z^2}{t^2}\right| d[\sigma(t)];$$

$\Sigma(z)$ is equal to $\log|\mathcal{S}(z)|$ for $\mathcal{S}(z)$ the entire function $e^{S(0)}\prod_{k=1}^\infty\left(1 - z^2/\sigma_k^2\right)$, where $0 < \sigma_1 < \sigma_2 < \cdots$ are the successive points t on $\mathbb{R}_+$ where $\sigma(t)$ *takes integral values.*

Our lack of knowledge about the local behaviour of $\sigma(t)$ makes it difficult to see the relation between $S(x)$ and $\Sigma(x)$ for $x \in \mathbb{R}$. That is why we compare those two functions on the line $z = x + i$.

Theorem 12.2 *For* $\Im z > 0$,

$$\Sigma(z) - S(z) \ \leq \ \log\frac{|z|}{\Im z}.$$

Proof Understanding here $\int_{-\infty}^\infty$ to mean $\lim_{A\to\infty}\int_{-A}^A$, we write

$$S(z) = \int_{-\infty}^\infty \log\left|1 - \frac{z}{t}\right| d\sigma(t) \ + \ S(0)$$

and

$$\Sigma(z) = \int_{-\infty}^\infty \log\left|1 - \frac{z}{t}\right| d[\sigma(t)] \ + \ S(0).$$

Here and in what follows, we are taking $[\sigma(t)]$ to be *odd* on $\mathbb{R}$ and equal to $\sigma(t)\mathrm{sgn}(t)$.

When $\Re z > 0$, the value of $\log|1 - z/t|$ *decreases* as $t \leq 0$ does, whilst in the passage from $d\sigma(t)$ to $d[\sigma(t)]$, mass is *moved away* from 0. This makes

$$\int_{-A}^0 \log\left|1 - \frac{z}{t}\right| d[\sigma(t)] \ \leq \ \int_{-A}^0 \log\left|1 - \frac{z}{t}\right| d\sigma(t)$$

when $\Re z > 0$ for each large A with $-A$ located at one of the places on the negative real axis where $[\sigma(t)]$ jumps (downwards, as $|t|$ decreases).

With $a > 0$ *small*, the *difference*

$$\int_a^\infty \log\left|1 - \frac{z}{t}\right| d([\sigma(t)] - \sigma(t))$$

is now integrated by parts. This yields an *integrated term* tending to 0 with a because $\sigma(a)\log 1/a \le \int_0^a (\log 1/t)\, d\sigma(t)$, which must $\to 0$ as $a \to 0$ since, as noted earlier in this section, $\int_{-1}^1 (\log 1/|t|)\, d\sigma(t) < \infty$. We thus have

$$\int_0^\infty \log\left|1 - \frac{z}{t}\right| d([\sigma(t)] - \sigma(t)) = \int_0^\infty \left(\frac{\partial}{\partial t}\log\left|1 - \frac{z}{t}\right| ([\sigma(t)] - \sigma(t))\right) dt.$$

In the second integral,

$$0 \le \sigma(t) - [\sigma(t)] \le 1.$$

And when t *increases* from 0 to ∞, the *distance* $|1 - z/t|$ first *decreases* to a certain value $d_0 < 1$ and then *increases again* to 1. This we can see from the following diagram:

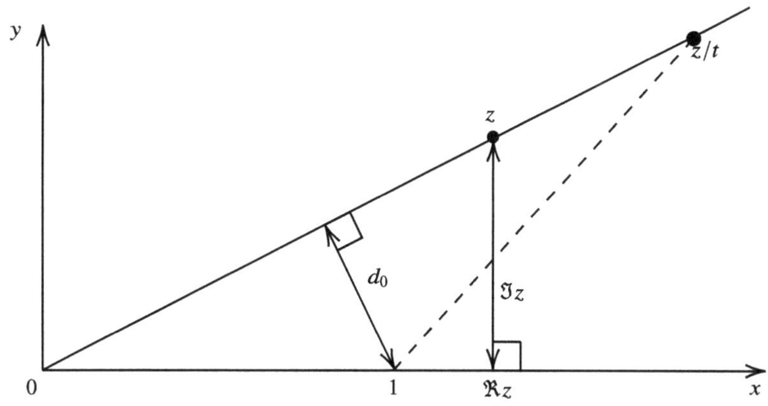

From it, we have, by similar triangles,

$$d_0 = \frac{d_0}{1} = \frac{\Im z}{|z|}.$$

Thence, and by the preceding observation,

$$\int_0^\infty \left(\frac{\partial}{\partial t}\log\left|1 - \frac{z}{t}\right|\right) (\sigma(t) - [\sigma(t)])dt \;\le\; \log\frac{1}{d_0} = \log\frac{|z|}{\Im z}$$

for $\Re z$ and $\Im z > 0$ which, added to the result for $\int_{-\infty}^{0}$, yields

$$\int_{-\infty}^{\infty} \log \left|1 - \frac{z}{t}\right| d([\sigma(t)] - \sigma(t)) \;\leq\; \log \frac{|z|}{|\Im z|},$$

a relation which of course holds just as well when $\Re z < 0$ or $\Im z < 0$.

The theorem is proved. □

Corollary 12.2 *For $n^3 \in \mathbb{K}$, we have*

$$\Sigma(n^3 + i) - S(n^3 + i) \;\leq\; 3 \log |n|.$$

And for $y > 0$,

$$\Sigma(iy) \leq S(iy).$$

(The last relation is also easily seen directly.)

We are going to use Lagrange interpolation with the entire function $\mathcal{S}(z)$ defined just after Theorem 11.1 and the points $n^3 \in \mathbb{K}$. We have $|\mathcal{S}(z)| = e^{\Sigma(z)}$ and, for $\mathcal{S}(z)$ with all its zeros real, we have

$$|\mathcal{S}(n^3)| \leq |\mathcal{S}(n^3 + i)| \;=\; e^{\Sigma(n^3+i)}.$$

Again, by the last corollary,

$$e^{\Sigma(n^3+i)} \leq |n|^3 e^{S(n^3+i)},$$

making

$$|\mathcal{S}(n^3)| \leq |n|^3 e^{S(n^3+i)},$$

By the first of the two preceding theorems, the right side is $\leq |n|^3 e^{S(n^3)} e^{O(1/n)}$, making finally

$$|\mathcal{S}(n^3)| \leq |n|^3 e^{S(n^3)} (1 + O(1/n)).$$

Now from the beginning of this section, we have

$$S(n^3) \leq \sqrt{3}\pi |n| - \mu(n\hat{G}(n^3) + J(n^3))$$

with

$$n\hat{G}(n^3) + J(n^3) + C \;\geq\; \log \underline{W}(n^3)$$

for a constant C, according to our use of the term *majorant*. But then, for $W_C(n^3) = e^{-C}W(n^3)$ (which makes $\underline{W}_C(n^3) = e^{-C}\underline{W}(n^3)$), the ratios $P(n^3)/W_C(n^3)$ are dense in $\mathscr{C}_\circ(\mathbb{K})$ *if and only if the* $P(n^3)/W(n^3)$ *are*. So we may as well *drop* the C here, and *do* so, in order to simplify the writing.

The last relations thus make

$$S(n^3) \leq \sqrt{3}\pi\,|n| - \mu \log \underline{W}(n^3),$$

where $\mu > 1$. And from this and a relation appearing above, we get

$$|\mathcal{S}(n^3)| \;\leq\; |n|^3 \frac{e^{\sqrt{3}\pi\,|n|}}{(\underline{W}(n^3))^\mu}\left(1 + O\left(\frac{1}{n}\right)\right).$$

Wishing to do Lagrange interpolation at the points $n^3 \in \mathbb{K}$, we must bring in the function

$$C_\circ(z) = \prod_{n=1}^{\infty}\left(1 - \frac{z^2}{n^6}\right)$$

from Sect. 4 and its derivative $C_\circ'(n^3)$ at the $n^3 \in \mathbb{K}$. For the latter we have (Sect. 4)

$$C_\circ'(n^3) \cong (-1)^n \frac{e^{\sqrt{3}\pi\,|n|}}{12\pi^3 n^5}$$

So Lagrange interpolation leads to the expression

$$C_\circ(z) \sum_{n^3\in\mathbb{K}} \frac{\mathcal{S}(n^3)}{C_\circ'(n^3)(z - n^3)},$$

where the coefficient

$$\gamma_n \;=\; \frac{\mathcal{S}(n^3)}{C_\circ'(n^3)} \;\cong\; \frac{12\pi^3(-1)^n n^5\,\mathcal{S}(n^3)}{e^{\sqrt{3}\pi\,|n|}}$$

is in absolute value

$$\leq\; \frac{12\pi^3|n|^5|n|^3 e^{\sqrt{3}\pi\,|n|}}{e^{\sqrt{3}\pi\,|n|}(\underline{W}(n^3))^\mu} \;=\; \frac{12\pi^3 n^8}{(\underline{W}(n^3))^\mu}$$

to within a factor of $1 + O(1/n)$, by a closely preceding estimate of $|\mathcal{S}(n^3)|$. So Lagrange interpolation leads to

$$C_\circ(z) \sum_{n^3 \in \mathbb{K}} \frac{\gamma_n}{z - n^3},$$

where

$$|\gamma_n| \;\le\; \frac{12\pi^3 n^8}{\underline{W}(n^3)^\mu},$$

with $\underline{W}(n^3)$ tending to ∞ *faster than any power of* $|n|$ as $n \to \pm\infty$.

And we say now that the above expression is *equal* to $\mathcal{S}(z)$, which is already clearly so for $z = n^3 \in \mathbb{K}$.

Indeed, we *in fact* claim that

$$z^k \mathcal{S}(z) \;=\; C_\circ(z) \sum_{n^3 \in \mathbb{K}} \frac{n^{3k}\gamma_n}{z - n^3}$$

for $k = 0, 1, 2, \cdots$. The summation on the right always converges absolutely thanks to the above-mentioned property of $\underline{W}(n^3)$.

That sum is clearly an entire function of order at most $1/3$. And the entire function $\mathcal{S}(z)$ also has that property.

Indeed,

$$\log|\mathcal{S}(z)| \;\le\; S(z) + \log\frac{|z|}{\Im z}$$

by the preceding theorem for $\Im z > 0$, so

$$\log|\mathcal{S}(z)| \le S(z) + \log|z|$$

for $|\Im z| > 1$. As stated at the beginning of the proof of the first lemma in this section,

$$S(z) < V_\circ(z) - \mu(v(z) + J(z)),$$

where $V_\circ(z)$ is described in Sect. 4. From its properties and those of the second term on the right, we deduce that $S(z)$ is $O(|z|^{1/3})$, so $\log|\mathcal{S}(z)|$ is $O(|z|^{1/3})$, at least for $|\Im z| > 1$. This then holds also for $|\Im z| \le 1$ because, with $\mathcal{S}(z)$ having all its zeros on the real axis, $|\mathcal{S}(x + iy)|$ is an increasing function of $|y|$ for each $x \in \mathbb{R}$.

It follows that for $k = 0, 1, 2, \cdots$, *the differences*

$$F_k(z) \;=\; z^k \mathcal{S}(z) - C_\circ(z) \sum_{n^3 \in \mathbb{K}} \frac{n^{3k}\gamma_n}{z - n^3}$$

are all entire functions of order $1/3$. Each one vanishes at the $n^3 \in \mathbb{K}$, which, together, constitute *all* the zeros of $C_\circ(z)$, so the *ratio*

$$\frac{F_k(z)}{C_\circ(z)} = z^k \frac{\mathcal{S}(z)}{C_\circ(z)} - \sum_{n^3 \in \mathbb{K}} \frac{n^{3k}\gamma_n}{z - n^3}$$

is also an entire function. Considering the Hadamard factorizations of the entire functions $\mathcal{S}(z)$, $C_\circ(z)$, and $F_k(z)$, all of order $1/3$, we see that the ratio $F_k(z)/C_\circ(z)$ must be of order *at most* $1/3$.

We look at the behaviour of that ratio on the imaginary axis. Considering the *second* term on the right, we observe that

$$\left| \frac{n^{3k}\gamma_n}{iy - n^3} \right| \leq \frac{12\pi^2 |n|^{3k+8}}{\sqrt{n^6 + y^2}(\underline{W}(n^3))^\mu}$$

by the preceding estimate of $|\gamma_n|$. So, since $\underline{W}(n^3) \to \infty$ faster than any power of $|n|$ as $n \to \pm\infty$, we see that the sum on the right converges absolutely for $z = iy$ and tends to zero for $y \to \pm\infty$.

We turn to the *first* term on the right. Here we use a relation mentioned above with $z = iy$ to get

$$S(iy) < V_\circ(iy) - \mu(v(iy) + J(iy)),$$

and another just preceding it, to thence obtain

$$\log \mathcal{S}(iy) \leq V_\circ(iy) + \log|y| - \mu(v(iy) + J(iy)).$$

We need also a *lower* bound on $C_\circ(iy)$. From Sect. 4 and the beginning of this section, we have

$$V_\circ(iy) = \int_0^\infty \log\left|1 + \frac{y^2}{t^2}\right| \frac{dt}{3t^{2/3}}$$

and again

$$\log C_\circ(iy) = \sum_{n=1}^\infty \log\left|1 + \frac{y^2}{n^6}\right| \geq V_\circ(iy) - \int_0^1 \log\left(1 + \frac{y^2}{t^2}\right) \frac{dt}{3t^{2/3}}.$$

The integral on the right is, for $y > 0$,

$$y^{1/3} \int_0^{1/y} \log\left(1 + \frac{1}{\tau^2}\right) \frac{d\tau}{3\tau^{2/3}},$$

and we need an estimate on it for *large* y. The function

$$\frac{1/\tau^2}{1+1/\tau^2} = \frac{1}{\tau^2+1}$$

decreases when $\tau > 0$ increases, so the above integral is (for $y \geq 1$)

$$< \; y^{1/3} \int_0^{1/y} \log \frac{2}{\tau^2} \cdot \frac{d\tau}{3\tau^{2/3}}.$$

Here

$$\begin{aligned} &\int_0^{1/y} \log \frac{2}{\tau^2} \cdot \frac{d\tau}{3\tau^{2/3}} \\ &= 2\int_0^{1/y} \log \frac{1}{\tau} \cdot \frac{d\tau}{3\tau^{2/3}} \; + \; \log 2 \int_0^{1/y} \frac{d\tau}{3\tau^{2/3}} \\ &= \frac{2\log y}{y^{1/3}} + \frac{6}{y^{1/3}} + \frac{\log 2}{y^{1/3}}, \end{aligned}$$

making

$$y^{1/3} \int_0^{1/y} \log\left(1+\frac{1}{\tau^2}\right) \frac{d\tau}{3\tau^{2/3}} \; < \; 2\log y + 6 + \log 2.$$

From this and a preceding relation, we now get

$$\log C_\circ(iy) \; \geq \; V_\circ(iy) - 2\log y - \log 2 - 6.$$

Whence, with the earlier estimate of $\log \mathcal{S}(iy)$, we obtain

$$\left|\frac{\mathcal{S}(iy)}{C_\circ(iy)}\right| \; \leq \; \frac{2e^6|y|^3 e^{V_\circ(iy)}}{e^{\mu(v(iy)+J(iy))}e^{V_\circ(iy)}} \; = \; \frac{2e^6|y|^3}{e^{\mu(v(iy)+J(iy))}}.$$

Now property v at the end of Sect. 7 states that

$$\frac{v(iy)+J(iy)}{\log y} \longrightarrow \infty$$

for $y \to \infty$. It follows then that for *any* k,

$$\frac{y^k \mathcal{S}(iy)}{C_\circ(iy)} \longrightarrow 0$$

as $y \to \infty$, and the same of course holds good for $y \to -\infty$.

With this and the previous observation about *the second term* in the formula for the ratio $F_k(z)/C_\circ(z)$ (an *entire function*, surely of order $\leq 1/3$), we see that $F_k(iy)/C_\circ(iy)$ is *bounded for* $-\infty < y < \infty$ and hence, by Theorem 1.4.1 in [12], p. 3, *bounded* in the *right half plane* and of course (here) in the *left* half plane also. So it is *constant*, and since $F_k(iy)/C_\circ(iy) \to 0$ for $y \to \pm\infty$, *the constant is zero.*

We have thus proved the following:

Theorem 12.3 *With*

$$\gamma_n \;=\; \frac{\mathcal{S}(n^3)}{C_\circ'(n^3)} \;\cong\; \frac{(-1)^n \cdot 12\pi^3 n^5 \mathcal{S}(n^3)}{e^{\sqrt{3}\pi|n|}}$$

(to within a factor of $1 + O(1/n)$ *in absolute value* $\leq 12\pi^3 n^8/(\underline{W}(n^3))^\mu$*, we have, for* $k = 0, 1, 2, \cdots$ *,*

$$\frac{z^k \mathcal{S}(z)}{C_\circ(z)} = \sum_{n^3 \in \mathbb{K}} \frac{n^{3k}\gamma_n}{z - n^3}.$$

And we have the following:

Corollary 12.3 *With the* γ_n *as in the theorem,*

$$\sum_{n^3 \in \mathbb{K}} n^{3k-1}\gamma_n \;=\; 0$$

for $k = 1$ (sic!), $2, 3, \cdots$.

Remark 12.1 The idea for part of the approach to this result goes back to Borichev and Sodin, in [2].

Referring to the statement of the theorem, we see that we can write

$$\gamma_n \;=\; \frac{12\pi^3 n^8 \rho_n}{(\underline{W}(n^3))^\mu},$$

where $0 < |\rho_n| < 1 + O(1/n)$. And now, for $l = 0$ (sic!), $1, 2, 3, \cdots$, we have, from the corollary,

$$\sum_{n^3 \in \mathbb{K}} \frac{n^{3l}}{\underline{W}(n^3)} \cdot \frac{12\pi^3 n^{11}\rho_n}{(\underline{W}(n^3))^{\mu-1}} \;=\; 0.$$

Here, for $n^3 \in \mathbb{K}$,

$$\frac{12\pi^3 n^{11}\rho_n}{(\underline{W}(n^3))^{\mu-1}}$$

is not zero, and also in $l_1(\mathbb{K})$ since $\underline{W}(n^3)$ increases faster than any power of $|n|$ as $n \to \pm\infty$ (which is *the main reason* for our having brought in the power $\mu > 1$!). So we have the following:

Theorem 12.4 *Under the conditions of this and the preceding section, the ratios* $P(n^3)/\underline{W}(n^3)$ *formed from polynomials* P *are not dense in* $\mathscr{C}_\circ(\mathbb{K})$.

Corollary 12.4 *Neither are the* $P(n^3)/W(n^3)$ *formed from polynomials* P *dense in* $\mathscr{C}_\circ(\mathbb{K})$.

Proof Refer to the first theorem in Sect. 2. □

13 Conclusion

Almost all of the preceding work pertains to *even* weights $W(n^3)$ on $\mathbb{K}$. But the results obtained for *these* have relevance to the approximation problem for general weights W on $\mathbb{K}$ according to the second theorem in Sect. 2. (See also the remark immediately preceding the *third* theorem there.)

Let us try to summarize what we have found out about weighted polynomial approximation on $\mathbb{K}$ with even weights.

The results are in terms of these weights' *Mergelian minorant* $\underline{W}(n^3)$ defined at the beginning of Sect. 2. They concern the possible existence of a particular kind of *majorant* for $\log \underline{W}(n^3)$, meaning that such a majorant *plus a constant* is $\geq \log \underline{W}(n^3)$ at the points $n^3 \in \mathbb{K}$. Stated in terms of $\hat{G}(x) = -G(x)$, such majorants are of the form $n\hat{G}(n^3) + J(n^3)$, where the functions $\hat{G}(x)$ and $J(x)$ have the following properties (according to the theorem at the end of Sect. 7):

i) $\hat{G}(x)$ is odd, bounded above and below, and continuous.
ii) The harmonic (Poisson integral) extension of $\hat{G}(x)$ to the upper half plane has a finite Dirichlet integral.
iii) $J(x)$ is even and, for $x > 0$, increasing and $< \sqrt{3}\pi x^{1/3}$.
iv) For $x > 0$, $J'(x) < \sqrt{3}\pi/3x^{2/3}$.
v) The harmonic extension $v(z) + J(z)$ of $x^{1/3}\hat{G}(x) + J(x)$ to $\Im z > 0$ is an increasing function of $y = \Im z$ for each z there, and

$$\frac{v(iy) + J(iy)}{\log y} \longrightarrow \infty$$

as $y \to \infty$.

Then:

1° If the $P(n^3)/\underline{W}(n^3)$ formed from polynomials P *are not dense* in $\mathscr{C}_\circ(\mathbb{K})$, a *majorant* of the form $x^{1/3}\hat{G}(x) + J(x)$ *exists* for $\log \underline{W}(n^3)$, and properties i–v of $\hat{G}(x)$ and $J(x)$ *hold*;

2° *If* such a majorant $x^{1/3}\hat{G}(x) + J(x)$ for $\log \underline{W}(n^3)$ *does* exist, but with iii) replaced by the *stronger* condition

iii)′ $J(x)$ is even and, for $x > 0$, increasing and $\leq \sqrt{3}\alpha\pi x^{1/3}$ for some α, $0 < \alpha < 1$; *then* the ratios $P(n^3)/\underline{W}(n^3)$ formed from polynomials P *are not dense* in $\mathscr{C}_\circ(\mathbb{K})$.

As stated at the end of the last section, iii)′ then implies that the ratios $P(n^3)/W(n^3)$ *are not dense* in $\mathscr{C}_\circ(\mathbb{K})$ (and vice versa, because $\underline{W}(n^3) \leq W(n^3)$).

One would like to improve iii)′ by allowing $J(x) < \sqrt{3}\pi x^{1/3}$ to (only sometimes, perhaps) be *closer* to $\sqrt{3}\pi x^{1/3}$ than $\sqrt{3}\pi\alpha x^{1/3}$, where $\alpha < 1$. I think that such an improvement cannot be obtained by use of the methods of the present article. This on account of the important role the quantity μ, larger than 1 (but for which, for $x > 0$, $\sqrt{3}\pi\alpha\mu x^{1/3}$ is *still* $< \sqrt{3}\pi x^{1/3}$) plays in the work leading to 2°.

14 Addendum

We prove the statement in Sect. 11, point iv, to the effect that a continuous even function $F(x)$ on $\mathbb{R}$ together with its Poisson extensions $F(z)$ to $\Im z > 0$ and $\Im z < 0$ has *a finite least superharmonic majorant provided that*

$$\int_E F(x) d\omega_D(x, 0) \leq M < \infty$$

for *all* finite collections $E = -E$, $0 \notin E$, of disjoint closed intervals on $\mathbb{R}$ (as usual, D is $\mathbb{C} \sim E$).

Without loss of generality, let $F(0) = 0$. Given $\epsilon > 0$, there is a $\delta > 0$ such that $F(x) < \epsilon$ for $-\delta < x < \delta$. Using the material on pp 391–395 of [10] (with l, δ of p. 365 for the notation), we show that the *restriction* $F_\delta^+(x)$ of $F^+(x)$ to the set $\{|x| \geq \delta\}$ together with the harmonic extension of $F_\delta^+(z)$ to $\mathbb{C} \sim \{|x| \geq \delta\}$ *has* a finite superharmonic majorant. Then $\epsilon +$ that majorant will be a superharmonic majorant of $F(z)$.

Let O be the set of x on $\{|x| \geq \delta\}$ where $F^+(x) > 0$; here $O = -O$. For any collection $E = -E$ of disjoint closed intervals in O, we have, by the above relation

$$\int_E F_\delta^+(x) d\omega_D(x, 0) = \int_E F(x) d\omega_D(x, 0) \leq M.$$

Assuming for the moment that $F(x)$ is of compact support, the boxed formula on p. 395 of [10] (with $A = 0$) shows, on making the ϵ in *that* formula (*not* the above ϵ!) tend to zero, that $F_\delta^+(z)$ has a superharmonic majorant $H_\delta(z)$ with $H_\delta(0) \leq M$.

A standard limiting argument now enables us to *remove* the condition that $F_\delta^+(x)$ be of compact support, and the result is that $F_\delta^+(z)$ *has* a superharmonic majorant $H(z)$ with $H(0) \leq M$ (and hence $H(z)$ being everywhere finite).

2002–2020

The author thanks the following people for encouragement and for help:

1. Michael Barr
2. Alexander Borichev
3. David Drasin
4. John Garnett
5. Javad Mashreghi
6. George Mauro
7. Mikhail Sodin
8. Suzie Synott

References

1. Koosis, P., Estimating polynomials and entire functions by using their logarithmic sums over complex sequences, *St. Petersburg Math. J.*, 13 (2002), N5, pp. 757–789.
2. Borichev, A. and Sodin, M., The Hamburger moment problem and weighted polynomial approximation on discrete subsets of the real line, *J. Anal. Math.*, 76 (1998), pp. 219–264.
3. Koosis, P., *The Logarithmic Integral*, Vol. I, Cambridge Studies in Advanced Mathematics 12, Cambridge University Press, 1988 and 1998.
4. Sodin, M. and Yuditskii, P., Another approach to de Branges' theorem on weighted polynomial approximation, in *Proceedings of the Ashkelon Workshop on Complex Function Theory* (May 1996), L. Zalcman, ed., *Israel Mathematical Conference Proceedings*, Vol. 11, Amer. Math. Soc., Providence, R.I., 1997, pp. 221–227.
5. Titchmarsh, E. C., *Introduction to the theory of Fourier integrals*, Second edition, Oxford, 1948.
6. Pedersen, Henryk L., Uniform estimates of entire functions by logarithmic sums, *Journal of Functional Analysis*, 146:2 (1997), pp. 517–555.
7. Benedicks, M., Weighted polynomial approximation on subsets of the real line. Preprint 1981:11, Uppsala University, Mathematical Department (1981). 12 pp.
8. Borichev, A. and Sodin, M., Krein's entire functions and the Bernstein approximation problem, *Illinois Journal of Mathematics*, 45:1, (2001), pp. 167–185.
9. Koosis, P. and Pedersen, H. L. Lower bounds on the values of an entire function of exponential type at certain integers, in terms of a least superharmonic majorant. *Algebra i Analiz*, 10 (1998) N3, pp. 31–44. Reprinted in *St. Petersburg Math. J.*, 10 (1999) N3, pp. 429–439.
10. Koosis, P., *The Logarithmic Integral*, Vol. II, Cambridge Studies in Advanced Mathematics 21, Cambridge University Press, 1992 and 2009.
11. Garnett, J. and Marshall, D., *Harmonic measure*, New Mathematical Monographs 2, Cambridge University Press, 2008. (Esp. pp. 447–451).
12. Boas, R. P., *Entire functions*, Academic Press, 1954.

Index

I. Binder et al. (eds.), *Functions Spaces, Theory and Applications*, Fields Institute Communications 87, https://doi.org/10.1007/978-3-031-39270-2

$$\begin{aligned} m_1, \dots, m_s &= \text{orders of } \beta_1, \dots, \beta_s \text{ as zeros of } B, \\ n_1, \dots, n_s &= \text{orders of } \beta_1, \dots, \beta_s \text{ as zeros of } \widetilde{C}. \end{aligned}$$

The forward problem in the complex plane that consists in the existence of a trace for a function in $H^p_\nu(\Omega)$ is solved: the operator $T : H^p_\nu(\Omega) \longrightarrow L^p_{\mathbb{R}}(\partial\Omega)$ given by $T(f) = \mathrm{Re}(tr_{\partial\Omega} f)$ is well defined, linear and bounded. It implies that the forward problem for $\mathrm{div}(\sigma\nabla u) = 0$ on Ω with $\mathrm{tr}_{\partial\Omega}\, u = b_D \in L^p_{\mathbb{R}}(\partial\Omega)$ and the operator $\widetilde{T} : \mathrm{Re}\,(H^p_\nu(\Omega)) \longrightarrow L^p_{\mathbb{R}}(\partial\Omega)$ given by $\widetilde{T}(u) = \mathrm{tr}_{\partial\Omega} u$ are well defined, linear and bounded.

Remark 3.8 Under the same hypothesis, there is a unique function $u \in \mathrm{Re}(H^p_\nu(\Omega))$ defined on Ω such that $\mathrm{tr}_{\partial\Omega} u = b$.

$$H^p_\nu(\Omega)/\left\{f \in H^p_\nu(\Omega) : \int_{\partial\Omega} \mathrm{tr}_{\partial\Omega}\, f d\mathcal{H}^1 = 0\right\}.$$

$$\mathcal{P}(\mathfrak{h}) = \mathcal{P}(\mathfrak{h}_{\upharpoonleft t}) \, [\dot{+}] \, \mathcal{P}(\mathfrak{h}_{t \upharpoonright}),$$

It is easy to see that $V \in \mathbb{K}_{\mathrm{Schr}}$ if and only if $(p; w) \in \mathbb{K}_{\mathrm{SL}}$ with p, w from (9.6); for the equivalence of the limit point property at b, see (9.15). Moreover, if $V \in \mathbb{K}_{\mathrm{Schr}}$, then $\Delta_{\mathrm{Schr}}(V) = \Delta_{\mathrm{SL}}(p, w)$.

(i) A large subclass of $\mathbb{K}_{\mathrm{Schr}}$ is the following. Let $b > 0$ or $b = \infty$, let $V_0 \in L^1_{\mathrm{loc}}(0, b)$ and let $l \in \big[-\frac{1}{2}, \infty\big)$. Moreover, set

Since $l = 0$ is allowed in (9.8), the class $\mathbb{K}_{\mathrm{Schr}}$ contains potentials where 0 is a regular endpoint. If $l \in \big[-\frac{1}{2}, \frac{1}{2}\big) \setminus \{0\}$, then (9.1) is in the limit circle case at 0 and $\Delta_{\mathrm{Schr}}(V) = 1$.

$$\mathcal{D}_\alpha^p = \mathrm{D}^{-1}\mathrm{A}_\alpha^p = \mathrm{D}^{-1}\mathrm{H}^{p,p,\frac{\alpha+1}{p}} = \mathrm{H}_1^{p,p,\frac{\alpha+1}{p}}.$$

$$\mathrm{B}^p = \mathrm{H}_{1+\frac{1}{p}}^{p,p,1}.$$

Consequently, if $\varphi \in L^\infty$ is such that φ or $\overline{\varphi}$ is of bounded type, then every subnormal Toeplitz operator must be either normal or analytic, since $\ker[S^*, S]$ is invariant under S for every subnormal operator S. We say that a Toeplitz operator T_Φ with matrix-valued symbol Φ is *analytic* if $\Phi \in H^\infty_{M_n}$. Evidently, any analytic Toeplitz operator with a normal symbol is subnormal because the multiplication operator M_Φ is a normal extension of T_Φ. As a first inquiry in the above reformulation of Halmos' Problem 5, the following question can be raised: *Is Abrahamse's Theorem valid for Toeplitz operators with matrix-valued symbols*? In general, a straightforward matrix-valued version of Abrahamse's Theorem is doomed to fail: for instance, if $\Phi := \left[\begin{smallmatrix} z+\overline{z} & 0 \\ 0 & z \end{smallmatrix}\right]$, then, clearly, both Φ and Φ^* are of bounded type and $T_\Phi = \left[\begin{smallmatrix} U+U^* & 0 \\ 0 & U \end{smallmatrix}\right]$ (where U is the shift on H^2) is subnormal, but neither normal nor analytic.

(a) We note that the assumption "Φ has a tensored-scalar singularity" is essential in Theorem 3.12. As we saw before, if $\Phi := \left[\begin{smallmatrix} \overline{z}+z & 0 \\ 0 & z \end{smallmatrix}\right]$, then T_Φ is neither normal nor analytic. But since $\ker H_\Phi = \ker H_{\left[\begin{smallmatrix} \overline{z} & 0 \\ 0 & 0 \end{smallmatrix}\right]} = \left[\begin{smallmatrix} z & 0 \\ 0 & 1 \end{smallmatrix}\right] H^2_{\mathbb{C}^n}$, it follows that $\Theta \equiv \left[\begin{smallmatrix} z & 0 \\ 0 & 1 \end{smallmatrix}\right]$ does not have any nonconstant diagonal-constant inner divisor, so that Φ does not have a tensored-scalar singularity.

If $c = \begin{pmatrix} b \\ a \end{pmatrix}$, then c is CE and $\|\boldsymbol{c}\|_c^2 = \|\boldsymbol{b}\|_b^2$, where $\boldsymbol{c} = L^* \otimes I_2\, c^{\mathrm{t}}$, $\boldsymbol{b} = L^* b^{\mathrm{t}}$.

$$A^{\mathrm{t}} := \sum_\beta \hat{A}_\beta^{\mathrm{t}} L^\beta 1, \quad \text{and} \quad D^{\mathrm{t}} = \sum_\alpha \hat{D}_\alpha^{\mathrm{t}} L^\alpha 1,$$

MIX
Papier aus verantwortungsvollen Quellen
Paper from responsible sources
FSC® C105338
www.fsc.org

If you have any concerns about our products,
you can contact us on
ProductSafety@springernature.com

In case Publisher is established outside the EU,
the EU authorized representative is:
Springer Nature Customer Service Center GmbH
Europaplatz 3, 69115 Heidelberg, Germany

Printed by Libri Plureos GmbH
in Hamburg, Germany